Invertebrate Palaeontology and Evolution

Invertebrate Palaeontology and Evolution

E.N.K. Clarkson
Professor of Palaeontology
Department of Geology
University of Edinburgh
Scotland

Fourth edition

Blackwell
Science

Published by Blackwell Science Ltd,
a Blackwell Publishing company

BLACKWELL PUBLISHING
350 Main Street, Malden, MA 02148-5020, USA
9600 Garsington Road, Oxford OX4 2DQ, UK
550 Swanston Street, Carlton, Victoria 3053, Australia

First published 1979 by Unwin Hyman Ltd
Second edition 1986
Third edition 1993 by Chapman & Hall
Fourth edition 1998 by Blackwell Science Ltd

10 2008

Library of Congress Cataloging-in-Publication Data has been applied for

ISBN 978-0-632-05238-7

A catalogue record for this title is available from the British Library.

Set by Wyvern 21 Ltd, Bristol
Printed and bound in Singapore
by C.O.S. Printers Pte Ltd

The publisher's policy is to use permanent paper from mills that operate a sustainable forestry policy, and which has been manufactured from pulp processed using acid-free and elementary chlorine-free practices. Furthermore, the publisher ensures that the text paper and cover board used have met acceptable environmental accreditation standards.

For further information on
Blackwell Publishing, visit our website:
www.blackwellpublishing.com

In memory of
Professor Peter Sylvester-Bradley
(1913–1978)

Contents

Preface

The first three editions of this textbook were published in 1979, 1986 and 1992, and I trust that this new one, necessitated by so many advances both in fact and theory, will retain its function as a course text for students of palaeontology from their second year onwards. I have made substantial changes to Chapters 1–3, 7, 8 and 12, and the Trace Fossils section has been transferred from Chapter 1 to the last chapter. All the other chapters have been revised to a greater or lesser extent. I have redrawn about half of the illustrations: a singularly congenial occupation for long winter evenings, and I hope that these will prove of value in helping students to understand the anatomy and terminology of fossil invertebrates.

Key words appear in bold at their first mention in the text. There is no specific index for them, but they appear in the General Index.

As before, I wish to thank all my friends, colleagues and family who have given me such assistance in preparing this new edition. Derek Briggs, Colin Scrutton and Rachel Wood gave excellent advice at the beginning of the project as to how I should proceed with the fourth edition, and I have taken on board most of their suggestions. As always, these colleagues, together with David Harper, Peter Sheldon, Susan Rigby, Liz Harper, Dick Jefferies, John Cope, Alan Owen and various others have helped me at all stages, and so have many others, too numerous to mention, in my own country and abroad, and my thanks are due to all.

I would like, above all, to record my heartfelt thanks to my true and stalwart friend Cecilia Taylor, for all the support and helpful suggestions she has given me at every stage. Had Cecilia not undertaken the vast job of rebuilding our teaching collections and preparing new course materials, I would never have had the time to complete this book before the deadline.

To Roisin Moran of University College, Galway I extend my grateful thanks for her two beautiful paintings for the Part Title pages.

I wish to thank also my editorial colleagues, Dr Ian Francis and Jane Plowman, who have given all possible assistance. From Ian came the suggestion, which we have followed up, that the new edition should be coupled with a CD-ROM of Fossil Images. This has been undertaken in conjunction with the Natural History Museum, London, and to Norman McLeod and Paul Taylor of that institution I owe a great deal. I am likewise grateful to David Hicks and Eve Daintith for their meticulous copy-editing and proofreading, respectively.

On a wet November day, over 50 years ago, I was taken into a museum in my native Newcastle-upon-Tyne, to escape from the rain. There, in a dusty glass case were two giant ammonites (probably *Titanites giganteus*), and I was given to understand that they had lived millions of years ago. They excited a fascination which has continued until now, and I am still glad to escape to the hills on a fresh summer morning, to search for fossils, and then to bring them back to the laboratory for study. I hope that this book will be of value to any student wishing to explore something of the richness and diversity of ancient life, and of the methods available for its scientific study. If so, I will have achieved what I set out to do.

Euan Clarkson
Edinburgh

PART ONE
General Palaeontological Concepts

Arnioceras cf. *hartmanni* (Oppel), an assemblage of immature ammonites from the Lower Jurassic of Black Ven, Charmouth, England (Lower Sinemurian). These specimens were probably catastrophically buried, since the soft parts must have been in place when they died preventing sediment penetrating the chambers. Painting by Roisin Moran; original specimen in the James Mitchell Museum, University College, Galway, Ireland.

1 Principles of palaeontology

1.1 Introduction

Once upon a time . . .

Some 4600 million years ago the Earth came into being, probably forming from a condensing disc of particles, dust and gas, which slowly rotated round the Sun. Larger particles, or planetismals, formed from this nebular disc, and as these collided they accreted, eventually forming the planets.

Of all the nine planets in the Solar System only Earth, as far as is known, supports advanced life, though at the time of writing much interest has been generated by the discovery of organic material on Jupiter's satellite Europa. It is, however, a striking fact that life on Earth began very early indeed, within the first 30% of the planet's history. There are remains of simple organisms (bacteria and 'blue-green algae', or cyanobacteria) in rocks 3400 Ma old, so life presumably originated before then. These simple forms of life seem to have dominated the scene for the next 2000 Ma, and evolution at that time was very slow. Nevertheless, the cyanobacteria and photosynthetic bacteria were instrumental in changing the environment, for they gave off oxygen into an atmosphere that was previously devoid of it, so that animal life eventually became possible.

Only when some of the early living beings of this Earth had reached a high level of physiological and reproductive organization (and most particularly when sexual reproduction originated) was the rate of evolutionary change accelerated, and with it all manner of new possibilities were opened up to evolving life. This was not until comparatively late in geological history, and there are no fossil animals known from sediments older than about 700 Ma. Needless to say, these are all invertebrate animals lacking backbones. All of them are marine; there is no record of terrestrial animals until much later. In terms of our understanding of the history of life, perhaps the most significant of all events took place about 543 Ma ago at about the beginning of the Cambrian Period, for at this stage there was a sudden proliferation of different kinds of marine invertebrates. During this critical period the principal invertebrate groups were established, and they then diversified and expanded. Some of these organisms acquired hard shells and were capable of being fossilized, and only because of this can there be any chance of understanding the history of invertebrate life.

The stratified sedimentary rocks laid down since the early Cambrian, and built up throughout the whole of Phanerozoic time, are distinguished by a rich heritage of the fossil remains of the invertebrates that evolved through successive historical periods; their study is the domain of invertebrate palaeontology and the subject of this book.

1.2 Occurrence of invertebrate fossils in Phanerozoic rocks

Hard-part preservation

Fossil invertebrates occur in many kinds of sedimentary rock deposited in the seas during the Phanerozoic. They may be very abundant in limestones, shales, siltstones and mudstones but on the whole are not common in sandstones. Sedimentary ironstones may have rich fossil remains. Occasionally they are found in some coarse rocks such as greywackes and even conglomerates. The state of preservation of fossils varies greatly, depending on the structure and composition of the original shell, the nature and grain size of the enclosing sediment, the chemical conditions at the time of sedimentation, and the subsequent processes of **diagenesis** (chemical and physical changes) taking place in the rock after deposition.

The study of the processes leading to fossilization is known as **taphonomy**. In most cases only the hard parts of fossil animals are preserved, and for these to be fossilized, rapid burial is normally a prerequisite. The soft-bodied elements in the fauna, and those forms with thin organic shells, did not normally survive diagenesis and hence have left little or no evidence of their existence other than records of their activity in the form of trace fossils. What we can see in the rocks is therefore only a narrow band in a whole spectrum of the organisms that were once living; only very rarely have there been found beds containing some or all of the soft-bodied elements in the fauna as well. These are immensely significant for palaeontology.

The oldest such fauna is of late Precambrian age, some 615 Ma old, and is our only record of animal life before the Cambrian. Another such 'window' is known in Middle Cambrian rocks from British Columbia, where in addition to the normally expected trilobites and brachiopods there is a great range of soft-bodied and thin-shelled animals – sponges, worms, jellyfish, small shrimp-like creatures and animals of quite unknown affinities – which are the only trace of a diverse fauna which would otherwise be quite unknown (Chapter 12). There are similar 'windows' at other levels in the geological column, likewise illuminating.

The fossil record is, as a guide to the evolution of ancient life, unquestionably limited, patchy and incomplete. Usually only the hard-shelled elements in the biota (apart from trace fossils) are preserved, and the fossil assemblages present in the rock may have been transported some distance, abraded, damaged and mixed with elements of other faunas. Even if thick-shelled animals were originally present in a fauna, they may not be preserved; in sandy sediments in which the circulating waters are acidic, for instance, calcareous shells may dissolve within a few years before the sediment is compacted into rock. Since the sea floor is not always a region of continuous sediment deposition, many apparently continuous sedimentary sequences contain numerous small-scale breaks (**diastems**) representing periods of winnowing and erosion. Any shells on the sea floor during these erosion periods would probably be transported or destroyed – another limitation on the adequacy of the fossil record.

On the other hand, some marine invertebrates found in certain rock types have been preserved abundantly and in exquisite detail, so that it is possible to infer much about their biology from their remains. Many of the best-preserved fossils come from limestones or from silty sediments with a high calcareous content. In these (Fig. 1.1) the original calcareous shells may be retained in the fossil state with relatively little alteration, depending upon the chemical conditions within the sediment at the time of deposition and after.

A sediment often consists of components derived from various environments, and when all of these, including decaying organisms, dead shells and sedimentary particles, are thrown together the chemical balance is unstable. The sediment will be in chemical equilibrium only after diagenetic physicochemical alterations have taken place. These may involve recrystallization and the growth of new minerals (**authigenesis**) as well as cementation and compaction of the rock (**lithification**), and during any one of these processes the fossils may be altered or destroyed. Shells that are originally of calcite preserve best; aragonite is a less stable form of calcium carbonate secreted by certain living organisms (e.g. corals) and is often recrystallized to calcite during diagenesis or dissolved away completely.

Calcareous skeletons preserved in more sandy or silty sediments may dissolve after the sediment has hardened or during weathering of the rock long after its induration. **Moulds** (often miscalled casts) of the external and internal surfaces of the fossil may be left, and if the sediment is fine enough the details these show may be very good. Some methods for the study of such moulds are described in section 7.2, with reference to brachiopods. If a fossil encloses an originally hollow space, as for instance between the pair of shells of a bivalve or brachiopod, this space may either be left empty or become filled with sediment. In the latter case a sediment **core** is preserved, which comes out intact when the rock is cracked open. This bears upon its surface an **internal mould** of the fossil shell, whereas **external moulds** are left in the cavity from which it came. In rare circumstances the core or the shell, or both, may be replaced by an entirely different mineral, as happens in fossils preserved in ironstones. If

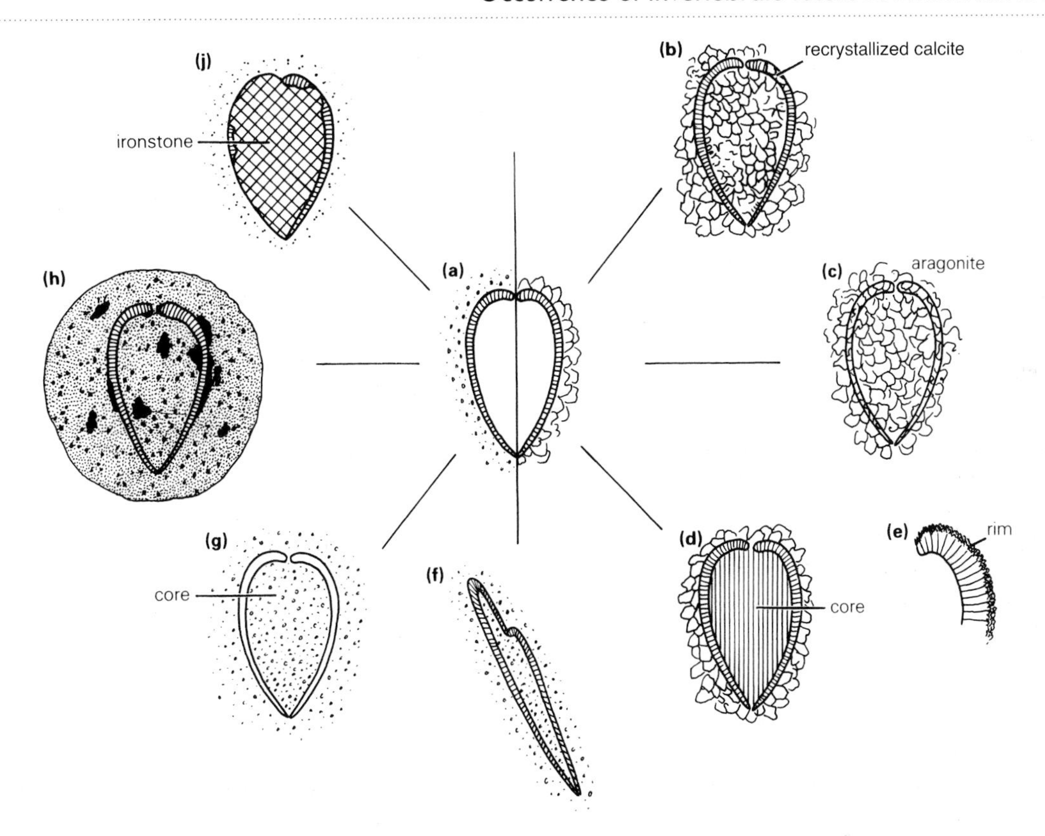

Figure 1.1 Possible processes of fossilization of a bivalve shell: (a) original shell, buried in mud (left) or carbonate (right); (b) the shell was calcite, was buried in a carbonate sediment and was preserved intact other than as a small crystallized patch; (c) shell originally of aragonite, now recrystallized to calcite which destroys the fine structure; (d) original calcite shell retained surrounding a diagenetic core of silica; (e) a silica rim growing on the outside of the shell; (f) tectonic distortion of a shell preserved in mudstone; (g) shell preserved in mud with original shell material leached away, leaving an external and an internal mould, surrounding a mudstone core; (h) a calcareous concretion growing round the shell and inside (if the original cavity was empty), with patches of pyrite in places; (j) ironstone replacement of core and part of shell.

the original spaces in the shell are impregnated with extra minerals, it is said to be **permineralized**, while the growth of secondary minerals at the expense of the shell is **replacement**. Cores may sometimes be of pyrite. Graptolites are often preserved like this, anaeorbic decay having released hydrogen sulphide, which reacted with ferrous (Fe^{2+}) ions in the water to allow an internal pyrite core to form. Sometimes a core of silica is found within an unaltered calcite shell. This has happened with some of the Cretaceous sea urchins of southern England. They lived in or on a sediment of calcareous mud along with many sponges, which secreted spicules of biogenic silica as a skeleton. In alkaline conditions (above pH 9), which may sometimes be generated during bacterial decay, the solubility of the silica increases markedly, and the silica so released will travel through the rock and precipitate wherever the pH is lower. The inside of a sea urchin decaying under different conditions would trap just such an internal microenvironment, within which the silica could precipitate as a gel. Such siliceous cores retain excellent features, preserving the internal morphology of the shell. On the other hand, silica may replace calcite as a very thin shell over the surface of a fossil as a result of some complex surface reactions. These siliceous crusts may retain a very detailed expression of the surface of the fossil and, since they can be freed from the rock by dissolving the limestone with hydrochloric acid, individual small fossils preserved in this way can be studied in three dimensions. Some of the most exquisite of all trilobites and brachiopods are known from material such as this.

A relatively uncommon but exquisite mode of preservation is **phosphatization**. Sometimes the external skeleton, especially of thin organic-shelled animals, may be replaced or overgrown by a thin sheet of phosphate, or the latter may reinforce an originally phosphatic shell. In the former situation the external form of the body is precisely replicated. Alternatively a phosphatic filling of the interior of the shell may form a core, picking out internal structures in remarkable detail. Such preservation is probably associated with bacterial activity directly after the death of the animal. Many small Cambrian fossils have been preserved by phosphatization (Chapters 3 and 12), but much larger fossils may be preserved also, for example crustaceans with a fluorapatite infilling and with all their delicate appendages intact.

Fossils are often found in **concretions**: calcareous or siliceous masses formed around the fossil shortly after its death and burial. Concretions form under certain conditions only, where a delicate chemical balance exists between the water and sediment, by processes as yet not fully understood.

Soft-part preservation

In very rare circumstances soft-bodied organisms can be preserved as fossils, and these provide otherwise unobtainable evidence of the diversity of metazoans living at particular periods; this is discussed in Chapter 12.

1.3 Divisions of invertebrate palaeontology

Invertebrate palaeontology is normally studied as a subdivision of geology, as it is within Earth science that its greatest applications lie. It can also be seen as a biological subject, but one that has the unique perspective of geological time. Within the domain of invertebrate palaeontology there are a number of interrelated topics (Fig. 1.2), all of which have a bearing on the others and which also link up with other sciences.

Three main categories of fossils may be distinguished: (1) **body fossils**, in other words the actual remains of some part, usually a shell of skeleton, of a once-living organism; (2) **trace fossils**, which are tracks, trails, burrows or other evidence of the activity of an animal of former times – sometimes these are the only guide to the former presence of soft-bodied animals in a particular environment; (3) **chemical fossils**, relics of biogenic organic compounds which may be detected geochemically in the rocks.

At the heart of invertebrate palaeontology stands **taxonomy**; the classification of fossil and modern animals into ordered and natural groupings. These groupings, known as **taxa**, must be named and arranged in a hierarchial system in which their relationships are made clear, and as far as possible must be seen in evolutionary perspective.

Evolution theory is compounded of various disciplines – pure biology, comparative anatomy, embryology, genetics and population biology – but it is only the palaeontological aspect that allows the predictions of evolutionary science to be tested against the background of geological time, permits the tracing of evolving lineages and illustrates some of the patterns of evolution that actually have occurred.

The rates at which animals have evolved have varied through time, but most animal types (**species**) have had a geological life of only a few million years. Some of these evolved rapidly, such as the ammonites, others very slowly (Chapter 2). A rock succession of marine sediments built up over many millions of years may therefore have several fossil species occurring in a particular sequence, each species confined to one part of the succession only and representing the time when that species was living. Herein lies the oldest and most general application of invertebrate palaeontology: **biostratigraphy**. Using the sequence of fossil faunas, the geological column has been divided up into a series of major geological time units (periods), each of which is further divided into a hierarchy of small units. The whole basis for this historical chronology is the documented sequence of fossils in the rocks. But different kinds of fossils have different stratigraphical values, and certain parts of the geological record are more closely subdivided than others. Some 'absolute' ages based on radiometric dating have been fixed at particular points to this relative scale, and these provide a framework for understanding the geological record in terms of real time

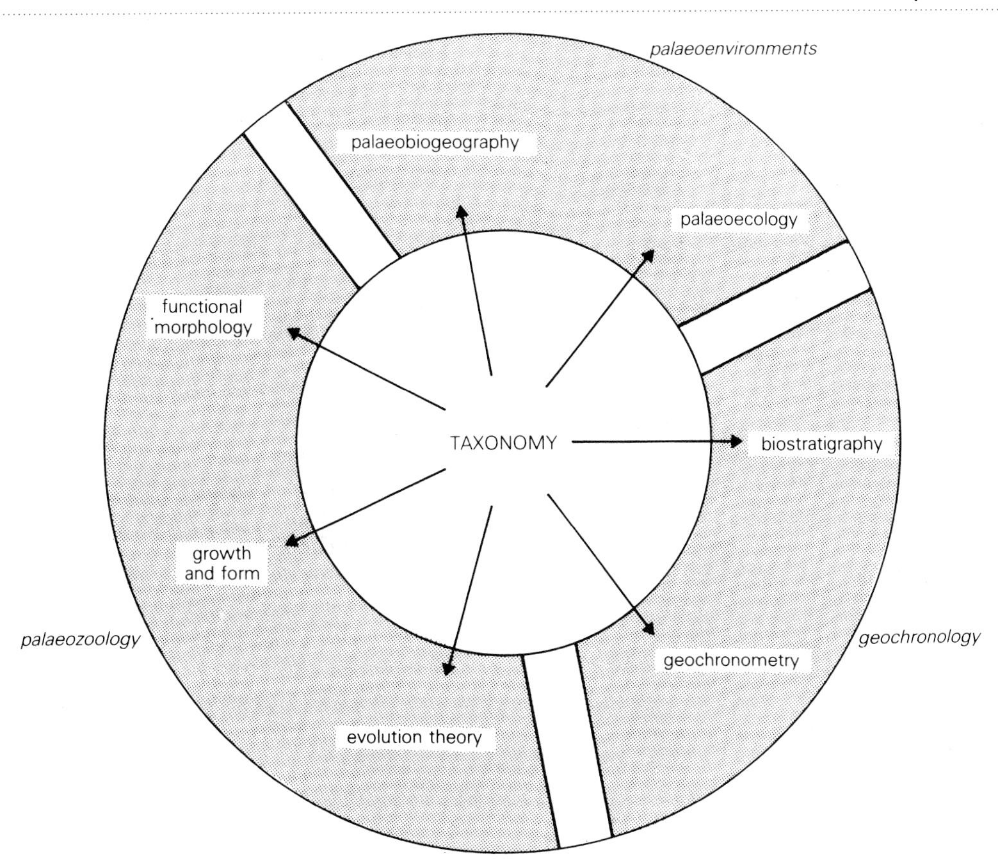

Figure 1.2 The various subdisciplines of palaeontology.

(i.e. known periods of millions of years) rather than as just a purely relative scale. This is only possible at certain horizons, however, and for practical purposes the geological timescale based on biostratigraphy is unsurpassed for Phanerozoic sediments.

Although stratigraphy is the basis of the primary discipline of **geochronology**, a small facet of palaeontological study has a bearing on what may be termed '**geochronometry**'. By counting daily growth rings in extinct corals and bivalves, information has been obtained bearing upon the number of days in the lunar month and year in ancient times. This has helped to confirm geophysical estimates on the slowing of the Earth's rotation (Chapter 5).

Since stratigraphical applications of palaeontology have always been so important, the more biological aspects of palaeontology were relatively neglected until comparatively recently. **Palaeoecology**, which has developed particularly since the early 1950s, is concerned with the relationships of fossil animals to their environment, both as individuals (**autecology**) and in the faunal communities in which they occur naturally; the latter is sometimes known as **synecology**.

Since the soft parts of fossil animals are not normally preserved, but only their hard shells, there are relatively few ways in which their biology and life habits can be understood. Studies in **functional morphology**, however, which deal with the interpretation of the biology of fossilized skeletons or structures in terms of their original function, have been successfully attempted with many kinds of fossils, restricted in scope though these endeavours may necessarily be. **Ichnology** is the study of trace fossils: the tracks, surface trails, burrows and borings made by once-living animals and preserved in sediments. This topic has proved valuable both in understanding the behaviour of the animals that lived when the sediment was being deposited and in interpreting the contemporaneous environment. Finally, it is only by the integration of taxonomic data on local faunas that the global distribution of marine invertebrates through time can be elucidated. Such studies of **palaeobiogeography** (or **palaeozoogeography** in the case of animals) can be used in conjunction with geophysical data in understanding the former relative positions and movements of continental masses.

All of these aspects of palaeontology are interrelated, and an advance in one may have a bearing upon any other. Thus a particular study in functional morphology may give information on palaeoecology and possibly some feedback to taxonomy as well. Likewise, recent refinements in taxonomic practice have enabled the development of a much more precise stratigraphy.

Chemical compounds of biological origin can now be recovered from ancient rocks and form the basis of **biomolecular palaeontology**. Such fossil molecules may help to diagnose which organisms they come from and their breakdown pathways may say something about the environment. Molecular phylogenetics based upon protein sequencing may show how far two or more related organisms have diverged from a common ancestor, and to some extent the available techniques can be applied to the recent fossil record. Immunological-determinant techniques can be used to detect proteins and polysaccharides in fossil shells but, for the moment, only shells younger than 2 Ma have proved amenable to analysis. There are also promising developments also in palaeobiochemistry and organic geochemistry which are applicable to the fossil record, though these are beyond the scope of this book.

Taxonomy

Taxonomy is often undervalued as a glorified form of filing – with each species in its folder, like a stamp in its prescribed place in an album; but taxonomy is a fundamental and dynamic science, dedicated to exploring the causes of relationships and similarities among organisms. Classifications are theories about the basis of natural order, not dull catalogues compiled only to avoid chaos. *The best monographs are works of genius . . .*

(S.J. Gould, 1990)

These words should make entirely clear the fundamental nature of taxonomy. For, as has often been said, to identify a fossil correctly is the first step, and indeed the key, to finding out further information about it. Sound classification and nomenclature lie at the root of all biological and palaeontological work; without them no coherent and ordered system of data storage and retrieval is possible. Taxonomy, or **systematics** as it is sometimes known, is the science of classification or organisms. It is the oldest of all the biological disciplines, and the principles outlined by Carl Gustav Linnaeus (1707–1778) in his pioneering *Systema naturae* are still in use today, though greatly modified and extended.

The species concept

The fundamental unit of taxonomy is the species. Animal species (e.g. Sylvester-Bradley, 1956) are groups of individuals that generally look like each other and can interbreed to produce offspring of the same kind. They cannot interbreed with other species. Since it is reproductive isolation alone that defines species it is only really possible to distinguish closely related species if their breeding habits are known. Of all the described 'species' of living animals, however, only about a sixth are 'good', or properly defined, species. Information upon the reproductive preferences of the other five-sixths of all naturally occurring animal populations is just not documented.

The differentiation of most living and all fossil species therefore has to be based upon other and technically less valid criteria.

Of these by far the most important, especially in palaeontology, is **morphology**, the science of form, since most natural species tend to be composed of individuals of similar enough external appearance to be identifiable as of the same kind. Distinguishing species of living animals by morphological criteria alone is not without hazards, especially where the species in question are similar and closely related. Supplementary information, such as the analysis of species-specific proteins, may be of help in some cases where there is good reason for it to be sought (e.g. for disease-carrying insects). For the rest some degree of subjectivity in taxonomy has to be accepted, though this can be minimized if enough morphological criteria are used.

Nomenclature and identification of fossil species

In the formal nomenclature of any species, living or fossil, taxonomists follow the biological system of Linnaeus, whereby each species is defined by two names: the **generic** and **specific** (or **trivial**) names. For example, all cats, large and small, are related, and one particular group has been placed in the **genus** *Felis*. Of the various species of *Felis* the specific names *F. catus, F. leo* and *F. pardus* formally refer to the domestic cat, lion and leopard, respectively. In full taxonomic nomenclature the author's name and the date of publication are given after the species, e.g. *Felis catus* (Linnaeus 1778) (see below for further discussion).

In palaeontology it can never be known for certain whether a population with a particular morphology was reproductively isolated or not. Hence the definition of species in palaeontology, as in most living specimens, must be based almost entirely on morphological criteria. Moreover, only the hard parts of the fossil animal are preserved, and much useful data has vanished. A careful examination and documentation of all the anatomical features of the fossil has to be the main guide in establishing that one species is different from a related species. In rare cases this can be supplemented by a comparison of the chemistry of the shell, as has proved especially useful in the erection of higher taxonomic categories. Within any interbreeding population there is usually quite a spread of morphological variation. On a broader scale there may be both geographical and stratigraphic variations, and all these must be carefully documented if the species is to be ideally established. Such studies may be very significant in evolutionary palaeontology.

When a palaeontologist is attempting to distinguish the species in a newly discovered fauna, say of fossil brachiopods, she or he has to separate the individual fossils out into groups of morphologically similar individuals. There may, to take an example, perhaps be eight such groups, each distinguished by a particular set of characters. Some of these groups may be clearly distinct from one another; in others the distinction may be considerably less, increasing the risk of greater subjectivity. These groups are provisionally considered as species, which must then be identified. This is done by consulting palaeontological monographs or papers containing detailed technical descriptions and illustrations of previously described brachiopod faunas of similar age, and comparing the species point by point. Some of the species may prove to be identical with already described species, or show only minor variation of a kind that would be expected in a local variant within the same species. Other species in the fauna may be new, and if so a full technical description with illustrations must be prepared for each new species, which should be published in a palaeontological journal or monograph. This description is based upon **type specimens**, which are always thereafter kept in a museum or research institute. Usually one of these, the **holotype**, is selected as the reference specimen and fully illustrated; comparative detail may be added from other specimens called the **paratypes**. There are various other kinds of type specimens; for example, a **neotype** may be erected when a holotype has been lost, or when a species is being redescribed in fuller and more up-to-date terms when no type specimen has previously been designated.

A new genus will be designated as gen. nov. by its author, this following the generic name, a new species as sp. nov. and a new subspecies as ssp. nov.

The new species must be named and allocated to an existing genus, or if there is no described genus to which it pertains then a new genus must also be erected. To appreciate the method let us consider the following historical tale. In the early nineteenth century brachiopods were poorly known and few distinct genera had been erected. One of these was the living *Terebratula*, named by O.F. Müller in 1776. When E.F. von Schlotheim first studied Devonian fossils from North Germany in 1820 he recognized that some of the shells were brachiopods, and he described one of the most abundant forms as the new species *Terebratula sarcinulatus*. By 1830, however, much more was known about brachiopods, and G. Fischer de Waldheim proposed a new genus for this species, so that it became correctly designated *Chonetes sarcinulatus* (Schlotheim). This is the 'type species' of *Chonetes*, a well-known Siluro-Devonian genus of the Class Strophomenata. Note that where a species was originally described under a different generic name, the original author's name is quoted in parentheses. In 1917 F.R. Cowper Reed, then working on Ordovician and Silurian brachiopods of the Girvan district, Scotland, recognized many new species. One of these had similarities in morphology to *Chonetes*, but it was sufficiently different to be regarded as a species of a new subgenus; this is written *Chonetes (Eochonetes) advena* Reed 1917. When in 1928 the taxonomic problems of *Chonetes* and similar forms were addressed by O.T. Jones, the new Superfamily Plectambonitacea was erected to accommodate *advena* and many other related brachiopods. At a later stage *Eochonetes* was elevated to the rank of a full genus. In the most recent treatment, D.A.T. Harper described a large fauna from the Girvan district of Scotland, and within this he recognized two subspecies of *E. advena*, of somewhat different ages and distinguished by minor differences in morphology. These, in Harper's (1989) monograph, are written *Eochonetes advena advena* Reed 1917 (designating Reed's original material), and *Eochonetes advena* Reed 1917 *gracilis* ssp. nov. Subsequent authors will refer to the latter, in full, as *Eochonetes advena gracilis* Harper 1989, or in abbreviated form as *E. advena gracilis*.

Where, due to indifferent preservation, or lack of an up-to-date monographic base, a species cannot be identified with certainty, it may be designated as aff. (related to) or cf. (may be compared with), an existing species (e.g. *Monograptus* cf. *vomerinus*). Where the fossil can be identified as belonging to a known genus, but cannot be ascribed to a species (as may be the case where preservation is poor or if only a fragment is preserved, the suffix sp. (plural spp.) is used (e.g. *Caloceras* sp.). If the specimen, in a more extreme case, can only tentatively be referred to a genus, one would write e.g. *?Kutorgina* sp. If only the species is dubious such an ascription as *Eoplectodonta ?penkillensis* might be used.

All taxonomic work such as this must follow a particular set of rules, which have been worked out by a series of International Commissions and are documented in full in the opening pages of each volume of the *Treatise on Invertebrate Paleontology* (a continuing series of volumes published by the Geological Society of America).

Taxonomic hierarchy

Although all taxonomic categories above the species level are to some extent artificial and subjective, ideally they should as far as possible reflect evolutionary relationships.

Similar species are grouped in **genera** (singular **genus**), genera in **families**, families in **orders**, orders in **classes** and classes in the largest division of the animal kingdom: **phyla** (singular **phylum**). There may be various subdivisions of these categories, e.g. superfamilies, suborders, etc., and in certain groups there is even a case for erecting 'superphyla'. There are only about 30 phyla in the animal kingdom, and only about a dozen of these, e.g. Mollusca and Brachiopoda, leave any fossil remains.

In taxonomy higher taxa are usually distinguished by their suffix (i.e. -ea, -a, etc.). As an example, the following documents the classification of the Ordovician brachiopod *Eochonetes advena gracilis*, referred to earlier, according to a taxonomic scheme in which the author of the taxon and the year of publication are quoted.

Phylum Brachiopoda Dumeril 1806
Subphylum Rhynchonelliformea Williams *et al.* 1996
Class Strophomenata Williams *et al.* 1996
Order Strophomenida Öpik 1934
Suborder Strophomenidina Öpik 1934
Superfamily Plectambonitacea Jones 1928
Family Sowerbyellidae Öpik 1930
Subfamily Sowerbyellinae Öpik 1930
Genus *Eochonetes* Reed 1917
Species *advena* Reed 1917
Subspecies *gracilis* Harper 1989

In the above section we have seen the divisions of the taxonomic hierarchy, but in defining these groupings how do taxonomists actually go about it? The basic principle is that morphological and other similarities reflect phylogenetic affinity (**homology**). This is always so unless, for other reasons, similarity results from convergent evolution. But in assessing 'similarities' how does one decide upon which characters are important? How should they be chosen to minimize subjectivity and produce natural order groups? There is no universally accepted method of facilitating these ends and taxonomists have used different methods. In recent years three sharply contrasting schools have emerged: these are the schools of (1) evolutionary taxonomy, (2) numerical taxonomy and (3) cladism.

Until the 1970s most palaeontologists, especially those working with fossil material which they have collected in the field, were evolutionary taxonomists. In erecting a hierarchical classification, such classical taxonomists used a traditional and very flexible combination of criteria. First there is **morpho-**

logical (or **phenetic**) **resemblance**, the extent to which the animals resemble one another. Second, **phylogenetic relationships** are along with phenetic resemblance considered important. By this is meant the way (as far as can be determined) in which animals are actually related to each other, i.e. in terms of recency of common origin, which of course grades into evolutionary taxonomy. The order of succession in the rock record and geographical distribution may play an important part in deciding relationships. This practical approach to taxonomy, which took all factors into consideration, has for a long time been the backbone of palaeontological classification, and is still considered to be the best method by stratophenetic palaeontologists, who place much emphasis on time in seeking ancestor–descendant relationships (Henry, 1984; Gingerich, 1990).

For some taxonomists, however, the uncertainties and subjectivity which are almost inescapable in any kind of classification seemed to be particularly acute in classical taxonomy, as did the limitations of the fossil record in terms of preservation. The numerical taxonomists tried to escape from this problem by opting for quantified phenetic resemblance as the only realistic guide to natural groupings. It was their view that if enough characters were measured, quantified and computed and represented by the use of cluster statistics, the distances between clusters could be used as a measure of their differences. **Numerical taxonomy** has been found very useful in some instances, but subjectivity cannot be eliminated since the operator has to choose (subjectively) how best to analyse the measurements made, and may need to 'weight' them, giving certain characters more importance (again subjectively) in order to obtain meaningful results. Hence the objectivity of numerical taxonomy is less than it might appear.

The third school relies upon phylogenetic criteria alone, emphasizing that features shared by organisms manifest, in nature, a hierarchial pattern, evident in the distribution of characters shared amongst organisms. It is known as **cladism** or **phylogenetic systematics** – a school founded by the German entomologist Willi Hennig (1966) and was soon applied vigorously to palaeontology (e.g. Eldredge and Cracraft, 1980). In the early days of cladistics there were many doubters (including, as I have to admit, the present author). But, as the method itself evolved, cladistics has come to be recognized as by far the most effective method for reconstructing phylogeny. Smith's (1994) explanation of cladistic methodology is so comprehensive that only brief comments are given here.

Hennig was of the opinion that recency of common origin could best be shown by the shared possession of evolutionary novelties or 'derived characters'. Thus in closely related groups we would see 'shared derived characters' (**synapomorphies**) which would distinguish this group from others. Hennig's central concept was that in any group characters are either 'primitive' (**symplesiomorphic**) or 'derived'. Thus all vertebrates have backbones; the possession of a backbone is a primitive character for all vertebrates and is not, of course, indicative of any close relationship between any group of vertebrates. What is a primitive character for all vertebrates is of course a derived character as compared to invertebrates – synapomorphy and symplesiomorphy thus delineate the relative status of particular characters with respect to a specific problem.

Hennig endeavoured to provide an objective methodology for determining recency of common origin in related taxa, based upon primitive and derived characters. Such relationships are expressed in a **cladogram** (Fig. 1.3), in which dichotomous branching points are arranged in nested hierarchies.

Here taxa A and B share a unique common ancestor. They are said to be **sister groups**. They both share an evolutionary novelty or synapomorphy, not possessed by taxon C. Now C is the sister group of the combined taxa A and B, and D is the sister group of combined taxa A, B and C. In performing a cladistic analysis, therefore, a taxonomist assumes that dichotomous splitting had occurred in each lineage and compiles an (unweighted)

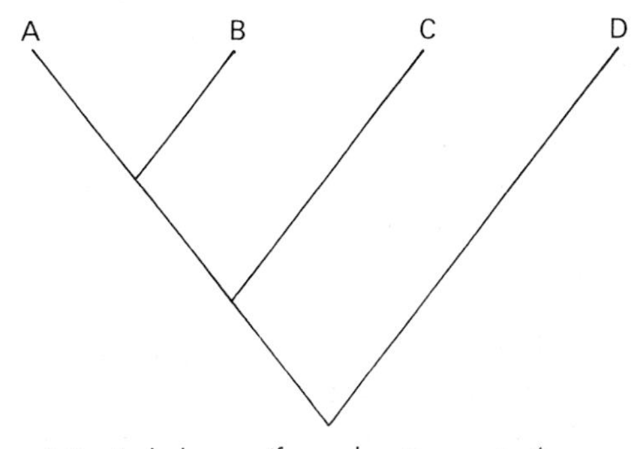

Figure 1.3 A cladogram (for explanation see text).

character data matrix. The more characters and character states there are available the larger the database, and large databases are often routinely processed and the construction of a cladogram speeded up by using one of several computer programs. The PAUP (Phylogenetic Analysis Using Parsimony) program, for example, is a technique which makes the fewest assumptions in ordering a set of observations.

How can we distinguish symplesiomorphic (shared primitive) from synapomorphic (shared derived) character states? The most useful way is 'outgroup comparison'. Here an 'ingroup' (of which the relationships are under investigation) is specifically designated, and compared with a closely related 'outgroup'. Any character present in a variable state in the 'ingroup' must be plesiomorphic if it is also found in the outgroup. Likewise, apomorphic characters are only present in the ingroup.

Hennig distinguished three kinds of cladistic groupings. **Monophyletic** groups contain the common ancestor and all of its descendants (D, C, B, A in Fig. 1.3); **paraphyletic** groups are descended from a common ancestor (usually now extinct and known as the stem group) but do not include all descendants (B and C, for example, in Fig. 1.3); **polyphyletic** groups on the other hand, are the result of convergent evolution. Their representatives are descended from different ancestors and hence, although these may look superficially similar, any polyphyletic group comprising them is artificial.

A cladogram is not an evolutionary tree; it is an analysis of relationships. As such it is a valuable and rigorous way of working out and showing graphically how organisms are related, and it forces taxonomists to be explicit about patterns and groups. The methodology of cladistics is especially good when dealing with discrete groups with large morphological and stratigraphic gaps and to these it brings the potential for real objectivity. A cladogram shows how sister groups are hierarchically related on the basis of shared–derived homologies, but although it portrays taxonomic relationships in terms of recency of common origin, the order of succession in the rock record is not taken into account (though implicitly cladograms have a time axis).

So where does that leave the potential contribution of stratigraphy in reconstructing phylogeny? Whilst a few 'transformed cladists' negate the value of the fossil record altogether (a view vigorously opposed by Ridley, 1985), the successive appearance of taxa in stratigraphy cannot be denied as an essential source of data, however imperfect the fossil record actually is. Thus as Gingerich (1990) comments, 'Time is a fundamental dimension in evolutionary studies, and a major goal of palaeontology should continue to be the study of the diversification of major groups in relationship to geological time'. So, having constructed an appropriate cladogram, the next stage in exploring relationships is to combine this information with biostratigraphic data. For this, Smith (1994) discusses methodology in detail. The result is a phylogenetic tree, which shows the splitting of lineages through time and is effectively 'a"best estimate" of the tree of life'. Very commonly there is an excellent correspondence between the cladogram and the rock record; on the other hand the combined cladistic and biostratigraphic approach may throw up unexpected patterns.

The ultimate problem, not only for cladism, but for all taxonomic methodology, results from convergent evolution. Resemblances in characters or character states may have nothing to do with recency of common origin, but from convergence, and it may not always be possible to disentangle the results of the two. Thus Willmer (1990) and Moore and Willmer (1997) in considering the relationships between major invertebrate groups argue that 'cladistic analysis based on parsimony will tend to minimize and thus conceal convergence', and they contend that convergence at all levels is far more important than has generally been believed. There is certainly a problem here. Even so, cladistic methodology, coupled with biostratigraphy seems to be the best way forward, and an essential prerequisite for drawing up meaningful phylogenetic trees.

Use of statistical methods

Inevitably palaeontological taxonomy carries a certain element of subjectivity since the information coded in fossilized shells does not give a complete record of the structure and life of the animals that bore them. There are particular complications that cause trouble. For instance, palaeontological taxonomy can do little to distinguish **sibling species**, which look alike and live in the same area but cannot interbreed. **Polymorphic species**, in which many forms are present within one biological species, may likewise be hard to speciate correctly. In particular, where **sexual dimorphism** is strong the males and females of the one species may be so dissimilar in appearance that they have sometimes been described as different species, and the true situation may be hard to disentangle (as with ammonites; Chapter 8).

When it comes to the distinction of closely related species, however, there are a number of statistical tests that may help to give a higher degree of objectivity. One simple bivariate test in common use, for example, can be used when a series of growth stages are found together. If a collection of brachiopods is made from locality A, the length/width ratios or some other appropriate parameters may be plotted on a graph as a scatter diagram. A line of best fit (e.g. a reduced major axis) may then be drawn through the scatter. This gives a simple $y = ax + b$ graph, where a is the gradient and b the intercept on the y axis. A similar scatter from a population collected from locality B may be plotted on the same graph, and the reduced major axis drawn from this too. The relative slopes and separations of the two axes may then be compared statistically. If these lie within a certain threshold the populations can then be regarded as being of the same species; if outside it then the species are different.

This is only one of a whole series of possible tests, and more elaborate techniques of multivariate analysis are becoming increasingly important in taxonomic evolutionary studies.

With the advent of microcomputers and the provision of specialist software packages designed specifically to meet the needs of palaeontologists (PALSTAT; Bruton and Harper, 1990; Ryan *et al.*, 1995), the use of numerical analysis is becoming standard. Statistical methods are likewise essential in defining palaeocommunities, in 'undeforming' populations of deformed fossils and comparing them to unaltered material.

Palaeobiology

Various categories may be included in palaeobiology: palaeoecology (here discussed with palaeobiogeography), functional morphology and ichnology, each of which requires some discussion.

Palaeoecology

Since ecology is the study of animals in relation to their environment, palaeoecology is the study of ancient organisms in their environmental context. All animals are adapted to their environment in all of its physical, chemical and biological aspects. Each species is precisely adapted to a particular ecological niche in which it feeds and breeds. It is the task of palaeoecology to find out about the nature of these adaptations in fossil organisms and about the relationships of the animals with each other and their environments; it involves the exploration of both present and past **ecosystems** (Schäfer, 1972)

Although palaeoecology is obviously related to ecosystem ecology, it is not and cannot be the same. In modern community ecology much emphasis is placed on energy flow through the system, but this kind of determination is just not possible when dealing with dead communities. Instead palaeontologists perforce must concentrate on establishing the composition, structure and organization of palaeocommunities, in attempting from here to work out linkage patterns in food webs and in investigating the autecology of individual species.

Many attempts have been made to summarize categories of fossil residue, to provide the background for interpreting original community structure. The scheme proposed by Pickerill and Brenchley (1975) and amended by Lockley (1983) is used here:

1. An **assemblage** refers to a single sample from a particular horizon.
2. An **association** refers to a group of assemblages, all showing similar recurrent patterns of species composition.
3. A **palaeocommunity** (or **fossil community**) refers to an assemblage, association or group of associations inferred to represent a once-distinctive biological entity. Normally this only represents the preservable parts of an original biological community, since the soft-bodied animals are not preserved. This definition corresponds more or less exactly to that of Kauffman and Scott (1976).

Palaeoecology must always remain a partial and incomplete science, for so much of the information available for the study of modern ecosystems is simply not preserved in ancient ones. The animals themselves are all dead and their soft parts have gone; the original physics and chemistry of the environment is not directly observable and can only be inferred from such secondary evidence as is available; the shells may have been transported away from their original environments by currents, and the fossil assemblages that are found may well be mixed or incomplete; post-depositional diagenetic processes may have altered the evidence still further. Despite this palaeoecology remains a valid, if partial science. Much is now known about the post-mortem history of organic remains (taphonomy; Chapter 12), of which an additional dimension involves burial processes (**biostratinomy**). This helps to disentangle the various factors responsible for deposition of a particular fossil assemblage, so that assemblages preserved *in situ*, which can yield valuable palaeoecological information, can be distinguished from assemblages that have been transported.

Biostratinomy or preservation history has both pre-burial and post-burial elements. The former include transportation, physical, chemical or biological damage to the shell, and the attachment of **epifauna**. Post-burial processes may involve disturbance by burrowers and sediment eaters (**bioturbation**), current reworking, solution and other diagenetic preservation changes.

Modern environments and vertical distribution of animals

Figure 1.4 shows the main environments within the Earth's oceans at the present day and the nomenclature for the distribution of marine animals within the oceans.

Modern marine environments are graded accordingly to depth. The **littoral** environments of the shore grade into the subtidal **shelf**, and at the edge of the shelf the continental **slope** goes down to depth; this is the **bathyal** zone. Below this lie the flat **abyssal** plains and the **hadal** zones of the deep-ocean **trenches**. There is often a pronounced zonation of life forms in depth zones more or less parallel with the shore. In addition there is a general decrease in **abundance** (number of individuals) but not necessarily **diversity** (number of species) on descent into deeper water from the edge of the shelf. The faunas of the abyssal and hadal regions were originally derived from those of shallow waters but are highly adapted for catching the limited food available at great depths. These regions are, however, impoverished relative to the shallow-water regions.

Animals and plants that live on the sea floor are **benthic**; those that drift passively or swim feebly in the water column are **planktic** (= **planktonic**) since they are the **plankton. Nekton** (**nektic** or **nektonic** fauna) on the other hand comprises active swimmers. **Neritic** animals belong to the shallow waters near land and include **demersal** elements which live above the continental shelves and feed on benthic animals thereon. **Pelagic** or **oceanic** faunas inhabit the surface waters or middle depths of the open oceans; **bathypelagic** and the usually benthic, abyssal and hadal organisms inhabit the great depths.

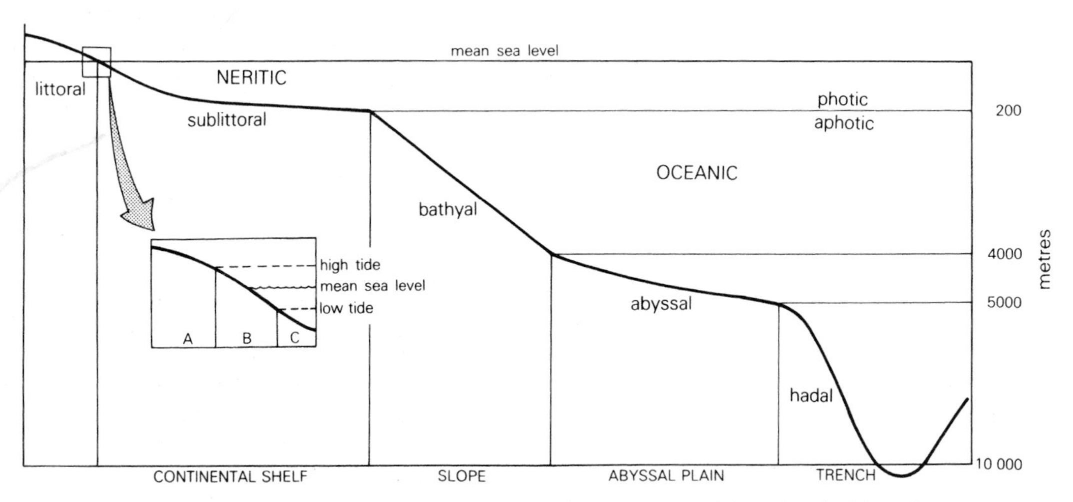

Figure 1.4 Modern marine environments. A, B and C in the inset refer to supralittoral, littoral and sublittoral environments. (Based on a drawing in Laporte, 1968.)

Only the shelf and slope environments are normally preserved in the geological record, the trench sediments rarely so. The abyssal plains are underlain by basaltic rock, formed at the mid-oceanic ridges and slowly moving away from them to become finally consumed at the subduction zones lying below the oceanic trenches; it moves as rigid plates. The ocean basins are very young geologically, the oldest sediments known therein being of Triassic age. These are now approaching a subduction zone and are soon to be consumed without trace. Hence there are very few indications of abyssal sediment now uplifted and on the continents.

What is preserved in the geological record is therefore only a fraction, albeit the most populous part, of the biotic realm of ancient times. The sediments of the continental shelf include those of the littoral, lagoonal, shallow subtidal, median and outer shelf realms. Generally sediments become finer towards the edges of the shelf, the muddier regions lying offshore. There may be reefs close to the shore or where there is a pronounced break in slope.

Horizontal distribution of marine animals

The main controls affecting the horizontal distribution of recent and fossil animals are temperature, the nature of the substrate, salinity and water turbulence. The large-scale distribution of animals in the oceans is largely a function of temperature, whereas the other factors generally operate on a more local scale. Tropical shelf regions carry the most diverse faunas, and in these the species are very numerous but the number of individuals of any one species is relatively low. In temperate through boreal regions the species diversity is less, though the number of individuals per species can be very large.

Salinity in the sea is of the order of 35 parts per thousand. Most marine animals are **stenohaline**, i.e. confined to waters of near-normal salinity. A few are **euryhaline**, i.e. very tolerant of reduced salinity. The brackish water environment is physiologically 'difficult', and faunas living in brackish waters are normally composed of very few species, especially bivalves and gastropods belonging to specialized and often long-ranged genera. These same genera can be found in sedimentary rocks as old as the Jurassic, and their occurrence in particular sediments which lack normal marine fossils is a valuable pointer to reduced salinity in the environment in which they lived.

Water turbulence may exercise a substantial control over distribution, and the characters of faunas in high- and low-energy environments are often very disparate. Robust, thick-shelled and rounded species are normally adapted for high-energy conditions, whereas thin-shelled and fragile forms point to a much quieter water environment, and it may be possible to infer much about relative turbulence in a fossil environment merely from the type of shells that occur.

MODERN AND ANCIENT COMMUNITIES

In shallow cold-temperature seas marine invertebrates are normally found in recurrent ecological communities or associations, which are usually substrate related. In these a particular set of species are usually found together since they have the same habitat preferences. Within these communities the animals either do not compete directly, being adapted to microniches within the same habitat, or have a stable predator–prey relationship.

Community structure is normally well defined in cold-temperature areas, but in warmer seas where diversity is higher it is generally less clear.

Petersen (1918), working on the faunas of the Kattegat, first studied and defined some of these naturally occurring communities. He also recognized two categories of bottom-dwelling animals: **infaunal** (buried and living within the sediment) and **epifaunal** (living on the sea floor or on rocks or seaweed). It was soon found that parallel communities occur, with the same genera but not the same species, on the opposite sides of the Atlantic. Since this pioneer work a whole science of community ecology has grown up, having its counterpart in palaeoecology. Much effort has been expended in trying to understand the composition of fossil communities, the habits of the animals composing them, community evolution through time and, as far as possible, the controls acting upon them (e.g. Thorson, 1957, 1971). This is perhaps the most active field of palaeoecology at present, as a host of recent original works testifies (e.g. Craig, 1954; Ziegler *et al.*, 1968; Boucot, 1975, 1981; Scott and West, 1976; McKerrow, 1978; Skinner *et al.*, 1981; Dodd and Stanton, 1990; Bosence and Allison, 1995; Brenchley and Harper, 1997).

As Fürsich (1977) makes it clear, however, most fossil assemblages 'lack soft-bodied animals. Where possible trace fossils can be used to compensate for this but they are no real substitute'. Hence assemblages, associations and even palaeocommunities must not be considered as directly equivalent to the sea-floor communities of the present day. Using biostratinomic and sedimentological data the 'degree of distortion' from the original community can in some cases be estimated.

FEEDING RELATIONSHIPS AND COMMUNITIES

All modern animals feed on plants, other animals, organic detritus or degradation products. The tiny plants of the plankton are the **primary producers** (**autotrophs**), as are seaweeds. Small planktic animals are the **primary consumers** (herbivores and detritus eaters); there are **secondary** (carnivores) and **tertiary consumers** (top carnivores) in turn. Each animal species is therefore part of a **food web** of **trophic** (i.e. feeding) relationships wherein there are a number of **trophic levels**. In palaeoecology it is rarely possible to draw up a realistic food web (though this is one of the more important aspects of modern ecology), but most fossil animals can usually be assigned to their correct feeding type and so the trophic level may be estimated reasonably.

Of primary consumers the following types are important:

filterers or suspension feeders, which are infaunal or epifaunal animals sucking in suspended organic material from the water;

epifaunal 'collectors' or detritus feeders, which sweep up organic material from the sea floor; some infaunal bivalves and worms are also collectors;

swallowers or deposit feeders, which are infaunal animals unselectively scooping up mud rich in organic material.

Secondary and tertiary consumers, the carnivores, may prey on any of these, but it is the communities of the primary trophic level that are most commonly preserved because of their sheer number of individuals.

In many living communities most of the '**biomass**' is actually contributed by very few taxa, usually not more than five (the **trophic nucleus**), but there may be representatives of a number of other species in small numbers. In this system competition between the species concerned seems to be minimized. It is thus mutually beneficial since the different taxa are exploiting different resources within the environment. Living communities are therefore generally well balanced, the number of species and individuals of particular species being controlled by the nature and availability of food resources. Fossil assemblages may be tested according to this concept. If they are 'unbalanced' then either (1) there may have been soft-bodied unpreservable organisms which originally completed the balance or (2) the assemblage has been mixed through transportation and thus does not reflect the true original community.

FAUNAL PROVINCES

Marine zoogeography (Ekman, 1953; Briggs, 1974; Hallam, 1996) is primarily concerned with the global distribution of marine faunas and with the definition of **faunal provinces**. These are large geographical regions of the sea (and most particularly the continental shelves) within which the faunas at the specific, generic and sometimes familial level have a distinct identity. In faunal provinces many of the animals are **endemic**, i.e. not found outside a particular province. Such provinces are often separated from neighbouring provinces by fairly sharp boundaries, though in other cases the boundaries may be more gradational.

Figure 1.5 shows the main zoogeographical regions of the present continental shelves as defined by Briggs (1974).

Tropical shelf, warm temperate, cold temperate and polar regions can be distinguished, whose limits are controlled by latitude but also by the spread of warmer or colder water through major marine currents. Tropical shelf faunas occur in four separate provinces. Of these the large Indo-West Pacific province, extending from southern Africa to eastern Australia, is the richest and most diverse and has been a major centre of dispersal throughout the Tertiary. Smaller and generally poorer provinces of tropical faunas are found in the East Pacific, West Atlantic and (least diverse of all) East Atlantic. These are separated both by land barriers (e.g. the Panama Isthmus) and by regions of cooler water (e.g. where the cold Humboldt Current sweeping up the western side of South America restricts the tropical fauna to within a few degrees south of the equator).

The shelf faunas of cooler regions of the world are likewise restricted by temperature, and again warm or cold currents exercise a strong control of their distribution. Some zoogeographic 'islands' can be isolated by regions of warmer or cooler water.

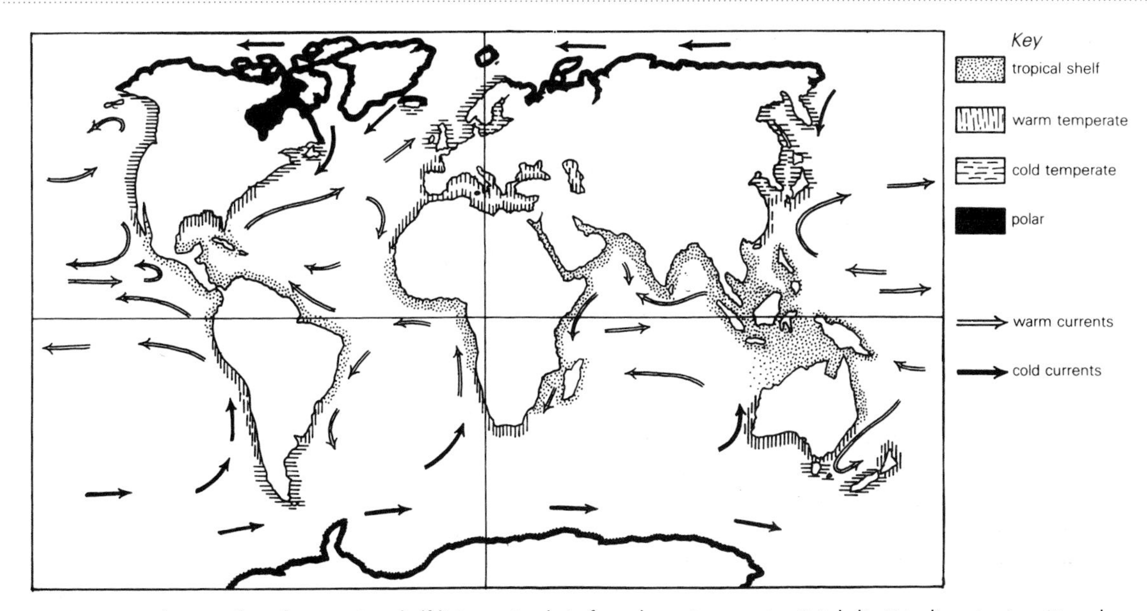

Figure 1.5 Distribution of modern marine shelf-living animals in faunal provinces, using Winkel's 'Tripel' projection. (Based on a drawing in Briggs, 1974.)

For example, the tip of Florida carries a tropical shelf fauna, isolated to the northeast and northwest by cooler water areas, but though this is part of the West Atlantic tropical shelf province it has been isolated for some time and therefore its fauna has diverged somewhat from that of the Caribbean shelf. Likewise some oceanic islands (e.g. the Galapagos) may be considered as part of a general zoogeographic region or province, but as their shelves have been initially colonized by chance migrants they may have very many endemic elements which have evolved in isolation. How many of these endemics there are may depend largely on how long the islands have been isolated.

The development of today's faunal provinces has been charted throughout most of the Tertiary. At present Mesozoic and Palaeozoic provinces are likewise being documented (Middlemiss *et al.*, 1971; Hallam, 1973; Hughes, 1975; Gray and Boucot, 1979; McKerrow and Scotese, 1990). When used in conjunction with palaeocontinental maps it may be possible to see how the distribution pattern of ancient faunas related to the position of ancient continental masses and their shelves. Sometimes palaeozoogeographical and geological evidence may have a bearing upon palaeotemperatures and even allow some inference to be made regarding ancient ocean current systems.

Modern and ancient reefs

Throughout geological time animals have not only become adapted to particular environments but also themselves created new habitats and environments. Within these there has been scope for almost unlimited ecological differentiation.

Perhaps the most striking examples of such **biogenic** environments are the reefs of the past and present. Reefs (Fig. 1.6) are massive accumulations of limestone built up by lime-secreting algae and by various kinds of invertebrates.

Through the activities of these frame-builders great mounds may be built up to sea level, with caves and channels within them providing a residence for

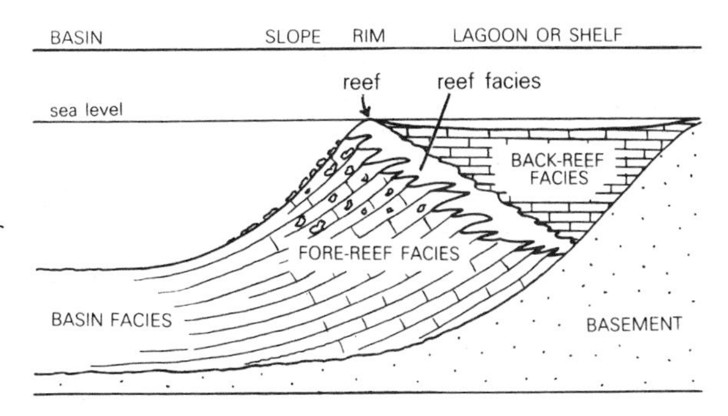

Figure 1.6 Generalized section through a reef (algal or coral).

innumerable kinds of animals, all ecologically differentiated for their particular niches.

In the barrier and patch reefs of the tropical seas of today, which grow up to the surface, the warm, oxygen-rich, turbulent waters allow rapid calcium metabolism and hence continuous growth. The principal frame-builders are algae and corals, but there are many other kinds of invertebrates in the reef community: sponges, bryozoans and molluscs amongst others. Some of these add in minor ways to the reef framework; others break it down by boring and grazing. The growth of the reef to sea level continually keeps pace with subsidence, but it is also being continually eroded. In a typical coral reef complex the reef itself is a hard core of cemented algal and coral skeletons facing seawards, and as the reef subsides it grows outwards over a forereef slope of tumbled boulders broken from the reef front. Behind it is a lagoon with a coral sand sediment and tidal flats along the shore colonized by cyanobacteria ('blue-greens'). Green algae are commonest in the back-reef facies; red algae are the main lime secreters of the reef itself.

Large reefs such as the above are known as **bioherms**; they form discrete mounds rising from the sea floor. **Biostromes**, on the other hand, are flat laminar communities of reef-type animals and barely rise above the sea floor.

Throughout geological history there have been various kinds of reef communities, which have arisen, flourished and become extinct. In all of these the frame-builders have included algae, but the invertebrate frame-builders have been of different kinds; the present corals are only the most recent in a series of reef-building animals (Newell, 1972).

The oldest known reefs are over 2000 Ma old, made up entirely of sediment-trapping and possibly lime-secreting cyanobacteria: the prokaryotic stromatolites. Some of these reefs reached considerable dimensions. One is reported from the Great Slave Lake region in Canada as being over 450 m thick, and separating a shallow-water carbonate platform from a deep-water turbidite-filled basin. There are no preserved metazoans in these reefs, however, and their ecological structure may have been very simple.

With the rise of frame-building metazoans in the Lower Cambrian a new kind of reef community made its appearance. Stromatolitic reefs were invaded by the sponge-like archaeocyathids, growing in clumps and thickets on the reef surface. When these earliest of reef invertebrates became extinct in late Middle Cambrian time there were no more reef animals until some 60 Ma later; the only reefs were stromatolitic. In Middle Ordovician time these algae were joined by corals and stromatoporoids (lime-secreting sponges; Chapter 4) as well as by red algae, which together formed a reef environment attracting a host of other invertebrates including brachiopods and trilobites. For unknown reasons this type of reef complex did not continue beyond late Devonian time. The reefs that arose in the Carboniferous were mainly algal, stromatoporoids and corals no longer playing such an important part in their construction.

The same kind of reef continued into the Permian, and many of these fringed the shrinking inland seas of that time. They rose at the edge of deeper basins in which the water periodically became more saline as it evaporated; there was likewise evaporation in giant salt pans behind the reef as water drawn through the reef dried out in the lagoons behind. In the Permian reefs of Texas and northern England, which are very large, the reef front rose as a vertical wall of laminated algal sheets, turning over at the reef crest where it reached sea level (Fig. 1.7).

The upper surface was intensely colonized by stromatolites, which died out towards the lagoonal back-reef facies. This kind of reef in general morphology therefore does not closely approximate the standard pattern of Fig. 1.6. When the Permian shelf seas had dried out completely the Permian reef-complex type vanished, and there were no more reefs until the slow beginnings of the coral–algal reef system that arose in the late Triassic. Coral reefs

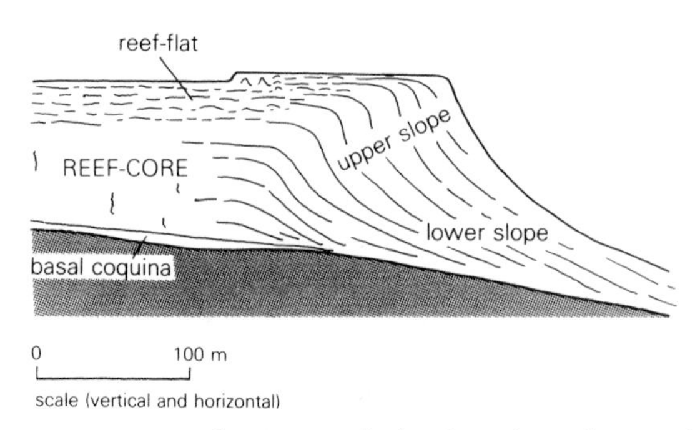

Figure 1.7 Crest of a Permian algal reef. (Redrawn from Smith, 1981.)

have expanded and flourished since then, other than during a catastrophic period in the early Cretaceous from which there are no reefs known (though corals must have been living somewhere at that time). When the corals recovered they were joined in many places by the peculiar rudistid bivalves, also reef formers, which at one period almost supplanted corals as the dominant reef frame-builders. Yet these too died out in the late Cretaceous, leaving corals the undisputed and dominant reef-building invertebrates.

There has been some decline in the spread of coral reefs and in the number of coral genera since the beginning of the Tertiary. They are now confined to the Indo-Pacific region and on a smaller scale to the West Atlantic. This decline may still be continuing, though the reefs were not significantly affected by the Pleistocene glaciation. The future of world reef communities, the most complex of all marine ecosystems, remains to be seen.

Functional morphology, growth and form

The functions of particular organs in fossils cannot be established by many of the methods available to zoologists, but it is still possible to go some way towards explaining how particular organs worked when the animal that bore them was alive. Such functional interpretation is, however, limited and in many ways it is hard to go beyond a certain point, even when the function of a particular organ is known (section 11.3).

Palaeontologists are often presented either with organs whose function is not clear and which have no real counterparts in living animals, or with fossils of bizarre appearance which are so modified from the normal type for the taxon that they testify to extreme adaptations. Some attempts can be made towards interpreting these morphologies in terms of adaptation and mode of life, which in turn may lead to a clearer understanding of evolutionary processes. If these problems are to be tackled, then a coherent methodological scheme is needed. One particular system of approach, the paradigm method of Rudwick (1961), has been much discussed, but it lies largely outside the scope of this work and so only some examples of its application are mentioned.

Two related aspects of palaeozoology which can be deduced from fossilized remains alone are growth and form. Following the classic work of d'Arcy Thompson (1917) it has been understood that the conformation of the parts of any organism is the result of interacting forces, dictated by physicomathematical laws, which have operated through growth. The central issue here (Thomas and Reif, 1993) is the balance perceived between 'accidents of history and the prescription of physical laws, as causes of organic design'.

Different marine invertebrate skeletons may be functionally convergent in the way they grow, and the same kinds of growth patterns turn up frequently in representatives of many phyla. This is because there are relatively few ways in which an animal can grow and yet can produce a hard covering. Invertebrate skeletons, by contrast with those of vertebrates, are generally external, and this narrows down the spectrum of growth possibilities still further. Some of these types of skeletons are as follows.

External shells growing at the edges only by accretion of new material along a particular marginal zone of growth. Very often such growth results in a logarithmic spiral shell as in brachiopods, cephalopods and in coralla of certain simple corals.

External skeletons of plates – disjunct, contiguous or overlapping – normally secreted along a single zone which may be but is not always marginal. A good example is the echinoderm skeleton in which the plates once formed are permanently locked into place, though they may thereafter grow individually by accretion of material in concentric zones.

External skeletons all formed at the one time. The arthropod exoskeleton is most typical. Growth here is difficult for the skeleton has to be periodically moulted, a process known as **ecdysis**. When the old exoskeleton is cast the arthropod takes up water or air and swells to the next larger size, and the new cuticle which underlies the old one then hardens. Growth is thus rapid and episodic and is only possible during moulting. The disadvantage of this system is the vulnerability to damage and predation during moulting.

Essentially, any kind of skeleton, internal or external is 'highly constrained by geometric rules, growth processes, and the properties of materials. This suggests that, given enough time and an extremely large number of evolutionary experiments, the discovery by organisms of "good" designs – those that are viable and can be constructed with available materials – was inevitable and in principle predictable . . . the recurring designs we observe are attractors, orderly and stable configurations of matter that must necessarily emerge in the course of evolution' (Thomas and Reif, 1993).

In point of fact the potential available has been very well exploited by living creatures, though specific constraints seldom allow the production of ideal organisms. Nature works as a tinkerer, rather than as an engineer (Jacob, 1977; see also Chapter 2) – a point to be borne in mind in considering all living and fossil organisms.

Stratigraphy

Sedimentary rocks have been built up by layer upon layer of sediment, which sometimes has been much the same for long periods of time but at other times has changed its character rapidly. The individual layers within sedimentary rocks are separated by bedding planes. These bedding planes are time horizons, and the history of a rock sequence is reflected in its layering. Each layer in the rock sequence must have been laid down on a pre-existing layer, so that the oldest rocks are at the bottom of an exposed sequence, the youngest at the top, unless the succession has been tectonically inverted. This is the **principle of superposition**, recognized as long ago as 1669 by the Danish scientist Nicolaus Steensen (Steno).

Stratigraphy is concerned with the study of stratified rocks, their classification into ordered units and their historical interpretation. It bears not only upon past geological events but also upon the history of life and is perhaps the most basic part of all geology (Harland and Armstrong, 1990; Benton, 1995).

Much of stratigraphy is concerned with chronology; the geological record has to be divided up into time periods which are standardized, as far as possible, all over the world. One of the primary aims of stratigraphy has been to produce an accurate chronology in which not only the order of events but also their dates are known. Stratigraphical classification is basic to all of this.

There are three principal categories of stratigraphical classification, **lithostratigraphy, biostratigraphy** and **chronostratigraphy**, all of which are ways of ordering rock strata into meaningful units (Hedberg, 1976; Holland *et al.*, 1978).

Lithostratigraphy

Lithostratigraphy is concerned with the erection of units based upon the characters of the rocks and differentiated on types of rock, e.g. siltstone, limestone or clay. It is useful in local areas and essential in geological mapping, but there is always the danger that even in a small area rock units cut across time planes. For instance, if a shoreline has been advancing in one direction a particular suite of sediments, probably of the same general kind, will be left in its wake. Though this bed will appear in the rock record as a single uniform layer, it will not all have been deposited at the one time; since it cuts across time planes it is said to be **diachronous**. Such diachronism is common in the geological record. Furthermore, many suites of sediments are laterally impersistent; different sedimentary facies may have existed at the same time within a small space – a sandstone layer, for instance, passing into a shale some distance away. Lithostratigraphy is thus only of real value within a relatively small region.

The divisions erected in lithostratigraphy are arranged in a hierarchial system: **group, formation, member** and **bed**. A bed is a distinct layer in a rock sequence. A member is a group of beds united by certain common characters. A formation is a group of members, again united by characters with features in common. It is the primary unit of lithostratigraphy and is most useful in geological mapping. Hence it is formations that are normally represented by different colours on geological maps and cross-sections, and a formation is normally defined for its mapping applications. Finally, a group ranks above a formation; it is composed of two or more formations and is often used for simplifying stratigraphy on a small-scale map.

Biostratigraphy

In biostratigraphy the fossil contents of the beds are used in interpreting the historical sequence. It is based upon the principle of the irreversibility of evolution. This means that at any one moment in the Earth's history there was living a unique and special assemblage of animals, characteristic of that period and of no other. As time went on these were replaced by others; each successive fossil assemblage is a pale reflection of the life at the time that the enclosing sediments were deposited. Thus during the early Palaeozoic trilobites and brachiopods were the most common fossils. By the Mesozoic the most abundant preservable invertebrates were the ammonites; they too became extinct, and snails and bivalves are the commonest relics of Cenozoic time. This is how it appears on a broad scale. However, when the time ranges of individual fossil species are examined it is evident that some of these lasted for only a fraction of geological time, characterizing very precisely a particular brief historical period.

In any local area, once the sequence of fossil faunas has been precisely established through assiduous collection and documentation from exposed sections, this known succession can be used for correlation with other areas. Certain fossil species have been found to be particularly good stratigraphical markers. They characterize short sections of the geological succession known as **zones**. To take an example, ammonites are particularly good zone fossils for Mesozoic stratigraphy. The Jurassic period lasted some 55 Ma, and in the standard British succession there are over 60 ammonite zones by which it is subdivided, so the zones are defined historical periods which have an average duration of less than a million years each.

The practical problems in biostratigraphy are, however, very complex, and some parts of the geological succession are much more closely zoned than others. The main problems are as follows.

1. Many kinds of fossils, especially those of bottom-dwelling invertebrates, are facies controlled. They lived in particular environments only, e.g. lime-mud sea floor, reef, sand or silty sea floor. They were often highly adapted for particular conditions of temperature, salinity or substrate and are not found preserved outside this environment. This means that they can only be used for correlating particular environments and thus are not universally applicable.
2. Some kinds of fossils are very long-ranged. Their rates of evolutionary change were very slow. They can only be used in a broad and general sense for long-period correlation and are of very little use for establishing close subdivisions.
3. Good zone fossils such as the graptolites are delicate and only preserved in quiet environments, being destroyed in more turbulent conditions.
4. Since fossil species could migrate following their own environment through time, there is always a possibility of diachronous faunas. The zone as defined in one area may not therefore be exactly time-equivalent to that in another region.

In the example of a graptolite, therefore, for the reasons outlined in (3) and (4), the total range or **biozone** of a species is not likely to be preserved in any one area, and it is therefore hard to draw ideal isochronous boundaries or time lines.

Ideally, zone fossils should have a particular combination of characters to make them fully suitable for biostratigraphy. These would be:

a wide horizontal distribution, preferably intercontinental;
a short vertical range so that they could be used to define a very precise part of the geological column;
enough morphological characters to enable them to be identified and distinguished easily;
strong, hard shells to enable them to be commonly preserved;
independence of facies, as would be expected from a free-swimming animal.

All of these conditions are seldom fulfilled in fossils used for zonation; perhaps the neritic ammonites come closest to it, and it is not surprising that the principles of really precise stratigraphical correlation were first worked out fully with these fossils, notably by the German palaeontologist A. Oppel in the 1850s.

It was Oppel too who first recognized that there are various ways of using fossils in stratigraphy which partially circumvent the difficulties mentioned, and hence different types of biozones. There are four main kinds of biozones generally used (Hedberg, 1976). **Assemblage zones** are beds or groups of beds with a natural assemblage of fossils. They may be based on all the fossils preserved therein or on only certain kinds. They are usually very much environmentally controlled and therefore of use only in local correlation. **Range zones** are perhaps of more general application. A range zone usually represents the total range of a particularly useful selected element in the fauna. One may therefore refer to the *Psiloceras planorbis* zone, based upon the eponymous ammonite that defines the lowest zone of the European Jurassic, above which is the *Schlotheimia angulata* zone. Each range zone is always named after a particular species which occurs within it. Where there are a number of zonally useful species, or where the ranges of individual species are long, a more precise time definition may be given by the use of overlapping stratigraphical ranges. Such zones are therefore called **concurrent range zones**. **Acme** or **peak zones** are useful locally. An acme zone is a body of strata in which the maximum abundance of a particular species is found, though not its total range. Such acme zones may be narrow but are often useful as marker horizons in geological mapping. Finally, an **interval zone** is an interval between two distinct biostratigraphical horizons. It may not have any distinctive fossils, or indeed any fossils at all, being simply a convenient way of referring to a group of strata bracketed between two named biostratigraphically defined zones.

Biostratigraphical units, unlike litho- and chronostratigraphical units, are not hierarchially arranged, apart from in the case of **subzones**, which are local divisions where a zone can be divided more finely in a particular region than elsewhere.

A different kind of stratigraphic concept, the **biomere** (Palmer, 1965, 1984) was defined as a regional biostratigraphic unit bounded by abrupt, non-evolutionary changes in the dominant elements of a single phylum. These changes are not necessarily related to physical discontinuities in the sedimentary record and they may be diachronous. The biomere concept has proved most useful in studies of late Cambrian trilobite faunas, where a repeated pattern of events is evident from the fossil record. In each biomere the shelf sediments contain an initial fauna of low diversity and short stratigraphical range (one or two species only). However, later faunas within the biomere become much more diversified and of longer stratigraphic range and suggest, by this stage, 'sound adaptive plans' and the zenith of the trilobite fauna. At the top of the biomere there is often a rather specialized fauna of short-lived trilobites, then all the groups become extinct abruptly.

The succeeding biomere begins in the same way as its predecessor, often with trilobites of similar appearance to those at the base of the first one. They may have migrated in from a stock of more slowly evolving trilobites in an outlying, possibly deeper-water area. The later development of the new biomere is as before: expansion and diversification followed by extinction. Several such biomeres have been defined in the intensively studied Upper Cambrian of North America. The pattern is invariably similar and could probably be discerned in other parts of the geological column as well.

Chronostratigraphy

Chronostratigraphy is more far reaching than either bio- or lithostratigraphy but has its roots in both of them. Its purpose is to organize the sequence of rocks on a global scale into chronostratigraphical units, so that all local as well as worldwide events can be related to a single standard scale. Hence it is concerned with the age of strata and their time relations. To do this a hierarchical classification of time-equivalent units must be employed. The conventional hierarchical system used is as shown in Table 1.1.

Table 1.1 Conventional hierarchical correlation between chronostratigraphical and geochronological units.

Chronostratigraphical units	Geochronological units
Eonothem	Aeon
Erathem	Era[a]
System[a]	Period[a]
Series[a]	Epoch[a]
Stage[a]	Age
Chronozone	Chron

[a] *These terms are in most common use.*

Chronostratigraphical units relate quite simply to geochronological units; thus the rocks of the Cambrian **System** were all deposited during the Cambrian **Period**. Most of these terms are self-explanatory, but it should be recognized that they are all, at least in theory, worldwide in extent.

The *Psiloceras planorbis* **chronozone** is a time unit equivalent to the time in which the said ammonite was in existence, even if it was confined to certain parts of the world only. It is hard indeed, however, to be able to delimit chronozones accurately, since most fossils were confined to certain geographical regions or provinces, as are most of the animals living today. There are relatively few well-established chronozones, or 'world instants' as they have been called, and so 'chronozone', though it has a real meaning, is not a term applicable to most practical stratigraphy. A **stage**, on the other hand, is a group of successive zones having great practical use, especially since it is normally the basic working time unit of chronostratigraphy, the narrowest that can actually be used on a regional scale.

It is usually at the stage level that rocks of widely different facies can be correlated. As an example there are some difficulties in making precise zonal correlations between Ordovician trilobite–brachiopod faunas and time-equivalent faunas with graptolites. Graptolites are rarely preserved in the siltstones and limestones favoured by the shelly fossils, and the latter, being benthic, could not inhabit the stagnant muds in which the graptolites were best preserved. In some areas, of course, the faunas do alternate in vertical sequence since the sites of deposition of these two facies fluctuated with oscillating shorelines, but though precise zone-to-zone correlations are possible at some levels it is found in practice that Ordovician graptolite zones correlate best with stages defined on shelly fossils.

Fossils give a relative chronology which can be used as the primary basis of chronostratigraphy. Nevertheless, it is often hard to correlate precisely beds of equivalent age in widely separated areas. The fossil sequences, though well documented with any one area, may contain very few elements in common, if indeed any at all, since they belong to different faunal provinces which are hard to correlate. Sometimes, however, the boundaries of such provinces may have oscillated to and fro. There may therefore appear elements of adjacent faunal provinces in vertical succession, thus facilitating stratigraphical correlation. At most stratigraphical horizons there are usually some ubiquitous worldwide fossils, so that intercontinental correlation is not impossible.

In chronostratigraphy the relative sequence given by the fossils is supplemented and enhanced by absolute dates which can be affixed at certain points wherever appropriate rocks occur. These are usually lavas bracketed between fossiliferous sediments, and their occurrence is not too common. It is most unlikely, therefore, that radiometric dating will supersede palaeontological correlation; the two are entirely complementary, and the great success of chronostratigraphy, in spite of its limitations, owes much to both. The use of automatic data-processing and retrieval systems is growing and may (Hughes, 1989) compensate for some of the constraints in present stratigraphical practice.

Bibliography

Books, treatises and symposia

Ager, D.V. (1963) *Principles of Palaeoecology*. McGraw-Hill, New York. (Useful basic text)

Barrington, E.J.W. (1967) *Invertebrate Structure and Function*. Nelson, London. (Invaluable zoology text)

Benton, M.J. (ed.) (1995) *The Fossil Record 2*. Chapman & Hall, London. (Invaluable summary of stratigraphical distribution of all known fossil genera)

Boardman, R.S., Cheetham, A.H. and Powell, A.J. (eds) (1987) *Fossil Invertebrates*. Blackwell, Oxford. (Exceptionally useful multi-author text)

Bosence, D.W.J. and Allison, P.A. (1995) (eds) *Marine palaeoenvironmental analysis from fossils*. Geological Society of London Special Publication No. 83. (12 useful papers)

Boucot, A. (1975) *Evolution and Extinction Rate Controls*. Elsevier, Amsterdam. (Advanced text, mainly dealing with brachiopod distribution)

Boucot, A.J. (1981) *Principles of Benthic Marine Palaeoecology*. Academic Press, New York. (Valuable if idiosyncratic)

Brenchley, P. and Harper, D.A.T. (1997) *Ecosystems, Ecology and Evolution*. Chapman & Hall, London. (New, vigorous and readable text)

Briggs, D.E.G. and Crowther, D. (1990) *Palaeobiology: a Synthesis*. Blackwell, Oxford, 583 pp. (The best and most useful palaeontological compendium of all)

Briggs, J.C. (1974) *Marine Zoogeography*. McGraw-Hill, New York. (Delimitation of faunal provinces)

Bruton, D.L. and Harper, D.A.T. (1990) Microcomputers in palaeontology. *Contributions of the Palaeontological Museum, Oslo* **370**, 1–105. (8 papers)

Cowen, R. 1994. *History of Life*, 2nd edn. Blackwell, Oxford. (Eminently readable)

Dodd, J.R. and Stanton, R.J. (1990) *Palaeoecology, Concepts and Applications*, 2nd edn. Wiley, New York, 502 pp. (Interesting approach)

Ekman, S. (1953) *Zoogeography of the Sea*. Sidgwick and Jackson, London. (An older but still useful text on marine life zones and faunal provinces)

Eldredge, N. and Cracraft, J. (1980) *Phylogenetic Patterns and the Evolutionary Process*. Columbia University Press, New York. (Explains cladistic methods)

Fairbridge, R.W. and Jablonski, D. (eds) (1979) *The Encyclopedia of Earth Sciences*, Vol. VII, *The Encyclopedia of Paleontology*. Dowden, Hutchinson and Ross, Stroudsberg, Penn. (Invaluable reference)

Goldring, R. (1991) *Fossils in the Field: Information Potential and analysis*. Longman, Harlow. (How palaeontological information is gathered in the field)

Gould, S.J. (1990) *Wonderful Life: the Burgess Shale and the nature of history*. Hutchinson Radius, London. (Comments on taxonomy)

Gray, J. and Boucot, A. (eds) (1979) *Historical Biogeography: Plate Tectonics and the Changing Environment.* Oregon State University Press, Corvallis, Ore. (Several papers on faunal provinces)

Hallam, A. (1973) *Atlas of Palaeobiogeography*. Elsevier, Amsterdam. (48 original papers on distribution of fossil faunas)

Hallam, A. (1996) *An Outline of Phanerozoic Biogeography*. Oxford University Press, Oxford.

Harland, W.B. (ed.) (1976) *The Fossil Record – a Symposium with Documentation*. Geological Society of London, London. (Contains time-range diagrams of all fossil orders and charts showing diversity fluctuations through time)

Harland, W.B. and Armstrong, R.L. (1990) *A Geological Time Scale*. Cambridge University Press, Cambridge, 253 pp. (Essential and up-to-date)

Hedberg, H.D. (1976) *International Stratigraphic Guide*. Wiley, New York. (Official guide to stratigraphical procedure)

Hedgpeth, J.W. and Ladd, J.S. (1957) *Treatise on Marine Ecology and Palaeoecology,* Vols 1 and 2, Memoirs of the Geological Society of America No. 67. Geological Society of America, Lawrence, Kan. (Standard text, with papers and annotated bibliography on ecology of all living and fossil phyla)

Hennig, W. (1966) *Phylogenetic Systematics*. University of Illinois Press, Urbana, Ill. (Original cladism; second edition published in 1979)

Hughes, N.F. (ed.) (1975) *Organisms and Continents Through Time*. Special Papers in Palaeontology No. 17. Academic Press, London. (23 original papers on distribution of faunas; many charted on global maps)

Hughes, N.F. (1989) *Fossils as Information: New Recording and Stratal Correlation Techniques*. Cambridge University Press, Cambridge, 144pp.

Joysey, K.A. and Friday, A.E. (1982) *Problems of Phylogenetic Reconstruction*, Systematics Association Special Volume No. 21, Academic Press, New York. (11 papers)

Laporte, L. (1968) *Ancient Environments*. Prentice-Hall, Englewood Cliffs, N.J.

McKerrow, W.S. (1978) *The Ecology of Fossils*. Duckworth, London. (Community reconstructions)

McKerrow, S. and Scotese, C.R. (eds) (1990) Palaeozoic palaeogeography and biogeography. Geological Society Memoir No. 12, pp. 1–240. (Many valuable papers including Palaeozoic world maps)

Middlemiss, F.A., Rawson, P.F. and Newall, G. (eds) (1971) *Faunal Provinces in Space and Time*. Seel House Press, Liverpool. (13 original papers on faunal distribution)

Moore, R.C. Teichert, C., Robison R.A. and Kaesler, R.C. (successive editors from 1953). *Treatise on Invertebrate Paleontology*. Geological Society of American and the University Kansas Press, Lawrence, Kan. (The standard reference work on invertebrate fossils – each phylum treated in separate volumes).

Murray, J.W. (ed.) (1986) *Atlas of Invertebrate Macrofossils*. Longman, Harlow, for the Palaeontological Association. (Excellent photographic coverage of main groups)

Paul, C.R.C. (1980) *The Natural History of Fossils*. Weidenfeld and Nicolson, London. (Simple but interesting)

Piveteau, J. (ed.) (1952–1966) *Traité de paléontologie*. Masson, Paris. (The French 'Treatise', slightly older than the American *Treatise on Invertebrate Paleontology*, but of very high quality)

Raup, D.M. and Stanley, S.M. (1978) *Principles of Paleontology,* 2nd edn. Freeman, San Francisco. (Excellent textbook emphasizing approaches and concepts, but not morphological or stratigraphical details)

Ridley, M. (1985) *Evolution and Classification. The Reformation of Cladism*. Longman, Harlow, 201 pp. (Clear treatment of different approaches to classification)

Ryan, P., Harper, D.A.T. and Whalley, J.S. (1995) *PALSTAT: User's manual and case histories. Statistics for palaeontologists*. 71 pp. and disc. Chapman & Hall, London and Palaeontological Association.

Schäfer, W. (1972) *Ecology and Palaeoecology of Marine Environments* (translated edn) (ed. G.Y. Craig). Oliver and Boyd, Edinburgh. (Standard work on Recent North Sea environments with applications for palaeoecology)

Schopf, T.J.M. (ed.) (1972) *Models in Paleobiology*. Freeman, Cooper, San Francisco. (10 original papers, many significant)

Scott, R.H. and West, R.R. (eds) (1976) *Structure and Classification of Paleocommunities*. Dowden, Hutchinson and Ross, Stroudsberg, Penn. (11 original papers)

Skinner, B.J. (ed.) (1981) *Paleontology and paleoenvironments: Readings from American Scientist*. Kaufmann, Los Altos, Calif. (21 collected papers)

Smith, A.B. (1994) *Systematics and the fossil record. Documenting evolutionary patterns*. Blackwell, Oxford (Lucid explanation of cladistic methodology)

Sylvester-Bradley, P.C. (ed.) (1956) *The Species Concept in Palaeontology*. Systematics Association, London. (Basic work of several papers on palaeontological taxonomy)

Teichert, C. (1975) *Treatise on Invertebrate Paleontology*, Part W (Suppl. 1), *Trace Fossils and Problematica*. Geological Society of America, Lawrence, Kan.

Thompson, d'Arcy W. (1917) *On Growth and Form*. Cambridge University Press, Cambridge. (Individualistic classic work on physical laws determining growth; an abridged edition of this book was edited by J.T. Bonner in 1961 and also published by Cambridge University Press)

Thorson, G. (1971) *Life in the Sea*. World University Library, London. (Simple, well illustrated text; deals with community structure)

Valentine, J.W. (1973) *Evolutionary Palaeoecology of the Marine Biosphere*. Prentice-Hall, Englewood Cliffs, N.J. (Valuable text on evolution and ecology)

Willmer, P. (1990) *Invertebrate Relationships*. Cambridge University Press, Cambridge, 400 pp. (Modern, very readable and up-to-date treatment)

Individual papers and other references

Craig, G.Y. (1954) The palaeoecology of the Top Hosie Shale (Lower Carboniferous) at a locality near Kilsyth. *Quarterly Journal of the Geological Society of London* **110,** 103–19. (Classic palaeoecological study)

Fortey, R.A. and Jefferies, R. (1981) Phylogeny and systematics – a compromise approach, in *Problems of Phylogenetic Reconstruction* (eds K.A. Joysey and A.E. Friday). Academic Press, New York, pp. 112–47 (Where cladistics is and is not useful)

Fürsich, F.T. (1977) Corallian (Upper Jurassic) marine benthic associations from England and Normandy. *Palaeontology* **20**, 337–85. (Palaeoecology, mainly of bivalve communities)

Gingerich, P.D. 1990. Stratophenetics, in *Palaeobiology; a Synthesis* (eds D.E.G. Briggs and P.R. Crowther). Blackwell, Oxford, 437–42.

Harper, D.A.T. (1989) *Brachiopods from the Upper Ardmillian succession (Ordovician) of the Girvan District, Scotland*, Part 2, 79–128. Palaeontographical Society Monographs (taxonomic procedure referred to in text)

Harper, D.A.T. and Ryan, P. (1990) Towards a statistical system for palaeontologists. *Journal of the Geological Society of London* **147**, 935–48. (Essential reading)

Henry, J.-L. (1984) Analyse cladistique et Trilobites; un point de vue. *Lethaia* **17**, 61–6. (Use and limitations of cladism)

Holland, C.H., Audley-Charles, M.G., Bassett, M.G. *et al.* (1978) A guide to stratigraphic procedure. Geological Society of London Special Report No. 10, pp. 1–43. (Invaluable modern treatment)

Jacob, F. (1977) Evolution and tinkering. *Science* **196**, 1161–7.

Kauffmann, E.G. and Scott, R.W. (1976) Basic concepts of community ecology and palaeoecology, in *Structure and Classification of Palaeocommunities* (eds R.W. Scott and R.R. West). Dowden, Hutchinson and Ross, Stroudsberg, Penn., pp. 1–28.

Lockley, M. (1983) A review of brachiopod dominated palaeocommunities from the type Ordovician. *Palaeontology* **26**, 111–45. (Terminology for palaeocommunity structure)

Moore, J, and Willmer, P. (1997). Convergent evolution in invertebrates. *Biological Reviews* **72**, 1–60.

Newell, N.D. (1972) The evolution of reefs. *Scientific American* **226,** 54–65. (Informative short paper)

Palmer, A.R. (1965) Biomere – a new kind of stratigraphic unit. *Journal of Palaeontology* **39**, 149–53.

Palmer, A.R. (1984) The biomere problem: evolution of an idea. *Journal of Paleontology* **39,** 149–53.

Petersen, C.C.J. (1918) The sea bottom and its production of fishfood. Reports of the Danish Biological Station **25**, 1–62. (The first definitive summary of benthic communities)

Pickerill, R.K. and Brenchley, P. (1975) The application of the community concept in palaeontology. *Maritime Sediments* **11,** 5–8. (Terminology and application)

Rudwick, M.J.S. (1961) The feeding mechanism of the Permian brachiopod *Prorichthofenia*. *Palaeontology* **3**, 450–71. (paradigm approach)

Smith, D.B. (1981) The Magnesian Limestone (Upper Permian) reef complexes of north-east England, in *European Reef Models* (ed. D.F. Toomey), Society of Economic Paleontologists and Mineralogists Special Publication No. 30, Tulsa, Oklahoma, SEPM, pp. 161–8.

Thomas, R.D.K. and Reif, W.-E. (1993) The skeleton space: a finite set of organic designs. *Evolution* **47,** 341–60.

Thorson, G. (1957) Bottom communities, in *Treatise on Marine Ecology and Palaeoecology*, Vol. 1 (ed. J.W. Hedgpeth), Memoirs of the Geological Society of America No. 67, Geological Society of America, Lawrence, Kan., pp. 461–534. (Standard work, invaluable, well illustrated)

Ziegler, A.M., Cocks, L.R.M. and Bambach, R.K. (1968) The composition and structure of Lower Silurian marine communities. *Lethaia* **1,** 1–27. (The first major work on Lower Palaeozoic communities; very well illustrated)

2 Evolution and the fossil record

2.1 Introduction

Amongst all the sciences concerned with organic evolution it is only palaeontology that has the unique perspective of geological time. It is the rock record alone that provides an historical perspective for the study of evolutionary events, and this time dimension could never have come from any other source. Accordingly, the input of palaeontology to evolution theory has been in understanding the history of life, in interpreting patterns of evolution (e.g. adaptive radiations) and lines of descent, and, importantly, in assessing rates of evolution. Now it has to be admitted that the fossil record is incomplete, and to interpret it can in some instances be 'like trying to read a diary with half the pages missing' (Sheldon, 1988). Yet it still remains an immensely rich source of primary data, and can give, if resolved finely enough, a fair picture of evolutionary events that actually took place, however long ago.

The student of palaeontology needs to have some background in biological evolutionary theory, otherwise he will be in the position of 'the man standing by the roadside and watching the cars go past but without any idea of how their engines work', to use Brouwer's (1973) graphic simile. For the fossil record cannot give much information on the mechanism of evolutionary change; this has to be provided by genetics, cytology, molecular biology and population dynamics, building upon the original conceptions of Charles Darwin. Many years ago the first real multidisciplinary amalgam of data was published as *Evolution: the Modern Synthesis* (Huxley, 1942). From this highly successful attempt at welding together information from various sources, the neo-Darwinian synthesis takes its name – and this is a useful starting point. But recent developments in molecular biology are transforming and adding to this synthesis, and new views on the nature of the gene seem to be changing our whole conception of how organisms evolve.

In the following text, therefore, I present firstly a simplified account of classic neo-Darwinian evolutionary theory, aimed particularly at students of Earth science who may only have a limited background in biology. This includes some information on the impact of new information from molecular genetics on evolution theory. This section is not intended to be comprehensive, nor does it pretend to be a guide to all recent developments and discoveries. Fuller treatments are readily available elsewhere viz., Simpson (1953), Maynard Smith (1975, 1982), Mayr (1963, 1976), Dawkins (1986), Dobzhansky *et al.* (1977), Gamlin and Vines (1987), Bonner (1988), Endler and McLellan (1988), Hoffmann (1989), Campbell and Schopf (1994), Maynard Smith and Szathmary (1995), Futuyma (1996), Strickberger (1996), Ridley (1996) and others listed in the bibliography, while Valentine (1973), Hallam (1977), Stanley (1979), Cope and Skelton (1985), Levinton (1988) and Skelton (1993) are more directly concerned with evolution and the fossil record. In the second part of this chapter there is a more extended account of what the fossil record can tell us about the nature of evolutionary change.

2.2 Darwin, the species and natural selection

The theory of evolution links together a multiplicity of biological phenomena and is underpinned by all the evidence of the geological record. It remains a theory, not a proven fact, for the immense timescale over which evolutionary changes have taken place does not permit their direct observation. There is, however, no other theory which encompasses so much and accords with the evidence of comparative anatomy, biochemistry and physiology, of genetics and cytology, and of the relationships between organisms perceived by taxonomy. Whereas the evidence in some of these fields is circumstantial, when brought together and interpreted it builds up to a theory of formidable consequence, and as Dobzhansky has said, 'Nothing in biology makes sense except in the light of evolution.'

The recent facile attempts to discredit evolution by self-styled 'creation scientists' have been so eloquently dispatched by Dott (1982), Kitcher (1982) and Stanley (1982) that no further comment is needed here.

Although Charles Darwin is generally regarded as the father of evolution theory (Darwin, 1859), there were many pre-Darwinian scientists who postulated that animals and plants had changed over long periods of time and that new types of 'species' had arisen from pre-existing ones. These early workers and Darwin himself identified many different points which could be regarded as evidence for the origin of modern species from pre-existing and more primitive ancestors. All these are accepted today, though refined and added to. Taxonomy, as always, lies at the heart of evolutionary thought, and closely intertwined with it is comparative anatomy. The five-fingered, or pentadactyl, limb of higher vertebrates, for example, has been modified in a variety of ways. The grasping hand of primates, the flippers of marine mammals, the wings of birds and bats; the hoofs of horses – they all look dissimilar, but are all variants on a common theme. Likewise the diversity in structure and function of the beaks of the Galapagos finches (*Geospiza*), which Darwin encountered during the voyage of the *Beagle* in 1833, led him to say 'Seeing this gradation and diversity in structure in one small, intimately related group of birds, one might really fancy that from an original paucity of birds in this archipelago, one species had been taken and modified for different ends'. Geographical distribution was seen as important to evolutionary thought in other ways too. Thus the existence of 'relict' and isolated species (e.g. lungfish) in different parts of the world surely indicated an original widespread ancestral type, a subsequent population collapse, and restriction of a few species to small areas only where each had become adapted to its own environment.

Darwin was particularly interested in selective breeding of animals and plants under domestication. He realized that the present great variety of dogs and cattle had been produced in only a few thousand years from only one or at most a few ancestral types. He concluded that a great potential for 'descent with modification' must exist in all animals and plants which could be speeded up by such artificial breeding. This led on to the belief that similar processes, though probably on a slower time scale, had operated in the wild state; in other words 'natural selection'. Thus breeding experiments are the foundation of classical genetics, where the mechanism of heredity is understood in terms of its effects.

Darwin was also concerned with palaeontology but he found that the fossil record was somewhat of a disappointment in supporting the case for evolutionary change. He had hoped to find evidence of gradational change between animal species, of 'infinitely numerous transitional links' connecting ancestors to descendants and of stratigraphically arranged series showing 'descent with modification'. In fact, he did not find what he had expected. Darwin assumed that the imperfections of the rock record and the limitations of knowledge at that time were the factors responsible. He was indeed partially correct, but to this point we shall return later.

Of the pre-Darwinian evolutionists the most prominent was the French naturalist Jean-Baptiste Lamarck (1774–1829), who proposed long before Darwin that all living organisms had originated from primitive ancestors and that in the slow process of such changes had become adapted for living in particular environments. The concept of such **adaptation** originated with Lamarck. He appreciated that in order to live, animals have to be efficiently adapted to all the physical and biological parameters of the environments they inhabit. In a sense, of course, an animal is 'all adaptation'; it has to be anatomically, physiologically and trophically adapted to its environment, and it is critical to evolution theory to understand how such adaptations came to be.

While appreciating the importance of adaptation, however, Lamarck linked his insight with some concepts no longer believed to be tenable. He believed that adaptation had come about through some kind of internal driving force, a 'vital spark' which made animals become more complex. He felt that new organs must arise from new needs and that these 'acquired characters' were inherited, as in his classic postulate that the neck of the giraffe had become longer in response to a 'need' to reach leaves up on the tree. The theory of inheritance of acquired characters is not highly regarded nowadays and is generally untestable (although there is some evidence of a kind of genetic feedback from the environment operating to produce apparent 'Lamarckian changes'). Darwin, on the other hand, provided a logical and testable theory for evolutionary change: one that has stood the test of time and provided a starting point for later developments.

The full title of Darwin's major work of 1859 was *On the origin of species by means of natural selection, or the preservation of favoured races in the struggle for life.* The main points of the theory are straightforward.

1. Animal species reproduce more rapidly than is needed to maintain their numbers. Animal populations, however, though fluctuating, tend to remain stable. (Here he was influenced by the Englishman Malthus, who had written on this subject some years earlier.)
2. There must therefore be competition within and between species in the 'struggle for existence', for food, for living space and (within members of the same species) for mates, if the characters that individuals bear are to be transmitted to the next generation.
3. Within species all animals vary, and this variation is inherited.
4. In the struggle for life those individuals best fitted to survive in a particular environment are the ones to live and to reproduce. The others are weeded out in the intense competition. The favourable characteristics that make such survival possible are inherited by future generations, and the accumulation of different favourable characters leads to the separation of species well adapted to particular environments. This is what Darwin called 'natural selection'.

All this seems logical enough, though Darwin's early critics, Mivart for instance, argued that Darwin had not really shown how favourable characters were actually accumulated, only how those animals less fitted to their environment failed to survive. In this they were not unsound, for the most serious weakness of the theory, as presented by Darwin, was that the nature of variation and heredity was largely unknown, so that his views on this were speculative and insufficient for the theory to be seen to work.

The pioneering work of the Austrian monk Gregor Mendel in 1865, and of the later school of T.H. Morgan which began in 1910, laid the foundation for genetic experiment and theory. It was this that supplied the necessary understanding of heredity essential to the amplification of Darwin's theory.

Inheritance and the source of variation

In the cells of all eukaryotes (all organisms except for viruses, bacteria and cyanobacteria) there is a **nucleus** (Fig. 2.1) containing elongated thread-like bodies, the **chromosomes**, which are made of protein and deoxyribonucleic acid (DNA).

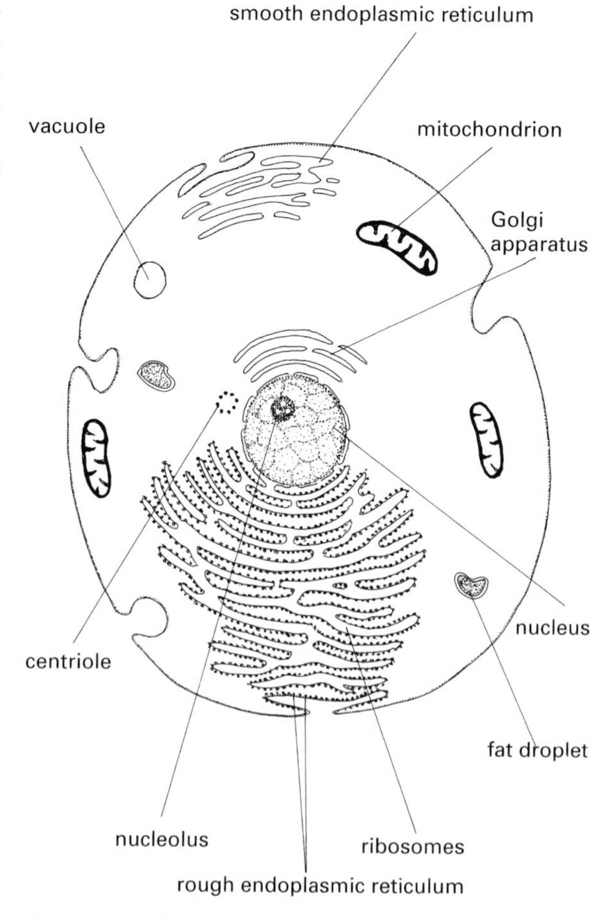

Figure 2.1 Eukaryote cell structure, with organelles.

The DNA molecule forms a long, twisted, spiral ladder of which the 'uprights' consist of alternate blocks of phosphoric acid and the pentose sugar deoxyribose, whereas the 'rungs' are matching pairs of relatively small units, the nucleotide bases. These bases, in DNA, are adenine, guanine, thymine and cytosine. They are attached to the pentose sugar units and project inwards, linking up together in pairs. Adenine can only pair with thymine, and cytosine with guanine.

In the chromosome the DNA strands are arranged in discrete units, the **genes**, which are strung together along the length of the chromosome, and of which there may be several hundred per chromosome. The **genome** is a term used to describe the whole genetic complex. These genes are the primary units of heredity, carrying the **genetic code**, which is involved in producing proteins and in directing the development and functioning of the whole organism. All the necessary information for these ends is carried in the DNA, and what is important is the sequence of the four nucleotide bases along its length. Though these form a kind of alphabet consisting only of four letters, the number of possible combinations in which they can exist, coding for specific proteins, is enormous. This sequence of nucleotide bases determines the sequence in which any of the 20 amino acids found in living organisms are strung together to make proteins. It has been shown that combinations of three bases acting together code for particular amino acids, and there are more than enough possible combinations in this 'base-triplet' system to make all the biogenic amino acids, and to link them up in the right order to make a specific protein.

Proteins are synthesized, not in the nucleus, but at the ribosomes in the cellular cytoplasm, and this means that all the information has to be transferred out of the nucleus to these sites of protein synthesis. For this to take place the nuclear DNA first produces a single-stranded copy of itself, nuclear RNA, but with uracil replacing thymine and ribose sugar replacing deoxyribose. Following this process of '**transcription**', a further molecule (messenger RNA) is formed, which moves out of the nucleus, presumably through pores in the nuclear membrane, and attaches itself to the ribosome. (This is not a simple process, however, for at this stage the genes themselves are modified. It has been shown recently that genes themselves are made of two kinds of components arranged in series. These are **exons**, which code for proteins, and **introns**, which do not. When the nuclear DNA is transcribed to nuclear RNA it retains the organization of the original DNA, but when nuclear RNA is reprocessed to messenger RNA the introns are lost and the exons are spliced together. This does not always take place in the original order, however, and such exon shuffling may be the basis of rapid evolution of proteins themselves in new combinations. Provided that these are functional, new gene systems may arise through comparatively few such shuffling events, and in short periods of time. This is a newly understood phenomenon, but may have important consequences for evolution theory.)

When messenger RNA arrives at the surface of a ribosome it does not form protein directly, but through yet another intermediate molecule, transfer RNA, and when this complex process is complete, the result is a protein sequence, coded for by the nuclear DNA, the transfer RNA meanwhile returning to the cytoplasm.

The understanding of protein synthesis has been a major triumph for molecular biology, but the molecular pathways that lead from genes to actual organs and characters are very complex, and at present largely unknown. How the proteins and other compounds which are produced are organized and built up into functional organs and whole bodies remains one of the main tasks for molecular biology for the future. Some progress has already been made; it is now known that some genes are 'structural', and concerned only with the synthesis of materials, others are 'regulator' genes, which control and organize the compounds and direct their building. Such genes release chemical products which start a whole host of complex reactions. In some kinds of development the genes are switched on and off in particular sequence, releasing products which react together in synthesizing complex molecules. Structural genes can be activated and deactivated when needed; evidently the initial stimulus for the switching on of structural genes is given when a sensor gene receives an appropriate stimulus.

Simple organisms such as bacteria and fungi have sets of adjacent genes known as **operons**, coding for a particular metabolic pathway. These can be switched off and on together. In higher organisms, however, genes which control different parts of a coordinated programme may be scattered around in small groups, but on different chromosomes.

Genetically identical organisms reared in different environments will not develop to exactly the same form, as is witnessed by the different appearance of vegetables grown in rich and poor soils, respectively. The inherited genetic material in any organism, known as the **genotype**, is reacted upon by the environment (probably through sensor genes) to create a developed individual. This individual, the product of both heredity and environment, is the **phenotype**.

Sometimes single genes control single characters. More often characters are **polygenic**; that is, many genes, each of small effect, contribute to the biochemical pathways that result in the formation of a particular character. Again, some genes are **pleiotropic**; that is, they affect several characters since the same gene products may be used in different biochemical pathways. A phenomenon which at first sight seems puzzling is that much of the genome consists of repeated DNA segments (multigene families) with anything from two to several thousand copies of the same gene. Though it may appear that much chromosomal material is redundant, there could well be important evolutionary implications here, as is discussed later.

Where does variation come from?

Within the nucleus of any cell in the body of most living organisms there is a specific number of chromosomes with their genes. The chromosome number is always constant for the species. In humans there are 23 pairs of chromosomes, while in the fruit fly *Drosophila* there are four pairs. The members of each pair are all **homologous**, i.e. similar in appearance and length, except for one pair, the sex chromosomes, upon which the genes regulating sexual characters are located. [The sex of an individual depends upon this pair of sex chromosomes. In one sex the chromosomes are identical (XX) in the other sex they are dissimilar (XY). The sex chromosomes carry other genes than those specifically responsible for sexual characters and thus certain physical characters are sex-linked. In mammals the female is XX and the male XY, but in birds it is the other way round].

When chromosomes are paired like this the organization is said to be **diploid**, the chromosome number being conventionally defined as $2n$.

When the body (**somatic**) cells of an animal divide as the organism grows, the chromosomes divide by longitudinal fission, and each '**daughter' cell** inherits an exact copy of the chromosomes and genes in the **parent cell**. This process, known as **mitosis** (Fig. 2.2), is effectively the same for all the somatic cells in any one body.

When a cell is not dividing, the chromosomes are largely invisible; this is known as interphase. As cell division begins, biological staining reveals the chromosomes, each already divided into two chromatids held together by a small spherical centromere, which does not stain. The first stage of cell division is prophase, where two centrioles move to opposite poles and a 'spindle' of protein threads appears between them. At metaphase the nuclear membrane vanishes and the chromosomes line up along the median plane, each splitting in two so that each chromatid becomes a daughter chromosome with its own centromere. Then, during anaphase they pull apart so that each bundle of daughter chromosomes moves to the opposite pole. Finally, at telophase, a new nuclear membrane appears round each set of chromosomes, the spindle vanishes and the cell divides into two identical daughter cells.

On the other hand, the formation of eggs or sperm (**gametes**) in the testes and ovaries respectively involves a very different process: **meiosis** (Fig. 2.3).

There are two cell divisions. The first of these is a reduction division in which each daughter cell inherits only one chromosome from an homologous pair. Here again there are four phases but some of these are significantly different from their mitotic counterparts. During prophase of the first meiotic division, when the chromosomes pair up, they divide longitudinally, forming groups of four chromatids (tetrads) but with only one centromere for each pair of chromatids. These four chromatids are often tangled at certain points known as **chiasmata**, and as they pull apart they break and re-form, each exchanging parts of the same length with its homologous chromosome. The result of such '**crossing over**' is that the unpaired chromosomes passed down to the daughter cell are not identical with those of the parent cell, as they are composed of bits of each of the homologous paired chromosomes. The second meiotic division is like a somatic cell mitosis, so that the end product of the two divisions

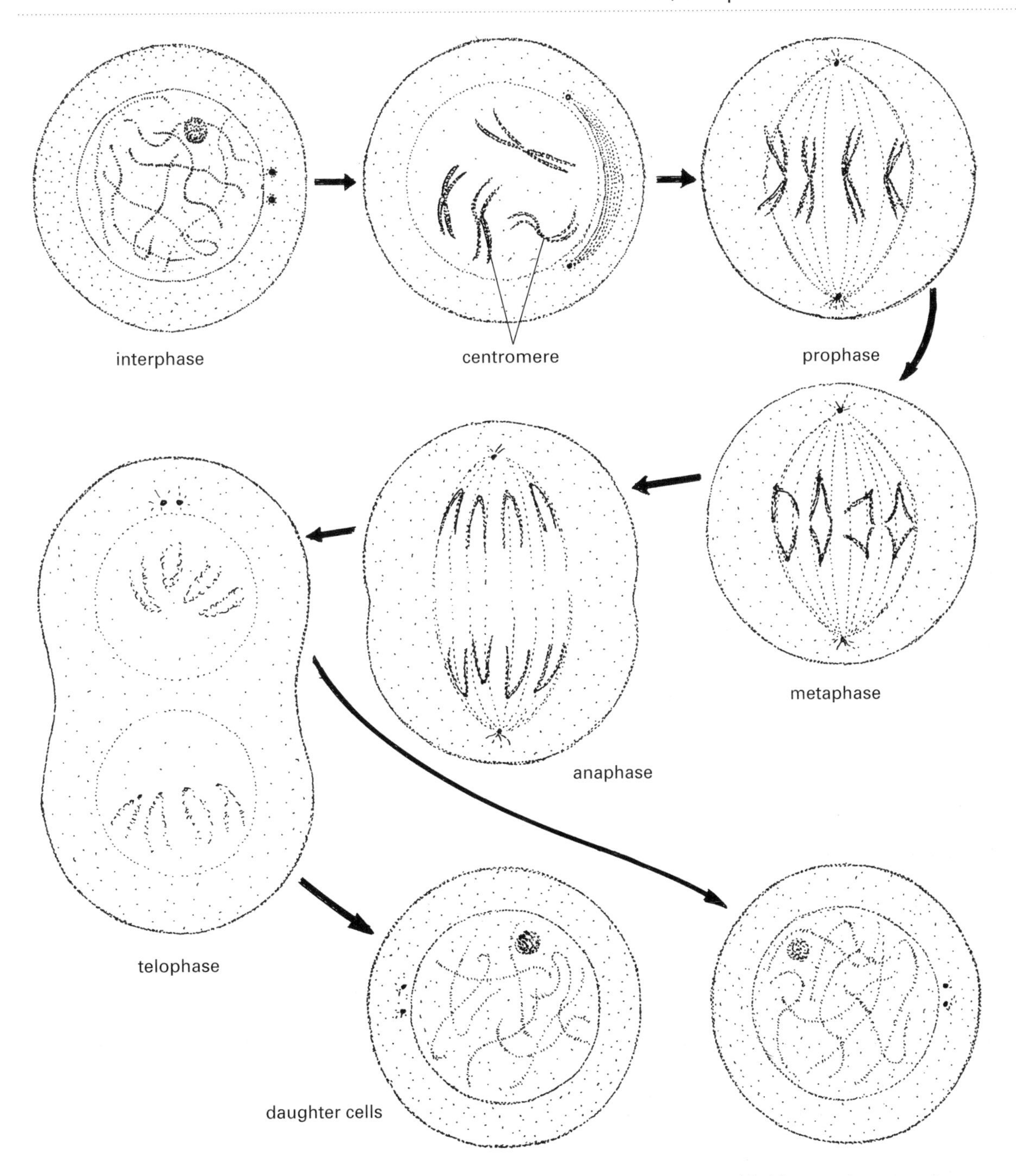

Figure 2.2 The stages of mitotic cell division in a eukaryotic cell. For explanation see text. (Modified from Kershaw, 1983.)

is four gametes from a single parent cell. Each gamete has half the number of parent chromosomes; the chromosomes are unpaired, and none are identical to any of the parent chromosomes since genetic material was exchanged during crossing over.

These gametes containing unpaired chromosomes are said to be **haploid**, and the chromosome number is *n*.

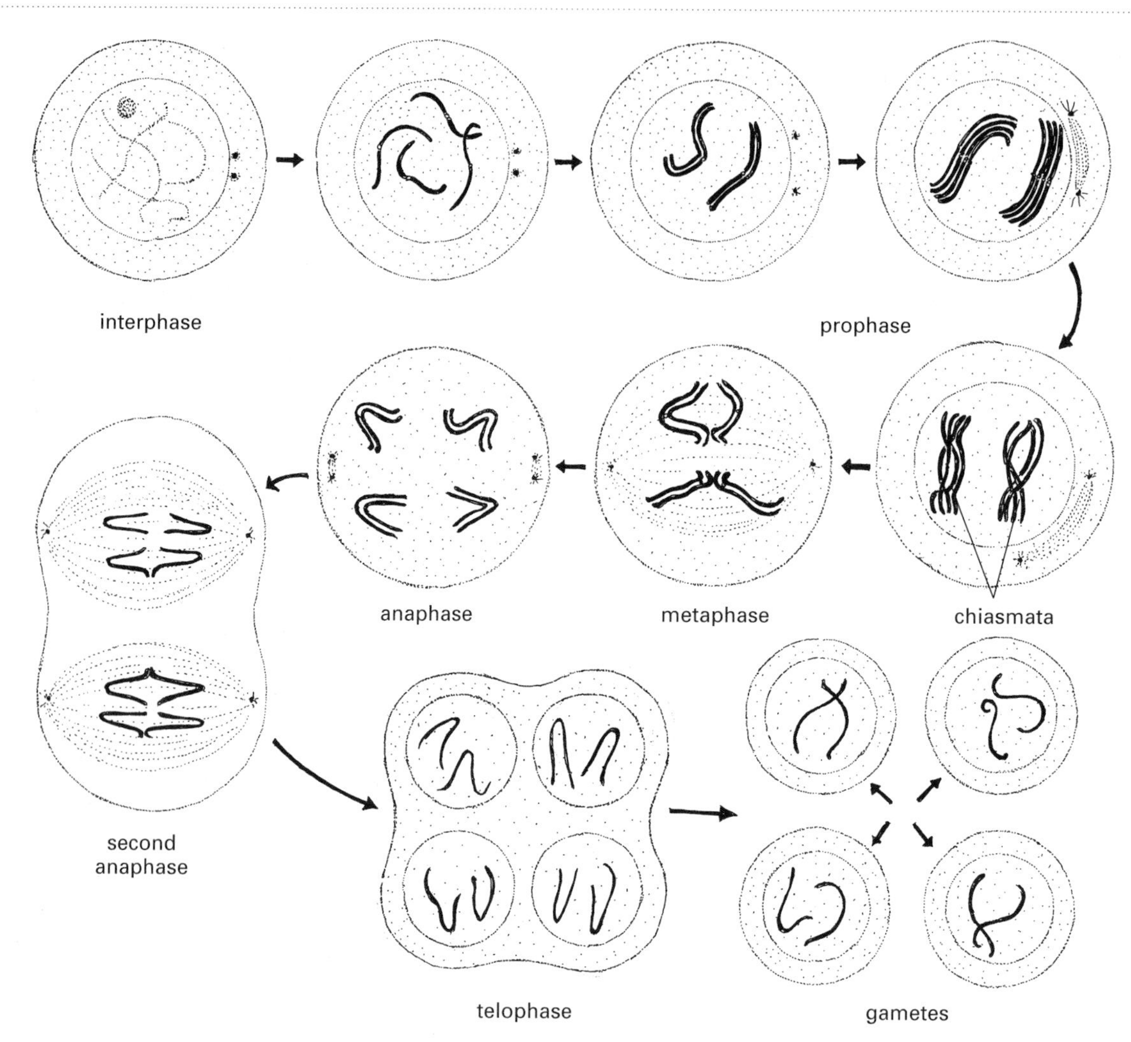

Figure 2.3 The stages of meiotic cell division in a eukaryotic cell. For explanation see text. (Modified from Kershaw, 1983.)

When eggs are fertilized by sperm the homologous chromosomes from different parents come together and pair up. In this process of **recombination** a new genotype results from already existing genes, deriving from different sources. The effect of both crossing over and recombination is that variation within the products of reproduction is high. It is mainly upon this variation that natural selection operates.

Significance of alleles

Homologous chromosomes lie side by side, matching gene by gene. The homologous genes, however, are not necessarily identical, for they may exist in a number of 'expressions', known as **alleles**, each of which is slightly different chemically and will lead to the development of differences in the characters it controls. Such differences may range from the minor to the very substantial.

Sometimes, as in the genes controlling human eye colour, there may be only two alleles, designated *R* and *r* in conventional notation, *R* controlling brown eye colour and *r* controlling blue eye colour. Alleles for eye colour may exist on homologous chromosomes in four possible combinations: *RR, Rr, rR* or *rr*. In the **homozygous** condition alleles on homologous chromosomes are of the same kind: *RR* or *rr*. In the **heterozygous** condition they are of different kinds: *Rr* and *rR*. An individual possessing *RR* will have brown eyes, whereas the possessor of *rr* will be blue eyed. But the combinations of *Rr* or *rR* will also lead to brown eyes, for the allele *R* is **dominant** over *r* and masks its effect. In this case *r* is said to be the **recessive** gene.

Sometimes alleles are multiple; some genes are known to exist in up to 11 alleles. Human blood types, for example, are controlled by only a single gene which occurs as three alleles designated *p*, *q* and *r*. Alleles *p* and *q* have equal dominance whereas *r* is recessive. Of the four human blood groups O has the recessive homozygote *rr*; group A may carry *pp* or *pr* and group B *qq* or *qr*, while the group AB has the dominant heterozygote *pq*.

The above examples deal with genes affecting single characters, but there are more complex patterns of variation when two or more genes control single characters. Thus plumage colours in parakeets are under the control of two genes which in different combinations may produce blue, green, yellow or white feathers – a wider combination of colours than would be possible with only a single pair of alleles.

These sources of variation – crossing over, recombination, homo- and heterozygosity, multiple alleles, etc. – are all consequences of diploidy. In addition, many plants increase their potential variability by **polyploidy**, i.e. duplication of chromosomes, allowing a greater number of loci for alleles. Polyploidy, however, is rarely found in animals.

Recessive genes might be expected to get lost during evolution, but in fact the proportions of alleles within populations do not normally change, unless there is a special survival advantage in possessing one particular type of allele. If there is no such advantage a genetic equilibrium is maintained in terms of the Hardy–Weinberg equation:

$$p^2 + 2pq + q^2 = \text{equilibrium}$$

where p^2 and q^2 represent the relative proportions of dominant homozygotes (*RR*) and recessive heterozygotes (*rr*), respectively. The equilibrium so defined will only be shifted when there is a decided selective advantage in possessing a particular allele. Thus if a particular phenotype is preserved by natural selection, the balance will alter and the proportion of the allele within the population will increase.

Evolution can thus be considered as based upon changes in gene frequencies through natural selection.

Mutation

Natural selection operates upon the great stock of variation inherent in the gene pool of a population. The genes are continually reshuffled by the mechanism discussed above in new combinations or genotypes. But if evolving organisms are ever to give rise to anything radically new there must be a source of new variation. Where does this variation come from?

It sometimes happens that a gene will suddenly and spontaneously change to a new allele. This kind of random change is known as a **gene mutation**. It is the result of a chemical change in the DNA helix: a substitution or rearrangement of nucleotide bases, or a duplication or loss of a small part of the sequence. The simplest kinds of mutation arise by one nucleotide base, or a pair of bases, being replaced by another base or base pair. This is substitution, but mutations can also arise by addition or deletion of base pairs and this may disrupt the function of the gene. Mutations on a larger scale can affect from tens to thousands of base pairs, and **chromosomal mutations**, which are the most extreme kind, may involve inversion, rearrangement, addition or deletion of long sequences of DNA. These tend to have larger-scale phenotype effects and are more often disadvantageous. Although it was formerly held that mutations are comparatively rare events, it is now apparent (Drake *et al.*, 1983) that spontaneous 'lesions' are continually occurring in chromosomes during replication; there may be many per cell each day. But such mispairings are continually being repaired by a complex enzyme system (DNA polymerase acting in conjunction with other enzymes). This acts as a 'proofreader' which picks up wrongly paired bases and puts them back in the right order. Even this proofreading system sometimes slips up, and mutations then are fixed in the chromosome, to have their effects upon the descendant.

While mutations have been traditionally understood as being due to DNA damage or misprocessing, it has recently been shown that many spontaneous mutations, especially those on a larger scale, may be due to mobile genetic elements (variously named '**jumping genes**', '**transposons**', or even '**molecular parasites**'), which are DNA sequences able to move along chromosomes or from one chromosome to another. They can alter the whole genetic structure of an organism through rearrangement or expansion of its genetic material. Some tend to attach themselves to particular sites; others seem able to fit in almost anywhere. The capacity for movement of these mobile or quasi-stable genetic elements allows a radical and rapid alteration of the pattern of gene expression by a single genetic event – a phenomenon which has clear implications for rapid evolution and for the origin of new types of organization.

A new allele, formed by mutation, is heritable and will produce changes in one or more characters in the organism. These changes are usually of small scale and of limited effect; if they are larger they are very often lethal. Most mutations are neutral, some are disadvantageous, but others may lead to new or somewhat different characters which may carry some selective advantage. An advantageous mutation, which may only have a marginally beneficial effect, will only be able to spread if (1) it is recurrent within the population so that new recruits are continually added to the mutant stock, (2) it carries some selective advantage so that the Hardy–Weinberg equilibrium is shifted in its favour and (3) the population is small enough for it to be carried and spread relatively quickly.

All alleles have resulted from mutation; multiple alleles have arisen at different times. Some mutant alleles may spontaneously mutate back again to the original gene expression from which they arose. The rates of mutation vary with the gene. Some are much higher than others, but they are generally low, one mutation in 10 000 for a particular gene being considered as high. The rate of mutation can be increased by radiation, including exposure to ultraviolet light, or by chemical means.

Spread of mutations through populations

Mutation is the primary source of variation, but it is only through the effects of diploidy in higher organisms that the mutations can spread. As mentioned, it is not likely that mutations will spread very fast in a large population, even if they are dominant and mutation rates are high. But most natural populations are not evenly distributed throughout their species range. They tend to be subdivided into relatively small units of population, known as **gamodemes**. Examples of such natural population groupings are rabbit warrens, schools of fishes, pond communities, the faunas of enclosed lagoons and, in the human sphere, the inhabitants of towns and villages. They tend to be partially isolated from other such gamodemes, but not completely so, for though each gamodeme carries its own gene pool there is some possibility of exchange and hence gene flow between the pools.

It is within such small natural populations that favourable mutations, especially those allowing better adaptations, are able to spread. A mutant condition established even in one family may, if it possesses selective advantage, soon spread through the population in the gamodeme within relatively few generations. It may then spread to other gamodemes in the vicinity, increasing the effective range of the useful mutant condition. Such a mutant condition, especially if occurring in a **peripheral isolate** (i.e. a single gamodeme, or small cluster of gamodemes on the fringes of a larger widespread population), will be the first step towards the formation of a new and distinct species. The accumulation of a few more mutations may then isolate it completely, as in the classic Darwinian mode. The origin of species from such peripheral isolates is known as **allopatric speciation** (section 2.3). When the new type is sterile with the parent stock, having diverged enough genetically, the 'parent' and 'daughter' species are then entirely distinct.

Recessive mutations spread comparatively slowly, and only in small populations with close inbreeding is there any likelihood of more rapid spreading. However, recessive mutations can become dominant under certain circumstances. Experiments have shown that changes in temperature alone may be enough to modify the dominance of a gene. In natural populations there are some genes which exist only as modifiers, changes in which may alter dominance to recessiveness in a gene. If a recessive allele is favourable it will be selected for dominance.

Isolation and species formation

It should now be clear that some kind of isolation is normally essential for species formation. It may be a geographical isolation, i.e. some part of a population being cut off from the rest by a barrier, as noted above. But there are other kinds of isolation which may be equally effective in certain cases. There is ecological isolation, in which species become adapted for microhabitats in the same general area. Physiological or biological factors may also isolate populations so that they speciate, e.g. changes in the timing of the breeding cycle or merely size differences precluding copulation. Again, there may be a simple genetic incompatibility; fertilization may take place but, because of different chromosome lengths or locations of alleles, recombination is impossible and the hybrids abort in embryo or are sterile.

There are other ways in which isolation might be achieved. A population 'explosion' or flush, derived from a founder population, may initially expand into many new areas in which competition or predation is minimal. After the first flush, however, may come an equally sudden crash, which leaves only a few small isolated populations, widely scattered and each with its own sample of the total gene pool. These in turn may act as founders of new species, at least some of which may survive. Where an initially large population has been all but wiped out by some catastrophe, the species may recover and expand again. However, it has passed through a 'genetic bottleneck' in which much of the potential variation has been lost. The loss of these alleles leads to homozygosity, and such lack of variation may ultimately prove detrimental.

Genetic drift: gene pools

If a population is very small it will contain only a limited and random sample of the total genetic variability within the whole species, i.e. only a fraction of the overall 'gene pool' of that species. The smaller the population, the fewer will be the alleles handed down to future generations.

At each meiosis only one chromosome of a pair, with its random sample of alleles, is passed down to the gametes; any alleles in a tiny population that are not recombined in fertilization are therefore lost, and a special evolutionary process, known as **genetic drift**, becomes important in these small populations. This is simply an evolutionary change in gene frequencies through random chance assortment, without selection being involved. Hence certain localized populations descended from an initial pair or few pairs may become quite distinctive or may even differentiate as new species merely because the founders of the population happened to carry a particular set of alleles.

In any gene pool the effects of all the factors considered so far are continually interacting, though many biologists believe that most mutations are neutral and occur at a more or less constant rate. Gene frequencies within the gene pool are related at the chromosome level to mutation, crossing over and recombination. They may be 'stabilized' in a particular kind of selection that actually prevents evolution in a large population, for if the species is successful, widespread and well adapted to its environment, the undesirables are weeded out and the only operational selection is **stabilizing selection**. This is analogous to what Mivart proposed in his critique of Darwin. On the other hand, gene frequencies begin to alter when the environment changes to become easier or harsher or when a population migrates into a new area and, finding little competition, begins to differentiate ecologically. The intensity of selection under these latter circumstances is much greater. At such times there may occur **directional** selection towards one extreme in response to a long-continued selection pressure, and this is seen in the fossil record as an **evolutionary trend** (though often it is hard to see the reason for the trend). Alternatively, **disruptive** selection may occur; the commonest phenotype in a population can be selected against, and the two extremes are positively selected, forming two different species from an original normally distributed population. One would therefore expect to find in the fossil record relics of long periods of species stability, separated by breaks representing times when speciation was very rapid, and then again stable periods. This is very often what is found, as will be explained at some length in the next section.

In any evolving system the more microenvironments there are, the greater will be the possibility of differentiation. Species thus can be said to arise in response to the environments actually available.

It is well known that natural selection is operating at the present time (Endler, 1986). Evidence for this comes from various sources. Topical examples could include the rise of myxomatosis-resistant strains of rabbits or antibiotic-resistant bacteria. The case of 'industrial melanism' in moths has been extensively studied and quoted. Before the Industrial Revolution in the north of England, the moths inhabiting the lichen-covered trunks of trees had a pale speckled colour harmonizing with the environment and rendering them inconspicuous. When records were first kept in the early 1850s these moths made up 99% of the population: the other 1% were dark-coloured types which were more easily picked out by predators. But by the end of the century the woodlands near the cities had become polluted and the lichens, which are very sensitive to unclean air, had been killed, exposing the blackened tree trunks below. The vast majority of the moths in these areas were then the dark-coloured forms, the pale varieties accounting for no more than 1%. Since it had become advantageous to be dark-coloured so that predators would less easily see the moths on the dark trunks, natural selection had been intense in the changing environment. Some 70 species of moth are known to have been affected by such industrial melanism within the very short timespan of 50 years. But only a few genes are affected here; it requires more changes to initiate speciation, involving a greater variety of genes controlling different characters.

A more recent example concerns the same Galapagos finches (*Geospiza*) studied by Darwin. One species is a specialist feeder on small seeds. Following a severe drought in 1977, the availability of small seeds became much reduced. Bigger seeds however, were more abundant, and those larger birds with stronger beaks that could cope with these were able to survive. It follows that the pressures of natural selection vary from place to place and from year to year and that there is no reason why small-scale evolution should not be reversible.

In the example quoted above the selective advantage of the gene is clear, but in many cases it is less easy to understand the selective advantages of particular characters.

Selection, then, is seen as a primary agent in evolution. It does not, however, produce ideal genotypes, but only the best available given the numerous constraints which operate at any time. Thus, as Jacob (1977) has well said, 'Natural selection does not work like an engineer. It works like a tinkerer'. The whole organism is a single system, and the morphology and function of its component parts have to be compromises amongst competing demands and based upon genetic and physical material which may itself be limiting.

For example, the lenses of a trilobite's eye are made of calcite (Chapter 11), as is the rest of the animal's cuticle. Presumably trilobites were unable to secrete material other than calcite for any part of the exoskeleton due to inherent genetic limitations, and although calcite has the virtue of transparency, its disadvantage for an optical system is that is a highly birefringent mineral, producing double images at different depths, for light travelling through it in any other directions than the mineralogical *c*-axis. Trilobites have overcome this disadvantage first by orientating the *c*-axis normal to the visual surface, and second by many ingenious modifications of the crystal lattice; they have also evolved lens doublets which bring light to a sharp focus. In spite, therefore, of an intrinsic limitation, selection pressure has 'forced' the best functional system that was possible.

Such anatomical and physiological adaptations are one result of natural selection, but animal populations are also adapted to their environment reproductively and trophically. In this respect, and in broad and general terms, two kinds of adaptive strategies have been distinguished. They are different ways of surviving and reproducing on a limited energy budget. What are known as '*r*-strategists' have very high fecundity so that great numbers of offspring are produced rapidly. Individuals may be short-lived, relying upon their superior reproductive abilities for species survival. '*K*-strategists', on the other hand, produce fewer offspring, but are usually slow growing and live a much longer time; they are better at competing for food and living space. This is an attractive concept and in some respects still useful. It has, however, been heavily criticized, since some species have both high fecundity and superior competitiveness, and all possible combinations of these end-points may occur (Stearns, 1977). Furthermore, the concept of the 'limited energy budget' is not invariably valid. It is therefore not very easy to categorize adaptive strategies very simply, and it seems rather that populations will tend to maximize whatever reproductive and survival potential they have, in the circumstances in which they find themselves.

Molecular genetics and evolution

Classical genetics makes use of breeding experiments so that the genes are understood, as far as possible, through their effects. Molecular genetics, on the other hand, concentrates upon the nature of the gene, and upon how genetic material is structured and organized. Now that we have covered the basics of classical evolution theory, it would seem appropriate to consider the impact of new discoveries in molecular genetics. Whereas these are revolutionary, the chemical pathways that lead from genes to fully developed organisms are still largely unknown, and until we have a clearer picture of the relationship of genes to development we shall still be far from understanding how evolution works at the molecular (and hence the most basic) level.

Even so, what has already been established is important for understanding hitherto elusive aspects of evolution theory, for it shows that the genome (the genetic apparatus) is a much more fluid and dynamic system than formerly supposed, with different levels of hierarchial complexity (e.g. Hunkapiller *et al.*, 1982).

Perhaps the most important of these observations is that many eukaryotes carry much more DNA than they need for coding proteins. This exists as thousands of copies of the same gene or sets of related genes. Such multiple repetitions seem to be meaningless, especially since there seems to be no clear relationship between the amount of excess DNA and the size or complexity of the animal; salamanders and primitive fishes carry much more DNA than most mammals. Such unessential genetic material has been called 'redundant' or 'junk' DNA. Some of it consists of many repetitions of the same gene, but most of the excess DNA exists as multigene families, i.e. groups of repeated sequences or multiple copies of the same gene. Such multiple copies of non-coding sequences exist too, and multiple copies of all types of gene sequences can be identical or quite widely variable. What does this excess DNA, which at first sight appears to be redundant, actually do? Some functions are reasonably clear. A multigene family, for example, can encode many closely related proteins at one and the same time, rather than just one. Moreover, the effects of a mutant gene, which otherwise might be deleterious, will be buffered by the other gene copies; it does no harm but stays in the system as a 'shielded' component. It may even, after much evolutionary divergence, turn out to be useful in some category, to an eventual descendant organism.

The 'amplified' amount of information carried in multigene families may, in ways not as yet fully understood, be necessary for the maintenance of the complex functions and interactions of the cells of metazoans. It had indeed been suggested that the Cambrian 'explosion' of metazoan life may be related to the initial emergence of such multigene families. When more is known of their functions, it may perhaps allow a further development of this concept. For the moment, multigene families seem to be primary organizational units within the genome, and an intrinsic part of a dynamic and interactive system of genetic control and regulation.

These new concepts have led to two possible challenges to neo-Darwinism. The first is Kimura's (1979) 'neutral theory of evolution'. This proposed that populations are much more diverse than would be expected from natural selection. The level of variation should be much lower because 'useful' differences should spread through the population, and 'harmful' variants should be eliminated. Since there is actually a high degree of evident variability, most of it is nothing to do with natural selection. It arises from chance differences at the molecular level, and most of the variation neither advances nor retards an individual's survival. It is purely neutral, and whether genes for particular characters persist or disappear is by chance alone. This is a debatable point, but conversely it could be argued that even if very many differences are neutral, those few that do confer a selective advantage are available to be moulded by the redoubtable power of natural selection.

Another challenge, deriving from the concept of multigene families is Dover's (1983) theory of molecular drive. It is based on the understanding that individual members of a multigene family show great similarity within a species. The individual members of a species do not evolve independently but are largely homogenized in a phenomenon known as '**concerted evolution**'. According to Dover this is due to fixation of mutations within a population because of molecular mechanisms within the genome. This he calls **molecular drive**. A number of such molecular mechanisms (e.g. gain or loss of sections of genome by transposition) can operate to allow genes to multiply out of phase with the organism that bears them. The genes controlling different characters can thus increase in frequency due to just such horizontal multiplication, and not to any effects of natural selection. Molecular drive can thus fix a variant in a multigene family, and such fixed variants, constantly turned over in a breeding population, give such homogeneity. Such a concerted pattern of fixation defines biological discontinuities from one species to the next and may indeed be one means of origin of biological novelty. Molecular drive in consequence may be another way in which species can arise, accidentally, and by non-Mendelian causes. Thus while natural selection is traditionally understood as operating at the level of the organism and is fundamentally adaptive, molecular drive is non-adaptive and operates at the level of the genome. It may also be contrasted with genetic drift, which operates at the level of the organism but is non-adaptive. Molecular drive may well prove to be the third cause of speciation; but it is hard to see how significant adaptation might arise therefrom.

Neither of the 'neutral' or the 'molecular drive' theories seriously undermines neo-Darwinism. They do, however, raise the question of how much of the variation we see is there because of chance alone and how much has accumulated because it actually confers a selective advantage.

Further discoveries at the molecular level include transposons, and collectively this new information must have a bearing upon rapid but coordinated evolutionary change, and the origin of new organizational patterns.

Gene regulation during development

When an adult organism develops from a fertilized egg, it does so according to a predetermined pathway, controlled by specific genes. These genes are switched on and off in regular sequence, and any error in this process could be lethal. The product of one gene is likely to have a sequential effect on other genes, and in long-term development we can envisage a 'cascade' of genetic controls. Thus a gene switched on, or turned off, at any one developmental stage controls the expression of other genes at a subsequent stage. The end result is a fully developed organism in which "there is a place for everything, and everything in its place' (Gee, 1996).

There are three groups of regulatory genes that have different functions during embryogenesis and operate in sequence. First, **maternal genes** establish the antero-posterior and dorso-ventral axes of the egg, before fertilization. Second, **segmentation genes** come into operation after the egg is fertilized, and they define the number and position of segments in the body. The third group consists of **homeotic genes**; these control the unique identity of a segment, i.e. how it is constructed and what it looks like. These seem to work on and integrate the signals produced by the segmentation genes. A mutation in a homeotic gene complex may produce a duplicate segment, resulting, for example, in a fly with four perfectly formed wings instead of two.

These three groups of genes act on each other in sequence and presumably code for proteins that in turn have their effect on other genes, including those controlling structure. Some genes (maternal, segmental and especially homeotic) contain a short, highly conservative DNA sequence (or motif) known as the **homeobox**. It is 180 base pairs long, and constant in size and shape. The homeobox was first discovered in homeotic genes in *Drosophila* in 1984, but has now been shown to be homologous in various invertebrate groups, including primitive metazoans, but also in vertebrates, including humans. Homeobox-containing genes must be very ancient, at least 500 Ma old, since they are found so diversely, and they always function in segmented animals in controlling the spatial expression of the body axis. They are, however, present in non-segmented phyla and evidently they do not always operate in the same way. It seems that these highly conservative homeobox genes can coopt other functions, and so, in the evolution of developmental processes 'the same regulatory elements are recombined into new developmental machines' (Raff, 1996). Clearly the homeobox concept has an evident bearing on one of the major areas of evolutionary palaeontology; how new body plans originate and how they can be modified. In this context palaeontology supplies the time dimension, which can be integrated with genetic and molecular data.

2.3 Fossil record and modes of evolution

We have already seen that palaeontology, with its unique perspective of geological time, can give special insights into the nature of evolutionary processes. In particular, descent of evolving lineages can be established and the history of life through time understood. This is an important task for palaeontology, and at present both phyletic evolution and community evolution are receiving much attention.

Of equal importance is the perception, gleaned from the fossil record alone, that rates of evolution have varied through time. In some instances change has been very rapid, but in other cases there has been very little evolutionary modification at all, even over long periods of time. It seems to be the case that evolutionary change in many instances does not proceed at a regular pace but in fits and starts. Sometimes we can discern rapid bursts of evolution in the fossil record, a single stock may suddenly diversify and its descendants quickly become adapted for life in various habitats. Such **adaptive radiations** are one of the most evident patterns of evolution revealed by palaeontology. Although the stratigraphic sequence is incomplete it still gives a sound, though in some instances fragmentary, guide to the course of evolutionary change. Although the nature of the patterns seen might have been predicted, they would be unlikely to be confirmed from observations on modern animals alone.

These various issues will be explored in greater detail in the following text. It is important to note at the outset, however, that palaeontologists and geneticists are working at quite different timescales. The changes that a geneticist can produce through selective breeding over 200–300 generations may take many years. Yet this same timespan may be represented geologically by no more than a single bedding plane! Bearing this in mind, we can proceed to examine the two main grades or modes of evolution: **microevolution** and **macroevolution**.

Microevolution refers to small-scale changes within species, and especially the transformation from one species to a new one. Macroevolution, on the other hand, concerns evolution above the species level, with the origins of major groups and with adaptive radiations. These two evolutionary modes will be considered in turn.

Microevolution

The origins of new species from existing ones, largely through changing interactions between the animal and its environment, are generally attributed to microevolution. Although the genetic and isolating processes by which species differentiate are well understood, it is only comparatively recently that full confirmation of how they have actually operated through time has been clearly seen. Simpson (1944, 1953), working mainly with vertebrates, linked genetic theory and palaeontological data in a timely manner, as did Carter (1954). It came to be understood at this time that species could arise either by lineage splitting (speciation) or by transformation of the whole population (anagenesis). Unfortunately, however, there were very few instances where fossil populations were actually seen to show a gradual modification through time. Indeed, many of the classic cases which were supposed to exhibit such unbroken series have not stood up well to modern analysis. Simpson and others argued, as had Darwin, that the rarity of fossil specimens intermediate in morphology between successive forms was due to the incompleteness of the rock record. There is, however, an alternative view, enunciated by Eldredge and Gould in 1972, that the record can be taken at face value, and the very fact that there are so few transitional series known has something fundamental to tell us about the nature of microevolutionary processes.

According to Eldredge and Gould, most palaeontological thinking had been dominated by the idea that evolutionary change must inevitably have been slow and steady. This model of gradual descent with modification seemed to be in tune with the majestically unfolding, long-term operation of the processes of natural selection as originally envisaged by Darwin. Was it not, after all, classic Darwinism? The idea of **phyletic gradualism** implies an even, slow, methodical change of the whole gene pool and hence of the characters of the entire population. This transformation would involve large numbers, usually the entire ancestral population, and would occur overall as a large part of the ancestral species geographical range. It seems a reasonable concept, and certainly some populations may have changed in just this way. Yet, as previously noted, continuous series of graded intermediate forms are seldom seen in stratigraphical sequence – what we normally see is a sequence of stable fossil populations separated by abrupt morphological breaks. To some degree this may indeed be due to incompleteness and bias in the fossil record, but to Eldredge and

Gould it suggested that evolutionary change occurs only in short-lived bursts (punctuations) in which a new species arises abruptly from a parent species, often with relatively large morphological changes, and thereafter remains more or less stable until its extinction. These authors had independently documented precise patterns of microevolutionary change in Devonian trilobites from eastern North America (section 11.3) and in Pleistocene land snails from Bermuda (section 8.4). All their evidence indicated the abrupt origins of new species, followed by stasis, and to this model of evolutionary change they gave the name '**punctuated equilibria**'.

They felt that this model seemed to accord with what is seen in the fossil record, and in addition would predict that intermediates would not be found. It was, moreover, not out of keeping with Darwinian thought, for Darwin himself had envisaged that not all evolutionary change need be gradual. Indeed in the fourth edition of the *Origin of Species* there is the following comment: 'Many species when once formed never undergo further change but become extinct without leaving modified descendants, and the periods during which the species have undergone modification, though long as measured in years, have probably been short in comparison with the periods during which they retain the same form'.

If the punctuated equilibrium model is a reasonably faithful reflection of what actually happened, we too have to consider explanations of how rapid changes occurred. Two of these will be pursued here.

Allopatric speciation

Mayr (1963) and others proposed that new species are only likely to arise by some kind of reproductive isolating mechanisms, for the process of speciation must create some kind of barrier to gene flow. One such kind of isolating mechanism involves **allopatric**, or geographic speciation. Let us suppose that a relatively homogeneous population exists within a particular area. Gene flow throughout the population is achieved by frequent interbreeding and stability is thus maintained. At some point a small part of this population migrates away from the parent population and becomes geographically isolated. Such a population carries its own gene pool and, if the population is very small, natural selection or genetic drift will modify it quite rapidly, especially if the population is adapting to a new local environment. To begin with, an isolated population such as this may develop as a 'race' with different ecological requirements, but (if put to the test) still capable of interbreeding with the original population. Within a relatively few generations, however, and with continued isolation, this population may become reproductively as well as geographically isolated from the parent population. The initial members of the peripheral isolate may well be few in numbers and are statistically likely to carry a set of genes somewhat different to the parent population. In these small populations evolutionary change is fast. These may be the founders of a new and flourishing species, which after the operation of further selection pressure will be sufficiently modified to be reproductively distinct from the original species.

The stepwise pattern of species change normally seen in the fossil record may now readily be interpreted. The sharp morphological discontinuities actually seen in so many sequences indicate that the descendant species, which originated as a peripheral isolate, has enormously expanded and has moved back into the area formerly occupied by the parent species. By this time the parent species may have become extinct, or may have broken down into smaller populations (and perhaps eventually to separate species), or may have moved away following a shifting habitat, or (though this is less likely) the new species may have competitively displaced it.

From the biological point of view, however, allopatric speciation is only one way in which reproductive isolation can occur, and in recent years there has perhaps been less emphasis on the allopatric model. Geographic isolation is not an essential factor for all modes of speciation though clearly it is an important factor.

The origin of species may be seen as a case of **quantum evolution** at the species level. The isolated population has to pass out of the set of ecologically precise parameters to which it and the parent population are adapted into another adaptive zone. The intervening conditions are unstable and inadaptive; most and indeed usually all populations undergoing such a transition will not survive to gain the next adaptive peak. It is convenient to consider the problem in terms of an 'adaptive landscape' (Wright, 1932) of high hills of adaptive fitness separated by low-lying inadaptive 'swamps', which have to be traversed and in which most adventurers will sink without trace. Those that do survive and gain the next adaptive peak will carry only a fraction of the total gene pool of the parent population. Thus this newly isolated population already starts with a biased sample of genes by comparison with the whole gene pool (Mayr's **founder principle**), and this in itself, subsequently reinforced by mutant alleles, may be a significant step towards speciation. On the other hand, if it has lost too many genes by 'genetic bottlenecking', referred to earlier, it may have little chance of renewal.

Heterochrony

When an organism develops from a fertilized egg the different organs appear and grow in a precise sequence, and the timing of these events is genetically controlled. The eventual size and shape of the organism are normally linked to the timing of the rate of development, but they can be decoupled from each other, and this is the basis of **heterochrony**, defined as 'changes through time in the appearance or rate of development of ancestral characters' (Gould, 1977; McNamara, 1990a,b, 1995; McKinney and McNamara, 1991). In the simplest case *size* alone can be affected. We may think, for example, of a trilobite in which coordinated development of size, shape and timing leads to an individual of normal size for the species. In descendant form, however, there may occur a mutation affecting rate of development. If so, the descendant trilobite may go on growing until it is much larger than the ancestral form. Then the genetic timing mechanism which puts an end to growth (and which has been relatively slowed down) catches up and growth terminates. The result is a giant version of the ancestral trilobite; equally a dwarf could result from the early onset of the growth-stopping programme.

Changes in shape or morphology result in different structures developing at different rates. There are two possibilities here, **paedomorphosis** and **peramorphosis**. In paedomorphosis the 'juvenile' characters of the ancestor are retained into the adult of the descendant. The whole organism may be affected, so that the descendant adult resembles a juvenile of the ancestral form. Alternatively, only certain characters change – the structures concerned have failed to pass through all the normal stages of development – their morphology has become 'frozen' in a more juvenile form, yet they are still functional.

McNamara (1990a,b, 1995) has summarized three separate paedomorphic processes. Progenesis occurs by precocious sexual maturation (and thus affects all structures leading to a small and juvenile version of the ancestor). Neoteny involves a reduced rate of morphological development (and may affect all or only some morphological structures; the descendant is usually the same size as the ancestor). Post-displacement occurs by delayed onset of growth of particular morphological structures relative to others (only single structures are involved and they usually resemble small and juvenile versions of the ancestral form).

Peramorphosis is the converse process: the descendant adult is morphologically further advanced than the ancestor. This process can likewise affect whole organisms or single characters only. Again there are three kinds of peramorphosis: hypermorphosis (delayed sexual maturity), acceleration (increased rate of morphological development) and pre-displacement (earlier onset of growth). The morphology of a descendant form may therefore be a mosaic of normal, paedomorphic and peramorphic characters.

Whereas heterochrony is more fully treated by McNamara (1990a,b, 1995) and McKinney and McNamara (1991), two important points can be summarized here. First, heterochronic changes are by nature 'instantaneous', and many of the microevolutionary 'stepwise' changes seen in the fossil record may have originated in this way. Second, heterochrony offers animals an escape from specialization. An animal species may be well adapted to a particular environment, but if the environment itself begins to alter, the only way out for the species is to change into something else, and heterochrony may enable it rapidly to do so.

While heterochrony is probably a primary control of microevolutionary change, it is perhaps even more important in understanding the origin of macroevolutionary novelties; it is indeed one of the most critical concepts in the whole of evolutionary palaeontology.

Testing microevolutionary patterns

The fossil record is undeniably incomplete, and it has been argued that there are so many gaps in it that it cannot really be used to interpret rates of evolutionary change. There is truth in this and we can only be fairly sure about processes in marine invertebrates if we sample on a fine enough stratigraphic scale. Ideally, to record rates and processes of change, we need to undertake very detailed microstratigraphic collection through an unbroken sedimentary sequence, using large volumes of material for statistical processing, with precise, independent stratigraphical control, and over a broad geographical area.

All these objectives are seldom obtainable, and even if they are, it has to be remembered that sedimentation in most depositional settings is either too slow or too intermittent to permit resolution of short-term, continuous biological processes in the fossil record. Even the most carefully sampled microstratigraphic sequences represent spans of time many times longer than human lifetimes (Schindel 1980, 1982). The details of local short-term processes (e.g. colonization or fluctuating population dynamics) cannot normally be resolved by palaeontological techniques.

This is one problem; another more serious one is that rates of sedimentation may change, often quite markedly, during the deposition of a fossiliferous sequence – which will impose a bias on interpretation. Suppose that a population of brachiopods living in the same environment had slowly changed in response to a continuous weak selection pressure. If the enclosing sediments had accumulated at a constant rate, then a gradual pattern of evolutionary change would be recorded. If, on the other hand, sedimentation had been intermittent and there had been erosional breaks, we should see an apparently stepwise pattern of evolution. An artificial and possibly undetectable bias is therefore imposed, making gradational events look as if they are punctuational. Our knowledge of small-scale evolutionary events, therefore, is only as good as the sedimentary record allows it to be.

Yet a further problem is how to recognize ancestor–descendant relationships. If we have a continuous sedimentary sequence with gradational transformations, then the older fossils would seem to be the ancestors and the younger ones the descendants. If, on the other hand, a sequence displays sedimentary changes coupled with sudden shifts in morphology, the relationships are far less clear; the younger elements may have been independently derived and have migrated in from elsewhere. The greater the morphological shift, the less can ancestor–descendant relationships be postulated with certainty. In such cases cladistic analysis is likely to prove useful.

Analysis of case histories

There are now a fair number of well-documented cases illustrating various patterns of microevolutionary change (summarized in Cope and Skelton, 1985; Clarkson, 1988; Sheldon, 1990a,b). There are good examples of long-term stasis and of the origin of new species by abrupt morphological leaps. Equally, there are several instances of an apparently gradual phyletic change. Both modes of evolution are possible, and there is no reason why they should be mutually exclusive.

Of the classic and often cited earlier studies which purported to show gradual change, many do not stand up to detailed analysis. They were not carried out on a sufficiently detailed scale and their authors were not uninfluenced by preconceptions. They found what they expected to find; gradual morphological shifts through time. Thus an analysis of the Jurassic oyster *Gryphaea* by Trueman in the 1920s seemed to show a gradual trend towards increased coiling. Hallam's (1982) more finely resolved studies, however, showed a pattern of stepwise change, with paedomorphosis as a control of speciation. The only indication of gradual evolution here is a phyletic size change through time.

Of course, the researches of Eldredge and Gould are not immune to criticism either. Eldredge's study of *Phacops* (section 11.3) covered a wide geographical area. It demonstrated that stasis was the norm and that speciation was rare and probably rapid. But the stratigraphy was not tightly constrained enough to allow the process of speciation to be pinpointed accurately. It could also be argued that the characters used, i.e. the number of files of lenses in the eyes of these trilobites, is itself discontinuous and would inevitably show stepwise changes. Likewise, although Gould's studies of evolution (section 8.4) covered only half a million years of geological time, snail shells are rather simple things with few characters that can be measured.

Williamson (1981) made a very detailed study of Late Tertiary molluscs in a lake basin in East Africa. In all his 13 lineages the pattern which emerged accorded with the pattern of punctuated equilibria and with long periods of stasis. Apparent speciation events were accompanied by a much higher than usual variability, and these seem to have taken no more than 5000–500 000 years. But it is always possible that some of the changes recorded are not truly genetic; some might have been ecophenotypic – a purely developmental response of individuals to a changing environment.

There are other examples, e.g. in the Cretaceous sea urchin *Discoidea* (Smith and Paul, 1985) where small-scale changes that can be followed through time were imposed by environmental causes alone. These can be correlated with alterations in sediment type – whenever the sediment was muddier, the test was higher. Such changes are ecophenotypic, not evolutionary. Yet other examples have been documented of truly genetic change, of which Sheldon's (1987, 1988) study of Ordovician trilobites from central Wales (section 11.3) is compelling. These fossils occur in great numbers and are superbly preserved in a virtually continuous sequence of black shales. The sequence studied spanned some 3 Ma and within it are eight common trilobite lineages, all of which showed a net increase in the number of ribs in the pygidium. It is a striking indication of gradual evolution occurring in parallel in the various genera and made all the more remarkable by character reversals, i.e. temporary decrease in rib number from time to time. Gradual evolution does not necessarily mean unidirectional changes. In another study spanning about the same length of time as Sheldon's (*c.* 3.5 Ma) trilobites from a Middle and Upper Devonian section in France were examined (Feist and Clarkson, 1989). Here conodont evolution was used as an independent control of biostratigraphy. These tropidocoryphine trilobites are found as two separate and successive lineages, and in both the eyes became reduced in size and finally non-functional. This may be coupled with the adoption of an infaunal, burrowing habitat. Such changes are interpreted as gradual and it is the precise stratigraphical control provided by the conodonts which enable this to be so.

Perhaps the clearest indication of both gradual and punctuated evolution occurring together is Fortey's (1985) work upon Lower Ordovician trilobites in Spitzbergen. Here there are many olenid trilobites. They are almost certainly benthic, and they show a clear pattern of stepwise evolution – the abrupt origin of new species followed by stasis. In the same beds one finds the large-eyed trilobite *Carolinites*, which swam in the upper waters of the sea (section 11.3). This pelagic trilobite, however, showed a continuous and gradual change through the same succession, and the fact that it does so demonstrates that the jumps between the olenid species are real, and not the result of hidden sedimentary breaks. This is the same succession in which we see both gradualistic and punctuational change. The two modes of evolution are real, and the apparently competing theories turn out, after all, to be complementary.

Phyletic changes seem to be important in the fine tuning of species to their habitats. We should consider however, the following question. What sort of environments are likely to favour punctuational change and stasis and which would favour gradualism? It has been suggested (Johnson, 1982) that punctuational change is most likely to occur in benthic, and gradual change in pelagic, environments. This would seem to be true of Fortey's trilobites. If evolution is largely a matter of adapting to changing environments, might we not expect to find abrupt morphological changes in environments where physical changes were likewise abrupt? Conversely, pelagic environments and those benthic regions less disturbed by violent changes would be expected to be more the locus for gradual adaptive change. Sheldon (1990a,b, 1996), with the deep-water environment of his Welsh trilobites in mind, would concur with the latter point, commenting that 'persistent phyletic evolution is more characteristic of narrowly fluctuating, slowly changing environments'. He suggests, however, that stasis is more likely to prevail in more widely fluctuating and rapidly changing environments. Quite possibly some morphologies are static and comparatively inert within quite wide environmental boundaries. They may be comparatively unaffected by 'normal' environmental changes and it is only when the broad bounds of these are exceeded that we might expect speciation events to happen.

Co-evolution

Many species are co-adapted to life in an intimate association with another species, usually of a totally different kind. Such associations have arisen by a kind of complementary process known as co-evolution and the adaptations of the two species concerned are essentially linked. A predator–prey system is one such example, as are symbiotic and commensal relationships, and of course there are the mutually necessary relationships between flowering plants and insects, or in some cases, birds.

In view of the close co-adaptation between species a modification of one will promote changes in others. This is clear, for example, in the evolution of molluscs, perhaps the most eminently edible of all invertebrates, in which 'evolutionary escalation', an 'arms race' between predator and prey has been a dominant control of their evolutionary history (Vermeij, 1987). Van Valen (1973) developed this theme in a different direction. He drew up plots of species survivorship, showing how long the components of an original sample of organisms survived through time. Rather surprisingly, he found that the likelihood of extinction of any species remained constant; a species might become extinct at any time, however long ago it had originated. In other words species do not necessarily have better chances of survival because they have existed for a long time. Consequently species must continually be in a state of evolutionary flux simply to keep up with each other to achieve a constant balance. Evolution is thus forced mainly by biological factors. This is the Red Queen hypothesis, based upon Lewis Carroll's formidable personage in *Alice through the Looking Glass* who told Alice 'Now here, you see, it takes all the running you can do to stay in the same place'.

The Red Queen hypothesis makes the assumption of progress, that later members of a lineage will be better at surviving than earlier ones. For molluscs this seems reasonable. For other animals this assumption is more difficult to test, even at the macroevolutionary level, where most of the current debate hinges (Benton, 1990). An alternative model is the Stationary hypothesis. This proposes that evolution is forced mainly by non-biological, environmental factors. In an unchanging environment there will be no change. Speciation, evolutionary bursts and extinctions only happen when the physical environment changes. Testable data remain fairly equivocal, but as Benton (1990) shows, continuous 'constant running' may just not be possible for many lineages in view of the constructional and developmental limits of the organisms themselves.

Macroevolution

Evolution is a hierarchical phenomenon, in the sense that changes of evolutionary importance may take place at any level: the genome, the individual, the species, or above. Moreover, processes such as selection and drift can operate at all these levels. We have already considered some of the mechanisms operating at the genome level, and we have seen that microevolution is concerned with changes within species or with minor taxonomic transitions. Macroevolution, on the other hand, refers to evolution at higher levels or, to use Levinton's (1983) definition, with 'the characteristic transitions that diagnose differences of major taxonomic rank'. It is in the study of macroevolution that palaeontology has, perhaps, its most direct contribution to make. Bearing in mind that several thousand generations may have lived and died during the time of deposition of only 1–2 mm of sediment, there is no way in which a truly fine focus can be obtained on events which took place during this time unless sedimentation was very rapid. Detailed palaeontological studies are better at resolving the more major features of evolution: species selection, the origin of new structural plans and higher taxa, adaptive radiations and extinctions, and rates of evolution.

Species selection

We begin with species selection, since it is here that the long-term differences between micro- and macroevolution are most clear. In microevolution the unit of selection is the individual, variations arise from mutation and recombination, and natural selection operates by favouring individuals with character combinations suitable for survival. At the macroevolutionary level, however, as Stanley (1979) has shown, the unit of selection is the species, the source of variation is in (rapid) speciation events, and selection acts upon species with favourable character combinations. This is species selection. It is, of course, analogous to natural selection. But while rate of reproduction and potential for individual survival operate at the individual level, rate of speciation (i.e. birth of new species) and species longevity are determinants in species selection.

So having defined species selection, how does it operate? On the punctuation model, speciation events produce discrete species which, after a brief and geologically instantaneous period of flux, have their characters defined from the start. Thereafter there is usually a long period of little change; virtual stasis except, perhaps, for a little phyletic fine-tuning. Within a plexus of related species, however, when followed through long time intervals, there is often a net evolutionary change or trend. There may be several coupled trends, some of which can be reversed. Such trends (e.g. the increase in complexity of the ammonite suture, or the numerous changes evident in graptolite phylogeny) have been known for a long time and have been interpreted in various ways. In modern macroevolutionary theory these trends result from species selection, those species with character combinations advantageous for survival being selected through time. How rapidly and how effectively they spread depends upon many factors, not the least of which is the rate of generation of new species which happen to have these characters and the tendency to survive for long periods. Variation in speciation rate and in extinction rate alone, or with selection operating, will affect the development of a new evolutionary plexus or clade.

An important concept, stressed by Stanley, is that macroevolution is decoupled from microevolution. This is because individuals are the units of selection at the microevolutionary level, and species are the equivalent selection units at higher levels. Hence large-scale phylogenetic trends are nothing to do with phyletic trends within a particular species. Though species selection is the basis of much macroevolutionary change, it is not the only agent; there is also what is known as directed speciation, where basically random changes move in the same general direction, and phylogenetic drift, which is comparable to genetic drift at a macroevolutionary level. These are less important, however, and species selection remains as 'microevolution in action' the primary mode of macroevolutionary change, over longer time periods.

Origins of higher taxa

New types of structure, and new grades and systems of animal organization, appear very suddenly in the fossil record, testifying to an initial and very rapid period of evolution. Such changes may be relatively small, defining new groups at lower taxonomic levels. Much larger changes, of great importance, result in the origination of higher taxa: orders, classes and phyla. Even in the first representatives of these groups, all the systems of the body, and all the anatomical, mechanical, physiological and biochemical elements therein, have to be precisely and harmoniously coordinated from the very beginning. The origination of higher taxa, in this sense, remains the least understood of palaeontological phenomena. It is sometimes referred to as **mega-evolution**, though the term is seldom used nowadays, as it seems to be another form of macroevolution operating at a higher level, and it is hard to know where to draw the boundary. When more is known about the functions of the genome, and how the information carried in the genetic code results in a fully developed individual, we shall have more of an insight into this most critical yet most elusive of all aspects of evolution.

The links between higher taxa are obscure, and are but poorly represented in the fossil record, as exemplified by the diversification of life in the early Cambrian, where transitional or linking forms are absent. The geological record gives no indication of such relationships, and our knowledge thereof rests entirely upon the traditional disciplines of comparative anatomy and embryology, supplemented, to some extent, by chemical taxonomy. But what the fossil record does give is many examples of the 'instantaneous' origin of new structural plans. The derivation of monograptids from diplograptids, the beginnings of irregularity in echinoids by the abrupt displacement of the periproct, the fusion of the mantle in bivalves to form siphons: all these are important, sudden key innovations which allowed great evolutionary diversification thereafter.

Such changes as these are now understood in terms of changes of regulatory genes which control the operation of structural genes. More importantly, evolutionary novelties may arise by relatively minor changes in the regulatory gene complex. This might affect a character or group of characters, but such regulatory gene changes might, on the other hand, affect the whole organism.

A well-known process that is presumably controlled by regulatory genes and seems to have been important in evolution is change in relative growth of different parts of the body giving the derivative stock different proportions. Such **allometric** changes might be sudden or gradual, and might well account for much evolutionary change.

Abrupt and major changes are less easy to understand, and have been the subject of much discussion. In 1940 Goldschmidt proposed a novel and certainly somewhat fanciful concept to account for the seemingly instantaneous origin of new species and indeed higher taxa. He held that major chromosomal mutations or rearrangements could in rare cases produce a 'hopeful monster', a viable new kind of organism which could be the progenitor of an entirely new taxon with its own character. Goldschmidt's views were never popular, largely because of the very low probability of such a 'freak', even if it happened to have new adaptive characters, being able to find a mate and breed. Yet there is still some support for a modified and toned-down version of the 'hopeful monster' concept, and it is not impossible to envisage a viable and 'not-too-freakish' freak, breeding initially with similar novel offspring of the same parent in a small population, as the founder of a new taxon. Few would accept Goldschmidt's idea in its original and extreme form, but it is quite possible that a comparatively few quantum events over several generations could have led to an entirely new type or organization. Moreover, the modern view of the genome, as a hierarchically organized and dynamically interactive system, allows for much more coordinated flexibility than would have seemed possible a few years ago.

We have already considered the importance of heterochrony in the origin of new species. It seems to be even more so in considering the origins of higher taxa and of new systems or organization. In many cases it provides an answer to the enigma of how a living organism can abruptly produce a descendant form, which may be quite different in some respects and yet still is a functional, coordinated organism. Here is one example. The eyes of most trilobites (section 11.3) are holochroal; that is, they have many contiguous lenses closely packed together upon the visual surface. This is the primitive type of eye which can be traced back to the early Cambrian. There appeared, however, in the early Ordovician a new kind of eye, confined to phacopid trilobites alone. These schizochroal eyes have relatively much larger lenses, not very numerous, and separated from each other by cuticular material. They look very different from holochroal eyes. It is at first sight very hard to see how the schizochroal eye could have been derived from any holochroal precursor. The answer lies in a study of the ontogeny. For the juvenile eye of an adult holochroal-eyed trilobite is, in effect a miniature schizochroal eye. It has relatively large, separate and few lenses. Arrest its development by paedomorphosis, and the juvenile 'schizochroal' eye becomes the adult form. A certain amount of reordering of the lens-packing system, a little fine tuning, and there we have a true phacopid eye. What we cannot know is the nature of the selection pressures which led to this fairly dramatic reorganization. We can be sure, however, that since this kind of eye was functional in the juvenile stage, it must have been equally functional when it was adopted through paedomorphosis, as the adult type.

This is just one example. There are many others. Paedomorphosis has often been considered as being responsible for the origin of early fish-like ancestors of vertebrates from the tadpole-like, free-swimming larvae of sea squirts. Likewise the small planktonic copepods may be products of the early onset of maturity in planktonic larvae of benthic crustaceans. The importance of heterochrony (Gould, 1977; McKinney, 1988; McKinney and McNamara, 1991) cannot be overstressed, as it can account for abrupt evolutionary changes at every level of organization.

Paedomorphosis may be seen as a special case of **pre-adaptation**. The concept of pre-adaptation, in its simplest form, means only that animals must already possess characters capable of being modified if they are going to be able to adapt to new environments. Sometimes these characters may be in the form of organs that have lost their original functions; they are therefore 'spare parts' and can readily become adapted without inhibiting an existing function. More often, however, they are already functional, but by becoming modified they may then become more flexible in the variety of functions they perform. There are many good examples drawn from the vertebrates, such as the use of a gut diverticulum as an air-breathing lung in Devonian and later lungfish living in rivers and ponds that are liable to dry up; the same structure became the swim bladder in ray-finned fish.

Numerous examples could also be cited in the invertebrates, the remarkable adaptation of already existing spines, cilia and tube feet in echinoids for deep burrowing being an ideal case.

However, not only may structures be pre-adapted before modification; they must also become **post-adapted** while their function – and probably form as well – are in the process of being transformed. Pre-adaptation has become recognized as a very critical concept, indeed Carter (1954) has written, 'No adaptation is possible without some grade of pre-adaptation in the structure to be adapted'.

An earlier concept relating to ontogeny as a tool for understanding evolutionary relationships was that of Haeckel, who believed that the ontogeny of any animal was a more or less direct recapitulation of its phylogeny. In other words, the young stages of development of any animal are very similar to what the ancestors of that species were like. In embryology and in palaeontology there is very little evidence for this (though one or two possible cases are discussed with reference to ammonites in section 8.4), and since the widespread acceptance of heterochrony, Haeckel's 'biogenetic law', as it was called, has been very largely discarded.

Rates of evolution, adaptive radiations and extinction

Evolutionary rates within evolving groups are far from constant, as has long been known (Campbell and Day, 1987). If we consider evolutionary rates in terms of production of new species, and how long these taxa persist, a common and general pattern emerges in most fossil groups. This is that in the initial stages of the life of a higher taxon, evolutionary change is much more rapid than it is later on. When a successful new animal group appears for the first time (usually with a new body plan or with a key adaptive innovation) there is a great initial evolutionary 'burst'. New taxa proliferate rapidly and expand their geographical range quickly, and the new groups arising from this burst become adapted to many environments. This process is known as **adaptive radiation** and is always a time of rapid evolutionary change (sometimes known as **tachytelic evolution**) and of high production rates of species, genera and possibly also families and orders.

Adaptive radiation may be virtually global or geographically quite localized. Moreover, it may occur at almost any taxonomic level, but it is most strikingly illustrated by such phenomena as the initial diversification of invertebrates in the early Cambrian, the radiation of Ordovician trilobites following the later Cambrian extinctions, or the sudden origin and rapid proliferation of higher crustaceans in the early Carboniferous. A further example, though on a more spread-out timescale, is the Jurassic evolutionary explosion of sea urchins, in which many highly successful and persistent orders diversified from an initial stock which had survived from the Palaeozoic. The initial impetus for an adaptive radiation is commonly a new structural plan. In marine organisms adaptive radiations are very often also linked with marine transgressions, for if new space is created and new environments are generated then opportunistic groups will arise to colonize them.

Once the initial burst of diversification is over and the main groups are differentiated, the rate of species production slows down, and the species tend to survive longer. Whereas the new taxa originated by quantum evolution during the early stages of the radiation, it is during the later stages that phyletic microevolutionary change can play a part, for it allows adaptive fine-tuning, such as slow change in size through time or alteration in the timing of breeding cycles. If a particular species becomes extinct, it will probably be replaced by a new one from a related stock and will become tuned to its environment in the same way. If the origin of new species more or less equates with the extinction of older ones (**horotelic evolution**), the lineage will survive and remain vigorous. If species abundance is plotted against time, then their differential survival or extinction will give rise to clades (i.e. clusters of branching lineages) of particular shapes. For example, when initial evolution has been rapid, and there is thereafter a slow and gradual decline, the resultant pattern will resemble a fir tree in outline. Conversely, though rarely, a more gradual increase in numbers of species, followed by a relatively abrupt extinction, produces a pattern of paintbrush shape. More commonly, clade shape resulting from a rapid initial burst, and expansion, followed by a fairly rapid decline, though with some members persisting thereafter, over long time periods, takes the form of a sprouting snowdrop bulb. In such a case the survivors continue as relicts long after the parent group which gave rise to them has disappeared. These survivors are living fossils, continuing in the virtual absence of evolutionary change (**bradytelic evolution**). Such genera as the brachiopod *Lingula* may be considered here. This genus was already in existence in the Upper Cambrian and since then has changed relatively little. It is able to survive because early on in history it colonized a 'difficult' but persistent environment, that of mudflats and shoals exposed at low tide (though specimens are known from deeper waters). There was relatively little competition for this since the physiological problems imposed by fluctuating salinity and periodic exposure are severe. But because *Lingula* was able to cross this adaptive threshold, which other invertebrates have rarely been able to do, there is no reason why it should not go on living in such an environment indefinitely. It is unlikely to be displaced by competitors.

Some 'bradytelic' animals now live in the quiet and stable waters of the deep sea. Hexactinellid sponges and the monoplacophoran mollusc *Neopilina* (whose most recent fossil relative is Middle Devonian) are amongst these. But the deep sea, though a stable enough environment and a haven for relict stocks, is not an ideal home because of the relatively limited food supply. Those animals which can cope with such unpopular environments have a fair chance of becoming bradytelic, under the influence of stabilizing selection.

Relicts of once important groups living thus may be restricted to certain habitats, but this does not necessarily mean that such 'living fossils' are declining: given the right conditions these taxa could go on for ever. Whereas, as mentioned above, they may inhabit environments for which there is little competition, it might be also that they have a wider environmental tolerance than do other animals. New views on the nature of living fossils are summarized in Eldredge and Stanley (1984). With the exception of living fossils, it seems to be the lot of all taxa to become extinct. In some cases there is a kind of relay of species, a new one coming in to occupy the vacant niche left by the extinction of a precursor. In particular groups there seems to be a definite correlation between a high rate of speciation and a high extinction rate. According to Stanley (1979) there are two reasons for this. One is that speciation and extinction are very high in small, local populations. If a population becomes widespread, it limits further speciation in the area it inhabits, and its broad geographical spread is a bulwark against extinction. Though parts of a widespread population living in local areas may die out, mainly due to accidental factors, the whole population is less likely either to become extinct or to allow the production of new species from the same stock. A second factor documented by Stanley is that rates of speciation and extinction vary with behaviour. Specialized behaviour patterns are characteristic of particular species, and if any species becomes too specialized (e.g. in food preferences) it may increase the vulnerability of that species to extinction through even minor environmental change.

Evolving lineages in the fossil record sometimes show patterns of directionality collectively known as 'evolutionary trends' (McNamara, 1990a,b). These may be anagenetic, occurring in a single non-branching lineage, or cladogenetic which involve speciation events. In some cases the changes come about through successive paedomorphic events so we have 'paedomorphoclines'. Some examples are summarized in Chapters 6, 8, 10 and 11, though more fully treated in McNamara (1990a,b). Many trends seem to be adaptive, arising from continued natural selection in a particular direction. Yet the role of adaptation has been called into question, and at least some may be the result of random sorting alone.

Throughout the history of life individual taxa have originated, flourished for a longer or shorter time, and then become extinct. The demise of individual species has left little trace upon the broad pattern of evolution. Much more important, however, have been periods of mass extinction which are dealt with from the geological and stratigraphic viewpoint separately in Chapter 3. But at this point it is worth stressing that when, through major environmental catastrophes of one kind and another, many kinds of organisms are killed off simultaneously, the whole evolutionary pattern is disrupted. When such catastrophes happen, as they have many times within the Phanerozoic, whole ecosystems collapse along with their component organisms. All the adaptations of individual species, the result of successful initial evolutionary bursts and of subsequent fine tuning over millions of years, are simply wiped out, together with all the delicate ecological balances between species.

When normal conditions are restored again after such a mass extinction period, and especially if there is a subsequent marine transgression opening up new space and habitats, then evolution resumes, based upon new adaptive radiations and on the development of new ecosystems. Mass extinctions, more than any other factor, control the whole pattern of the evolution of life, by resetting it from time to time in new and different combinations. It seems indeed to be the case that major physical or environmental catastrophes have caused more large-scale change than has competition between organisms.

So we should now consider what the role of competition has been.

2.4 Competition and its effects

Darwin emphasized the role of competition as an important factor in natural selection. It is clear that competition for food, living space and mates is a primary control within a single species or population. This is competition in simple ecological terms. Such competition may be direct, in terms of aggression and the establishment of a hierarchical order within certain kinds of higher animal societies. On the other hand it may be indirect, in terms of too many individuals passively exploiting the available food supply. But what are the effects of competition between species, especially if one invades the territory of another or if the geographical ranges of two species overlap? Undoubtedly interspecific competition is highly important, as is witnessed by the catastrophic reduction of the native British population of red squirrels after the introduction of the American grey squirrel in the nineteenth century. There are now very few red squirrels left in Britain, after only about 100 years.

An invading species may be expected to fail to displace an established species, to take over part of its range and set up a territorial balance, to infiltrate and take up a previously unoccupied niche, or to eliminate the native species entirely. How often the latter happens is not clear, for in the fossil record abrupt faunal breaks may signify the migration of an allopatrically derived species into an area after the native stock has become extinct through environmental or other causes. Quite commonly neither species becomes extinct and competitive elimination between **sympatric** species (i.e. species living in the same area) may be avoided by a variety of adaptive strategies. One way is spatial segregation; the animals come to inhabit and feed only within strictly localized geographically or vertically (depth) restricted habitats. Hence the two sympatric species are not directly competing for food or living space.

Another strategy is character displacement, either in sympatric species or where the ranges of neighbouring species overlap geographically. In the latter case some rather curious reactions between the two species with their similar ecological requirements are set up. Either individuals in the range of overlap may come to look very similar to those of the competing species, or alternatively they may diverge in characters to a much greater degree than normal. Evolutionary convergence may be interpreted as a strategy for coexistence which would minimize the effects of selection, so that it would operate upon all the individuals in the range of overlap as if they were just one species and not two competing ones. Divergence of characters, on the other hand, suggests selection within the two species for slightly different habitats, so that competition is avoided by habitat proliferation.

Cases illustrating these are known from the fossil record. In Cambrian agnostid trilobites there is evidence of depth segregation, as well as geographical separation in closely related species, and at the same time size displacement seems to operate in several species. It has been predicted that, in order to be an effective strategy for avoiding competitive elimination, size displacement would need to be in the ratio of about 1:1.28 for sympatric species. The fossils investigated, including these agnostids, seem to come remarkably close to this theoretical figure.

Eldredge (1974a) has shown how during Devonian times in North America there lived two species of the large trilobite *Phacops*. *P. iowensis* inhabited the western area, whereas successive allopatrically derived subspecies of *P. rana* (probably originating as a European invader) occupied the terrain to the east. The species range of the two did not normally overlap – a case of non-competitive geographical separation – but where the two species are found together, *P. rana* and *P. iowensis* mutually reacted; *P. iowensis* diverged in morphology from *P. rana* whereas *P. rana* converged, diverged or stayed 'neutral' in different characters, exhibiting a 'mixed' reaction. On the basis of these studies Eldredge has suggested that character displacement may be no more than a 'magnified microcosm of the general pattern of interaction between two species even when allopatric', provided that allopatry is sustained through competitive exclusion.

Even though such mechanisms exist, competition does operate to limit species and perhaps to cause extinction, at the interspecific, ecological level to which we have referred. But does it also operate at major extinction periods, and if so, how?

Since competition has been identified as an evident agent in evolution, there was for a long time a generally, if lightly held view that competition alone was perhaps the most important cause for the extinction of some groups and their replacement by others. Although competition undoubtedly is important, the older view that particular taxa somehow became outdated and were gradually replaced by more efficient competitors is no longer seen as an adequate explanation for all types of apparent replacement. Though in some respects, for example, trilobites are biologically limited, lacking strong jaws and grasping equipment, these animals were not just for their own time, but surely for all time, remarkably well organized and highly functional creatures. Recent studies upon their functional morphology, for example, their complex mechanisms for enrollment as a defence against predators, hardly support the view that trilobites died out simply 'because they weren't much good', as one of the author's students memorably phrased it some years ago. The history of trilobites, like that of so many other groups, is punctuated by a series of major and minor extinctions, some of which can be directly related to physical events such as marine regressions and not to competition. At each of the major events many trilobite groups became extinct more or less simultaneously. Those that were left continued to flourish for long periods of time. Some of these were then removed during the next critical event, and finally the last trilobites disappeared during the great Permian catastrophe, during which over 90% of the total shallow-water marine invertebrate biota became extinct.

A case which may seem, perhaps, more directly related to the effects of competition, was the replacement of brachiopods at the end of the Palaeozoic by bivalves as the dominant marine sessile filter-feeding benthos. Brachiopods dominated the level-bottom nearshore and shelf habitats during Palaeozoic time, but they were able to share a number of habitats (especially in the nearshore region) with bivalves and a long-term equilibrium was set up. From the Mesozoic onwards, however, this dominance was lost to the bivalves and brachiopods became restricted to certain habitats alone. Was this a direct consequence of competition? Or were physical factors involved here too?

There is no doubt that bivalves are much more effective filter feeders than are brachiopods. The pumping and filtering system of the bivalves is more efficient than the lophophore of brachiopods, though both systems are composed of filaments with beating cilia. However, the organization of the bivalve feeding gill produces a relatively stronger current, and the net energy gain for bivalves is decidedly higher than in a brachiopod filtering apparatus of the same length. Moreover, processes of particle trapping, sorting and digestion in bivalves are highly effective by comparison with those in brachiopods, and bivalves have in any case a greater environmental tolerance. These, and other factors, collated by Steele-Petrovic (1979), all indicate the relative physiological superiority of bivalves over brachiopods and suggest that competition cannot be dismissed as a factor in faunal replacement.

Even so, the replacement of brachiopods by bivalves was not directly due to a competitive takeover, but the result of one major episode, the Permian extinction crisis (Gould and Galloway, 1980). This affected brachiopods more severely than bivalves – the latter 'weathered the storm' better, due no doubt to their physiological superiority. Competition was undoubtedly intensified during the Permian and the brachiopod-dominated communities were disrupted. When the crisis was over, the bivalves, moving from their original nearshore habitats, were able to colonize the now vacant ecospace, previously occupied by brachiopods. It was here that the physiological superiority of bivalves was so important, for they responded differently to the mass extinction period than did brachiopods; they were not too seriously affected by the Permian crisis. Subsequently they recovered and underwent a major radiation, whereas the brachiopods lost their former dominance and remained of low diversity thereafter. Here, of course, direct competition for sea-floor space and food may have had a 'tuning' effect on the brachiopod–bivalve bias, but the faunal replacement was in the first place set in motion by a major physical catastrophe.

As Benton (1983) has shown, competition on its own may increase the likelihood of extinction in a lineage, but it is unlikely to have been the only cause. Major catastrophes or physical changes, as demonstrated so often in the fossil record, have usually by themselves been sufficient.

2.5 Summary of palaeontological evolution theory

Evolution can be defined as a change in gene frequencies through time. In any gene pool most variation results from recombination of dominant and recessive alleles, but new variation is introduced by mutation and under certain circumstances mutant alleles can become dominant. In very small populations much evolutionary change is non-adaptive because of the effects of genetic drift, but in larger populations selection acting upon the variation present in the gene pool and leading to adaptation is the primary force in evolutionary change. Molecular drive may be a third mechanism of evolutionary change but, like genetic drift, is non-adaptive.

New species may arise by the 'classic' pathway of slow continuous change in a large population, and there are some indications of this in the fossil record. Equally, and perhaps more commonly, new species may arise abruptly. Some may originate through heterochrony, especially in small populations. Other new species arise as peripheral isolates on the fringes of the parent population, migrating wherever there is a vacant niche. Once established over a wide area, populations tend to remain static, being stabilized by selection unless there is a marked environmental change which increases the selection pressure. Such stasis is the norm for populations, and the origin of new species is a rather special and relatively rare event. Competition between species in a new environment may be direct in the initial stages before a balance is achieved, but in a more balanced community of longer standing a stable predator/prey ratio is maintained and competition is usually avoided, character displacement between similar types being one of the mechanisms. Some types of biotic association are remarkably persistent over long periods of time, and these include co-evolutionary relationships.

Microevolution refers to the origins of new species. This seems to take place either by allopatric speciation in accordance with the predictions of population dynamic theories, or through heterochrony. The overlay of the sedimentary record, however, obscures to a greater or lesser extent the actual processes of speciation. Macroevolution refers to the processes of evolution at higher levels, leading to major taxonomic discontinuities.

Macroevolutionary species selection leads to the evolutionary trends observed in the fossil record; these are decoupled from microevolutionary trends. Adaptive radiation is a standard macroevolutionary phenomenon referring to the descendants of an original type (often with a key adaptation) undergoing a great burst of evolution and becoming adapted for specific habitats and niches. Such radiations are only possible where there is vacant exospace. They usually take place where a formerly stable ecosystem has collapsed, and when mass extinctions have 'cleared the field' for new kinds of animals and new ecosystems to arise. Such mass extinctions, based upon major physical changes, are probably the most important factors underlying large-scale evolutionary changes, and are explored more fully in the next chapter.

Neo-Darwinian theory remains a vital background for the understanding of palaeontological evolution. By responding flexibly to justified criticism and assimilating new data and concepts, it has grown and developed while still retaining its integrity as a theory. It is believed by some (Dawkins, 1986) that the creative force of natural selection is adequate to explain all of the diversity of life and its evolution through time. Yet there are others who do not feel that this reductionist approach is the only possible one, and who prefer to study organisms as integral wholes. A new evolutionary paradigm may be emerging (Ho and Saunders, 1984). There is, for example, some evidence, in certain cases for the inheritance of acquired characters (Pollard, in Ho and Saunders, 1984) and a plausible genetic mechanism has been proposed. There is some support for a semantic theory of evolution (Barbieri, 1985) suggesting a plurality of evolutionary laws resembling grammatical rules in language. There is a new appreciation of genes having their own dynamic life, and the possibility that genes can organize themselves independently of the environment and that organisms can evolve without any selection pressure (Campbell, 1987). The new paradigm may balance exogenous (neo-Darwinian) against endogenous (internal) causes of evolutionary change. Perhaps, in 20 or 30 years, we may be reading the fossil record in a very different light.

Bibliography

Books, treatises and symposia

Ayala, R.J. and Valentine, J.W. (1979) *Evolving: the Theory and Processes of Organic Evolution*. Benjamin, Cummings, Menlo Park, Calif. (Very good)

Barbieri, M. (ed.) (1985) *The Semantic Theory of Evolution*. Harwood, London, 188 pp. (Very clear presentation)

Bonner, J.T. (1988) *The Evolution of Complexity by Means of Natural Selection*. Princeton University Press, New Jersey.

Brouwer, A. (1973) *General Palaeontology*. Oliver and Boyd, Edinburgh.

Campbell, J.H. and Schopf, J.W. (1994) *Creative Evolution?!*, Jones and Bartlett, Boston, Mass. (6 papers)

Campbell, K.S.W. and Day, M.H. (eds) (1987) *Rates of Evolution*. Allen and Unwin, New York, 314 pp. (16 papers)

Carter, G.S. (1954) *Animal Evolution: a Study of Recent Views of its Causes*. Sidgwick and Jackson, London. (Classic textbook)

Cherfas, J. (ed.) (1982) *Darwin Up to Date*. *New Scientist*/IPC, London. (24 papers – most useful)

Cope, J.W. and Skelton, P.R. (1985) *Evolutionary Case Histories from the Fossil Record*. Special Papers in Palaeontology No. 33, Academic Press, London. (Many recent classic papers)

Cowen, R. (1994) *The History of Life*. Blackwell, Oxford. (Eminently readable)

Darwin, C. (1833) *The Voyage of the Beagle*. John Murray, London.

Darwin, C. (1859) *On the Origin of Species by Means of Natural Selection, or the Preservation of Favoured Races in the Struggle for Life*. John Murray, London.

Dawkins, R. (1986) *The Blind Watchmaker*. Pelican, Harmondsworth. (Natural selection as creative force)

Dobzhansky, T., Ayala, F.J., Stebbins, G.L. and Valentine, J.W. (1977) *Evolution*. Freeman, San Francisco. (Comprehensive advanced text)

Eldredge, N. (1991) *Fossils, the Evolution and Extinction of Species*. Aurum, New York. (Clearly written; superb colour photographs)

Eldredge, N. and Stanley, S.M. (1984) *Living Fossils*. Springer Verlag, New York. (34 papers)

Endler, J.A. (1986) *Natural Selection in the Wild*. Princeton University Press, New Jersey. (Major review, but technically difficult)

Futuyma, D.J. (1996) *Evolutionary Biology,* 2nd edn. Sinauer, Sunderland, Mass.

Futuyma, D.J. and Slatkin, M. (1984) *Coevolution*. Sinauer, Sunderland, Mass. (Valuable text)

Gamlin, L. and Vines, G. (1987) *The Evolution of Life*. Collins, London. (Fine treatment, colour illustrations)

Gee, H. (1996) *Before the Backbone. Views on the Origin of the Vertebrates*. Chapman & Hall. London.

Gould, S.J. (1977) *Ontogeny and Phylogeny*. Belknap/Harvard, Cambridge, Mass. (Excellent historical and philosophical treatment of ideas on recapitulation, neoteny and paedomorphosis)

Gould, S.J. (1978) *Ever Since Darwin – Reflections on Natural History*. Deutsch, London. (Very readable)

Gould, S.J. (1980) *The Panda's Thumb – More Reflections on Natural History*. Norton, New York . (Very readable essays)

Hallam, A. (1977) *Patterns of Evolution as Illustrated by the Fossil Record*. Developments in Palaeontology and Stratigraphy No. 3. Elsevier, Amsterdam.

Ho, M.-W. and Saunders, P.T. (eds) (1984) *Beyond Neo-Darwinism. An Introduction to the New Evolutionary Paradigm*. Academic Press, New York, 376 pp.

Hoffman, A. (1989) *Arguments on Evolution*. Oxford University Press, Oxford.

Huxley, J. (1942) *Evolution: the Modern Synthesis*. Allen and Unwin, London. (Classic text)

Jepsen, G.L., Mayr, E. and Simpson, G.G. (1949) *Genetics, Palaeontology and Evolution*. Princeton University Press, New Jersey. (Many individual papers; classic, if a little dated)

Kershaw, D.R. (1983) *Animal Diversity*. Unwin Hyman, London. (Very clear, excellent illustrations)

Kitcher, D. (1982) *Abusing Science: the Case Against Creationism*. Open University Press, Milton Keynes. (Excellent)

Laporte, I. (1982) *The Fossil Record and Evolution: Readings from Scientific American*. Freeman, San Francisco. (17 collected papers)

Levinton, J. (1988) *Genetics, Palaeontology and Macro-evolution*. Cambridge University Press, Cambridge. (A most important synthesis)

Maynard Smith, J. (1975) *The Theory of Evolution*, 3rd edn. Penguin, Harmondsworth. (Invaluable)

Maynard Smith, J. (ed.) (1982) *Evolution Now: a Century after Darwin*. Macmillan, London. (Many key papers with introduction)

Maynard Smith, J. and Szathmary, E. (1995). *The major transitions in evolution*. Freeman, New York (Genetic approach)

Mayr, E. (1963) *Animal Species and Evolution*. Harvard University Press, Cambridge, Mass.

Mayr, E. (1976) *Evolution and the Diversity of Life – Selected Essays*. Belknap Press, Cambridge, Mass.

Mayr, E. (ed.) (1978) *Evolution Readings from Scientific American*. Freeman, San Francisco. (Nine articles)

McKinney, M.L. (ed.) (1988) *Heterochrony in Evolution: a Multidisciplinary Approach*. Plenum, New York, pp. 1–348. (16 papers)

McKinney, M.L. and McNamara, K.J. (1991) *Heterochrony – the Evolution of Ontogeny*. Plenum, New York, pp. 1–437. (The most useful synthesis available)

McNamara, K.J. (ed.) (1990a) *Evolutionary Trends*. Belhaven, Cambridge, Mass., pp. 1–368. (14 papers, discussion and case histories)

McNamara, K.J. (ed.) (1995) *Evolutionary Change and Heterochrony*. Wiley, New York. (13 valuable papers)

Raff, R.A. (1996) *The Shape of Life. Genes, Development, and the Evolution of Animal Form*. University of Chicago

Press, Chicago. (Links developmental and evolutionary biology)
Raff, R.A. and Kaufman, T.C. (1983) *Embryos, Genes and Evolution – the Developmental Genetic Basis of Evolutionary Change*. Macmillan, New York. (Advanced integrative text)
Ridley, M. (1996) *Evolution*, 2nd edn. Blackwell, Oxford. (Clear and up-to-date)
Ruse, M. (1982) *Darwinism Defended – a Guide to the Evolution Controversies*. Addison Wesley, Reading, Mass. (Excellent and readable)
Simpson, G.G. (1944) *Tempo and Mode in Evolution*. Columbia University Press, New York.
Simpson, G.G. (1953) *The Major Features of Evolution*. Columbia University Press, New York. (Classic work combining many aspects of evolution theory)
Stanley, S.M. (1979) *Macroevolution: Pattern and Process*. Freeman, San Francisco. (Already a classic)
Skelton, P.W. (1993). *Evolution. A Combined Biological and Palaeontological Approach*. Addison Wesley/Open University. (Admirable fusion of approaches)
Stanley, S.M. (1982) *The New Evolutionary Timetable – Fossils, Genes and the Origin of Species*. Basic Books, New York. (Simple but very good)
Strickberger, M.W. (1996) *Evolution*, 2nd edn. Jones and Bartlett, Boston, Mass. (Excellent, comprehensive treatment)
Valentine, J.W. (1973) *Evolutionary Palaeontology of the Marine Biosphere*. Prentice-Hall, Englewood Cliffs, New Jersey. (See Chapter 1)
Vermeij, G.J. (1987) *Evolution and Escalation: an Ecological History of Life*. Princeton University Press, New Jersey.

Individual papers and other references

Benton, M.J. (1983) Large-scale replacements in the history of life. *Nature* **302**, 16–17. (Competition is not all)
Benton, M.J. (1990) Red Queen Hypothesis, in *Palaeobiology: a Synthesis* (eds D.E.G. Briggs and P.R. Crowther), Blackwell, Oxford, pp. 119–23.
Campbell, J.H. (1987) The new gene and its evolution, in *Rates of Evolution*, (eds K.S.W. Campbell, K.S.W. and M.H. Day), Allen and Unwin, New York, pp 283–309.
Clarkson, E.N.K. (1988) The origin of marine invertebrate species a critical review of microevolutionary transformations. *Proceedings of the Geologists' Association* **99**, 153–71.
Dott, R.H. (1982) The challenge of scientific creationism. *Journal of Paleontology* **56**, 267–70. (Presidential address)
Dover, G. (1983) Molecular drive: a cohesive model of species evolution. *Nature, London* **299**, 111–17.
Drake, J.W., Glickman, B.W. and Ripley, L. (1983) Updating the theory of mutation. *American Scientist* **71**, 621–30.
Eldredge, N. (1971) The allopatric model and phylogeny in Palaeozoic invertebrates. *Evolution* **25**, 156–67. (Very important work)
Eldredge, N. (1974a) Character displacement in evolutionary time. *American Zoologist* **14**, 1083–97. (Advanced)
Eldredge, N. (1974b) Stability, diversity and speciation in Palaeozoic epeiric seas. *Journal of Paleontology* **48**, 540–8. (Important work stressing allopatric evolution)
Eldredge, N. and Gould, S.J. (1972) Punctuated equilibria an alternative to phyletic gradualism, in *Models in Palaeobiology* (ed. T.J.M. Schopf), Freedman, Cooper, San Francisco, pp. 82–115. (Allopatric evolution)
Endler, J.A. and McLellan, T. (1988) The process of evolution towards a newer synthesis. *Annual Review of Ecology and Systematics* **19**, 395–421. (Excellent summary of evolutionary processes)
Feist, R. and Clarkson, E.N.K. (1989) Environmentally controlled phyletic evolution blindness and extinction in late Devonian tropidocoryphine trilobites. *Lethaia* **22**, 359–374.
Fortey, R.A. (1985) Gradualism and punctuated equilibria as competing and complementary theories, in *Evolutionary Case Histories from the Fossil Record* (eds J.W. Cope and P.R. Skelton), Special Papers in Palaeontology No. 33, Academic Press, London, pp. 17–28.
Gould, S.J. (1969) An evolutionary microcosm Pleistocene and Recent history of the land snail *P. (Poecilizonites)* in Bermuda. *Bulletins of the Museum of Comparative Zoology* **138**, 407–532. (Allopatric evolution)
Gould, S.J. and Galloway, C.B. (1980) Bivalves and clams: ships that pass in the night. *Paleobiology* **6**, 383–407. (Why did bivalves take over?)
Hallam, A. (1982) Patterns of speciation in Jurassic *Gryphaea*. *Paleobiology* **8**, 334–66.
Hunkapiller, T., Huang, H., Hood, L. and Campbell, J.H. (1982) The impact of modern genetics on evolutionary theory, in *Perspectives in Evolution* (ed. R. Milkman), Sinauer, Sunderland, Mass., pp. 164–89.
Jacob, F. (1977) Evolution and tinkering. *Science* **196**, 1161–7.
Johnson, J.G. (1982) Occurrence of phyletic gradualism and punctuated equilibria through time. *Journal of Paleontology* **56**, 1329–31. (Sedimentary overlay obscures events)
Kimura, M. (1979) The neutral theory of molecular evolution. *Scientific American* **241**, 94–104.
Kitts, D.B. (1974) Paleontology and evolutionary theory. *Evolution* **28**, 458–72.

Levinton, J.S. (1983) Stasis in progress: the empirical basis of macroevolution. *Annual Review of Ecology and Systematics* **14**, 103–37.

Mayr, E. (1954) Change of genetic environment and evolution, in *Evolution as a Process* (eds J. Huxley, A.C. Hardy and E.B. Ford), Allen and Unwin, London, pp. 157–80. (Allopatric speciation)

McNamara, K.J. (1982) Heterochrony and phylogenetic trends. *Paleobiology* **8**, 130–42. (Paedomorphosis and peramorphosis)

McNamara, K.J. (1990b) Heterochrony, in *Palaeobiology: a Synthesis* (eds D.E.G. Briggs and P.R. Crowther), Blackwell, Oxford, pp. 111–18.

Pollard, J.W. (1984) Is Weismann's barrier absolute?, in *Beyond Neo-Darwinism*, (eds M.-W. Ho and P.T. Saunders), Academic Press, New York, pp. 291–314.

Schindel, D.E. (1980) Microstratigraphic sampling and the limits of paleontologic resolution. *Palaeobiology* **6**, 408–26. (See text)

Schindel, D.E. (1982) Resolution analysis a new approach to the gaps in the fossil record. *Paleobiology* **8**, 340–53. (See text)

Sheldon, P.R. (1987) Parallel gradualistic evolution of Ordovician trilobites. *Nature* **330**, 561–63.

Sheldon, P.R. (1988) Making the most of the evolution diaries. *New Scientist* **21**, January, 52–54.

Sheldon, P.R. (1990a) Microevolution and the fossil record, in *Palaeobiology: a Synthesis*, (eds D.E.G. Briggs and P.R. Crowther), Blackwell, Oxford, pp. 106–10.

Sheldon, P.R. (1990b) Shaking up evolutionary patterns. *Nature* **345**, 772.

Sheldon, P.R. (1996) Plus ca change: a model for stasis and evolution in different environments. *Palaeogeography, Palaeoclimatology, Palaeoecology* **127**, 209–27.

Smith, A. and Paul, C.R.C. (1985) Variation in the irregular echinoid *Discoidea* during the early Cenomanian, in *Evolutionary Case Histories from the Fossil Record* (eds J.W. Cope and P.R. Skelton), Special Papers in Palaeontology **33**, 29–37.

Stearns, S.C. (1977) The evolution of life-history traits. *Annual Review of Ecology and Systematics* **8**, 148–71. (*r*- and *K*-selection too simplistic)

Steele-Petrovic, M. (1979) The physiological differences between articulate brachiopods and filter-feeding bivalves as a factor in the evolution of marine level bottom communities. *Palaeontology* **22**, 101–34. (Is competition important?)

Van Valen, L.M. (1973) A new evolutionary law. *Evolutionary Theory* **1**, 1–30. (Red Queen hypothesis)

Williamson, P.G. (1981) Palaeontological documentation of speciation in Cenozoic molluscs from the Turkana Basin. *Nature* **293**, 437–43.

Wright, S. (1932) The roles of mutation, inbreeding, cross-breeding and selection in evolution. *Proceedings of the Sixth International Congress of Genetics*, 356–66. (Adaptive landscapes)

3 Major events in the history of life

3.1 Introduction

Life began on Earth over 3500 Ma ago; that is, within the first 30% of the planet's history. At this time the Earth's atmosphere was reducing and devoid of free oxygen. Just how reducing it was is a matter for debate, as is its actual composition. It may have consisted of CH_4, NH_3, H_2O and H_2, though alternative models suggest N_2 and CO_2, with H_2O, or various other mixtures. Experimental work has shown that many kinds of simple organic molecules could have been synthesized abiotically from any of these gaseous mixtures, provided that a suitable energy source was present to power the reactions. In particular amino acids have been produced experimentally, as have at least some nucleotide bases and carbohydrates, all vital to the origins of living material. Some simple organic molecules have also been spectroscopically detected in interstellar space and complex ones have been found in meteorites.

The energy required for the synthesis of such molecules was continually available in the form of electrical discharges, ultraviolet light and ionizing radiations, hot springs, lava outpourings and solar heat. After several hundred million years the surface of the Earth was covered by an ocean, or a series of ocean basins, containing a 'hot dilute soup' of organic material (a classic phrase due to J.B.S. Haldane). The next stages in organic evolution are less clear, since the amino acid molecules had to be brought together and polymerized as proteins, eventually to replicate themselves through the agency of DNA. At this point the origin of the cell, albeit in a primitive form, was essential for further evolution, for the cell membrane acts as a buffer between the living material and the outer environment, and allows an enclosed internal environment of its own.

Some of the early protein-like substances probably polymerized in spherical bodies, such as have been produced experimentally in conditions simulating those of the early Earth. These may be regarded as cell precursors. From these, by processes still poorly understood, came the earliest true organisms. Since the present state of knowledge of the origins of life and the processes of chemical evolution have been summarized elsewhere (e.g. Awramik, 1982; Woese and Wächtershaüser, 1990), these will not be followed further here, nor will a very detailed picture be given of the later stages of pre-biotic and early biotic evolution which Cloud (1976), Schopf (1983, 1992a,b), Lipps and Signor (1992) and others have treated.

3.2 Prokaryotes and eukaryotes

The earliest living organisms were undoubtedly **prokaryotes**: cells devoid of nuclei and with their DNA not arranged in chromosomes. There are two main groups, the Archaebacteria (halophiles, methanogens and thermophiles, able to live in singularly hostile anoxic conditions) and Eubacteria (which include, amongst other groups, purple and green bacteria, and the living and fossil cyanobacteria, colloquially if incorrectly known as 'blue-green algae' (though it is legitimate to call them 'blue-greens'). These Eubacteria were the first true photosynthesizers. Prokaryotic communities can be traced back to about 3500 Ma, and dominated the scene for a long time thereafter.

By about 2500 Ma there existed teeming communities of bacteria, including cyanobacteria; these were of many kinds and metabolized in different ways. Moreover, they were able to exchange genetic material in the way that their modern counterparts do. Indeed 'the only major form of genetic exchange yet to be evolved was the complicated kind invented by eukaryotes' (Dyer and Obar, 1994).

All living organisms, other than prokaryotes and viruses, are eukaryotic. **Eukaryote** cells are appreciably larger than those of prokaryotes and of a much higher grade of organization (Fig. 2.1). Each cell has a membrane-bound nucleus, within which reside the chromosomes, and the cytoplasm contains various 'organelles' specialized for sundry functions, including sexual reproduction. The mitochondria, for instance, are the sites of cell respiration, and they have their own DNA, distinct from that of the chromosomes. So do the chloroplasts in the cells of green plants; these are strikingly similar to certain kinds of free-living cyanobacteria. The cilia and flagella (collectively undulipodia) have a constant structure, identical in all eukaryotic cells. They have a central pair of contractile microfibrils, surrounded by a ring of nine other pairs. Such striking similarities, along with much other evidence, strongly support the theory of endosymbiosis, which postulates that the organelles were once independent, free-living bacteria. They became incorporated successively in other eukaryotic cells, and lost their autonomy. The eukaryotic cell is therefore a 'community of micro-organisms', working together as a 'marriage of convenience' (Margulis and McMenamin, 1990). Even the acquisition of motility, and the complex system of mitotic and meiotic cell division probably arose by the invasion of spirochaete bacteria, which have now undergone such modification that only their microtubules remain.

The first endosymbionts were ancestral bacteria which incorporated other bacteria, were able to respire aerobically, and which subsequently became mitochondria. At this time free oxygen was an active poison for most bacteria, and not only did the symbiotic union of mitochondria impart an effective way of getting rid of it, but it also gave them a new way of acquiring energy. The build-up of free oxygen in the oceans from photosynthesis may thus have been the trigger for the origin of eukaryotes, the timing of which can be linked to this event. It is probable that the nuclear membrane came into being at about the same time, and plastids were acquired in turn. It is very likely that all this took place rather quickly (Dyer and Obar, 1994), between about 2.8 and 2.1 Ga, an hypothesis which can be tested against the rich fossil record of the later Precambrian. Such fossils occur in three kinds of environment (Strother, 1989).

(1) Chert–carbonate facies. Prokaryotic life-forms were first found in 1954 by studying thin sections of the 2000 Ma old Gunflint chert from the Precambrian Iron Formation of southern Ontario. This work revealed a rich and diverse assemblage of prokaryotes, presumably photosynthetic and preserved as three-dimensional moulds in the chert. Some of these are very similar to living cyanobacteria. Further work revealed that prokaryotes are common in Precambrian chert–carbonate horizons. The oldest known microfossils in chert come from the Warrawoona Group of Western Australia showing filamentous and coccoid structures associated with probable stromatolites in cherts some 3400 Ma old, and an active ecosystem must have been present then. From about 2000 Ma ago onwards, rich microbiotas are often found in a typical chert–carbonate facies. Very rich assemblages occur in the later Proterozoic (e.g. the 850 Ma Bitter Springs Formation in central Australia). These are dominated by prokaryotes, and in general terms shed comparatively little light upon the time of origin of prokaryotes. One particular coccoid genus, *Glenobotrydion*, has attracted much attention in this respect, for within each cell there is a dark spot originally interpreted as a nucleus (Schopf, 1968). If this were so, *Glenobotrydion* would be a eukaryote. Although the cell wall is ultrastructurally compatible with a eukaryote affinity, other evidence may indicate that this organism is actually a prokaryote. The 'nuclei' may be no more than degradation products of cellular cytoplasm. For as Knoll and Barghoorn (1974) showed, the cell contents in living cyanobacteria can contract to the centre of the cell after death, giving a spurious impression of being a nucleus. There are some 700 known examples of later Precambrian chert–carbonate assemblages, and although possible eukaryotes have been reported in other diverse and well-preserved assemblages [e.g. the 1.2–1.4 Ga Beck Spring Dolomite in California (Licari, 1978)], it has recently become clear that chert–carbonate facies are not the right place to look for eukaryotes. They are equivalent, instead, to the microbial mats of today, which are almost exclusively prokaryotic. These live in narrowly defined ecological conditions which are often harsh enough to exclude eukaryotic organisms altogether, often

associated, for example, like their Precambrian predecessors, with evaporites.

(2) Stromatolites. Precambrian prokaryotes are often found associated with stromatolites. These are laminated structures built up layer by layer, normally through the sediment-binding and carbonate-secreting activities of the cyanobacteria which formed microbial mats on the surface (some 'stromatolites' may actually be chemical precipitates). They may be flat, but are often columnar or form hemispherical mounds. The stromatolite record may be traced back to 3000 Ma. Stromatolites (Walter, 1976) were the first true reef-formers, but were badly affected by the rise of grazing organisms in the early Cambrian. The best modern examples of columnar stromatolites are found living in hypersaline waters in Shark Bay, Western Australia.

(3) Shale–siltstone facies. Precambrian microfossils occur also in organic rich shales. These are all **acritarchs** (reproductive or encystment stages of eukaryotic algal plankton). They were first reported by Timofeev in 1959. The acritarchs are usually found as flattened and folded compressions, and most of the earlier ones are simple spheroids. They can be traced back to some 1400 Ma, from which time diversity rose to some 30 taxa, about 750 Ma ago, and some of these later acritarchs were spiny and of more complex form. These were dominant in coastal waters. By the later Vendian, some 630 Ma BP, some 70% of this rich biota had become extinct, a major event which coincided with the late Precambrian (Varangian) glaciation.

When the climate ameliorated there was a further diversification of eukaryotic planktic algae, but high diversity levels were not reached again until well into the Early Cambrian.

A conception of ecological diversity in the later Proterozoic is given by an 800 Ma (Riphean) sequence in Svalbard (Knoll and Calder, 1983). Here there are three distinct assemblages. The first is a stromatolite–mat assemblage, with six or seven prokaryote genera preserved in silica; the second is a planktic assemblage dominated by large acritarchs living in the water column above the coastal sediments. There also occurs here the much-discussed *Glenobotrydion*. Third, there is quite different assemblage of vase-shaped microfossils (VSMs), probably pelagic heterotrophic **protozoans**, which have turned up independently in many other localities. While differential preservation must limit our view of late Riphean life, these microbiotas show something of the diversity and evolutionary complexity of both prokaryotes and eukaryotes just before the rise of the first metazoans. All eukaryotes reproduce sexually, and bear the potential for rapid evolution, since their potential for variation, upon which natural selection can act, is much higher than in prokaryotes. Yet this potential does not seem to have been exploited immediately. The oldest known eukaryote, *Grypania*, is a coiled unbranched filament up to 30 mm long, and it comes from rocks over 2.1 Ga old. There are likewise spherical microfossils from China, somewhat younger, and the earliest eukaryotic cells known to belong to any modern taxon are red algae, around 1 Ga old.

The Late Precambrian ocean can be understood as harbouring a dynamic and vital ecosystem of autotrophs (organisms that make their own food) and filter feeders, but with very short food chains. Its collapse was abrupt and the subsequent Cambrian 'explosion' of animal phyla, in a new, competitive and multitrophic ecosystem was perhaps (McMenamin, 1989) the most decisive event in the history of life.

3.3 Earliest metazoans

The diversification of life which began in the later Proterozoic resulted, by the early Palaeozoic, in the establishment of the main groups of living organisms which persist today. Of the many attempts to classify the essential major groups or kingdoms, that of Whittaker (1969) and Whittaker and Margulis (1978) has been generally adopted and is followed here. These authorities suggest five major kingdoms as follows: (1) **Monera** (all prokaryotes, i.e. bacteria, cyanobacteria and green prokaryotic algae), (2) **Protoctista** (protozoans, nucleated algae and slime moulds), (3) **Fungi** [which cannot make their own food but which must live as parasites, saprophytes or symbionts (lichens)], (4) **Plantae** (all eukaryotic plants) and (5) **Animalia**, including the sponges (parazoans) and multicellular heterotrophs (metazoans).

All of these were certainly established by the later Proterozoic. **Protozoans**, which may be photosynthetic or predatorial, remained small and evolved, often to elaborate form, within the confines of a single cell. They are represented in the fossil record by various kinds of microfossils, such as the Foraminiferida, Radiolaria and Chitinozoa.

Animals, all of which are multicellular, may have begun as clustered aggregates of protozoans which remained together after cell division. One line led to the sponges (Chapter 4), retaining in general terms such an 'aggregate' organization, another to the **metazoans**, in which different parts of the body became specialized for different functions and in which tissues (muscle, gut, nerve, etc.) developed.

Once the earliest multicellular animals were established, metazoan life underwent successive radiations. Of these the first was the soft-bodied Ediacara fauna of the late Proterozoic, then came a radiation of small shelly fossils at the base of the Cambrian, and there followed a dramatic 'explosion of life' in which calcareous algae, 'shelled' protozoans, sponges and archaeocyathids, conodontophorids, arthropods (especially trilobites), pogonophorans, molluscs and echinoderms first made their appearance, as did many short-lived 'problematic' groups of dubious affinity. Different kinds of skeletons appeared, made of both agglutinated and secreted material, of various kinds of chemical composition. Although most of these early groups were suspension or possibly deposit feeders, there were some predators and an ecological balance was soon achieved. This 'Cambrian explosion' was a real event, and not an artefact of the record; it probably lasted no more than a few million years.

The Proterozoic–Cambrian transition is now remarkably well calibrated (Knoll, 1996), due in part to the recognition that carbon isotope variations of that time were globally synchronous. There is now a refined stratigraphy which along with new fossil discoveries have given a fresh perspective on the 'Cambrian explosion' and the evolution of the earliest metazoans.

The base of the Cambrian (which in my scientific lifetime has hovered between 500 and 590 Ma) now seems firmly fixed at 543 Ma. New records of possible metazoans, the oldest known, are dated from about 615 Ma, just prior to the Varangian Ice Age (*c.* 610–590 Ma), and the oldest Ediacaran faunas from about 575 Ma. The Early Cambrian, in ascending sequence comprises the Nemakit–Daldynian (543–530 Ma), the Tommotian, followed by the Adtabanian (together 530–525 Ma), then the Botomian and finally the Toyonian (together 525–520 Ma), all named after type areas in Siberia. The following Middle and Late Cambrian lasted a remarkably short time (520–505 Ma). With this timescale in mind we can now consider the various faunas in turn.

Ediacara fauna: two viewpoints

Before 1947 almost nothing was known of the nature of Precambrian metazoan life. The term 'Proterozoic' coined for the later Precambrian (2500–543 Ma) was based on the understanding that there were organisms (e.g. stromatolites) living then, but whether or not there was animal life was then unknown.

The discovery in that year of a rich and well-preserved fauna of soft-bodied animals in the Ediacara Hills of South Australia (Fig. 3.1) was the first acceptable faunal record from the Precambrian (Glaessner, 1961, 1962, 1984).

The fossils came from a shallow marine to littoral sequence with flaggy and cross-bedded sandstones; the occasional emergence of shoals is shown by polygonal mudcracks at certain horizons. The fossils are preserved along the interface of fine argillaceous laminae and hard sandstones. Apparently the dead animals came to be stranded upon the tidal mudflats or in the bottom of tidal pools and were preserved when covered by sand. The sand is of unusual texture, rather like a foundry sand, which is a factor in this uncommon mode of preservation.

Several thousand specimens of Ediacara fossils are now known, belonging to some 30 genera. Traditionally they have been considered as early representatives of modern phyla, especially Cnidaria (Glaessner and Wade, 1966; Wade, 1972; Glaessner, 1984). A contrary, and admittedly heretical view, however, is that they are relics of a 'failed' evolutionary experiment, quite unrelated to any modern organisms and 'almost like life on another planet' (Seilacher, 1989).

The traditional view

According to the traditional interpretation the Ediacara fossils fall into four main categories, and before considering alternative hypotheses, they are first described under these headings.

The majority are jellyfishes (**medusoids**) and soft corals (**pennatulaceans**) belonging to Phylum Cnidaria; there are also segmented worms

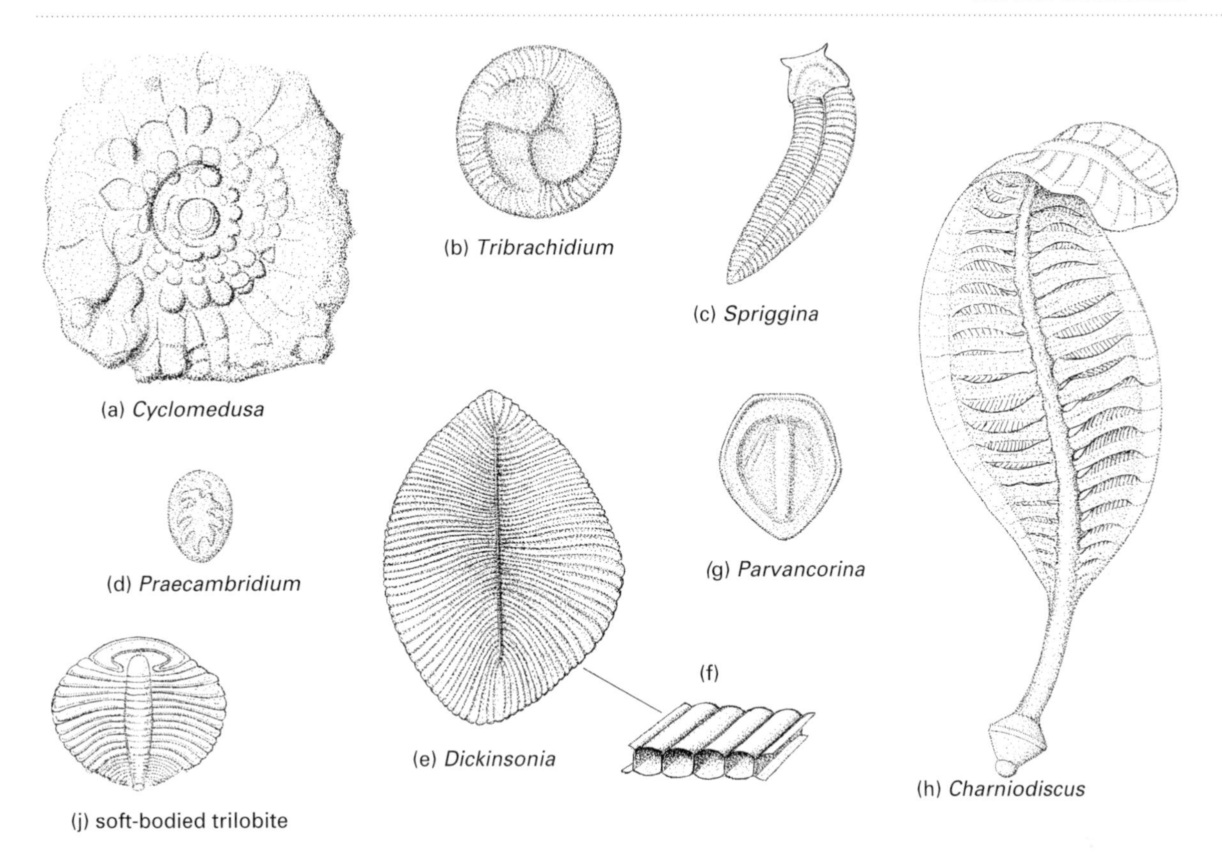

Figure 3.1 Elements of the Precambrian Ediacara fauna of Australia. (a) *Cyclomedusa* (× 0.5); (b) *Tribrachidium* (×1); (c) *Spriggina* (× 0.75); (d) *Praecambridium* (×1); (e) *Dickinsonia* (× 0.75); (f) same, three-dimensional reconstruction; (g) *Parvancorina* (× 0. 75); (h) *Charniodiscus* (× 0. 25) ; (j) 'soft-bodied trilobite' (×1). [(a)–(d), (g) redrawn from Glaessner and Wade, 1966; (e), (f) based on Seilacher, 1989; (j) from Jenkins, 1992.]

(**annelids**) and some peculiar organisms of unknown affinities. Sometimes one surface of the fossil alone is preserved, but there are also composite moulds with both surfaces compressed together.

Medusoids

Fifteen species of jellyfish have been described. *Mawsonites* is up to 125 mm across, is dome-shaped and has a strongly sculptured surface with large radially arranged lobes. *Cyclomedusa* is much smaller and has a central cone with concentric rings and a wide flat disc. *Kimberella* (Fig. 5.3) probably had a much longer, bell-shaped dome with distinct longitudinal zones. The tentacles of *Eoporpita* are preserved, but the sculpture of the other nine jellyfish genera is less clear.

Pennatulaceans

Charniodiscus and *Pteridinium* are elongated feather-shaped bodies. In *Charniodiscus* there is a basal disc with a central shaft about 150 mm long from which springs primary branches at about 45° to the shaft, and some 20 or so secondary branches are produced from each of these. *Pteridinium* is somewhat similar, but poorly known, because of elastic deformation during preservation. The affinities of these have been much debated, but the morphological comparisons they bear with sea pens, i.e. alcyonarian corals (Chapter 5) with polyps ranged along the secondary branches, are so close that they are generally believed to be related to this group.

Annelids

Three species of broad, flat, bilaterally symmetrical 'worms' are known, referred to as *Dickinsonia*. They have a thin central axis and narrow segments normal to it, arranged in a 'quilted' pattern radially around the front and rear. Some specimens bear gut lobes, but no mouth or eyes have been preserved. A close resemblance to the modern segmented worm *Spinther*, which lives on and eats sponges, has been noted, so *Dickinsonia* has been considered to be an annelid. It differs from *Spinther* only in that the distal claws (parapodia) situated on the ends of the segments and used for crawling are lacking. *Dickinsonia* specimens usually lie flat on the bedding planes, but some of the larger ones are deformed by traction, allowing details of the quilted structure to be clarified. In these, sharp creases crossing the quiltings show that the cuticle must have been leathery and flexible. Evidently the upper and lower cuticular surfaces were not knitted together at their outer circumference but held apart by stiff, biradially arranged segmental septa.

The elegant *Spriggina* has a horseshoe-shaped 'head' with long projecting setae. Some 40 segments follow, each, according to some authorities, terminating in a short parapodium, and hence it has been considered to be an annelid, with some points in common with modern planktonic worms.

Fossils of unknown affinities

Praecambridium is an ovoid discoidal form some 5 mm long. The 13 known specimens show three or four pairs of raised lobes, seemingly arranged in segments, behind a horseshoe-shaped anterior part. Some authorities (Glaessner and Wade, 1971) have interpreted this as an arthropod with a soft, non-mineralized integument. It has some resemblance to a trilobite protaspis or larval stage, but the analogy is not a close one. A similar though larger form, *Vendia*, has been found in Siberia.

Parvancorina is a shield-shaped animal up to 26 mm long, with a broad, anchor-like ridge on the upper surface and a narrow, incised rim within the margin. Fine striae extend obliquely backwards from the main ridge. Some authors have suggested arthropodan affinities in view of a supposed resemblance to the 'shell' of the little crustacean *Triops*, but this is speculative.

Tribrachidium is discoidal with a central raised platform. Upon this three 'arms' radiate from the centre, and as they near the edge of the platform they are sharply bent, each giving rise to a series of marginal striae. In many ways *Tribrachidium* has a superficial similarity to the edrioasteroids (Chapter 9), especially to a few genera which have triradial symmetry. Some specimens seem to have a Y-shaped central mouth and bristle-like appendages on the surface, conceivably precursors of tube feet. The possibility that this ancient fossil was a kind of 'proto-echinoderm' is therefore not to be disregarded. Some specimens of rather trilobite-like appearance have recently been recorded (Jenkins, 1992; Fig 3.1j) from the Ediacara Hills, at various stages in growth. Might these be some kind of soft-bodied trilobite precursor? We await further evidence.

Vendozoan hypothesis

Seilacher (1989) rejects the view that the Ediacaran organisms are ancestors of modern metazoan phyla. He categorizes all these organisms as Vendozoa, arguing from the standpoint of constructional morphology. He comments that in spite of an apparent diversity in the fauna, nearly all the genera have a striking basic uniformity; they are thin and flattened, round or leaf-like, and possess a ridged or quilted upper surface. Moreover, such forms as *Dickinsonia* (Fig. 3.1,e,f) are constructed of biradially arranged hollow tubes of flexible material, open to the outside. Here, as with all Ediacaran organisms, there is no trace of a gut. Such features suggest that all these animals fed by passive absorption of dissolved food and metabolites over their whole surface. If this were the case, and if internal digestive, vascular and transportation systems were lacking, then these animals had no relationship with any modern group (except, in feeding mode, with pogonophorans), and even *Spriggina* may be envisaged, not as a crawling worm, but as an upright frond, held in the sand by the 'head'.

If this interpretation is followed, it is possible to envisage the Vendozoa as lying on the surface of bacterial or cyanobacterial mats, and many may have had endosymbionts within their bodies. The vision of an endless spread of passive, flat green quilts is quite appealing. Yet the 'fall from the Garden of Ediacara' (McMenamin, 1989; McMenamin and McMenamin, 1990) came abruptly. With the rise of the first predators in the early Cambrian the pattern of global ecology changed from primarily autotrophic to the heterotrophic system that was to dominate until the present time.

Faunas of Ediacaran age and type have been recorded from various parts of the world: Australia, Newfoundland, Siberia, England and Wales, Scandinavia and Namibia. They are generally of marine shallow-water origin. The diverse Australian fauna (Glaessner, 1984; Jenkins, 1992) dates from about perhaps 560 Ma. A somewhat older fauna of about 565 Ma from the Avalon Peninsula in Newfoundland occurs in sediments of relatively deeper-water origin, at various horizons. They are pectinate (comb-like), bushy, spindle-shaped and star-shaped forms belonging to about 30 genera, and on the traditional interpretation might be largely referred to the Cnidaria, as might be expected from the age of the faunas, but 10 of the genera are of unknown affinity. There are no worm-like or trilobite-like forms present. The fossiliferous Charnian rocks of central England are now known to be of similar age to the lower Avalon suite and like them contain abundant specimens of the long-ranged genus *Charniodiscus*.

Until recently there seemed to be a substantial gap between the last Ediacaran fauna and the base of the Cambrian, reinforcing Seilacher's concept of an isolated 'first experiment', which died out long before the Cambrian began. But there is no longer a temporal gap between the youngest Ediacaran fauna and the first small shelly fossils; this is purely a preservational artefact. In Namibia (Grotzinger *et al.*, 1995) there has been described a richly fossiliferous sequence with an abundant Ediacara fauna, especially *Pteridinium*. Several tuffs within the sequence provide reliable U–Pb dates, and the youngest Namibian Ediacara-type assemblage extends right up to the base of the Cambrian (543 Ma) and is contemporaneous with the first small shelly fossil assemblages. If the Middle Cambrian *Thaumaptilon* from the Burgess Shale, which closely resembles *Charniodiscus* (Conway Morris, 1993) is truly an Ediacaran survivor, and if more is known about the 'soft-bodied trilobites' mentioned by Jenkins (1992) from the Ediacara Hills, then the distinction between at least some of the Ediacaran taxa and Cambrian faunas becomes less clear. This is sustained by Seilacher's (1992) comments on the curious Ediacaran Psammocorallia which seem to have been true anthozoan polyps, i.e. 'sand corals' with skeletons formed by sand grains which entered the gastric cavity and lodged within the mesogloea (Chapter 5). This would give some rigidity and would anchor the coral in the substrate. Nevertheless the body plans of most Ediacaran organisms are very different from those of other metazoans, and there remains much support for Seilacher's Vendozoan hypothesis.

The Ediacaran fauna ranges throughout much of the Vendian, a time span of some 40 Ma, and the diversity of types increases through time. There is only one undoubted skeletalized fossil of similar age, the pipe-like annulated calcareous tube known as *Cloudina*, which can be up to 4 cm long; in Namibia these tubes are so abundant, along with undescribed goblet-shaped microfossils, as to form bioclastic sheets. By the end of the Vendian, a few small shelly fossils are present (*Anabarites* and *Cambrotubulus*), presaging their dramatic expansion in the early Cambrian.

Small shelly fossils

In recent years rich and varied faunas of small shelly fossils (SSFs; Figs 3.2, 3.3) have been documented from the base of the Cambrian and just into the Precambrian (e.g. Matthews and Missarzhevsky, 1975; Raaben, 1981; Rozanov and Zhuravlev, 1992).

The best-known sequences come from the Siberian Platform, where late Precambrian calcareous beds with dolomitic, oolitic and stromatolitic horizons are overlain by early Cambrian sediments.

The Tommotian stage does not contain trilobites but has archaeocyathids and abundant SSFs, many with phosphatic shells. Some of these (e.g. *Aldanella*) are helically coiled gastropods; there are also **hyolithids** (an extinct molluscan group) and common sponge spicules. Tommotian faunas are replete with short-lived problematic groups of which two are of most interest: the **hyolithelminthids** (laminated phosphatic tubes open at both ends) and **tommotiids** (phosphatic conoidal shells usually occurring in symmetrical pairs). These latter have been compared with plates of barnacles (cirripede crustaceans), though they are phosphatic rather than calcareous. If so, it would suggest a very early origin for these crustaceans. The resemblances, however, are only superficial, and the tommotiids probably belong to an extinct phylum (Bengtson, 1977). Some of these microfossils may actually be

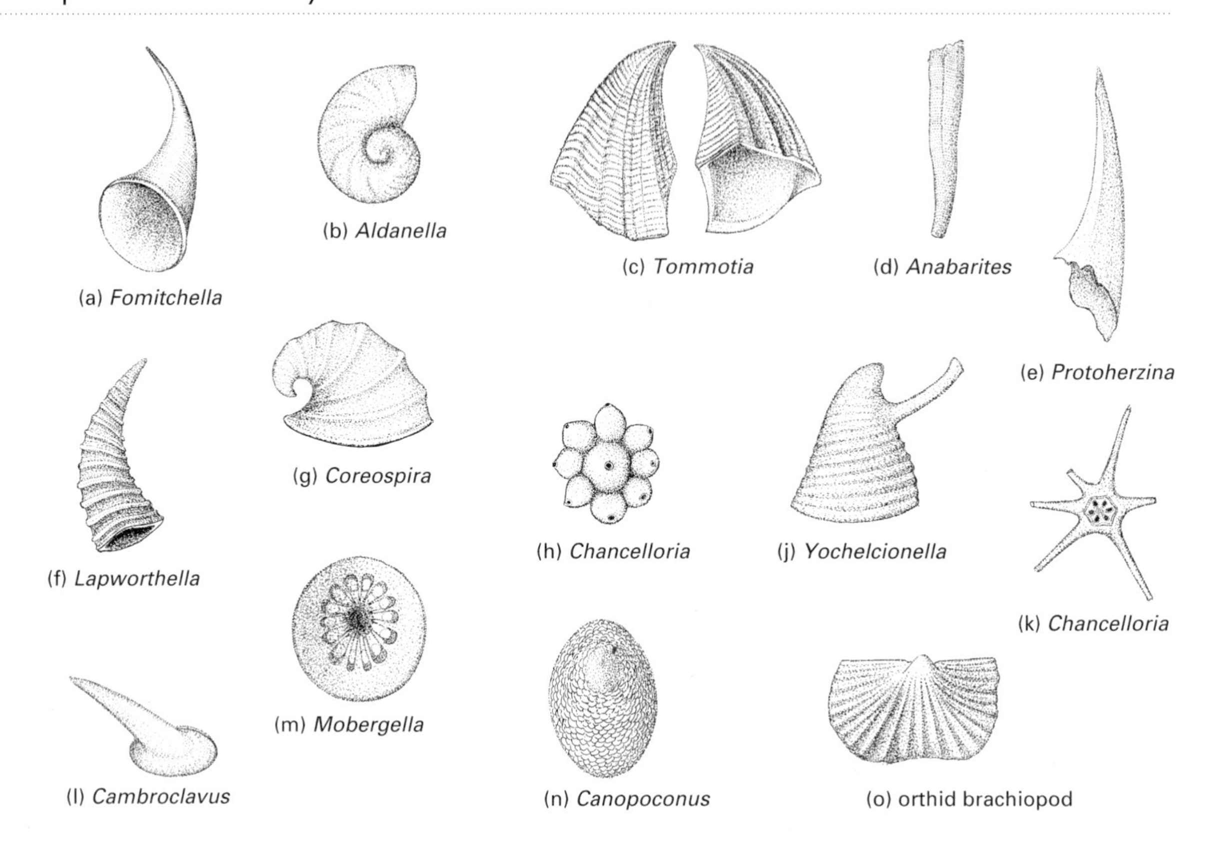

Figure 3.2 Latest Precambrian (*Anabarites*) and earliest Cambrian (others) small, shelly fossils. (a) *Fomitchella* (× 40); (b) *Aldanella* (× 20); (c)*Tommotia* in obverse and reverse views (×15); (d) *Anabarites* (×15); (e) *Protoherzina* (×15); (f) *Lapworthella* (× 20); (g) *Coreospira* (× 20); (h) *Chancelloria* (× 30); (j) *Yochelcionella daleki* (× 8); (k) *Chancelloria ramifundus* (× 30); (l) *Cambroclavus* (× 50); (m) *Mobergella* (× 5); (n) *Canopoconus* (× 50); (o) orthid brachiopod (×10). (Mainly redrawn from Rozanov and Zhuravlev, 1992; others from Qian Yi and Bengtson, 1989; Bengtson *et al.*, 1990.)

components of larger organisms. Thus small plates and spines with a net-like ornament (*Microdictyon*) have now been shown to be part of the armour of a large slug-like form (Chapter 12).

An exceptionally complete and important section through the latest Precambrian and early Cambrian has been described from Meishucun in Yunnam, China (Qian Yi and Bengtson, 1989). Within the Meishucunian stage, directly equivalent to the Tommotian, three chronozones and eight subzones have been erected on the basis of SSFs alone. The first trilobites do not appear until the overlying Qionzhusian stage. The abundance of trace fossils and acritarchs make this an ideal section for studying the sequence of early Cambrian events, and a candidate for the Precambrian–Cambrian global boundary stratotype.

Similar fossils of Tommotian age (Bengtson *et al.*, 1990) are known from Australia, England and Scandinavia. Many of the species continue well up into the Lower Cambrian, where they are found with trilobites. By contrast, there are relatively few microfossils in the latest Precambrian stage: the Yudomian. There are only a few small tubular organisms, e.g. *Anabarites*, a phosphatic tube, *Cloudina*, an annulated calcareous tube, and *Sabellidites*, an organic-walled tube, thought to be the skeleton of a pogonophoran (a simple unsegmented 'worm' that has no gut but absorbs dissolved organic matter over its surface).

In the type area in southeastern Siberia and elsewhere, the Tommotian rests unconformably on the late Precambrian Yudomian, and the Nemakit-Daldynian stage is missing. Since, in the Tommotian,

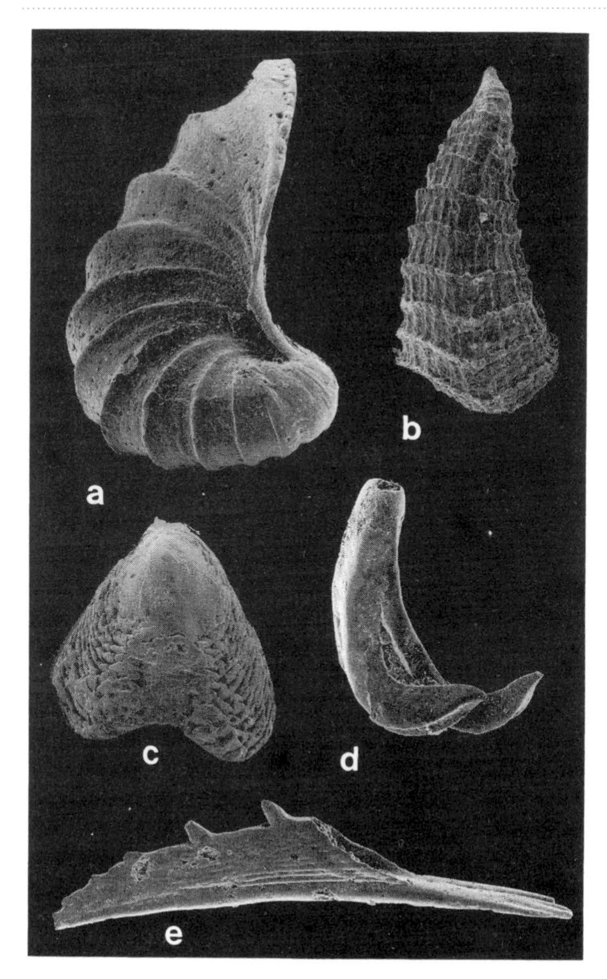

Figure 3.3 (a) *Archaeospira ornata* (× 35); (b) *Lapworthella fasciculata* (× 70); (c) *Maikhanella pristinis* (× 35); (d) *Drepanochites dilatatus* (× 35); (e) *Scoponodus renustus* (× 35). (Photographs reproduced by courtesy of Dr Stefan Bengtson.)

SSFs appear en masse, the impression is given that their origin was very abrupt. Recent discoveries in northeastern Siberia, however (Knoll, 1996; Kaufmann *et al*., 1996) record a continuous succession with the basal Cambrian Nemakit-Daldynian stage resting conformably on the Yudomian boundary. Here we see SSFs appearing sequentially and diversifying through a period some 3–6 Ma long, and the great diversity of SSFs in the Tommotian is only the culmination of a long evolution. Accordingly, the origin of SSFs was a stepwise process, and the evolutionary burst thereof a much less dramatic event than it formerly seemed to be.

Precambrian trace fossils

The earliest appearance of trace fossils may help to date the origins and initial radiation of the metazoans more readily than do the soft-bodied organisms themselves, with their limited prospects of preservation.

Trace fossils, however, do not extend far back into the Proterozoic. The oldest so far described, which at first seemed to be the burrow of a worm-like organism in rocks 1000 Ma old, is now believed to be a termite burrow. Trace fossils achieved some degree of diversity in the Vendian, and their makers flourished in shallow seas with sandy floors (Crimes, 1992a,b). There are scattered records of trace fossils in rocks of over 750 Ma, but trace fossils do not become abundant or diverse until near the Precambrian–Cambrian boundary.

Most of the early trace fossils probably resulted from the burrowing activities of soft-bodied infaunal worms, but the number of types represented increased greatly in the uppermost Vendian; horizontal sinuous feeding burrows, surface trails, resting marks and vertical pipes are not uncommon some little distance below the Precambrian–Cambrian boundary. It is not always easy, however, to date these in absolute terms, nor is it clear what kinds of animals made them.

However, in many sequences traces of a kind normally attributable to trilobite resting traces (*Rusophycus*) appear some considerable stratigraphical distance below the trilobites themselves. They are mainly simple forms with paired chevron-like markings, evidently the scratch marks of the appendages. They may have been made by late Precambrian trilobite precursors with thin, non-mineralized shells, propelled by 'a simple musculature of low efficiency' (Crimes, 1976).

In general, trace fossils are more common than body fossils in the later Precambrian, and while they go further back in time they do not take the presumed origin of motile metazoans very far back into the Precambrian, certainly no more than 1000 Ma before the present time and probably much less. The first good trace fossils postdate the widespread late Precambrian glaciation. Following the first Vendian diversification, a second radiation of trace fossils took place in the Tommotian/Adtabanian, and three trace fossil biozones have been established for this interval. Curiously, there was very little change in the trace fossil biota for the rest of the Lower Palaeozoic, other than a possible migration of some types into deeper water.

Causes of the Cambrian 'explosion of life'

The proliferation of skeletalized fossils in the Early Cambrian was a real phenomenon and not an artefact of the geological record. The appearance of the main invertebrate phyla was phased through some 30–40 Ma, but it was a critical turning point in the history of life. But can we establish why it happened? Specifically, did changes in external factors drive it? Or was it merely that life had reached a particular evolutionary threshold? Quite evidently, both physicochemical and biological factors were involved, together forcing the most dramatic event in the history of life.

Physicochemical factors

During the Vendian and Early Cambrian there were major changes in atmospheric composition, ocean chemistry, plate tectonic activity and climate. To begin with there was a slow build-up of free atmospheric oxygen as a result of photosynthesis which evidently attained a level sufficient to support animal life (Rhoads and Morse, 1971), and likewise a decrease in the concentration of CO_2 in the atmosphere (Tucker, 1992). There is still much doubt, however, as to how much free O_2 was actually present in the Proterozoic – there may in fact have been a fair amount (McMenamin and McMenamin, 1990). Other factors include a worldwide series of marine transgressions, the first following the Varangian glaciation, successively flooding the continental shelves and opening up new shallow-water ecological niches. Most of the continental masses at that time lay in low latitudes, and the warming, shallow shelf seas were ripe for colonization. It has been suggested (Brasier, 1979, 1992a–c) that evolutionary patterns of diversification may be linked with successive stages of flooding of the shelves. Likewise, the availability of extra food, especially phytoplankton, may be related to a major periods of upwelling. This latter hypothesis seems to be borne out by the remarkable synchroneity of large phosphorite deposits, on a global scale, clustered around the Precambrian–Cambrian boundary (Cook and Shergold, 1984; Cook, 1992). It is likely that during the later Precambrian, phosphorus-rich sediments accumulated on the deep-ocean floor as this element was depleted from surface waters. During an extended period of enhanced oceanic overturn, in the earliest Cambrian (and possibly associated with the marine transgression) phosphorus was brought up to surface waters to give increased nutrient levels in the photic zone. Phosphorus is essential to life and this event may have provided the trigger for the sudden diversification of living organisms (as well as the accumulation of phosphorites on the sea floor). It is significant that many early Cambrian organisms (e.g. elements of the Tommotian fauna, and linguliform brachiopods) have shells of calcium phosphate, which was very abundant at the time. High phosphorus levels inhibit calcification in living organisms, which may explain why calcareous skeletons did not appear commonly in the fossil record until later, when the early Cambrian phosphogenic event was largely over.

The spread of nutrient-enriched waters over low-latitude shelves, as recorded in C-isotope fluctuations (Brasier, 1992a–c), may well be linked to a global rise in temperature as the Earth emerged from a late Precambrian 'icehouse' to the 'greenhouse' phase of the later Cambrian.

Finally, there were certainly four, probably synchronous, events of global regression and emergence during the time in question. The first, the Kotlinian crisis, marked the end of the Proterozoic, and the virtual demise of the Ediacara fauna. The last, the Toyonian crisis of the latest early Cambrian, is synchronous with the end of the SSF radiation. The effects of these several crises are currently a subject of detailed study.

Biological factors

The most familiar representatives of the early Cambrian fauna are trilobites, archaeocyathids, certain primitive molluscs, small brachiopods (chiefly with phosphatic shells) and the earliest echinoderms (edrioasteroids, helicoplacoids and others). Some of these are found in a few isolated localities alone and did not spread until later; others are widespread from the start. The archaeocyathids, for instance, appeared first and spread from their Siberian centre of origin to other parts of the world, only to become virtually extinct by the end of the Toyonian. More than any other group, they dramatically chart the rise and decline of the Early Cambrian biota. While contemporaneous phosphatic-shelled SSFs proliferated and diversified, reaching their acme in the Tommotian, the first trilobites, echinoderms and brachiopods did not appear until the Adtabanian. There occur, too, a good number of peculiar and problematic fossils, often geographically very localized. All of these appear very abruptly and are fully organized and differentiated on their first appearance, which at first sight may seem difficult to understand.

There are two factors by which the very abrupt appearance of these early fossils can be explained. One is that evolution at this time was very rapid, perhaps due in part to intense selection pressures operating at that time. The second point is that the acquisition of hard shells as opposed to merely an unmineralized integument must have brought the bearers thereof to a 'fossilization threshold', a potential not available even to their immediate ancestors. Though the mineralized shell which made fossilization possible must have been acquired in all the above groups over a period of a few million years – it was spread out in other taxa throughout the rest of the Cambrian and the early Ordovician. But how did a mineralized external covering originate in the first instance? One suggestion (Glaessner, 1962) is that the calcareous or phosphatic material of which it is composed originated as an excretory product, possibly accumulating over the skin and hardening it. Once it was formed quite accidentally, the possession of a shell which could be used as a base for muscle attachment (giving enhanced locomotory prospects) and as a protection would give a singular selective advantage to the animal possessing it. Any invertebrates with hard coverings formed in such a way would then have great potential for evolutionary development. Moreover, once hard mouthparts had originated, the selection pressure on other phyla to develop hard coverings for protection would be significantly increased.

Eukaryotic cells in general are able to pump ions and produce protein matrices capable of mineralizing. They are already pre-adapted for the processes of biomineralization (Simkiss, 1989), but they did not realize their full mineralizing potential until metazoan cellular systems were well differentiated, specialized organs had originated and extensive surfaces for crystal growth had become available. The incorporation of the citric acid cycle into the cellular metabolism of many organisms provided a vast increase in available energy for biomineralization.

Something of the vast diversity that life had reached by Atdabanian times is shown by the Chengjiang 'window' (Chapter 12). So diverse and 'advanced' is this fauna that it would seem to signify either (1) extremely rapid evolution of most types of animals in the early Cambrian or (2) an origin for **ceolomates** (animals with an internal cavity in which the gut is suspended) some time before the Precambrian–Cambrian boundary. It is highly probable that both factors were involved.

Biological evidence on metazoan relationships

There are several levels or grades of organization in multicellular animals living today, the comparative study of which sheds light upon the probable course of their evolution (e.g. Clark, 1979; Willmer, 1990; Nielsen, 1995; Gee, 1996).

Undoubtedly the earliest multicellular animals were aggregates of protozoans, loosely integrated but with cells all of the one kind. Some of today's simplest sponges would serve as models for this grade. Sponges (q.v.), however, developed along their own evolutionary pathway and are not really metazoans.

All other multicellular organisms have their cells organized into tissues, i.e. cell layers or masses specialized for different functions, and it is this that defines the metazoans proper.

In the simplest metazoan grade, represented by Phylum Cnidaria, the organization is **diploblastic**; that is, the body wall consists of two layers: an **ectoderm** and an **endoderm** (Fig. 3.4a).

These together form the bag-like body which encloses a single cavity, serving as a gut, with a single opening which acts as both mouth and anus.

The next level, which includes all higher types of animals, is the **triploblastic** grade. In this condition there is a third layer, the **mesoderm**, sandwiched between the endoderm and the ectoderm. In its most rudimentary form, as represented by the unsegmented flatworms (the Platyhelminthes), the body is bilaterally symmetrical and the mesoderm forms a more or less solid mass of tissue

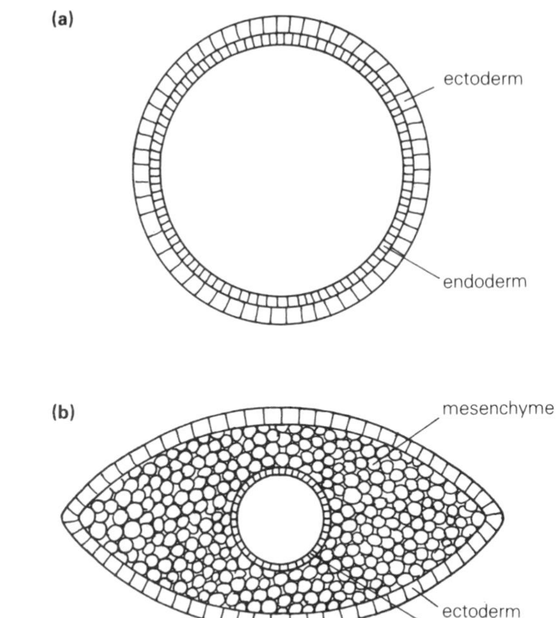

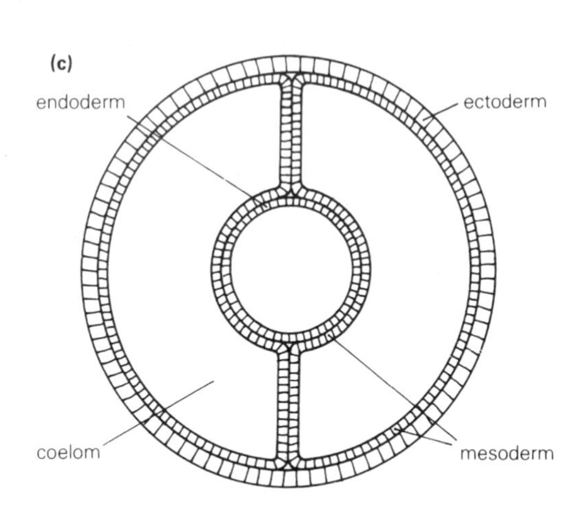

Figure 3.4 Grades of organization in metazoans: (a) diploblastic, (b) acoelomate triploblastic and (c) coelomate triploblastic.

(mesenchyme) between the endoderm and the ectoderm (Fig. 3.4b). The endoderm forms the gut, the ectoderm the outer skin. In all other invertebrates, as well as in the vertebrates, the mesoderm forms a lining to the ectoderm and overlies the endodermal gut also, enclosing a fluid-filled cavity known as the **coelom** (Fig. 3.4c). The gut lies freely within the coelom, and since it may be looped or coiled it does not have to follow the contours of the ectoderm. The coelom may have functioned primitively as a hydrostatic skeleton (the contained fluid giving some rigidity), and this function is retained in the annelid worms. These move by waves of contractions running forwards from the rear. Each contraction forces the incompressible fluid forwards in the direction of movement, and it is against the fluid itself that the muscles contract. Many other uses have been found for the coelom in higher animals, especially as a space for storage of internal organs and materials. The development of the various organs in animals can be followed in embryology. Organs of ectodermal origin include the skin and nervous system, most of the gut and some associated organs are endodermal, while respiratory and excretory organs and all muscles are mesodermal.

What can be established about the relationship between the main invertebrate phyla? The fossil record is of little value here since the phyla, of which at least 32 are already defined on their first appearance. It should be possible to establish how the phyla themselves are grouped in terms of common ancestry in 'mega-groups' or 'superphyla'.

Those schemes that have been drawn up are highly variable. Some authors favour a more-or-less orderly progression from simple to advanced organisms. Rather more envisage some kind of tree-like pattern, with dichotomous branchings, others again envisage a polyphyletic origin for most invertebrates, the various phyla or superphyla being derived independently from protistan ancestors. The problem, as Willmer (1990) puts it, is 'attempting to reconstruct a tree with evidence only from the most recent leaves and a few twigs".

The likely ancestor of later groups (itself derived from a protozoan precursor) is considered to have been a 'planula', a small, ciliated, ovoid mass of solid cells, floating in the sea. From this were independently derived cnidarians, ctenophores and possibly sponges. The next stage was the origin of acoelomate flatworms likewise originating, probably several times over, from a planula-type ancestor. This probably had solid mesenchyme and may well have resembled living platyhelminth flatworms. These gave rise, again perhaps polyphyletically, to the coelomates.

Two coelomate 'superphyla' are traditionally agreed as distinct, on embryological grounds, though some authorities claim that the apparent differences are less clear than they might appear. These are the **Protostomia** (annelids, molluscs and arthropods) and the **Deuterostomia** (chordates, echinoderms and a few minor phyla). In Protostomia (1) the early sequence of cell division (cleavage), follows a spiral course and the fate of each cell is determined from an early stage; (2) when the embryo has reached the 'blastula' stage, to become a hollow ball of cells, it invaginates to form a double-walled sphere with a hole at one end. This point of invagination becomes the mouth (protostomy); (3) the coelom arises by schizocoely, i.e. by the mesoderm splitting to produce a coelomic cavity; and (4) the swimming, ciliated larva is a trochophore. In the Deuterostomia on the other hand (1) cleavage follows a radial pattern and the fate of each cell remains flexible for a long time; (2) the point of invagination in the blastula becomes the anus and the mouth develops separately (deuterostomy); (3) the coelom arises by enterocoely, i.e. as invaginated pouches from the intestine; and (4) the larva is a 'tornaria'.

[A controversial view of the use of larvae as indicators of relationship, however, has been put forward by Williamson (1992). Radially symmetrically echinoderm adults have bilaterally symmetrical larvae. Why should this be so? Other incongruities between larvae and adults abound. Williamson experimentally fertilized eggs belonging to ascidians (sea squirts) with the sperm of echinoids; these are very distantly related groups. Some of these matured to become apparently normal sea-urchin larvae (plutei), and a few survived as adult sea urchins. Such experiments led Williamson to conclude that 'a larval form can be transferred from one animal to a very distantly related one by cross-fertilization'. If it is indeed possible, through naturally occurring hybridization, for one kind of marine invertebrate to 'hijack' the larval form of another, and that the amalgamation of different kinds of organisms actually occurred in the past, then the importance of larvae as phylogenetic indicators is much diminished.]

Within the deuterostomes a separate 'lophophorate' superphylum can be distinguished, including the Brachiopoda and Bryozoa, and the small 'worms' known as Phoronida. These all have the same kind of food-gathering apparatus, known as the **lophophore**. None of these phyla are segmented but they could have been derived from an animal with a very few segments.

Many authorities believe that the proto- and deuterostomes were derived independently from an acoelomate 'platyhelminth'. Bergström (1989, 1997), however, considers that deuterostomes were derived from protostomes by loosening developmental constraints; they are not really very far different. He envisages a stem group of metazoans, about the platyhelminth grade, which was pseudosegmented and with a true mouth and anus, creeping along on a ciliated locomotory sole. These, the Phylum Procoelomata, inhabited the later Proterozoic shallow seas, their larval stage being a trochophore.

They gave rise directly to Mollusca which retain many protostomian characters and variously to the many phyla that differentiated during the Cambrian radiation. Tommotiids and machaeridians could be procoelomates that developed external 'scales' and survived into the Palaeozoic. Willmer (1990) presents a slightly different view in her exhaustive review of invertebrate relationships, envisaging a great proliferation of coelomates in the early Cambrian, independently derived from acoelomate 'platyhelminths' and, although superphyla such as the protostome lophophorates can be distinguished, the overall impression is one of dramatic polyphyly, the majority of metazoan phyla being derived independently from an ancestral stock. The quite remarkable diversity of the Middle Cambrian Burgess Shale, British Columbia (Chapter 12), which includes many animals impossible to place in modern groups, lends further support to a polyphyletic origin for the coelomates.

The advent of molecular genetics on the one hand and cladistic methodology on the other has revivified the age-old debate on metazoan relationships. While much of this is beyond the scope of the present text, the following works give something of the flavour of the modern debate. Raff's (1996) synthesis of palaeontological data with developmental genetics puts special emphasis on the role of regulatory genes. Likewise Davidson *et al.* (1995) explore the origin of bilateral body plans in terms of developmental regulatory mechanisms. Gee's (1996) analysis of the origins of deuterostomes (echinoderms and chordates) relies greatly on cladistic methods, as does Nielsen's (1995) survey of relationships within the animal kingdom. Moore and Willmer (1997), on the other hand, believe convergent evolution to be so widespread that cladistic analysis based on parsimony (the choice of the simplest cladogram) will minimize and actually conceal convergence, leading to false results. They believe that the evaluation of morphological characters, coupled with molecular data, is the best way forward. The debate will evidently continue for a very long time.

It is possible that the earliest metazoans inhabited the **meiobenthos**, that is the poorly oxygenated sulphide-rich layer just below the surface of many marine sediments. Whereas the meiobenthos today has been invaded by some secondarily adapted benthos, at least some of its present-day small organisms (protistans and metazoans) might well be the relics of original Precambrian metazoan fauna, which originated in the sulphide-rich layers, and some of which invaded the oxygenated layers above as the progenitors of the later true benthos. If this were the case, then the early evolution of small metazoans might well have begun long before the beginning of the Cambrian. Unfortunately such meiobenthos is most unlikely to be fossilized and this we shall never know.

3.4 Major features of the Phanerozoic record

Diversification of invertebrate life

Once the major groups of invertebrates were established they were able to expand and differentiate, though their relative dominance has fluctuated significantly though time. The great Cambrian 'explosion' gave rise to many groups which were short-lived, but those phyla which survived the Cambrian have continued to the present day. Taxa of lesser rank have generally shorter time ranges. From most groups it has been possible to work out some of the details of their phylogeny, depending on how complete their fossil record is. To some extent, through functional studies, it is also possible to understand something of the biological quality of the fossil organisms even if these are of great antiquity, and it is very clear that even from the beginning of the fossil record invertebrate life was highly organized and well adapted, even if some of the later representatives of a stock apparently 'improved' on the earlier ones.

Trilobites, for example, which have been extinct for some 250 Ma, were undoubtedly highly functional invertebrates, as is witnessed by the complex morphology of their eyes and other environmental sensors. In a sense they could be described as 'primitive', since their appendages are undifferentiated and they lack some of the elaborations of other arthropod groups. The term 'primitive', however, should not be construed as meaning biologically inferior or ill-adapted, for though in an evolutionary sense they were unable to escape from the limitations of being trilobites they were able to adapt to many specialized niches and were amongst the most successful life forms of their time.

In trilobites, as in all of the invertebrate fossil groups studied here, once the thematic pattern of organization of each taxon was established it tended to remain very conservative. Thus, within any phylum each hierarchical level of organization can be related to an archetypal pattern of construction – a heritage which though highly functional was at the same time confining and restrictive. Every archetypal pattern, however much evolutionary plasticity was possible within it, remained an archetype for all the phylum. At taxonomic levels lower than the phylum, new kinds of animals could sometimes originate, breaking away from the functional system of their ancestors but immediately setting up new archetypes, which were likewise confining but which allowed, within certain limits, all manner of new evolutionary possibilities.

The molluscs, one of the most diverse of all invertebrate phyla, show very well how the potential inherent in a single archetypal plan was realized. A hypothetical 'archimollusc' (see Fig. 8.1) has a single shell covering the mass of viscera, a rasping mouth and a posterior cavity in which the gills were located. From this at least six new and independent groups emerged. Some, such as the monoplacophorans and polyplacophorans, remained quite close to the ancestral type; others – the bivalves, gastropods, scaphopods and most of all the cephalopods – diverged away in various degrees and became adapted to various habitats and modes of life. Each of these classes had an archetypal 'master plan' in which all the features of the archimollusc were present, though modified, added to and altered in a particular functional combination, allowing in each case a focus for new evolutionary potential.

In the molluscs and in most other invertebrate groups the initial differentiation took place very late in the Precambrian or in the earliest Phanerozoic. The earliest evolutionary stages are not preserved and can only be reconstructed by inference and with subjectivity. Likewise, though the formation of new invertebrate archetypes went on at lower taxonomic levels throughout the whole of the Phanerozoic, there is normally little record of the types that carried the potential for further development. This is probably because most of them lived in very small and rapidly evolving populations confined to very localized areas. The chances of their preservation and discovery are therefore very small, and such critically important intermediates are rarely found.

Changes in species diversity and habitat

Do more species live now than in former times? Or has the number of species remained more or less constant since, say, the late Cambrian? This has not been an easy problem to solve because of the imperfections and bias inherent in the fossil record, and until now there has been no clear agreement. Latterly, however, much data deriving from various sources has been pooled and shows a consistent pattern of change – a real increase in species diversity through time (Sepkoski *et al.* 1981). This is based upon analysis of (1) trace fossil diversity, (2) species per million years, (3) species richness, i.e. numbers of invertebrate species within fossil communities through time, (4) generic diversity through time and (5) diversity of families through time.

All five estimates, statistically tested, showed a single underlying pattern. Low diversity prevailed in the Cambrian, and although diversity was appreciably higher during the rest of the Palaeozoic it levelled off and did not go on increasing. After the late Permian extinctions there was an initial period of low diversity, followed by a real and continuous increase in diversity until the present day. There are, indeed, at least twice the number of species living in today's oceans as there were at any time during the Palaeozoic.

Modern seas abound in gastropods, bivalves, crabs, echinoids and fishes. In each of these modern groups the diversity of species exceeds that attained by any Palaeozoic phylum (other, perhaps, than trilobites in the Ordovician and crinoids in the Carboniferous). Conversely, however, the number of body plans or defined groups in the Palaeozoic was much higher. There are fewer kinds of animals, i.e. groups of high taxonomic rank, today than there were formerly, and this net reduction in types of organic design is a manifest trend in the fossil record. The highly successful groups of today, which include so many species, seem to be 'winnowed survivors' of what Gould has termed 'early experimentation and later standardization'. The early experimental groups, such as so many Cambrian Problematica, the Burgess Shale faunas and some of the more bizarre Lower Palaeozoic echinoderms, did not last long, but to what extend it was competition from other groups that weeded them out, or purely random physical processes, must remain conjectural.

Not only has there been a shift in diversity and biomass through time, but there has also been an overall displacement in habitat. In the earliest Palaeozoic the immobile suspension feeders (e.g. brachiopods, dendroids, bryozoans, corals, archaeocyathids and some kinds of echinoderms) occupied soft sediment surfaces. Throughout time, however, the soft sediment environment came to be occupied by deposit and infaunal suspension feeders such as protobranch bivalves, irregular echinoids, some crustaceans, holothurians and worms, as well as a number of scavengers and predators (Thayer, 1979). The great increase of such sediment-disturbing ('bulldozing') animals rendered the soft sediment surface less habitable to its former immobile denizens, and those that do survive (endobyssate bivalves and stalked crinoids) tend to favour relatively unbioturbated sediment. There was, through time, a corresponding increase in the diversity of mobile taxa in soft sediments, and of immobile suspension feeders on hard surfaces. Some initially immobile taxa, such as the crinoids, have mobile representatives today. It is quite possible that the rise of reef-forming and dwelling animals is at least partly due to the destruction of the original habitat by bioturbation.

Problematic early Palaeozoic fossils

The early Cambrian faunas represent a major metazoan radiation. Along with the 'standard' trilobites, molluscs, inarticulate brachiopods and others that were the progenitors of long-lived stocks, there were many short-lived, often geographically localized and frequently bizarre forms which cannot be related to any other groups, but which must be classified as extinct phyla. It was a time during which many independent lineages originated, but only some of which survived (Stanley, 1976), for as previously noted there were many more body plans in the early Palaeozoic than thereafter.

The rapid diversification of metazoans around the beginning of the Cambrian was a real evolutionary phenomenon, and the appearance of mineralized exoskeletons was only one aspect of this diversification (Stanley, 1976; Bengtson, 1977). There was surely an equivalent radiation of soft-bodied animals at phylum level and, although most traces of these have been lost, there are indeed a few isolated occurrences. For example, some curious soft-bodied pentameral organisms up to 10 cm in diameter and known only from a single bedding plane in the Scottish Highlands (Campbell and Paul, 1983) may have echinoderm affinities, but are more likely to belong to an unknown phylum. Similarly, organisms which formed their shells by agglutinating particles together are found for the first time in the Lower Cambrian.

The genus *Salterella*, for example, of Lower Cambrian age, is a small, conical fossil with its walls diverging at some 20° and with a central tapering bore. The walls consist of small agglutinated grains arranged in nested conical layers, each of which makes an angle of about 45° to the axis. *Salterella* bears a superficial resemblance to cephalopods with which it was originally confused, but it is now referred to the extinct, short-lived phylum Agmata (Yochelson, 1977, 1983). *Salterella* is widespread, ranging through much of North America, and is found also in the Northwest Highlands of Scotland.

Many of the 'small shelly fossils' (Fig. 3.2) from the Tommotian are of problematic origin. Some may well have been parts of larger organisms. Bengtson (1968, 1977; Qian and Bengtson, 1989). For example, the round, discoidal shell *Mobergella*, known from Scandinavia, carries some seven pairs of markings (which may be either elevations or depressions) in its concave inner surface (Fig. 3.5).

These may have been muscle attachments, and one might envisage the animal as a limpet-like creature, protected by the shell. Small specimens of such a shell could not have afforded any protection to its bearer unless it was a burrower. Likewise, the small button-shaped *Lenargyrion*, some 0.45 mm in diameter and consisting of a dense tuberculated lapping over a porous core, is of unknown affinity. Might these have been **dermal sclerites** – a kind of body armour of an extinct group? For the moment, and perhaps indefinitely, our knowledge of many of these bizarre, early Cambrian 'Problematica' must remain tantalizingly incomplete.

(a)

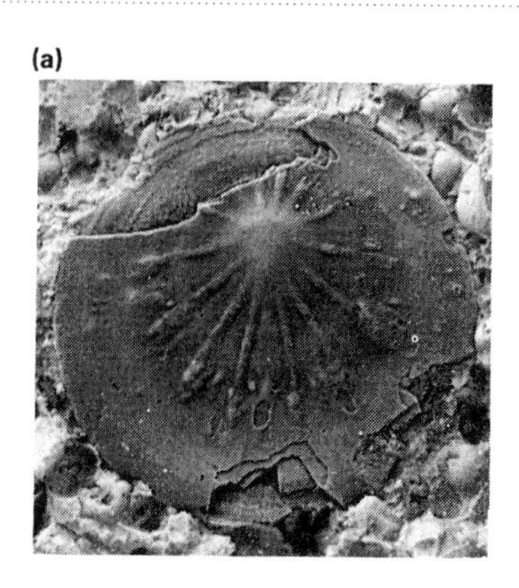

(b)

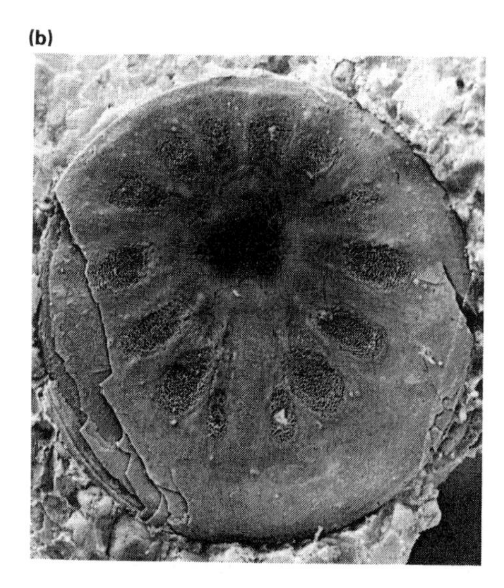

Figure 3.5 *Mobergella*, an enigmatic Lower Cambrian fossil from Sweden: (a) young stage with convex inner surface; (b) older example with concave inner surface (×15). (Photographs reproduced courtesy of Dr Stefan Bengtson.)

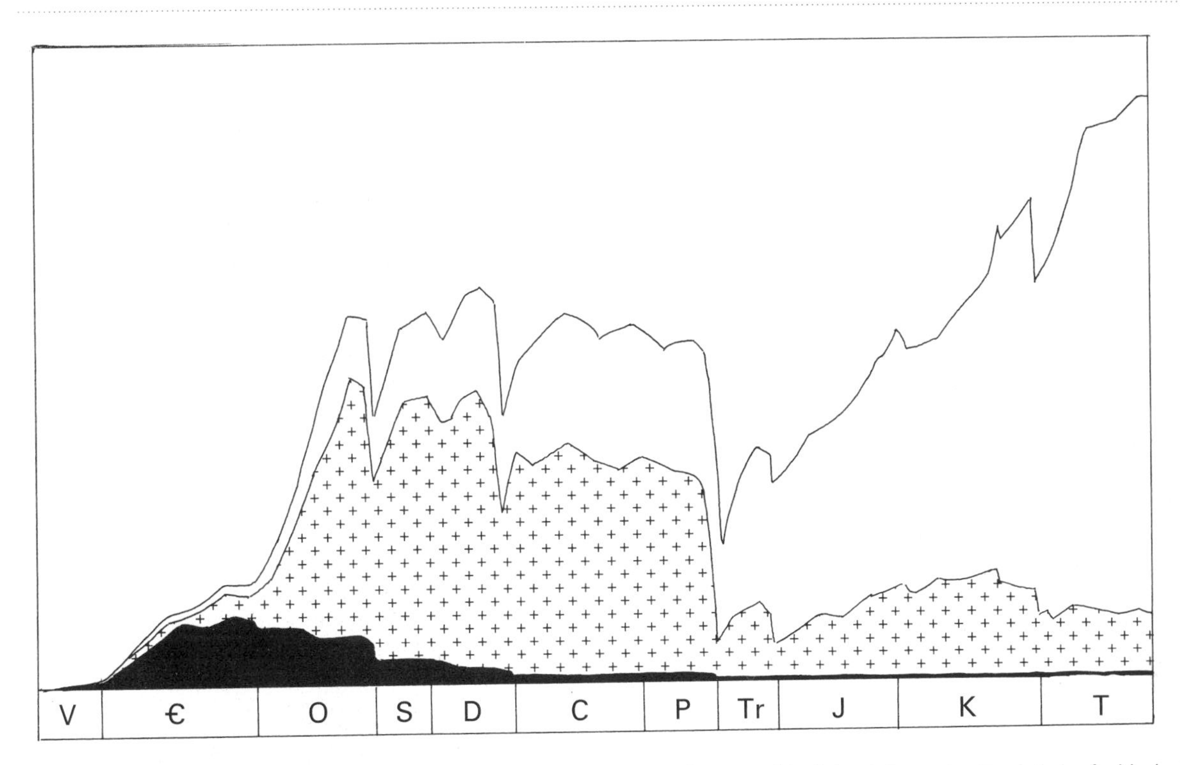

Figure 3.6 The three successive, but overlapping marine evolutionary faunas (e.f.) of the Palaeozoic. Cambrian e.f., black; Palaeozoic e.f., crosses; Modern e.f., clear. (Based on Sepkoski, 1990.)

Marine evolutionary faunas (Fig. 3.6)

The Ediacara fauna of 'vendozoans' and the Tommotian fauna of small shelly fossils and archaeocyathids record distinct early episodes in the history of marine life. Following this, and from the early Cambrian onwards, there has been a continual relay of successive invertebrate faunas, punctuated by a series of extinction periods of greater or lesser effect. This Phanerozoic record can be resolved into three successive 'evolutionary faunas' (Sepkoski, 1984, 1990). These are the Cambrian, Palaeozoic and Modern evolutionary faunas; their component elements seem to diversity rapidly and to decline slowly together, and they share other characteristics too. Each evolutionary fauna is not seen therefore as simply composed of randomly associated taxa, but are effectively linked in some way.

The Cambrian evolutionary fauna is dominated by trilobites, along with inarticulated brachiopods, hyolithids, monoplacophorans and early echinoderms, and the soft-bodied or organic shelled Burgess Shale animals belong here too. This evolutionary fauna diversified rapidly and its members were generalized deposit feeders and grazers. It began to decline in the Ordovician and became progressively restricted to deep-water environments. By the Permian it had all but vanished, apart from inarticulated brachiopods and monoplacophorans which survive until today.

Meanwhile the Palaeozoic evolutionary fauna had arisen with an 'explosion' of calcareous-shelled invertebrates in the early Ordovician. Rhynchonelliform brachiopods, corals, molluscs, bryozoans and echinoderms rapidly diversified, new trilobite families arose and graptolites with collagen skeletons dominated the plankton. Much of the benthos consisted of epifaunal suspension feeders, with more complex levels of ecological tiering than the Cambrian faunas had shown. The Ordovician radiation of this evolutionary fauna established the pattern of invertebrate life on the continental shelves at least until the end of the Permian (despite sharp changes at the end of the Ordovician and in the late

Devonian). Although the Palaeozoic fauna was severely affected by the end-Permian extinction event, many representative families survived through the Mesozoic to the present day.

The Modern, or Mesozoic–Cenozoic evolutionary fauna of bivalves, arthropods, echinoids, foraminiferids, crustaceans, fish and other groups had some representatives going back to the Cambrian and had slowly diversified through the Palaeozoic. When, following the Permian event, vacant ecospace became available again on the continental shelves it was the Modern evolutionary fauna which took over. There are many evident changes at this time, for example, the brachiopods losing their dominance to bivalves, and the complete replacement of coral and echinoderm faunas. Very diverse ecological types are present amongst representatives of the Modern fauna.

The three great evolutionary faunas overlapped but they are distinct and they provide a useful way of understanding the expansion of the biosphere. One characteristic feature that emerges clearly is that each successive evolutionary fauna diversified more slowly than that which preceded it, but has a higher level of maximum diversity, and a greater range of ecological strategies too. The stepwise pattern of increase in global diversity seen in Fig. 3.6 has led to twice as many families living today than there were at any time during the Palaeozoic. Parallels may also be sought in the expansion of terrestrial and freshwater plants and animals (Gray, 1987).

Despite the coherence of each of the three evolutionary faunas, each has undergone substantial internal turnover. This is well illustrated by the Modern fauna, and especially by the rise and fall of the ammonoids. Whereas the rise of the ammonoids in the Devonian was slow, by the Triassic they had become amongst the most common and characteristic of all preservable invertebrates and very diverse. Yet they too nearly became extinct again at the end of the Triassic, having already suffered a late Permian setback. All the ammonites of the Jurassic and Cretaceous were derived from a single small group of late Triassic ammonoids which alone survived the rigours of the end of the Triassic.

Ammonites flourished in the Mesozoic, along with bivalves, gastropods, echinoderms, corals and brachiopods, which were still locally common. In late Cretaceous time the ammonites declined. Over a period of some 10 Ma they became reduced in numbers and diversity and confined to only certain parts of the world, finally dying out at the end of the Cretaceous.

When, following the final demise of the ammonites, the earliest Tertiary faunas were established they were not unlike those of today. The hard-shelled faunal assemblages in the Tertiary consist of bivalves, gastropods, echinoderms, bryozoans, scleractinian corals and other forms, with brachiopods becoming a very minor component. Indeed shell debris on an early Tertiary shoreline would not, in terms of groups represented, be very different from that on a modern beach. It would contrast singularly with a Mesozoic beach, which would be covered by ammonite shells.

Climatic and sea-level changes

The Earth's climate has dramatically fluctuated during the Phanerozoic, oscillating between alternate 'icehouse' and 'greenhouse' conditions (e.g. Frakes *et al.*, 1992). The late Precambrian (Varangian) ice age gave way to a warmer, though still cool climate in the early Cambrian. By the late Cambrian the Earth had entered a hot 'greenhouse' phase, lasting until the end of the Devonian and punctuated only by the short, sharp, late Ordovician glaciation. Then from the Carboniferous to the late Triassic there was another 'icehouse' with extensive ice sheets covering the southern continent of Gondwana during the late Carboniferous and Permian. A warming episode in the late Triassic led to a second 'greenhouse' phase, peaking in the Cretaceous, and thereafter there was a general decline in temperature, leading to the present 'icehouse'.

Such large-scale climatic fluctuations can be related to changes in the internal workings of the Earth. 'Icehouse' phases occur when plate tectonic movements are minimal. At such times ocean ridges subside, thereby accommodating vast amounts of water in the ocean basins, so that the continental masses are left high and dry. Lowered vulcanicity minimizes atmospheric CO_2, and heat loss increases. With the onset of renewed plate tectonic activity, continents move, split and collide, ocean ridges rise, continents are flooded, and increased vulcanicity throws dust and CO_2 into the atmosphere so that heat is kept in. Thus begins a new 'greenhouse' phase. Such megacyclicity inevitably has profound effects on living organisms, since when due to global warming the continental shelves are flooded, vast new marine habitat areas are opened up for colonization. These disappear during a later 'icehouse' phase. Moreover, climatic change by itself, especially when rapid, is likely to generate a biological crisis.

Upon these two 'greenhouse' and three 'icehouse' megacycles of the Phanerozoic are imposed progressively shorter second-, third- and fourth-order cycles generated by other causes (Hallam, 1992). These too had their biological effects and these varying cycles, operating at different scales, are part of the restless stage upon which the history of life has been acted out.

Extinctions

It seems to be the fate of virtually all organisms eventually to become extinct. In the same way that new taxa have arisen through time so they decline and disappear and most genera and species have limited geological ranges. There is thus a continuous level of background faunal changeover, consisting of innumerable minor extinction and replacement episodes. Sometimes, however, many taxa become extinct more or less simultaneously. Such larger-scale extinction episodes terminate each of the geological periods and the systems themselves are defined on the basis on the faunas they contain. For example, at the end of the Cambrian the trilobite faunas that had dominated the scene for a long time underwent a worldwide change, the majority of families becoming extinct and being replaced by entirely new stocks (Whittington, 1966; Westrop, 1989). This abrupt change is a convenient stratigraphical marker, but there must have been reasons for it. Some kind of catastrophe seems to have taken place, though not necessarily a sudden one.

Such events are obviously destructive, but in a different sense they are also creative. For once new ecosystems are set up they tend to become rather rigid in an evolutionary sense. There is not much prospect within them for innovative change. Once such an ecosystem has broken down, however, the potential then becomes available for opportunistic replacement and it is at such times that new and promising evolutionary developments can prosper. Thus according to Boucot (1983) the marine Phanerozoic fossil record can be divided into 12 level-bottom 'ecological–evolutionary' units punctuated by extinction events of a greater or lesser magnitude. Within each of these units, communities were remarkably stable and their structure was tightly constrained within ecologically dictated limits. During these long periods of stability there were few evolutionary innovations and stabilizing selection operated both at the population and the community level. When, however, major extinction events disrupted the structure of ecosystems, not only were new structural and physiological innovations established during the subsequent adaptive radiations, but community types were completely reconstituted in new and different patterns.

Possible causes of mass extinctions

Of all the extinction periods which have been documented, by far the most severe was the late Permian crisis, which reduced the number of marine invertebrate families by 57% (with perhaps 95% of all species disappearing).

There were other crises of intermediate severity. Up to 22% of all families died out in the late Ordovician (end-Ashgillian), and crises of comparable magnitude took place in the late Devonian (Frasnian–Famennian; 21%) and late Triassic (Norian; 20%), and finally in the late Cretaceous (Maastrichtian; 15%). The last-mentioned of these is of particular interest as it affected land animals as well as marine faunas. None of these episodes, nor the less important extinction events down to the biomere level, were part of a regular cycle of events but were irregularly spaced throughout geological time.

The view that extinction events are coupled with a 26 Ma periodicity (which would suggest a single underlying causal mechanism, and probably extraterrestrial) has now rather fallen from favour (Patterson and Smith, 1987; McGhee, 1989a,b).

What is difficult, and rather more speculative, is to find causes. Since Alvarez *et al.* (1980) proposed that the terminal Cretaceous extinction event resulted from an asteroid colliding with the Earth, mass-extinction analysis has become something of a growth industry in palaeontology (Silver and Schulz, 1982; Larwood, 1988; Albritton, 1989; Chaloner and Hallam, 1989; Donovan, 1989; Alvarez, 1997).

Whereas there has been a natural, if misguided tendency to look for a common cause for all extinction events, many authorities would prefer to seek explanations in random combinations of more ordinary factors (Teichert, 1988). Indeed, as Holland (1989) has rightly pointed out, the use of the term 'mass extinction' can overdramatize the patterns that the fossil record actually has to show.

Potential causes of the deterioration of global ecosystems were determined by McGhee (1989a,b) as follows.

Earthbound mechanisms

Global temperature decline. This would particularly affect shallow-water tropical (and reef) ecosystems. Marine organisms living at higher latitudes would track migrating temperature belts and would be less severely affected, and likewise inhabitants of deep water.

Marine regression. Fluctuations in sea level have taken place hundreds of times during the Phanerozoic. Some have been no more than minor localized incursions or regressions of the sea; others on a much more widespread scale. Regressions would reduce habitable area on continental shelves (though habitat space around oceanic islands would increase). A flooded continental shelf or epicontinental sea provides a great area of living space and very many habitats. There will be especially many habitats if the temperature gradient across the sea is high. If the shelf seas are contracted, then the living space will disappear, many of the habitats will vanish and the ecological disturbance will be profound. A fall in sea level of only a few metres could do untold damage. What have been termed 'perched faunas' (Johnson, 1974), i.e. faunas that have evolved rapidly and have colonized a shelf or continental sea during a time of maximum flooding, are especially vulnerable when the sea retreats.

Furthermore, if any 'key species', plant or animal, occupying a critical position in a food web becomes extinct, the whole network of feeding relationships dependent upon it will be immediately disrupted, and those animals at higher trophic levels which depend on it for food may well become extinct too. Climatic buffering effects might likewise vanish. Marine regression would, of course, occur during glaciation, with water being locked up in polar ice. The combination of shrinking living space and cooler climate could be acute.

Volcanism. The production of masses of volcanic dust into the atmosphere would undoubtedly result in global cooling and result in environmental stress which could be long continued.

Oceanographic effects. Patterns of oceanic circulation may change through time; new systems of upwelling could be set up, and possibly large-scale overturn of stagnant bottom water could occur, with lethal effects on the pelagic and shelf fauna. On the other hand, a diminution of upwelling systems could, over an extended time period, deplete the plankton of essential nutrients.

Extraterrestrial mechanisms

Supernova radiations. Lethal blasts of radiation from a nearby supernova could well have adversely affected the Earth. It is not possible to tie any such event in with the fossil record, however, and most marine organisms would have been protected by sea water.

Bolide impact. Rare but devastating meteorite or asteroid collisions with the Earth could have totally disrupted the pattern of life. If a bolide some 10 km in diameter hit the Earth's surface, the first effect would be a shock wave with initial devastation. Second, there would be a period (lasting several months) of darkening by atmospheric dust. Following the initial blast the cessation of photosynthesis, acid rains and lowered temperatures due to sunlight being unable to penetrate the dust would have had lethal effects on sensitively balanced ecosystems and could have precipitated a major extinction. There is no doubt that the Earth has been bombarded by extraterrestrial bodies through time, for there are up to 100 known impact craters of various sizes on the Earth's surface which have been made during the Phanerozoic. The 26 km Ries crater in the Schwäbische Alb, in southern Germany, one of the best studied, is an impact structure some 15 Ma old; the meteorite must have been 1–2 km in diameter. A 70 km diameter Triassic impact crater is known from Ontario, and many of the lunar impact craters are much larger still. It has been estimated that meteorites of more than 10 km in diameter, producing craters of 200 km diameter, should occur about once every 40 Ma, and the likelihood of major extinction resulting from the impact of celestial projectiles is actually quite high.

In analysing particular extinction events, sedimentological and geochemical data are used in conjunction with information from the fossil record. It has become evident from this that no single mechanism can be invoked to account for all extinction events, and indeed some may have resulted from several coincident causes.

Late Ordovician (Ashgillian) extinction event

Towards the end of the Ordovician, sea level stood high, productivity was vigorous and environments were fairly stable. Shelf faunas were diverse and black graptolitic shale accumulated in deeper water. Then, at the beginning of the Hirnantian (the topmost stage in the Ordovician), the climate deteriorated rapidly, and the Gondwanan ice sheets expanded greatly, extending to quite low latitudes, and drastically lowering sea level. The resulting ecological disruption came in two abrupt phases with a more diffuse time of extinctions between (Brenchley, 1989). The first phase (late Rawtheyan) affected the plankton (notably graptolites and conodonts) and also the deeper-shelf benthos, where there was a great drop in diversity. This may be linked to oceanic overturn of anoxic or toxic deep water (Owen and Robertson, 1995). Shallow-shelf brachiopods were also affected but less severely, and there arose the unique, short-lived, but virtually worldwide *Hirnantia* fauna, dominated by brachiopods, which persisted until the end of the Ordovician. While this fauna is normally of low diversity (<10 genera), some variants may have up to 25 taxa, including local relicts of earlier faunas, and there is evidence of bathymetric zonation within the Hirnantian fauna too (Rong and Harper, 1989).

The second phase occurred during the glacial maximum. Sea level was drastically lowered and habitable shelf area reduced, causing further disruption extinction of the shallow-shelf biota.

The third and final phase of extinction, some hundreds of thousands of years later, occurred in late Hirnantian times, at the very end of the Ordovician. By this time the glaciation was largely over, and sea level rose quickly bringing with it anoxic waters which spread onto the shallow shelves and killed off the *Hirnantia* fauna. Only later did warm-water faunas re-establish themselves, from the beginning of the Silurian.

In North America the Late Ordovician brachiopod faunas became abruptly extinct, and when the seas rose again following the melting of the glaciers, European invaders colonized the North American shelf areas (Sheehan, 1973). The original North American communities had enjoyed a long period of stability, but were narrowly adapted to particular niches in defined community structures. The European species, however, had a greater flexibility and were less closely adapted. These were able to cope with adverse environmental conditions, but it was some 3–5 Ma before persistent community structures were re-established.

Late Devonian (Frasnian–Famennian) extinction event

This event is well attested (with some 21% of marine families lost) but less well understood. It seems to have been spaced out over some 3 Ma (McGhee, 1989a,b), but as a series of separate events, and with a particularly sharp drop at the end of the Frasnian. The worst effects were concentrated on the tropical reef ecosystems and warm-water shallow marine communities. The various components, however, were affected in different ways. Tabulate corals never recovered, though rugose corals did better, and although 75% of brachiopod genera became extinct, they staged a remarkable recovery, as did the ammonoids. High-latitude faunas and deeper-water associations (as well as terrestrial biotas) were much less badly affected. These effects are consistent with a model of drastic global cooling, but in this case sea level stood high throughout the whole biotic crisis, and glaciation was certainly not involved. Another factor, however, is the global spread of euxinic bottom conditions into shallow seas during the later Devonian, and this might itself have been promoted by a sea-level high stand. One or more major oceanic overturns of this stagnant bottom water would have poisoned the upper waters, and have led to severe faunal extinctions. Anoxic overturns of this kind have been considered as side effects of global cooling. At present (McGhee, 1996) a search continues for evidence of meteorite bombardment. Some evidence does exist in the form of several microtektite layers (melted glass spherules), soot horizons and a possible iridium anomaly, but these are scattered at various times in late Devonian history. More dramatically, several craters attributed to bolide impact are of late Devonian age. One of these, the 52 km diameter Siljan crater in central Sweden, may have formed in the late Frasnian, others at different times. The concept of multiple impacts, spaced out over an extended time period, is beginning to look appealing.

Late Permian extinction event

This acute crisis seems to have spanned at least 10 Ma, and was also coupled with extinction of terrestrial faunas. Those marine faunas which suffered the greatest reduction were tropical stenohaline groups, both benthic and pelagic, and the crinoid–brachiopod- and bryozoan-dominated ecosystems of the Late Palaeozoic were totally destroyed. Yet bivalves, nautiloids, gastropods and conodonts suffered comparatively little, and Permian gastropods indeed have proved an excellent group for studying controls on extinction and survival.

Faunal and geochemical data support a gradual, and not a catastrophic end-Permian event, though with successive sharp pulses within it. Many factors were involved, and no one single cause is envisaged (Erwin, 1993). One important factor was the palaeogeography of the time, since all the continents were then assembled into the giant supercontinent of Pangaea, with the triangular Tethyan ocean indenting its eastern margin. Such a configuration would certainly have affected climate and ocean circulation, and knock-on biological effects would have been inevitable. Second, there was the most extensive known marine regression of any time during the Phanerozoic. This regression began in the early Permian and by the end of the Permian the overall drop in sea level has been calculated as 210–280 m. The loss of marine habitat was very great, and this, together with marked seasonal climate swings, led to a singular ecological instability and explains also why there are so few good boundary sections against which these faunal changes can be calibrated. Warm shallow seas became reduced to certain parts of the world only and much of the Permian marine sedimentation was deposited in inland seas connected to the ocean by narrow channels and bordered by enormous algal reefs. Of these the best known are the seas of western Texas and northwestern Europe. Evaporites were deposited both behind the reefs as salt water washed over them and in the basins themselves as they periodically dried out. The low-lying regions formerly occupied by these seas were sometimes flooded, only to dry out again; in the Zechstein sea of northwestern Europe five such desiccation cycles have been recognized. While the hypersaline waters of these seas were generally hostile to life, except for the algal reefs themselves and a few specialized invertebrates, the amount of salt taken out of the upper water of the sea and permanently locked away in evaporite sediments must have been considerable. Whether this was enough to reduce the salinity of surface waters so as to affect adversely the life of stenohaline marine organisms is debatable.

It was formerly thought that this drop in sea level was the result of global cooling, but the Permo-Carboniferous glaciation was finished by the late Permian; sinking oceanic ridges is a more likely cause.

A further factor seems to have been massive vulcanicity, centred in Siberia, causing a dramatic increase in atmospheric CO_2 levels, which may have led to oceanic anoxia. These multifarious causes collectively led to previously unparalleled environmental instability and so to collapsing ecosystems.

By the end of the Permian the last reef environments had vanished, and all associated ecosystems had collapsed. Normal marine conditions do not seem to have been restored for several million years. The Permian extinction therefore (Maxwell, 1989) seems to have involved global cooling, marine regression, knock-on effects of these and perhaps several other factors.

Late Triassic (Carnian–Norian) extinction event

In the late Triassic there were heavy losses in many invertebrate groups and the ammonoids and bivalves only just survived. Conodonts, conulariids and strophomenide brachiopods became extinct. Land faunas also suffered dramatically at this time. Extinctions of marine faunas were spread through the Scythian, Carnian and subsequent Norian (Benton, 1990), and an extraterrestrial cause thus seems remote. Although stratigraphic resolution is less good in the Triassic than it is, for example, in the Jurassic, data from European pectinid bivalves and crinoids especially (Johnson and Simms, 1989) show substantial reductions in the Carnian, contemporaneously with widespread facies changes. Simms and Ruffell (1990) couple these estimations with synchronous climatic changes, particularly to the onset of increased rainfall during an otherwise arid time. This may have led to a change in ocean surface temperature, pH or salinity changes or loss of carbonate habitats.

Cretaceous–Tertiary boundary extinction

The K–T boundary is marked by the more or less simultaneous extinction of many groups of animals and plants, ranging from phyto- to zooplankton, through ammonites and belemnites to the terrestrial, marine and flying reptiles and land plants. Even though this crisis was less severe than that of the late Permian, some estimates suggest that 15% of all marine families and up to 75% of all Cretaceous species were eliminated.

There may have been no single cause for these events, indeed the slow decline of the ammonites during the late Cretaceous, and their restriction to only a few parts of the world prior to their final extinction, may suggest that gradual environmental change was a primary control. This view has been reinforced recently (McLeod *et al.*, 1997) by refined calibration of the time ranges of several Cretaceous marine organisms, which show that ostracods, bryozoans, and bivalves, as well as land reptiles were also declining during this time. Yet there was a pronounced sharp extinction at the very end of the Cretaceous, and in recent years much support has been given to 'catastrophic' theories to explain this and other severe extinction periods. There are two main contenders here; bolide impact and intense vulcanicity. The present debate dates from the discovery of an iridium-rich layer in marine clays in Italy and Denmark by Alvarez *et al.* (1980). This layer lies precisely at the Cretaceous–Tertiary boundary, as recognized by faunal changes. Iridium is depleted in the Earth's crust but is 10^3–10^4 times more abundant in meteorites, and it was postulated that it was derived from a large asteroid or meteorite some 10 km in diameter which collided with the Earth at the end of the Cretaceous.

In support of the asteroid impact hypothesis the iridium layer has now been found worldwide (up to 75 localities), often with shocked quartz grains in non-marine as well as marine sequences. Whereas stratigraphic resolution does not permit a precise correlation between marine and terrestrial sequences, data from plant fossils over the K–T boundary points to a dramatic and sudden environmental disturbance. Sections in western North America (Wolfe and Upchurch, 1986; Boulter *et al.*, 1988; Upchurch, 1989) are particularly instructive in this context. Here a high-diversity evergreen broadleaved flora, indicative of rather dry conditions, abruptly terminates at the iridium-rich K–T boundary clay layer. Above this lie a few millimetres of rock devoid of spores or plant cuticles, but often with sapropel or fusain, indicating rotting or burned vegetation (and soot layers at this level indicating huge wildfires have been found worldwide). There follows a few centimetres of coal or mudstone with only fern spores and megafossils, and thereafter angiosperm remains increase with evidence of greater precipitation, showing a phased and gradual recovery until a tropical rainforest was established. The timescale here is about 1.5 Ma.

Evidence of this kind is compelling, and is consistent with a drastic and sudden environmental trauma, followed by a slow recovery. The likelihood that an asteroid impact was responsible has been much accentuated by the discovery of a 200–300 km wide crater of end-Cretaceous age at Chicxulub in the Yucatan Peninsula. It is buried by later sediments but has now been geophysically mapped, and shows at least three vast concentric rings of the kind to be expected from the impact of a bolide over 10 km in diameter. (Alvarez, 1997). Drillings in the Caribbean made in 1997 give evidence of great disturbance in sediments of the K–T boundary, which are surely the result of shock waves generated by this collision. Some geologists, however, believe that intense volcanic activity around this time could equally be responsible for the iridium layer (or layers), global wildfires, and the dramatic floral and faunal changes documented, and indeed, the 65 Ma Deccan Trap flood basalts in India coincide exactly with the K–T boundary. These flood basalts erupted for a million years or so. Possibly India was migrating over a hot mantle plume at the time. Alternatively the volcanism might have been triggered off by another asteroid impact. There are iridium layers spanning up to 20 Ma on either side of the K–T boundary, and more needs to be known about the distribution of iridium in sediments generally. Likewise, a more refined correlation of marine and terrestrial events is needed as a test for synchroneity. Even though the relative effects of asteroid impact and vulcanicity remain unclear, and although this has to be set against a background of other changes, the available evidence shows unequivocally that a great and far-reaching catastrophe, or indeed multiple catastrophes, devastated the Earth 65 Ma ago.

Smaller, but nevertheless important stepwise extinctions took place in the Eocene and Oligocene (Prothero, 1989) and dominantly affecting land animals in the late Pleistocene.

Bibliography

Books, treatises and symposia

Albritton, C.C. (1989) *Catastrophic Episodes in Earth History*. Topics in the Earth Sciences No. 2. Chapman & Hall, London, 221 pp. (Clear, simple presentation)

Alvarez, W. (1997) *T. rex and the Crater of Doom*. Princeton University Press, Princeton, New Jersey, 185 pp. (Popular account of meteorite impact hypothesis)

Berggren, W.A. and Van Couvering, J. (eds) (1984) *Catastrophes and Earth History*. The New Uniformitarianism. Princeton University Press, Princeton, New Jersey. (18 papers, deals with mass extinctions)

Chaloner, W.G. and Hallam, A. (1989) Evolution and extinction. *Philosophical Transactions of the Royal Society of London B* **325**, 241–490. (16 symposial papers; most important)

Clark, R.B. (1964) *Dynamics in Metazoan Evolution*. Clarendon Press, Oxford. (Classic text)

Donovan, S.K. (ed.) (1989) *Mass Extinctions: Processes and Evidence*. Belhaven, London, 266 pp. (Exceptionally useful compilation of several papers)

Dyer, B.P. and Obar, R.A. (1994) *Tracing the History of Eukaryotic Cells; the Enigmatic Smile*. Columbia University Press, New York. (Clear, easily understood)

Erwin, D. (1993) *The Great Paleozoic Crisis: Life and Death in the Permian*. Columbia University Press, New York (Very good and readable)

Frakes, L.A., Francis. J.E. and Sytkus, J.I. (1992) *Climate Modes of the Phanerozoic*. Cambridge University Press (Excellent synthesis)

Gee, H. (1996) *Before the Backbone: Views on the Origin of Vertebrates*. Chapman & Hall, London. (Deuterostome relationships)

Glaessner, M.F. (1984) *The Dawn of Animal Life – a Biohistorical Study*. Cambridge University Press, Cambridge, 244 pp. (New, very clear summary)

Hallam, A. (1977) *Patterns of Evolution as Illustrated by the Fossil Record*. Elsevier, Amsterdam. (17 original papers)

Hallam, A. (1992) *Phanerozoic Sea-level Changes*. Columbia University Press, New York. (Readable synthesis)

House, M.R. (ed.) (1979) *The Origin of Major Invertebrate Groups*, Systematics Association Special Volume No. 12. Academic Press, London. (Many essential papers)

Larwood, G.P. (ed.) (1988) *Extinction and Survival in the Fossil Record*. Systematics Association Special Volume No. 34. Clarendon, Oxford, 365 pp. (15 papers, excellent quality)

Lipps, J.H. and Signor, P.W. (1992) *Origin and Early Evolution of the Metazoa*. Plenum Press, New York and London. (16 papers)

Margulis, L. (1970) *Origin of Eukaryotic Cells*. Yale University Press, Newhaven. (Symbiotic origin of eukaryotes; biology, biochemistry, palaeontology)

Margulis, L. (1981) *Symbiosis in Cell Evolution – Life and its Environment on the Early Earth*. Freeman, San Francisco. (Up-to-date statement of endosymbiosis)

McGhee, G.R. (1996) *The Late Devonian Mass Extinction: the Frasnian/Famennian Crisis*. Columbia University Press, New York. (The best treatment available)

McMenamin, M.A.S. and McMenamin, D.L.S. (1990) *The Emergence of Animals – the Cambrian Breakthrough*. Columbia University Press, 217 pp. (Straightforward, readable text)

Nielsen, C. (1995) *Animal Evolution*. Oxford University Press. (Cladistics emphasized)

Raaben, M.E. (ed.) (1981) *The Tommotian Stage and the Cambrian Lower Boundary Problem* (translated from Russian), Amerind Publishing, New Delhi. (Illustrated)

Raff, R.A. (1996) *The Shape of Life: Genes, Development, and the Evolution of Animal Form*. University of Chicago Press. (links palaeontology and molecular genetics)

Schopf, J.W. (1983) *Earth's Earliest Biosphere – its Origin and Evolution*. Princeton University Press, Princeton, New Jersey. (15 sections on origin of life and first ecosystems)

Schopf, J.W. (ed.) (1992a) *Major Events in the History of Life*. Jones and Bartlett, Boston, Mass. (6 synthetic papers)

Schopf, J.W. and Klein, C. (eds) (1992) *The Proterozoic Biosphere*. Cambridge University Press, Cambridge. (41 papers)

Silver, L. and Schultz, P. (eds) (1982) *Geological Implications of the Impact of Large Asteroids and Comets on the Earth*, Geological Society of America Special Paper No. 190, Geological Society of America, Lawrence, Kan. (Many papers)

Walter, M.R. (ed.) (1976) *Stromatolites*. Elsevier, Amsterdam. (Many original papers)

Ward, P.D. (1992) *On Methuselah's Trail: Living Fossils and the Great Extinctions*. Freeman, New York. (Excellent light reading)

Williamson, D.I. (1992) *Larvae and Evolution: Towards a New Zoology*. Chapman & Hall, London. (Unorthodox, but interesting)

Willmer, P. (1990) *Invertebrate Relationships*. Cambridge University Press, Cambridge, 400 pp. (Modern, very readable and up-to-date treatment)

Individual papers and other references

Alvarez, L.W., Alvarez, W., Asaro, F. and Michel, H.V. (1980) Extraterrestrial cause for the Cretaceous–Tertiary extinction. *Science* **208**, 1095–108. (The meteorite impact hypothesis)

Awramik, S.M. (1982) The origins and early evolution of life, in *The Cambridge Encyclopaedia of Earth Sciences* (ed. D.G. Smith), Cambridge University Press, Cambridge, pp. 349–62. (Useful synthesis)

Barghoorn, E.S. (1971) The oldest fossils. *Scientific American* **224**(5), 30–42. (Fully illustrated work on Precambrian evolution)

Bengtson, S. (1968) The problematic fossils in the early Palaeozoic. *Acta Universitatis Upsaliensis* **415**, 1–71. (Lower Cambrian problematica)

Bengtson, S. (1977) Early Cambrian button-shaped microfossils from the Siberian platform. *Palaeontology* **20**, 751–62.

Bengtson, S, Conway Morris, S., Cooper, B.J. *et al.* (1990) Early Cambrian fossils from South Australia. *Memoirs of the Association of Australasian Palaeontologists* **9**, 1–364.

Benton, M. (1990) End-Triassic, in *Palaeobiology: a Synthesis* (eds D.E.G. Briggs and P.R. Crowther), Blackwell, Oxford, pp. 194–8.

Bergström, J. (1989) The origin of animal phyla and the new phylum Procoelomata. *Lethaia* **22**, 259–69.

Bergström, J. (1997) Origin of high-rank groups of organisms. *Palaeontological Research* **1**, 1–14.

Boucot, A.J. (1983) Does evolution take place in an ecological vacuum? *Journal of Paleontology* **57**, 1–30. (Very stimulating)

Boulter, M.C., Spicer, R.A. and Thomas, B.A. (1988) Patterns of plant extinction from some palaeobotanical evidence, in *Extinction and Survival in the Fossil Record*, Clarendon, Oxford, pp. 1–36.

Brasier, M.D. (1979) The Cambrian radiation event, in *The Origin of Major Invertebrate Groups* (ed. M.R. House), Systematics Association Special Volume No. 12, Academic Press, London, pp. 103–59.

Brasier, M.D. (1992a) Background to the Cambrian explosion. *Journal of the Geological Society of London* **149**, 585–7.

Brasier, M.D. (1992b) Nutrient-enriched waters and the early skeletal fossil record. *Journal of the Geological Society of London* **149**, 621–9.

Brasier, M.D. (1992c) Palaeoceanography and changes in the biological cycling of phosphorus across the Cambrian–Precambrian boundary, in *Origin and Early Evolution of the Metazoa* (eds J.H. Lipps and P.W. Signor), Plenum Press, New York and London, pp. 483–524.

Brenchley, P. (1989) The late Ordovician extinction, in *Mass Extinctions: Processes and Evidence* (ed. S.K. Donovan), Belhaven, London, pp. 104–32.

Campbell, N. and Paul, C.R.C. (1983) Pentameral fossils from the Lower Cambrian pipe rock of the North-West Highlands. *Scottish Journal of Geology* **19**, 347–54. (Soft-bodied problematica)

Clark, R.B. (1979) Radiation of the Metazoa, in *The Origin of Major Invertebrate Groups* (ed. M.R. House), Systematics Association Special Volume No. 12, Academic Press, London, pp. 55–101.

Cloud, P. (1976) Beginnings of biospheric evolution and their biogeochemical consequences. *Paleobiology* **2**, 351–87. (Recent work on late chemical–early biological evolution)

Conway Morris, S. (1993) Ediacaran-like fossils in Cambrian Burgess Shale-type faunas of North America. *Palaeontology*, **36**, 563–635.

Cook, P.J. (1992) Phosphogenesis around the Proterozoic–Phanerozic transition. *Journal of the Geological Society of London* **149**, 637–646.

Cook, P. and Shergold, J. (1984) Phosphorus, phosphorites and skeletal evolution at the Cambrian–Precambrian boundary. *Nature* **308**, 231–6.

Crimes, T.P. (1976) The stratigraphical significance of trace fossils, in *The Study of Trace Fossils* (ed. R.W. Frey), Springer-Verlag, Berlin, pp. 109–30. (Trace fossil evidence on beginnings of metazoan life)

Crimes, T.P. (1992a) Changes in the trace fossil biota across the Proterozic–Phanerozoic transition. *Journal of the Geological Society* **149**, 637–46.

Crimes, T.P. (1992b) The record of trace fossils across the Proterozoic–Cambrian boundary, in *Origin and Early Evolution of the Metazoa* (eds J.H. Lipps and P.W. Signor), Plenum Press, New York and London, pp. 177–204.

Davidson, E.H., Peterson, K.J. and Cameron, R.A. (1995) Origin of bilateralian body plans: evolution of developmental regulatory mechanisms. *Science* **270**, 1319–25.

Gehli, J.C. 1987. Earliest known echinoderm – a new Ediacaran fossil from the Pound subgroup of South Australia. *Alcheringa* **11**, 337–43.

Glaessner, M.F. (1961) Precambrian animals. *Scientific American* **204**, 72–8. (Ediacara fauna)

Glaessner, M.F. (1962) Precambrian fossils. *Biological Reviews* **37**, 467–94. (Ediacara fauna)

Glaessner, M.F. and Wade, M. (1966) The late Precambrian fossils from Ediacara, South Australia. *Palaeontology* **9**, 599–628. (Ediacara fauna; technical descriptions)

Glaessner, M.F. and Wade, M. (1971) *Praecambridium*: a primitive arthropod. *Lethaia* **4**, 71–8. (Ediacara fauna)

Gray, J. (1987) Evolution of the freshwater ecosystem: the fossil record. *Palaeogeography, Palaeoclimatology, Palaeoecology* **62**, 2–214.

Grotzinger, J. P, Bowring, S.A., Saylor, B.Z. and Kaufman, A.J. (1995) Biostratigraphic and geochronologic constraints on early animal evolution. *Science* **270**, 598–604.

Holland, C.H. (1989) Synchronology, taxonomy and reality, in *Evolution and Extinction* (eds W.C. Chaloner and A. Hallam). *Philosophical Transactions of the Royal Society B* **325**, 263–77.

Jenkins, R.J.F. (1992) Functional and ecological aspects of Ediacaran assemblages, in *Origin and Early Evolution of the Metazoa* (eds J.H. Lipps and P.W. Signor), Plenum Press, New York and London, pp. 131–176.

Johnson, A.L.A. and Simms, M.J. (1989) The timing and cause of the late Triassic marine invertebrate extinctions: evidence from scallops and crinoids, in *Mass Extinctions: Processes and Evidence* (ed. S.K. Donovan), Belhaven, London, pp. 174–94.

Johnson, J.C. (1974) Extinction of perched faunas. *Geology* **2**, 479–82. (Faunas colonizing flooded shelves are most vulnerable to extinction when the sea retreats)

Kaufman, A.J., Knoll, A.J., Semikhatov, M.A. *et al.* (1996) Integrated chronostratigraphy of the Proterozoic–Cambrian boundary in the western Anabar region, Northern Siberia. *Geological Magazine*, **133**, 509–33.

Knoll, A.H. (1996) Daughter of time. *Paleobiology* **22**, 1–7 (update on Precambrian–Cambrian boundary; excellent)

Knoll, A.H. and Barghoorn, E.S. (1974) Precambrian eukaryotic organisms: a reassessment of the evidence. *Science* **190**, 52–4. (Casts doubt upon fossil eukaryotic algae)

Knoll, A.H. and Calder, S. (1983) Microbiotas of the late Precambrian Rysskö Formation, Nordaustlandet, Svalbard. *Palaeontology* **26,** 467–96. (Diverse prokaryotic flora)

Knoll, A.K. and Swett, K. (1987) Micropalaeontology across the Precambrian-Cambrian boundary in Spitzbergen. *Journal of Paleontology* **61**, 898–926.

Licari, C.R. (1978) Biogeology of the late pre-Phanerozoic Beck Spring Dolomite of eastern California. *Journal of Paleontology* **52,** 767–92.

Margulis, L. and McMenamin (1990) Marriage of convenience. *The Sciences* **30**, 31–6. (Eukaryote origins).

Matthews, S.C. and V. Missarzhevsky (1975) Small shelly fossils of late Precambrian and early Cambrian age: a review of recent work. *Quarterly Journal of the Geological Society of London.* **131**, 289–304. (Microfossil evidence on explosion of life in early Cambrian)

Maxwell, W.D. (1989) The mid Permian mass extinction, in *Mass Extinctions: Processes and Evidence* (ed. S.K. Donovan), Belhaven, London, pp. 152–73.

McGhee, G. (1989a) Catastrophes in the history of life, in *Evolution and the Fossil Record* (eds K.C. Allen and D.E.G. Briggs), Belhaven, London, pp. 26–50.

McGhee, G. (1989b) The Frasnian–Famennian extinction event, in *Mass Extinctions: Processes and Evidence* (ed. S.K. Donovan), Belhaven, London, pp. 133–51.

McLeod, N. *et al.* (22 authors) (1997) The Cretaceous–Tertiary biotic transition. *Journal of the Geological Society of London* **154**, 265–92.

McMenamin, M.A.S. (1989.) The origins and radiation of the early Metazoa, in *Evolution and the Fossil Record* (eds K.C. Allen and D.E.G. Briggs), Belhaven, London, pp. 73–98.

Moore, J. and Willmer, P. (1997) Convergent evolution in invertebrates. *Biological Reviews* **72**, 1–60.

Newell, N.D. (1963) Crises in the history of life. *Scientific American* **208**, 1–16. ('Catastrophism' as a determinant for major extinction periods)

Owen, A.W. and Robertson, D.B.R. (1995) Ecological changes during the end-Ordovician extinction. *Modern Geology* **20**, 21–39.

Patterson, C. and Smith, A.B. (1987) Is the periodicity of extinctions a taxonomic artefact? *Nature* **330**, 248–51.

Prothero, D.R. (1989) Stepwise extinctions and climatic decline during the later Eocene and Oligocene, in *Mass Extinctions: Processes and Evidence* (ed. S.K. Donovan), Belhaven, London, pp, 217–34.

Qian Yi and Bengtson, S. (1989) Palaeontology and biostratigraphy of the Early Cambrian Meishucunian stage in Yunnan Province, South China. *Fossils and Strata* **24**, 1–156.

Rhoads, D.C. and Morse, J.W. (1971) Evolutionary and ecologic significance of oxygen-deficient marine basins. *Lethaia* **4,** 413–28. (Was early evolution limited by O_2 deficiency?)

Rong, J.-Y. and Harper, D.A.T. (1989) A global synthesis of the latest Ordovician Hirnantian brachiopod faunas. *Transactions of the Royal Society of Edinburgh: Earth Sciences* **79**, 383–402.

Rozanov, A. Yu. and Zhuravlev. A. Yu. (1992) The Lower Cambrian fossil record of the Soviet Union, in *Origin and Early Evolution of the Metazoa* (eds J.H. Lipps and P.W. Signor), Plenum Press, New York and London, pp. 205–282.

Runnegar, B. (1992) Evolution of the earliest animals, in *Major Events in the History of Life* (ed J.W. Schopf), Jones & Bartlett, Boston, Mass., pp. 41–62.

Schopf, J.W. (1968) Microflora of the Bitter Springs Formation, late Precambrian, Central Australia. *Journal of Paleontology* **42**, 651–8. (*Glenobotrydion* - possible eukaryote?)

Schopf, J.W. (1992b) The oldest fossils and what they mean, in *Major Events in the History of Life* (ed J.W. Schopf), Jones & Bartlett, Boston, Mass., pp. 29–63.

Schopf, J.W. and Oehler, D.Z. (1976) How old are the eukaryotes? *Science* **193**, 47–9. (Defends Precambrian algae as eukaryotic)

Seilacher, A. (1989) Vendozoa: Organismic construction in the late Proterozoic biosphere. *Lethaia* **22**, 229–39

Seilacher, A. (1992) Vendobionta and Psammocorallia: lost constructions of Precambrian evolution. *Journal of the Geological Society of London* **149**, 607–13.

Sepkoski, J. (1984) A kinetic model of Phanerozoic taxonomic diversity III. Post-Palaeozoic families and mass extinctions. *Paleobiology* **10**, 246–67.

Sepkoski, J. (1990) Evolutionary faunas, in *Palaeobiology: a Synthesis* (eds D.E.G. Briggs and P.R. Crowther), Blackwell, Oxford, pp. 37–41.

Sepkoski, J., Bambach, R.K., Raup, D.M. and Valentine, J.W. (1981) Phanerozoic marine diversity and the fossil record. *Nature,* London **293**, 435–7. (Change in diversity through times)

Sheehan, P.M. (1973) The relation of Late Ordovician glaciation to the Ordovician–Silurian changeover in North American brachiopod faunas. *Lethaia* **6**, 147–84.

Simkiss, K. (1989) Biomineralisation in the context of geological time. *Transactions of the Royal Society of Edinburgh: Earth Sciences*. **80**, 193–9.

Simms, M.J. and Ruffell, A.H. (1990) Synchroneity of climatic change and extinctions in the Late Triassic. *Geology* **17**, 265–8.

Stanley, S.M. (1976) Fossil data and the Precambrian–Cambrian evolutionary transition. *American Journal of Science* **276**, 56–76.

Strother, P. (1989) Pre-metazoan life, in *Evolution and the Fossil Record* (eds. K.C. Allen and D.E.G. Briggs), Belhaven, London, pp. 51–72.

Sylvester-Bradley, P.C. (1975) The search for protolife. *Proceedings of the Royal Society B* **189**, 213–33. (Chemical and early biological evolution)

Tappan, H. (1982) Extinction or survival: selectivity and causes of Phanerozoic crisis, in *Geological Implications of Impacts of Large Asteroids and Comets on the Earth* (eds L. Silver and P. Schultz), Geological Society of America Special Paper No. 190, Geological Society of America, Lawrence, Kan., pp. 265–76.

Teichert, C. (1988) Extinctions and extinctions. *Palaeobiology* **2**, 411.

Thayer, C.W. (1979) Biological bulldozers and the evolution of marine benthic communities. *Science* **203**, 458–61. (Change in habitat through time)

Tucker, M.E. (1992) The Precambrian–Cambrian boundary: seawater chemistry, ocean circulation and nutrient supply in metazoan evolution, extinction and biomineralisation. *Journal of the Geological Society of London* **149**, 655–68.

Upchurch, G.R. (1989) Terrestrial environmental changes and extinction patterns at the Cretaceous–Tertiary boundary, North America, in *Mass Extinctions: Processes and Evidence* (ed. S.K. Donovan), Belhaven, London, pp. 195–216.

Vidal, G. and A.H. Knoll (1982) Radiations and extinctions of plankton in the late Proterozoic and early Cambrian. *Nature* **297**, 57–60. (Early history of algae)

Wade, M. (1968) Preservation of soft-bodied animals in Precambrian sandstone at Ediacara, South Australia. *Lethaia* **1**, 238–67. (Why the Ediacara fauna was preserved)

Wade, M. (1972) Hydrozoa and Scyphozoa and other medusoids from the Precambrian Ediacara fauna. South Australia. *Palaeontology* **15,** 197–225. (Technical descriptions)

Westrop, S. (1989) Trilobite mass extinctions near the Cambrian–Ordovician boundary in North America, in *Mass Extinctions: Processes and Evidence* (ed. S.K. Donovan), Belhaven, London, pp. 89–103.

Whittaker, R.H. (1969) New concepts of kingdoms of organisms. *Science* **163**, 150–60. (Five kingdoms)

Whittaker, R.H. and Margulis, L. (1978) Protist classification and the kingdoms of organisms. *Biosystems* **10**, 3–18.

Whittington, H.B. (1966) Phylogeny and distribution of Ordovician trilobites. *Journal of Paleontology* **40,** 696–737.

Woese, C.F. and Wächtershaüser, G. (1990) Origin of life, in *Palaeobiology: a Synthesis* (eds D.E.G. Briggs and P.R. Crowther), Blackwell, Oxford, pp. 3–9. (Valuable, up-to-date synthesis)

Wolfe, J.A. and Upchurch, G.R. (1986) Vegetation, climate and floral changes at the Cretaceous–Tertiary boundary. *Nature* **324**, 148–52.

Yochelson, E.L. (1977) Agmata, a proposed extinct phylum of early Cambrian age. *Journal of Paleontology* **51**, 437–54.

Yochelson, E.L. (1983) *Salterella* (Early Cambrian; Agmata) from the Scottish Highlands. *Palaeontology* **26**, 253–60.

PART TWO
Invertebrate Phyla

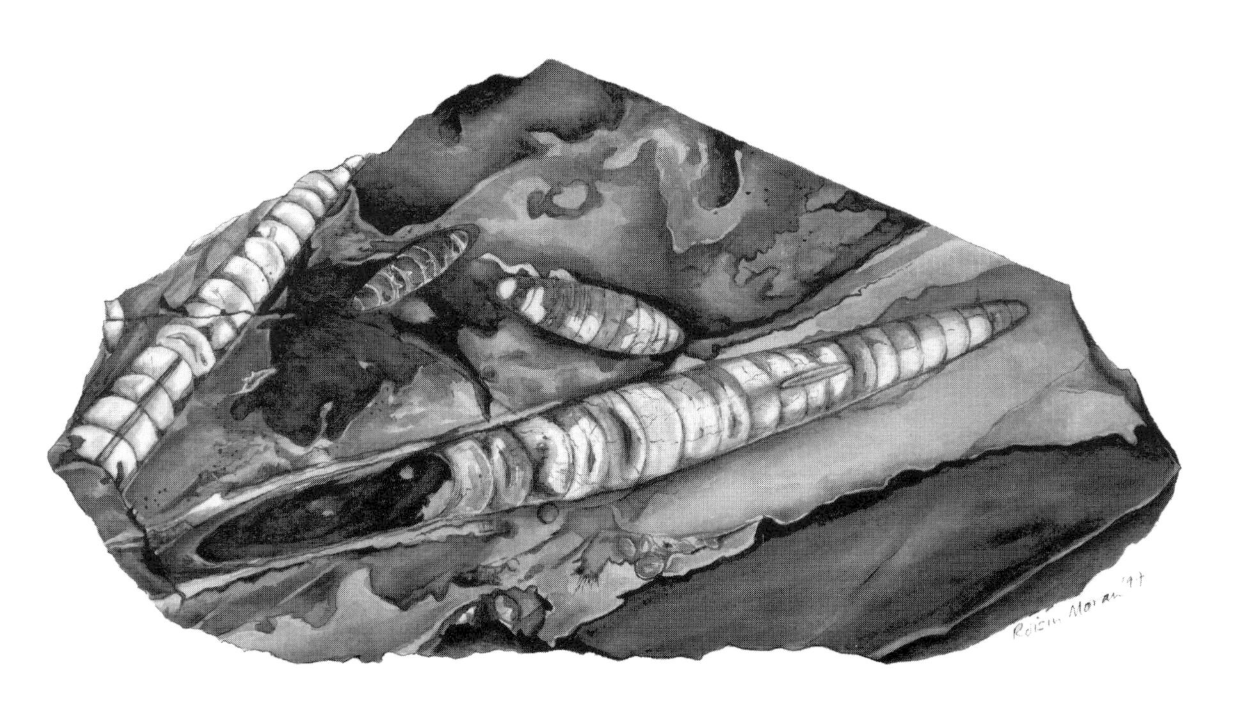

Orthocone cephalopods cut tangentially, showing internal septa and siphuncular structures, in red Ordovician limestones from Sweden. Painting by Roisin Moran; original specimen in the James Mitchell Museum, University College, Galway, Ireland.

4 Sponges

4.1 Phylum Porifera: sponges

The sponges are multicellular organisms but are not regarded as metazoans. They have only a few types of cell, and these are not really organized in tissues. There is no nervous system. In many ways they are of a grade of organization in between protozoans and metazoans; hence they are sometimes known as **parazoans.** Apparently they were derived from a protozoan ancestor but were an evolutionary blind alley, not ancestral to any metazoan.

The fine structure of sponges can often be used for identifying the species, but the external morphology may vary within fairly broad limits depending upon the environment. Sponges have remarkable powers of regeneration, and a living sponge squeezed through a silk net will re-form again on the other side. But individual cells removed from sponges, even though they may look just like some protozoans, will not live long; they are only viable as part of the sponge.

Sponges are generally sessile benthic animals and are all filter feeders (Bergquist, 1978). A typical sponge (Fig. 4.1a) has an upright bag-shaped body with a central cavity (**paragaster**) opening at the top via an **osculum**. The outer surface of the sponge is perforated by numerous tiny holes (**ostia**), which lead to **incurrent canals** and thence to **chambers** within the sponge body.

These chambers are lined by **collar cells** or **choanocytes** (Fig. 4.2), each of which faces into the chamber.

Exhalant passages lead from these to the central cavity. The collar cells are the most important elements in the organization of the sponge. Furthermore, these cells demonstrate the relationship between sponges and protozoans, since the single-celled or sometimes colonial protozoans known as the Choanoflagellida are in form virtually identical with sponge collar cells. Some planktonic marine choanoflagellids, like sponges, construct basket-like capsules of geometrically organized rods of silica – a further indication of biological affinity.

Each collar cell is a small globular cell with a cylindrical collar projecting from it, composed of fine pseudopodia. This collar encircles a solitary central flagellum. In life all the flagella beat continually, the tip of each whirling in a spiral motion. The current generated by all the flagellae draws water in through the ostia to the chambers and out to the central cavity and the osculum. Particulate organic matter then adheres to the sticky outside of the collar, which ingests it. Though the whirling of each flagellum is comparatively slow, the combined effect of all the flagellae operating together produces an efficient though low-pressure pump which can pass through the sponge every minute a volume of water equal to the sponge's own volume. Sponges generally have other kinds of cells than the collar cells. In one major group, Subphylum Gelatinosa (Fig. 4.4), the outer surface is covered with flattened epithelial cells (**pinacocytes**) of which those in the pore regions (**porocytes**) are perforated and can close off the pores by contraction if necessary. Each porocyte is stimulated to close individually by the presence of noxious substances in the water; the conduction of such stimulation to neighbouring cells may take place, but in the absence of a nerve net is very limited. There are also amoeboid cells (**amoebocytes**) which wander through the sponge body and transfer nourishment from the choanocytes to other parts of the sponge.

Most sponges have a skeleton; this may simply be a colloidal jelly, but in most living kinds it consists of a horny material (**spongin**) or of calcareous or siliceous **spicules** (Figs 4.6, 4.10, 4.11) or of both. Some living poriferans (sclerosponges) and several fossil groups

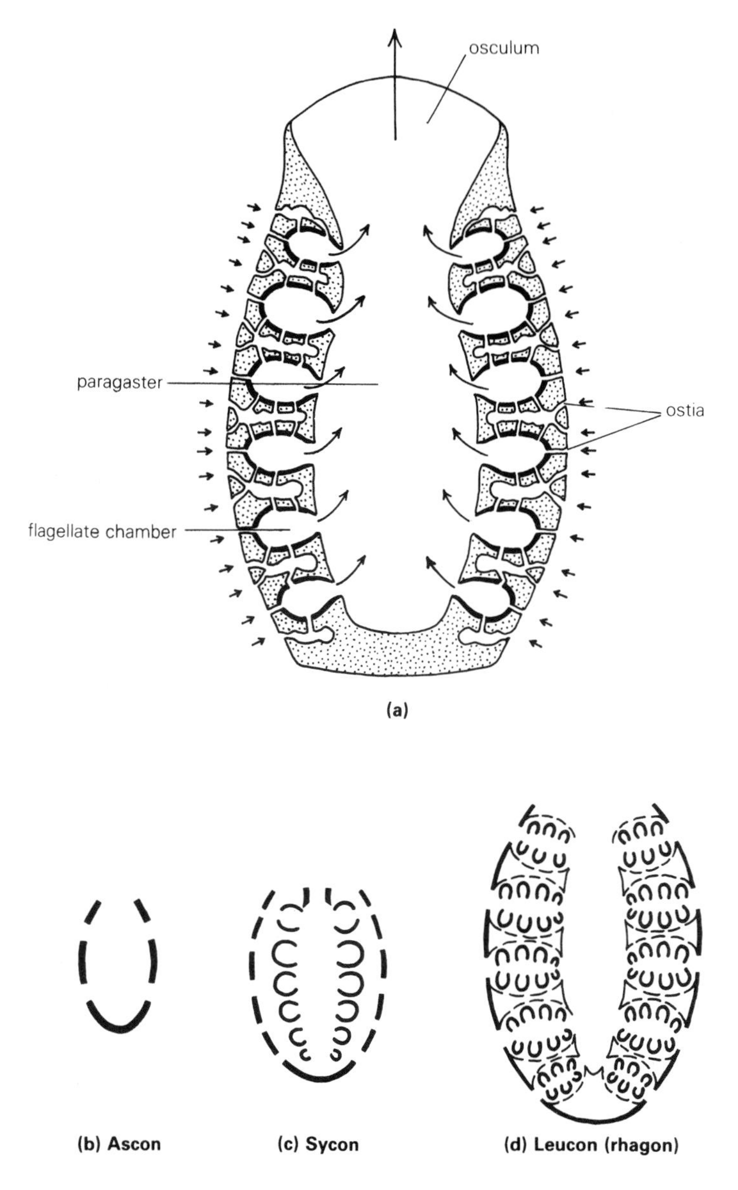

Figure 4.1 Elements of sponge morphology: (a) leucon (rhagon) type showing passage of water currents; (b)–(d) ascon, sycon and leucon grades of organization. [(a) based on drawing in Marshall, 1978.]

have a calcareous skeleton in addition to the spicules, and the extinct archaeocyathans have a skeleton consisting of calcium carbonate alone. Spicules can be fossilized, especially when they are united so as to hold the body of the sponge together after death. Thus complete or partial sponge bodies, mats of spicules or isolated spicules are found in rocks extending back to the Cambrian and possibly further. Sometimes sponges use shell debris, sand or even the spicules of dead sponges to strengthen their skeletons.

There are three grades of organization in living sponges. In the simplest (**ascon**) grade (Fig. 4.1b) the sac-like body is merely a single chamber lined with choanocytes, with attendant epithelial cells and amoebocytes. **Sycon** sponges (Fig. 4.1c) have a number of grouped ascon-like chambers with a cen-

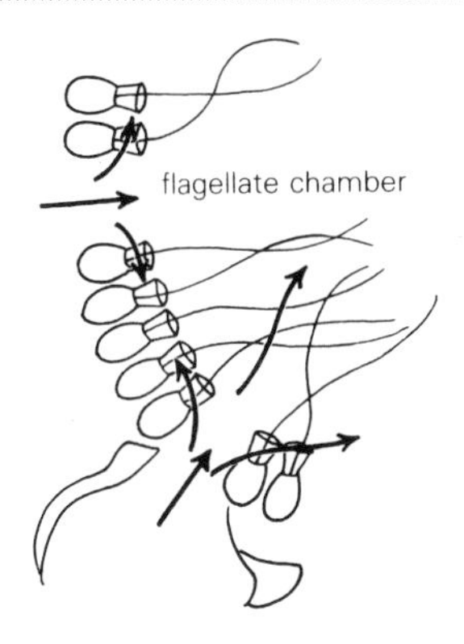

Figure 4.2 Wall of chamber, showing incurrent pores with arrows indicating current directions and collar cells (choanocytes).

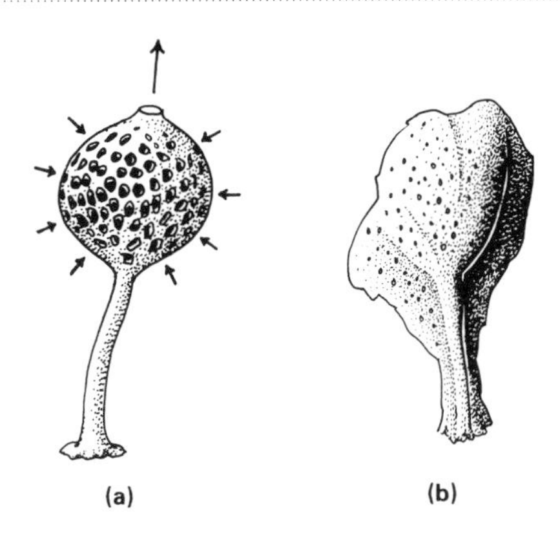

Figure 4.3 (a) Life form of a stalked, still-water sponge and (b) life form of a 'rheophilic' sponge with a wide lateral osculum directed away from the current. (Redrawn from an illustration in Carter, 1952.)

tral opening, but the vast majority of sponges are of **leucon** (**rhagon**) type (Fig. 4.1a,d), in which a number of sycon-like elements open into the large central cavity (paragaster). In the larger modern sponges, shrimps, ophiuroids or other commensal animals may live in the paragaster, and such associations may be found in fossil forms. Ascons are relatively small and not much larger than 10 cm high. Sycon and leucon sponges with their folded chambers exhibit obviously greater filtering efficiency and grow to a larger size.

There are about 1500 genera of modern sponges, of which some 80% are marine (the rest are freshwater). The shallow marine genera show a remarkable tolerance of intertidal conditions. Sponges abound in the deep sea, especially the Hexactinellida or siliceous glass sponges, which are confined to quiet, still waters. Sponge morphology is well adapted to the external environment, especially in such functional necessities as the separation of incurrent and excurrent water (Bidder, 1937; Hartman and Reiswig, 1973). Thus many deep-water sponges are **stalked,** the incurrent water coming in near the stalk attachment and the excurrent wastewater passing through the osculum (Fig. 4.3a).

Sponges living in shallow water in which current directions are constant may have the osculum down one side, so that the broader fan-like surface faces into the current and the lateral osculum away from it (Fig. 4.3b). This not only makes use of the prevailing current but also prevents reuse of excurrent water. A parallel with this is found in bryozoans and crinoids.

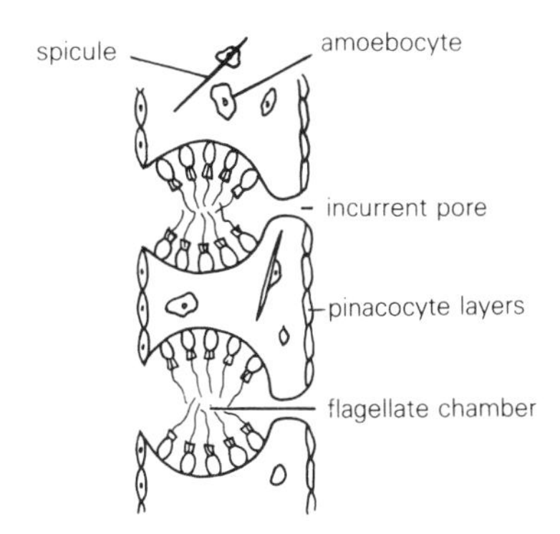

Figure 4.4 Wall structure of an advanced gelatinosan sponge. (Redrawn from Reid, 1958–1964.)

4.2 Classification

Sponge taxonomy at the highest level is based upon the soft tissues, though only the skeleton is preserved in the fossils. Two subphyla have been distinguished (Reid, 1958–1964) on the structure of the sponge wall. In Subphylum Gelatinosa (Fig. 4.4) the outer epthelial layer of flattened pinacocytes overlies a gelatinous middle layer (**mesenchyme**) in which the spicules are secreted by **scleroblasts** and wherein the amoebocytes wander.

The innermost layer, in sponges of ascon grade, is that bearing the choanocytes; in more advanced grades these are invaginated within chambers.

The other subphylum, Nuda (Fig. 4.5), has neither pinacocyte layer nor mesenchyme; the

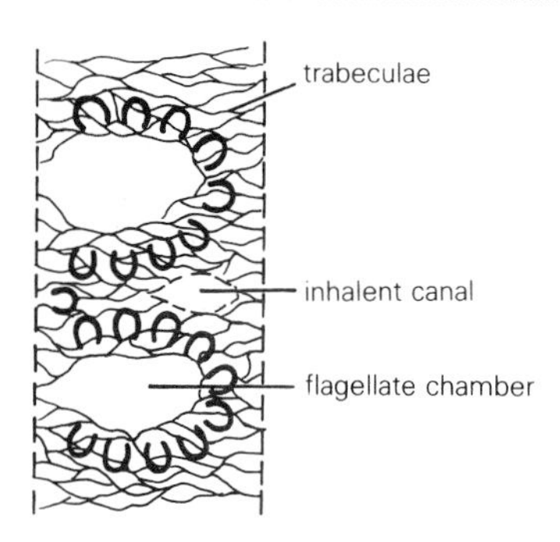

Figure 4.5 Wall structure of a hexactinellid (Nuda) of leucon grade. (Redrawn from Reid, 1958–1964.)

choanocytes are borne in a network of syncitial filaments (**trabeculae**) but are likewise organized in chambers.

The majority of living Porifera have a skeleton of spicules, and until recently it was imagined that the only sponges in the fossil record were likewise spicular. Since the discovery of Recent sclerosponges, with both a spicular and a calcareous skeleton, the classification and relationships of fossil and living sponges has been dramatically revised, more so, perhaps, than in any other fossil group. It has now been recognized that the skeleton of stromatoporoids, chaetetids and sphinctozoans likewise possesses both spicules and calcareous material. Though the aragonite skeleton is often recrystallized, obscuring the primary structure, acid etching and SEM micrography has enabled spicules to be detected. These groups, together with the entirely calcareous archaeocyathids, are now all regarded as true sponges. It has also become clear that solid, calcareous skeletons developed independently within at least two of the main poriferan classes (Demospongea and Calcarea), and that 'stromatoporoids', for example, are thus best regarded as 'grades of organization' rather than as of separate taxonomic rank, and they can be accommodated within modern sponge classes (Zhuravlev *et al.*, 1990). Because of such convergent evolution, many problems remain in trying to produce a satisfactory classification of the Porifera. A provisional classification is given below: this no doubt may well be modified in the future.

PHYLUM PORIFERA (Cam.–Rec.): Multicellular animals with choanocytes and a skeleton of spongin and/or calcareous or siliceous spicules and/or massive carbonate.

SUBPHYLUM 1. GELATINOSA: Porifera with pinacocytes and mesenchyme; choanocytes on the inner mesenchyme surface.

CLASS 1. DEMOSPONGEA (Cam.–Rec.): Gelatinosa of leucon grade with siliceous spicules and/or spongin and sometimes with foreign inclusions. Spicule rays usually diverge at 60 or 120°. Living sclerosponges, chaetetids, most sphinctozoans and most stromatoporoids are included here.

CLASS 2. CALCAREA (CALCISPONGEA) (Cam.–Rec.): Ascons, sycons or leucons with a skeleton of calcareous spicules. A few sphinctozoans and stromatoporoids belong to this class.

SUBPHYLUM 2. NUDA: Porifera without pinacocyte cells or mesenchyme; choanocytes set in a network of protoplasmic threads.

CLASS HEXACTINELLIDA (HYALOSPONGEA) (L. Cam.–Rec.): Nuda, usually leucons, with spicular rays diverging at 90°. Spicular skeleton only.

INCERTAE SEDIS. ARCHAEOCYATHA (L.–M. Cam.): Calcareous perforated skeleton of two 'nested' inverted cones joined by radial partitions. A few sphinctozoans and stromatoporoids are included here.

4.3 Class Demospongea

Spicular demosponges

Most fossil demosponges are represented by siliceous spicules only, the skeleton having collapsed. Such spicules may have either one single ray (**monaxon**) or four rays (**tetraxon**) diverging at 60 or 120° (Fig. 4.6c).

Ancient spicules, even of Cambrian age, can normally be related to modern families. Only in Order Lithistida (Cam.–Rec.) are the spicules of different form. These knobbly or tubercular **desmas** (Fig. 4.6a,b) are often so interlocked that the skeleton holds together after death. Lithistid genera are common from the Jurassic onwards, and such types as the Cretaceous *Siphonia* show details of the canals, chambers, paragaster and osculum when sectioned. Clionid sponges can penetrate into hard shell or limestone material, spreading out below the surface and secreting a spicular skeleton within the shell. Such sponges contribute greatly to the destruction of hard shell tissues and thus are of some geological importance.

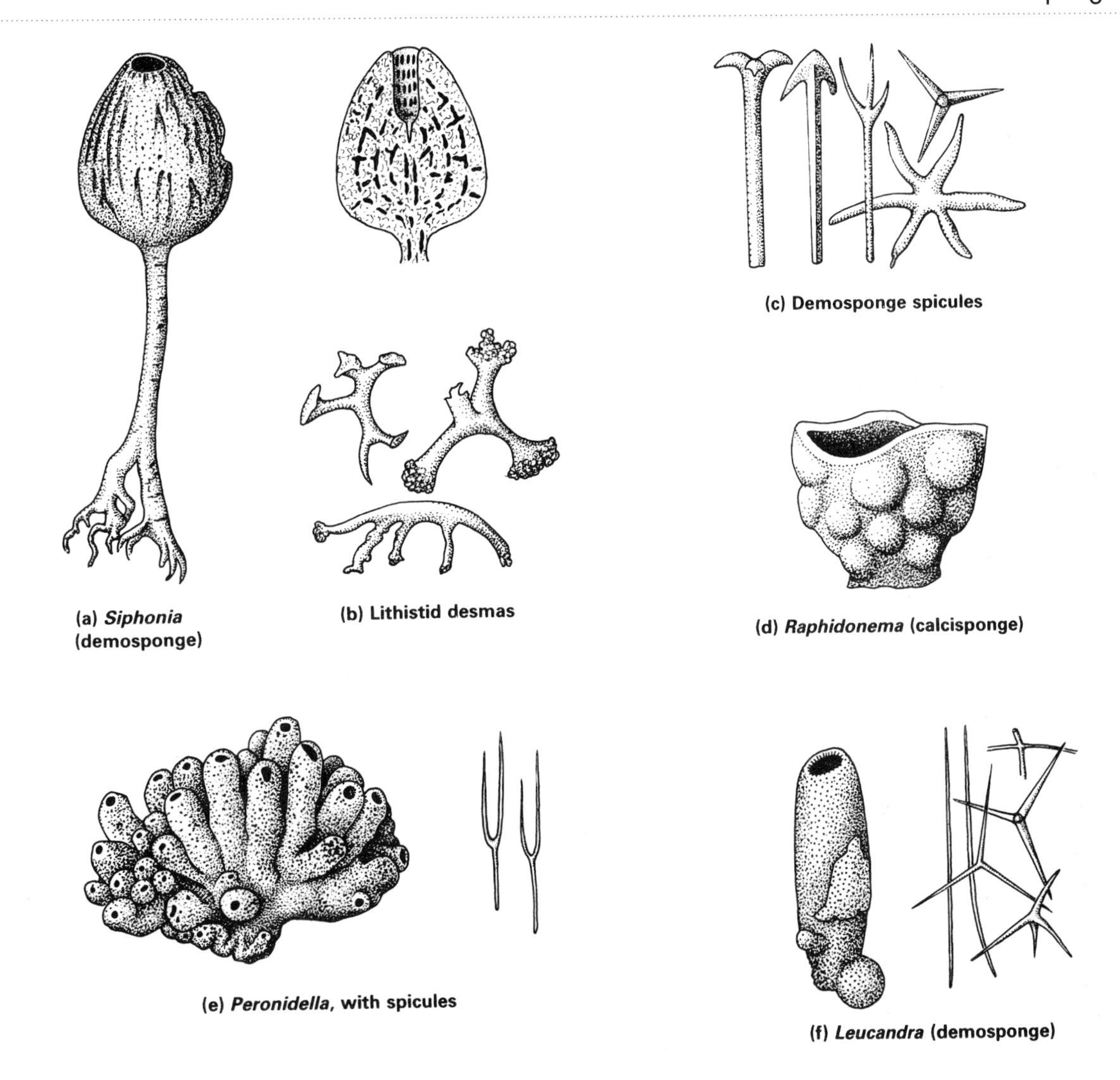

Figure 4.6 Demospongea and Calcarea: (a) *Siphonia* (Cret.), a stalked example (left) cut in section (right), showing the internal canals and paragaster (× 0.5); (b) lithistid desmas (× 25); (c) demosponge spicules (× 30); (d) *Raphidonema*, a Cretaceous calcareous sponge (× 0.25); (e) *Peronidella*, a Cretaceous calcareous sponge (× 0.5), with 'tuning fork' spicules (× 30); (f) *Leucandra*, a demosponge (× 0.5), with tetraxon spicules (× 30). (Mainly redrawn from Hinde, 1887–1893.)

Sclerosponges

Some 13 living species of 'coralline sponges' have been described since 1970, inhabiting submarine caves and other dark cryptic recesses (Hartman and Goreau, 1970). These secrete a massive, encrusting, aragonitic skeleton in addition to a skeleton of spongin and spicules. The calcareous skeleton may consist of both aragonite and calcite and it progressively engulfs the spicular skeleton during growth. The spicules then become incorporated in the calcareous material.

The outer surface of the living tissue investing the skeleton is pierced with two sizes of pores; small incurrent ostia, and larger, widely spaced excurrent oscula. These join with a system of radiating star-shaped canals, often running in grooves within the outer layer of the skeleton. Water is sucked in

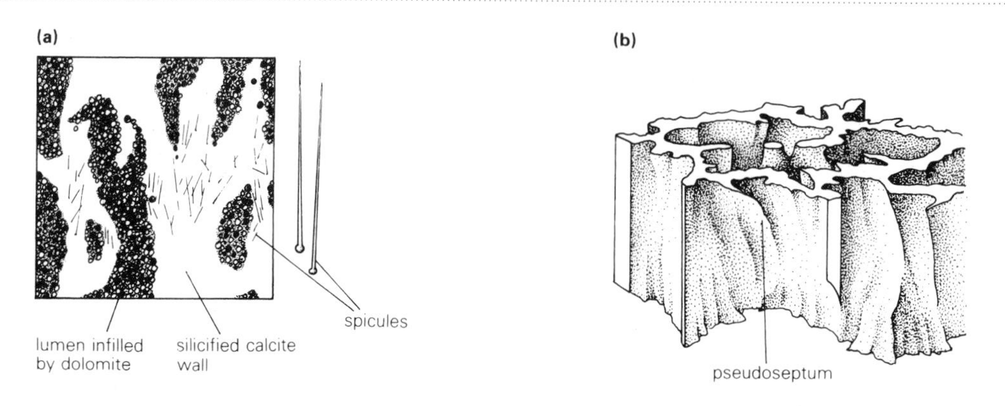

Figure 4.7 *Chaetetes*: (a) wall structure and (b) drawing showing appearance in thin section, with spicule pseudomorphs. (Based on a drawing in Gray, 1980.)

through the ostia by choanocytes lying in linked chambers and expelled through the excurrent oscula.

Fossil sclerosponges are known from the Tertiary of California.

Chaetetids

These are laminar, hemispherical or encrusting organisms, often of quite large size. They were long believed to be tabulate corals, but they are now recognized as closely related to sclerosponges.

Spicules were first discovered in the Carboniferous *Chaetetes* (Fig. 4.7) (Gray, 1980).

In this genus the colonies display rhythmic growth bands, and on sectioning they show an irregular to subpolygonal cellular structure, the cells being up to 0.5 mm across and known as calicles. The calicle walls vary in thickness, often extending into ridges, which may in turn develop into a pseudoseptal structure and occasionally into isolated columns. These structures commonly give the calicle walls a scalloped appearance. The spicules of *Chaetetes* are found within the calicle walls. In this material from Wales these walls are often diagenetically silicified, a process which has preserved the calcite spicules in patches. The spicules themselves are straight or slightly curving tylostyles up to 0.3 mm long, sharply pointed distally, and with a distinct boss at the proximal end. The presence of such calcareous spicules seems to settle the affinities of the Chaetetida and they are now recognized as being related to sclerosponges. In the Cretaceous *Stromatoaxinella* there are also spicules (Wood and Reitner, 1988), but these are of quite a different kind, and characteristic of a different group of demosponges. Indeed, in many ways this genus is intermediate between chaetetids and stromatoporoids. From such evidence, it is now clear that 'chaetetids' are polyphyletic, and that this grade of organization is convergent, having arisen more than once in geological history.

Stromatoporoids

Stromatoporoids are calcareous masses of layered and structured material found in carbonate sequences of Cambrian to Oligocene age and dominant in the Silurian and Devonian. Palaeozoic stromatoporoids, especially those of the Ordovician to Devonian, were important reef formers. Though common enough and greatly studied, their zoological nature has only recently been clarified; they are likewise demosponges with both spicules and a calcareous skeleton.

They bear some resemblance to compound tabulate corals, since the calcareous skeleton is of irregular layered form and forming rounded masses or thin flat sheets, occasionally cylindrical or discoidal (Fig. 4.8).

Where the upper surface is preserved it normally shows a pattern of polygonal markings. It also may have small swellings (**mamelons**) at intervals and frequently a most characteristic feature of stromatoporoid organization, stellate grooves known as **astrorhizae**. These are strikingly similar to the star-

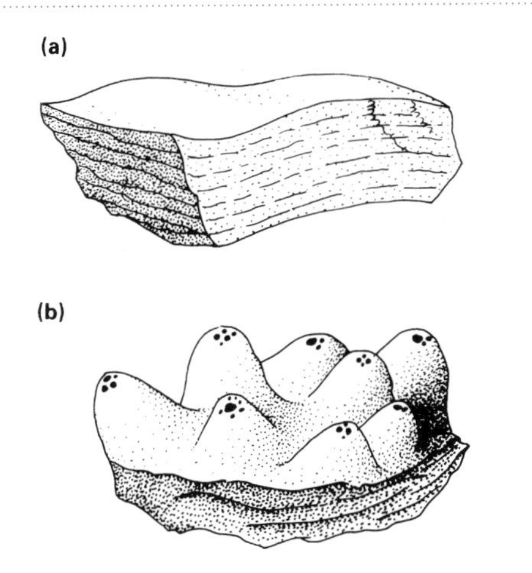

Figure 4.8 (a) *Stromatopora* (Dev.), showing laminae in a fractured part of a colony (×0.75); (b) upper surface of *Stromatopora*, showing mamelons with openings of astrorhizae (×1). (Based on a drawing by Lecompte in *Treatise on Invertebrate Paleontology*, Part F.)

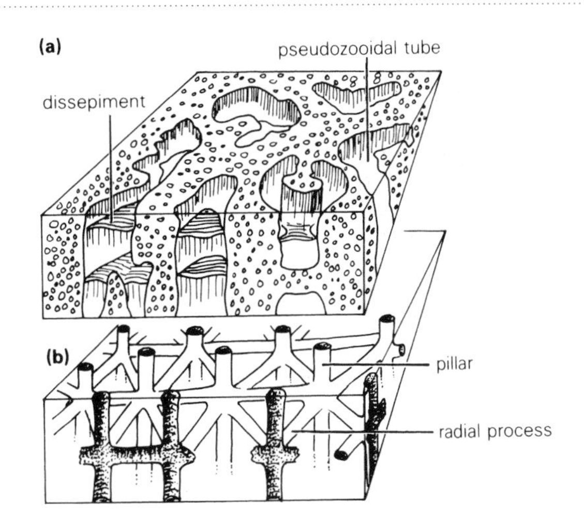

Figure 4.9 Wedge diagrams showing hard tissue in (a) *Stromatopora* and (b) *Actinostroma*. (Redrawn from Stearn, 1966.)

shaped exhalent canal systems on the surface of sclerosponges.

The structure of stromatoporoids may be studied in vertical and tangential sections. It appears at first sight to be less defined than in corals, but the dominant structures are clear enough. A few selected types serve to illustrate the range in morphology. In *Stromatopora* (Ord.–Perm.; Figs 4.9, 4.10) the **coenosteum** (colonial skeleton) in vertical section shows stout, upright **pillars,** joined at intervals and traversed by thin horizontal **laminae,** though the differentiated structure is often lost and only a reticulate pattern of anastomosing elements remains.

There may also be **tabulae** making partitions between chambers. With upward growth (Fig. 4.9) new materials were secreted by the cells in contact with the upper coenosteal surface, thus roofing over the successive astrorhizal canals. In *Stromatopora* the calcareous material is constructed of carbonate spherulites in contact with one another, but this is unusual.

Actinostroma (Cam.–L. Carb.; Figs 4.9–4.11) is somewhat similar, though the pillars are straighter and more slender and joined by horizontal radial processes in a kind of laminar network. Astrorhizae are not common and are absent altogether in some species.

Labechiella (Sil.–Dev.) has strong pillars and thin horizontal tabulae and possesses rare astrorhizae, whereas *Labechia* (Fig. 4.9; Ord.–Carb.) has very

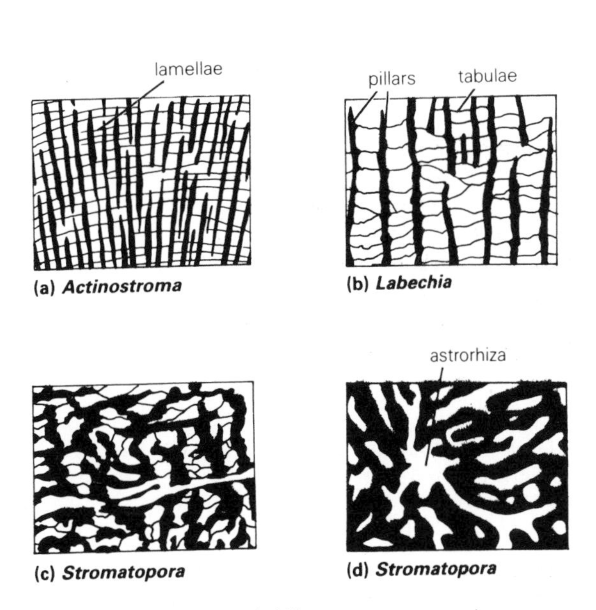

Figure 4.10 Structure of different stromatoporoids in section (×10 approx.): (a)–(c) vertical sections; (d) horizontal section. (Based on a drawing by Lecompte in *Treatise on Invertebrate Paleontology*, Part F.)

prominent tabulae between the pillars, convex upwards. In other genera horizontal laminae, often with cellular partitions, are well defined. These structural elements – pillars, laminate, tabulae, astrorhizae, etc. – are found variously developed in different stromatoporoids; sometimes one kind

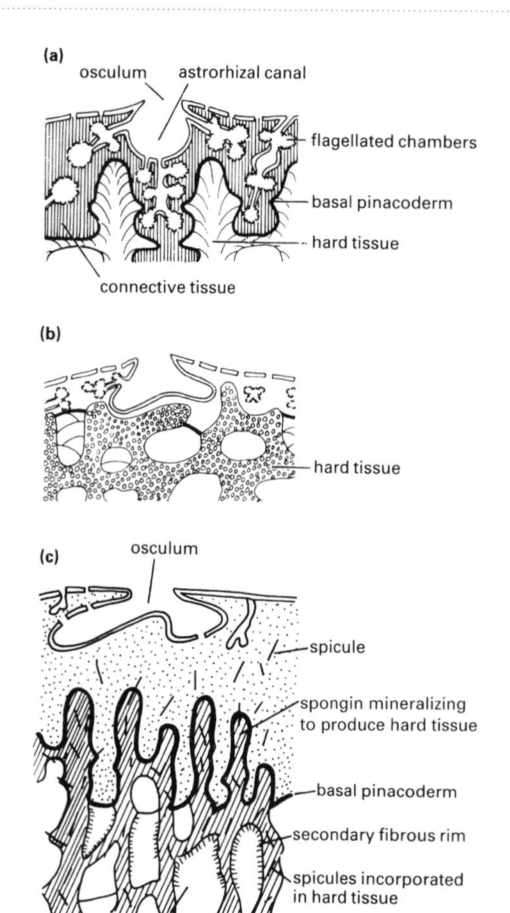

Figure 4.11 Restored morphology of stromatoporoids modelled upon sclerosponges: (a) *Actinostroma* (Cam.–L. Carb.); (b) *Stromatopora* (Ord.–Perm.) in which the hard tissue is composed of carbonate spherulites.; (c) *Actinostromarinina* (U. Jur.) showing how spicules are trapped by sponges which then becomes mineralized to form a primary calcareous skeleton. This skeleton then grows a secondary fibrous rim. [(b) based on Stearn, 1972; (c) based on Wood, 1987.]

of structure develops much at the expense of others.

Spicules have been found in the skeletons both of late Mesozoic (Wood, 1987) and Carboniferous stromatoporoids. Most of the spicules are siliceous, and thus the majority of these stromatoporoids are regarded as demosponges. Apparently the primary skeleton was of spicules. These then became trapped by a coating of spongin, and it was upon this framework that the calcareous skeleton (usually aragonitic) became secondarily precipitated.

The original composition of stromatoporoid coenostea is unknown: it may have been calcite or

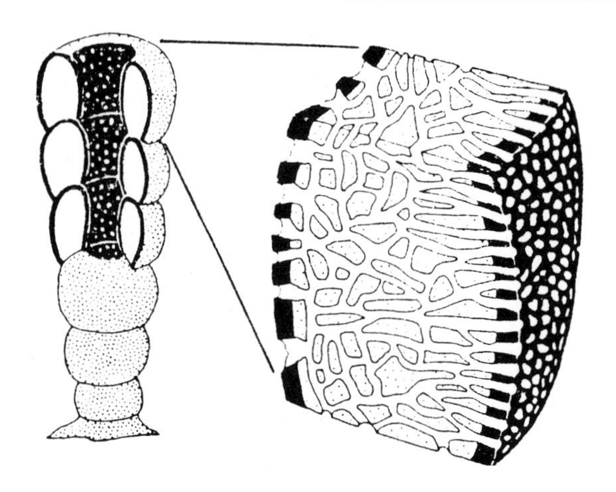

Figure 4.12 A generalized sphinctozoan sponge, showing wall structure. (Redrawn from Wood, 1990.)

aragonite. Some authorities (e.g. Stearn, 1966) have suggested an originally aragonitic skeleton which is usually much altered differentially by diagenesis. Stearn attempted to distinguish primary from secondary microstructures and by so doing helped to remove the source of much taxonomic confusion. Figure 4.11 shows the reconstructed soft-part morphology, modelled upon that of living sclerosponges.

Stromatoporoids grow as laminar sheets accreted one after the other – a pattern which has proved amenable to computer simulation, as pixels on a raster array (Swan and Kershaw, 1994). Many growth forms corresponding to actual stromatoporoid morphologies were produced, varying according to simulated sedimentation patterns; domal forms, merging and branching types and columnar stacked sheets with ragged margins analogous to real types. Curiously, conical-based morphologies, easily generated by the computer program, are rarely found in nature. These experiments suggest that each stromatoporoid 'growth unit' was largely autonomous, like each pixel in the model, and there is no clear evidence of centralized control of growth.

Sphinctozoans

Sponges of sphinctozoan grade have segmented, irregularly proliferating chambers arranged round a central cavity (Fig. 4.12).

The walls are perforated and may be of complex

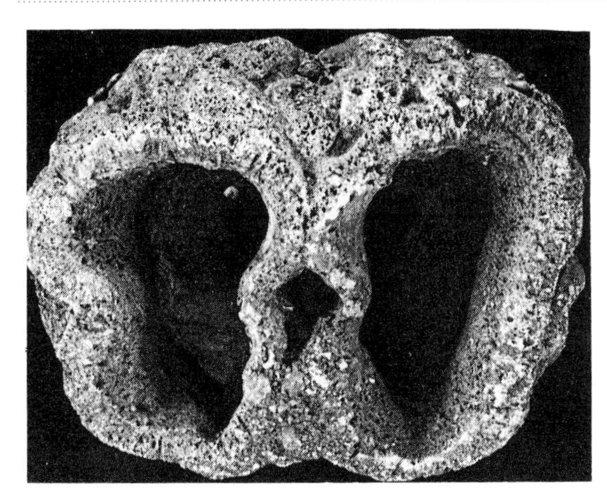

Figure 4.13 External appearance of an unusual twinned sponge. *Raphidonema*, from the Lower Cretaceous of Faringdon, England (× 0.75). (Photograph reproduced by courtesy of C. Chaplin.)

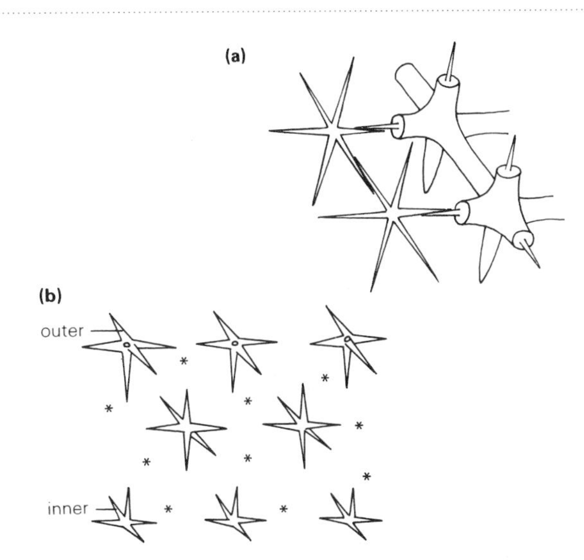

Figure 4.15 (a) Part of a skeleton of a dictyonine hexactinellid, showing overgrowth of contiguous hexact spicules by a siliceous envelope and (b) construction of the hexactinellid skeleton with outer, middle and inner megascleres and scattered microscleres. (Redrawn from Reid, 1958–1964.)

Figure 4.14 *Valospongea gigantea* (M. Cam.) from Utah. The surface of this keg-shaped hexactinellid is shown, with low, round mounds covering the entire surface. The oscular margin is uppermost and spicular impressions are clear (× 5). (Photograph reproduced by courtesy of Dr J.K. Rigby.)

structure. They flourished mainly in the late Palaeozoic and early Mesozoic, but several Cambrian genera are known, and a Recent form, *Vaceletia*, has also been discovered, living in cryptic habitats. There are similarities between sphinctozoans and some archaeocyathids, suggesting a possible biological relationship between the two grades.

4.4 Class Calcarea

The skeleton of calcareous sponges normally consists entirely of calcite spicules; there is neither spongin nor silica. The spicules are often of tuning-fork shape (Fig. 4.6e). In the large Order Pharetronida the spicules form a closely packed mesh of different-sized 'tuning forks', giving a rigid and easily fossilized skeleton. Rich Jurassic calcareous sponge faunas are known, sometimes associated with reefs. Well-known Cretaceous genera include the vase-like *Raphidonema* (Figs 4.6d, 4.13) and the digitate *Peronidella* (Fig. 4.6e) which together form extensive sponge beds in the Lower Cretaceous of southern England.

A few sphinctozoans seem to belong to the Calcarea.

4.5 Class Hexactinellida

Hexactinellids are of the normal sponge shape and may be anchored by an expanded flange or by a tuft of glassy fibres, especially in the deep-sea forms (Fig. 4.16).

Bag-shaped, vase-like and dendritic forms are known, some with lateral oscula. The skeleton is of opaline silica, consisting of large (**megasclere**) and small (**microsclere**) spicules (Figs 4.15, 4.16).

These are well ordered, expressing cubic symmetry. The megascleres line the outer and inner walls and are here five-rayed, with the odd ray (**axon**) pointing inwards. Six-rayed megascleres are found in the central trabecular and choanocytic layers, together with the much smaller stellate or dumb-bell-shaped microscleres, not often found fossilized. Often the megascleres unite as they grow together by complete fusion or by the formation of siliceous links between adjacent spicules. In such deep-sea glass sponges as *Euplectella* the whole skeleton becomes rigidly united, usually by the growth of a common sheath of silica enveloping adjacent parallel rays. Here the skeleton forms a rigid lattice protected against torsion by spiral siliceous girders running round the outside at 45° to the axis.

Hexactinellids are classified mainly on the basis of their skeletal characters – the structure of the mega- and microscleres, whether they unite and how they link, but since microscleres are seldom fossilized there are some difficulties with the taxonomy of fossil forms. Hexactinellids can be traced back to the

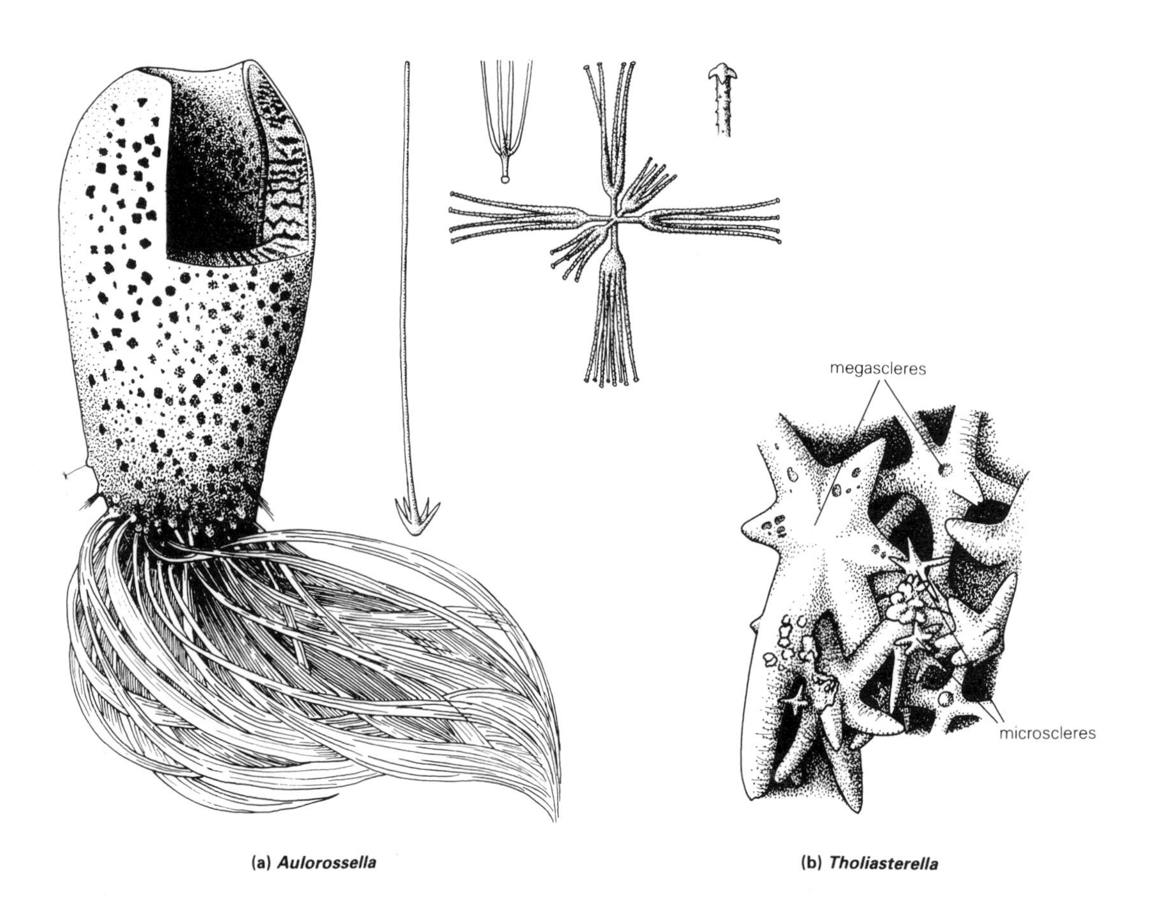

Figure 4.16 (a) A deep-sea hexactinellid glass sponge, *Aulorossella* (× 1) and (b) spicules of *Tholiasterella* (Hexactinellida) with micro- and megascleres (× 10). [(a) redrawn from Schulze and Kirkpatrick, 1904; (b) redrawn from Hinde, 1887–1893.]

Cambrian, and many now-extinct groups flourished in the Upper Palaeozoic. These sponges were affected severely by the Permian extinctions but had recovered by the Jurassic, when many new taxa arose. Order Hexactinosida was very important in the Cretaceous but is now nearly extinct. Hexactinellids were abundant in the Tertiary on the continental shelves, but since most hexactinellids today live in the bathyal and abyssal zones either there has been a shift of habitat or the shelf forms have become extinct and only descendants of former deep-sea faunas survive.

4.6 Incertae sedis: Archaeocyatha

The archaeocyathids are a very early group of calcareous fossils, usually in the form of a porous inverted cone, found mainly in lower Cambrian carbonate facies and persisting only until the upper Cambrian. Their acme was during the lower Cambrian, but they declined thereafter, becoming very rare and confined to cryptic habitats before their final extinction at the end of the Cambrian. They lived in shallow waters, often associated with stromatolites and forming thickets, localized bioherms or biostromes, or as isolated individuals. They have been considered to be the first 'reef-forming' animals, but they are usually subordinate in these reefs to stromatolitic algae, growing in clumps or scattered on the surface.

Archaeocyathids are found in all continents, but the best-known faunas are in Russia, South Australia and western North America. They are absent from the British Isles and northern Europe. Where they occur in continuous sequence, as in Russia, they have proved to be of considerable stratigraphical value.

Archaeocyathids (Fig. 4.17) are normally solitary organisms bearing a superficial resemblance to corals or sponges.

The skeleton or cup in a typical example such as *Ajacicyathus* (Class Regulares) is no more than a pair of inverted cones, one inside the other, forming **outer** and **inner walls** which are connected by vertical radial partitions (**septa**) and separated by an annular cavity (the **intervallum**). A large central cavity seems analogous to the paragaster of sponges. The lower part of the cup is usually expanded into a basal flange with root-like holdfasts.

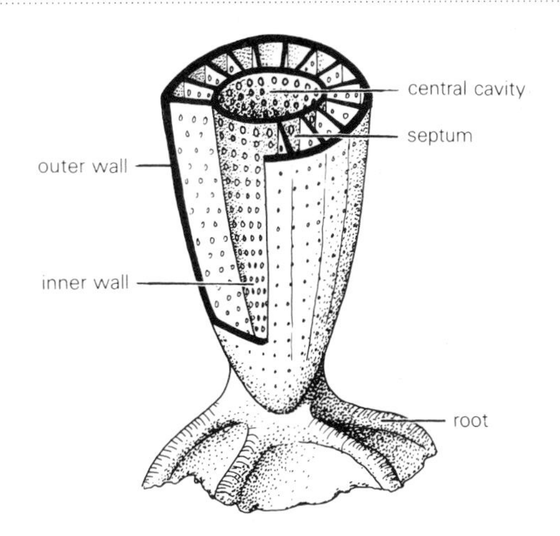

Figure 4.17 Archaeocyathid morphology modelled on *Ajacicyathus* (Cam.; × 1.5). (Modified from Hill in *Treatise on Invertebrate Paleontology*, Part E.)

The outer and inner walls are perforated by numerous holes arranged in longitudinal rows; the septa are sparsely perforate. In most archaeocyathids the outer wall has small pores whereas those of the inner wall are much larger. The microstructure of the cup is of polymicrocrystalline calcite, usually altered by diagenesis.

While *Ajacicyathus* is of relatively simple construction, other archaeocyathids may have perforated walls and septa, and also accessory structures crossing the intervallum (Fig 4.18).

These include radial rods, perforate transverse plates (**tabulae**) and imperforate arched plates (**dissepiments**). These elements tend to be concentrated towards the base of the cup. A few genera do not have an inner wall; some of these are very flattened cones, nearly discoidal in shape. Outgrowths from the outer wall are quite common and include tubular rods, tubercles or simply dense irregular masses; the latter apparently grow as a defensive mechanism where neighbouring specimens contact the individual in overcrowded conditions. Though most archaeocyathids are conical and solitary some rare colonial forms are known, with individuals being either arranged in chains or branched and shrub-like.

The usual size of archaeocyathids is 10–25 mm in diameter and up to 50 mm high, though specimens can reach a height of 150 mm and a few giants exceed this range.

Most archaeocyathids, like *Ajacicyathus*, belong to

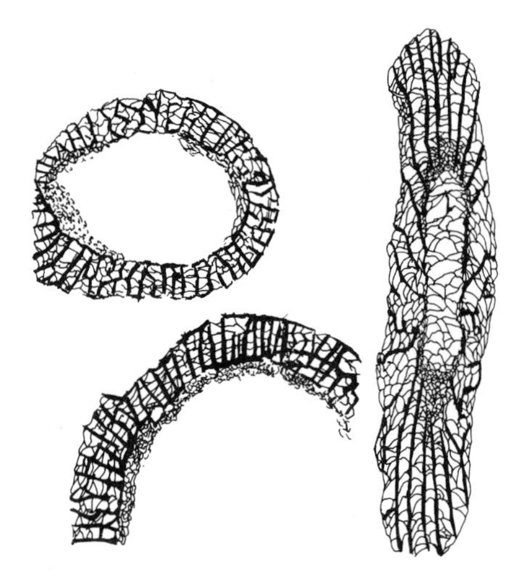

Figure 4.18 Structure of the archaeocyathid *Metaldetes* shown in cross-section (left) and oblique longitudinal section (right) with radial septa. (Drawn by E. Bull from photographs in Debrenne and James, 1981.)

Class Regulares. The other class, Irregulares, has fewer members, distinguished by irregular pore structures on both walls and often an irregular outline of the cup as well.

Soft parts, organization and ecology

The soft tissues of archaeocyathids are quite unknown, though they presumably possessed choanocytes, like those of other sponges. One useful pointer to their biological 'grade', however, is their capacity to regenerate the cup where damaged, which points to a quality of organization and 'individual integrity' at least as good as that of the living Porifera.

Most authorities agree that archaeocyathids were filter feeders, actively pumping water through the cup and straining off the food particles, perhaps with choanocyte-like cells, but the filtering function is not obvious in Irregulares. Flume tank experiments using model archaeocyathids (Balsam and Vogel, 1973) suggest that 'passive flow' without flagellar pumping may have been important. The models were fixed in a flume tank in a laminar current whose strength could be varied. In this current a velocity gradient was established, generating a secondary flow from the outer wall through the intervallum and inner wall to the central cavity and out through the osculum. Short, stout archaeocyathids were found to be better adapted to rapid current velocity than tall, slender ones. Apparently the conical shape of archaeocyathids, and in particular the large excurrent opening, was well adapted for the generation of such passive flow. Though the passive generation of currents may well have been important, it is unlikely that the archaeocyathids abandoned the useful flagellar pumping system devised by their presumed protozoan ancestors, and feeding may have been a dominantly active process economically boosted by the passive flow component.

Archaeocyathids have been regarded by many as 'nature's first attempt to make a multicellular skeletal organism'. But the success of other kinds of sponges may well have been an important factor in their early extinction.

Nearly all archaeocyathids come from carbonate shelf sediments deposited in warm seas. They commonly adhered to the substrate by holdfasts or similar devices, though in some of the cup-like forms found in soft sediments the holdfasts were very small or absent. They are most commonly associated with algal stromatolites; however, they also occur between and around the algal reefs and in bedded limestones with trilobites, brachiopods and hyolithids, but very rarely with other sponges.

There is evidence from their occurrence in wave-smashed, tumbled blocks and their common association with algae that they were depth limited, the optimum being 20–30 m, individuals becoming smaller and fewer down to 100 m. Their mode of growth is shown from a life assemblage of mid Lower Cambrian age from Australia (Fig 4.19; Brasier, 1976).

In this fauna many species of Regulares were found clustered together, growing in the same general direction and reaching heights of up to 90 mm. Irregulares are present but fewer. The larvae always settled on dead archaeocyathids and usually on the outer wall near the top. Where the juveniles attached they fixed themselves by **exothecal outgrowths**. Irregular masses of exothecal material formed where individuals of the same and different species come into contact suggest competitive interactions – perhaps the oldest known case in the fossil record. In this fauna a micrite layer surrounds each cup, the work of the problematic encrusting organism *Renalcis*, which bound the archaeocyathid community together and so enabled it to be preserved as a life assemblage.

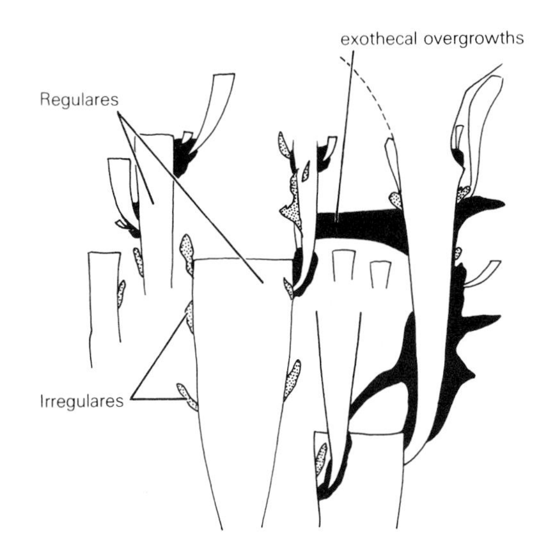

Figure 4.19 Archaeocyathid relationships in Cambrian life assemblages from South Australia (reconstructed from serial sections). (After Brasier, 1976.)

There are various examples of reef-associated archaeocyathids. One interesting fauna described by Debrenne and James (1981) from Labrador and the facing coast of Newfoundland is of late Lower Cambrian age. It consists only of Irregulares, and though only a very few species are present these are densely packed. The paucity of species may be because by this stage the archaeocyathans were becoming fewer in any case. The archaeocyathans occur either with tabular biostromes or with small patch reefs (bioherms), where they may comprise up to 50% of the total volume of the rock. Subsidiary fossils include trilobites and brachiopods together with echinoderm and hyolithid debris. In this rock the genus *Metaldetes* (Fig. 4.18) may be of stick-like or cup-like form and individuals are surrounded and welded together by an extensive '**coenenchyme**', so that the cups have a massive habit, resembling that of corals, and from the coenenchyme numerous buds are found to swarm.

Many archaeocyathids are solitary, and these are especially common in the early part of the Cambrian. Subsequently the modular forms became more common and successful. In these the organism as a whole consists of repeated functional units or individuals with soft-part continuity (Wood *et al*., 1992). Such modules may be added to any part of the whole organism, which is then able to grow in various directions. Modularity has proved especially appropriate for reef-building archaeocyathans; they can grow larger, they have improved regenerative powers and they can encrust substrates. Indeed, archaeocyathans were the first of all skeletonized metazoans to have developed modularity. It was, however, only the forms with perforated septa which became modular, presumably because soft-tissue connections are essential for such to develop. The degree of integration varies in archaeocyathans, as with corals. The less well-integrated forms are 'pseudocolonial', i.e. they consist of many relatively similar linked elements, and the branching forms of these were the most successful and long-lasting of all archaeocyathans. They were common in early Cambrian reefs, where turbulence, sedimentation rates and nutrient supply were fairly high. Other archaeocyathans, however, became increasingly well integrated, especially the massive laminar or columnal types, so the whole organism may be considered as a single individual, rather than an aggregate of separate modules. This is particularly the case in the Irregulares, some of which bear a striking resemblance to stromatoporoids.

Distribution and stratigraphic use

Archaeocyathids are found in all continents, though their distribution is patchy; none are reported from the British Isles. Their abundance in the Lower Cambrian of the USSR has enabled very precise time ranges to be worked out for over 100 genera and used stratigraphically most successfully. The oldest known species come from the Tommotian stage in Russia (Raaben, 1981). They diversified in the subsequent Adtabanian, reached an acme in the Botomian, and declined greatly in the Toyonian stage which terminated the Lower Cambrian. Species are few in the Middle Cambrian, and only one Upper Cambrian survivor has been recorded from Antarctica. There is some evidence of provinciality, with three or more provinces in the Botomian, limiting intercontinental correlation using archaeocyathids. Even so, the Russian work has shown the potential stratigraphical value of these remarkable extinct poriferans, whose zonal use will undoubtedly increase as the faunas become better known.

4.7 Geological importance of sponges

Sponges are of little stratigraphical value, though they have been used as markers at specific horizons. Their main importance geologically lies in that they have been a source of biogenic silica and that they are an element, though not often an important one, in reef fabrics.

Rich faunas of varied middle Cambrian demosponges and hexactinellids are known from the Burgess Shale and from Utah. In the latter region they are often preserved flattened but complete (Fig. 4.14), and they have recently been described in elegant detail by Rigby (1983, 1986).

Hexactinellids through time seem to have been mainly confined to the 'off-reef' facies in deeper waters, whereas demosponges and calcareous sponges preferred shallower waters. Ordovician lithistid sponges, especially in the carbonate facies of North America, were quite important as frame builders in stromatoporoid, bryozoan (Rigby, 1971) and algal reefs, sometimes being as much as half the total reef volume. From time to time in the Palaeozoic there were quite extensive localized developments of sponges, such as the rich Silurian sponge faunas of Tennessee and the Devonian and Mississippian glass sponges of New York and Pennsylvania, but these were not associated with reefs.

In the Permian the sponge faunas associated with the Texan reef complex are very well known. Here the calcisponges derived from Carboniferous shelf-living faunas became increasingly important on the patch reefs and finally on the barrier reefs, whereas hexactinellids as usual preferred the off-reef marginal facies. Some well-developed Triassic sponge reefs occur in the Alps; in contrast, the Jurassic sponge reefs of southern Germany, which are very well known, contain sponges as primary frame-builders. Here the sponges are mainly siliceous thin-walled forms, often preserved in life position, forming small sponge mounds, which occasionally grew to great size and united to produce large masses. But these sponges did not raise themselves far above the substratum, forming thin flat biostromes rather than true reefs. Other organisms are rarely associated with these Jurassic biostromes, being unable to provide suitable anchorage. Many modern sponges, e.g. *Cliona* (Cobb, 1969), bore into hard substrates and may contribute quite significantly to the destruction of shells.

4.8 Sponge reefs

Spicular sponge reefs

In the Jurassic there developed various and much-studied kinds of sponge reefs. In Normandy, for example, small Middle Jurassic bioherms were built up by the lithistid *Platychonia* (Palmer and Fürsich, 1981), individuals of which cemented to each other to form a rigid framework. These probably grew in the lower photic zone, below normal wave base but above storm base since some of them show signs of storm damage. A low-diversity community dominated by the oyster *Atreta* grew on the upper surface, while the quieter-water lower surfaces supported a very rich assemblage of encrusting calcisponges, cyclostome bryozoans and small thecideacean brachiopods, all of which were suspension feeders. Small nesting brachiopods, vagile echinoids and starfish and boring worms and bivalves also colonized the reef, whereas erect bryozoans lived on the inter-reef mud surface. Thus a diverse but balanced ecosystem based upon the frame builder *Platychonia* was able to develop.

These small reefs contrast with the much larger Upper Jurassic sponge algal reefs of southern Germany, which grew to some considerable height above the sea floor. In these (e.g. Flügel and Steiger, 1981) the dominant frame builders are hexactinellid and lithistid siliceous sponges, but calcareous algae also played a dominant role. The sponges may be cup-shaped or discoidal, depending on whether they grew on hardgrounds or soft surfaces, respectively. In the best-studied examples growth was cyclical and the reef built up in stages. An initial sponge layer was covered by marl and killed, then more sponges grew on top. These died in turn and thus growth of further sponges continued, sponge and marl layers alternating. Periodically, cyanophycean algae grew on top of the dead sponge layers and towards the top of the reefs these became dominant. There was a great deal of postmortal recrystallization and dissolution in the reefs so that the original form of the sponge is not always preserved. Some micromorphic faunas of cephalopods and brachiopods are sometimes associated with the sponge reefs, but often no suitable anchorage could be provided and other organisms are rare. Stick bryozoans in places formed meadows between the sponge beds.

Most Cretaceous and Tertiary sponges, though locally abundant, do not seem to have been reef related. Jackson *et al.* (1971) have described Recent 'coralline' sponge brachiopod communities, which are also of palaeontological interest.

Calcareous sponge reefs

Throughout geological time, archaeocyathids (stromatoporoids, chaetetids and sphinctozoans) have contributed in various ways to reef formation.

Some archaeocyathids were able to fix themselves to hard substrates. Others tended to root in soft sediments. They are often found associated with stromatolite reefs but their contribution is minor. They flourished, however, in 'cryptic' habitats within the early Cambrian reefs, in grottoes, cavities within the framework, and on the undersides of blocks of debris and spreading skeletal organisms (Zhuralev and Wood, 1995). Such crypts offered an attractive habitat, with reduced exposure and environmental stress and housed solitary chambered archaeocyathans, calcified cyanobacteria and microborers, spiculate sponges and other organisms. Much of the biomass of the reefs was actually in the form of crypt-dwellers. From the earliest Cambrian, archaeocyathans appeared simultaneously, but in separate associations both in crypts and on the open surface, and it is unlikely that the crypts served as 'safe havens' for former denizens of the exposed sea floor.

The immense Permian Capitan Reef of Texas and New Mexico rims the margin of the ancient Delaware Basin. Some parts are characterized by enormous platy sponges, *Gigantospongia*, forming the ceilings of great cavities housing a rich cryptic fauna, including pendant sphinctozoans and bryozoans. Other regions have a founder bryozoan–sponge community with the smaller platy sponge *Guadelupia*, but likewise with a cryptic fauna (Wood *et al.*, 1996). As with the Cambrian reefs, most of the preservable benthos was housed in the crypts.

Chaetetids and sphinctozoans are often associated with reefs, but the major sponge reef formers were the stromatoporoids. Many Palaeozoic stromatoporoids were reef builders and, since they had geological requirements similar to those of reef-building tabulate corals, the two are often found together as frame builders. Stromatoporoids, in general terms, tend to displace tabulates in high-energy environments.

Four basic growth forms are found amongst stromatoporoids: laminar, domical, bulbous, and dendroid (Kershaw, 1984, 1990). In Devonian reefs the large domical kinds ('ballstones') and laminar forms are associated with high-energy conditions on the reef crust and are major frame builders. Dendroid and small bulbous forms preferred quiet water conditions in back-reef lagoons. There is a direct parallel here between these and living corals.

Very large reef-building stromatoporoid masses are often found to have marginal invaginations of sediment and inclusions of sediment within the coenosteum, indicative of cessations of growth for a time. Tongues of marginal sediment may be found on one side of the stromatoporoid mass, suggesting banking up of sediment on the lee side. The growth of such masses is often asymmetrical, and they often seem to lean over into the current, from which, of course, their food came. The most spectacular Palaeozoic reefs built largely by stromatoporoids are those of the Silurian of Gotland, where they form huge reef masses up to 20 m high and 200 m in diameter. In these bioherms stromatoporoids were dominant and corals and other elements subsidiary. The stromatoporoids in these reefs developed various growth strategies and a variety of responses to environmental pressures. Abrupt changes in the direction of growth are due to the shifting or overturning of the coenosteum, hence the stromatoporoids record the various events that have taken place while they lived (Kershaw, 1984). Other such Silurian reefs occur in the Great Lakes Region of North America (Heckel and O'Brien, 1976). In the Wenlock of the Welsh Borders a giant coral stromatoporoid reef, with subsidiary bryozoans, formed a marginal barrier between an outer deeper-water zone in which reefs were absent and an inner shallow-water region where smaller reefs abounded (Scoffin, 1971). The acme of coral–stromatoporoid reef development was in the Devonian, and immense reefs of this age are found in southwestern England and in Alberta. Mesozoic stromatoporoids, which were in many ways dissimilar to those of the Palaeozoic, were not generally reef formers.

It seems that sponges of various kinds were able to secrete strong calcareous skeletons in response to environmental opportunities, at different times throughout geological history. Such large colonial forms are very good at colonizing hard substrates. Yet living sponges with calcareous skeletons are few, and of low diversity, and are confined to cryptic habitats. The decline of coralline sponges, and especially the stromatoporoids, after the early Mesozoic is probably correlated with the rise of reef-building scleractinian corals. Whereas living calcified demosponges have much tougher skeletons than do scleractinians, the latter grow very much more quickly. Such rapid growth is a consequence of their symbiotic relationship with zooxanthellae. These corals can thus colonize available space very rapidly: they are ideally constituted for building reefs.

Coralline sponges, however, lack zooxanthellae. Since the rise of scleractinians they have been unable to compete as reef formers, and have lost their former dominance. They now exist only as relicts living in dark and cryptic habitats where hermatypic corals cannot penetrate. Yet it is these few survivors, the modern sclerosponges, undiscovered until recently, which have enabled a substantial reinterpretation of the Phylum Porifera to be effected.

Bibliography

Books, treatises and symposia

Bergquist, P.R. (1978) *Sponges*. University of California Press, Berkeley, Calif. (Living faunas, illustrated)

Carter, G.S. (1952) *General Zoology of the Invertebrates*. Sidgwick and Jackson, London.

Heckel, P. and O'Brien, G.D. (1976) *Silurian Reefs of the Great Lakes Region of North America*. AAPG Reprint Series No. 14, American Association of Petroleum Geologists, Tulsa, Okl. (Numerous papers)

Marshall, N.B. (1978) *Aspects of Deep Sea Biology*. Blandford, London.

Moore, R.D. (ed.) (1955) *Treatise on Invertebrate Palaeontology*, Part E, *Archaeocyatha and Porifera*. Geological Society of America and University of Kansas Press, Lawrence, Kan.

Raaben, M.E. (ed.) (1981) *The Tommotian Stage and the Cambrian Lower Boundary Problem*. Amerind Publishing, New Delhi.

Reid, R.E.H. (1958–1964) *The Upper Cretaceous Hexactinellida*. Palaeontographical Society, London. (Classification, growth, morphology)

Reitner, J. and Keupp, H. (1991) *Fossil and Recent sponges*. Springer-Verlag. Berlin. 595 pp. (Many valuable papers)

Teichert, C. (ed.) (1972) *Treatise on Invertebrate Palaeontology, Part E (revised volume), Archaeocyatha*. Geological Society of America and University of Kansas Press, Lawrence, Kan.

Individual papers and other references

Balsam, W.L. and Vogel, S. (1973) Water movement in archaeocyathids: evidence and implications of pressure flow in models. *Palaeontology* **47**, 979–84. (Functional morphology)

Bidder, G.P. (1937) The perfection of sponges. *Proceedings of the Linnean Society of London* **149**, 119–46.

Brasier, M. (1976) Early Cambrian intergrowths of archaeocyathids, *Renalcis* and pseudostromatolites from South Australia. *Palaeontology* **19**, 223–45. (Successive generations of archaeocyathids in life position; bibliography)

Cobb, W.R. (1969) Penetration of calcium carbonate substrates by the boring sponge *Cliona*. *American Zoologist* **9**, 783–90.

Debrenne, F. and James, N.P. (1981) Reef-associated archaeocyathids from the Lower Cambrian of Labrador and Newfoundland. *Palaeontology* **24,** 343–78. (See text)

Debrenne, F. and Wood, R. (1990) A new Cambrian sphinctozoan sponge from North America, its relation to archaeocyaths, and the nature of early sphinctozoans. *Geological Magazine* **127**, 435–43.

Flügel, E. and Steiger, T. (1981) An Upper Jurassic sponge–algal build-up from the northern Frankenalb, West Germany, in *European Reef Models* (ed. D.F. Tooney), Society of Economic Palaeontologists and Mineralogists Special Publication No. 30, Society of Economic Palaeontologists and Mineralogists, Tulsa, Okl., pp. 37–98. (Typical sponge reef)

Gray, D.I. (1980) Spicule pseudomorphs in a new Palaeozoic chaetetid and its sclerosponge affinities. *Palaeontology* **23**, 803–20.

Hartman, W.F. and Goreau, T.E. (1970) Jamaican coralline sponges: their morphology, ecology, and fossil representatives. *Symposia of the Zoological Society of London* **25**, 205–43. (Sclerospongidean demosponges)

Hartman, W.D. and Reiswig, H.M. (1973) The individuality of sponges, in *Animal Colonies* (eds R.S. Boardman, A.H. Cheetham and W.A. Oliver), Dowden, Hutchinson and Ross, Stroudsburg, Penn., pp. 567–84.

Hinde, G.J. (1887–93) A monograph of the British fossil sponges. *Palaeontographical Society Monographs*,1–254.

Jackson, J.B.C., Goreau, T.F. and Hartman, W.B. (1971) Recent brachiopod coralline sponge communities and their palaeoecological significance. *Science* **173**, 623–5.

Kershaw, S. (1984) Patterns of stromatoporoid growth in level-bottom environments. *Palaeontology* **27**, 113–30.

Kershaw, S. (1990) Stromatoporoid palaeobiology and taphonomy in a Silurian biostrome in Gotland, Sweden. *Palaeontology* **33**, 681–706.

Palmer, T. and Fürsich, F. (1981) Ecology of sponge-reefs from the Upper Bathonian of Normandy. *Palaeontology* **24,** 1–23.

Rigby, J.K. (1971) Sponges and reef and related facies through time. *Proceedings of the North American Paleontological Convention*, Chicago, 1969(J), pp. 1374–88. (A recent treatment of sponge reefs)

Rigby, J.K. (1983) Sponges of the Middle Cambrian Marjum Limestone from the House Range and Drum Mountains of Utah. *Journal of Paleontology* **57**, 240–70. (Excellent illustrations).

Rigby, J.K. (1986) Sponges of the Burgess Shale (Middle Cambrian), British Columbia. *Palaeontographica Canadiana* **2,** 1–105.

Scoffin, T.P. (1971) The conditions of growth of the Wenlock reefs of Shropshire (England). *Sedimentology* **17**, 173–219. (Classic).

Schulze, F.E. and Kirkpatrick, R. (1904) Die Hexactinelliden der Deutschen Südpolar-Expedition 1901–03. *Deutsch Südpolar Expedition* **12**, Zool. 4, 1–62. Reimer, Berlin.

Stearn, C.W. (1966) The microstructure of stromatoporoids. *Palaeontology* **9**, 74–124. (Distinction of primary from secondary microstructure)

Stearn, C.W. (1972) The relationship of the stromatoporoids to the sclerosponges. *Lethaia* **5**, 369–88. (See text)

Stearn, C.W. (1975) The stromatoporoid animal. *Lethaia* **8,** 89–100. (Sponge affinities of stromatoporoids)

Swan, A.R.H, and Kershaw, S. (1994) A computer model for skeletal growth of stromatoporoids. *Palaeontology* **37**, 409–23.

Wood, R. (1987) Biology and revised systematics of some late Mesozoic stromatoporoids. *Special Papers in Palaeontology* **37**, 1–89.

Wood, R. (1990) Reef-building sponges. *American Scientist* **78**, 224–235.

Wood, R. and Reitner, J. (1986) Poriferan affinities for Mesozoic stromatoporoids. *Palaeontology* **29**, 469–75.

Wood, R. and Reitner, J. (1988) The Upper Cretaceous 'chaetetid' demosponge *Stromataxinella irregularis* n.g. (Michelin) and its systematic implications. *Neues Jahrbuch für Paläontologie Abhandlungen* **177**, 213–224.

Wood, R., Dickson, J.A.D. and Kirkland, B.L. (1996) Observations on the ecology of the Permian Capitan Reef, Texas and New Mexico. *Palaeontology*, **39,** 733–62.

Wood, R., Reitner, J. and West, R. (1989) Systematics and phylogenetic implications of the haplosclerid stromatoporoid *Newellia mira* nov. gen. *Lethaia* **22**, 85–93.

Wood, R., Zhuralev, Yu. A, and Debrenne, F. (1992) Functional biology and ecology of the Archaeocyatha. *Palaios* **7**, 131–56.

Zhuralev, Yu. A. and Wood, R. (1995) Lower Cambrian reefal cryptic communities. *Palaeontology* **38**, 443–70.

Zhuravlev, Yu. A., Debrenne, F. and Wood, R.A. (1990) A synonymised nomenclature for calcified sponges. *Geological Magazine* **127**, 587–9.

5 Cnidarians

5.1 Introduction

Corals, sea anemones, jellyfish and the small colonial hydroids are all representatives of Phylum Cnidaria. They were formerly grouped with the ctenophores, 'sea gooseberries' or 'comb jellies' (Fig. 5.3a) in Phylum Coelenterata, but cnidarians and ctenophores are now regarded as separate though related phyla. Ctenophores are globular or elongated gelatinous organisms swimming by serried combs of fused cilia arranged in rows. They catch food with their long tentacles, which generally are armed with adhesion cells for catching prey; one species alone has stinging cells. Until recently they were believed to be the only phylum with no fossil record, but two specimens found in the Lower Devonian of Hunsrückschiefer (Chapter 12), and revealed by X-radiography (Stanley and Stürmer, 1983), extends the known range far back in time.

Cnidarians are the simplest of all true metazoans, but they are of an evolutionary grade higher than sponges since their cells are properly organized in tissues which are normally constructed on a radial plan. The cell wall is **diploblastic**, having cells organized in two layers only (Fig. 5.1a–c).

These are the outer **ectoderm** and inner **endoderm**, which have no body cavity between them but only a jelly-like structureless layer. In this **mesogloea** runs a simple nerve net. The mesogloea is sometimes invaded by cells from the two primary layers. The single body cavity (**enteron**) has only one opening, the mouth, which also serves as an anus and is normally surrounded by a ring of tentacles. It is lined by endoderm which is sometimes infolded to form radial partitions or **mesenteries**, increasing the area over which digestion may take place, for the primary function of the endoderm is digestive. The cells of the endoderm ingest food by means of amoeboid pseudopodia: alternatively they can extend flagella into the enteron to stir its contents.

The cells of the ectoderm are more highly differentiated. Of these the principal types are the **musculoepithelial** cells – columnar cells with contractile fibres – and there are also sense cells leading to the nerve net below. In addition there are ectodermal stinging cells, the **nematocysts** (Fig. 5.1c,e). These are large cells with a sealed central cavity containing poisonous fluid, within which lies a tightly coiled elongated tubular thread carrying barbs and stylets inside it. A small sensory hair on the outside of the nematocyst, the **cnidocil**, is sensitive to vibrations in the water, and when a small organism passes close to the cnidarian it triggers the nematocyst to discharge. When this happens the central cavity is intensely compressed by strong muscle fibres and the seal breaks; the thread is shot out with great force, turning inside out as it extends so as to expose the spiral barbs, which penetrate the prey so that it is injected and paralysed by the poison. Batteries of nematocysts are found on the tentacles, which all discharge together. The captured prey, held fast by the threads, is conveyed to the mouth as the tentacle then bends into it. Further transport of food is by mucus strings.

Cnidaria are characterized by a life cycle in which successive generations are of different kinds. This **alternation** of **generations** or **temporal polymorphism** is typical of the less specialized Cnidaria but may be suppressed entirely in the more 'advanced' kinds. Two types of individual, the normally fixed **polyp** (Fig. 5.1a) and the free **medusa** (Fig. 5.1b), alternate successively so that the polyp gives rise asexually to medusae, which reproduce sexually so that their zygotes produce polyps, and so on (Fig. 5.1d).

A polyp is a sedentary animal with a cylindrical

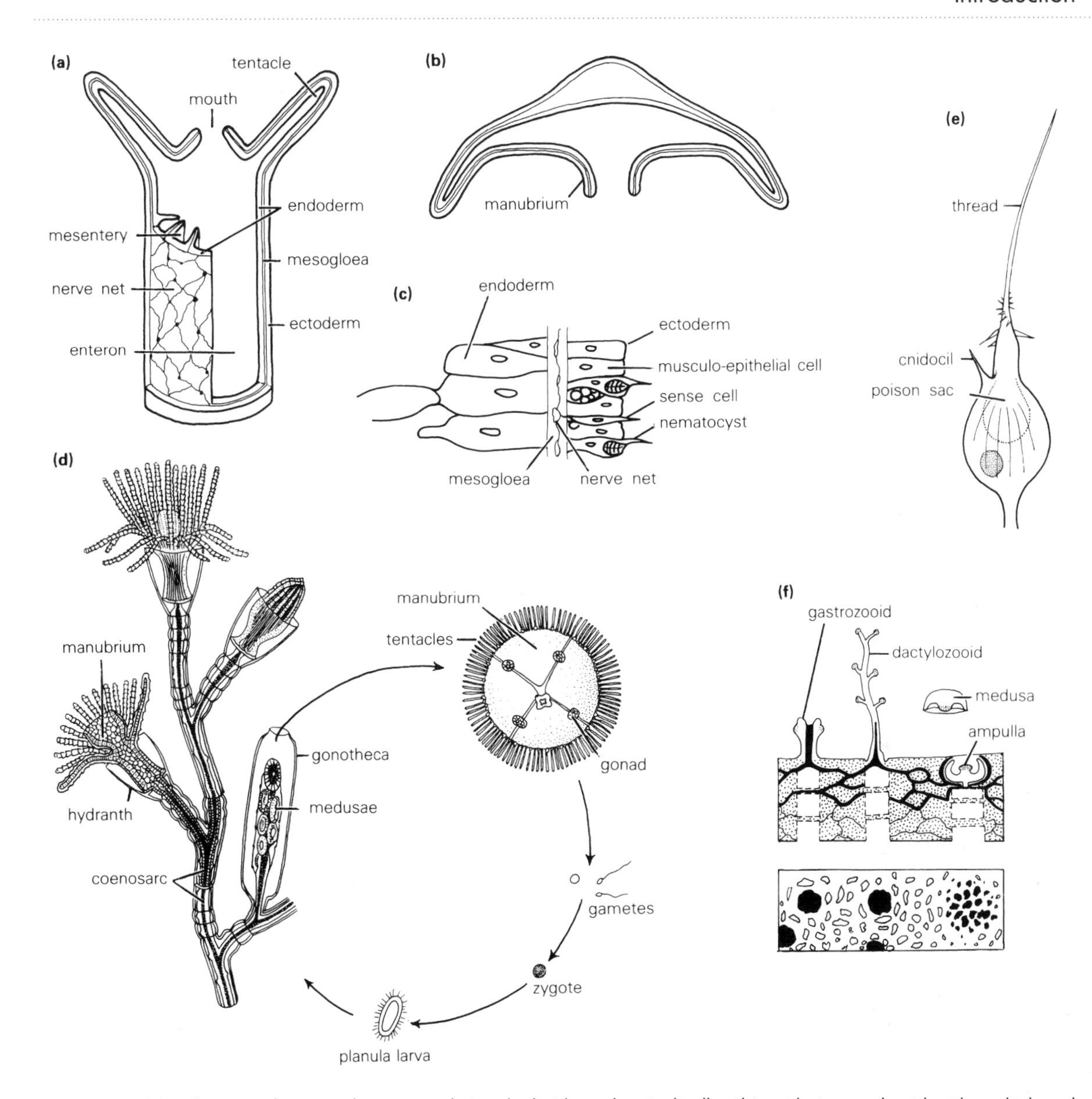

Figure 5.1 (a) Schematic diagram showing cnidarian hydroid cut longitudinally; (b) cnidarian medusoid with a thickened mesogloea above the manubrium; (c) body wall of a cnidarian, showing details of the diploblastic structure; (d) *Obelia* – morphology and life cycle (× 25): hydroid phase (left), medusoid (right) from below; (e) discharged nematocyst, showing spiral barbs near the base of the thread; (f) *Millepora* (Cret.–Rec.) – longitudinal section (above) and surface view (below) (× 10): the small medusa has recently been budded off from the ampulla. [(d) Modified from an illustration in Borradaile *et al.*, 1956; (f) based on a drawing by Boschma in *Treatise on Invertebrate Paleontology*, Part F.]

body and an upward-facing mouth surrounded by a ring of tentacles. The mesogloea is comparatively thin. Medusae are free-swimming inhabitants of the plankton, often very small; their orientation is inverted relative to that of polyps. The medusoid body is discoidal, with the mouth prolonged centrally into a downward-facing funnel or **manubrium**. The mesogloea is very thick, and below it the enteron extends through four or more radial canals which join a peripheral circular canal, forming a kind of circulatory system for dissolved food material. The tentacles hang down freely beneath the outer rim, each usually carrying at its base an organ of balance or statocyst.

Since medusae are the sexual as well as the dispersal phase of the cycle, they carry the four gonads, one of which is located along each of the primary radial canals. When the gametes are ripe they are shed by rupture of the gonad wall, and fertilization takes place in the sea water. The zygote then develops into a small ciliated **planula larva**. At first this has a solid core of endoderm, which eventually develops a central cavity, and settles down to become the next polyp generation.

Medusae move by rhythmic pulsation of the bell-like body through highly modified musculo-epithelial cells. They normally die after reproducing.

Though alternation of generations seems to have been basic to the more generalized Cnidaria, such as most of Class Hydrozoa, it has been modified or suppressed in others. In Class Scyphozoa the medusoid phase is dominant and the polypoid phase very reduced, whereas in Class Anthozoa the medusoid has been eliminated entirely and the polypoid phase has become the sexual generation.

A description of the major divisions of the Cnidaria (simplified) follows.

5.2 Major characteristics and classes of Phylum Cnidaria

The Cnidaria (Precam.–Rec.) are metazoans with radial or biradial symmetry. They have body walls of ectoderm and endoderm separated only by mesogloea; a single body cavity (enteron) with a mouth but no separate anus; a simple nerve net; and no separate excretory or circulatory systems. Nematocysts are present. They may be solitary or colonial, and they are often polymorphic with alternate polyps and medusae. Many have calcareous or organic skeletons and they are frequently colonial. There are three classes:

CLASS 1. HYDROZOA (Precam.–Rec.): See below.
CLASS 2. SCYPHOZOA (Precam.–Rec.): Most large jellyfish.
CLASS 3. ANTHOZOA (Precam.–Rec.): Corals, sea anemones, gorgonians, sea pens.

5.3 Class Hydrozoa

The Hydrozoa are usually polymorphic, with radial and often tetrameral symmetry. Several hydrozoan orders embrace a wide range of forms:

ORDER 1. HYDROIDA (Cam.–Rec.): Have colonial sessile polyps and planktonic medusae. e.g. *Obelia*.
ORDER 2. TRACHYLINA (?Jur.–Rec.): Free medusae only, the polyp stage having been suppressed entirely. e.g. *Petasus*.
ORDER 3. HYDROCORALLINA (Tert.–Rec.): These secrete a calcareous skeleton and are often reef formers. e.g. *Millepora*, *Stylaster*.
ORDER 4. CHONDROPHORA (Precam.–Rec.): The hydroid phase is a single large polyp, floating at the surface by a gas-filled float. Reproductive phase is a free medusa. e.g. *Velella*, *Porpita*.
ORDER 5. SIPHONOPHORA (Rec.): Large and complex floating colonies, not known as fossils; extreme polymorphism of zooids, no free sexual medusae. e.g. *Physalia*.
ORDER 6. SPONGIOMORPHIDA (Trias.–Jur.): Form massive colonies with radial pillars united by horizontal bars, and structures resembling the astrorhizae of stromatoporoids

Order Hydroida

Hydroids are rather generalized hydrozoans in which there is a chitinous external skeleton (**perisarc**). The order includes three suborders: Eleutheroblastina (free solitary naked forms, the pond *Hydra* being one), Calyptoblastina and Gymnoblastina. Representatives of the latter two suborders are occasionally found in fossil form. *Obelia* (Suborder Calyptoblastina; Fig. 5.1d), a standard example of a colonial hydroid, has a hollow root-like structure for attachment from which branching tubes arise, giving rise to polyps (**hydranths**) alternately on each side. The perisarc is annular in places, especially below the cups (**hydrothecae**) which surround each polyp. Though each polyp is an individual, the polyps are all connected together by a tubular system (**coenosarc**) through which the enteron continues. Each polyp has many tentacles and a prominent bulbous funnel or manubrium upon which the mouth is set. Towards the base of the colony are cylindrical **gonothecae**: flask-shaped structures with a constricted aperture. Within each of these is a central stem arising from the coenosarc, and from this the medusae form, being budded off continually from the upper end of the stem as they mature. These escape through the aperture and thereafter swim freely and grow until they are old enough to reproduce sexually. Gonothecae are typical of the Calyptoblastina; in the Gymnoblastina the medusae form directly from the coenosarc without being enclosed, and there is no hydrotheca.

Fossil chitinous hydroids are well known in the Ordovician and have been recorded from several localities. The calcareous tubes of several Mesozoic and Tertiary serpulid tube-worms (e.g. *Parsimonia)* are often found infested by a colonial organism *(Protulophila)* which has been interpreted as a hydroid buried in the outer part of the calcareous tube (Scrutton, 1975; Fig. 5.3b). This has two components: polyp chambers, and stolonal networks which connect them in a regular scale-like pattern. The polyps grew around the rim and were probably incorporated into the tube as the serpulid grew, forming polyp chambers opening through semicircular apertures. This hydroid–serpulid association was probably symbiotic or commensal, in the same way that several species of the modern hydroid *Proboscidactyla* are symbiotic with tubular sabellid worms.

Protulophila would have been able to feed on some of the supply of food particles drawn in by the feeding arms of the worm, whereas the serpulid in turn probably derived some protection from the nematocyst batteries of the hydroid. It must have been a successful association, for the degree of infestation is sometimes as high as 95% and infested serpulids are found in beds ranging from the Middle Jurassic to the Pliocene, 170 Ma in all.

Order Hydrocorallina

The hydrocorallines (milleporines and stylasterines) are hydrozoans which superficially resemble corals and may be quite important in some modern reefs. *Millepora* (Figs 5.1f, 5.2) has a thick laminar calcareous skeleton of many vertical tubes with cross-partitions connected by thin horizontal ramifying canals.

The soft tissues invest only the upper part of the skeleton, degenerating in the older lower layers as the skeleton builds up. There are three types of vertical tube, each with its own zooid arising from the surface and capable of retracting into the tube: the **gastrozooids**, which are purely manubrium-like mouths; the **dactylozooids**, which are elongated tentacles with batteries of nematocysts; and the **ampullae**, which produce medusae. All these are connected by soft tissues running through the horizontal canals. This kind of colony illustrates an effective 'division of labour' and consequent modification of the zooids.

Figure 5.2 Millepora, a Recent specimen from Barbados (× 0.35). (Photograph reproduced by courtesy of C. Chaplin.)

Millepora in its natural reef habitat is very variable in form. The millepores of Barbados (Stearn and Riding, 1973) either encrust dead corals or gorgonians in thin sheets less than 1 cm thick, or form 'boxwork', i.e. thick near-vertical blades joined along their edges to form box-like, subangular, upward-opening cavities up to 30 cm high. Alternatively they may be bladed, i.e. with erect blades buttressed by other blades, making colonies up to 50 cm high, or branching, up to 20 cm high, with delicate or stubby branches.

It is possible that the boxwork, bladed and branching forms are distinct non-intergrading biospecies, but the encrusting form grades into the others and is probably an environmental growth form of the other species, being most common in shallow water or encrusting gorgonians. The habitat preferences of the other species seem to be governed by water turbulence, boxwork forms being the strongest and bladed the most delicate.

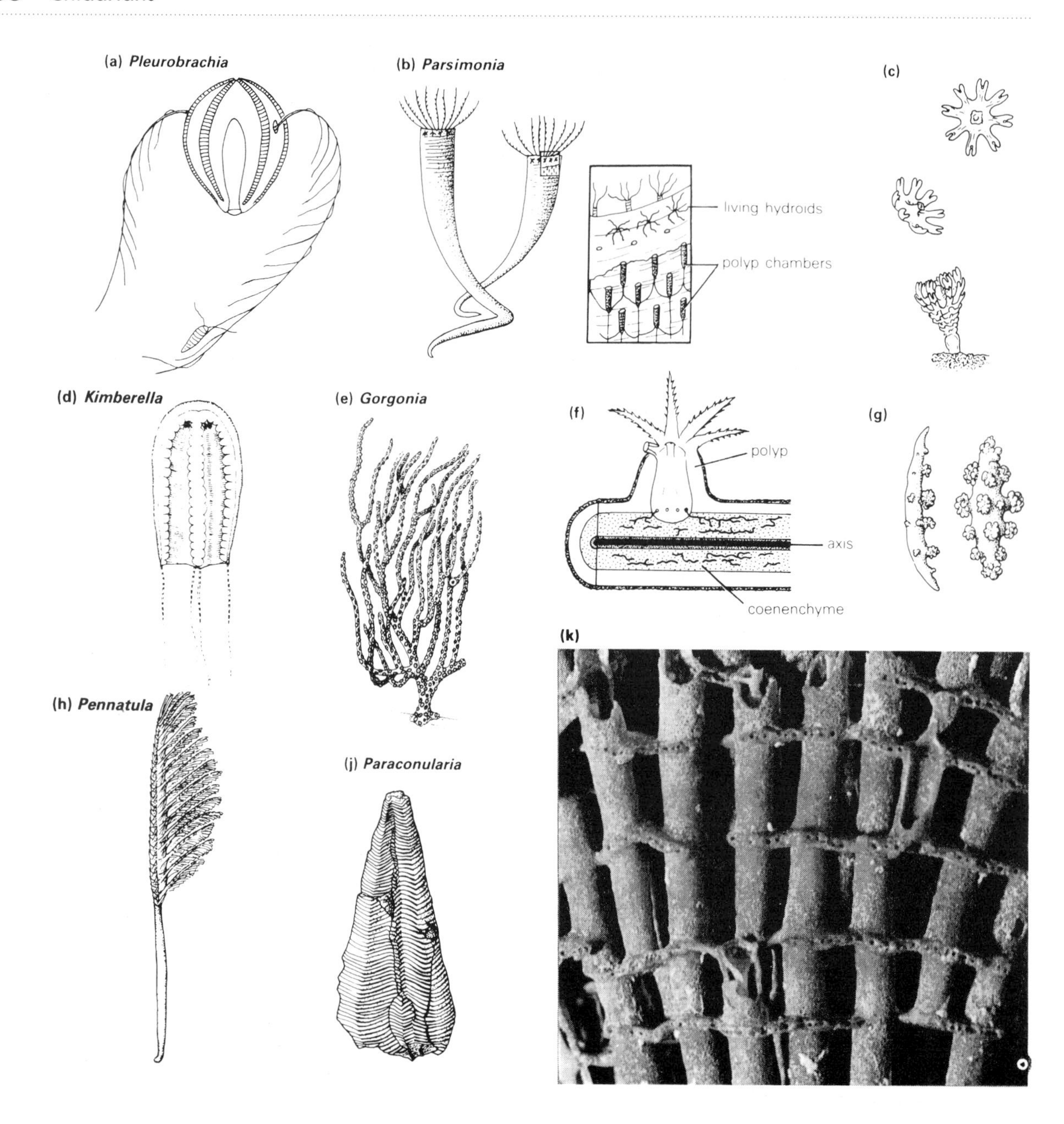

Figure 5.3 (a) A ctenophore, showing four of the comb-like, fused plates of cilia enteron and tentacles, which have just caught a copepod (× 1); (b) Cretaceous serpulid worm infested by the symbiotic hydroid *Protulophila* (× 1) – to the right is an illustration showing detail of the rim of the serpulid with living hydroids. The worm tube below is exfoliated to show the arrangement of the polyp chambers linked by connecting stolons; (c) strobilating scyphozoan scyphistoma releasing ephyra larva; (d) internal mould of *Kimberella*, a probable Precambrian scyphozoan (× 0.75); (e) a Recent *Gorgonia* with visible polyps [see (f), (g)]; (f), (g) section through gorgonian stem with eight-armed polyps set into organic coenenchyme with central axis and spicules (× 200) right; (h) *Pennatula*, a Recent sea pen (× 0.25); (j) *Paraconularia*, a conulariid (Cam.–Perm.) – possibly of cnidarian affinities; (k) *Tubipora*, a Recent red octocoral (× 4). [(b) Based on Scrutton, 1975; (d) redrawn from Wade, 1972; (f), (g) based on illustrations by Bayer in *Treatise on Invertebrate Paleontology*, Part F; (j) redrawn from Slater, 1907.]

5.4 Class Scyphozoa

The Scyphozoa are free-swimming medusae, entirely marine and usually with radial symmetry based on a tetramerous plan. The sessile polyp stage (Fig. 5.3c) is called a **scyphistoma**, which buds by **strobilation**.

This means that towards its upper end it is continually growing and being divided by transverse grooves, so that a pile of small medusae forms, one on top of the other, and are released in turn to break free as **ephyra larvae**: the juveniles of the adult jellyfish.

Few jellyfish are found in the fossil record. Of these, some are found as compressions in very fine sediment; others, including at least some of the Ediacaran fauna, are preserved as sand infillings of the gut cavity (but see Chapter 3). The Precambrian *Kimberella* from Ediacara is preserved as positive composite moulds. It is a primitive scyphozoan of quadrate form (Wade, 1972), possibly related to the living cubomedusans, and the gonads adhered to the radial canal and projected into the central cavity of the bell.

One group of problematic fossils, the Conulata (conulariids) (Cam.–Trias.) appears to be of scyphozoan affinities. These are steeply pyramidal phosphatic fossils (Fig. 5.3j), which have a quadrate cross-section and often marked herringbone ridges down the sides. Whereas they have been referred to various groups (chordates, annelids or an extinct phylum; Babcock, 1991), reconstructed cross-sections through *Eoconularia loculata* from the Silurian of Gotland, show a four-branched internal partitioning virtually identical to that of living coronated scyphozoans (Jerre, 1993, 1994). Likewise in three-dimensional conulariids from the USA (van Iten, 1991; van Iten *et al*., 1996) internal structures are very similar, and moreover the ultrastructure reveals a striking resemblance to the thecae of coronated scyphozoans, and an evident similarity in mode of growth and injury repair. There is some evidence of strobilation and, in three-dimensional specimens from Spain, of longitudinal fission. Possible tentacles have been reported below the pyramidal structure in rare instances and one rather flattened genus (*Conchopeltis*) resembles a medusa. Conulariids, therefore now seem to be related to coronate scyphozoans, independently acquiring a phosphatic exoskeleton.

Conulariids are not often preserved because they broke up easily after death, but the presence of very numerous fragments dissolved out of limestones in Gotland testifies to their former importance as a common component of Palaeozoic faunas. When they are found whole, they may be solitary, or occur in monospecific (possibly clonal) assemblages attached to tubes or nautiloid shells.

5.5 Class Anthozoa

The Anthozoa are polyps with no trace of a medusoid stage. Anthozoans (corals, sea anemones, gorgonians and sea pens) are solitary or colonial, entirely marine cnidarians. The polyps produce gametes which develop directly into planulae larvae after fertilization and thence to a new generation of polyps; the medusoid generation has been eliminated entirely. They resemble hydrozoan polyps in many respects, though they are often very large. They always have a tubular **gullet** or **stomodaeum** leading down into the enteron, which hydrozoan polyps do not have. The interior itself is divided by radial partitions (mesenteries) whose number and morphology is important in the subdivision of the class.

Those that secrete hard parts, and especially the corals, are of great geological importance, often forming thick beds in the Palaeozoic and sometimes, in association with stromatoporoids, reef-like masses. From Tertiary time onwards true coral reefs of vast thicknesses have formed, and corals living as reef formers are probably more important now than at any other time.

Anthozoans are grouped in three subclasses: Ceriantipatharia, Octocorallia and Zoantharia. Only the last of these, which includes the corals, will be considered in detail.

Subclass Ceriantipatharia

Ceriantipatharia are solitary or colonial polyps which for morphological reasons are placed apart from other groups. They are virtually unknown as fossils.

Subclass Octocorallia

Octocorallia (?Precam., Ord.–Rec.), poorly known as fossils, are especially represented by the gorgonians (Order Gorgonacea) common in many modern coral reefs (Fig. 5.3e,f,h). The colony forms a flat fan of anastomosing branches. The skeleton consists of horny branching tubes, each of which may have a central calcified core. From the outer surface of the branching tubes project the many small polyps (**autozooids**), each of which has eight tentacles, characteristic of the subclass. These autozooids are set in a thick gelatinous outer coating over the branching tubes, within which are set innumerable calcareous **spicules** which help to support the autozooids. These are the only part to preserve as fossils, and only from them can the former presence of gorgonians be inferred.

The fossil record of octocorals is sparse, but recently organic skeletons believed to be of gorgonians have been found in rocks as early as the Lower Ordovician (Lindström, 1978). Furthermore, abundant spicules of Lower Silurian age, usually isolated, but sometimes found in nest-like aggregations, have been shown to belong to octocorals of Order Alcyonacea (Bengtson, 1981). These warty, spindle-shaped spicules of the genus *Atractosella*, once thought to pertain to sponges, are virtually indistinguishable from those of modern alcyonaceans, hence this group has much more ancient origins than was formerly believed. *Atractosella* probably lived on a sand–gravel sea floor; it was not a reef former.

Heliopora (Order Coenothecalia) has a massive aragonitic skeleton, bright blue in colour. It is an important reef former in the Indo-Pacific province. Its similarity to the tabulate *Heliolites* has suggested that the latter is a Palaeozoic octocoral, but this resemblance is far more likely to result from convergent evolution.

The red octocoral *Tubipora*, the 'organ pipe' coral (Order Stolonifera), has polyps inhabiting long horny tubes (Fig. 5.3k) and often forming large colonial masses. The sea pens (Order Pennatulacea; Fig. 5.3h) are another kind of octocoral. Here the colony has the form of a feather, the base of which is set in mud while the feathery upper branches are lined with autozooids. The Ediacaran *Charniodiscus* has usually been interpreted as a sea pen (though see Chapter 1 for alternative views) and if so indicates early success for this group.

Subclass Zoantharia: corals

The structure of the Zoantharia is exemplified by the Upper Jurassic to Recent solitary coral *Caryophyllia* (Order Scleractinia). Recent species are of cosmopolitan distribution, living on various substrates and at depths of 0–2750 m. *C. smithii* lives in cool, temperate waters off western Scotland (Wilson, 1975, 1976). In regions of strong tidal currents it is found attached to pebbles or boulders, but in weaker currents on the outer shelf it can live in a variety of substrates; the most common form of attachment in sand patches is to calcareous tubes of the worm *Ditrupa* (Fig. 5.4a).

The polyp of *Caryophyllia*, which is some 2–3 cm across (Fig. 5.4b), is many-tentacled and has a stomodaeum leading down to the enteron, which is divided biradially by numerous mesenteries. Each of these has frilled free edges. The soft basal tissues secrete an aragonitic cup or **corallum** which is short and horn-shaped. It has an outer wall (**epitheca**) within which are numerous radially arranged **septa** which lie between the paired mesenteries. The first-formed **protosepta** (Fig. 5.4c) are larger and more pronounced than the **metasepta** intercalated between them.

When preserved as a fossil, the aragonite is often altered to calcite or dissolved completely leaving a negative mould in the matrix. Aragonitic skeletons are known as far back as the Triassic, and almost all Scleractinia are aragonitic. In *Caryophyllia*, as in all Scleractinia, the septa are laid down cyclically in multiples of six. When the coral is very young the corallum is simply a **basal plate** with six small prosepta inserted between the mesenteries. Later the metasepta appear, and there may be two or three generations of these so that the whole becomes multiseptate. A slight shift of their spacing throughout growth as the corallum becomes elliptical imparts the characteristic biradial symmetry. Throughout growth new material is added to the exposed edges of the septa and to the epitheca so that the coral expands as it grows.

In broad and general terms the simple structure of *Caryophyllia* applies to most living and fossil corals, but there are major differences in skeletal structure, septal arrangement and mode of colony formation. Eight orders are now defined (Scrutton, 1997b):

ORDER 1. RUGOSA (M. Ord.–U. Perm.)
ORDER 2. TABULATA (L. Ord.–U. Perm.)
ORDER 3. SCLERACTINIA (Trias.–Rec.)
ORDER 4. HETEROCORALLIA (U. Dev.–L. Carb.)
ORDER 5. COTHONIIDA (M. Cam.)
ORDER 6. KILBUCHOPHYLLIDA (U. Ord.)
ORDER 7. NUMIDIAPHYLLIDA (Perm.)
ORDER 8. TABULACONIDA (L. Cam.)

Order Rugosa

Morphology

The Rugosa are an exclusively Palaeozoic group of solitary and colonial corals. They show bilateral symmetry in a primitive manner, arising because the numerous metasepta are inserted in four loci alone. Such bilaterality is clear in some of the Rugosa, though in others it is obscured by the proliferation of septa, and in genera such as *Hexagonaria* no pattern at all is visible. Two examples are chosen to show the basic structure of rugose corals. The widespread Carboniferous *Zaphrentites* (Superfamily Hapsophyllidae; Fig. 5.4d–f) is a small, solitary 'horn coral', so called because of its shape.

The outer horn-shaped part of the corallum is covered by a thin calcareous skin, the **epitheca**, which extends from the tip to the upper (distal) surface or **calice**, where the skeletal elements filling the inside of the corallum are exposed. These internal elements form the basis of classification of the Rugosa and are of two main types: the vertical elements (**septa** and **axial structure**) and the horizontal elements (**tabulae**), which will be considered separately. (**Dissepiments**, which are important horizontal elements in some Rugosa, are lacking in *Zaphrentites*.)

The septa are thin vertical plates arranged in a characteristic biradial pattern, which develops to maturity throughout the ontogeny of the coral. Their manner of insertion can be studied by investigating the ontogeny of the coral from the early stages; this is usually done by making serial sections normal to the axis, from the tip to the calice, and arranging them in a successional series. This provides a full record of how the coral grows, since structures once formed by accretion are not resorbed but retain their original form through the life of the coral.

When *Zaphrentites is* very young a single proseptum divides it (Fig. 5.4f), soon becoming separated into **cardinal** (C) and **counter** (K) prosepta. Two other pairs of prosepta follow: the **alar** (A) adjacent to the cardinal septum, and the **counterlateral** (KL). Even at this stage some bilaterality is evident, which becomes more pronounced when the metasepta are inserted serially in four quadrants only, on the 'cardinal' side of the alar and counterlateral septa. Through differential growth the counterlateral and alar septa move towards the counterseptum to make room for new metasepta. Short minor septa are laid down between the metasepta. Around the cardinal septum there is a **cardinal fossula** where metasepta are not inserted, and especially during the intermediate growth stages there are lateral (**alar**) fossulae as well. These are visible in the adult *Zaphrentites*, but in many large Rugosa, where the major and minor septa have proliferated greatly, the fossulae are so compressed as to be hard to distinguish on account of the increasing radial distribution of septa. Many Rugosa, e.g. *Phillipsastraea*, have no trace at all of a fossula.

In *Zaphrentites*, as in most other Rugosa, two other skeletal elements are (1) the epitheca, and (2) **tabulae**, which are flat horizontal plates forming a floor to the cavity in which the polyp resided. The major septa join together in a central boss formed of updomed tabulae.

Heliophyllum (Subfamily Zaphrentidae; Fig. 5.4n–q), a large North American rugosan, is sometimes found as poorly developed colonies. In this coral there are so many septa that it is not easy to see the four main lines of emplacement. The spacing of the septa is more or less equidistant, and the symmetry is virtually radial. There are some structural elements not found in *Zaphrentites*, especially the dissepiments, which are concentrated in a broad marginal zone or **dissepimentarium**. These are small curving plates located between the septa and set normal to them, inclined downwards at about 45° to the epitheca. The major septa may reach the centre, in which updomed tabulae are the most prominent components. Another characteristic feature is the presence of **carinae**: short 'yard-arm' bars projecting laterally from the septa.

Though bilateral symmetry has been masked by radial structure, the cardinal fossula can still be seen where the dissepimentarium narrows and the axial ends of neighbouring septa are shortened and curve round it.

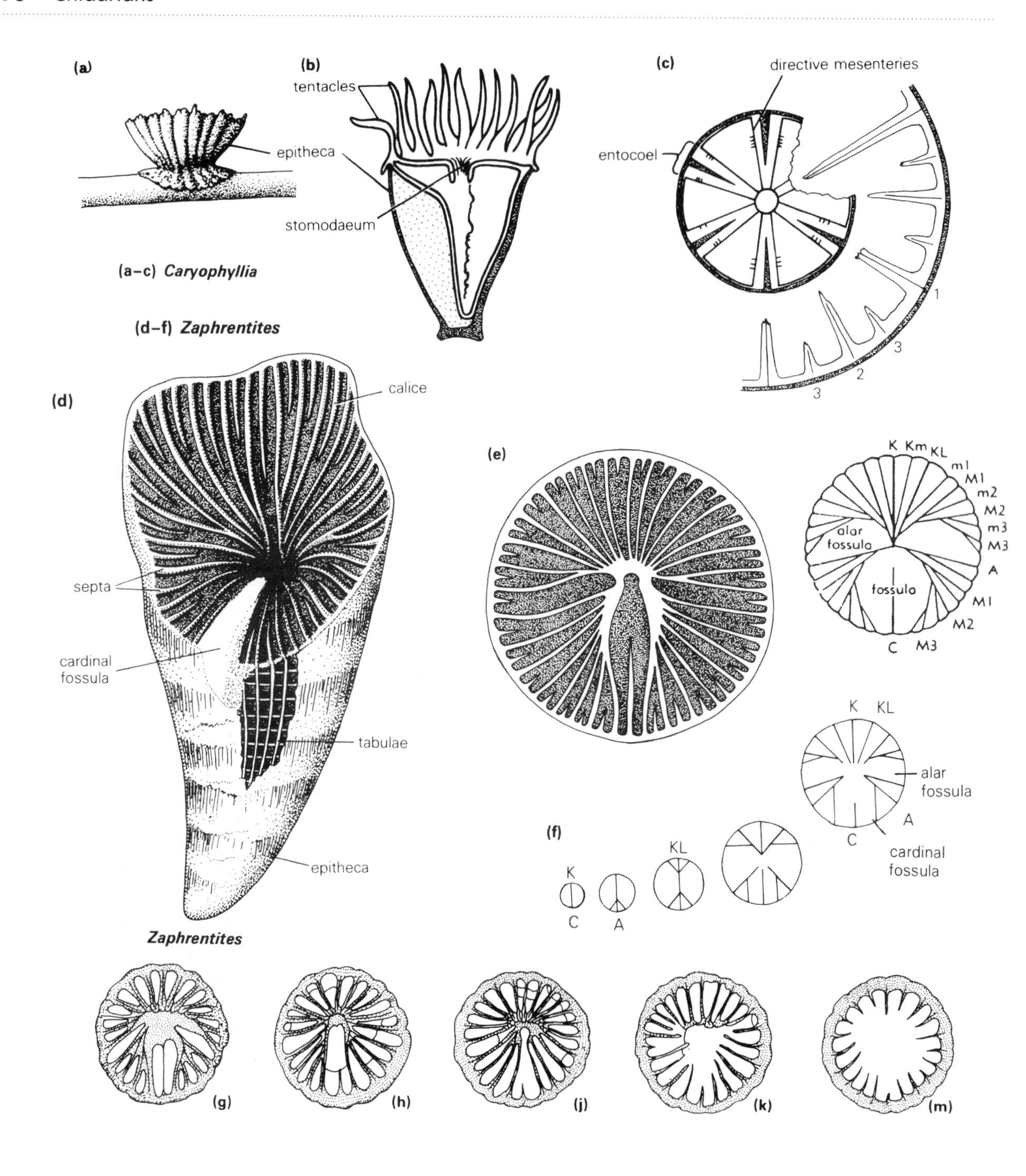
(a)
(b)
(c)
tentacles
epitheca
stomodaeum
directive mesenteries
entocoel
1
2
3
3
(a–c) Caryophyllia
(d–f) Zaphrentites
(d)
calice
septa
cardinal fossula
tabulae
epitheca
Zaphrentites
(e)
K Km KL
m1
M1
m2
M2
m3
M3
A
M1
M2
C M3
alar fossula
fossula
K KL
alar fossula
A
C
cardinal fossula
(f)
K
C
A
KL
(g)
(h)
(j)
(k)
(m)

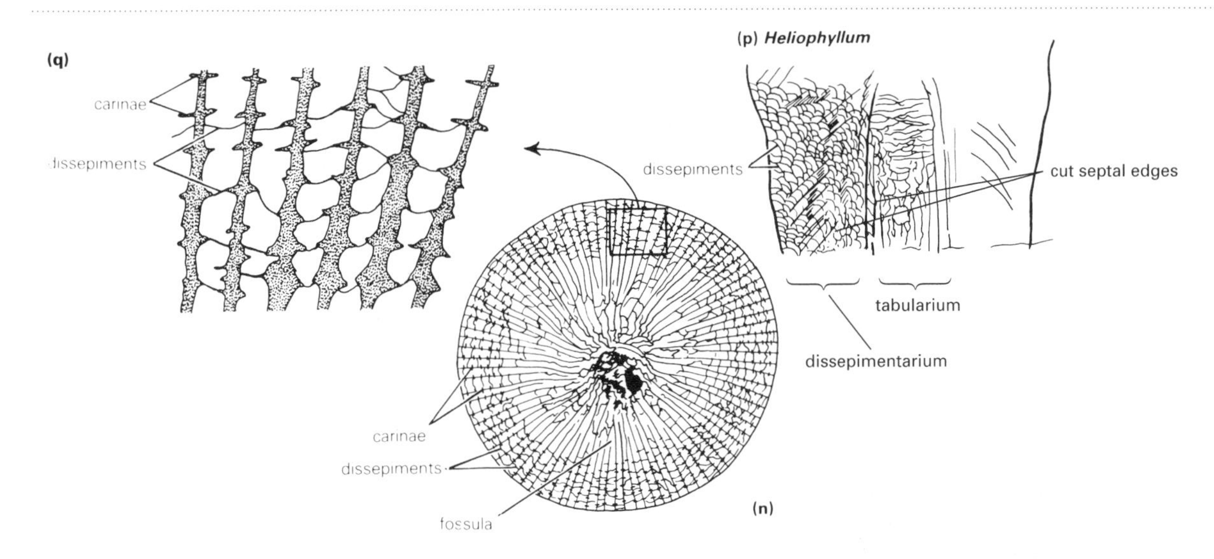

Figure 5.4 (a) Young *Caryophyllia*, a Recent scleractinian coral attached to the tube of the worm *Ditrupa*; (b) *Caryophyllia* vertical section; (c) transverse section through young (centre) and older (right) scleractinian coral (1, protoseptum; 2, 3, metasepta); (d) *Zaphrentites* (Carb.) with part of the epitheca removed, showing bilateral symmetry; (e) same, in transverse section; (f) septal emplacement shown by serial sections from tip (C, cardinal septum; KL, counter-lateral; K, counter-septum; A, alar); (g)–(m) examples of the *Zaphrentites delanouei* group (Carb.), including (g) *Z. delanouei*, (h) *Z. parallela*, (j) *Z. constricta*, (k) *Z. disjuncta* (early) and (m) *Z. disjuncta* (late) (not to scale); (n) *Heliophyllum* (Dev.), transverse section showing near-radial symmetry (× 1.5); (p) vertical section (× 1.5); (q) enlargement of peripheral zone, showing relationships of septa and yardarm carinae. [(a) Redrawn from Wilson, 1976; (d), (e) based on Milne-Edwards and Haime, 1850; (g)–(m) from Carruthers, 1910; (n)–(q) redrawn from Hill in *Treatise on Invertebrate Paleontology*, Part F.)

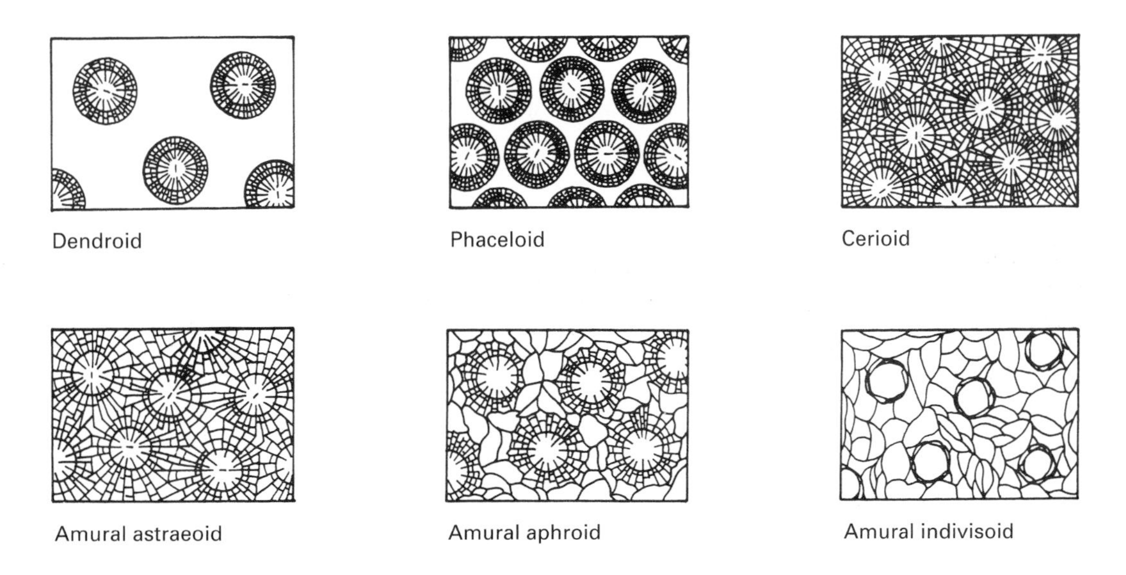

Figure 5.5 Transverse sections through compound rugose coral colonies, with terminology of different types.

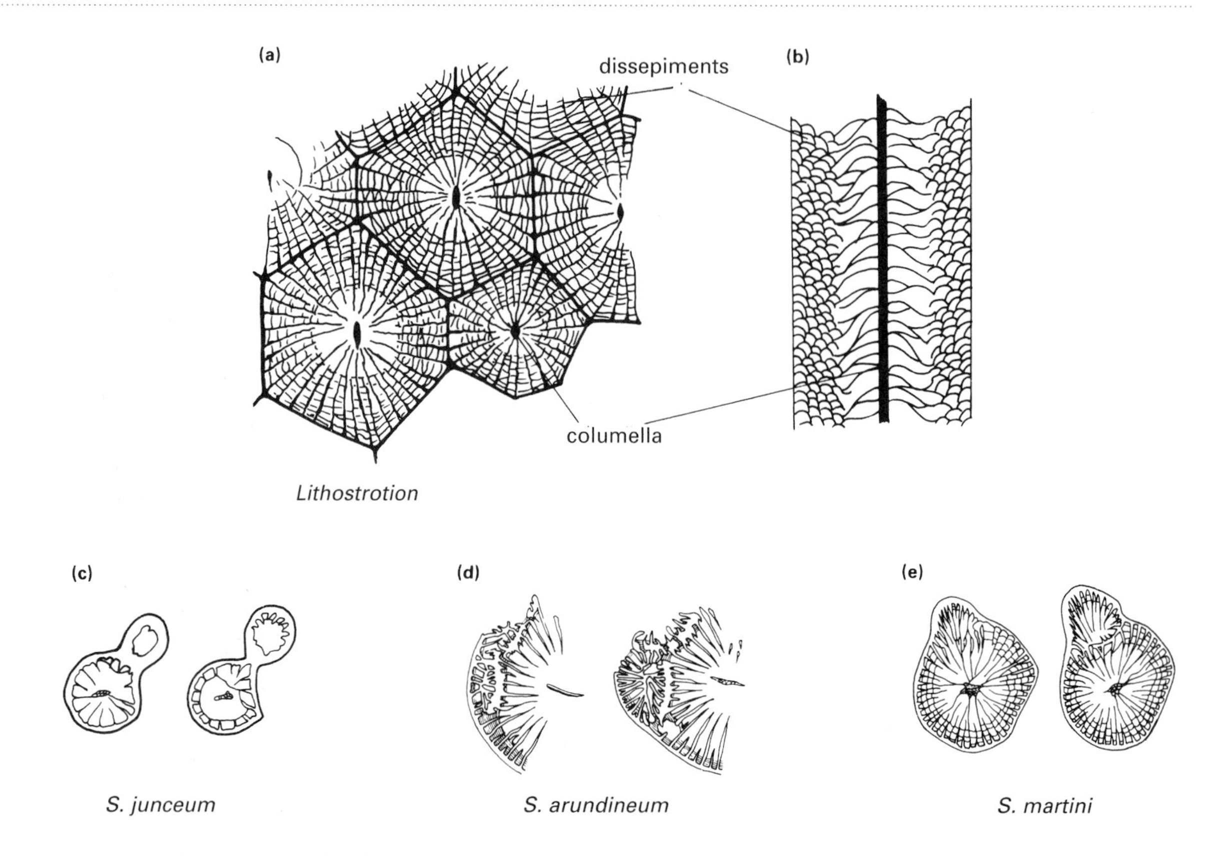

Figure 5.6 *Lithostrotion*, a Carboniferous compound rugose coral: (a) transverse; (b) longitudinal section; (c)–(e) lateral increase in three *Siphonodendron* species, *S. junceum*, *S. arundineum* and *S. martini*. (Redrawn from Jull, 1965.)

Range of form and structure in Rugosa

FORM, TYPE AND HABIT OF CORALLUM

Rugosa are either solitary or colonial (compound), and either form may exist even within closely related genera or even species of the same family. In solitary Rugosa (Fig. 5.4d) the usual shape is a curved horn: the ceratoid shape of *Zaphrentites*. If the 'cone' has expanded very fast, a flat discoidal shape may result (Fig. 5.7c); with progressively less abrupt expansion **patellate**, **turbinate** (Fig. 5.7d), **trochoid** and **ceratoid** forms will be the product. **Cylindrical** corals are virtually straight-sided, except for the first-formed part; **scolecoid** forms are irregularly twisted cylinders; **pyramidal** types have sharply angled sides; whereas **calceoloid** genera, oddest of all (Fig. 5.10q), possess a curved corallum with one flattened side and a lid or **operculum**.

Colonial Rugosa (Figs 5.5–5.8) are those in which a single skeleton (corallum) is produced by the life activities of numerous adjacent polyps each contributing a **corallite** to the whole.

The habit of the colony may be variably influenced by the environment, but colony type, which includes the morphology and relations of the corallites, is more rigidly defined genetically. In the **fasciculate** type the corallites are cylindrical but not in contact. There are two kinds of fasciculate morphologies: **dendroid**, with irregular branches, and **phaceloid**, in which the corallites are more or less parallel and sometimes joined by connecting processes. **Massive** corals are those in which the corallites are so closely packed as to be polygonal in section. Several kinds of massive corals are distinguished (Fig. 5.5):

cerioid (e.g. *Lithostrotion)*, in which each corallite retains its wall;

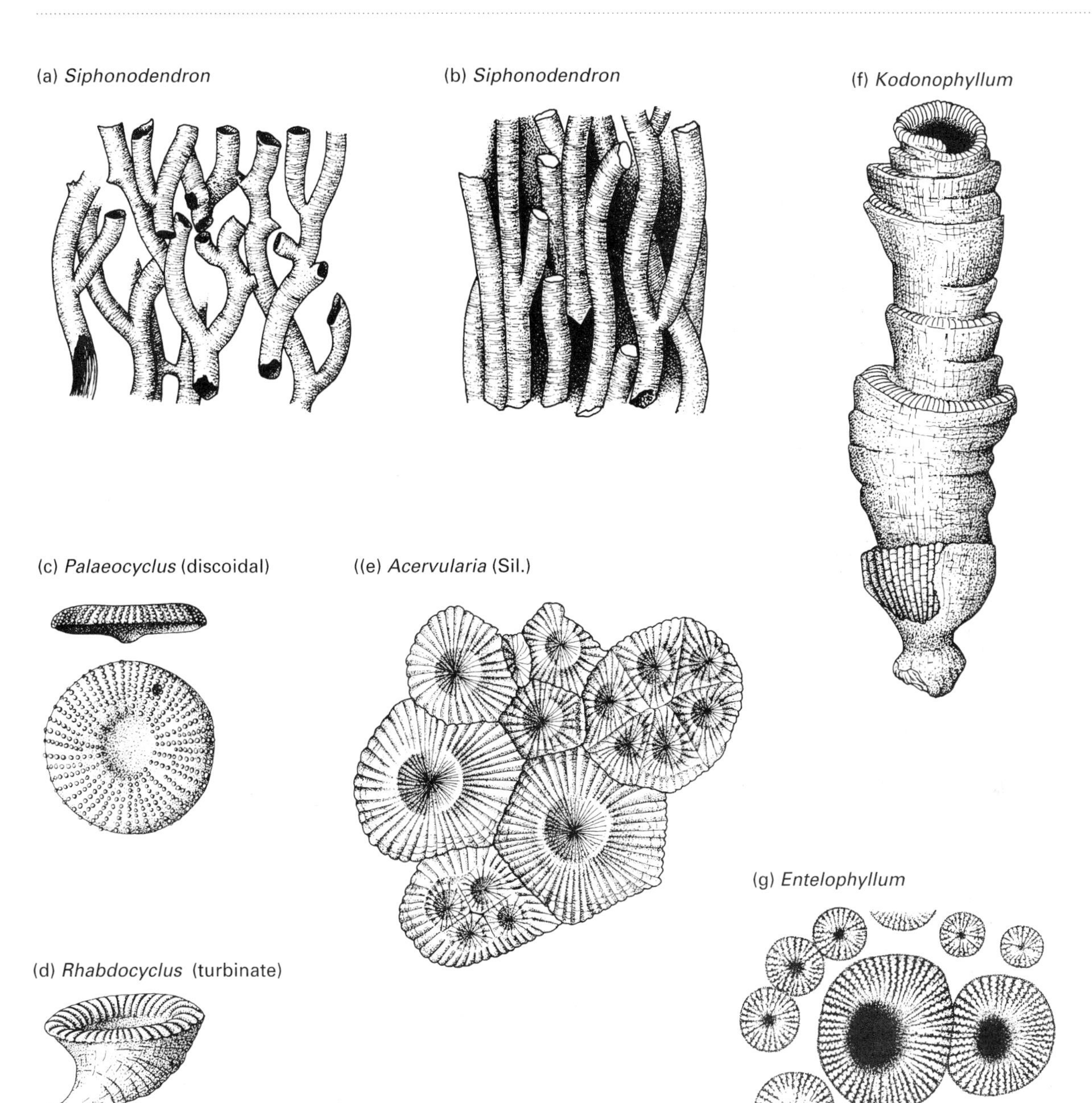

Figure 5.7 Form of corallum in Rugosa: (a) *Siphonodendron*, fasciculate dendroid; (b) *Siphonodendron*, phaceloid; (c) *Palaeocyclus* (Sil.), discoidal; (d) *Rhabdocyclus*, turbinate form; (e) axial increase in ceriod Silurian *Acervularia*; (f) rejuvenescence in Silurian *Kodonophyllum*; (g) offsets in Silurian *Entelophyllum*. (Mainly redrawn from Milne-Edwards and Haime, 1850.)

astraeoid (e.g. *Phillipsastraea*), in which the epitheca is wholly or partially lost but the septa stay unreduced;
thamnasterioid (e.g. *Haplothecia*), in which the septa of adjacent corallites are confluent and often sinuous or twisted;
aphroid (e.g. *Arachnophyllum*), where the septa are reduced at their outer ends so that neighbouring corallites are united by a zone of dissepiments alone;
indivisoid (e.g. *Orionastraea*), in which septa are absent and dissepimental material is dominant; this type is very uncommon.

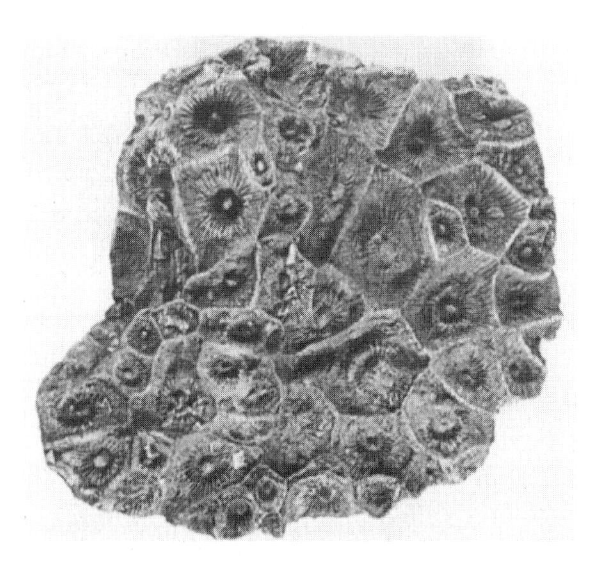

Figure 5.8 *Actinocyathus floriformis* (L. Carb.) from Bathgate, Scotland: a small colony showing lateral increase with a few small encrusting *Syringopora* (× 0.75). (Photograph reproduced by courtesy of Jeremy Jameson.)

FINE SKELETAL STRUCTURE OF SEPTA

The septa are composed of serially arranged radiating whorls of tiny fibres (sclerodermites), grouped in cylindrical bundles known as trabeculae (Fig. 5.9).

In some corals the axes of these bundles are simpler and parallel (monacanthine trabeculae), while in others there are in addition second-order trabeculae, radiating upwards and outwards (rhabdacanth and rhipidacanths). The sclerodermites are secreted in linear series from a continually producing centre of calcification: the trabecula grows as long as calcification goes on. Trabecular tissue is primary in Rugosa, and other kinds of tissue that have been described (fibro-normal and lamellar) may be the result of growth increments or diagenesis and therefore secondary (Sorauf, 1971).

The skeleton of Rugosa probably originally consisted of high-Mg calcite. This generally retains its original microstructures after death, though the high-Mg calcite usually converts to low-Mg calcite at the earliest stage of diagenesis, after which the original mineralogy may alter. Aragonite, however, seems to have been the primary skeletal mineral in

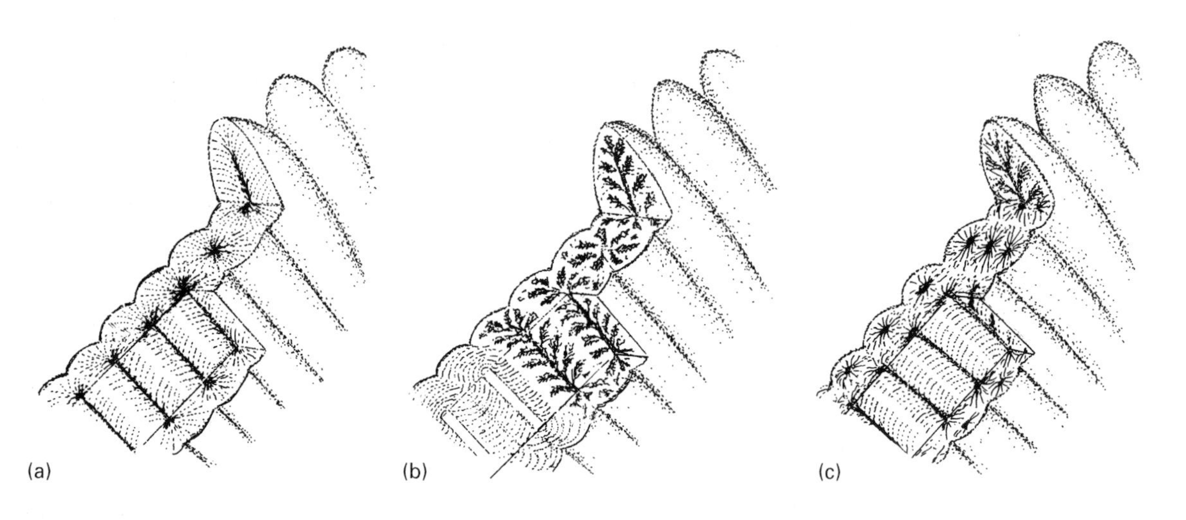

Figure 5.9 Fine structure of septa in Rugosa: (a) monacanthine trabeculae; (b) rhabdacanthine trabeculae (holacanths are probably recrystallized rhabdacanths or monacanths); (c) rhipidacanthine trabeculae. (Redrawn from Scrutton, 1986.)

some Permian Rugosa shortly before their extinction (Wendt, 1990)

Since the gross morphology of corals is so highly variable and has been subject, in so many independent stocks, to repeated convergent evolution, there is a clear need in taxonomy to find more 'stable' characters to use as the basis of classification. Although microstructure has proved useful here along with other characters, its interpretation is still the subject of much discussion and it is by no means a universal key.

TYPES OF SEPTA

In the Rugosa septal microstructure and arrangement may vary greatly, and those differences are of high taxonomic value. Septa are normally laminar, but perforated septa do occur in some genera where the trabeculae are not closely joined. Usually they are straight, sometimes zigzag. Commonly septa are not of uniform thickness throughout their length. In *Kodonophyllum* (Fig. 5.10k), for instance, the outer parts of the septa are so greatly thickened that they are contiguous, forming a well-marked stereozone.

In *Aulophyllum* (Fig. 5.10a) only half of the coral usually has medially thickened septa. The sides of rugosan septa are usually smooth but are often provided with flanges or small denticulations where the trabecular tips show through. In *Orionastraea* and some other aphroid genera the outer parts of the septa are replaced by dissepiments, whilst adjacent corallites of thamnasteroid types have confluent septa. *Amplexus* illustrates a type of coral in which septa are imperfectly developed, only being properly formed immediately above each tabular surface.

AXIAL STRUCTURE

Very many Rugosa have a central structure (Fig. 5.10a–f), which may be one of three basic kinds. An axial vortex is formed by the joined axial ends of the major septa, slightly twisted as in *Ptychophyllum*. A columella is simply a dilated end of the counter-septum and is representative of *Lithostrotion*, *Siphonodendron*, *Lophophyllidium* and other similar genera. An axial column, as in *Aulophyllum*, *is* a broad axial zone formed of **tabellae**, i.e. small incomplete tabulae within the axial complex. In *Lonsdaleia*, *Actinocyathus* and *Dibunophyllum* the axial column appears web-like in section and is likewise composed largely of tabulae with septa crossing it; in the latter genus it is divided by a central septal plate formed by the counter-septum. *Aulina* has a very simple axial column, the **aulos**, which is a vertical tube truncating the inner ends of the septa and traversed by horizontal tabulae.

TABULAE AND DISSEPIMENTS (FIG. 5.10F–H)

Tabulae are transverse plates which may be flat, convex or concave. They usually occupy a central space or tabularium, and if there is an axial complex they join with it. Tabulae may be replaced by a number of smaller plates called tabellae. Tabulae in the Cyathaxoniicae, the most primitive superfamily, generally extend right across the corallum, but in most Rugosa they terminate externally in a marginarium, which is either a septal stereozone or a dissepimentarium, and hence do not reach the edge.

Dissepiments, the small plates usually found towards the edge of the corallum, lie peripheral to the tabularium and like the tabulae are constructed of fibro-normal tissue. They are of various kinds. They may be simply inclined or slightly swollen plates dipping towards the axial region, but in *Phillipsastrea*, *Thamnophyllum* and relatives, for instance, there evolved an extraordinary range of dissepimental types, including horseshoe-shaped ones, which have proved useful in classification.

Dissepiments are not found in the earliest rugose genera. On their first appearance in late Ordovician streptelasmatinids and columnariids they are confined to a thin peripheral dissepimentarium. In later genera, however, they become more important structural components, most noticeably in types such as *Lonsdaleia*. Rarely, as in *Cystiphyllum*, the dissepimentarium together with tabellae with which dissepiments may intergrade, make up virtually the whole skeleton, since the septa are reduced to separate trabeculae developed only on the upper surface of the tabellae and dissepiments.

CALICE

The **calice** is that part of the coral which, in life, was in contact with the basal ectoderm of the polyp. It is in effect a mould of the secretory surface. It is not commonly seen in specimens preserved in limestones unless these have been weathered, but in individuals collected from shales calicular surfaces can often be clearly seen. Usually there is an annular calicular platform surrounding a deep central depression (calicular pit). Some genera with an axial complex have a

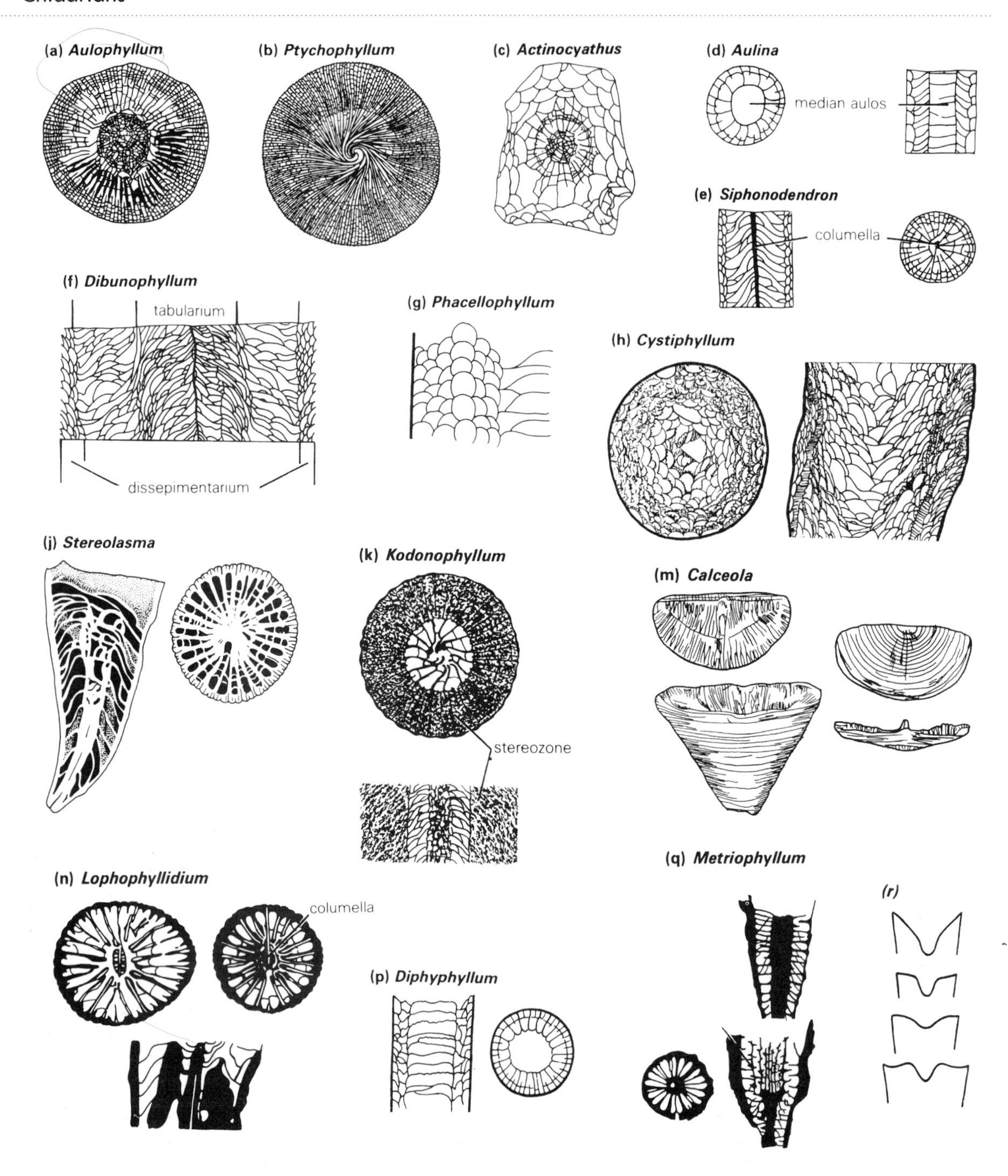

Figure 5.10 Morphology of Rugosa: (a) *Aulophyllum* (Carb.) with axial structure and semicircular zone of thickened septa (× 1); (b) *Ptychophyllum* (Sil.) with axial vortex (× 0.75); (c) *Actinocyathus* (Carb.) with lonsdaleoid dissepiments or presepiments (× 1.25); (d) *Aulina* (Carb.) with median aulos (× 1.6); (e) *Siphonodendron* (Carb.), vertical section showing dissepimentarium and tabularium (× 1.5); (f) *Dibunophyllum* (Carb.), showing axial structure and location of central tabularium and dissepimentarium; (g) *Phacellophyllum* (Dev.), vertical section of dissepimentarium with horseshoe-shaped dissepiments (× 4); (h) *Cystiphyllum* (Sil.), showing broad dissepimentarium (× 0.75); (j) *Stereolasma* (Dev.), vertical and transverse sections (× 1.5); (k) *Kodonophyllum* (Sil.), with septal stereozone (× 1.5); (m) *Calceola*, a Devonian solitary rugose coral with operculum (× 0.5); (n) *Lophophyllidium* (Carb.–Perm.), two transverse and one vertical sections; (p) *Diphyphyllum* (Carb.), showing downturned outer edges of tabulae forming a tube (× 1.5); (q) *Metriophyllum* (Dev.), with very thickened axial structure and epitheca; (r) diverse calycal shapes in Spongophyllidae (diagrammatic). (Mainly redrawn from illustrations in Hill in *Treatise on Invertebrate Paleontology*, Part F.)

marked calicular boss where it emerges to the surface, but this is not always present. The form of the calice is quite variable even within families, as shown in Fig. 5.10m, which illustrates various forms found in the Spongophyllidae in which there are conical, bell-shaped, saucer-shaped and everted types.

BUDDING AND THE PRODUCTION OF NEW CORALLITES (FIG. 5.6)

When a compound coral grows the first-formed **protocorallite** gives rise asexually to offsets, which may form in a number of ways. The term **budding** is used for the soft parts, **increase** for the skeleton. Detailed studies have been made, in the Lithostrotionidae and many other groups, of the three main methods of increase and budding. There are two kinds of increase, parricidal, in which the parent is destroyed (1, 2), and non-parricidal, where it is not.

1. Axial increase. In this mode the corallite splits by fission into two or more daughter polyps within the calice. Such budding is inevitably parricidal. The upper surfaces of many compound corals can often be seen with several corallites undergoing axial increase at approximately the same time. As the colony continues to grow the parent corallite is buried and the daughter corallites grow to full size. This is the least common of modes of increase in the Rugosa, though normal in the Silurian *Acervularia* (Fig. 5.7e).
2. Peripheral increase. Here small offsets arise round the parent corallite (Fig. 5.7g). This is likewise parricidal; it is quite common in fasciculate genera such as *Entelophyllum*.
3. Lateral increase. This is the most common method of increase in the Rugosa and is not parricidal. In *Siphonodendron* it has been well investigated (Jull, 1965) but has proved to be so variable as to be of limited taxonomic value at the generic level. Three distinct types of lateral increase have been distinguished in *Siphonodendron* using serial sections (Fig. 5.6c–e). In one type (e.g. *S. junceum*) the lateral bud arises fully external to the parent calice and is initially aseptate; septa only form later. In a second type, typical of species with a narrow dissepimentarium (e.g. *S. arundineum*), the new corallite arises from near the periphery and, though septate from the start, does not inherit septa from the parent. In a third type, in species such as *S. martini* with a wide dissepimentarium, the daughter arises from well within the parent calice and inherits its initial septa, especially those of the outer wall, from the ends of the parent septa. Cerioid species of *Siphonodendron* increase in a manner analogous to the third type, but the daughter remains attached to the parent and is polygonal from the start. Nudds (1981) has extensively reviewed colony form in *Siphonodendron*.

There are some rugose corals (notably *Actinocyathus*) which show a curious feature associated with offset production. A pair of narrow tubular ducts may be seen extending from the base of the offset obliquely upwards through the dissepimentarium of the parent. The ducts may then merge at higher levels in the parent corallite. They can only have served to prolong gastric and nervous communication between the parent corallite and the daughter until the latter was well established (Scrutton, 1983). Oliver (1976a) has reviewed many aspects of colony formation in Rugosa and other corals.

REJUVENESCENCE (FIG. 5.7F)

In solitary cylindrical forms the corallite is often found to be constricted at irregular intervals, leaving a broad or narrow shelf where the septate older calice is exposed. Above this it may expand again, before once more being constricted. These constricted bands represent times of stress. Thus in periods of famine the polyp resorbed its own tissues in order to stay alive and shrank away from the edges, becoming smaller while retaining its form. The next period of increased food supply permitted growth to begin once more; this is **rejuvenescence**. Starvation and regrowth, however, are only one explanation; partial smothering of the coral by sediment is another alternative. In some species rejuvenescence may be especially strong, so that sharp rims are formed with deep and almost slit-like contractions between. In some instances small buds are found on the upper surface of a calice. These probably result from a late rejuvenescence of a coral which almost died, from small areas of still-living tissue.

Evolution in the Rugosa

The Rugosa have relatively few phylogenetically useful characters. Furthermore, they were very plastic in their evolutionary potential and notoriously subject to homeomorphic trends in separate stocks. Yet however common such iterative trends may have been, the guiding principle controlling their evolution was always towards a strong and firm skeleton. Thus many of the observed structures can be interpreted functionally; for instance, the complication of the trabeculae increased skeletal strength, the carinae held the polyp more firmly and the axial column gave greater support in the central region.

There is thus some functional meaning in all the trends that have been observed, though it is not always clear what the function was.

OVERALL PATTERN OF EVOLUTION

The earliest known of all corals are the Lower Cambrian Tabulaconida and the Middle Cambrian Cothonionidae. Undoubted Rugosa do not appear until the Middle Ordovician (Blackriveran) in North America. They spread rapidly and evolved in a series of successive episodes. During each of these one particular faunal group was dominant, and subsidiary elements at the time were (1) small persistent bradytelic stocks (2) early members of a later dominant stock and (3) remnants of a formerly dominant stock.

The earliest rugose corals appear just after the first Tabulata and the overall patterns of evolution in the two groups are remarkably similar (Scrutton, 1988).

In the Mid-Ordovician to Lower Silurian small solitary corals were dominant, mainly without dissepiments and with only feeble development of trabecular tissue. Thereafter there was a steady increase in numbers of Rugosa up until the Middle Devonian and with colonial corals, fasciculate cerioid and amural becoming increasingly important. The reef habitat was colonized, though in most reef habitats the Rugosa were subordinate to the stromatoporoids and tabulate corals. The succession is clear from Fig. 5.11, and particular orders come into dominance in turn. Then in the Frasnian (Late Devonian) came an almost total collapse of rugose coral faunas, paralleled by an equally dramatic drop in the tabulates.

Recovery from this major extinction episode was rapid, and the coralline ecosystems were so quickly restocked so that by the later Lower Carboniferous the Rugosa had nearly achieved their Lower Devonian maximum. By this time the Lithostrotionidae, Aulophyllidae and other 'classic' Carboniferous rugosans were dominant, but there was then a slow decline into the Permian. The last rugosans included genera such as *Plerophyllum* and *Waagenophyllum*. No rugosans are known after the end of the Permian. On the whole the ratio of solitary to colonial forms remains relatively constant, the solitary forms being dominant, but there is a weak trend towards a higher percentage of amural forms.

EVOLUTIONARY TRENDS

Rugose corals had only a limited evolutionary potential; hence it was inherently likely that the same kind of structure would be evolved several times over. According to Lang (1923), who first tried to establish such repetitive 'trends', in Carboniferous corals, 'each character follows in varying degrees and at varying rates one of a comparatively few possible developments pointing to the existence . . . of limited tendencies . . . repeated in each lineage'. In taxonomy, which becomes confused by such trends, the only way to separate and identify the genera is by very detailed study of morphology, fine structure and ontogeny.

In general terms the development of a marginarium (stereozone or dissepimentarium) is perhaps the most persistent and universal of all trends in the Rugosa, and since this happened in several groups independently in the Upper Ordovician and Lower Silurian just prior to the first success of the Rugosa as reef builders it was clearly important in allowing increase in colony size while firmly supporting the polyp at the margin. A number of other trends fundamental to all the Rugosa have also been distinguished.

The phylogeny of the Rugosa, however, is still too poorly known to allow trends to be properly documented except in a few specific cases.

Evolutionary studies of trends within Family Lithostrotionidae have shown that the parallels are so close in several descendent stocks that species originally lumped in the same genera were actually derived polyphyletically. Thus '*Orionastraea*', which has a weak or absent columella and a tendency for septa to be withdrawn from both the axis and the periphery, is evidently polyphyletic and has had to be split since it included morphs that are of similar form but derived from different ancestral species.

Apart from these trends, there is a cerioid trend in some Lithostrotionidae and in addition a particularly interesting trend which involves the loss of dissepiments through time. In the early forms (e.g.

Siphonodendron praenuntius) there are six circlets of dissepiments, reducing in the descendent *S. martini* to four and then to two. In addition the septa generally become fewer and the tabulae more widely spaced. The evolutionary principle governing the origin of the new types is probably paedomorphic since the descendent adults tend to resemble ancestral juveniles. Such morphology would seem to be especially appropriate for an environment in which rapid growth with minimal metabolic expenditure on skeletal elements was desirable. Likewise, the advantage of food sharing between the linked polyps in the 'orionastraeoid' aphroid morphologies, and the fact that in these all the available surface space is covered by polyps, would lead to greater efficiency of the corallum as an integrated unit.

Each of these trends therefore, expressed polyphyletically, has its own advantage suitable for particular ecological conditions.

MICROEVOLUTION

There are relatively few well-documented accounts of small-scale evolutionary change in the Rugosa. One of the best known is that of Carruthers (1910), who worked on the *Zaphrentites delanouei* group in Scotland and England. The species group actually ranges as far as Belgium. His collections were made from thin limestones in an alternating limestone–shale–sandstone–coal sequence, all well located stratigraphically. In studying the populations in ascending sequence he noted that most characters (shape, size, tabular spacing, etc.) tended to remain constant but that there seemed to be successive changes in the shape of the cardinal fossula and the length of the major septa (Fig. 5.4g–m). The oldest beds contain *Z. delanoeui* ssp., which has septa meeting in the centre and a large cardinal fossula expanded axially. A variant of this, *Z. parallela*, with a parallel-sided fossula, occurs in the same beds. It was probably derived paedomorphically.

A short-lived side branch led to the small *Z. lawstonensis*. In younger beds *Z. delaneoui* and *Z. parallela* are rare and replaced largely by *Z. constricta* in which the axial end of the cardinal fossula is constricted to a keyhole shape. In the youngest limestones there occurs *Z. disjuncta;* this form passes through a *constricta*-type morphology in its juvenile development, but thereafter septa retreat away from the centre.

Though this has been regarded as a classic study it is severely limited in two respects. One is that since the corals occur only at certain horizons within a highly variable sedimentary sequence, it was not possible to establish continuity of variation. Second, the study was done within a limited area alone, and it is not clear how far the sequence can be substantiated in other regions. Also, since different 'species' coexisted in time (*Z. delanouei* and *Z. parallela)* and there is some evidence of the same species being found at different horizons in different areas, the pattern of evolution was probably very complex and is perhaps best regarded as the expression of a trend within a species group rather than as a matter of linear descent involving separate species. In functional terms the morphology of *Z. constricta* makes for a stronger axial region than that of *Z. delanouei*, but that of *Z. disjuncta* is clearly weaker. Perhaps the need for rapid growth, as would be allowed by septal reduction, may in this case have offset the axial weakness.

So many factors seem to have been involved in the evolution of the *Z. delanouei* group – including trending, response to local environmental conditions, paedomorphosis, allopatry and polyphyly – that the evolutionary pattern now seems to be so complex as to elude analysis until better data are available.

Classification of the Rugosa (Fig. 5.11)

The following classification, erected upon the various structural features already described, is based upon that in the *Treatise on Invertebrate Paleontology*. It is largely founded on megascopic features; knowledge of microstructure is not yet advanced enough to be of much use. The inevitable effect of parallel trends and convergent evolution, however, creates taxonomic problems which if resolved may lead to a radical rearrangement of the group. Classification of the Rugosa is in a state of some flux; many authorities (e.g. Scrutton, 1997a) no longer use Cystiphyllida and Stauriida.

ORDER RUGOSA (?Cam./Ord.–U. Perm.): Solitary or colonial corals with the major septa inserted in four quadrants only relative to the protosepta. Epitheca normally present, and up to three orders of septa. Dissepiments occur in more advanced groups; tabulae normally present. First- and second-order septa well developed; third-order septa less common.

SUPERSUBORDER CYSTIPHYLLIDA (M. Ord.–?U. Dev.): Solitary or sometimes fasciculate or massive. Septa usually spiny and well-developed but may be much reduced.

Tabulae and dissepiments variably developed and of vesicular form, not clearly distinct from each other, floor of calice forms a shallow bowl; e.g. *Palaeocyclus, Rhabdocyclus, Tryplasma, Holmophyllum, Goniophyllum, Calceola, Cystiphyllum, Mesophyllum*.

SUPERSUBORDER STAURIIDA (M. Ord.–U. Perm.): Solitary or colonial with laminar septa usually composed of trabeculae. The septa are usually well developed but many retreat from the periphery. A marginal stereozone may develop, in which case dissepiments may be absent. Tabulae usually present, though may be incomplete. There are at least 16 suborders (Fig. 5.11), common genera include *Stauria, Streptelasma, Calostylis, Cyathaxonia, Entelophyllum, Arachnophyllum, Acervularia, Phillipsastraea, Heliophyllum, Zaphrentites, Caninia, Aulophyllum, Palaeosmilia, Lithostrotion, Siphonodendron, Lonsdaleia*.

Ecology

Solitary Rugosa had no really effective means of anchorage on the sea floor. They did not normally cement themselves to the substratum, though some have cicatrices of attachment or root-like 'talons' which may have helped to fix them. They seem to have preferred soft substrates, but since relatively few 'hardgrounds' are found preserved it is not known how far they could colonize such hard substrates. The horn-like shape would give a reasonable degree of support if the coral were to be half sunk in mud, convex side down. This would allow the polyp to be exposed above the sea floor and would enable it to stay in the same general attitude whilst the coral grew. There is some evidence that Rugosa such as *Aulophyllum* could grow like this, orientated with respect to currents. Commonly, however, in any population some or all of the coralla are twisted; they must have toppled over and then grown up again.

The solitary *Calceola* (Fig. 5.10m) lay on the sea floor, convex side down with its operculum

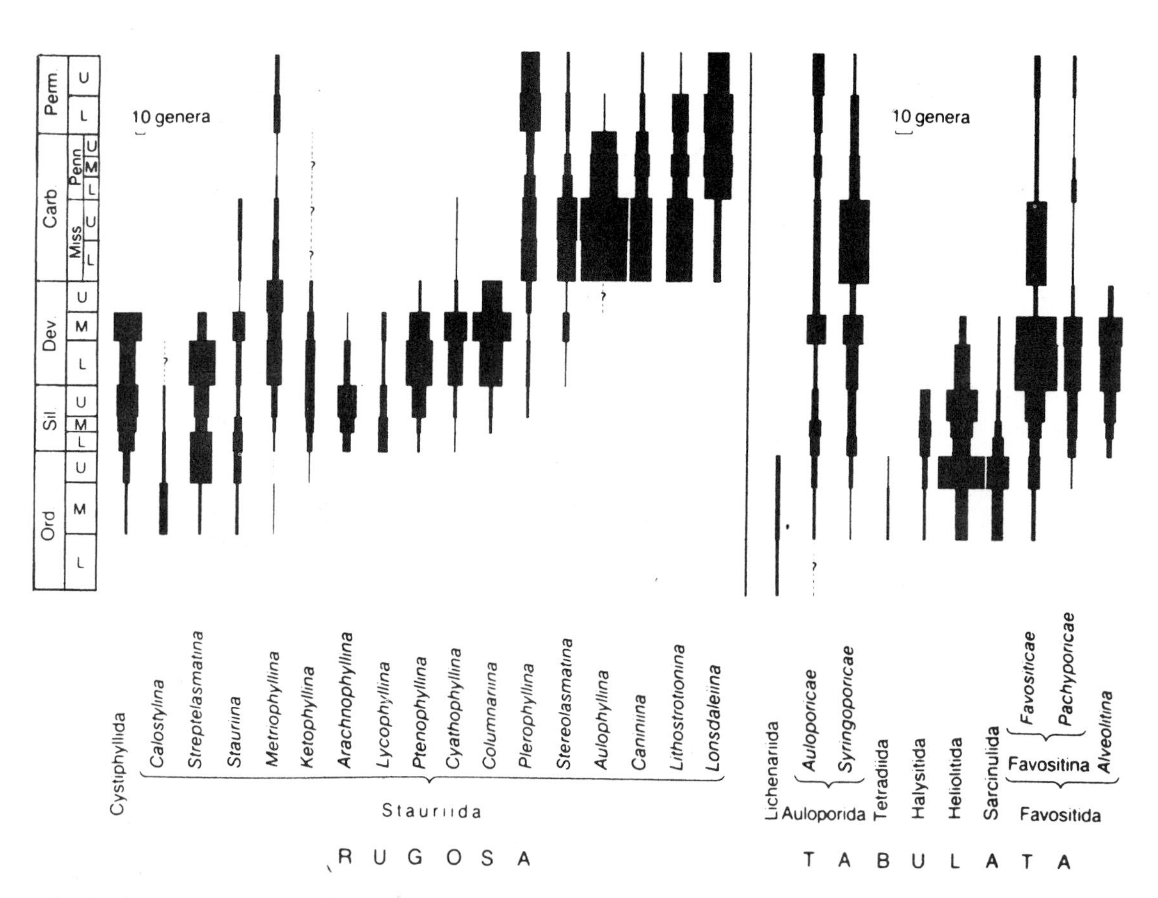

Figure 5.11 Time ranges of supersuborders (and some suborders and families) of Rugosa and Tabulata. (Redrawn from Scrutton, 1986.)

open to 90°. Presumably it could shut it in case of emergency.

Compound Rugosa did not normally fix themselves to the sea floor either, though rare examples of hardgrounds encrusted by Rugosa are known. In softer substrates the weight of the colony in the mud would no doubt give some stability. Since they were uncemented they were environmentally restricted and could not build proper reefs. This seems to have been their most severe evolutionary limitation, from which, in spite of their 230 million year history, they were never able to escape. Thus they are very rarely found within the limits of strong wave action and so were never really involved as frame builders in the algal and coral–stromatoporoid reefs of the Palaeozoic (Wells, 1957).

It is not known for certain whether rugose corals possessed endosymbiotic algae (zooxanthellae) in their tissues or not. Judging by the low level of morphological integration in rugose corallites (Coates and Jackson, 1987), however, it is probable that they did not, since modern corals lacking zooxanthellae are likewise poorly integrated.

Rugose coral faunas were very sensitive to environmental conditions, and the form of both solitary and colonial species was strongly influenced by ecological factors (Wells, 1937). Various life strategies have been distinguished in solitary Lower Palaeozoic Rugosa that lived on a soft sea floor (Neuman, 1988; Fig. 5.12).

Most were liberosessile, i.e. initially attached for a short time to a sediment grain and thereafter lying recumbent (e.g. *Holophragma*). This is evident from the shape of the coral and its attachment scar. Ambitopic forms such as *Phaulactis* were attached for a longer period, to a shell, for example, before falling over. *Rhegmaphyllum* seems to have grown upright, its base embedded in soft sediment. Rhizosessile genera have root-like holdfasts emanating from the corallite at several levels, and fastened to sediment grains (e.g. *Dokophyllum*). These forms grew upright. The small discoidal *Palaeocyclus*, which incidentally forms a globally widespread marker horizon in the Upper Llandovery, may have been vagile, able to move around by its peripheral tentacles on the sediment surface. Finally, several species of *Grewinkia* and other genera had a fixosessile mode of life, firmly attached through life to patches of hard bottom and to other fossils.

In large, strongly curved rugosans such as *Aulophyllum*, the horn-like form was adapted for resting in the mud, convex downwards and partially embedded. If it toppled over it regrew upwards again, which accounts for the convoluted shapes often seen in such corals.

Separate facies faunas can be distinguished,

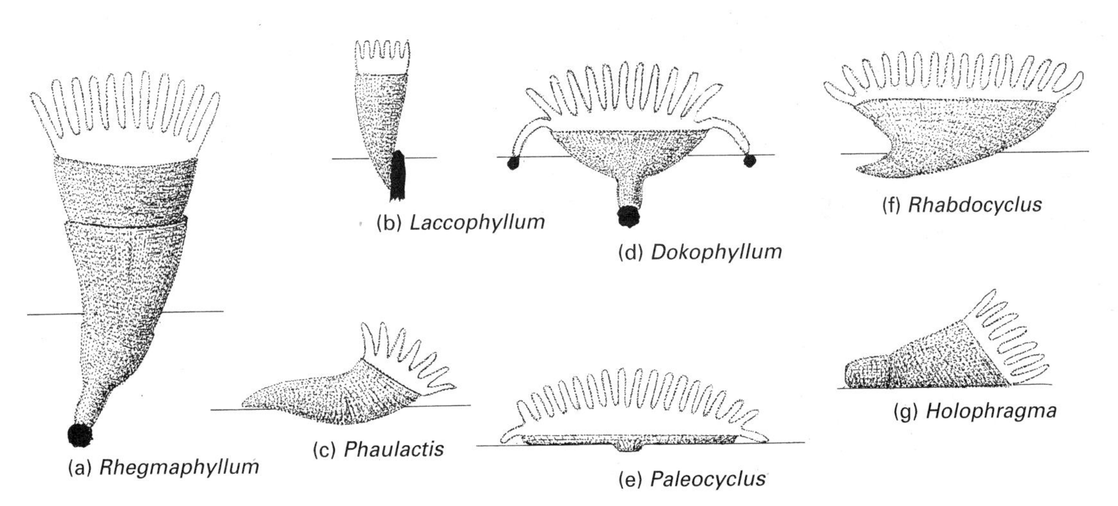

Figure 5.12 Life strategies of some Lower Palaeozoic solitary Rugosa: (a) *Rhegmaphyllum* growing on an originally unstable sedimentary grain; (b) *Laccophyllum*, attached to a vertical shell fragment; (c) *Phaulactis*, small specimen adopting a recumbent life style; (d) *Dokophyllum*, a rhizosessile form with tentacle-like appendages producing tubular holdfasts attaching to sand grains; (e) *Palaeocyclus*, adult form possibly capable of vagile movement by lateral tentacles; (f) *Rhabdocyclus*, adult, turbinate form with recumbent lifestyle; (g) *Holophragma*, recumbent lifestyle. (All redrawn from Neuman, 1988.)

adapted for particular circumstances and normally found with characteristic associations of brachiopods and tabulates. Three such facies faunas were noted by Hill (1937) working mainly with Scottish Carboniferous successions. These are as follows.

1. *Cyathaxonia*—laccophyllid fauna. Usually found in black or greenish calcareous shales, though sometimes in bryozoan reefs also, these corals are all small and solitary with dissepiments poorly developed. This facies fauna is very long ranged and certainly antedates the Carboniferous; indeed the earliest of all rugose coral faunas seem to have been of this type, though occurring in mudstones. It is usually found with tabulates, small brachiopods and trilobites. As represented typically in the Carboniferous the sea floor was generally muddy, with decaying organic matter, but must have been fairly well oxygenated.
2. *Caninia*–clisiophyllid fauna. This fauna, found first in the Tournaisian but probably older, consists of large solitary genera with dissepiments. Clisiophyllids gradually took over from caniniids with time. These faunas were adapted to shallow

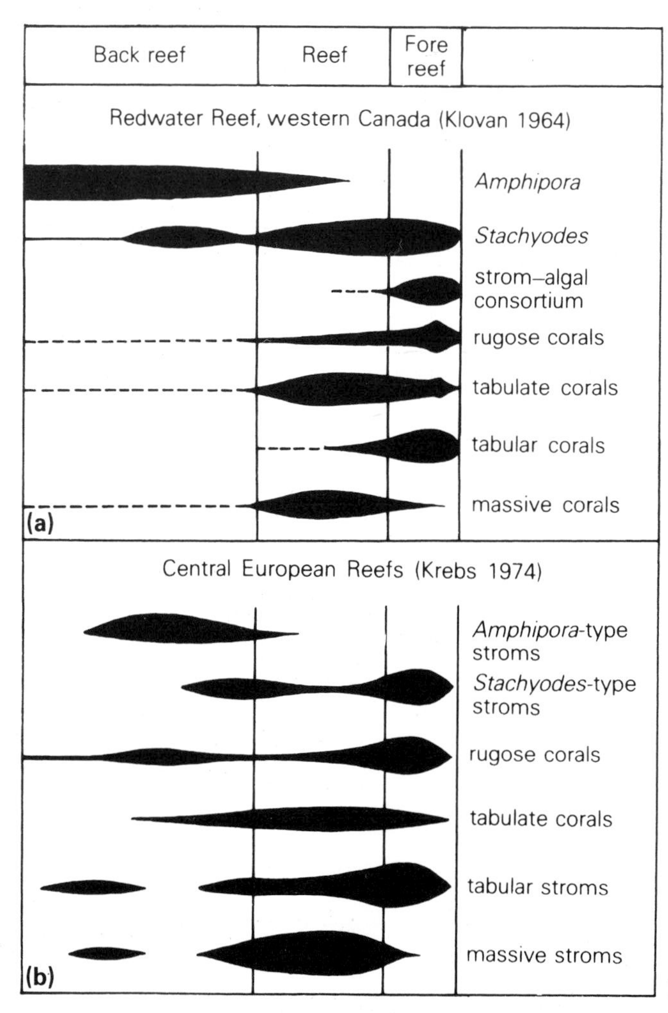

Fig. 5.13 cont. on p. 123

Figure 5.13 (a)–(d) Distribution of rugose corals, stromatoporoids and other reef-building organisms in various Devonian reefs. (a) Redwater reef, Western Canada; (b) Central European reefs; (c), (d) Frasnian (c) and Famennian (d) reefs in the Canning Basin, Western Australia; (e) palaeoecological succession in a Devonian reef from pioneer to climax stages. [(a) Redrawn from Klovan, 1964; (b) redrawn from Krebs, 1974; (c), (d) redrawn from Playford, 1980; (e) redrawn from Copper, 1974.]

limestone seas, well lit and oxygenated but with little terrigenous material.

3. Colonial or 'reef' coral faunas. Compound rugose corals with dissepiments may occur in small and scattered 'bioherms'; they are represented in bedded limestones as well. These were first considered to be reef faunas analogous to modern reef corals, but since the Rugosa were not really reef formers they are best considered as just a group of colonial corals. These colonial corals evolved from the *Caninia*–clisiophyllid fauna.

Fig. 5.13 cont.

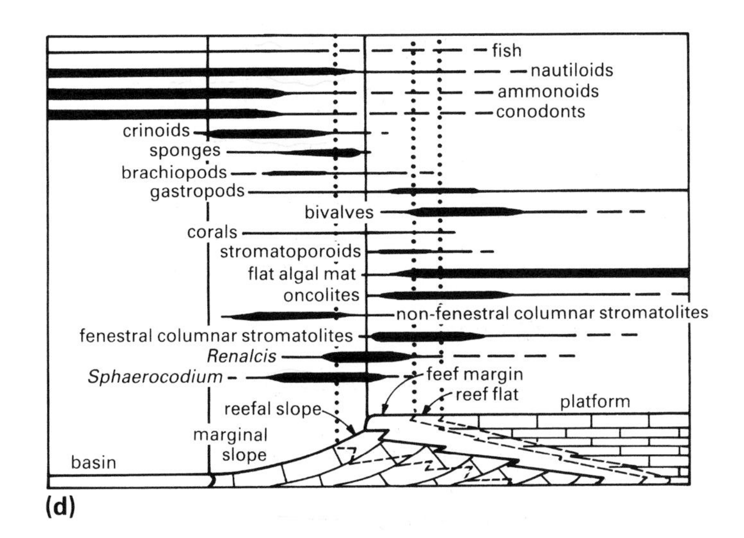

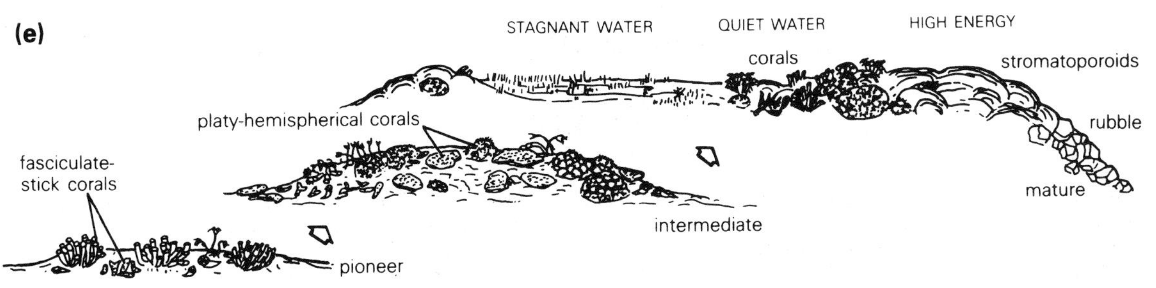

Hill's three facies associations have generally been well substantiated, though the *Cyathaxonia* fauna is very much separate whereas the other two are sometimes intergrading and ecologically less well defined. Thus it is not uncommon to see rapid alterations of faunas (2) and (3) in vertical sequence, as in the Carboniferous of the west of Ireland where, in addition, thickets of fasciculate lithostrotionids are found with the large solitary *Aulophyllum* growing within the thickets.

More recent work has concentrated upon the distribution of rugose and other corals in Palaeozoic reefs. Klovan (1964) and Krebs (1974) have charted the facies associations of corals and other reef organisms in Devonian reefs of western Canada and central Europe respectively (Fig. 5.13).

In these reefs colonial Rugosa tend to be confined to the quieter forereef facies around the flanks of rigid frame reefs; they do not normally occur within the reef itself nor, except in localized patches, within the back-reef facies. Solitary Rugosa are likewise confined but are numerically insignificant. Probably one of the best examples to show a changing reef-related coral ecology is in the Devonian Canning Basin in Western Australia. Here there is an exhumed Middle to Upper Devonian barrier reef belt some 350 km long (Playford, 1980); the reef complexes formed limestone platforms standing tens to hundreds of metres above the sea floor and were flanked by marginal slope and basin deposits. In the Frasnian, rugose corals, together with stromatoporoids and algae are characteristic of the reef margin, reef flat and platform, their numbers diminishing shorewards and also down the reef slope. The corals were abundant at depths of down to 10 m and rare at greater depths. Following the great global extinctions at the end of the Frasnian and the temporary drowning of the reefs, the character of the reef complex changed dramatically, and the few corals that persisted were confined to the reefal and marginal slope, the reef margin and the reef flat.

Tabulate corals, especially those with a creeping or foliaceous habit, were somewhat less restricted but could not compete with stromatoporoids and algae in the turbulent zone of most active reef growth. The hard skeletons of both rugose and tabulate corals, however, seem to have contributed to establishing a firm foundation for later reef growth of algae and stromatoporoids. Although the broad pattern of distribution of rugose and tabulate corals with respect to ancient reefs has been generally confirmed (e.g. Jamieson, 1971), a more detailed picture of the various coral–stromatoporoid facies associations is emerging. Scrutton (1977), for example, who worked upon an evolving reef complex in southwestern England, has shown how the genera in stromatoporoid reef limestones differ from those in biohermal, bioclastic and bedded dark limestones.

Copper (1974; Fig. 5.10e) has described a generalized palaeoecological succession of a Devonian reef from pioneer through intermediate to a mature reef system. Devonian reef systems were evidently more extensive and elaborate than those of the Silurian, but by the end of the Devonian had collapsed. A fuller treatment of the theme has been given by Walker and Alberstadt (1975).

On a global scale, evidence is also accumulating on the biogeographical distribution of rugose corals (Oliver, 1976b).

Order Tabulata (Fig. 5.14)

The tabulate corals are entirely Palaeozoic, and though they appear a little earlier than the Rugosa they have an otherwise similar time range. They are always colonial, never solitary, and usually their corallites are small. Invariably they have prominent tabulae, but other skeletal elements, in particular the septa, are reduced or absent.

Having relatively few structural elements, the Tabulata are of comparatively simple construction. *Favosites* (Sil.–Dev.; Fig. 5.14a), for example, has elongated, thin-walled, prismatic, polygonal corallites with horizontal tabulae extending right across the corallum (Stel, 1979). The septa are reduced to short and somewhat irregular spines. All the walls are perforated by numerous mural pores, connecting the corallites. In *Favosites*, as in tabulates generally, elements of 'skeletal microstructure' (tabulae and walls) are fibro-normal. Trabeculae are scattered in tabulates. In *Favosites* they are normal to the surface of the tabulae but inclined upwards and outwards along the axis of the septa. In other genera septal spines are not necessarily trabeculate. Some genera (e.g. *Trabeculites*) have trabeculate walls, and *Halysites* has trabeculae in the walls.

Range in form and structure

FORM OF CORALLUM

The corallum (colonial skeleton) is built up by individual polyps which may or may not be directly

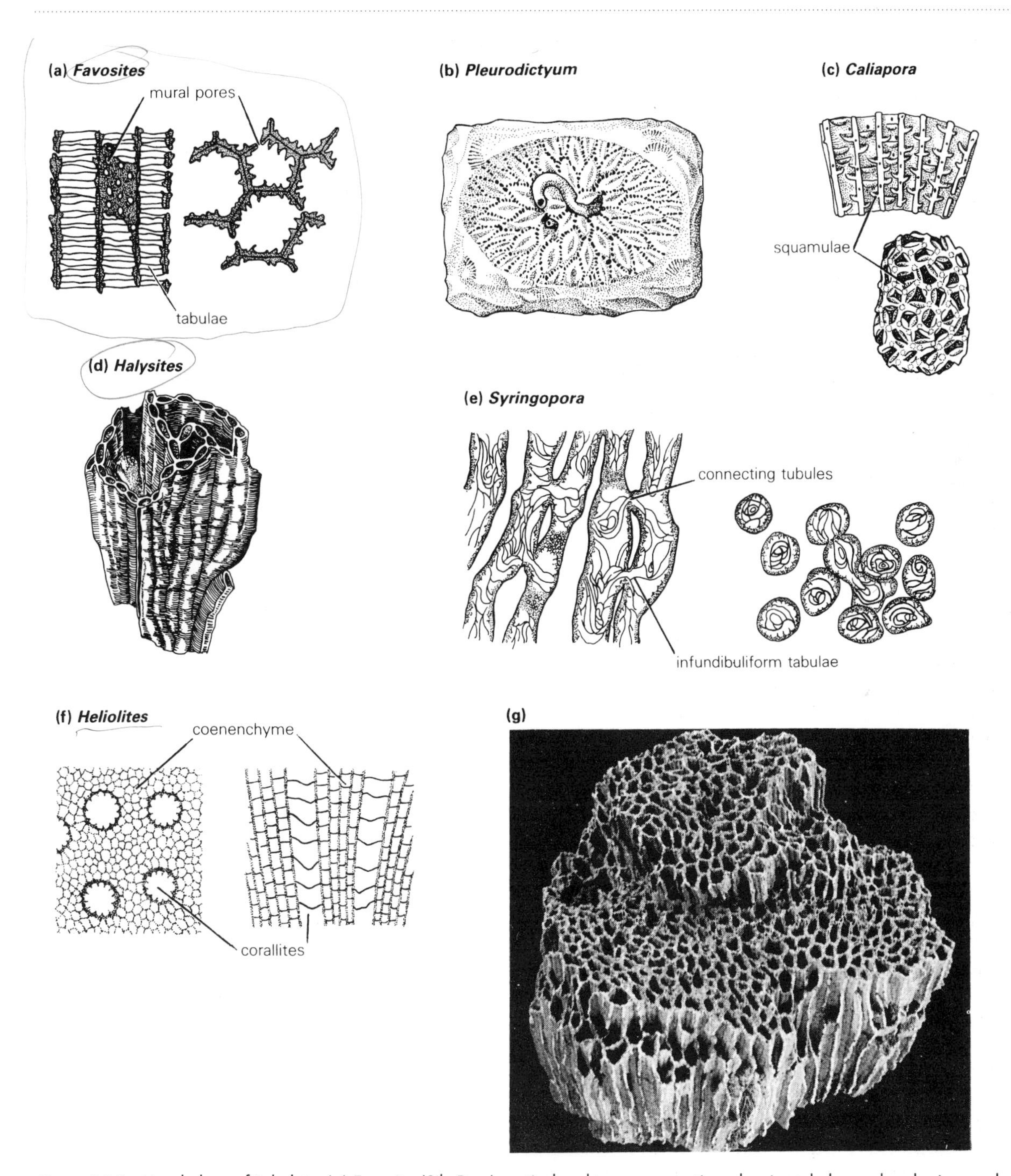

Figure 5.14 Morphology of Tabulata: (a) *Favosites* (Sil.–Dev.), vertical and transverse sections showing tabulae, reduced spines and mural pores (× 4); (b) *Pleurodictyum* (Dev.; internal mould) with commensal worm (× 1.5); (c) *Caliapora* (Dev.), vertical and transverse sections showing shelf-like squamulae (× 4); (d) *Halysites* (Ord.–Sil.), an external view of a colony showing cateniform growth (× 0.75); (e) *Syringopora* (Sil.–Carb.), vertical and transverse sections showing infundibuliform tabulae and connecting tubules (× 3.5); (f) *Heliolites* (Sil–Dev.), transverse and vertical sections with corallites embedded in coenenchyme; (g) *Halysites*, a Silurian colony cleared of matrix and showing cateniform growth (× 1). [(a)–(e) Redrawn from illustrations by Hill and Stumm in *Treatise on Invertebrate Paleontology*, Part F; (f) redrawn from an illustration in Woods, 1896; (g) reproduced by courtesy of C. Chaplin.)

connected to each other. **Cerioid** forms (e.g. the Favositina; Fig. 5.14a) have polygonal corallites all in contact. **Cateniform** colonies (e.g. the Halysitina; Fig. 5.14d,g) have elongated corallites joined end to end in wandering palisades. **Fasciculate** tabulates (e.g. *Syringopora)* (Fig. 5.14e) have cylindrical corallites which may be dendroid or phaceloid and may be provided with connecting tubules. **Auloporoid** genera have a branching (**ramose**) tubular structure and often an encrusting creeping (**reptant**) habit. They usually encrust hard substrates but are sometimes free living. One of the latter *Aulopora* species has been reconstructed, using a computer program for serial sections (Scrutton, 1989). Finally, **coenenchymal** (Fig. 5.14f) types, characteristic of the more advanced tabulates (e.g. the Heliolitina), have no dividing walls between the corallites but instead a common mass of complex tissue, the coenenchyme, deposited by a common colonial tissue (coenosarc) in between the polyps and forming a dense calcareous mass in which the corallites are embedded. As in the Rugosa, colony form, which refers to the relations of the corallites, must be distinguished from colony habit; thus *Thamnopora* is a ramose but cerioid favositid. Other tabulates may be **massive**, **foliaceous** (in which habit the coral forms thin overlapping laminar sheets) or **creeping**.

SKELETAL ELEMENTS

Tabulae, the most important skeletal elements, are always present and commonly traverse the corallite horizontally. Sometimes they are replaced by smaller tabellae. In the fasciculate genus *Syringopora* (Fig. 5.14e) the tabulae are funnel-shaped (**infundibuliform**) and run through the tubules that connect the branches. Septa in tabulates are rarely more than short spines, often 12 in number, but in some cases they do reach the centre of the corallite. In *Caliapora* (Fig. 5.14c) and other genera of the Favositina they may be replaced by small shelves (**squamulae**) which jut inwards at intervals from the corallite wall. Septal insertion in a sequence similar to that of the Rugosa is seen in *Palaeacis* and a few other tabulates but its significance is uncertain.

Where there is a marginal zone, as in the more advanced tabulates, it can be of two kinds. It may be only a thickened zone of annular lamellae as in *Striatopora*, constricting the aperture. In the other kind the growth of the marginarium is accompanied by the loss of the corallite walls, resulting in a coenenchyme. This can be constructed in various ways and evidently evolved independently in several stocks; tabulae, trabeculae borne on dissepiments and trabeculae closely packed or sometimes organized into tubes may all have a part in it. The production of a coenenchyme is one of the more important evolutionary trends in the Tabulata, another being the development of mural pores or connecting tubules for interconnection of the corallites.

BUDDING AND GROWTH OF THE CORALLITES

Tabulate coral colonies have two end-member adaptive growth strategies (Scrutton, 1997a). In peripheral growth, the new corallites arise at the edge of the colony; in medial growth they emerge between corallites which formed earlier. Some tabulates could only grow in one of these two ways, others could change their growth strategies as they developed.

Evolution and ecology

Tabulate-like corals are recorded from the Cambrian, but the resemblances are superficial. The earliest undoubted Tabulata are known in the Canadian (Lower Ordovician) of North America, somewhat antedating the arrival of the first Rugosa, and by the Middle Ordovician the full range of tabulate colonial types had arisen. Their overall pattern of evolution is remarkably similar to that of the Rugosa (Scrutton, 1988; see also Fig. 5.11). Tabulates were actually more numerous than Rugosa prior to the Silurian, but they were more acutely affected by the Late Ordovician extinction event than were the latter group. Thereafter they climbed to a new peak of diversity in the Middle Devonian, but then underwent a catastrophic decline, contemporaneously with the Rugosa, in the Frasnian. They never recovered fully from this, though they continued until the end of the Permian when they finally became extinct. It is considered likely that their limited success in the later Palaeozoic may have been partially due to the late Devonian extinction of the stromatoporoids, which had created the reefal and other niches in which tabulates had formerly flourished.

Through time different kinds of colonial structures dominated in turn. Thus heliolitids were commonest up to the early Silurian (massive coenenchymal imperforate structure), followed by favositids in the Devonian (massive cerioid perforate) and finally the fasciculate syringoporoids after the Devonian. This represents an overall decline from the range of types established in the early Ordovician, but also a general trend away from massive types and towards a more open, erect structure.

Larger tabulates are found in coral–stromatoporoid reefs and were relatively important, though they were not really frame builders since they had no proper means of attachment. Smaller tabulates tended to occur in deeper waters, and fasciculate genera usually lived in quieter environments. Ordovician and Lower Silurian tabulates of small size are often found with early solitary Rugosa, forming a characteristic association in calcareous mudstones. In later rocks also the two are frequently associated, and evidently the Rugosa and Tabulata had similar habitat preferences. Generally they lived in conditions similar to those of modern corals, but they are not normally found in very high-energy environments.

A study of Silurian patch reefs in the Welsh borderland, which are dominated by tabulates (Scrutton, 1997a) shows how different growth strategies and colony forms prevailed through successive stages of reef development. Water depth was less than 30 m (Scoffin, 1971). In the initial stages, peripheral growth led to a dominance of tabular form. As the reef grew, a core developed in which halysitids were abundant, and both peripheral and medial growth strategies in other tabulates gave rise to domal and bulbous colonies. In the reef margin, however, medial growth was prevalent and nodular colonies were common. In deeper waters elsewhere (<50–60 m) peripheral growth dominated and hence tabular and low-domal colonies were the norm.

There has been much dispute about the affinities of *Favosites* and related genera. Some organisms formerly believed to be tabulate corals (e.g. *Chaetetes*) have now been shown to be sclerosponges, and a case has been made for transferring favositids to the sclerosponges too. Structures claimed as spicules in *Favosites*, however, seem to be endolithic borings (Scrutton, 1987) and the mural pores and calcite skeleton of fibro-normal tissue of *Favosites* has no counterpart amongst the sponges. In addition fossilized polyps, each with 12 radiating tentacles have been found in Canadian *Favosites* (Copper, 1985), and the cumulative evidence is heavily in favour of favositids being corals.

Palaeofavosites colonies from the Silurian of Quebec can be produced both sexually and, more commonly, by fragmentation (Lee and Noble, 1990). Sexually produced colonies always have conical bases, showing where the colony has been produced from a single protocoralite. Such colonies are found randomly dispersed. Populations produced by fragmentation, however, are found in clumps, consisting of many colonies with densities of up to 50 m^{-2}. These never show a protocorallite, but have circular and concave bases and limited variation in corallite shape. They probably originated by fragmentation of a parent colony, as an adaptation to a muddy substrate. Sexual reproduction here may have been comparatively rare for the purposes of dispersal and gene flow only.

Tabulates are not of great stratigraphical value, but they do sometimes occur in useful marker bands. The Lower Devonian *Pleurodictyum* (Fig. 5.14b), a small domed tabulate always found with a commensal tubiculous worm in the centre of the colony, is characteristic of sandy facies, by contrast with other corals. Along with other fossils of the same facies it has been used successfully to determine directions of faunal migration across Europe.

Classification

ORDER TABULATA is divided into seven or eight suborders which are quite clearly circumscribed and easily distinguished. The following classification is simplified and a slightly more complex arrangement is shown in Fig. 5.11.

ORDER TABULATA (Ord.–Perm.): Invariably colonial. Generally composed of small corallites with tabulae; septa (usually 12 in number) reduced to vertical rows of spines or absent.

SUBORDER 1. LICHENARIINA (L. Ord.–L. Sil.): Massive colonies, corallites prismatic with 16 or more septa and with mural pores lacking or rare; e.g. *Lichenaria*.

SUBORDER 2. SARCINULINA (M. Ord.–M. Dev.): Massive coenenchymal colonies, with coenenchyme enclosing horizontal spaces and formed largely of tabular and septal extensions. Up to 24 septa; e.g. *Sarcinula*.

SUBORDER 3. FAVOSITINA (M. Ord.–Perm.): Variform colonies with slender corallites having mural pores and short spinose septa; e.g. *Favosites*, *Alveolites*, *Michelinia*, *Pleurodictyum*.

SUBORDER 4. AULOPORINA (Ord.–Perm.): Corals with creeping or erect habit, fasciculate, with tabulae widely spaced or absent; e.g. *Aulopora*.

SUBORDER 5. SYRINGOPORINA (M. Ord.–Perm.): Large

erect coralla, dendroid or phaceloid, with cylindrical corallites connected by horizontal tubules. Septa often absent; tabulae horizontal or funnel-shaped; e.g. *Syringopora*.

SUBORDER 6. HALYSITINA (M. Ord.–U. Sil.): Cateniform colonies of imperforate coralla having in some cases small vertical tabulate tubules between the large tubes; e.g. *Halysites*.

SUBORDER 7. HELIOLITINA (M. Ord.–M. Dev.): Massive colonies with 12 septate corallites embedded in coenenchyme. (These are sometimes classified apart and have some similarity to certain octocorals.) e.g. *Heliolites*.

Order Scleractinia (Fig. 5.15)

All post-Lower Triassic corals are included in Order Scleractinia. Like *Caryophyllia* (q.v.) they secrete an aragonitic exoskeleton in which the septa are inserted between the mesenteries in multiples of six (Fig. 5.4a–c). After the first six protosepta grow, successive cycles of six, 12 and 24 metasepta are inserted in all six quadrants. Through their pattern of septal insertion the Scleractinia are immediately distinguished from the Rugosa. Each of the earlier cycles is complete before the next cycle grows, though in some cases this system breaks down in the higher cycles.

The skeleton originates as a thin basal plate from which the septa arise vertically. As this skeleton grows upwards the lower margin of the polyp hangs over it as an **edge zone**, Within the cup there may be dissepiments as well as septa. In general, the structure of both simple and compound scleractinians is light and porous, rather than solid as in the Rugosa. In several respects other than the primary septal plan the Scleractinia differ from Palaeozoic groups, as mentioned briefly below; independent evidence (Oliver, 1980, 1996; Scrutton and Clarkson, 1990; Scrutton, 1997b) now shows that Scleractinia arose from a group of sea anemones rather than from any Palaeozoic coral.

Range in form and structure

TYPE AND HABIT

Scleractinians may be solitary or compound. In solitary scleractinians the form of the corallum depends on the relative rates of vertical and peripheral growth once the basal plate has been secreted. Where peripheral growth has been dominant a discoidal form results (Fig. 5.16a), often with everted septa in cases where these have grown rapidly.

Perhaps the commonest kinds are those with turbinate or conical coralla in which the axis is straight (Fig. 5.15a), though horn-shaped and cylindrical types are common.

In colonial Scleractinia, as with other colonies, the corallites are interconnected as a result of repeated asexual division by the polyps. Colony type is genetically defined and refers mainly to the relationships of the corallites. As with the Rugosa there are dendroid, phaceloid, thamnasteroid and, rarely, aphroid types. In cerioid scleractinians the thecae (i.e. walls) of adjacent polygonal corallites are closely united and of dissepimental or septal origin, by contrast with the walls of cerioid rugosans, which are epithecal. **Plocoid** types of corallites have distinct walls but the corallites are united to each other by dissepiments. Ramose types of creeping habit, paralleling the tabulate *Aulopora*, also exist.

In addition there are two other types which do not correspond to any rugosan type: (1) the **meandroid** type (Figs 5.15b, 5.16b) in which corallites are arranged in linear series with the cross-walls absent, and confined within the lateral walls which run irregularly over the surface like the convolutions of a human brain; (2) the **hydnophoroid** type (Fig. 5.15c) in which the centres of the corallites are arranged around little hillocks or **monticules**.

In habit scleractinian colonies may be branching (ramose), massive, encrusting and creeping (reptant) or foliaceous. Sometimes adjacent colonies of the same species fuse to form a single colony.

SEPTA AND ASSOCIATED STRUCTURES

Septa are formed of aragonitic trabeculae (simple or compound) normally arranged in a fan-like system and often with a denticulate upper surface, but laminar tissue is unknown (Fig. 5.15). Usually the trabeculae are united, though sometimes, paralleling the Palaeozoic Family Calostylidae, the trabecular framework is loosely united and may be perforated. Such **fenestrate septa** are more important in scleractinians than rugosans. The grouping of trabeculae and their structure are important stable characters for taxonomy. They form initially from aragonitic spherulites in an organic matrix (Sorauf, 1972).

The septa originate between the mesenteries, which are also the site of digestion, excretion and gonad development. These mesenteries have a double layer of endoderm separated by a thin sheet of mesogloea. Muscles are arranged on one side of the mesentery only (Fig. 5.4c), and the mesenteries are grouped in pairs with the pleated muscle blocks

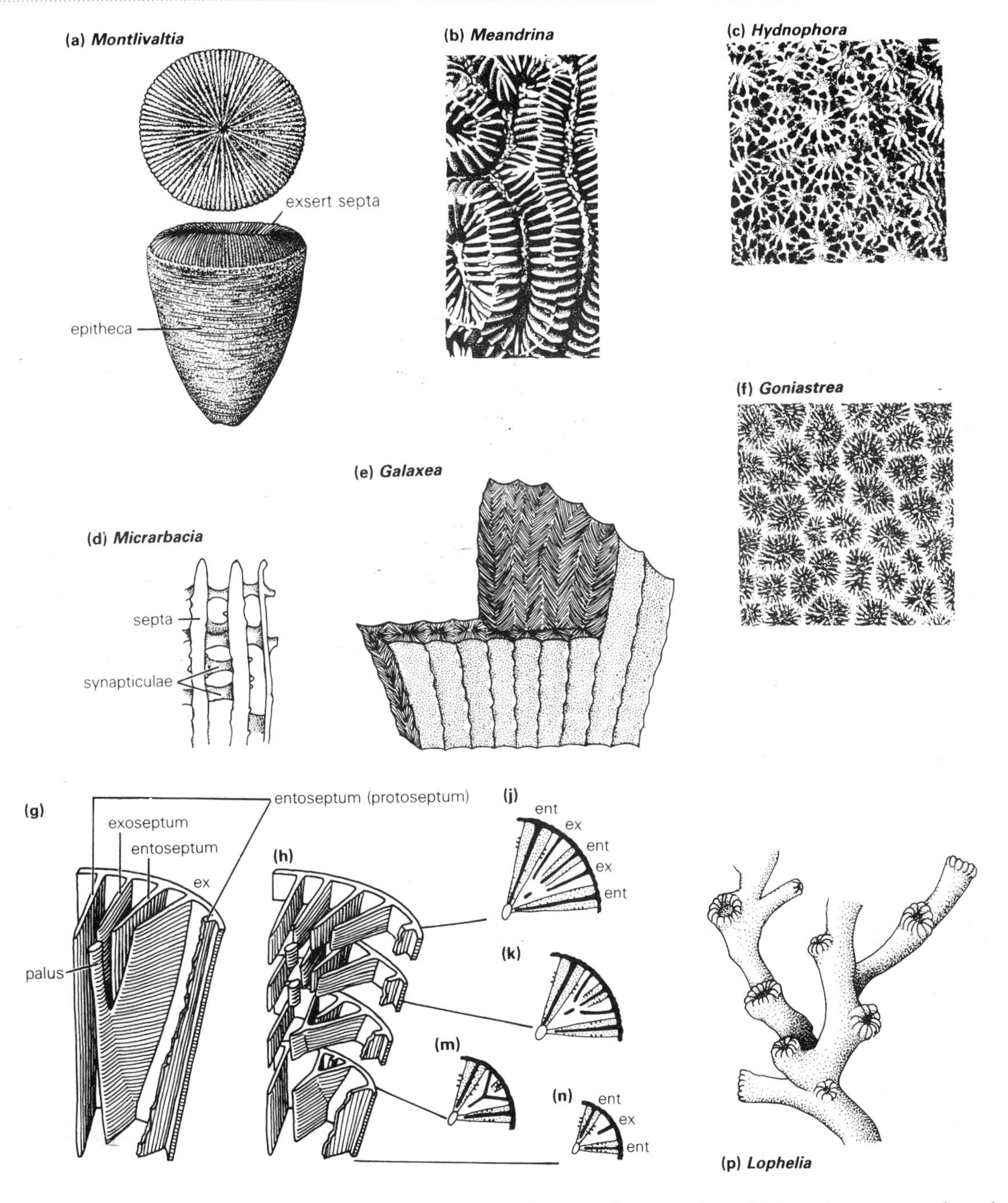

Figure 5.15 Scleractinian morphology: (a) *Montlivaltia* (Jur.), a well-preserved specimen (× 1); (b) *Meandrina*, upper surface of a Recent meandroid coral, showing a growth pattern resulting from intratentacular budding (× 1); (c) *Hydnophora* (Rec.), upper surface of a hydnophoroid coral, showing a growth pattern resulting from extratentacular (circum-mural) budding (× 2); (d) synapticulae in *Micrarbacia* (Cret.–Rec.); (e) trabecular structure in *Galaxea* (Mio.–Rec.); (f) *Goniastrea* (Eoc.–Rec.), showing distomodaeal budding (i.e. division of a single stomodaeum into two) on the upper surface (× 2); (g)–(n) the origin of pali, showing (g) a sextant of solitary scleractinian, (h) the same divided into layers and (j)–(n) transverse sections of polyp and corallum at each of these levels; (p) *Lophelia* (Oligo.–Rec.), a dendroid deep-water caryophilline (× 1). (Redrawn from Wells, 1956 in *Treatise on Invertebrate Paleontology*, Part F, with the exceptions of (a) and (p); (p) is original, (a) is redrawn from *British Mesozoic Fossils*, British Museum, London, 1962.)

(a)

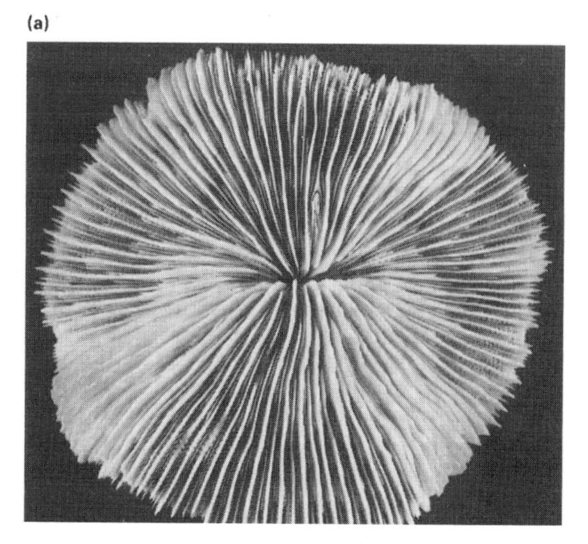

(b)

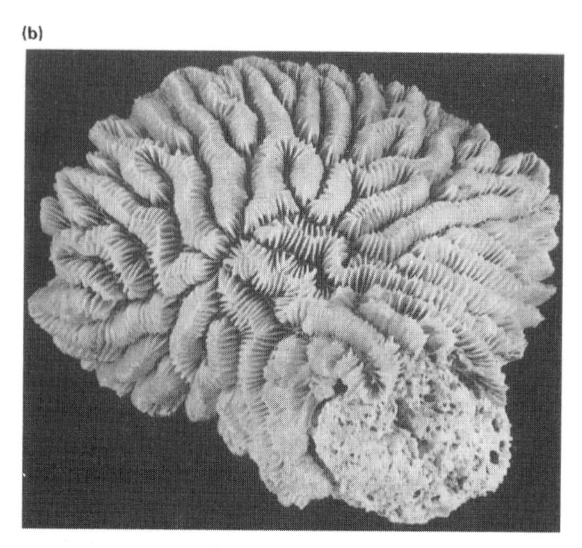

Figure 5.16 (a) *Fungia*, a large Recent solitary scleractinian, showing several orders of septa (× 0.5); (b) *Meandrina*, a Recent colonial coral from Barbados with typically meandroid structure resulting from intratentacular budding (× 0.5; cf. Fig. 5.12b). (Photographs reproduced by courtesy of C. Chaplin.)

facing each other. The space within such a pair is the **entocoel**, while in between (the muscles facing away from each other) is the **exocoel**. But two pairs of mesenteries on opposite sides, the **directive mesenteries**, have the muscle pleating reversed, giving a bilateral symmetry to an otherwise radial coral. If the coral is elliptical the long axis extends in the directive plane.

Septa are thus entocoelic (**entosepta**) or exocoelic (**exosepta**) in origin; usually the first two cycles are of entosepta, the rest of exosepta. In some scleractinians there may be a peculiar growth system producing vertical pillars (**pali**) along the inner edges of some of the entosepta (Fig. 5.15g–n).

If a columella forms it is always of septal origin and may begin as pali; it is usually a central rod or a dividing plate formed from a proseptum.

The septa are often connected by cross-bars called **synapticulae** (Fig. 5.15d) which grow towards each other from the walls of adjacent septa and eventually fuse, perforating the mesenteries in the process.

Dissepiments, like those of the Rugosa, are secreted by the base of the polyp. As the latter grows it moves upwards, detaching a small part of the base from the calice at a time and forming a dissepiment to enclose each space. Where the fine skeletal structure of dissepiments have been studied (Sorauf, 1972), dissepiments have been shown to have two layers: a first-formed primary layer, overlain by vertically aligned clusters or aragonitic spherulites. Endothecal dissepiments are confined within the corallite, but in some colonial scleratinians the corallites are united by a common spongy tissue, the coenosteum, which may be formed partially of exothecal dissepiments and partially by rods and pillars (**costae**), as in *Galaxea* (Fig. 5.15r).

Tabular dissepiments are flattish plates extending across the whole width of the corallite or confined only to its axial part (thus resembling tabulae in rugosans). **Vesicular dissepiments** are small arched plates, convex upwards and overlapping, and usually inclined downwards and inwards from the edge of the corallite.

OTHER PRIMARY STRUCTURES

The thin basal plate is semitransparent and firmly adherent to the substratum; it may later be thickened by secondary deposits. The epitheca is not developed in many scleractinians, but if present may consist of chevron-like crystallites of aragonite.

SECONDARY STRUCTURES

Stereome is an adherent layer of secondary tissue which may cover the septal surface. It is composed of transverse bundles of aragonitic needles. Stereome normally thickens the epitheca internally as well, but in some cases its function is taken over by primary thickening of the septa, by dissepiments or by synapticulae.

In compound scleractinians individual corallites are usually separated by a complex perforated tissue: the **coenosteum**. This may consist entirely of exothecal dissepiments (tabular) or other material, but it provides a support for numerous canals linking the individual corallites and binding the living tissues together in a functionally cohesive mass.

BUDDING AND COLONY FORMATION

The form of the colony in the Scleractinia is largely determined by the mode of budding combined with relative growth rates. The following types have been distinguished:

Intratentacular budding. Here the polyp divides by simple fission across the stomodaeum, each daughter (of which there may be two, three, four or more) retaining part of the original stomodaeum regenerating the rest. The products of such **polystomodaeal** budding may remain linked or become separate. Meandroid corals (Fig. 5.15b) are clearly the results of incomplete intratentacular budding.

Extratentacular budding (Fig. 5.15f). In this type new stomodaea are produced outside the tentacular ring of the parent. These extratentacular buds soon separate and do not remain linked. The colony that results from such budding (whether branching or massive) thus consists simply of numerous separate individuals and is not integrated functionally in the same way as the products of intratentacular budding. There may, however, be some nervous or chemical linkage if the soft tissues are in contact, even if the enterons are not united.

Evolution

The oldest known scleractinians are Middle Triassic in age. They derived from a group of soft anemones. On their first appearance they are already differentiated into a number of important families. **Hermatypic** in type, they formed small patch reefs, may have possessed zooxanthellae and were confined to warm, shallow, well-lit waters. By the late Triassic they were expanding fast in many parts of the world, and many new families arose. The first small coral reefs date from this time. This pattern persisted through the Lower Jurassic and the first true deeper-water **ahermatypic** corals (Suborder Caryophylliina) emerged at about that time, though these were of limited importance. By the Middle Jurassic scleractinians were on the increase everywhere, and patch reefs erupted wherever conditions were suitable. In the Tethyan Ocean the first large reefs developed, persisting until a general setback in the early Cretaceous. Later the large reefs and coral banks developed, and although some of the older families died out others arose. The success of reef corals fluctuated in the Cretaceous, and the ahermatypes became really important for the first time. By the late Cretaceous both hermatypic and ahermatypic corals were becoming of distinctly modern type, though like many other groups they suffered a major extinction event at the end of the Cretaceous. By the end of the Eocene the dominant groups were much like those of the present since the formerly important Mesozoic families (e.g. the Montlivalitiidae) had disappeared. During the late Tertiary the two main coral faunal provinces of the present – the Indo-Pacific and western North Atlantic provinces – had become distinct, and reef-building corals flourished on an unprecedented scale. Many reefs were killed by the lowering of sea level in the Pleistocene, but their remains have provided stable surfaces for the regrowth of coral since their submergence. Comparatively few genera were affected by the conditions of the Pleistocene, so that Pliocene coral faunas are effectively the same as those of the present.

Most families of scleractinians (Veron, 1995) show evolutionary trends, especially from solitary to compound coralla and, in colonial types, from phaceloid through plocoid and cerioid to meandroid form. This probably increased the degree of colonial integration.

Ecology

Scleractinians fall into two main physiological categories: **zooxanthellate** and **non-zooxanthellate** (approximately equivalent to the ecological terms hermatypic and ahermatypic). In zooxanthellate corals the endodermal cells are replete with symbiotic algae (zooxanthellae, which are a kind of dinoflagellate). These symbionts benefit the corals in two respects. Firstly, up to 95% of the organic carbon produced by the zooxanthellae is taken up by the polyp and used as food. Secondly, zooxanthellate corals are able to grow up to three times more quickly than if they had possessed no symbionts due to the extra O_2 production by the algae. These algae exist in enormous numbers, and they are so essential to the metabolism of the coral supplying it with nutrients and oxygen that such corals can only flourish in the photic zone, at depths normally less than 50 m (rarely 90 m). Since the corals have firm anchorage they grow successfully and most abundantly within the zone of wave action, at depths less than 20 m.

Reef-building or hermatypic corals are nearly all zooxanthellate and generally need a minimum water temperature of 18°C and flourish best between 25 and 29°C and at normal salinity. Hence most reef corals are restricted to shallow, well-lit, warm water. In addition, they need a good supply of oxygen and generally prefer a firm, non-muddy substrate for otherwise the planulae will not settle. There are, however, hermatypes living and building reefs in muddy regions. Light is perhaps the most important of all the limiting factors, though a reef can be killed by tidal emergence during heavy rain, during a hurricane for instance, and may take 20 years to re-establish itself.

Non-zooxanthellate corals are largely ahermatypic (i.e. non-reef formers). Some do live in coral reefs, though others are found at depths of up to 6000 m. They are, however, most abundant down to 500 m, and though they can survive temperatures below 0°C they flourish best between 5 and 10°C. Furthermore, they can live in total darkness. Over two-thirds of these 'deep-sea corals' are solitary and do not form reefs, though the colonial dendroid caryophyllid *Lophelia* (Fig. 5.15p) forms deep-water banks or thickets at the edge of the western European continental shelf.

The earliest (Triassic) scleractinians were evidently hermatypes, though perhaps not so adapted for the reef habitat as those of the present since they formed only small reefs. Ahermatypic corals developed from many hermatypic stocks by the slow spread of certain types into colder, deeper waters. The first known ahermatypes were the Jurassic caryophyllids whose descendants retained this habit, but representatives of many other groups also become ahermatypic, so that the hermatype–ahermatype division today cuts across many families and even genera.

It is likely that the great expansion of the hermatypic reef builders was directly linked with their adoption of zooxanthellae. Such symbiotic algae are found in other organisms too. The ecology of modern corals has been much studied; useful references are Yonge (1968), Muscatine and Lenhoff (1974), Lewis and Price (1975), Gill and Coates (1977), Goreau *et al.* (1979) and Vosburgh (1982).

Classification

In classifying the Scleractinia, as with the Rugosa and Tabulata, it has been necessary to find stable taxonomic characters. Septal structure can be used for defining subordinal categories whilst mode of colony formation is of value only at the familial level. Habitat and dimensions at any level of classification are of little value.

ORDER SCLERACTINIA (M. Trias.–Rec.): Solitary or colonial corals with mesenteries paired and the septa arranged in multiples of six. Basal plate gives rise to septa and epitheca.

SUBORDER 1. ASTROCOENIINA (M. Trias.–Rec.): Normally colonial hermatypes with small corallites. Septa of up to eight trabeculae, spinose to laminar; e.g. *Acropora*, *Thamnasteria*.

SUBORDER 2. FUNGIINA (M. Trias.–Rec.): Solitary or colonial with perforate septa linked by synapticulae. Corallites usually large. Mainly hermatypic; e.g. *Fungia*, *Cyclolites*, *Isastraea*.

SUBORDER 3. FAVIINA (M. Trias.–Rec.): Laminar septa with fan-like trabeculae. Dissepiments present but synapticulae rare. Mainly hermatypic; e.g. *Favites*, *Montlivaltia*, *Thecosmilia*.

SUBORDER 4. CARYOPHYLLINA (Jur.–Rec.): Normally solitary and ahermatypic. Septa laminar with simple trabeculae in fan system; rare synapticulae; e.g. *Parasmilia*, *Caryophyllia*.

SUBORDER 5. DENDROPHYLLINA (U. Cret.–Rec.): Solitary or colonial. Septa laminar but irregularly perforate. Wall of swollen synapticulae. Mainly ahermatypic; e.g. *Dendrophyllia*.

Coral reefs

Recent coral reefs (Yonge, 1963; Stoddard, 1969; Ladd, 1971; Cameron *et al.*, 1974; Sheppard, 1983; Veron, 1986; Wells, 1988) are confined to tropical seas and are best developed in the Indo-Pacific area and in the western North Atlantic. The Indo-Pacific reefs, where maximally developed (Melanesia to Southeast Asia), have 92 recorded genera and subgenera and over 700 species (see Veron, 1986, for illustrations and distribution maps), while the coral fauna of the North Atlantic is smaller with only 34 genera and 62 species. The North Atlantic fauna seems to have been derived initially from a Pacific source, since Oligocene reefs of the West Indies have many of the same corals, but it has since developed in isolation, particularly since the Americas were linked by the Panama isthmus in the Pleistocene.

In **structural reefs** the scleractinians have built up massive wave-resistant structures, often to great thicknesses, where the bulk of the material is contributed by corals, even though calcareous algae may play an important part in reef structure. Such a large rigid structure built up by generations of corals is the habitat of a host of diverse organisms, the whole forming one of the most complex ecosystems known. In coral reefs productivity, calcium metabolism and carbonate fixation (3500 $g^{-2}a^{-1}$) are very high. The range of habitats is such that there is a high degree of specialization in the associated fauna, especially fishes, molluscs and arthropods; these are found in zoned and localized niches in various parts of the reef.

In **non-reef communities**, however, though equally confined to tropical seas, the corals merely live together without building up a rigid structure.

Types and genesis

Structural coral reefs today belong to three major types: **fringing reefs**, **barrier reefs** and **atolls**. Fringing reefs develop along shorelines, especially those of volcanic islands. Barrier reefs are formed some distance from the shore. Atolls are circular or horseshoe shaped and usually orientated with respect to the prevailing wind, with a central shallow lagoon in which small patch reefs may develop. The Great Barrier Reef (Maxwell, 1968; Bennett, 1971) off eastern Australia is a special case, being a linear feature parallel to the fault-bounded margin of the continent, where coral growth was initiated on the edges of an upraised fault.

It is now well known that the three major types occur as successive stages in the growth of coral around a sinking volcanic island (Fig. 5.17a).

The growth of corals has kept pace with subsidence, while broken reef talus has formed cascading fans going down to depth on the outside. This theory, originally suggested by Darwin, has been confirmed by deep borings into atolls. At Eniwetok atoll, bored in 1951, basalt was encountered at about 1.25 km below the surface. The limestone overlying the basalt was Eocene in age. Thus Eniwetok is a limestone pillar 1.25 km thick standing on a deeply drowned volcanic island some 3.2 km high on the ocean floor. The maximum rate of subsidence has been about 50 cm per year. Confirmatory seismic and drilling evidence from other atolls suggests that this is the common pattern. The now sunken volcanic islands arose at 'hot spots' on a submarine ridge crest and have been carried away into deeper waters (as on a conveyor belt) as the ocean floor spread away from the ridge (Scoffin and Dixon, 1983).

Darwin's theory of atoll formation, though now vindicated, did not take into account Pleistocene changes of sea level, which were far reaching since the reefs were exposed and the corals killed off. The exposed surfaces were eroded to some extent and often fretted into a karst topography. Hence the veneer of corals which has grown since the end of the Pleistocene has re-established itself on an anomalous topographic surface, and it is because of this that the form of a modern reef is not simply a reflection of constructional activity.

Growth and zonation of corals

Corals on reefs have few natural enemies other than coral-eating fishes and the starfish *Acanthaster*. The latter has recently wreaked havoc on some Pacific reefs, for after the depredations of the starfish the dead coral becomes covered with an algal film which prevents coral planulae from settling. Repeated *Acanthaster* 'invasions', however, appear to follow a cyclical pattern waxing and waning over many thousands of years, and the damage to the reef is eventually repaired.

Because of the steep environmental gradients within the reef complex, the corals themselves are ecologically zoned with specific reference to the shoreline (Fig. 5.17b,c). Each species occupies a particular habitat and is not found elsewhere. Within particular habitats there is a kind of 'pecking order' in corals; dominant species may actually eat other species of the same genus by mesentery extrusion and digestion. Different species inhabit the windward and leeward sides of the island. There is, furthermore, a great change both in species and in their growth forms, from the upper part of the forereef slope across the reef flats to the central lagoon with its shell sand floor and patch reefs. In some cases massive rounded corals such as *Diploria* tend to inhabit the surf zone, while the stout branching *Acropora* (the 'stag's horn' coral) lives at greater depths out of the range of surf action. In other areas

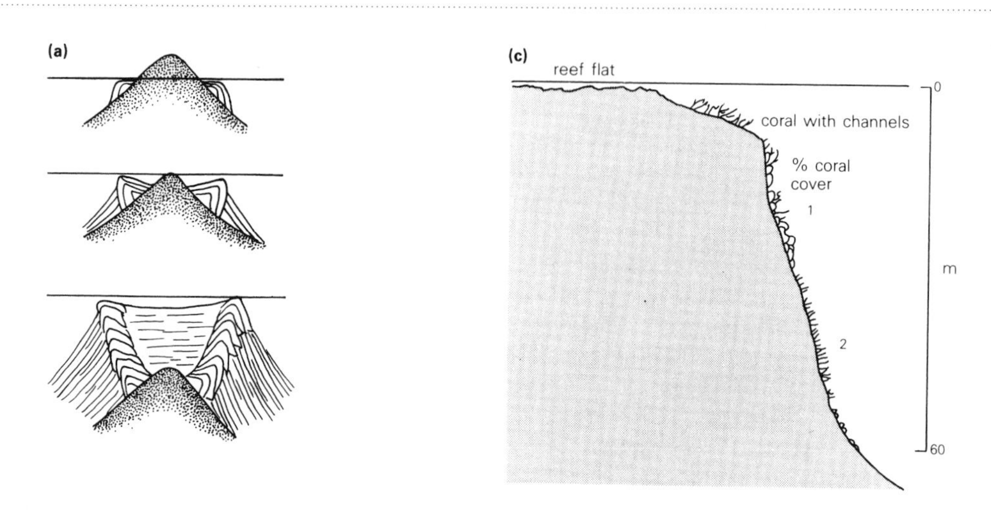

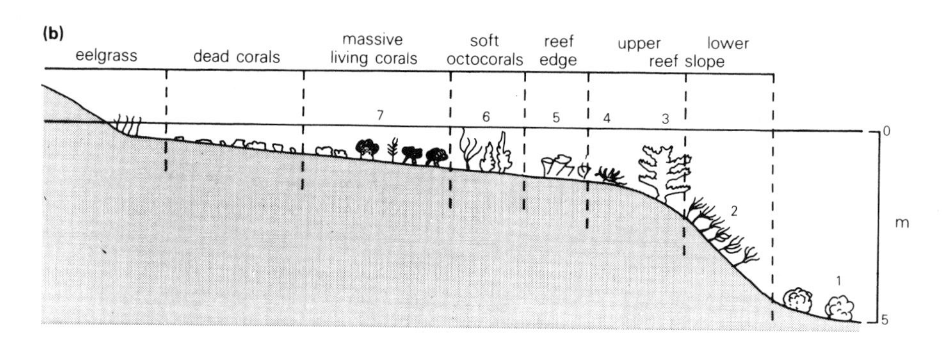

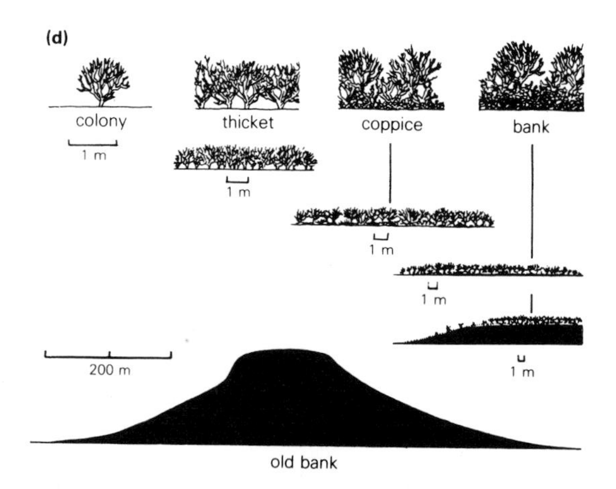

Figure 5.17 Coral reefs: (a) growth stages in the development of an atoll by volcanic subsidence according to Darwin, showing a fringing reef (top) a barrier reef (middle) and an atoll (bottom); (b) section through the edge of a growing reef, showing zonation and based on Manauli reef off the northwest of Sri Lanka (coral genera are represented diagrammatically: 1, *Symphyllia recta;* 2, *Acropora formosa;* 3, *A. hyacinthus;* 4, *A. humilis;* 5, *Montipora foliosa;* 6, the soft corals *Alcyonaria, Gorgonaria* and *Antipatharia;* 7, *Glavia, Pocillopora, Goniostrea* and *Porites*); (c) coral zonation on the growing edge of Arno Island (Marshall Islands; 1, *Porites* dominant; 2, *Pachyseria* dominant); (d) stages in the development of a deep-water coral bank. [(b), (c) Simplified from Mergner and Scheer, 1974, and Dahl *et al.*, 1974, respectively; (d) is redrawn from Squires, 1964.]

massive corals are commoner in quiet waters, and *Acropora* occurs in high-energy zones (Vosburgh, 1982). The apparent structural strength of corals is therefore only one factor amongst many that control ecological zonation; ability to compete for space and to cope with sediment influx may likewise be important.

Zonation varies from reef to reef depending on various factors. The presence of resistant algal ridges, for instance, can greatly affect the pattern of zonation. Not only the scleractinians but also the gorgonians (in the Caribbean reefs) and octocorals (in the Indo-Pacific reefs) are ecologically zoned, and so are the various bivalves and other molluscs that form specialized communities in the very varied reef habitats (Yonge, 1974).

The growth rate of corals varies greatly with the environment and the season but may reach 26 cm per year in branching corals and 1 cm per year in the more massive colonies. Reef growth as a whole may reach a maximum of 1.2 cm per year, but in the past it has generally been less.

Within reefs there is not only continual construction by the scleractinians and encrusting algae, but also a constant attrition through wave action and, equally important, destruction by boring organisms and coral-eating fishes. These produce a great quantity of fine comminuted coral sand, which fills up the lagoon and is transported to deeper waters.

Deep-water coral banks

Deeper-water corals (ahermatypes such as *Lophelia* and *Dendrophyllia)* may form structures differing from those of their shallow-water counterparts. Some of these deep-water corals are solitary; others are colonial and dendriform.

The development of these structures (Squires, 1964) takes place in four stages (Fig. 5.17d). At first there is a single **colony.** When this colony is joined by others it becomes a **thicket** a few metres across, in which the members may be all of the same species or of different species. This newly developed environment attracts other organisms, e.g. fishes, molluscs and crustaceans, giving it a new ecological character. As the thicket spreads and matures, skeletal debris accumulates from broken and bored coral, providing a new substratum which again attracts other animals. This stage is now a **coppice** and may be several metres across. Eventually, with the accumulation of more debris and further coral growth, a **bank** develops: a topographic entity with a core of solid skeletal debris, a covering mat of more open debris and a capping of live coral. The proportion of living to dead coral decreases with time.

The exposed sequences studied by Squires were Tertiary in age, but similar banks are forming today. The study of fossil thickets, coppices and banks can yield valuable palaeoecological data.

Geological uses of corals

Corals as stratigraphical indicators

Corals are generally too long-ranged for use as zone fossils, though the widely distributed *Palaeocyclus* provides a fine example of a useful acme biozonal indicator for the late Llandovery–early Wenlock. Nevertheless, corals have been used where no shorter-ranged fossils are available, especially in the Carboniferous. This work began in Belgium and in 1905 Vaughan, working on the Lower Carboniferous of Bristol, England, erected a zonal scheme based upon the first appearance of corals and brachiopods. This scheme was later extended to the North of England and proved, though crude and inexact, to be applicable, with minor modifications, to much of Europe.

Vaughan believed that the faunal sequence he erected reflected evolutionary lineages, but in fact the corals really occur as assemblage biozones with each of the main cycles bringing in a new migrant fauna.

Recently Mitchell (1989) has plotted the time ranges for 68 rugose coral species against standard Visean stages, showing that a now-refined coral stratigraphy is possible for the Lower Carboniferous, supplementing other stratigraphic data based on goniatites, conodonts and foraminiferids. New stages based upon transgressive cycles in the Lower Carboniferous are now standard, but much refinement is still needed.

Coral zonation in the European Devonian, once elaborated in great detail by Wedekind, has proved to be overoptimistic.

Corals as geochronometers

In well-preserved specimens of many rugose and scleractinian corals the epithecal surface shows fine growth ridges, some 200 per cm. Each of these represents a growth increment: the former position of the rim of the calice. These fine growth ridges are often grouped in prominent bands or annulations between which the epitheca is constricted. It is normally recognized that the fine growth ridges represent daily growth increments, whereas the banding is monthly and the broader and more widely spaced annulations are yearly. Wells (1963) first established in the scleractinian *Manicina* that the growth ridges were diurnal, and since the Recent *Lophelia pertusa*, from Norwegian fjords, has 28 growth ridges (presumably diurnal) per band, each band corresponding to a lunar cycle, the monthly banding hypothesis seems at least possible.

Wells (1963), working on Devonian corals with annual rather than monthly groupings, counted the number of growth ridges per annulation in a number of species. Though the results were limited by the preservation of the corals, he concluded, mainly from observations on *Heliophyllum*, *Eridophyllum* and *Favosites*, that there were an average of 400 days in the Devonian year. This figure corresponded well to astronomical estimates that the Earth's rotation has been slowing down through tidal friction by about 2 s per 100 000 years. Assuming that the Earth's annual circuit round the Sun was the same length, which seems to be fairly well substantiated, there must have been more days in the Devonian year, but they were shorter (Scrutton and Hipkin, 1973).

Annual banding is not often found in corals, having only been recorded in specimens that originally lived in waters where there were seasonal fluctuations. It is normally easier to work with the Rugosa that show the growth ridges grouped in monthly bands. These are more abundant and do not have to have a complete epitheca. Many Recent corals have monthly breeding cycles during which carbonate deposition is inhibited, and this may have been the case with the Rugosa.

The monthly banded Middle Devonian corals studied by Scrutton (1965; Fig. 5.18) had an average of 30.6 growth ridges per monthly band, and if the figure of 399 (the astronomical estimate for length of the Devonian year) is divided by the average growth-ridge count per band, the result is consistently 13.

Hence it seems that 13 bands each of about 30.6 growth ridges were laid down by these corals in the Middle Devonian year.

The 'coral clock' provides a consistent check on estimates of the slowing of the Earth's rotation based upon a biological rather than a physical system. It has, however, become increasingly clear that corals are not ideal tools for this, for while diurnal banding is clear enough on all unworn corals, clear monthly and annual cycles are harder to detect and may not be present in many corals (Scrutton, 1978). Tidal rhythmites (ebb-tidal deposits whose variable lamina thickness encodes a full spectrum of semidiurnal to lunar modal palaeotidal events; Williams, 1989) offer excellent potential for geochronometry, especially for the unfossiliferous Precambrian. Random events may in any case have left a masking record in the growth of the skeleton. Periodic growth features in bivalves are now better understood and may prove to be more suitable for geochronometry.

Figure 5.18 Banding on the epitheca of the Middle Devonian rugose coral ?*Heliophyllum* sp. (× 3). (Copyright photograph reproduced by courtesy of Dr C.T. Scrutton.)

Corals as colonies: the limits of zoantharian evolution

Colonies have been defined as 'groups of individuals structurally bound together in varying degrees of skeletal and physiological integration; all genetically linked by descent from a single founding individual' (Coates and Oliver, 1973).

Such colonies (e.g. cnidarians, bryozoans and graptolites) can exhibit a wide range of organization and integration. At one extreme the zooids budded off from the parent may be completely independent, whereas at the other end of the scale the individuals of the colony may become linked and coordinated so that the whole colony can function as a single unit. In some cases zooids may become highly specialized, promoting division of labour amongst the members of the colony.

Advantages of the colonial state presumably are mainly connected with greater efficiency in protection and stability as well as in reproduction, feeding and respiration. Thus a colony of asexually reproducing individuals can soon develop a large and effective biomass, with a strong and stable skeleton in which there is the potential for integration and cooperation between zooids.

In the octocorals (Bayer, 1973) there are two distinct functional types: (1) those which secrete a massive scleractinian-like skeleton with very little integration, and (2) those (the vast majority), including pennatulaceans, in which integration varies from the very simple, through numerous intermediate stages, to very complex colonies with the various zooids specialized for different functions. In the most extreme cases not only are interzooidal canals linking the enterons retained, so that food captured by a few zooids can be shared by the rest of the colony, but also there are specialized 'siphonozooids' organized for pumping water through the coenenchymal canals, to ensure that adequate supplies of oxygenated water for respiration reach all members of the colony. Such **polymorphic** zooids function together but cannot function apart, and the colony acts as a 'superindividual'. Despite the fact that this level of integration in the octocorals evidently originated very early in their history (if the interpretation of *Charniodiscus* as a Precambrian pennatulacean is correct), the zoantharians have never fully reached the same levels of colonial organization since their zooids are not polymorphic.

During the Palaeozoic the ratio of solitary to colonial Rugosa remained relatively constant, there being a somewhat higher percentage of solitary corals. Following the Devonian extinction there was a slightly enhanced trend towards coloniality, and a fuller integration of the colony. Tabulate corals were all colonial, and in *Favosites* the mural pores suggest enteron connections, indicating at least a degree of colonial integration. They were severely affected by successive extinction episodes, however, and the diversity established in the Ordovician was actually reduced through time (Scrutton, 1988).

Although cerioid and phaceloid forms are known amongst scleractinians, a generally higher level of integration is apparent in the preponderance of meandroid and coenosteoid types. Meandroid genera have a **polystome** system in which hundreds of mouths with their adjacent tentacles connect in each 'valley' with a single elongated enteron. In coenosteal Scleractinia the enterons of adjacent polyps are also all connected, and skeleton building has become a cooperative venture. But even the most advanced scleractinians, with the possible exception of a very few rare genera, do not exhibit polymorphism.

Despite the fact that rugose and tabulate corals were able to make some headway towards an integrated colonial system, they were generally less successful in this than the Scleractinia. But their comparative failure as reef builders was probably due not so much to this as to certain other features of their organization. The Rugosa and to some extent the Tabulata had slow-growing and rather solid and heavy skeletons in which a great deal of calcium carbonate was deposited in internal structures. They lacked an edge zone and could not attach themselves firmly to the substratum. It is not known whether Palaeozoic corals had zooxanthellae in their tissue; if they did not, one of the great advantages possessed by reef-building scleractinians was denied to them, and this would have affected their metabolic efficiency, their ability to get rid of waste material and other factors.

However, the strong, yet light and porous, stable, fast-growing and at the same time increasingly integrated skeletons of the scleractinians permitted the development of the coral reef environment, so laying down the framework for some of the most complex and productive of all ecosystems that have ever evolved.

Minor orders

In recent years many 'coralline' taxa have been described from the Cambrian, and informally placed in the Coralomorpha. Of these, two orders are true zoantharians, Tabulaconida and Cothonioniida. Neither of these appears to be ancestral to any post-Cambrian group. These and other invertebrates and other minor orders are discussed in full by Scrutton (1997b).

Order Tabulaconida

Tabulaconus (L. Cam.) is a solitary or dendroid coral, with large (>16 mm) corallites which are usually aseptate but with evident tabulae. *Moorowipora* from the Lower Cambrian of Australia (Sorauf and Savarese, 1995) is a cerioid form with septa and tabulae, and *Arrowipora* may also belong here. None of these have mural pores, their mode of growth is unlike that of tabulates, and they are classified separately.

Order Cothonioniida (Fig. 5.19b)

This monospecific subclass is known only from the early Middle Cambrian of New South Wales where individuals are abundant.

The corallites of *Cothonion* are small and conical and up to 10 mm in diameter. They have weak septa internally but no horizontal elements. There is, however, an operculum or lid for each corallite upon the underside of which are strong biradial septa. Jell and Jell (1976), who first described these corals, suggested that they might be allied to the Rugosa, but they are most probably a quite independent group, now elevated to the rank of order.

Order Heterocorallia (Fig. 5.19a)

The heterocorals are a small group of elongated, pipe-like corals, confined to the Carboniferous of Europe and Asia and very short-lived. They have septa and tabulae but often lack an epitheca. Their pattern of septal insertion is unusual and characteristic. Some authorities believe that they led an epiplanktonic life, attached to floating seaweeds.

Orders Kilbuchophyllida (Fig. 5.19c) and Numidiaphyllida

A single genus of scleractinian-like coral has recently been described from mass-flow deposits in the Upper Ordovician of Southern Scotland (Scrutton

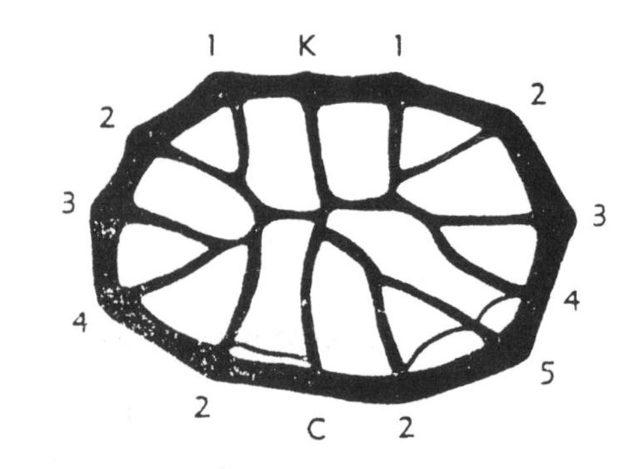

(a) ***Heterophyllia***

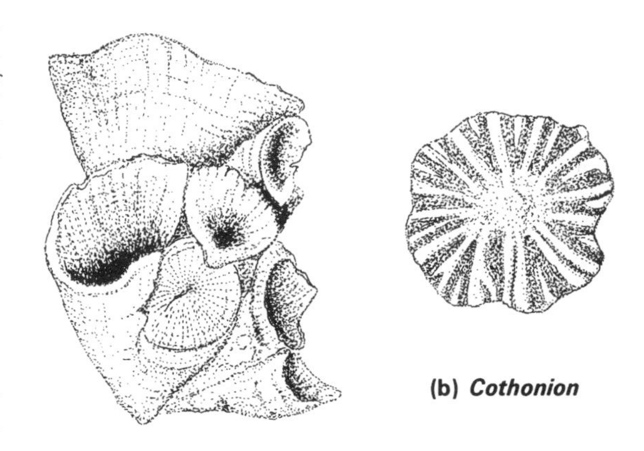

(b) ***Cothonion***

(c) ***Kilbuchophyllia***

Figure 5.19 Representatives of minor orders: (a) *Heterophyllia* (× 5); (b) *Cothonion* (M. Cam.; × 2) (c) adolescent *Kilbuchophyllia* (× 3). [(a) Redrawn from Hill, 1962; (b) redrawn from Scrutton, 1979; (c) redrawn from Scrutton and Clarkson, 1990.]

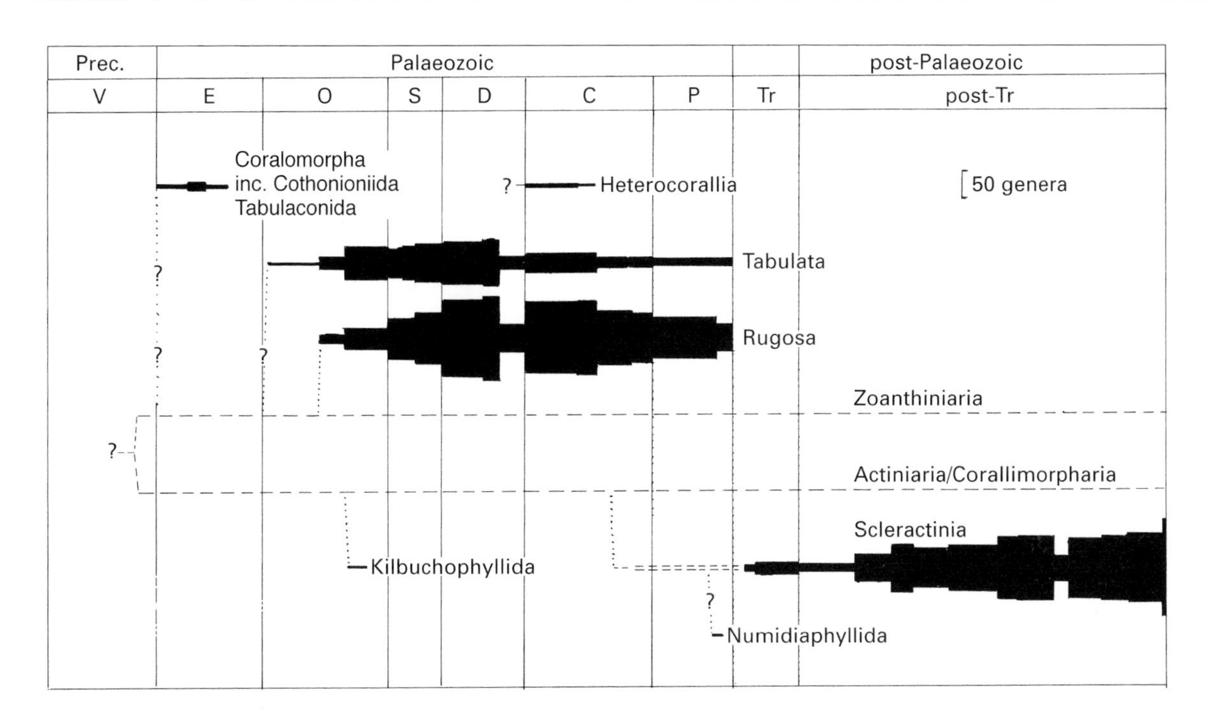

Figure 5.20 Relationships of the various groups of zoantharian corals. The ranges of the uncalcified sea anemones from which they were derived (Zoantharinaria, Actinaria, Corallimorpharia) are shown by dashed lines. (Simplified from Scrutton, 1997b.)

and Clarkson, 1991). In its appearance, pattern of septal insertion and septal microstructure, *Kilbuchophyllia* is remarkably similar to recent scleractinians, and is now known to have been aragonitic in composition.

It is not likely that this genus was actually ancestral to living scleractinians on account of the Ordovician–Triassic time gap. What is probable is that *Kilbuchophyllia* represents an earlier 'attempt at skeletonization' by the same group of sea anemones (Corallimorpharia) that gave rise to true scleractinians, some 220 Ma later (Fig. 5.20).

This early example of skeletal acquisition was short-lived and very localized, in Southern Scotland and Northern Ireland. The contemporaneous rugosans, which originated at about the same time, were derived from a different group of sea anemones (Zooanthiniaria), became dominant for the duration of the Palaeozoic, and the corallimorpharian sea anemones did not, other than the short-lived Permian *Numidiaphyllum*, which independently developed an aragonitic skeleton, calcify again until after the rugosans had become extinct.

Bibliography

Books, treatises and symposia

Bennett, I. (1971) *The Great Barrier Reef.* Lansdowne, Dee Why West.

Borradaile, L., Potts, F.A., Eastham, L.E.S. and Saunders, J.T. (1956) *The Invertebrata.* Cambridge University Press, Cambridge.

Cameron, A.M. *et al.* (eds) (1974) *Proceedings of the Second International Symposium on Coral Reefs*, Vols 1 and 2. Great Barrier Reef Committee, Brisbane.

Hill, D. (1937) *A Monograph of the Carboniferous Rugose Corals of Scotland.* Palaeontographical Society, London. (Taxonomy and coral community structure)

Maxwell, W.G.H. (1968) *Atlas of the Great Barrier Reef.* Elsevier, Amsterdam. (Illustrated atlas)

Moore, R.C. (ed.) (1956) *Treatise on Invertebrate Palaeontology*, Part F, *Coelenterata.* Geological Society of America and University of Kansas Press, Lawrence, Kan.

Muscatine, L. and Lenhoff, H.M. (1974) *Coelenterate Biology.* Academic Press, New York. (Excellent compilation)

Sheppard, C.R.C. (1983) *A Natural History of the Coral Reef.* Blandford, Poole.

Stel, J.H. (1979) *Studies on the Palaeobiology of Favositids*. Stabo, Groningen.

Teichert, C. (ed.) (1981) *Treatise on Invertebrate Paleontology*, Part F (Supplement 1), *Rugosa and Tabulata*. Geological Society of America and University of Kansas Press, Lawrence, Kan..

Veron, J.E.N. (1986) *Corals of Australia and the Indo-Pacific*. Angus and Robertson, North Ryde, NSW, Australia, 644 pp. (Superbly illustrated with colour photos and distribution maps of all species)

Veron, J.E.N. (1995) *Corals in Space and Time: Biogeography and Evolution of the Scleractinia*. Cornell University Press, Ithaca.

Wells, S. (ed.) (1988) *Coral Reefs of the World* (3 vols). UNEP IUCN (United Nations Environment Programme: International Union for Conservation of Nature and Natural Resources).

Individual papers and other references

Babcock, L. (1991) The enigma of conulariid affinities, in *The Early Evolution of Metazoa and the Significance of Problematic Taxa* (eds A.M. Simonetta and S. Conway Morris), CUP, Cambridge, pp. 133–43.

Bayer, F.M. (1973) Colonial organisation in octocorals, in *Animal Colonies* (eds R. Boardman, A. Cheetham and W.J. Oliver), Dowden, Hutchinson and Ross, Stroudsburg, Penn., pp. 69–93.

Bengtson, S. (1981) *Atractosella*, a Silurian alcyonacean octocoral. *Journal of Paleontology* **55**, 281–94.

Carruthers, R.G. (1910) On the evolution of *Zaphrentis delanouei* in Lower Carboniferous times. *Quarterly Journal of the Geological Society of London* **66**, 523–36. (Classic, though now dubious, evolutionary study)

Coates, A.G. and Jackson, J.B.C. (1987) Clonal growth, algal symbiosis, and reef formation by corals. *Palaeobiology* **13**, 363–78.

Coates, A.G. and Oliver, W.A. (1973) Coloniality in zoantharian corals, in *Animal Colonies* (eds R. Boardman, A. Cheetham and W.J. Oliver), Dowden, Hutchinson and Ross, Stroudsburg, Penn., pp. 3–217. (Integration of colonies).

Copper, P. (1974) Structure and development of early Palaeozoic reefs, in *Proceedings of the Second International Symposium on Coral Reefs*, (eds A.M. Cameron *et al.*), Great Barrier Reef Committee, Brisbane, pp. 365–86. (Changes in reef structure from incipient to maturity)

Copper, P. (1985) Fossilised polyps in 430-Myr-old *Favosites* corals. *Nature* **316**, 142–4.

Dahl, A., Macintyre, I.G. and Antonius, A. (1974) A comparative survey of coral reef research sites. *Atoll Research Bulletin* **172**, 37–120.

Garwood,. E.J. (1913) The Lower Carboniferous succession in the northwest of England. *Quarterly Journal of the Geological Society of London* **68**, 449–596. (Zonation using coral and brachiopod fauna)

Gill, G.A. and Coates, A.G. (1977) Mobility, growth patterns and substrate in some fossil and recent corals. *Lethaia* **10**, 119–34.

Goreau, T.F. and Yonge, C.M. (1968) Coral community on muddy sand. *Nature*, **217**, 421–3. (Soft-bottom habitat)

Goreau, T.F., Goreau, N.I. and Goreau, T.J. (1979) Corals and coral reefs. *Scientific American* **241**(5), 110–21.

Jamieson, E.R. (1971) Paleoecology of Devonian reefs in western Canada. *Proceedings of the North American Paleontological Convention*, Chicago, 1969 (J), pp. 1300–40. (Reef environments)

Jell, J.S. (1969) Septal microstructure and classification of the Phillipsastraeidae, in *Stratigraphy and Palaeontology* (ed. K.S.W. Campbell), Australian National University Press, Canberra, pp 50–73. (Composition of skeletal elements and chemistry)

Jell, P.A. and Jell, J.S. (1976) Early Middle Cambrian corals from western New South Wales. *Alcheringa* **1**, 181–95. (The oldest known corals)

Jerre, F. (1993) Conulariid microfossils from the Silurian Lower Visby beds of Gotland, Sweden. *Palaeontology* **36**, 403–24.

Jerre, F. (1994) Anatomy and phylogenetic significance of *Eoconularia loculata* (Wiman 1895), a conulariid from the Silurian of Gotland. *Lethaia* **27**, 97–110.

Jull, R.K. (1965) Corallum increase in *Lithostrotion*. *Palaeontology* **8**, 204–25. (Various methods of growth)

Kato, M. (1963) Fine skeletal structures in Rugosa. *Journal of the Faculty of Science of Hokkaido University Series* **411**, 571–630. (Microstructure)

Klovan, J.E. (1964) Facies analysis of the Redwater Reef Complex, Alberta, Canada. *Bulletin of Canadian Petroleum Geology* **12**, 1–100. (Distribution of corals within a reef)

Krebs, W. (1974) Devonian carbonate complexes of central Europe, in *Reefs in Time and Space* (ed. L.F. Laporte), Society of Economic Mineralogists and Paleontologists, Tulsa, Okl., pp. 155–208. (Distribution of corals within reefs)

Ladd, H.S. (1971) Existing reefs: geological aspects. *Proceedings of the North American Paleontology Convention*, Chicago, 1969 (J), pp. 1273–1300. (Good summary and bibliography)

Lang, W.D. (1923) Trends in British Carboniferous corals. *Proceedings of the Geological Association* **34**, 120–36. (An early analysis of evolutionary trends; actually of limited value)

Lee, J. and Noble, I.P.A. (1990) Colony development and formation in halysitid corals. *Lethaia* **23**, 179–94.

Lewis, J.B. and Price, W.S. (1975) Feeding mechanisms and feeding strategies of Atlantic reef corals. *Journal of the Zoological Society of London* **176**, 527–44.

Lindström, M. (1978) An octocoral from the Lower Ordovican of Sweden. *Geologica et Palaeontologica* **12**, 41–52. (Oldest known)

Mergner, H. and Scheer, G. (1974) The physiographic zonation and ecological conditions of some south Indian and Ceylon coral reefs, *Proceedings of the Second International Symposium on Coral Reefs*, (eds A.M. Cameron *et al.)* Great Barrier Reef Committee, Brisbane, pp. 3–30.

Milne-Edwards, H. and Haime, J. (1850) A monograph of the British fossil corals. *Palaeontographical Society Monograph*, pp. 1–156.

Mitchell, M. (1989) Biostratigraphy of Visean (Dinantian) rugose coral faunas in Britain. *Proceedings of the Yorkshire Geological Society* **47**, 233–47.

Neuman, B. (1984) Origin and early evolution of rugose corals. *Paleontographica Americana* **54**, 119–26.

Neuman, B. (1988) Some aspects of life strategies of Early Palaeozoic rugose corals. *Lethaia* **21**, 97–114.

Nudds, J.R. (1981) An illustrated key to the British lithostronid corals. *Acta Palaeontologica Polonica* **22**, 301–404.

Oliver, W.A. (1976a) Some aspects of colony development in corals. *Memoirs of the Paleontological Society 2 (Journal of Paleontology* **42**, Suppl.), 16–34.

Oliver, W.A. (1976b) Presidential address: biogeography of Devonian rugose corals. *Journal of Paleontology* **52**, 365–73. (Coral faunal provinces)

Oliver, W.A. (1980) The relationship of the scleractinian corals to the rugose corals. *Paleobiology* **6**, 146–60. (Favours origin of scleractinians from sea anemones)

Oliver, W.A. (1996) Origins and relationships of Palaeozoic coral groups and the origins of Scleractinia, in *Paleobiology and Biology of Corals* (ed. G.D. Stanley), The Paleontological Society Papers 1. Lawrence, Kan., pp. 107–34.

Playford, P.E. (1980) Devonian 'Great Barrier Reef' of Canning Basin, Western Australia. *American Association of Petroleum Geologists Bulletin* **64**, 814–40.

Sandberg, P.A. (1975) Bryozoan diagenesis: bearing on the nature of the original skeleton of rugose corals. *Journal of Paleontology* **49**, 587–606. (Detailed microstructural study with good bibliography)

Scoffin, T. P. (1971) The conditions of growth of the Wenlock reefs of Shropshire. *Sedimentology* **17**, 173–219

Scoffin, T.P. and Dixon, J.E. (1983) The distribution and structure of coral reefs; one hundred years since Darwin. *Biological Journal of the Linnean Society* **20**, 11–38. (Reefs and plate tectonics)

Scrutton, C.T. (1965) Periodicity in Devonian coral growth. *Palaeontology* **7**, 552–8. (Monthly banding and the length of the Devonian year)

Scrutton, C.T. (1975) Hydroid–serpulid symbiosis in the Mesozoic and Tertiary. *Palaeontology* **18**, 225–74. *(Parsimonia* and *Protulophila)*

Scrutton, C.T. (1977) Reef facies in the Devonian of eastern South Devon, England. *Memoires du Bureau de Recherches Gèologiques et Miniéres* **89**, 125–35. (Faunal variation with facies).

Scrutton, C.T. (1978) Periodic growth features in fossil organisms and the length of the day and month, in *Tidal Friction and the Earth's Rotation* (eds P. Broche and J. Sundermann), Springer-Verlag, Berlin, pp. 154–96. (Corals less valuable than bivalves)

Scrutton, C.T. (1979) Early fossil cnidarians, in *The Origins of Major Invertebrate Groups* (ed. M.R. House), Systematics Association Special Volume No. 12, Academic Press, London, pp. 161–207.

Scrutton, C.T. (1983) New offset-associated structures in some Carboniferous rugose corals. *Lethaia* **16**, 129–44. (Remarkable growth structures)

Scrutton, C.T. (1984) Origin and early evolution of tabulate corals. *Palaeontographica Americana* **54**, 110–18.

Scrutton, C.T. (1986) Cnidaria, in. *Atlas of Invertebrate Macrofossils* (ed. J.W. Murray), Longman, Harlow, pp. 11–46.

Scrutton, C.T. (1987) A review of favositid affinities. *Palaeontology* **30**, 485–92.

Scrutton, C.T. (1988) Patterns of extinction and survival in Palaeozoic corals, in *Extinction and the Fossil Record* (ed. G.P. Larwood), Systematics Association Special Volume No. 34, pp. 65–86.

Scrutton, C.T. (1989) Ontogeny and astogeny in *Aulopora* and its significance illustrated by a new non-encrusting species from the Devonian of southwest England. *Lethaia* **23**, 61–75.

Scrutton, C.T. (1997a) Growth strategies and colonial form in tabulate corals. *Boletin Real Sociedad Espánola de Historia Natural (Section Geologia)* **91**, 177–89.

Scrutton, C.T. (1997b) The Palaeozoic corals, 1: Origins and relationships. *Proceedings of the Yorkshire Geological Society* **51**, 177–208.

Scrutton, C.T. and Clarkson, E.N.K. (1991) A new scleractinian-like coral from the Ordovician of the Southern Uplands, Scotland. *Palaeontology* **34**, 179–94.

Scrutton, C.T. and Hipkin, R.G. (1973) Long-term changes in the rotation rate of the Earth. *Earth Science Review* **9**, 259–74. (Partially based on coral banding)

Slater, I.L. (1907) A monograph of British Conulariae. *Palaeontographical Society Monograph*, pp. 1–40.

Sorauf, J.E. (1971) Microstructure in the exoskeleton of some Rugosa (Coelenterata). *Journal of Paleontology* **45**, 23–32. (Scanning electron micrographs)

Sorauf, J.E. (1972) Skeletal microstructure and microarchitecture in Scleractinia (Coelenterata). *Palaeontology* **15**, 88–107. (Scanning electron micrographs)

Sorauf, J.E. and Savarese, M. (1995) A Lower Cambrian coral from South Australia. *Palaeontology* **38**, 757–70.

Squires, D.F. (1964) Fossil coral thickets in Warrarapa, New Zealand. *Journal of Paleontology* **38**, 904–15. (Development of thickets, coppices and banks)

Stanley, D.G. and Stürmer, W. (1983) The first fossil ctenophore from the Lower Devonian of West Germany. *Nature* **303**, 518–20.

Stearn, C.W. and Riding, R. (1973) Forms of the hydrozoan *Millepora* on a Recent coral reef. *Lethaia* **6**, 187–200. (Living milleporines)

Stoddard, D.C. (1969) Ecology and morphology of Recent coral reefs. *Biological Reviews* **44**, 433–98. (Valuable reference work with bibliography)

Taylor, J.D. (1968) Coral reef and associated invertebrate communities, mainly molluscan, around Mahe, Seychelles. *Philosophical Transactions of the Royal Society B* **293**, 129–206. (Coral/molluscan ecology)

van Iten, H. (1991) Anatomy, patterns of occurrency and nature of the connulariid schott. *Palaeontology* **34**, 939–54.

van Iten, H., Fitzke, J.A. and Cox, R.S. (1996) Problematical fossil cnidarians from the upper Ordovician of the north-central USA. *Palaeontology* **39**, 1037–64.

Vaughan, A. (1905) The palaeontological sequence in the Carboniferous limestone of the Bristol area. *Quarterly Journal of the Geological Society of London* **61**, 181–307. (Classic study using coral brachiopod zonation)

Vosburgh, F. (1982) *Aeropora reticulata:* structure, mechanics, and ecology of a reef coral. *Proceedings of the Royal Society B* **214**, 481–99.

Wade, M. (1972) Hydrozoa and Scyphozoa and other medusoids from the Precambrian Ediacara fauna, South Australia. *Palaeontology* **15**, 197–225.

Walker, K.R. and Alberstadt, L.P. (1975) Ecological succession as an aspect of structure in fossil communities. *Palaeobiology* **1**, 238–57. (Reef structural development and ecology with increasing maturity)

Wang, H.C. (1950) A revision of the Zoantharia Rugosa in the light of their minute skeletal structure. *Philosophical Transactions of the Royal Society B* **234**, 175–246. (Classic study, now rather superseded)

Wells, J.W. (1937) Individual variation in the rugose coral species *Heliophyllum belli*. *Palaeontologia Americana* **2**, 1–22. (Environmental factors in variation)

Wells, J.W. (1957) Annotated bibliography: corals, in *Treatise on Marine Ecology and Palaeoecology*, Vol. 2 (ed. J. Hedgpeth), Memoirs of the Geological Society of America No. 67, Geological Society of America, Lawrence, Kan., pp. 773–82. (Invaluable source work)

Wells, J.W. (1963) Coral growth and geochronometry. *Nature* **197**, 948–50. (Annual banding and the slowing of the Earth's rotation)

Wendt, J. (1990) The first aragonitic rugose coral. *Journal of Paleontology* **64**, 335–40.

Williams, G.E. (1989) Late Precambrian tidal rhythmites in South Australia and the history of the Earth's rotation. *Journal of the Geological Society of London* **146**, 97–111.

Wilson, J. (1975) The distribution of the coral *Caryophyllia smithii* S and B on the Scottish continental shelf. *Journal of the Marine Biological Association of the United Kingdom* **55**, 611–25. (Precise documentation)

Wilson, J.B. (1976) Attachment of the coral *Caryophyllia smithii* S and B to tubes of the polychaete *Ditrupa arictina* (Muller) and other substrates. *Journal of the Marine Biological Association of the United Kingdom* **56**, 291–303. (Documentation)

Yonge, C.M. (1963) The biology of coral reefs, in *Advances in Marine Biology*, Vol. 1, (ed. F.S. Russell), George Allen and Unwin, London, pp. 209–60. (Valuable summary and bibliography)

Yonge, C.M. (1968) Living corals. *Proceedings of the Royal Society B* **169**, 329–44. (Invaluable reference)

Yonge, C.M. (1974) Coral reefs and molluscs. *Transactions of the Royal Society of Edinburgh* **69**, 147–66. (Environmental niches in reefs)

6 Bryozoans

6.1 Introduction

Bryozoa, because they are small and often delicate, tend to be less familiar than most invertebrate groups preserved in fossil form. Yet they are common in sedimentary rocks and abundant in the sea today; at least 3500 living and 15 000 fossil species are known. The fronds of flat seaweeds such as *Laminaria* are often covered with the lacy calcareous skeletons of the bryozoan *Membranipora*, each colony being composed of hundreds of individuals.

All bryozoans are colonial, most are marine, and in the majority of cases the units or zooids of the colony secrete tubes or boxes of lime partially encasing the soft parts (e.g. Fig. 6.2). Each zooid is basically cylindrical, has a ring of tentacles and at first sight seems to resemble a small cnidarian polyp. However, the zooids are coelomate, having a freely suspended gut with both mouth and anus, and are unquestionably of a higher grade of organization. They were recognized as being distinct from cnidarians in 1820, and the name Bryozoa was given formally to them in 1831 by Ehrenberg, 1 year after the name Polyzoa had already been given by the Irish naturalist J. Vaughan Thompson. However, since both Thompson and Ehrenberg included in the phylum, as understood by them, certain other invertebrate groups, the true bryozoans are sometimes (and especially in American usage) known as Phylum Ectoprocta. The term Bryozoa, being in more common use, is retained here.

Two genera, one simple in construction and one more complex, are described here to show the basic morphology. Both of these belong to the largest and most successful marine class: the Gymnolaemata.

6.2 Two examples of living bryozoans (Figs 6.1–6.5)

Bowerbankia (Fig. 6.1)

Bowerbankia (Order Ctenostomata; Fig. 6.1a–e) has a creeping cylindrical stolon from which numerous bottle-shaped **zooids** arise in clusters.

The body walls of each living zooid (membranous and unpreservable in this case) are called the **cystid**. (The term **zooecium**, which has sometimes been used as an alternative to cystid, is now restricted to the calcified zooidal skeleton.) The calcified parts of a whole colony may be called a **zoarium**. When feeding, the zooid extends its **lophophore**, a ring of 10 ciliated tentacles, into the surrounding water. These tentacles converge on a central mouth from which the gut descends into the body of the zooid. This gut is U-shaped, hanging down in the coelom, and has an oesophagus, stomach and intestine which opens by an anus close to the mouth but outside the ring of tentacles. The tentacles attach to the body by an eversible tentacle sheath. In *Bowerbankia* and its relatives the sheath is protected by a collar, but this is not represented in other Bryozoa. Between the mouth and anus is a major ganglion from which nerves run to all parts of the body. From the base of the stomach a thread, the **funiculus**, connects to a main **funicular tube** running along the stolon and connecting all the zooids.

The tentacles can be either extruded for feeding (Fig. 6.1c) or retracted entirely within the body for protection (Fig. 6.1d). In the resting zooid the tentacle sheath becomes invaginated, with its surface facing inwards and surrounding the tentacles rather

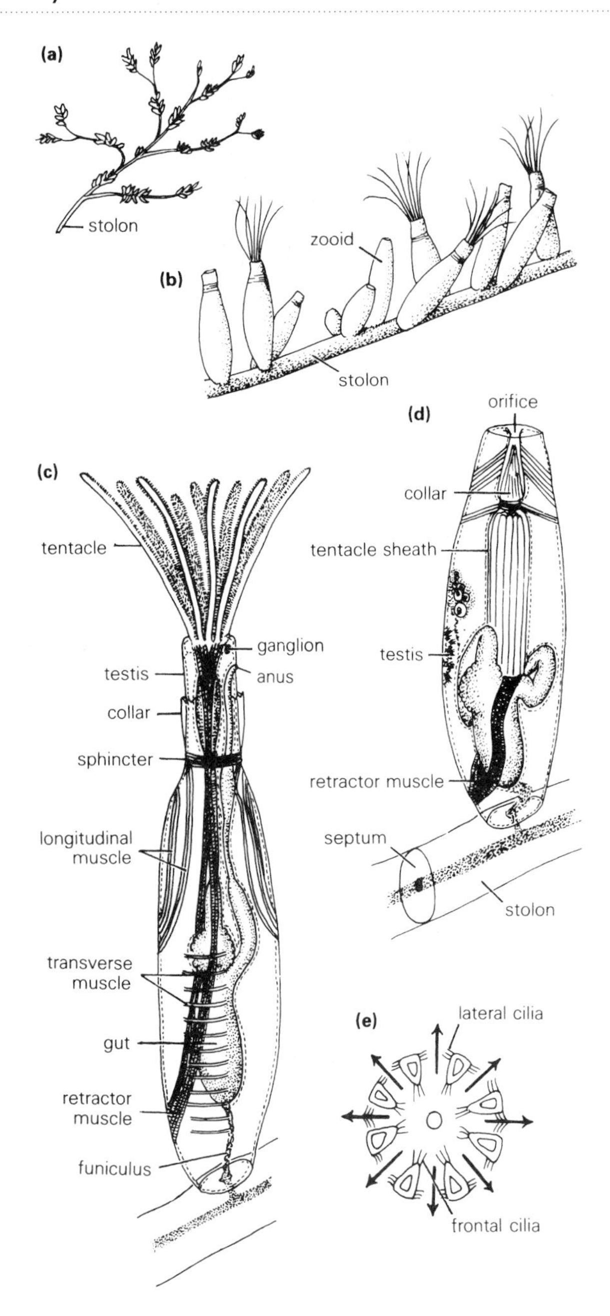

Figure 6.1 *Bowerbankia*, a Recent ctenostome: (a) appearance of colony (× 1); (b) section of zoarium with extended and retracted zooids (× 7); (c) zooid with tentacles extended (× 35 approx.); (d) zooid with tentacles retracted (× 35 approx.); (e) section through tentacle crown, with arrows showing current directions. [(a) Redrawn from Bassler in *Treatise on Invertebrate Paleontology*, Part A; (b)–(e) redrawn from Ryland, 1970.]

than facing outwards below them. Eversion of the tentacles is accomplished quite simply by compression of the body wall by transverse muscles; this raises the hydrostatic pressure of the coelomic fluid so that the tentacles have to emerge. A large retractor muscle, attached to the tentacular base and accompanied by longitudinal muscles, pulls the tentacles within the body when danger threatens, and the zooid is finally closed off by a circular sphincter muscle just below the collar. The collar then folds inwards in a series of pleats and completes the closure of the zooid.

When the zooid feeds, the tentacles are extended in an erect funnel by hydrostatic pressure of the coelom. They are quadrate in cross section and have **ciliary tracts** on each side, those of adjacent tentacles nearly touching. Another tract of 'frontal cilia' on each tentacle faces the mouth, being more strongly developed towards the base. When the lateral cilia beat downwards in a coordinated metachronal rhythm, currents are generated which pass straight down the funnel towards the mouth and out between the tentacles (Fig. 6.1e). The food particles (mainly phytoplankton) in the incoming stream are ingested by the mouth, but precisely how this is done is not entirely clear. The flow rates of currents leading directly to the mouth are much higher than peripherally, and possibly only the particles in central streams are captured. The operation of the lophophore and its structure are decidedly similar to those of brachiopods; hence the Bryozoa and Brachiopoda are assumed to be related and are grouped together with phoronids in 'Superphylum' Lophophorata.

There are no separate excretory, circulatory or respiratory organs. The colony grows from an initial zooid, the **ancestrula,** by growth of the stolon (itself a series of modified zooids) and by asexual budding of new zooids (Fig. 6.3). New colonies, however, are produced sexually. The zooids are hermaphrodite, but the ovary and testis may develop at different times. The ovary is a cluster of several egg cells, which are released one at a time into the tentacle sheath where they are fertilized. Each then develops into a **trochophore larva** while still in the tentacle sheath. When fully mature, the larva swims away; meanwhile the zooid degenerates. A new egg is released only when the last fertilized egg has developed into a larva. Sperm

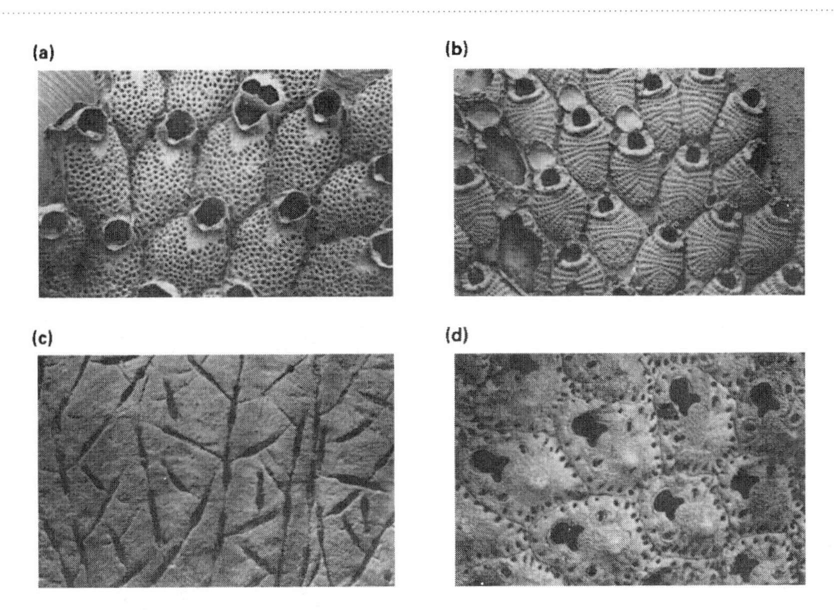

Figure 6.2 Bryozoan structure: (a) *Smittina* sp. (ascophoran cheilostome), Holocene, Antarctica (× 16); (b) *Castanopora magnifica* (Anascan cheilostome), Cretaceous, England (× 14); (c) *Orbignyopora* (cheilostome borings on brachiopod), Silurian, Pennsylvania (× 16); (d) *Crepidacantha* (ascophoran cheilostome), Recent, New Zealand. (All SEM photographs reproduced by courtesy of Dr P.D. Taylor.)

developed within the testis is seen in many bryozoans to make its way out into the sea via tiny pores in the tentacles.

Smittina (Figs 6.2a, 6.4, 6.5)

Smittina (Order Cheilostomata, Suborder Ascophora) is an example of one of the more complex bryozoans. The colony is encrusting, and the individual zooids are arranged like flat elliptical boxes radiating away from the ancestrula or first-formed zooid. They are all constructed of calcium carbonate. Mature zooids may develop distal **ovicells**, swollen spherical structures in which the fertilized eggs develop into larvae. In different cheilostomes ovicells may lie within the zooid, overlap onto the next one distally or be embedded in its posterior wall; alternatively they may be wholly or partially separated. Directly behind the ovicell is the **orifice** or opening to the zooid. This is keyhole-shaped (though this is not immediately evident, since the primary orifice is hidden beneath the upper surface) and closed by an **operculum**, hinged at its narrowest points on **cardelles**, so that when the operculum

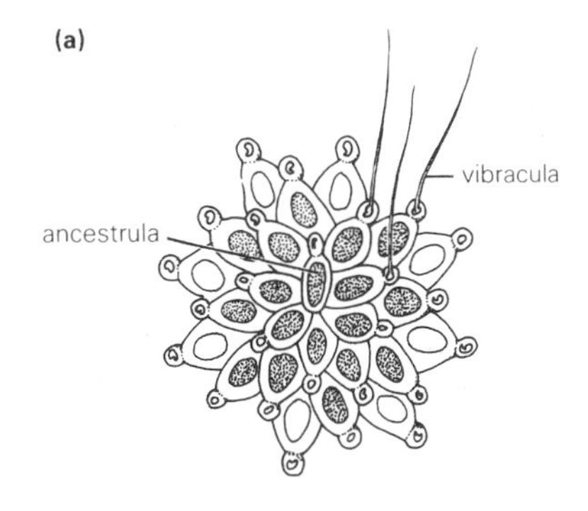

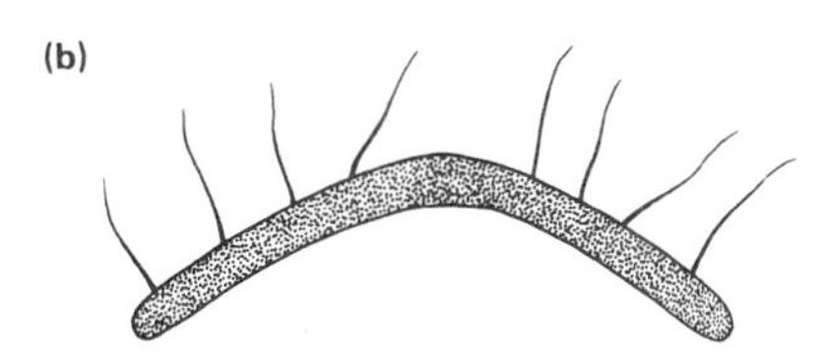

Figure 6.3 (a) *Cupuladria* (Mio.–Rec.), a young zoarium showing growth from central ancestrula – blank zooecia are the last formed (×10); (b) probable life position of zoarium. (Modified from Lagaaij, 1963.)

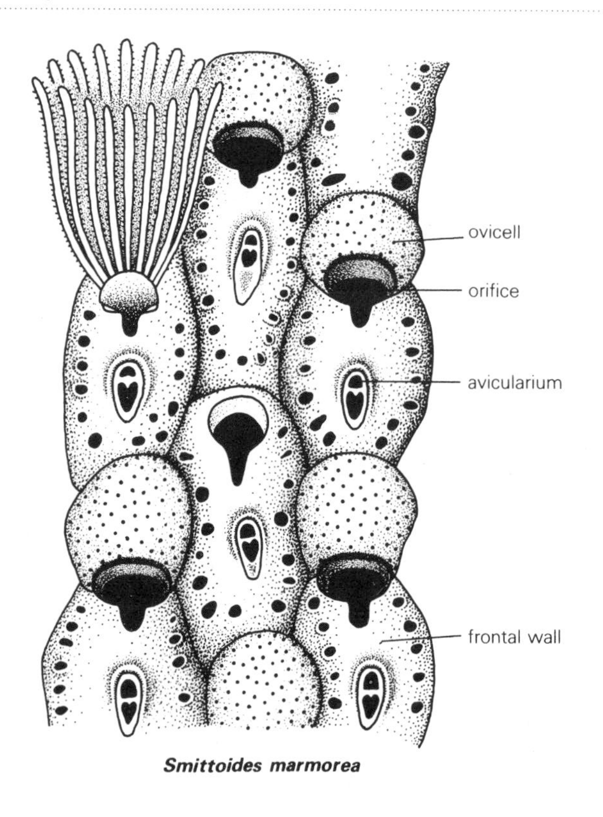

Figure 6.4 Structure of the ascophoran cheilostome *Smittoidea marmora*, showing mature individuals with ovicells (right).

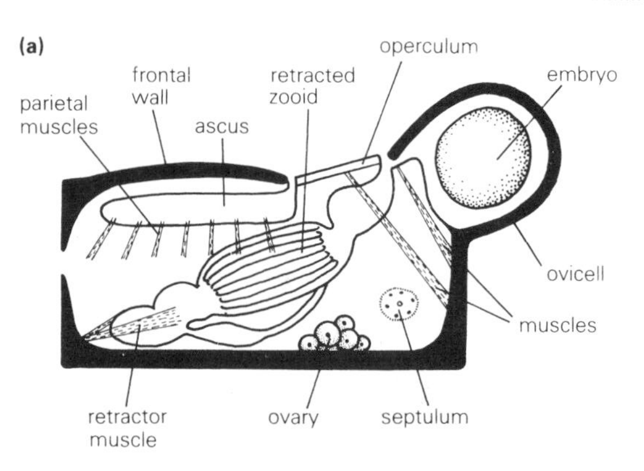

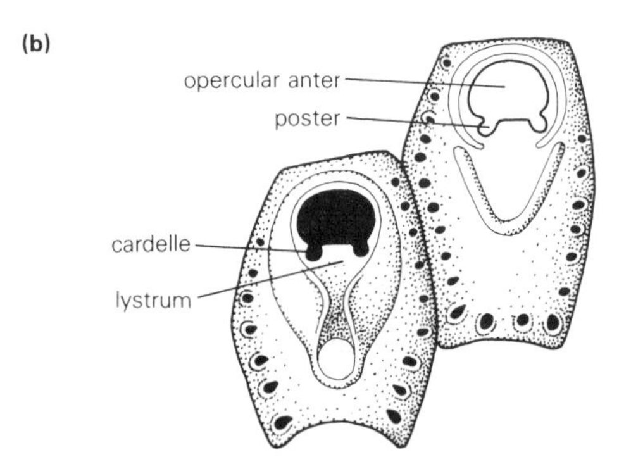

Figure 6.5 (a) Longitudinal section through a generalized ascophoran with zooids retracted; (b) *Smittoidea marmorea*, showing progressive development of the peristomial region in young zooids. (Redrawn from Hayward and Ryland, 1979.)

opens about this fulcrum the distal part (**anter**) rises and the proximal part (**poster**) sinks. When the **polypide** is retracted, it lies entirely within the cystid and the orifice is closed. When it emerges, it comes through the distal part of the orifice as the anter lifts up. Connected to the proximal part of the orifice below the poster is a sac, the **ascus**, sometimes known as a **compensatrix**, which is suspended from the body wall by many **parietal muscles**. This compensatrix is concerned with polypide extrusion. When the radial muscles contract, the compensatrix expands and the poster is depressed so that water enters the sac. As the compensatrix swells, the polypide is displaced and has to emerge from the anter because of the hydrostatic pressure. The polypide can be pulled back in again when the retractor muscle contracts. This causes the compensatrix to be evacuated hydrostatically (since the radial muscles are by this time relaxed) and the operculum shuts.

The **frontal wall** (i.e. upper surface) is complex in structure with a regular sculpture, secreted in several layers. Cheilostomes (Fig. 6.2) are strongly polymorphic, as is shown by the specialized structures **avicularia** and **vibracula**, which are both modified zooids. The avicularia (Fig. 6.6a) are attached to the upper surface at species-specific locations.

One kind, found in *Bugula*, resembles a bird's head in shape and contains a single modified polypide. It has a hinged chitinous mandible which opens and snaps shut in a constant motion, thought to discourage both predators and settling larvae.

The polypide is reduced to a rudiment, and the main internal soft-part structures are the paired antagonistic muscle sets with which the mandible snaps. The mandible is infrequently preserved fossil, but since its edge generally fits the avicularium, its

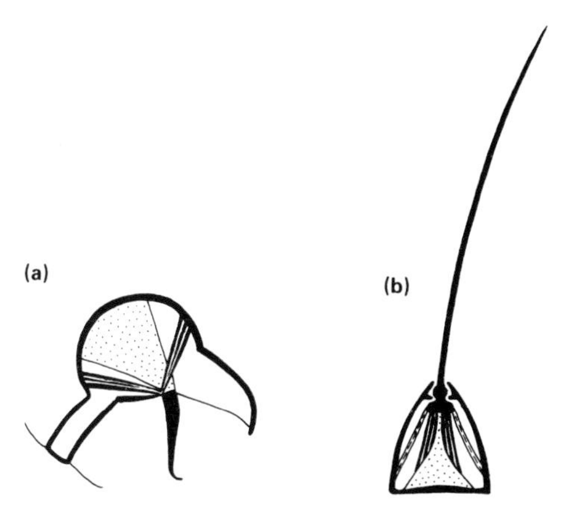

Figure 6.6 (a) Avicularium in lateral view with different sets of muscles to open and close; (b) vibraculum, with musculature for lashing shown. (Redrawn from Ryland, 1970.)

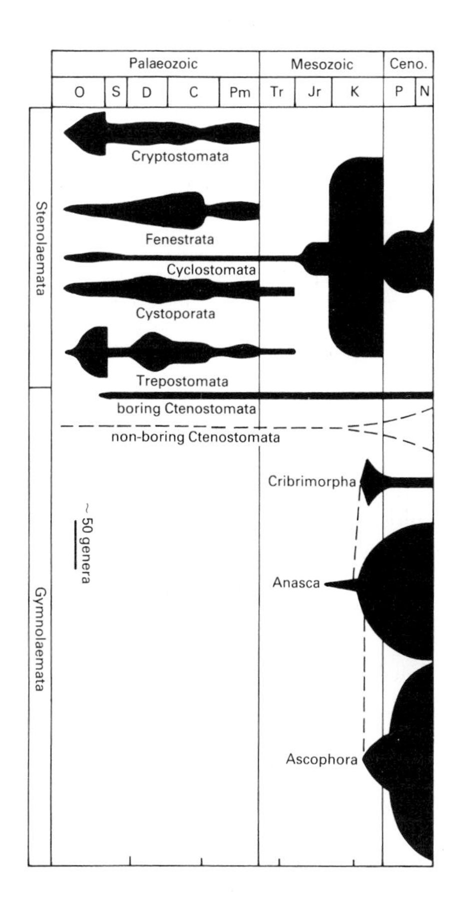

Figure 6.7 Time ranges and relative abundance of main bryozoan groups. (Redrawn from McKinney and Jackson, 1989.)

form is quite well reflected by the preservable skeleton. Avicularia may be sessile or, in the most extreme form, pedunculate, i.e. mounted on a short stalk. The least-modified sessile avicularia are said to be **vicarious**; these are slightly smaller than normal zooecia and replace them in the colony at regular intervals. **Interzooidal** avicularia occur between zooids and are reduced in size, whereas the much smaller **adventitious** and sometimes pedunculate avicularia are found mainly in cheilostomes in which they may occur anywhere on the frontal wall. There is a continuum between the two kinds of avicularium. Ryland (1970) has discussed their possible evolution in some detail.

Vibracula (Fig. 6.6b) possess a long whip-like bristle (**seta**) projecting from a sessile basal chamber which contains only the muscles. The seta swings on a pair of opposing pivots just above its lower end. The muscles are attached below the pivots. When triggered into action the contraction of the muscles causes the seta to lash violently. This stimulates neighbouring vibracula, and the whole ensemble will strike hard against any alien object, discourage settling larvae or winnow away sediment. In lunulitiform bryozoans, vibracula may be modified as 'legs' on the underside of the colony, enabling it to move.

6.3 Classification

PHYLUM BRYOZOA (Polyzoa; Ord.–Rec.): Sessile colonial coelomates, normally marine, rarely freshwater, consisting of small linked zooids, usually with a calcareous or more rarely an organic skeleton. Zooids have a tentacle-bearing retractile lophophore and a U-shaped gut with the anus outside the tentacular ring. Colonies arise from an ancestrula (or rarely a **statoblast**), and are encrusting, creeping, erect or in chains, polymorphic in some groups.

The classification adopted here is that of Boardman *et al.* (1983) in Robison, *Treatise on Invertebrate Paleontology* and McKinney and Jackson (1989; Fig. 6.7).

CLASS 1. PHYLACTOLAEMATA (Tert.–Rec.): Non-calcareous freshwater bryozoans. Zooids with horseshoe-shaped lophophores; statoblasts produced as resting buds. Twelve genera only. *Plumatella, Cristatella.*

CLASS 2. STENOLAEMATA (Ord.-Rec.): Calcified marine bryozoans, usually non-operculate, with an extensive fossil record. Zooids are cylindrical and elongated zooecia continuing to grow throughout the life of the colony and set at an angle to the direction of colony growth. Each polypide is surrounded by a membranous sac. Tentacle extrusion in living forms is brought about by muscular

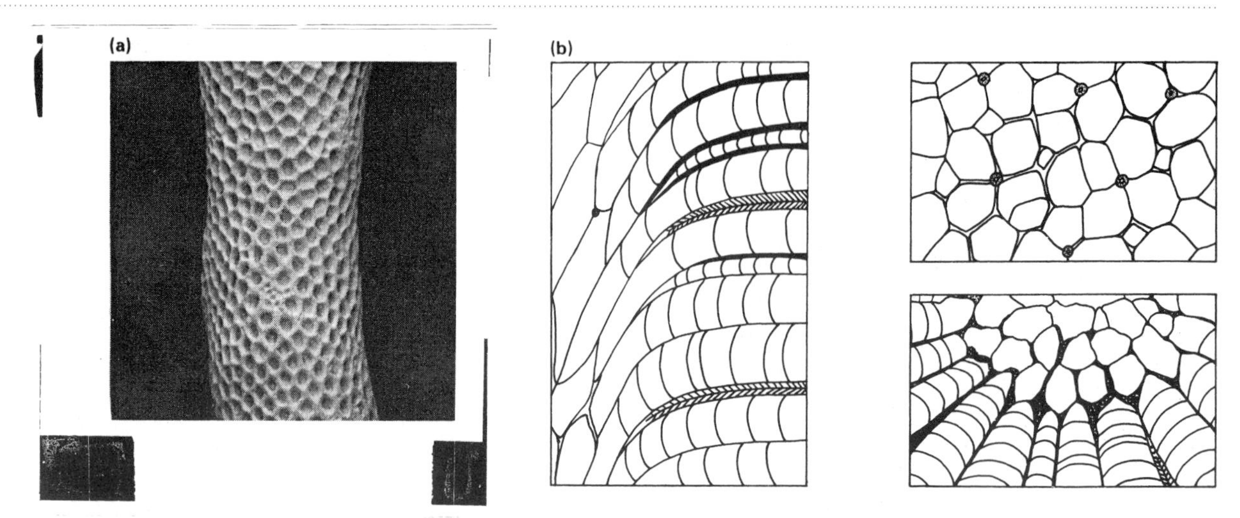

Figure 6.8 Trepostomes: (a) *Anisotrypa* (Carb.), Alabama (× 6); (b) *Dekayia* (Ord.) showing a longitudinal section (left), a tangential section with acanthostyles (above right) and a transverse section (below right). (Drawn from a photograph by J.P. Ross in *Journal of Paleontology*, 1962.)

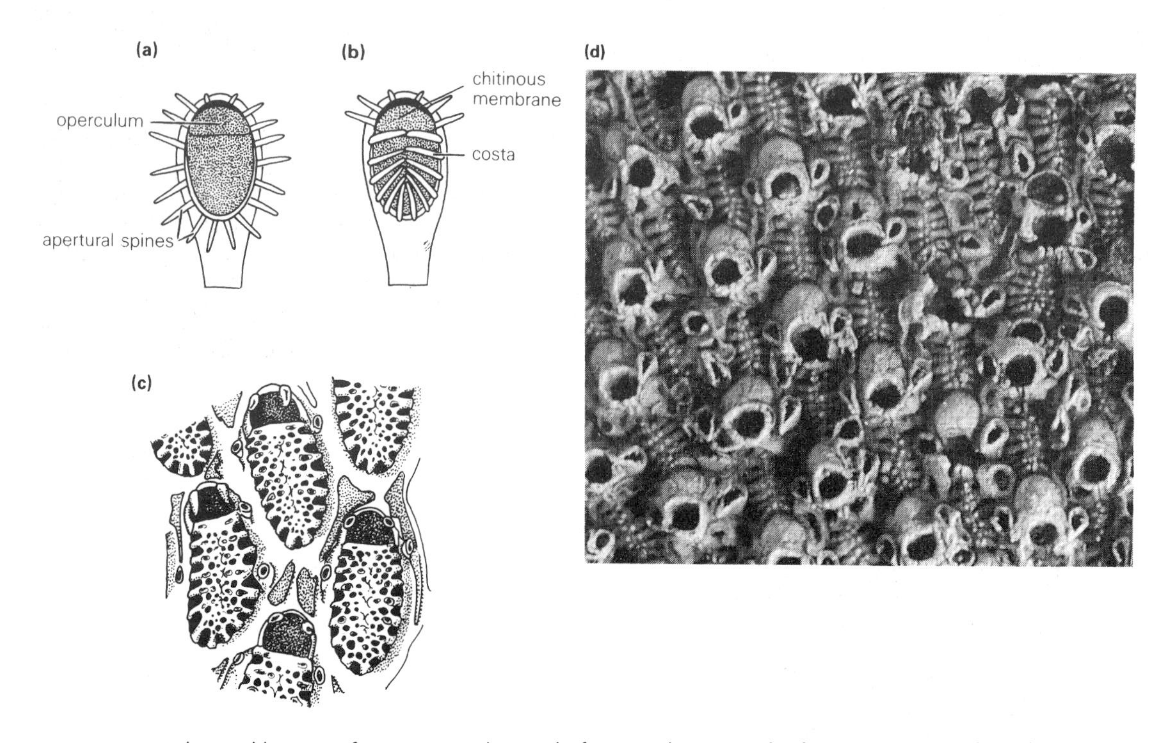

Figure 6.9 (a), (b) Possible origin of Cretaceous cribrimorphs from membraniporoids, showing (a) ancestral membraniporoid type with radiating apertural spines and (b) spines to form costae covering frontal chitinous membrane; (c) *Pelmatopora gregoryi*, a Cretaceous cribrimorph (frontal view) (× 25); (d) *Tricephalopora pustulosa*, a Cretaceous cribrimorph (× 25 approx.). [(a)–(c) Redrawn from Larwood, 1962; (d) photograph reproduced by courtesy of Dr G.P. Larwood.]

action forcing coelomic fluid into the proximal part of the zooid displacing the tentacles distally, 550 genera.

ORDER 1. CYCLOSTOMATA (better known as Tubuliporata, the usage adopted in the *Treatise on Invertebrate Paleontology*; Ord.-Rec.): Erect or encrusting zoaria of tubular zooecia possessing either circular apertures separated by pseudoporous frontal walls, or contiguous polygonal apertures. Interzooecial walls usually pierced by **mural pores**; e.g. *Crisia, Berenicea* (Fig. 6.11), *Stomatopora* (Fig. 6.11), *Mesenteripora* (Fig. 6.12).

Figure 6.10 *Theonoa diplopora* (Cyclostomata), Jurassic, England, showing radial ridges (×10).

ORDER 2. CYSTOPORATA (L. Ord.–U. Perm.): Similar to cyclostomes but with regions of curved **cystiphragms** separating zooecia and/or have crescentic projections (**lunaria**) around zooecial apertures; e.g. *Fistulipora, Ceramopora*.

ORDER 3. TREPOSTOMATA (Ord.–Trias.): 'Stony bryozoans' forming massive zoaria with elongate **autozooecia** which are initially thin walled but become thick walled close to the zoarial surface where small **mesozooecia**, filled by closely spaced **diaphragms**, may intervene between autozooecia. Mural pores absent; e.g. *Monticulipora, Anisotrypa* (Fig. 6.8a), *Dekayia* (Fig. 6.8b), *Hallopora*.

ORDER 4. CRYPTOSTOMATA (L. Ord.–U. Perm.): Zoaria erect and tree-like or forming bilaminar sheets, autozooids short, with basal diaphragms or incomplete partitions (hemisepta); e.g. *Arthrophragma*.

ORDER 5. FENESTRATA (L. Ord.–U. Perm.): Zoaria erect and net-like with unilaminate branches, autozooids short and commonly with hemisepta; e.g. *Fenestella*, *Rhabdomeson*, *Archimedes*.

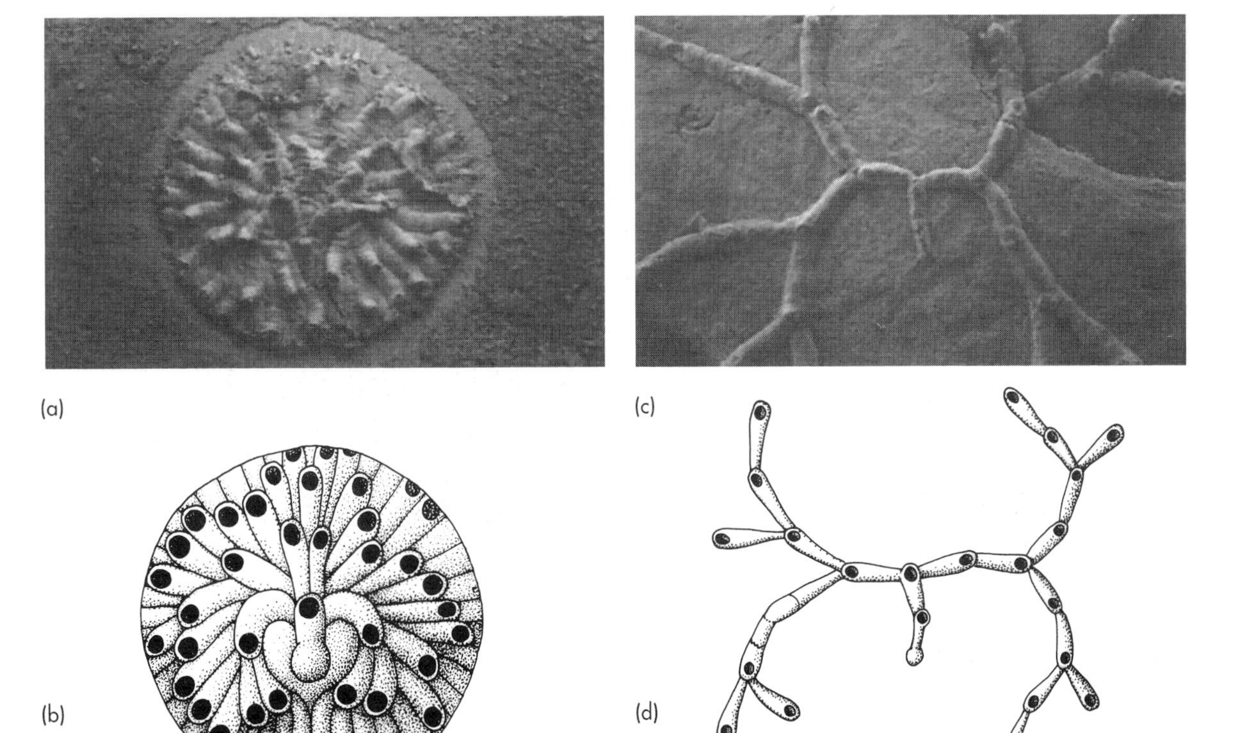

Figure 6.11 (a), (b) *Berenicea* and (c), (d) *Stomatopora*, two Jurassic cyclostomes with contrasting adaptive morphologies. (Redrawn from Taylor, 1979.)

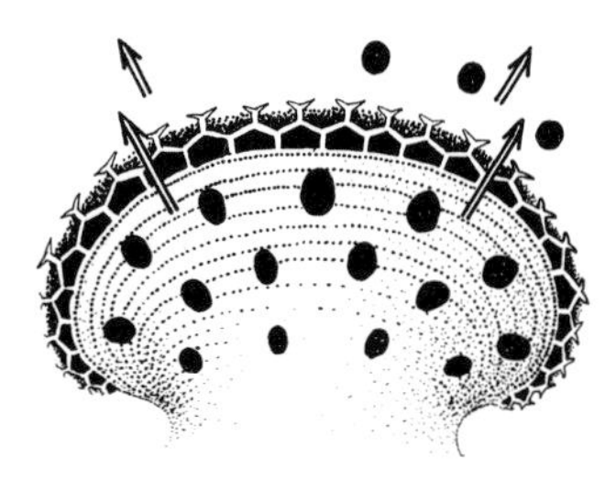

Figure 6.12 Mesenteripora, showing lunate structure by means of which overgrowth of old zooecia has taken place. (Redrawn from Taylor, 1976.)

CLASS 3. GYMNOLAEMATA (Ord.–Rec.): Marine, occasionally brackish or freshwater bryozoans which may be calcified. Zooids are box-like or may form short cylinders. Their size is fixed early in development and their long axis coincides into the local direction of growth of the zoarium. Zooids connected by a funicular network. Lophophores everted by muscular deformation of part of the body wall. Strongly polymorphic. 650 genera.

ORDER 1. CTENOSTOMATA (Ord.–Rec.): Zooids uncalcified, walls membranous or gelatinous, lacking ovicells, frequently **penetrant**. Examples include *Bowerbankia* (Fig. 6.1a–e), *Alcyonidium*, *Orbignyopora* (Fig. 6.2c).

ORDER 2. CHEILOSTOMATA (Jur.–Rec.): Zooids calcareous, with short box-like zooecia having a distal orifice closed by a hinged operculum. Avicularia and vibracula common. Embryos brooded in ovicells. The four suborders are defined on calcification of the frontal wall and according to how the lophophore is protruded. Examples include *Smittina* (Figs 6.2, 6.4, 6.5), *Membranipora*, *Cribrilina*, *Pelmatopora* (Fig. 6.9c), *Cupuladria* (Fig. 6.3a,b), *Crepidacantha* (Fig. 6.2d), *Castanopora* (Fig. 6.2b).

6.4 Morphology and evolution

Shortly after their first appearance in the early Ordovician the Bryozoa underwent a great burst of evolution, resulting in the establishment of the early Stenolaemata, which by the late Ordovician were very abundant and diverse. The Stenolaemata are all extinct now apart from the Cyclostomata, but were the dominant class of Palaeozoic bryozoans. Of all the Stenolaemata the most important and abundant in Lower Palaeozoic times were the 'stony bryozoans', the Trepostomata (Fig. 6.8a,b). They formed stick-like or globular calcareous zoaria up to 50 cm across. The zooecia are tubular and closely packed with cross-partitions and thickened distal parts. Sometimes the zooecial apertures are clustered in groups at the summit of small mounds (**monticules**). The zooecial walls are constructed of thin laminae, with the individual laminae at intervals forming meniscus-like diaphragms or cross-partitions. There is no communication through autozooecial walls. Between the large zooecial apertures (autozooecia) there are often smaller openings (mesozooecia): these suggest the former presence of some kind of smaller polymorphic zooid. Sometimes **acanthostyles** (rod-like spines of cone-in-cone calcite) are visible at the intersections of zooecial walls.

The Trepostomata, which after their initial evolutionary burst became so abundant in the Ordovician and Silurian, declined thereafter as cryptostomes became dominant, and they finally died out in the Triassic.

The Cryptostomata appeared in the early Ordovician and, declining somewhat thereafter, held their own until the end of the Permian. The Fenestrata however, to which they are related, reached their acme in the later Palaeozoic.

Fenestellid colonies (Fig. 6.13a–d) grew from an ancestrula forming first a ring of zooecia and then a circlet of upright branches; these bifurcated at intervals giving the colony a cup-shaped form or in some species the form of a half-cup or fan.

Whatever the form of the colony, the branches are subtriangular in cross-section and have rows of zooecia opening onto only one face of the colony. The branches are connected by cross-bars with rectangular spaces or **fenestrules** between (Fig. 6.13e). The skeleton was probably secreted by a thin epithelium which extended over the whole surface and, as in some living cyclostomes, the zooids probably shared a common coelom.

Since the growth of the bryozoan skeleton is a 'colonial effort', the colony may become highly integrated rather than remain as no more than an aggregate of individuals. The functional morphology of bryozoan colonies is potentially a fruitful field for research. Fenestellid colonies, to take an example, have been analysed functionally by Cowen and Rider (1972). The arrangement of the zooecia and the form of the zoarium are clearly important in understanding how the colony operated as a whole. It has been postulated that the 'operational subunits' of the fenestellid colony are the fenestrules. In the main branches, the lophophores of equally spaced zooids on opposite sides of a median ridge are

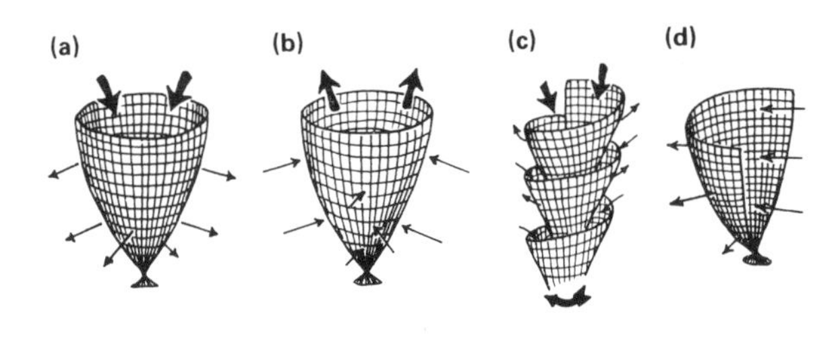

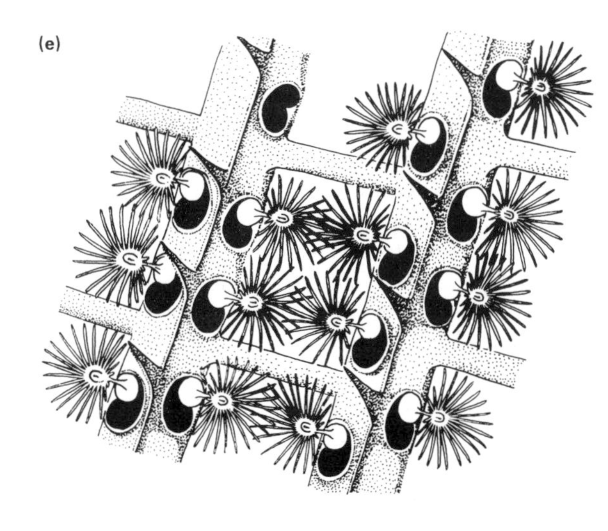

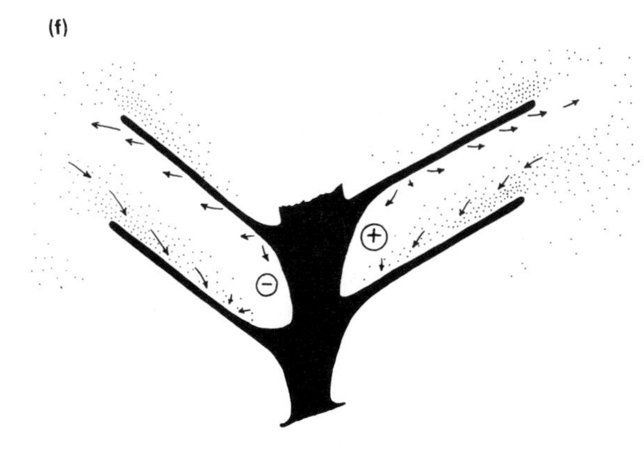

Figure 6.13 Morphology and direction of currents inferred in fenestellids: (a) Carboniferous fenestellid with inward-facing zooecia; (b) Silurian fenestellid with outward-facing zooecia; (c) *Archimedes* (Carb.–Perm.), a spiral colony; (d) fan-shaped fenestellid, using unidirectional current; (e) *Fenestella* (Carb.–Perm.), part of the zoarium showing zooids reconstructed as extending into fenestrules (× 8 approx.); (f) *Archimedes*, interpretation of feeding by spiralling currents. (Redrawn from Cowen and Rider, 1972.)

considered to have extended into the fenestrule forming a filtering net, the cross-bars giving support and dividing up the fenestrules. A combined feeding current set up by the zooids would draw water in through the fenestrules, enabling them to strain off all the food material with their lophophores, and would thus give high filtering efficiency.

The growth forms of fenestellid colonies may have been adapted for maximizing filtering efficiency at the colonial level. Two kinds of cup-shaped colonies are known: those with zooecia facing outwards (common in the Silurian), and those with inward-facing zooecial apertures (more common in the Carboniferous). There are also fan-shaped (half-cup) and spiral colonies. The cup-shaped colonies would probably be equally efficient whether they drew in water from the top and exhaled it laterally through the fenestrules, or whether they sucked it in at the sides (like a sponge) and sent out an excurrent stream from the central cavity. Fan-shaped colonies (again as with certain sponges) would have the best shape for taking in water in a regime where a weak current was constantly flowing normal to the fan surface.

The peculiar *Archimedes* has a typically fenestellid network structure, but this describes a helical spiral around a thick central calcareous column. Such a bryozoan standing upright on the sea floor would be functionally efficient if the combined action of the lophophores generated a current stream coming in at the top and running spirally along the 'deck' to the base, as down a fairground helter-skelter (Fig. 6.13f). As the main stream travelled spirally downwards, some of it would be sucked through the innumerable fenestrules and passed away in a centrifugal stream just below each deck. Since the fenestellid feeding current reconstructions were conceived, based upon fossil material alone, some very similar functional analogues have been found in the modern cheilostome *Retepora*, which resembles and feeds in precisely the manner which had been predicted for fenestellids.

Not uncommonly, new spiral *Archimedes* colonies originated from pre-existing, ageing branches. Possibly hundreds or even thousands of genetically identical colonies were produced in this way from an original founder (McKinney, 1983). Likewise the living cheilostome *Bugula turrita* has erect spiral colonies, the result of equivalent growth parameters, as computer modelling has shown.

Some fenestellids were lyre-shaped, e.g. *Lyropora*. They have a thickened marginal rim supporting a fenestrate network in the shape of a bowed fan. These bryozoans (McKinney, 1977) evidently lay flat upon the sea floor so that there was an open vault below the fan, the zooids being on the upper surface. The fan was probably orientated with the proximal end on the upstream side, so that water was sucked through the fenestrules, filtered and discharged through the open end of the vault.

Cryptostomes and fenestrates outclassed the trepostomes in the late Palaeozoic, but by the end of the Permian they too were extinct.

Many kinds of Palaeozoic bryozoans exhibit homeomorphy, since there are only a limited number of 'ways of being a bryozoan' (Blake, 1980; Taylor and Badve, 1995). Such evolutionary convergence can operate either at the level of the colony or the zooid. Thus there are lyre-shaped fenestrates and cyclostomes of virtually identical form, operating functionally in the same way. In rarer instances both zooid and colony form are strongly homeomorphic, as in the Cretaceous *Chiplonkarina*, a cheilostome with long tubular zooecia just like those of cyclostomes, and a colony type of remarkably similar form.

The Cyclostomata (Fig. 6.11) were the only group of calcareous bryozoans to cross the Palaeozoic–Mesozoic boundary (though the uncalcified ctenostomes did also and survived into the Jurassic). Since there was little competition they were able to dominate the Mesozoic scene. They had become very diverse and important by the Lower Cretaceous; indeed according to Ryland (1970) 'their zenith during this period constitutes one of the highlights in the history of the Bryozoa'. They have declined (relative to the cheilostomes) since, though some survive to the present, and since they are the only living representatives of an ancient stock, most of our conceptions of soft-part morphology in Palaeozoic bryozoans are based upon them. The cyclostome zooid is quite like that of *Bowerbankia*, though with a few characteristic differences, and there is some degree of polymorphism.

Many cyclostome colonies seem to have been quite well integrated, and the colony rather than the individual, as with fenestellids, is the functional unit in terms of effective feeding.

It is possible to infer the former presence and direction of extrazooidal feeding current systems here, both from the differential spacing of autozooecial apertures and from their orientation (Taylor, 1979). Some bryozoans have regularly spaced autozooecial apertures, often arranged in a system of hexagonal close packing. But in many kinds of bryozoan, the autozooecia are unevenly distributed over the colony surface. For example, the Jurassic cyclostome *Spiropora*, an erect branching genus, has autozooecial apertures arranged in bands arranged crosswise to the axis of the branch. In *Theonoa* (Fig. 6.10) the discoidal apertures are arranged in radial bundles (fascicles) which can be regarded as subcolonies. In the large, subspherical colonies of the Pliocene *Meandropora* the surface displays circular fascicles, each bounded by an exterior wall. During the growth of the colony these fascicles alternately differentiate and anastomose (Balson and Taylor, 1982).

In all these cases, the regions of clustered autozooecia – whatever their arrangement – must be sites of inhalent flow. The closely packed zooecia cooperated to produce a stronger current, filtering more water in unit time, than would have been possible if they had operated individually. The exhalent currents were channelled away through regions devoid of autozooids. Another such system is exhibited by the many bryozoan colonies in which the surface is raised into mounds or **monticules**. The autozooids are located in the depressions between these monticules, and because of their setting on the concave surface their tentacle crowns leaned together. These depressions must have been in the sites of cooperative inhalant currents, while the monticule summits, where autozooecia are sparse or absent, acted as exhalant chimneys (Banta *et al.*, 1974; Taylor, 1978). The powerful exhalant currents, in their turn, were able to clear sediment and to disperse larvae or excreta.

Homeomorphy, as previously mentioned, is common. Such evolutionary convergence can operate either at the level of the colony or the zooid. Thus there are lyre-shaped fenestrates and cyclostomes of virtually identical form. In rarer instances both zooid and colony form are strongly homeomorphic, as in the Cretaceous *Chiplonkarina*, a cheilostome where the zooids, long and tubular, are virtually identical with those of cyclostomes, as is the type of colony (Taylor and Badve, 1995).

During Jurassic time the genera *Stomatopora* and *Berenicea* (Fig. 6.11) were the dominant cyclostomes.

They are both encrusters and whereas they are often found together they are quite unlike each other in colony form, and their adaptive strategies were very different (Taylor, 1979). *Stomatopora* colonies consist of dichotomously branching uniserial rows of zooecia, by means of which the colony spreads rapidly and radially. The angle between successive bifurcations diminishes as the colony grows and, as computer simulations have shown, this delays the overlapping of branches until the colony is very large. *Berenicea* colonies, on the other hand, are initially fan-shaped, later becoming discoidal. They have much more closely packed zooecia, linked by numerous pores. The colony was well integrated, generating extrazooidal feeding currents, which flowed radially inwards and discharged upwards above the colony centre.

Stomatopora has been interpreted as an **opportunistic** species capable of rapidly colonizing new areas. The advantage of its uniserial dichotomous growth is that the colony could spread quickly from its point of origin. Some parts of the expanding colony might therefore be able to locate safe 'refuges', e.g. recesses in the substrate, where they would be protected from predation or other hazards. *Berenicea,* on the other hand, although more slowly growing, used space more economically and was much more fully integrated. *Berenicea* normally replaced *Stomatopora* in ecological succession and, competing more effectively for substrate space, is regarded as an **equilibrium** species. In evolutionary terms, therefore, *Stomatopora* is an *r*-strategist, whereas *Berenicea* is a *K*-strategist.

The relative numerical decline of the Cyclostomata in the later Cretaceous and Tertiary relates to the great contemporaneous expansion of the Cheilostomata, the last and perhaps the most successful of the bryozoan orders to arise. They probably originated not from cyclostomes but from ctenostomes, which were then in existence but whose fossil record is poor since they are always uncalcified.

Order Cheilostomata is divided by Ryland into four suborders: Ascophora, Anasca, Gymnocystidea and Cribrimorpha. Ascophoran structure has been described above in relation to *Smittina,* though within this suborder there is much structural variation in the position of the ovicell and the orifice and in the sculpture of the upper surface.

The Anasca lack the ascus, and the polypide extrudes through the action of internal muscles on the flexible frontal membrane. Both the Anasca and the Ascophora are very important today (Ryland and Hayward, 1977; Hayward and Ryland, 1979). Cribrimorpha (Fig. 6.9), however, after a brief though substantial expansion in the late Cretaceous (e.g. Larwood, 1962) have now greatly declined. These are usually unilaminar encrusting forms with calcified side walls, though the primary frontal wall in which the aperture lies is of chitin. This membranous wall is overarched by calcareous **costae** or ribs which form a secondary frontal wall and meet in the midline. These fuse, making a kind of porous cage over the primary wall – perhaps the most elaborate wall structures ever evolved in the bryozoans. The apertural region is usually protected by a semicircle of oral spines.

Cheilostomes, by their marked polymorphism of the avicularia and vibracula and by the connections between zooids, express a high degree of integration, which may have involved some modification of the structure of the colony to make maximum effective use of the feeding currents.

The geological history of bryozoans is incompletely known because of the poor record of the non-calcified forms. Nevertheless, it is clear that particular groups were dominant at certain times (Fig. 6.7). In the early Palaeozoic the trepostomes and cryptostomes were especially important, whereas the acme of the fenestrates was in the late Palaeozoic. After the Permian extinctions the cyclostomes, which had been present throughout the Palaeozoic, expanded vastly in the Jurassic and late Cretaceous, declining in relative abundance only when the cheilostomes arose to become the dominant bryozoans of the latest Cretaceous and Tertiary. In Recent seas they are perhaps the most numerous lophophorates. In each of these groups the degree of colonial integration varies. It is claimed indeed that the colony rather than the zooid has been the unit of natural selection. Thus in Recent seas the specialization of avicularia and other polymorphs can only be seen in terms of benefit to the colony as a whole. In addition, polymorphism occurs in 75% of all living cheilostomes, especially in those species living in predictable environments where sufficient food resources would allow the 'luxury' of non-feeding zooids.

Several trends, indicating progressive adaptive evolution in bryozoans, have been defined

(McKinney and Jackson, 1989). Examples are increased calcification of zooids, increased integration of zooids in cheilostomes, and a shift of rigidly erect species into deeper water. Such long-term, often polyphyletic adaptations can usually be related to the improvement of the colony as a living mechanism, or to defend itself against predators.

6.5 Ecology and distribution

Bryozoans are abundant in all oceans with a maximum in the western Pacific. They are found at all depths from the shoreline down to the abyssal zone – the deepest record is over 8500 m – but they decrease in numbers and importance in the fauna with depth.

The controls of distribution are as follows.

Temperature, for though a few species are eurythermal most have restricted temperature ranges.

Wave action, since the colonies are liable to damage.

The availability of a hard substrate upon which the larvae can settle. This is especially important in limiting depth range, for deeper sea sediments are much finer than those of the continental shelves and deep-sea oozes offer little prospect of a firm anchorage. There are, however, some gymnolaemates which can settle on fine particles, and anchor themselves by roots which penetrate the substratum.

Salinity, which is a fairly important control. For example, since waters off large river mouths have reduced salinity as well as much suspended sediment, few bryozoans are found there. Even so, there are a few euryhaline bryozoans which can withstand salinities of only 20 per mille, including the ubiquitous *Bowerbankia*.

Shallow-water bryozoans

Relatively few Recent bryozoans are intertidal since the high environmental energy and the problems of desiccation between tides are too great for such delicate organisms. The sublittoral zone, however, has a wealth of bryozoan colonies. They feed largely on the abundant phytoplankton of this zone and are especially common at depths between 20 and 80 m. The depth ranges of many bryozoan species from shallower waters are well known and they tend to form characteristic associations with other organisms, some of which have been remarkably persistent through time (Hayward *et al.*, 1994; Smith, 1995).

Bryozoans may be encrusters, of erect colony form, free living, or may live rooted in soft sediment. Such growth forms are a reflection of adaptive strategies, and as McKinney and Jackson (1989) comment, 'growth pattern and form of a bryozoan colony is an expression of its ecological niche more than its phylogenetic history'.

Encrusting forms tend to be common in shallow waters, owing to the availability of suitable substrates, but are more common on ephemeral than long-lasting substrates, e.g. dead shells, and seaweeds. Although they do occur also on stable substrates, they are usually outlasted by dominant competitors. Some Mesozoic species were able to produce frontal or peripheral subcolonies, and most, though not all, seem to have reproduced once only, and died soon afterwards (McKinney and Taylor, 1997) They may be runners (cf. the Jurassic *Stomatopora*), or form sheets consisting of a single layer or several layers; the latter kind may form mounds and thick sheets. Multilayered sheets are common on substrates of limited dimensions. Some Jurassic encrusting bryozoans are known to have made the best use of limited substrates by repeated overgrowth of old zooecia by young autozooecia, orientated in the same direction. During such 'multilamellar' growth (Taylor, 1976), new zooecia develop along a characteristic C-shaped growth margin (Fig. 6.12). The two ends of the 'C' remained stationary while successive increments expanded away from them, retaining the 'C' shape. As the growth margins expanded, they usually came in contact with other C-shaped growth margins, of which there were many per colony, and the resultant interactions were often of complex form. The great advantage of this system is that the colony as a whole thus has a longer lifespan than normal, by repeated incrustation of the same area, whereas each zooid is allowed an equal length of life before being covered by the next layer.

Another kind of multilamellar growth has been shown in the Mesozoic tubular bryozoan *Reptomulticava* (Nye and Lemone, 1978). Here a parental zooecium gave rise to two or more new intrazooecial buds (founding zooecia). These then coalesced as they grew up and over the apertural rims of the parent zooecia and merged with adjacent founding zooecia. Thus there originated a new layer of overgrowths which, as in the genera mentioned previously, reused existing space with a new layer of living zooecia.

Erect bryozoans may form unilaminar or bilaminar sheets, or arborescent colonies where the growing tips of the branches have repeatedly divided to produce a tree-like form (adeoniform when bilaminar). There are several advantages in erect growth, including the increase of tissue area relative to the substrata, the raising of the colony above the sea floor giving improved access to feeding grounds above the sea floor, and protection from predators and swamping by sediment. Of the erect forms found in shallow waters, many, e.g. the bilaminar *Flustra*, have flexible zoaria which are better able to withstand current actions. Stoutly constructed erect genera can withstand moderate current strengths and are able to absorb the stresses, while the more delicately branched forms prefer quiet waters. McKinney and Jackson (1989) discuss various adaptive strategies in erect bryozoans with respect to problems of breakage and flow.

A free-living habit has been common in bryozoans since the Ordovician, and while many of the 'stony' trepostomes, for example, may have rolled on the sea floor, other, large colonies lay passively. Since the late Cretaceous, however, many kinds of small discoidal or cap-shaped free-living colonies have arisen. These are the dominant free-living forms today, successfully colonizing loose, moving rippled sands.

One such species of particular ecological interest is the widespread anascan cheilostome *Cupuladria canariensis* (Fig. 6.3a,b), for its habitat and distribution in Recent seas and from the Miocene onwards has been very thoroughly researched (Lagaaij, 1963). It lives at depths between about 5 and 500 m, but it is most common on continental shelves on a sand substrate in the Atlantic and East Pacific. It can tolerate temperatures between 12 and 31°C, though it is normally confined within the 14°C isocryme and to salinities between 27 and 37 per mille.

The larvae have strong sediment preferences, so that colonies are always found where the particles (e.g. quartz grains, foraminiferids, glauconite pellets and shell fragments) are large enough to permit settlement, but not too large. *Cupuladria* (Fig. 6.3) can tolerate a certain amount of mud since the vibracula whip constantly and prevent the settlement of clay particles.

Colonies are always lunulitiform, i.e. have a concavo-convex lensoidal form, and rest on the bottom raised on vibracular setae. Many lunulitiform types (such as the related *Selenaria,* though not *Cupuladria)* can 'walk' on these setae across coarse unconsolidated sediments. Since the temperature limits of Recent colonies are well known, and assuming no change of habitat preference through time, it has been possible to use *C. canariensis* as a good palaeotemperature gauge. Since it is common in Miocene and Pliocene sediments of the North Sea Basin, the temperature of the water during deposition of these sediments must have been at least 9°C higher than it is today. The first appearance of this species is a good stratigraphical marker for the Lower Miocene, and on this criterion several suites of Tertiary sediments have been assigned to their correct system.

Rooted bryozoans (most of which are cheilostomes) are erect colonies attached by long tubes from the proximal end to grains of sediment. Some of these (e.g. *Sphaeropora*) project above the sediment surface like small fungi, though some may reach 15 cm in height. Rooted bryozoans are common in soft sediments in deep waters (>1000 m).

It may be possible to infer depth relationships in fossil bryozoan-bearing assemblages from colony form, but only in the broadest and most general sense and in conjunction with other criteria, since by itself the shape of a colony is not an unequivocal palaeoecological indicator.

Reef-dwelling bryozoans

Bryozoans have been quite significant as frame builders or as sediment binders in various kinds of reefs through geological time. In Ordovician to Devonian coral–stromatoporoid reefs they formed a subordinate part of the reef fabric and assisted in binding the sediment. Commonly they bridged gaps and allowed cavities to form below them, these often becoming filled with fine sediment. There is some evidence of vertical zonation of bryozoans in some Silurian reefs.

In the large Permo-Carboniferous algal reefs they

contribute in a minor way, or on a localized scale sometimes more importantly, to the reef framework. Fenestellids have been found in life position, projecting outwards from a steeply dipping reef face. They also occur in patch reefs as frame builders.

Modern coral reefs may carry abundant faunas of encrusting bryozoans, which are in places significant frame builders. Strongly built, thick-walled encrusters are found in regions of turbulent water, whereas the more delicate cribrimorphs live in sheltered cavities. The larvae of these types have strong habitat preferences and will only settle on particular substrates, usually coralline algae or dead skeletal material. Water turbulence seems to be the primary control of distribution within the multifarious habitats provided by the coral reef and bryozoans of various types, though almost all encrusters will flourish almost anywhere that is free of suspended sediment. Most species seem to have particular functional adaptations for such habitats, though many of these have not been investigated in detail.

Deep-water bryozoans

Most of the abyssal and bathyal Bryozoa of Recent seas are cheilostomes. Many species taken from depths of over 1000 m were attached to shells, pebbles and other hard surfaces, and they have been found at some 25% of deep-sea stations. Rooted species, with long root-like threads capable of holding them securely in soft sediment, are now known to be very abundant and diverse in deep waters, and are dominant below 1000 m.

6.6 Stratigraphical use

Since Palaeozoic bryozoan genera tend to be long ranged and facies controlled, their stratigraphical applications are usually poor. Species assemblages within given facies are regionally very useful for zonal purposes, however, especially in widespread carbonate shelf sediments. Some of the Cretaceous and Tertiary bryozoans seem to have limited vertical distribution, and as their time ranges become better known so their stratigraphical potential increases.

Bibliography

Books, treatises and symposia

Bigey, F.P. (ed.) (1991) Bryozoaires actuels et fossiles. *Bulletin de la Societé Scientifique d'histoire Naturelle de l'ouest de la France*, Mémoire 1. (55 symposium papers)

Busk, G. (1859) A Monograph of the Fossil Polyzoa of the Crag. Palaeontographical Society, London. (Early description of a bryozoan-rich bed)

Hayward, P.J. and Ryland, J.S. (1979) *British Ascophoran Bryozoans*,

Synopses of the British Fauna, Vol. 14. Academic Press, London. (Excellent illustrations)

Hayward, P.J., Ryland, J.S. and Taylor, P.D. (eds) (1994) *Biology and Palaeobiology of Bryozoans*. Olsen & Olsen, Fredensborg. (Many valuable papers)

Larwood, G.P. (ed.) (1973) *Living and Fossil Bryozoans*. Academic Press, New York. (Many original papers; a standard reference work)

Larwood, G.P. and Abbott, M.B. (1979) *Advances in Bryozoology*, Systematics Association Special Volume No. 13. Academic Press, London. (Many papers)

Larwood, G.P. and Neilsen, C. (eds) (1981) *Recent and Fossil Bryozoans*. Olsen & Olsen, Fredensborg. (Many valuable papers)

Larwood, G.P. and Rosen, B.R. (eds) (1979) *Biology and Systematics of Colonial Organisms*. Academic Press, London. (Bryozoans, corals, graptolite, etc.)

McKinney, F.K. and Jackson, J.B.C. (1989) *Bryozoan Evolution* (eds C.T. Scrutton and C.P. Hughes), *Special Topics in Palaeontology 2*. Unwin Hyman, London. (The best synthesis available: excellent value)

Moore, R.C. (ed.) (1953) *Treatise on Invertebrate Paleontology*, Part G, *Bryozoa*. Geological Society of America and University of Kansas Press, Lawrence, Kan.

Robison, R. A. (ed.) (1983) *Treatise on Invertebrate Paleontology*, Part G (revised) *Bryozoa, Vol. 1, Introduction, Order Cystoporata, Order Cyptostomata*. Geological Society of America and University of Kansas Press, Lawrence, Kan.

Ryland, J.S. (1970) Bryozoans. Hutchinson, London. (Invaluable account of Recent and fossil bryozoans)

Ryland, J.S. and Hayward, P.J. (1977) *British Anascan Bryozoans*, Synopses of the British Fauna, Vol. 10. Academic Press, London. (Excellent illustrations)

Woollacott, R.M. and Zimmer, P.L. (eds) (1977) *Biology of Bryozoans*. Academic Press, New York. (Numerous papers)

Individual papers and other references

Balson, P.S. and Taylor, P.D. (1982) Palaeobiology and systematics of large cyclostome bryozoans from the Pliocene Coralline Crag of Suffolk. *Palaeontology* **25**, 529–54. (Well worth reading)

Banta, W.C., McKinney, F.K. and Zimmer, P.L. (1974) Bryozoan monticules: excurrent water outlets? *Science* **185**, 783–4 (Important concept)

Blake, D. (1980) Homeomorphy in Palaeozoic bryozoans: a search for explanations. *Palaeobiology* **6**, 451–65.

Cowen, R. and Rider, J. (1972) Functional analysis of fenestellid bryozoan colonies *Lethaia* **5**, 147–64. (See text)

Lagaaij, R. (1963) *Cupuladria canariensis* (Busk): portrait of a bryozoan. *Palaeontology* **6**, 172–217. (Morphology, distribution and eco1ogy)

Larwood, G.P. (1962) The morphology and systematics of some Cretaceous cribrimorph Polyzoa (Pelmatoporinae). *Bulletin of the British Museum of Natural History* **6**, 1–281. (Excellent illustrations)

McKinney, F.K. (1977) Functional interpretation of lyre-shaped bryozoans. *Paleobiology* **3**, 90–7.

McKinney, F.K. (1983) Asexual colony multiplication by fragmentation: an important mode of genetic longevity in the Carboniferous bryozoan *Archimedes*. *Paleobiology* **9**, 35–43.

McKinney, F.K, and Taylor, P.D. (1997) Life histories of some Mesozoic encrusting bryozoans. *Palaeontology* **40,** 515–56.

Nye, O.B. and Lemone, D.V. (1978) Multilaminar growth in *Reptomulticava texana,* a new species of cyclostome bryozoa. *Journal of Paleontology* **52**, 830–45.

Schopf, T.J.M. (1977) Patterns and theories of evolution among the Bryozoa, in *Patterns of Evolution* (ed. A. Hallam), Elsevier, Amsterdam, pp. 159–207.

Smith, A.M. (1995) Palaeoenvironmental analysis using bryozoans: a review, in *Marine Palaeonvironmental Analysis from the Fossil Record* (eds D.W.J. Bosence, D.W.J and P. Allison), Geological Society of London Special Publication No. 83, pp. 231–44.

Taylor, P.D. (1976) Multilamellar growth in two Jurassic cyclostomatous bryozoa. *Palaeontology* **19**, 298–306.

Taylor, P.D. (1978) Functional significance of contrasting colony form in two Mesozoic encrusting bryozoans. *Palaeogeography, Palaeoclimatology, Palaeoecology* **26**, 151–8.

Taylor, P.D. (1979) The inference of extrazooidal feeding currents in fossil bryozoan colonies. *Lethaia* **12**, 47–56.

Taylor, P.D. and Badve, R.M. (1995) A new cheilostome bryozoan from the Cretaceous of India and Europe, a cyclostome homeomorph. *Palaeontology* **38**, 627–58.

7 Brachiopods

7.1 Introduction

Brachiopods are benthic marine animals whose soft parts are enclosed within a two-valved shell. They have some resemblance to bivalves in that they possess a hinged pair of **valves** and feed by drawing water into the shell and filtering off the food particles, but zoologically they are quite separate. The two valves are of different sizes but symmetrical about a **median plane**, by contrast with the equal-sized but inequilateral valves of the bivalves. Brachiopods are first found in the Cambrian and are very abundant in the fossil record, often being the commonest and most ubiquitous fossils in any shallow-water deposit. In the Palaeozoic they were a very important phylum and, though they were decimated in the Permian, some genera continued as the dominant benthos in localized areas during the Mesozoic where they may be vastly abundant, and by far the commonest fossils. Although their importance has since declined, they are, however, commoner today than formerly thought, especially in deep and cold waters. This may represent a real shift in environment through time.

The shallow waters off New Zealand contain abundant brachiopods (12 species of five genera), and this is the classic area for research on living brachiopods. There are, however, more diverse faunas elsewhere; the British Isles have 21 species of 17 genera (Brunton and Curry, 1979), though these are normally found in deeper waters and are hence rarely seen.

Over 4500 fossil genera are now known, though no more than 100 are living today, albeit these are widely distributed and found at all depths in the sea. Many of the fossil genera have been found to be stratigraphically useful at various horizons.

7.2 Morphology

Certain fundamental characters are common to all brachiopods. They all have a shell of two valves, usually fixed to the sea floor by means of a stalk or **pedicle** (though some are cemented, attached by spines or free-lying), and a complex food-gathering organ called the **lophophore**. Until recently, (and as in the third edition of this book), brachiopods were grouped in three classes: Lingulata, Inarticulata and Articulata. These are broadly equivalent to the present Subphyla Linguliformea, Craniiformea and Rhynchonelliformea, respectively, in the new classification of Williams *et al.* (1996). This has been established on a cladistic basis and tested against the stratigraphic record. While the original terms are now superseded, it is still proper to refer colloquially to 'inarticulated brachiopods' for the first two subphyla, and 'articulated brachiopods' for the third. A detailed classification is set out in section 7.4.

SUBPHYLUM 1. LINGULIFORMEA (L. Cam.–Rec.): Brachiopods in which the valves are not hinged by teeth and sockets and in which the shell is chitinophosphatic in composition. The pedicle is fleshy and muscular, and emerges between the valves or from the apex of one of the valves.

SUBPHYLUM 2. CRANIIFORMEA (L. Cam.–Rec.): Brachiopods likewise lacking teeth and sockets, but with a calcareous shell. Pedicle reduced or absent.

SUBPHYLUM 3. RHYNCHONELLIFORMEA (L. Cam.–Rec.): Calcareous-shelled brachiopods in which the valves are hinged by teeth in one valve and sockets in the other. The pedicle is made of a dead horny material and in some fossil genera appears to have atrophied. The rhynchonelliformes are much more diverse and abundant than the linguliformes or craniiformes. They appear at the same level in the Lower Cambrian; however, the main early radiation of the rhynchonelliformes was not until the early Ordovician.

Subphylum Rhynchonelliformea

Morphology of three genera

Magellania

The recent brachiopod *Magellania* (Order Terebratulida, Suborder Terebratellidina; Figs 7.1, 7.2) has been well chosen by many authors as a typical example of a modern rhynchonelliform brachiopod.

It shows clearly how the hard parts relate to the living anatomy, which is useful in interpreting the ubiquitous terebratulides of the Mesozoic.

Magellania lives epifaunally, attached by its pedicle

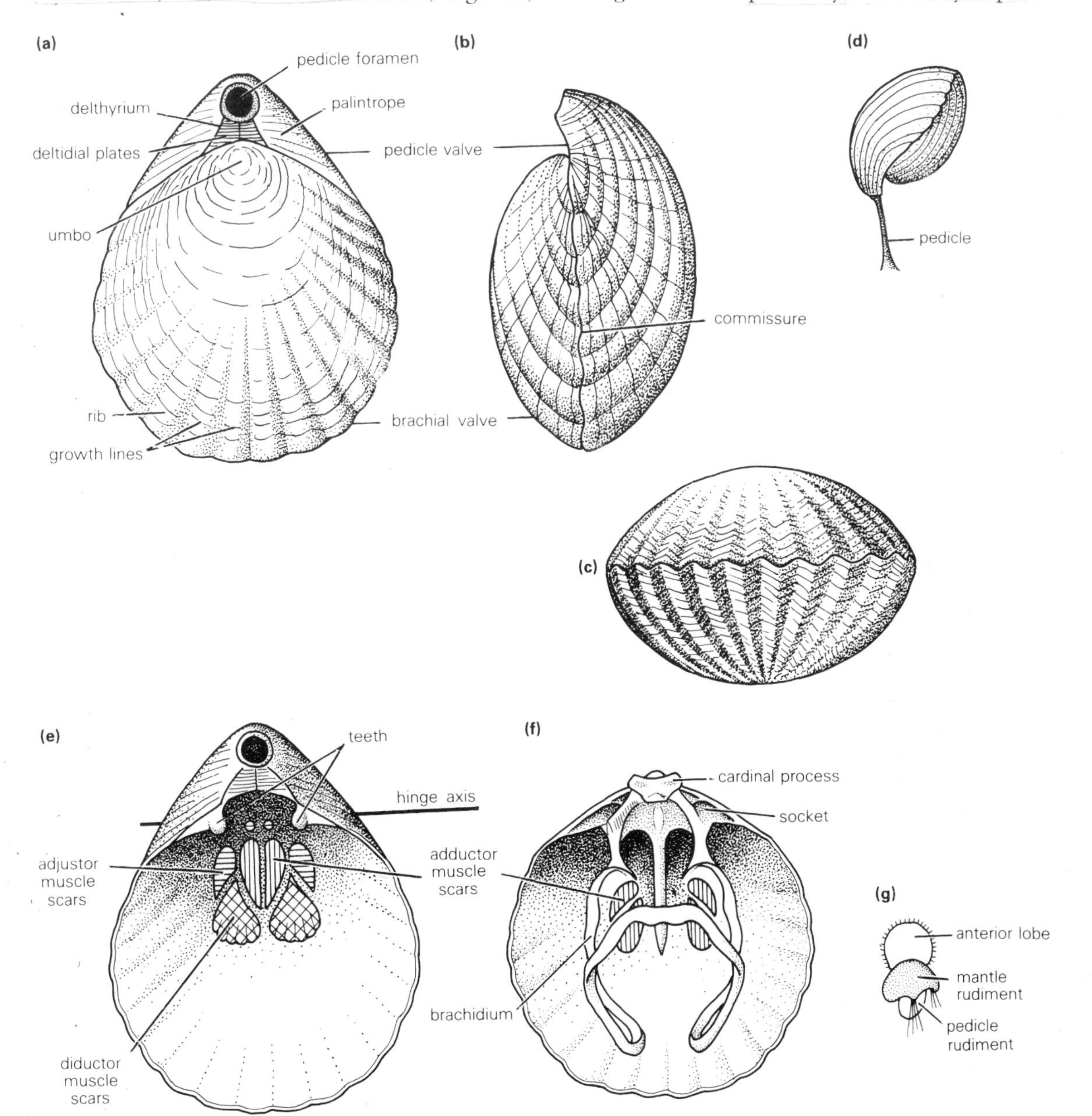

Figure 7.1 *Magellania flavescens*: (a) upper surface with brachial valve (× 2 approx.); (b) lateral view (× 2 approx.; (c) anterior view (× 2 approx.); (d) in life position, showing pedicle attachment; (e) internal view of pedicle valve (× 2 approx.); (f) internal view of brachial valve (× 2 approx.); (g) larva. [(a)–(f) Based on Davidson, 1851; (g) based on illustration in Percival, 1944.)

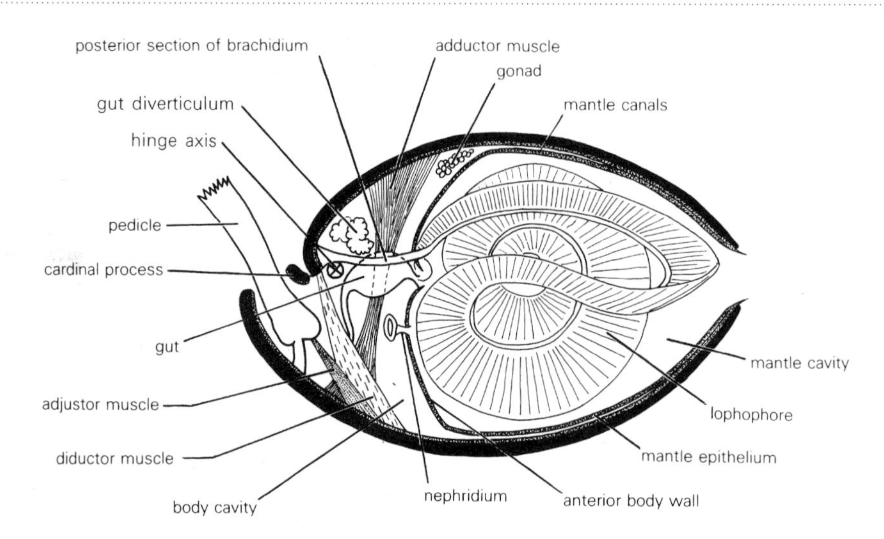

Figure 7.2 Magellania flavescens: median section, somewhat stylized.

to rocks and stones at depths of 12–600 m in Australian and Antarctic waters.

The two oval calcareous valves differ in size but are equilateral and divided by a single plane of symmetry. In standard orientation the 'upper' valve is smaller than the 'lower'. There has been some controversy as to what these two valves should be called; the 'upper' valve is variously known as the **brachial** or **dorsal valve**, the 'lower' as the **pedicle** or **ventral valve**. The terms 'brachial' and 'pedicle' refer to structures contained within the valves and thus are more technically correct. Furthermore, the conventional dorsal or ventral orientation is not necessarily a life orientation; in fact the brachiopod, when fixed by its pedicle, has both valves vertically held or may have the pedicle valve uppermost. Hence the terms 'brachial' and 'pedicle' are adopted here as standard practice to designate the valves though the other system is admittedly simpler.

EXTERNAL MORPHOLOGY

The superficial resemblances of such brachiopods as *Magellania to* Roman oil lamps has long been drawn, hence the vernacular term 'lamp shells'. The two valves are unequally biconvex. Each grows from the first-formed part, the **protegulum**, which later becomes part of the umbo. It grows in a logarithmic spiral, as does the shell of a bivalve, and at each growth increment new shell material is accreted round the growing edge. The valves meet along a line of junction or **commissure**. In conventional orientation the umbones are located at the **posterior** end of the brachiopod and, since the **hinge line** is in front of them, the valves gape at the **anterior** end.

Each valve is externally sculptured with faint **ribs** radiating from the umbo. These become stronger anteriorly so that the anterior part of the commissure is crenulated where the ribs meet it. The **growth lines** are subconcentric round the umbo and likewise become rather crenulated anteriorly where the ribs are stronger. Each of these growth lines represents a record of the former position of the edge of the shell, and since they are especially prominent where growth has ceased for a short time the shell retains a permanent record of its own ontogeny.

The pedicle valve is the larger of the two and is inturned posteriorly so that the most convex part of it can be seen from the upper surface. This part of the shell is the **palintrope**. The umbo is perforated by a round **pedicle foramen** through which the pedicle emerges. The latter is a stalk of horny tissue, cemented at its distal end to rocks or pebbles, and by it the brachiopod is raised above the bottom. Between the umbo and the inner edge of the valve is a triangular cavity, the **delthyrium**, closed by a pair of **deltidial plates**. They are marked by straight, parallel growth lines, transverse to the brachiopod's plane of symmetry.

Though the commissures of the two valves fit exactly nearly all the way round the shell, the smaller brachial valve has its umbo tucked just into the pedicle valve anterior to the delthyrium. Through this region runs the transverse **hinge axis** operating on internal teeth (pedicle valve) and sockets (brachial valve).

INTERNAL ANATOMY

In Fig. 7.2 the internal organs are displayed in a section cut through the plane of symmetry. These organs relate to structures made of calcium carbonate which can be seen in the separated valves. The shell is secreted by a **cellular epithelium** which secretes calcareous material mainly at the valve margin. This epithelium underlies every part of the shell. It is two layers thick in the anterior part of the brachiopod, where the inner layer forms the **mantle** enclosing a large **mantle cavity**. Towards the posterior end of the shell the mantles of the two valves abruptly leave the internal valve surface and join, forming a single sheet of tissue, the **anterior body wall**, which crosses between the valves and separates the mantle cavity from the posterior **body cavity**. The only organ contained within the mantle cavity is the lophophore, which itself is invested with a continuation of the mantle epithelium. This is the main food-gathering and respiratory mechanism of the brachiopod. In *Magellania* the lophophore has a hydrostatically supported fluid-filled canal (the **brachial canal**) as its axis. This is supported by a long, loop-shaped calcareous ribbon (**brachidium**) which is symmetrical about the median plane and attached at two points to the inside of the brachial valve at the rear of the mantle cavity. The brachidium may remain attached to the shell after death when the investing tissue has rotted away, but more often it is not preserved. From the strong brachidium-supported axis of the lophophore there spring a large number of slender parallel filaments. These are sticky and lined with cilia. The beating of these cilia generates the currents that bring in and exhale water. Inhalant currents come in laterally, and all the filaments are arranged so as to provide an effective net for trapping food particles; they are strained off while the filtered water is exhaled anteriorly in a single stream. The food particles caught are passed down the filament in a mucus belt to a food groove running along the brachial axis and so to the mouth.

The mouth is situated in the anterior body wall; it leads to a small gut with an oesophagus, stomach and blind-ended intestine, but no anus. There are also digestive glands or diverticulae associated with the gut (cf. Fig. 7.3c).

Excreta are voided through the mouth into the mantle cavity and disposed of by the exhalant current. There are no other perforations in the anterior body wall other than the paired funnel-shaped **nephridia** ('kidneys'), which remove nitrogenous waste taken up by wandering cells (**coelomocytes**) and also act as passages for the escape of gametes from the gonads. Sometimes the latter become so swollen with eggs or sperm as the breeding season approaches that they can expand into the **mantle canals**. These canals are tubular branching extensions of the body cavity (coelom) which run between the mantle and the inner epithelium. Within them coelomic fluid circulates, being used mainly for respiration. In dead specimens of *Magellania*, divested of soft tissue, the scars of these canals are clearly seen on the inside of the shell.

The only other organs found in the body cavity are the pedicle and related structures, and the muscles for opening and closing the shell. The pedicle is a cylindrical stalk having a thick external cuticle, within which is a thin epithelium and a central core of connective tissue. Though the shell is fixed to the pedicle, it can be turned in any direction by two sets of **adjustor muscles** near its base. These allow the upright brachiopod to swing into or away from the current. Frequently the pedicle splits into several strands for attachment (cf. Fig. 7.3e).

Brachiopods open and close the shell using two sets of paired muscles. The **adductor muscles** which close the shell are analogous, though not homologous, to the adductors of bivalves. They join the two valves somewhat obliquely, and their points of attachment are both anterior to the hinge axis, so that when they contract the shell must close. By contrast, the **diductor muscles** which open the shell have a quite different line of action, with their bases on opposite sides of the hinge. They are attached to the pedicle valve just outside the adductors, but they are fixed to the brachial valve by a calcareous boss, the **cardinal process**, on the posterior side of the hinge axis. If the diductor muscles contract while the adductors correspondingly relax, the cardinal process swings downwards, and as the rest of the shell lies anterior to the hinge

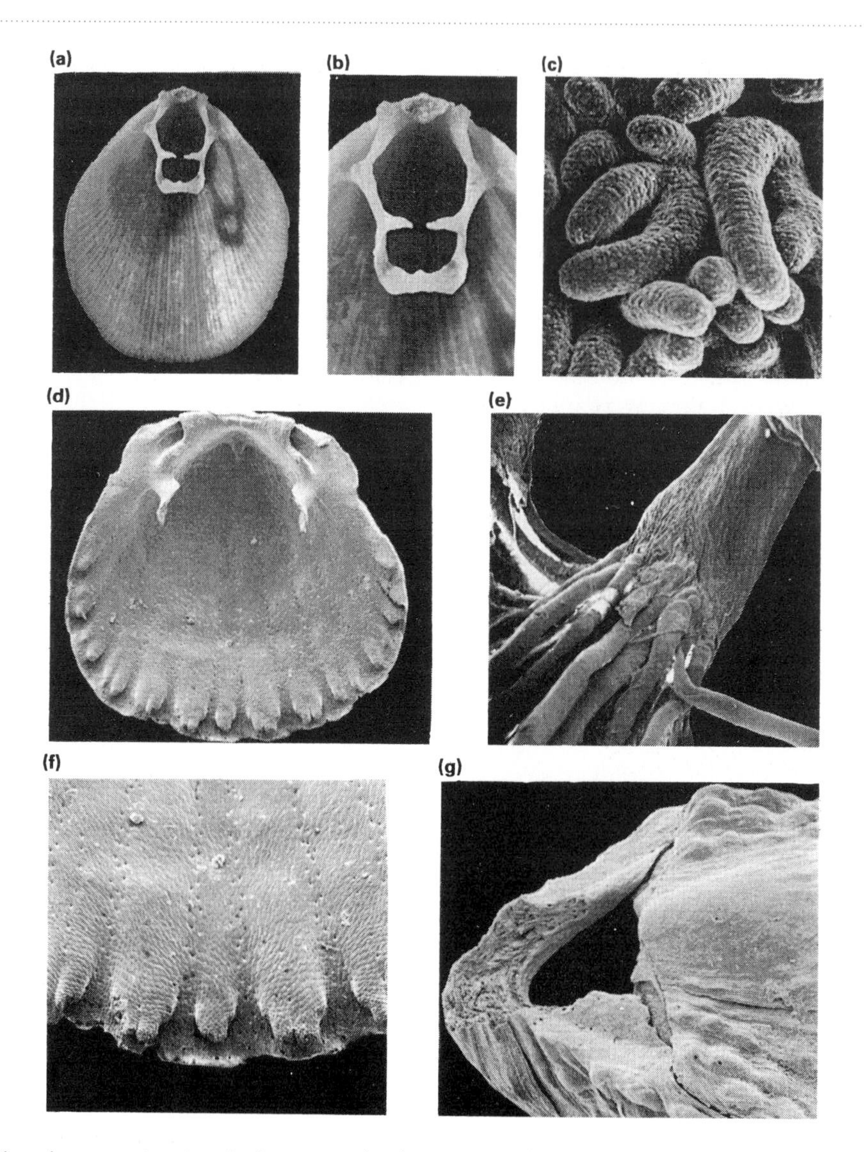

Figure 7.3 *Terebratulina retusa* (Rec.), Firth of Lorne, Scotland: (a) interior of adult (18 mm long) brachial valve (× 3); (b) enlargement of lophophore (× 6): (c) digestive diverticulae (× 100); (d) juvenile brachial valve showing alignment of punctae (× 25): (e) pedicle showing separate attachment strands (× 75); (f) enlargement of growing edge (× 50); (g) lateral view of pedicle foramen in complete juvenile shell (× 70). (Photographs reproduced by courtesy of Dr Gordon B. Curry.)

the valves gape open by a few degrees. Feeding may then begin.

These two sets of muscles, and also the adjustors, leave well-defined **muscle scars** inside the valves, visible in isolated valves of dead individuals and sometimes in fossil valves also. Of the scars in the pedicle valve, the larger outer pair are diductor scars, and the smaller inner pair the adductor scars. There are likewise two main pairs of muscle scars in the brachial valve, but these all belong to the adductor muscles since the latter divide in two.

Separated valves from which the soft tissues have been removed thus show the following characters.

Brachial (dorsal) valve. Sockets, cardinal process, adductor muscle scars, brachidium, mantle canals; there may also be a **median septum** or partition in the plane of symmetry.

Pedicle (ventral) valve. Palintrope, teeth, pedicle foramen, delthyrium closed by deltidial plates, adductor and diductor muscle scars, adjustor scars and mantle canals.

The delthyrium, incidentally, is the ancestral site of the pedicle foramen. In brachiopods that retain the more 'primitive system' the delthyrium is open and through this the pedicle emerges. In the early ontogeny of *Magellania* the pedicle is first located in an open delthyrium, but the foramen gradually migrates to its adult position as the brachiopod grows, and the delthyrium closes off behind it by the growth of the deltidial plates. The shell and associated structures in the Recent *Terebratulina*, similar in many ways to *Magellania*, though with a shorter brachidium, are shown in Fig. 7.3.

Magellania is an example of a brachiopod with a **non-strophic shell**. The hinge axis passes through the teeth and sockets, which are its only fulcra. The shell edges adjacent to this are curved and not coincident with the hinge line. Most living brachiopods (e.g. the Terebratulida, Terebratellida and Rhynchonellida) are of this type. Many groups of Palaeozoic brachiopods (e.g. the Orthida and Spiriferida), however, have **strophic shells**. In these the hinge line is straight, often extends the full width of the shell and is coincident with the hinge axis.

Visbyella

Visbyella (Order Orthida, Suborder Orthidina) is a Silurian brachiopod with a strophic shell (Fig. 7.4).

It belongs to one of the earliest of all rhynchonelliform brachiopod stocks to have evolved, the

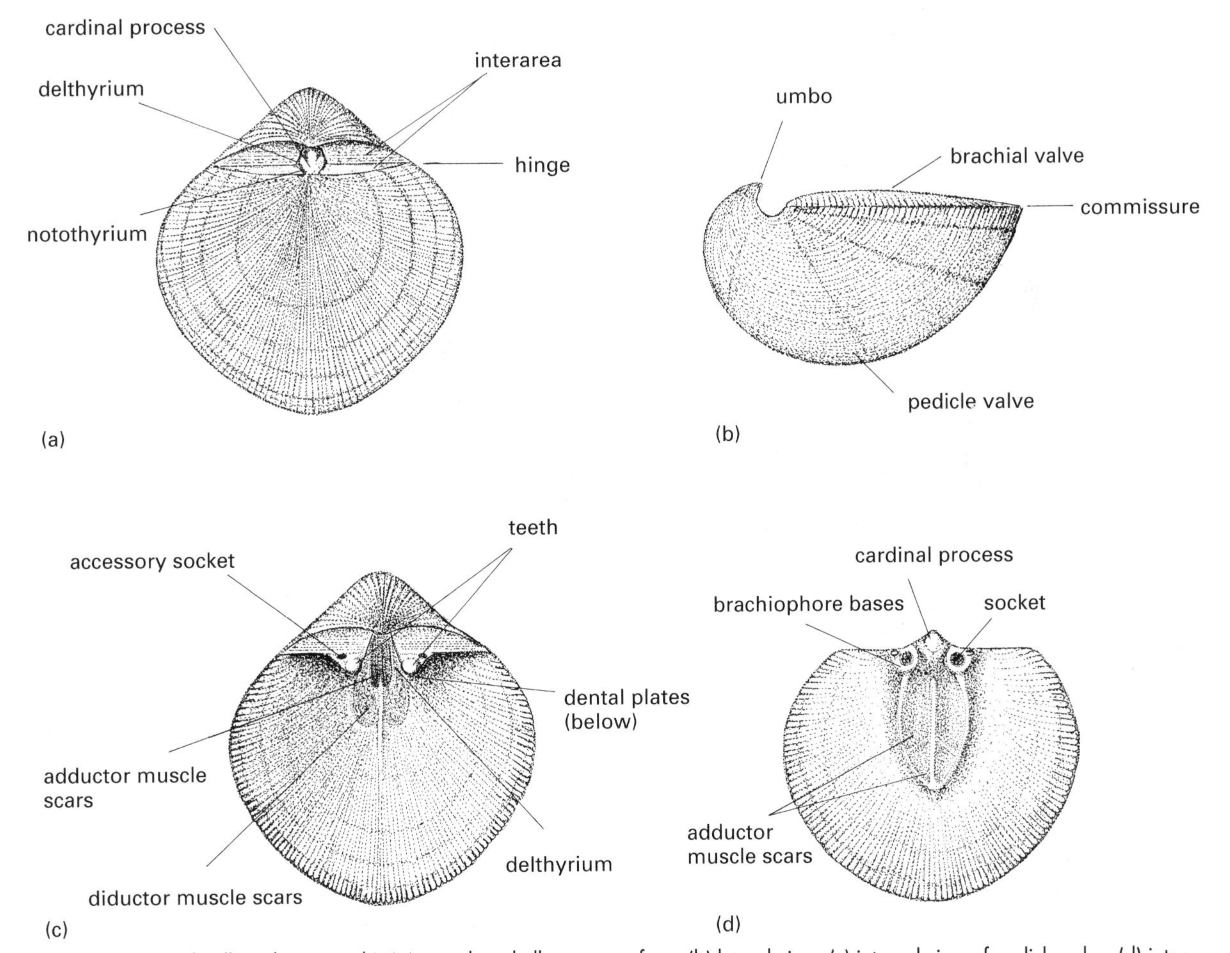

Figure 7.4 *Visbyella visbyensis* (Sil.): (a) complete shell, upper surface; (b) lateral view; (c) internal view of pedicle valve; (d) internal view of brachial valve (all $\times$ 5).

Orthida, first known from the Cambrian. The two valves are dissimilar in size and convexity. The large pedicle valve is very deep, the brachial only slightly plano-convex, but both have the same surface sculpture of fine ribs or **costellae** radiating from the umbo and curving outwards laterally to the commissure. A pronounced umbo characterizes the pedicle valve, anterior to which is the large open delthyrium without deltidial plates. This is flanked by a triangular **interarea**, which forms a flat, slightly sloping shelf whose anterior edge is the straight hinge line. The interarea has closely spaced growth lines parallel with the hinge. If the valves are separated, the large crenulate teeth are visible, situated anterior to the hinge line where the delthyrium reaches its greatest width. Joining the teeth to the floor of the pedicle valve are the vertical **dental plates**, which enclose a deep **umbonal cavity** continuous with the delthyrium. Flooring this cavity and extending some way anteriorly are the recessed muscle scars; the diductor scars are outside those of the adductors and the whole muscle field is heart-shaped. The adjustor scars are not recessed and are barely visible.

In the brachial valve there is an interarea, though it is narrower than that of the dorsal valve. Between them there is a small triangular opening, the **notothyrium**, which lies directly opposite the delthyrium so that the **pedicle opening** is diamond-shaped and is formed by the delthyrium and notothyrium together. A bilobed or trilobed **cardinal process** projects posteriorly and almost fills the notothyrium when the valves are closed. The adductor scars are very large. Supports for the lophophore are represented only by a pair of divergent **brachiophore bases**, which are simply a pair of oval knobs into which the sockets are recessed; nothing else is known about the lophophore. In both valves there are mantle canals of distinctive form. The costellae are more pronounced towards the commissure so that they interlock along a crenulated edge.

Eoplectodonta

Eoplectodonta (Order Strophomenida, Suborder Strophomenidina; Figs 7.5, 7.6) is a Silurian genus in which, as in many brachiopods of this order, the two valves are concavo-convex.

They fit inside one another, the brachial valve having a concave exterior, the pedicle valve being deeper and convex externally. The shell is strophic, and the hinge extends the full width of the semicircular shell, being bordered by very narrow interareas. There is a large cardinal process which blocks the pedicle opening when the valves are closed. Externally the valves have a sculpture of very thin ribs, with somewhat more pronounced single ribs spaced at intervals and dividing the shell into radial segments. There are no teeth. These have been lost in the evolutionary history of strophomenides but their function has been taken over by many small **denticles** which run along the hinge line, those of the two valves interlocking. The ventral muscle field has a bilobed diductor scar within which the smaller adductor scars are located. In the brachial valve there is a large V-shaped cardinal process which projects posteriorly carrying the point of attachment of the diductor muscles well to the rear of the hinge. This is necessary in view of the very narrow body space inside the shell and the restricted line of action of the muscles. A series of **vertical plates** supported on a raised platform (**bema**) hangs down from the roof of the brachial valve, almost touching the floor of the pedicle valve when the shell is closed (Fig. 7.5b,d,e). Presumably this structure was in some way connected with the lophophore which may have adhered to it. Lateral to this lie the adductor scars. An interesting feature of the inside of the shell are the numerous small projections (**taleolae**) which are calcite rods obliquely set in the shell and radially arranged parallel with the costellae. These are rod-like units lodged within the shell itself and projecting internally, and they are found only in the Strophomenida.

The delthyrium and notothyrium are largely blocked by the large cardinal process, and the small supra-apical foramen is often sealed. If there was a pedicle for this strange, thin shell, it cannot have had a supporting function and the apparent absence of adjustor muscle scars suggest that it was absent. Individuals must therefore have lain flat upon the sea floor (ambitopic habit), as is generally common in strophomenides, which lost their pedicles during ontogeny.

Preservation, study and classification of articulated brachiopods

The brachiopod species described above, one living and two fossil, belong to successful and important groups and give some indication of the variety

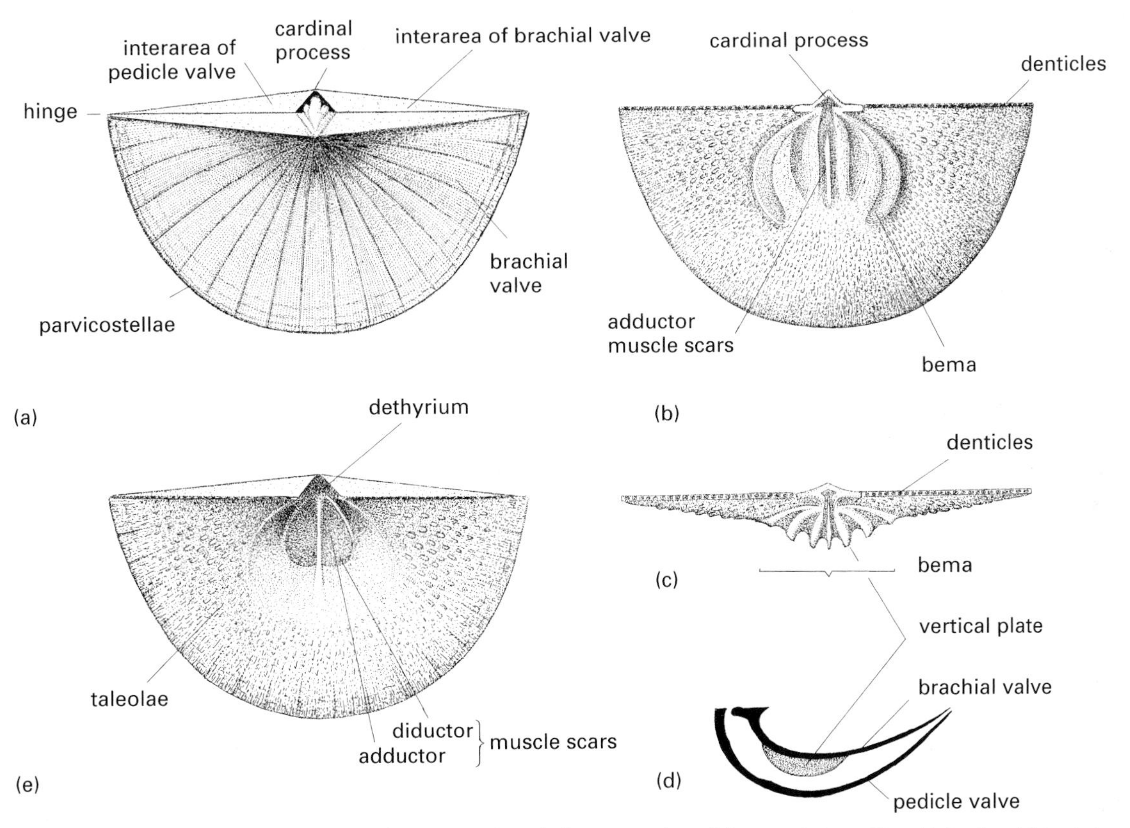

Figure 7.5 *Eoplectodonta penkillensis* (Sil.): (a) complete shell upper surface; (b) internal view of brachial valve; (c) posterior view of brachial valve; (d) vertical median section, showing disposition of vertical plates; (e) internal view of pedicle valve (all × 6)

of form and function in Subphylum Rhynchonelliformea. Yet they by no means show the full structural range, even though most brachiopods do not depart too radically from the kind of morphology shown in these. Nevertheless, there are some peculiar genera amongst the Brachiopoda, especially some Permian Productida which are so modified, at least externally, that they are hardly recognizable as brachiopods.

The majority of brachiopods can be assigned to their correct order and usually referred to their family on the basis of external morphology alone, but other features – the lophophore, muscle scars, hinge, dentition, cardinal processes, etc. – may have to be used in identifying a brachiopod at genus and species levels. Unfortunately the brachidium is found only in some of the many fossil groups. Amongst Palaeozoic orders the brachidium is

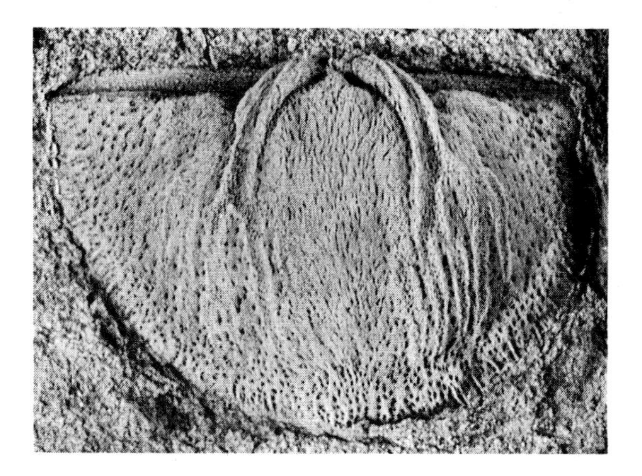

Figure 7.6 *Eoplectodonta penkillensis* (Sil.) from the Pentland Hills near Edinburgh: an internal mould showing the pedicle valve with mantle canals (× 8).

present only in the Spiriferida, Atrypida, Athyrida and Terebratulida, and possibly some Pentamerida (Fig. 7.7).

In these the brachidium may be preserved by mineral incrustation, if the shell is not infilled, or buried in matrix where it is. In the latter case the structure can be revealed by dissection or sectioning.

Many rhynchonelliform brachiopods are preserved with their shell material intact. Since the shell is calcitic it is not subject to the potential diagenetic transformation of an aragonitic shell. The external morphology of such shells is often very well preserved and, provided that they can be freed cleanly from the matrix, may readily be examined. Since it is very often the internal morphology that is really diagnostic, isolated valves are of particular value. If separated valves do not occur in the fauna, it may be necessary to undertake serial grinding in order to reveal internal structure. A parallel grinding machine that can grind to fixed distances is used for this, but each successive ground face must be photographed or drawn before regrinding. Though this is time consuming, such serial sections are often the only source of information on internal morphology. A complete internal reconstruction can be prepared from them by building a wax model or, more rapidly, by computer modelling.

If the brachiopod is preserved in a silty matrix but still retains its shell, the internal structures may be studied after removal of the shell by acid or by calcining, i.e. by burning off the outer shell so that the mould in matrix is visible.

Many Palaeozoic brachiopods and some more recent ones may be preserved in siltstone or mudstone from which all the calcareous material, including the shell, has been leached away by acidic groundwater. The resultant internal moulds often appear very different from moulds of the external surface. Hence in the *Eoplectodonta* specimens illustrated in Fig. 7.6 the umbonal cavity of the pedicle valve seems very pronounced, the mantle canal grooves are represented by ridges and the muscle fields are in negative relief. In the brachial valve of *Visbyella* (Fig. 7.4d) deep indentations would represent the brachiophore bases and the cardinal process. Much of the relief represents, of course, variations in thickness of the shell. The internal structures, such as the muscle scars, are often strikingly well preserved and can be studied with minimal preparation. It is in fact very often easier to work with and identify brachiopods from internal and external moulds than to use specimens with the calcitic shell preserved.

Furthermore, latex replication allows 'positives' to be taken which show all the features of shell morphology just as they were in life. Here a thin film of

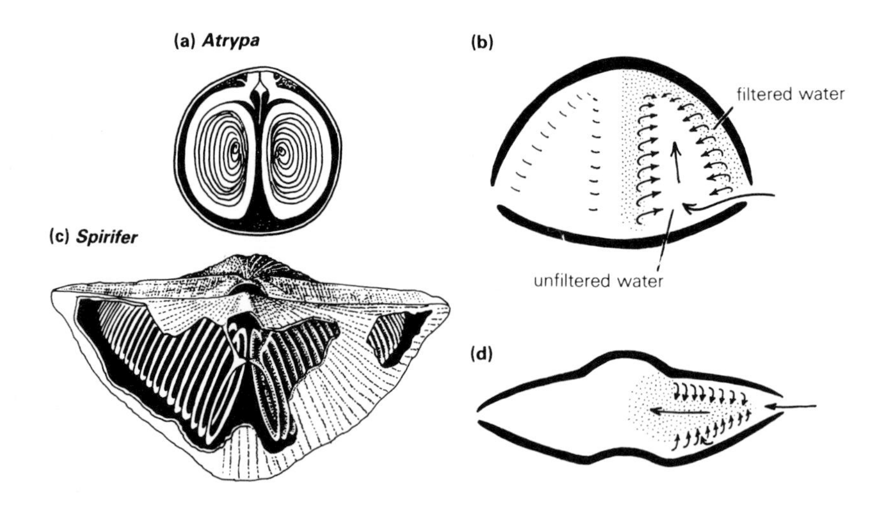

Figure 7.7 Spiral brachidia – their form and function: (a) *Atrypa reticularis*, interior of brachial valve from below (× 1.25); (b) same, vertical section showing spiral brachidia with unfiltered and filtered water; (c) *Spirifer striata*, silicified specimen with shell partially removed, showing spiral brachidia (× 1); (d) same, vertical section showing laterally directed spires with filtering system the reverse of that in *Atrypa*. [(a), (c) After Davidson, 1851; (b), (d) after Rudwick, 1970.]

rubber solution is poured on to a shell surface and allowed to dry; then a latex block is built up on top of this from successive thicker layers, and when fully dry may be stripped off. It is like producing a photographic print from a negative, and the result is normally just as clarifying.

Major features of brachiopod morphology

Form of shells

Brachiopod shells, as noted before, grow by accreting new material from the mantle at the valve edges. The ultimate form of any developing brachiopod shell is a product of the relative growth rates of the different parts of the valve edges (Rudwick, 1959). At any point of the valve margin, growth may be resolved into radial, anterior, lateral and vertical components. If all of these keep pace throughout development the result will be a **rectimarginate** shell, i.e. one with a planar commissure (Fig. 7.8f,g).

On the other hand, if the vertical component grows more rapidly than the others there will be a localized growth anomaly, resulting in a vertical deflection of the commissure. Whether the deflections are median, paired or serially arranged (Fig. 7.8h) depends on their locality, whereas their amplitude and whether they are sharp or gradual is a product of the rapidity of change of relative growth.

Relatively few shell shapes have actually been adopted by brachiopods, presumably for good functional reasons. Biconvex brachiopod shell shapes fall within a fairly narrow actual range, out of a much

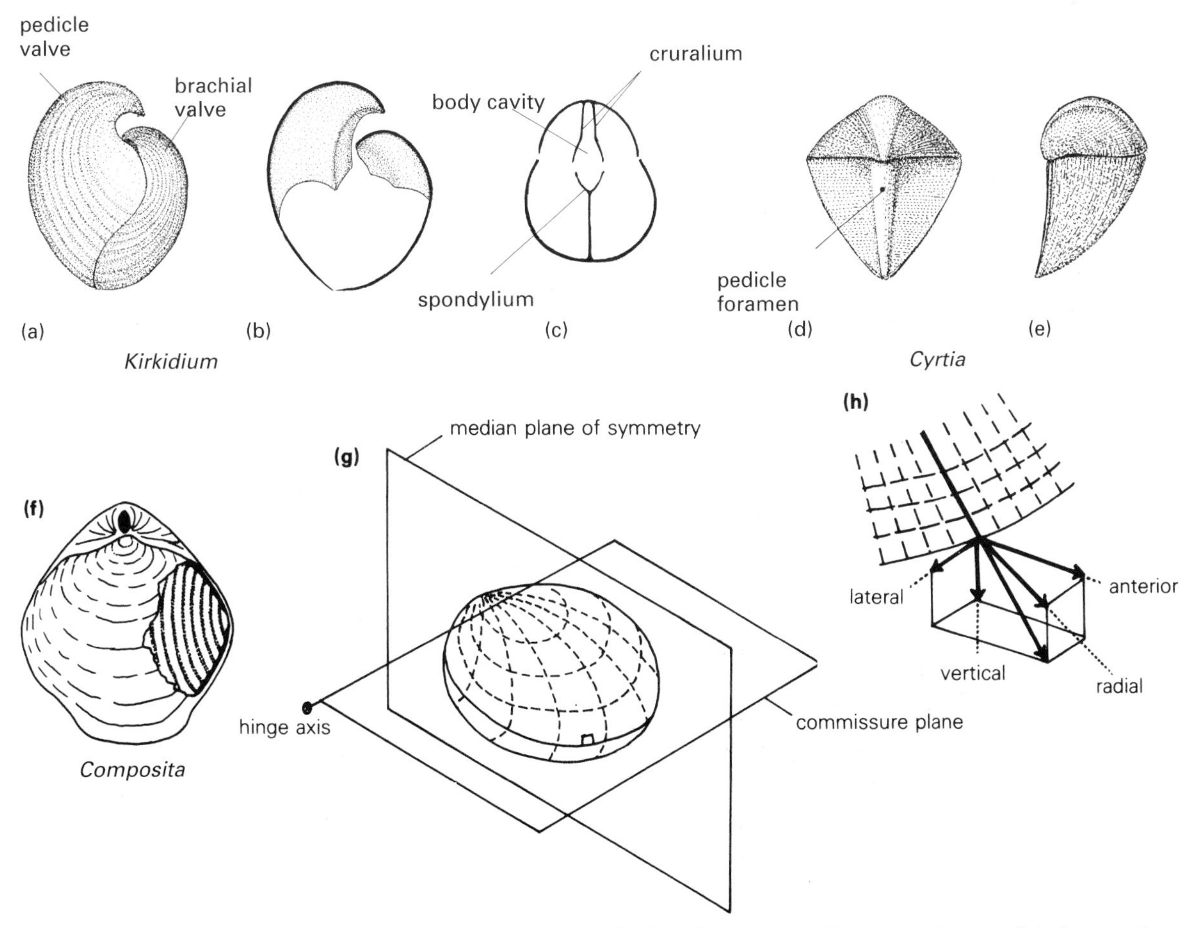

Figure 7.8 Brachiopod growth and form: (a) *Kirkidium knightii* (Sil.), lateral view (× 0.5); (b) same, with external shell removed to expose spondylium and cruralium; (c) same, in section, with body cavity enclosed by spondylium and cruralium; (d) *Cyrtia exporrecta* (Sil.) in frontal view and (e) *C. exporrecta* (Sil.) in lateral view, showing greatly expanded pedicle valve interarea (× 1.25); (f) *Composita* (Carb.) an athyrid resembling a terebratulide but with spiral brachidium; (g) radial and concentric growth elements of a rectimarginate shell, showing radial vectors, concentric zones, median plane of symmetry, commissure plane and hinge axis; (h) growth vectors operating at any point on the edge of a shell. [(f)–(h) Modified from Rudwick, 1959.]

broader range of possible shapes. Many theoretical shell shapes that might have been used were never, in fact, adopted, and the shell types that can be observed in nature seem to represent an optimization of areal and volume constraints (McGhee, 1980). Furthermore, many of the shell shapes that are found in nature, even if they depart markedly from the biconvex norm, tend to appear over and over again in unrelated stocks through time. In such **homeomorphy**, the result of convergent evolution, the descendants of a defined stock come to have shell shapes (or indeed other characters) closely reminiscent of those of other and separate groups.

Composita (Order Athyridida; Figs 7.8f, 7.16) is in external appearance strikingly like many terebratulides of the same and different ages. Both are non-strophic, though the majority of spiriferides are strophic. The only real external difference is in the style of perforation of the umbo, though the internal morphology is radically dissimilar of course. This is but one example of a ubiquitous phenomenon. There are **isochronous** and **heterochronous homeomorphs** (i.e. present at the same time or at different times, respectively), and they may occur either within the same taxon or in different taxa.

Some brachiopods, such as the rhynchonellides and spiriferides, have pronounced serial zigzag deflections all along the anterior part of the commissure, with a concave central **sulcus** on the brachial valve fitting into a pronounced **fold** in the pedicle valve. Such structures probably resulted from differentiation of growth rates at localized sites, not only in the vertical components but in the radial ones too. These sharp and serially arranged zigzags gave a relatively much greater effective length of commissure over which food particles could be taken in, for no extra gape. When the shell gaped slightly, small particles could be drawn in while large and harmful sand grains were excluded. Only at the sharp angle of the zigzag could larger particles enter, and these angles are normally protected by **setae** (which also act as early warning sensors) or occasionally by spines. The fold and sulcus system of rhynchonellides seems to be instrumental in separating lateral inhalant currents from a median exhalant stream. Spiriferides also have a median fold and sulcus, but they are relatively smaller and rarely ornamented by zigzags and probably mark the exhalant system.

The costellae or ribs may sometimes be prolonged into hollow or solid spines. One function of such spines appears to have been for anchorage. The large *Productus* has many spines on the base of the pedicle valve which seem to have had this function, and the thin-shelled strophomenide *Chonetes*, which has spines along the hinge margin, could conceivably have used them for fixing itself upright on the sea floor with the spines held vertically. But in the rhynchonellide *Acanthothiris* the spines of the pedicle and brachial valves when first formed at the edge of the mantle were hollow and have been interpreted as sensory tubes for extending the sensitive tip of the mantle well away from the body: another kind of early warning sensor (Rudwick, 1965b). These spines, initially hollow, became filled up with calcite as the mantle edges migrated away and eventually took up an anchoring function. External ornamentation in some silicified brachiopods is shown in Fig. 7.9.

Microstructure of shells (Fig. 7.10)

In all rhynchonelliform brachiopods the shell is multilayered, and the various layers can usually be distinguished in fossil brachiopods as well as in living ones. Shell structure and development has been extensively studied in the Recent rhynchonellide *Notosaria nigricans* (Fig. 7.10a), which serves as a standard model (Williams, 1968). There are three shell layers: an outer non-calcareous **periostracum**, a middle calcareous **primary layer** and an inner **secondary layer** of calcareous and inorganic material.

Within the periostracum there are three main proteinaceous layers which underlie an outer gelatinous sheath. This sheath protects the growing edge of the shell and is the first element to be formed, but being gelatinous it is soon rubbed off and does not extend far beyond the edge of the shell.

The primary shell layer beneath the periostracum is of rather structureless crystalline calcite, whereas the very distinctive secondary layer consists of elongated calcitic rods (fibres) with rounded ends, inclined at about 10° to the shell surface and having a regular system of stacking and a characteristic trapezoidal cross-section. Although the primary layer is of constant thickness, the secondary layer keeps on growing throughout the life of the brachiopod and so is thickest nearest the umbones.

A section through the growing edge of the shell shows how these several layers are secreted. The

Figure 7.9 Silicified brachiopods from the Visean (L. Carb.) of County Fermanagh, Ireland, showing details of surface sculpture and morphology: (a), (b) an adult shell of the spiriferacean *Tylothyris laminosus* (McCoy) viewed posteriorly (showing high ventral interarea with delthyrium almost completely covered by deltidium) and dorsally (× 1.5); (c) *Cleothyridina fimbriata* (Phillips), an athyrid spire bearer, showing part of the spire and having typical spinose lamellae externally (× 2); (d) *Productidina margaritacea* (Phillips), a juvenile specimen of a productoid, showing the pedicle valve posteriorly (× 7); (e) *Dasyalosia panicula* (Brunton), an adult shell showing two-directional spines on the pedicle valve (× 1); (f) *Overtonia fimbriata* (J. de C. Sowerby), a young pedicle valve with characteristic interdigitating ridged and spine pattern (× 2); (g) *Dasyalosia panicula* (Brunton), an adult shell with a well-developed interarea. (Photographs reproduced by courtesy of Dr Howard Brunton.)

mantle, which secretes the shell, is infolded into a groove under the shell edge; here there is a **generative zone** from which all the shell-secreting cells are produced. As each new cell is formed it moves towards the edge of the shell, producing the various layers in turn from its outer surface. First the cell secretes the gelatinous sheath, then the three layers of the periostracum in turn. Thus, by the time the cell has reached the outer edge of the shell, it has already produced four different kinds of material in succession. It then begins to grow calcite rhombs from its upper surface, embedded in a protein cement; these then grow together and amalgamate as the crystalline primary layer. The cell then swivels round an axis of rotation, so that its distal surface now faces outwards, and becomes permanently fixed in its final place as the growing edge of the shell moves away from it, all the while accreting new material in conveyor belt fashion. When the secretion of primary cell material is complete, the secretory function of the cell changes for the sixth and last time and it produces a single long calcitic

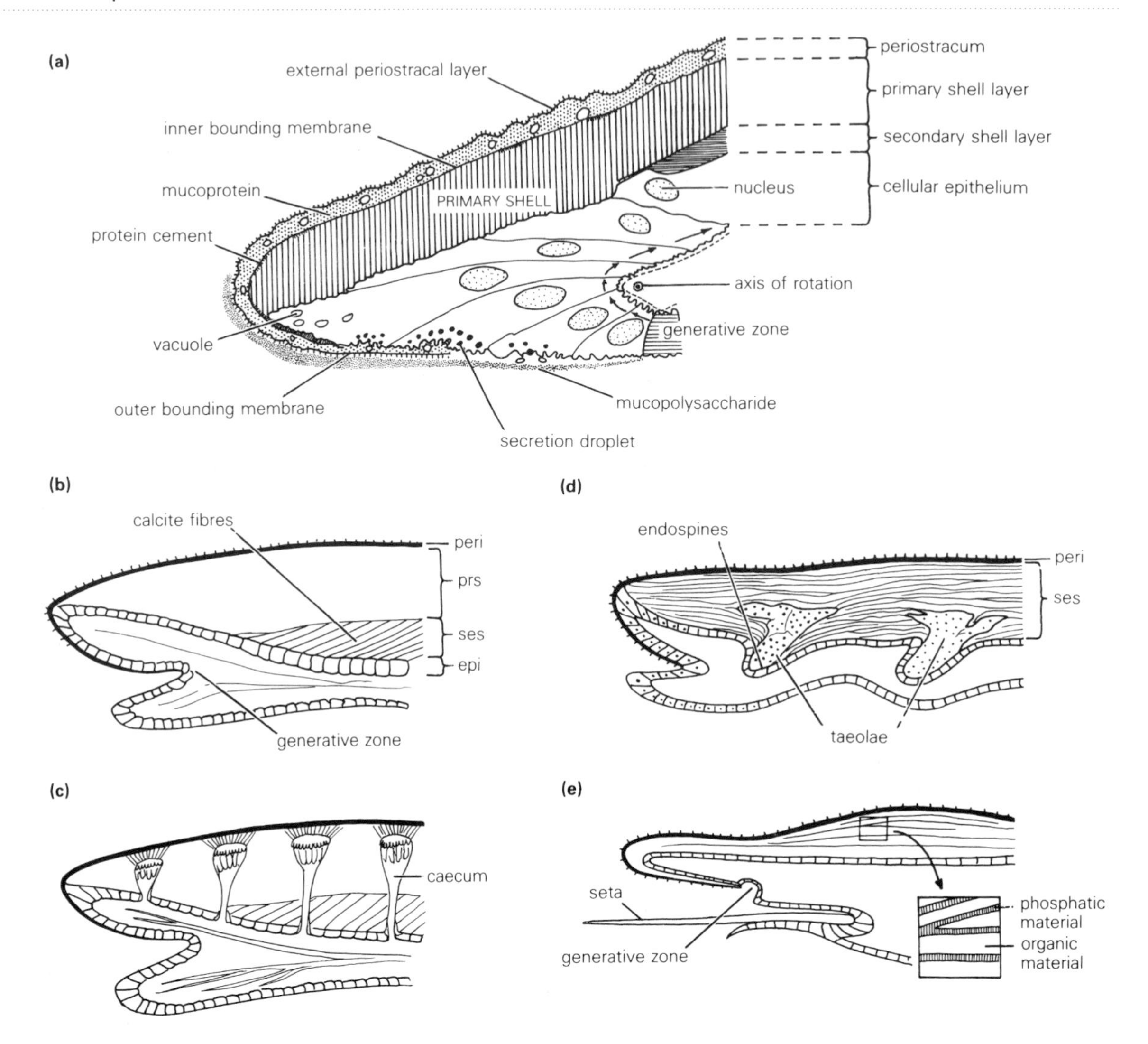

Figure 7.10 Shell structure in brachiopods: (a) standard secretory regime in *Notosaria nigricans* (for explanation see text); (b) impunctate shell (e.g. rhynchonellide); (c) endopunctate shell (e.g. terebratulide) with caecae possessing core cells freely suspended in the cavity – the secondary shell fibres are shown as cut in section; (d) pseudopunctate shell (e.g. strophomenide) with no primary layer but with taleolae prolonged internally as endospines; (e) shell of *Lingula*, with primary shell constructed of alternating layers of phosphatic and organic material – a marginal seta is shown. (Mainly based on illustrations by Williams and Rowell in *Treatise on Invertebrate Paleontology*, Part H.)

fibre, one of many identical elements in the secondary layer. In this layer the regular pattern of the many long inclined fibres naturally reflects the ordered arrangement of the individual cells that produce them. Since new material is continually added to the lower end of the fibre by the surface of each cell, it is not surprising to find that the secondary layer is thickest at the umbones, which are the oldest part of the shell.

The modern terebratulide *Waltonia* produces its shell according to the same 'standard secretory regime' as *Notosaria*, but the periostracum is much thicker and of a more complex labyrinth-like structure.

Most rhynchonelliform brachiopods, living and fossil, probably secreted their shells in much the same way as *Notosaria*. Differences in shell structure of some groups, however, indicate that the secretory programme must have been modified. In the 'advanced' Strophomenida (Fig. 7.10d) the shell below the periostracum is only a single layer of laminar calcite traversed by inclined calcitic rods (**taleolae**). But in the ancestral strophomenide group, Superfamily Plectambonitacea (Suborder Strophomenidina), to which *Eoplectodonta* belongs, there is a secondary layer of inclined calcitic fibres like that of *Notosaria*.

Recent Terebratulida and Thecideidina have an endoskeleton of calcareous spicules, secreted within the living tissue and assisting in its support. Such spicules are found in some Cretaceous terebratulides.

Endopunctation and pseudopunctation in shells

In rhynchonelliform brachiopods such as *Notosaria* the shell is **impunctate**, i.e. has no perforations or cavities within the shell structure (Fig. 7.10b). Other kinds of rhynchonelliformes (e.g. terebratulides) have **endopunctate** shells (Fig. 7.10c) in which the shell structure is penetrated from the inside by large, regularly arranged, elongated cavities known as **punctae**, normal to the shell surface. These contain tubular outgrowths of the mantle, the **caecae**. Caecae are formed at the edge of the shell during the standard secretory regime by small knots of cells which behave independently of the rest of the mantle. As the growing edge of the shell moves away the caecae are permanently locked in position but, since each caecum retains its contact with the mantle throughout the deposition of the primary and secondary layers, the lower end may become very long.

The caecum nearly reaches the outer shell surface and is connected to it by a 'periostracal brush' of tiny tubes filled with mucopolysaccharide. Below the brush are core cells filled with glycogen and proteins; these hang down freely into an empty cavity below, so that the only connection of the caecum with the mantle is by the flattened peripheral cells that line the wall of the punctae. Caecae are primarily storage chambers; they are also respiratory, and they inhibit boring organisms. In addition, the periostracal brush can release an organic 'glue' (for the repair of injury) if the periostracum is accidentally ruptured.

Since microborings in the shells of Recent brachiopods clearly avoid punctae (Curry, 1983a), the suggestion that caecal contents inhibit boring organisms would appear to be substantiated.

Impunctate and endopunctate shells in rhynchonelliform brachiopods seem to cut across established systematic boundaries, and it is probable that endopunctation evolved more than once. Thus most orthides are impunctate, apart from one superfamily, the Enteletacea. All atrypides and pentamerides are impunctate. Most spiriferides and rhynchonellides are also impunctate, but there is one mainly punctate spiriferide superfamily, the Spiriferinacea, and one punctate rhynchonellide superfamily of three genera, the Rhynchoporacea. All terebratulides are punctate.

In the shells of most Strophomenata (Fig. 7.10d) the thin and often irregularly formed taleolae give a spurious impression of punctation, especially to weathered specimens. Hence the term **pseudopunctate** is used to distinguish them, perhaps unfortunately, for in other respects their microstructure is clear enough.

Hinge and articulation

Strophic and non-strophic shells are distinguished by their hinge structure. There are no apparent intermediates between them; indeed it would be hard to imagine a brachiopod of intermediate kind. The various structures associated with the hinge region are dealt with in turn.

PEDICLE OPENING

The delthyrium and notothyrium are more closely seen on strophic shells since the hinge is straight. Primitively, as in orthides, enteletides and pentamerides, they together form a diamond-shaped opening for the emergence of the pedicle. However, this is not their sole function, for in brachiopods with high interareas they also allow a space for the diductor muscles when these are contracting. Since the diductor muscles are attached to the cardinal process, their line of action is brought close to the hinge, and if there were no opening the muscles would catch on the hinge line and fray.

Where there is no pedicle the delthyrium and notothyrium may be protected by various kinds of plates. In the pedicle valve there may be a single plate (the **deltidium**; Fig. 7.11j) or, as in *Magellania*, a pair of plates (**deltidial plates**), united by a median suture and arched so that they do not interfere with the line of action of the muscles.

The notothyrium may likewise be plated by a pair of **chilidial plates** or by a single **chilidium**. Rarely

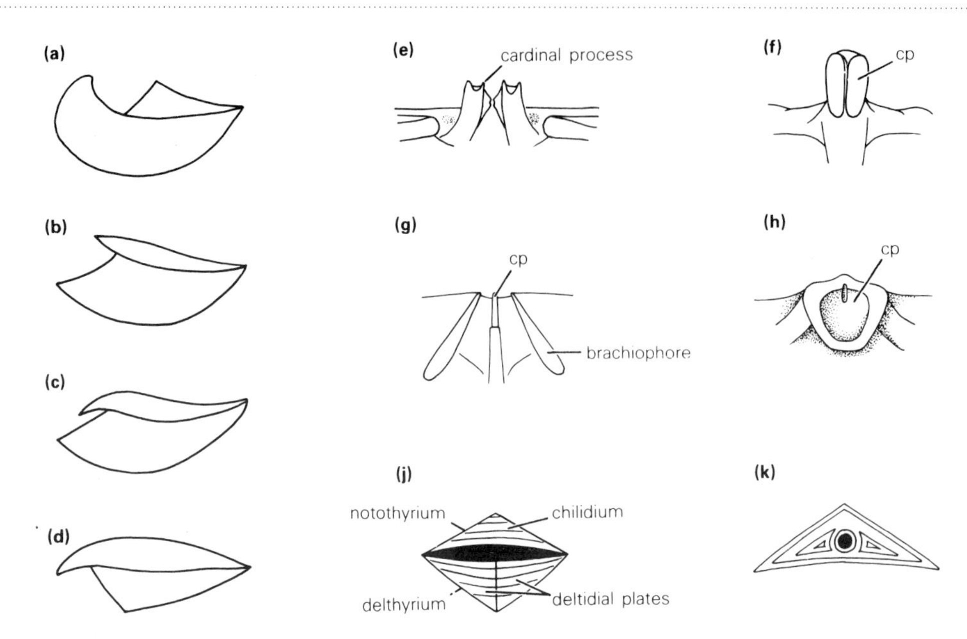

Figure 7.11 (a)–(d) Inclination of interareas: (a) brachial valve hypercline, pedicle valve anacline; (b) brachial valve anacline, pedicle valve apsacline; (c) brachial valve apsacline, pedicle valve apsacline; (d) brachial valve apsacline, pedicle valve procline; (e)–(h) cardinal processes: (e) *Strophomena;* (f) *Pustula;* (g) *Hesperorthis;* (h) *Leptellina;* (j) pedicle foramen with closing structures; (k) triangular stegidial plates closing delthyrium. (Redrawn from Williams and Rowell in *Treatise on Invertebrate Paleontology,* Part H.)

in spiriferides there may be a **stegidium** (Fig. 7.11k). This special type of delthyrial covering is a series of concentric triangular accretions formed during growth which successively close the delthyrial cavity from the outside in.

Certain strophomenides have only a very tiny hole in the umbo for the emergence of the pedicle. Likewise in the spiriferide *Cyrtia* (Fig. 7.8d,e), which has a much expanded pedicle valve interarea, the pedicle opening resides in the centre of the interarea. Any pedicle emerging from such an opening could not have enabled the brachiopod to be held upright on the sea floor, but could certainly have acted as a tether to hold it so that the shell was not swept away by currents.

Non-strophic shells such as the large Silurian *Kirkidium* (Fig. 7.8a–c) have a large, long pedicle valve with a strongly curved umbo but an open delthyrium some distance anterior to the umbo. In juvenile shells of *Magellania* and other terebratulides the pedicle opening is located in a normal delthyrium. As the shell grows this opening migrates posteriorly towards the umbo by resorption of the shell. Meanwhile deltidial plates form anteriorly and eventually join in the midline, closing off the pedicle opening entirely from the delthyrium; this can then be distinguished as a pedicle foramen.

INTERAREAS

Interareas proper are found only in strophic shells. Their attitude relative to the hinge line is very important in taxonomy, and the nomenclature of some of the various attitudes is illustrated in Fig. 7.11a–d. Of these **apsacline** and **anacline**, respectively, are the most common. Some shells have the interareas of the pedicle valve attenuated into a pronounced beak-like **rostrum,** which in *Uncites is* curiously twisted to one side.

CARDINALIA

The structures at the posterior end of the brachial valve, collectively known as **cardinalia**, are highly differentiated and serve various functions. Of these the medially placed cardinal process is often the most prominent (Fig. 7.11e–h). In its simplest form, as found in the more primitive rhynchonelliformes, it is no more than a pair of undifferentiated muscle bases behind the notothyrium (Fig. 7.11h). A more elaborate kind of structure is a **median ridge**; this primitively separated the two areas of muscle attachment, but in more advanced forms it actually became the site of diductor attachment through the inward migration of the muscle bases (Fig. 7.11g).

A further development of this kind of structure led to a shaft with a head (**myophore**; Fig. 7.11f), which increases the area of muscle attachment and may be variously crenulated, forked, multilobate or comb-like.

Other cardinalia include **socket walls** and **brachiophores**, expanded knobs which may have been involved in support of the lophophore.

TEETH, SOCKETS AND ACCESSORY STRUCTURES

Teeth are always found in the pedicle valve, sockets in the brachial. This system apparently evolved only once and then remained relatively stable. The teeth consist of knobs of secondary calcite with smooth or crenulated surfaces. In many brachiopods they may be supported by a pair of vertical dental plates which join them to the floor of the pedicle valve.

The teeth may be supplemented or even replaced (e.g. *Eoplectodonta)* in many strophomenides by denticles which grow along the margins of the hinge line. Members of Order Productida generally have neither teeth nor denticles.

Sockets, forming part of the cardinalia, often have large and thick interior walls, frequently as large as the teeth themselves. In such cases the interior socket wall may even indent a **secondary socket** close to the tooth on the pedicle valve. Socket walls can be swollen and, as in *Visbyella*, closely associated with the brachiophores marking the lophophore bases (Fig. 7.4d).

Teeth and sockets are constructed of secondary shell. As the shell grows, the teeth move away from the hinge line, taking up successive though closely spaced positions and leaving a 'track', which is the cumulative effect of the position that these structures had throughout ontogeny.

Muscle attachments

Adductor, diductor and adjustor muscles are normally attached directly to the insides of the two valves, forming scar patterns which are genus or species specific.

Sometimes, however, all the muscles may be raised off the floor by **muscle platforms**, such as those clearly displayed in the pentameraceans (Fig. 7.8a–c). In these the pedicle valve has a vertical wall divided in two distally so that it appears Y-shaped in cross-section. This structure, the **spondylium**, lies directly opposite a **cruralium**: a pair of near-vertical plates which are outgrowths of the brachial processes, hanging down from the roof of the pedicle valve. Together these structures enclose a narrow vertical cavity, open anteriorly about a third of the way from the umbo, in which the muscles and probably all the other organs of the body cavity were enclosed as well.

Muscle platforms such as these are not uncommon in rhynchonelliform brachiopods, and also in craniiformes such as *Trimerella* (Fig. 7.14g), and seem to have arisen several times during evolution. Rudwick (1970) has noted that it is not necessary to have a contractile muscle running the full height of the shell in order to open and close it; a shorter muscle could do the job more economically and with greater control. Muscle platforms tend to occur in those shells which are highly biconvex; they allow for the attachment of a short but optimally effective muscle system. Not all strongly biconvex shells have this system; the surviving rhynchonellides and terebratulides have part of the muscle strand replaced by a non-contractile **tendon**, which is another way of operating with equal efficiency.

Lophophore (Fig. 7.12)

The brachiopod lophophore, whose structure in *Magellania* has been described, is functionally similar in all brachiopods.

In the brachiopods, though the filamentous structure of the lophophore is constant, there are several distinct ways in which the lophophore itself is arranged within the mantle cavity. Furthermore, though some brachiopods have a calcareous support for the lophophore (spiriferides, atrypides and terebratulides), others do not, and in the orthides, as in most strophomenides, pentamerides, and fossil linguliformes and craniiformes, it is hard to tell what the original lophophore structure must have been. Living rhynchonellides likewise have no calcified lophophore support, and in these, as in living linguliformes and craniiformes, the lophophore is held up purely by the hydrostatic pressure of fluid in the great brachial canal (which early in life becomes closed off from the coelom). In many Strophomenata there are traces of possible lophophore structures impressed upon the inside of

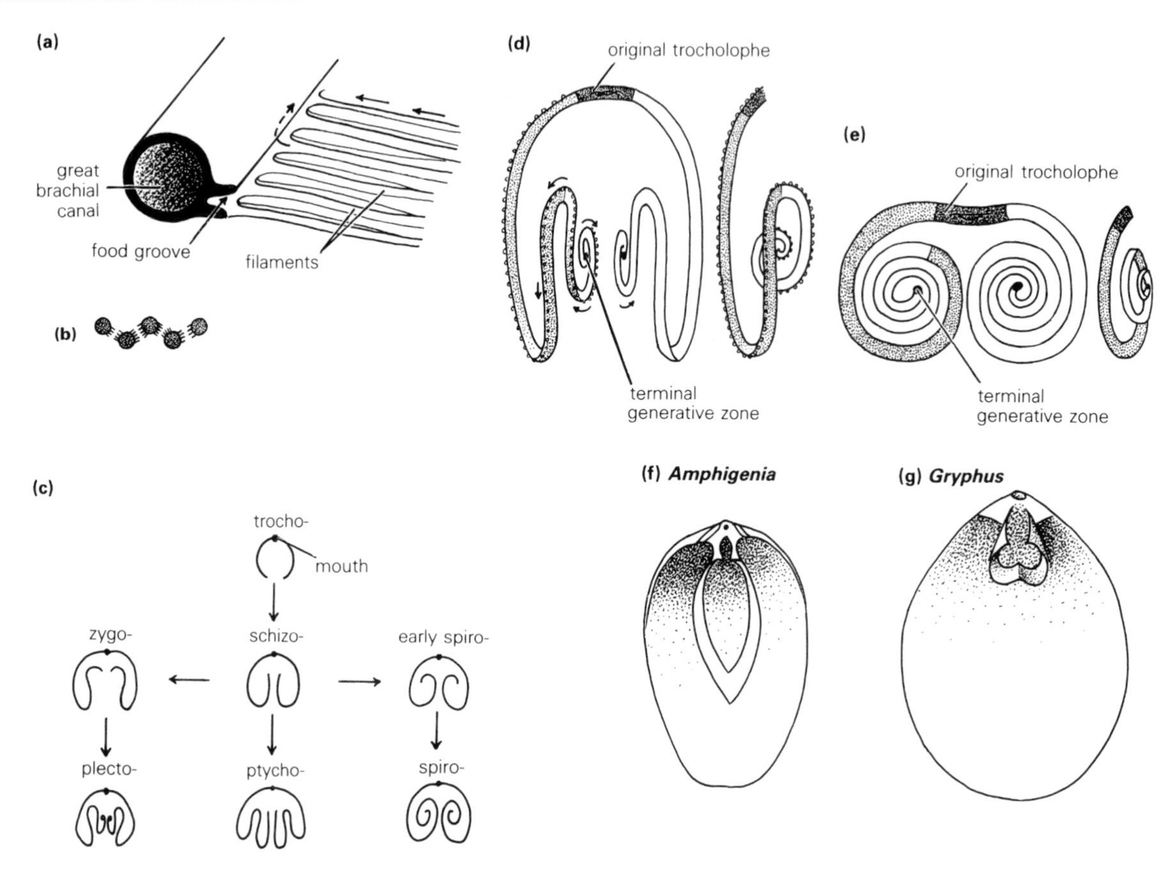

Figure 7.12 Brachiopod lophophores: (a) sectional view of lophophore with fluid-filled great brachial canal, food groove and filaments – arrows show the direction of movement of food particles; (b) arrangement of filaments in section; (c) the main types of lophophores showing various developmental pathways; (d) *Terebratula* plectolophe, showing mode of growth from a terminal generative zone; (e) *Rhynchonella* spirolophe, same; (f) *Amphigenia* (Dev.), primitive loop structure of early terebratulide (Centronellidina); (g) *Gryphus* (Rec.) short loop of Terebratulidina. [(a)–(c) Redrawn from Rudwick, 1970; (d), (e) redrawn from Williams and Wright, 1961; (f), (g) redrawn from *Treatise on Invertebrate Paleontology*, Part H.)

the brachial valve, while the arrangement of the vertical plates in *Eoplectodonta* and related genera suggests the form that the lophophore must have had; other possible lophophore structures give little information on the original form.

The **lophophore support** is represented in rhynchonellides and some pentamerides only by a pair of short struts (**crura**) projecting from the cardinalia. In groups where the brachidium is a calcareous ribbon (terebratulides, atrypides and spiriferides), it always grows (or grew) by accretion on the (anterior) growing edge and by resorption on the (posterior) trailing edge as the shell enlarges. Invariably new filaments are added at the growing ends of the lophophore alone, never intercalated within the filament series.

Changes in the form of the brachidium during ontogeny provide a basis for the classification of lophophores into different types (Fig. 7.12c). All lophophores pass through the same early stages in development. The first stage, whether supported or not, is a **trocholophe**. Here the lophophore is merely a pair of curving 'horns' projecting horizontally on either side of the mouth and forming an incomplete ring. Only one genus of micromorphic adult brachiopods has this kind of structure and, since the trocholophe forms the initial developmental stage of all brachiopods, this particular case is probably a neotenous development. The next stage in development is the **schizolophe**, where the two original horns have become larger but are bent back and run parallel, directed towards the mouth. All brachiopod lophophores go through this stage too; a few micromorphic terebratulides have schizolophes in the adult. From here there are three possible developmental pathways:

one starting with a **zygolophe**, which leads to a **plectolophe**: the long or short loop of many terebratulides;
one leading to the multilobed **ptycholophe**, as found in the modern *Lacazella*, a thecidean, and in some fossil productides;
one leading to the twin spiral system of the **spirolophe**, perhaps the commonest and most widespread functional system, possessed by Recent rhynchonellides and lingulides, as well as by fossil spiriferides athyrides and atrypides and possibly other fossil groups.

In both the plectolophe of terebratulides and the spirolophe of many groups the twin ribbons may be joined by a stout rod, the jugum, near the posterior end.

The mode of operation of spirolophes is well known in some modern rhynchonellide brachiopods, and some inferences can be made about how it functioned in fossils. As in all brachiopods, the ciliated filaments borne on the brachia create currents which normally enter the shell laterally (Fig. 7.7). When the water passes into the shell, it first of all lies outside the lophophore in an **inhalant chamber**. It then has to pass through the cilia from the outside to the space enclosed within the spires, i.e. into an **exhalant chamber**. As it does so, food particles are strained off. Brachiopod feeding currents are very precisely controlled. Inhalant and exhalant currents remain intact and there is neither mixing nor eddying once the currents have entered the shell. The water from the exhalant chamber comes out medially, and the marked deflections of the anterior margin of the rhynchonellides appear to separate inhalant from exhalant currents.

Notosaria is a Recent rhynchonellide whose twin conical spires point downwards. These spires form the inhalant chamber and contain unfiltered water which it passes through the filaments to an exhalant chamber outside the spires. In Order Atrypida (Fig. 7.7a,b) the morphology of the lophophore is strikingly similar, with the sole difference that the spires point upwards rather than downwards, and there is a calcified brachidium. It seems reasonable to assume that current systems functioned in a similar manner and that the unfiltered water lay inside the spires while the exhalant chamber was outside them. In *Spirifer*, however, the spires point laterally (Fig. 7.7c,d), and water must have filtered from the outside to the inside of the spires.

Plectolophes, as developed in modern terebratulides, are highly efficient structures of complex three-dimensional form. Each arm develops as a curving horn (the trocholophe stage), then recurves sharply as a bar running parallel with the original arm and lying above it (the zygolophe stage). In the fully developed plectolophe this bar turns inwards to terminate in a vertical spiral coil. The two spiral median coils developed from the paired arms lie side by side and with their filaments occupy much of the space in the centre of the shell; between them is inhalant water.

Members of Superfamily Terebratellacea (Suborder Terebratellidina) have plectolophes supported along most of their length by a long calcareous loop in which the two ends have joined up, though the median coil is supported by hydrostatic pressure alone. However, in terebratulaceans only the basal parts are supported by a much shorter loop; the rest remains erect through hydrostatic pressure, though some additional support is given by the calcareous spicules within the lophophore itself.

Ptycholophes are generally rather flat structures and do not form the complex three-dimensional coils found in the other two lophophore types. Presumably they are less efficient filtering systems. They are uncommon today, occurring only in Family Megathyrididae (Superfamily Terebratellacea) and thecideans, but they may have been the normal lophophore type in strophomenids, as is witnessed by the brachial ridge system of *Eoplectodonta*. Most fortunately, intact calcified ptycholophes have been found (Grant, 1972) in the small Permian productacean *Falafer* (Fig. 7.17a–c).

Subphylum Linguliformea

This subphylum of organophosphatic-shelled brachiopods is considered to be more closely related to bryozoans and phoronids than to other brachiopods. The best-known representative is *Lingula*.

Lingula

Lingula (Order Lingulida) is a phosphatic-shelled inarticulate brachiopod which has persisted since the Cambrian with relatively little change: a type example of bradytelic evolution. Modern *Lingula* lives successfully in a brackish to intertidal environment, normally in fine sand (Craig, 1951). The early invasion of *Lingula* into such a 'difficult' environment for which there were few competitors is probably one of the primary reasons for its virtual evolutionary stasis. Nevertheless, there is reason to believe that some fossil *Lingula* species lived in environments other than the intertidal, including shelf and basinal regions (Cherns, 1979), so the use of *Lingula* as an environmental indicator is limited. Some differences in *Lingula*-bearing 'palaeocommunities' through time have been documented by West (1976).

Modern *Lingula* (Fig. 7.14a–c) has a pair of almost identical, bilaterally symmetrical valves. They are gently convex, and from between them projects an elongated pedicle. The animal lives in a vertical position in a burrow, and its pedicle extends deep into the sand, with the upper edges of the two valves just at the level of the surface. In life the distal part of the pedicle has little roots which anchor the animal firmly, and if *Lingula* is disturbed the pedicle instantly contracts, withdrawing the shell deep within the burrow. By analogy with other brachiopods, the pedicle end is described as posterior; the slightly larger valve is called the pedicle valve and the smaller the brachial valve. Horny setae round the exposed edges of the shell mark off three apertures leading to the inside, the outer two for incurrent water, the central for excurrent water.

The shell is lined by a mantle, and the anterior body wall, the lophophore and other internal organs are as in the rhynchonelliformes. The lophophore is a spirolophe with the two spires directed inwards, each being an 'inhalant spire' carrying unfiltered water which is strained and exhaled medially. The gut, however, has an anus discharging posteriorly, and there are four nephridia. The circulatory system is simple, and there is a nervous system with a ganglion near the mouth, as in the articulates. One of the greatest differences between *Lingula* and articulated brachiopods lies in the musculature for opening and closing the shell (Fig. 7.14c). Since there is neither a hinge line nor teeth and sockets, the shell cannot open by leverage, and so the muscular system is unlike that of articulates. The two valves close by adductor muscles: a single posterior muscle and a pair of anterior ones. When these relax, the valves move apart slightly through the elasticity of the muscles, but otherwise the valves can only rotate, shear or slide against one another through the action of various **oblique muscles** which are large and well developed. The pedicle has a leathery external cuticle within which is an epithelium, embryologically part of the mantle. Inside it there are muscles enabling the pedicle to contract, so that when the end is firmly fixed in the sand the brachiopod can be withdrawn inside its burrow. As in articulates, the shell is secreted by cells produced in a generative zone in the mantle, under the valve rim. In *Lingula* (Fig. 7.10e) the periostracum is underlain by a primary shell with many alternating stratified layers of organic and phosphatic material. They are all subparallel with the shell margin, and rather than being continuous layers they are elongated lenses whose mode of formation is incompletely understood. The phosphatic layers are thicker where they overlie the body cavity. There is no secondary layer; nor are there punctae.

It is interesting to note how *Lingula* burrows (Thayer and Steele-Petrovic, 1975). The lingulid *Glottidia*, which is the best studied, enters the sediment anterior end first with the pedicle initially used as a stiff prop and later trailing behind. The brachiopod then burrows downwards through the sediment using scissor-like and rotary sawing movements of the valves. In addition, periodic rapid valve closure ejects water from the mantle cavity and the lateral setae pass mucus-bound sand upwards. When the brachiopod has thus burrowed, anterior end downwards, for some distance, it then turns upwards, describing a U-shaped track. Within a few hours the animal comes to the surface in standard feeding position, the correct way up. This remarkable burrowing system does not at any time involve the use of the pedicle as a digging 'foot' (as in bivalves); it is a method unique in the animal kingdom.

Lingula is widely reputed to be a euryhaline genus, but the euryhalinity is in fact only moderate. Thus the Australian *L. anatina* can tolerate salinity levels from 20 to 50 per mille, but only for about 4 weeks, and such features as pedicle regeneration and avoidance of predators by rapid withdrawal are impaired (Hammond, 1983) in a much shorter time. Other lingulids survive at salinities ranging from 13 to 42 per mille, and are well adapted to waters of fluctuating salinity, usually in waters of 8–10°C, and although they show a preference for shallow waters they have been recorded at depths of down to 500 m (Emig *et al.*, 1978).

Other Linguliformea (Fig. 7.14)

There are two classes within this subphylum, Lingulata and Paterinata. Class Lingulata consists of four orders (Holmer, 1989). Order Lingulida, in which the shells are mainly elongated ovals with growth lines, concentric round a near-marginal umbo, seems to represent a natural grouping. Though *Lingula* and its relatives seem to be typical 'living fossils', having persisted for long time periods, there are, however, clear differences in shell microstructure between living and Lower Palaeozoic *Lingula* species, which casts doubt upon whether the latter do in fact pertain to this genus. Family Obolidae belongs to this order, its representatives have suboval or rounded shells (Fig. 7.13e) in which the pedicle valve often has an internal flange divided into two **propareas**, which resemble the interareas of articulated brachiopods and have a triangular **pedicle groove** in between them. The musculature of *Obolus* can be reconstructed with reference to the muscle attachments and, though there are clearly defined adductor and oblique muscles, some evidence suggests that the leverage system could have functioned more like that of a rhynchonelliform than that of *Lingula*.

Order Acrotretida (L. Cam.–Rec.) often has circular valves of differing convexity, some possessing high, conical pedicle valves and flat, disc-shaped brachial valves (e.g. *Conotreta*, *Numericoma*) and have median septa in the pedicle valve, often of complex form. Order Discinida has rather flat, round or oval shells, with a submarginal umbo and usually strong subconcentric growth lines; the living *Discinisca* has a subcentral pedicle foramen from which emerges a short stout pedicle; the foramen of the Ordovician *Orbiculoides* is in a similar position. Order Siphonotretida (e.g. *Nushbiella*; Fig. 7.15) differs from the acrotretids mainly in shell structure, ontogeny and the development of the pedicle foramen.

Class Paterinata (L. Cam.–late Ord.), unlike Lingulata, has strophic shells and some other features suggestive of incipient rhynchonelliform structures, including a rudimentary notothyrium, delthyrium and paired adductor muscles. *Micromitra* is typical; in the genus *Dictyonites* the brachial valve has a remarkable surface sculpture of open pores.

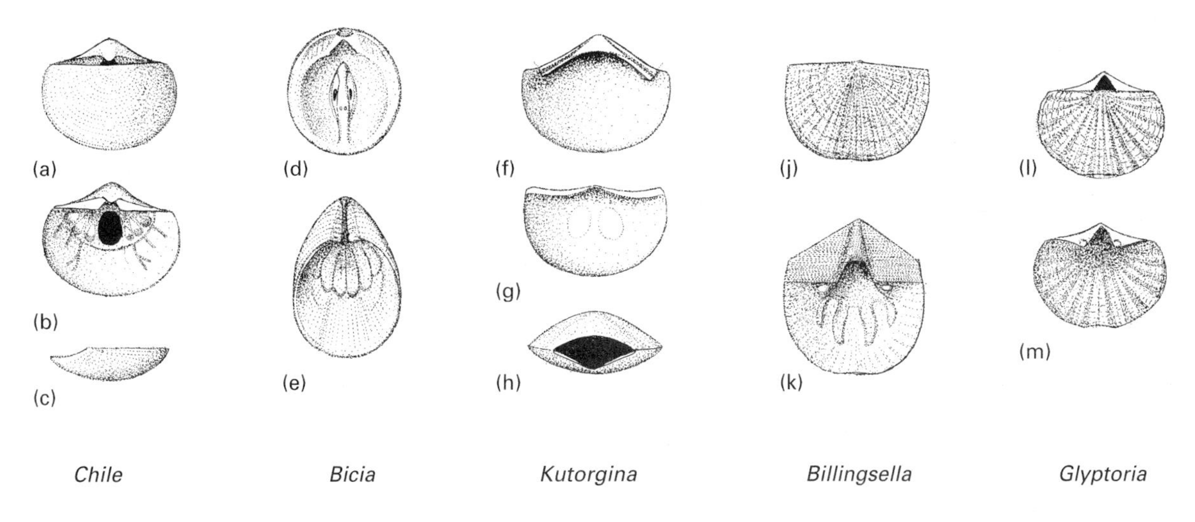

Figure 7.13 Cambrian brachiopods: (a) *Chile* (Chileida), complete shell; (b) same, interior view of pedicle valve, showing central perforation; (c) same, lateral view of pedicle valve (× 15): (d) *Bicia* (Obolellida), interior of brachial valve; (e) same, interior of pedicle valve (× 5): (f) *Kutorgina* (Kutorginidae), interior of pedicle valve; (g) same, interior of brachial valve; (h) same, posterior view of complete shell showing pedicle opening (× 8); (j) *Billingsella* (Billingselloidea), exterior view of brachial valve; (k) same, interior view of pedicle valve (× 3); (l) *Glyptoria* (Protorthida), complete shell; (m) same, interior of pedicle valve (× 3). [(d), (e), (j), (k) Redrawn from *Treatise on Invertebrate Paleontology*, Part H; others from Holmer, 1996, *Geologiskt Forum* **9**, 10–14.]

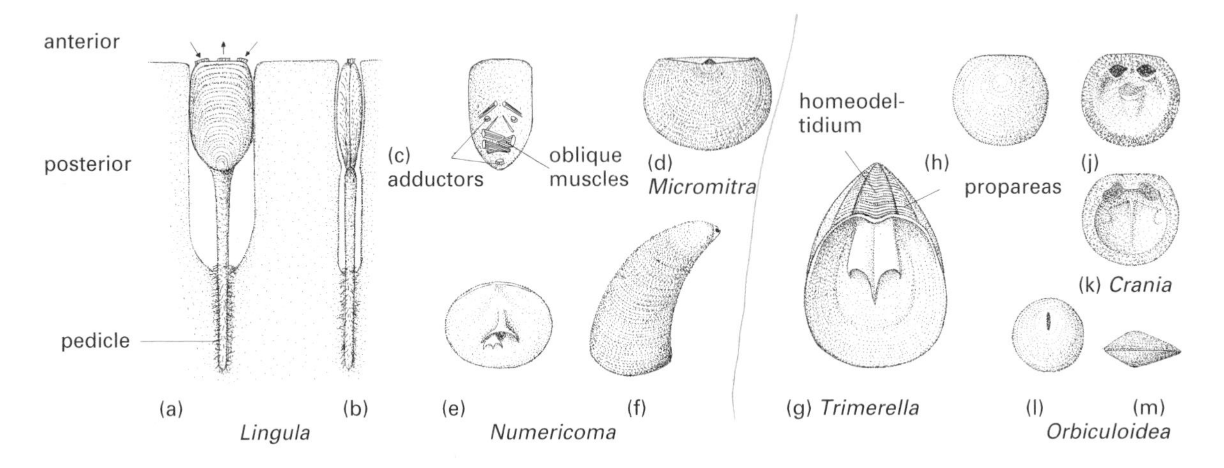

Figure 7.14 Linguliformes [(a)–(f), (l), (m)] and Craniiformes [(g)–(k)]. (a), (b) *Lingula* (Rec.) in its burrow, frontal and lateral views, showing inhalent and exhalent currents; (c) same, showing internal muscles in the brachial valve (× 1); (d) *Micromitra* (Paterinida) (L. Cam), complete shell (× 5): (e) *Numericoma* (Acrotretida; Ord.), interior of brachial valve; (f) same, exterior of pedicle valve (× 4); (g) *Trimerella* (Ord.), interior of pedicle valve (× 4); (h) *Crania* (Rec.), exterior and (j) interior of brachial valve; (k) same, interior of cemented pedicle valve (× 2); (l) *Orbiculoidea* (Acrotretida; Sil.) pedicle valve; (m) same, complete shell in lateral view (× 2). [Based mainly on photographs by Rowell in *Treatise on Invertebrate Paleontology*, Part H, except for (e), (f) from Holmer, 1989, *Fossils and Strata* **26**, 1–176.]

Subphylum Craniiformea

In this subphylum, with its single Class Craniata, the shells are of calcium carbonate, articulation is absent and the pedicle is absent altogether. The alimentary canal has a true anus. This class is regarded as closer to Rhynchonelliformea than to Lingulata. Three orders have been defined. Members of Order Craniida (e.g. *Crania*) are unusual craniiformes in having a shell of protein and calcium carbonate. They survive today as the most successful of calcareous-shelled craniiformes. In the brachial valve the shell is triple layered. Externally there is an impersistent mucopolysaccharide sheath with the periostracum below. Below this is the primary shell made up of needle-like calcite crystallites, and finally there is a secondary layer of calcitic laminae sheathed by protein sheets. The secondary layer is found only in the brachial valve; it is absent in the pedicle valve, which cements itself to the surface by its outer mucopolysaccharide film. Craniaceans are punctate with branching caecae in both valves, used for storing polysaccharides and protein. Where the muscles are inserted, the calcite of the primary layer forms blades normal to the surface. Craniids have no pedicle and the pedicle valve cements itself to an appropriate substratum.

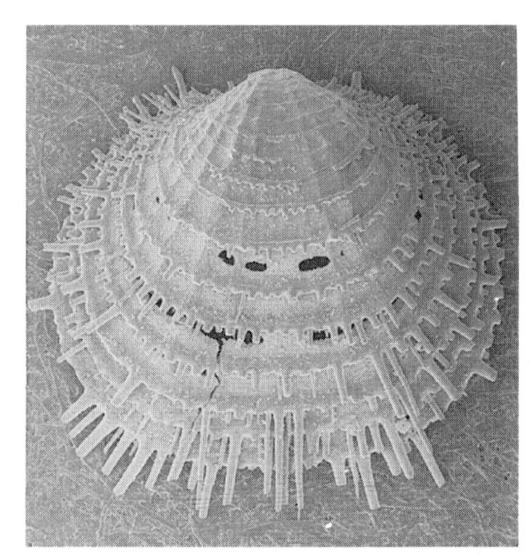

Figure 7.15 *Nushbiella*, (Lingulata, Siphonotretida), brachial valve (×10). (Photograph by courtesy of Dr Lars Holmer.)

The small Order Craniopsida (e.g. *Craniops*) includes forms with an impunctate shell having crystalline primary and laminar secondary layers and platforms in the central part of each valve. Order

Trimerellacea (Ord.–Sil.; Fig. 7.14g) consists of large brachiopods with thick biconvex shells, probably aragonitic, having in the pedicle valve a large **pseudointerarea** and strongly defined propareas with a **homeodeltidium** in between. A peculiar feature is the common presence of muscle platforms supported by a central buttress and often turned down at the edges to enclose paired chambers. This development of muscle platforms is, of course, quite independent of that in rhynchonelliformes.

7.3 Ontogeny

The sexes are separate in brachiopods. Fertilization of the eggs takes place in the sea water after the copious genital products have been shed via the nephridia through the mantle cavity and out into the sea. Breeding habits are poorly known, being only well documented in *Lingula* and *Terebratulina*. Apparently some species have a single annual breeding time; others produce multiple broods during the year.

A species in which spawning patterns are now well documented is *Terebratulina retusa* (Fig. 7.3) from waters 100–300 m deep off western Scotland (Curry, 1982). These brachiopods normally live attached to bivalves. Their biannual spawning cycles in late spring and late autumn are initiated at temperatures of 10–11°C, and the reproductive cycle is very precisely timed. Growth is at first rapid, decreasing throughout the (maximum) 7-year lifespan. There is a slowing down or cessation of growth in winter, producing clear growth banding. Size distribution patterns recorded as histograms show regularly spaced peaks corresponding to the biannual settlement 'cohorts'; these, incidentally, turned out to be identical with computer-based and theoretically predicted simulations.

In a few Recent brachiopods the development of the fertilized eggs takes place in special invaginated **brood pouches** in the shell of the female, and this seems to have been the case in certain fossil brachiopods as well.

The development of the zygote and larval stages is quite different in linguliform, craniiform and rhynchonelliform brachiopods, providing good embryological evidence that the three classes are natural divisions. In *Lingula* the free-swimming early larva has two ciliated segments: one becomes the body, the other the mantle. From the mantle segment there develops a mound-like ridge, which starts to grow lophophore filaments while the larva is still free-swimming. At the same time the larval shell (**protegulum**) develops as a single plate (though this is unusual in any brachiopod), later splitting into two separate plates for the brachial and pedicle valves, respectively. The pedicle too grows from the mantle segment, and when it is large enough the brachiopod settles upon it, usually by the time some 10 or so mantle filaments have been produced. All other structures (e.g. muscles, gut and nephridia) originate from the body segment.

The rhynchonellide *Notosaria* (cf. Fig. 7.1g) shows the very different character of rhynchonelliform development. The larva has three segments; the upper is the globular body segment or anterior lobe, the middle the presumptive mantle, and the lower the pedicle segment. Initially the mantle segment hangs down freely over the pedicle segment as a cylindrical 'skirt'. This then inverts itself to cover the body segment, leaving the pedicle exposed. Shell secretion then begins on the now reversed outer surface of the mantle, but the brachial and pedicle valve protegula are separate from the beginning. After a very brief larval life (shorter than in most inarticulated brachiopods) the young individual settles on the pedicle and completes its development.

7.4 Classification (Fig. 7.16)

Brachiopods have never proved very easy to classify, and to some extent the difficulties have been compounded by homeomorphy. Some of the early schemes were retained for far too long, simply for want of anything better.

The basis of a new classification of brachiopods was laid by Muir-Wood (1955), who made it clear that the concept of working down from 'orders' predetermined by a few key characters had to be abandoned in favour of a scheme of building up from the generic level to higher units, thus recognizing but avoiding the problems arising from homeomorphy. This defined the main orders, which are clearly set out in the *Treatise on Invertebrate Paleontology* of 1965. However, there remained a fundamental problem, which was how the defined groups were actually related to each other, and a

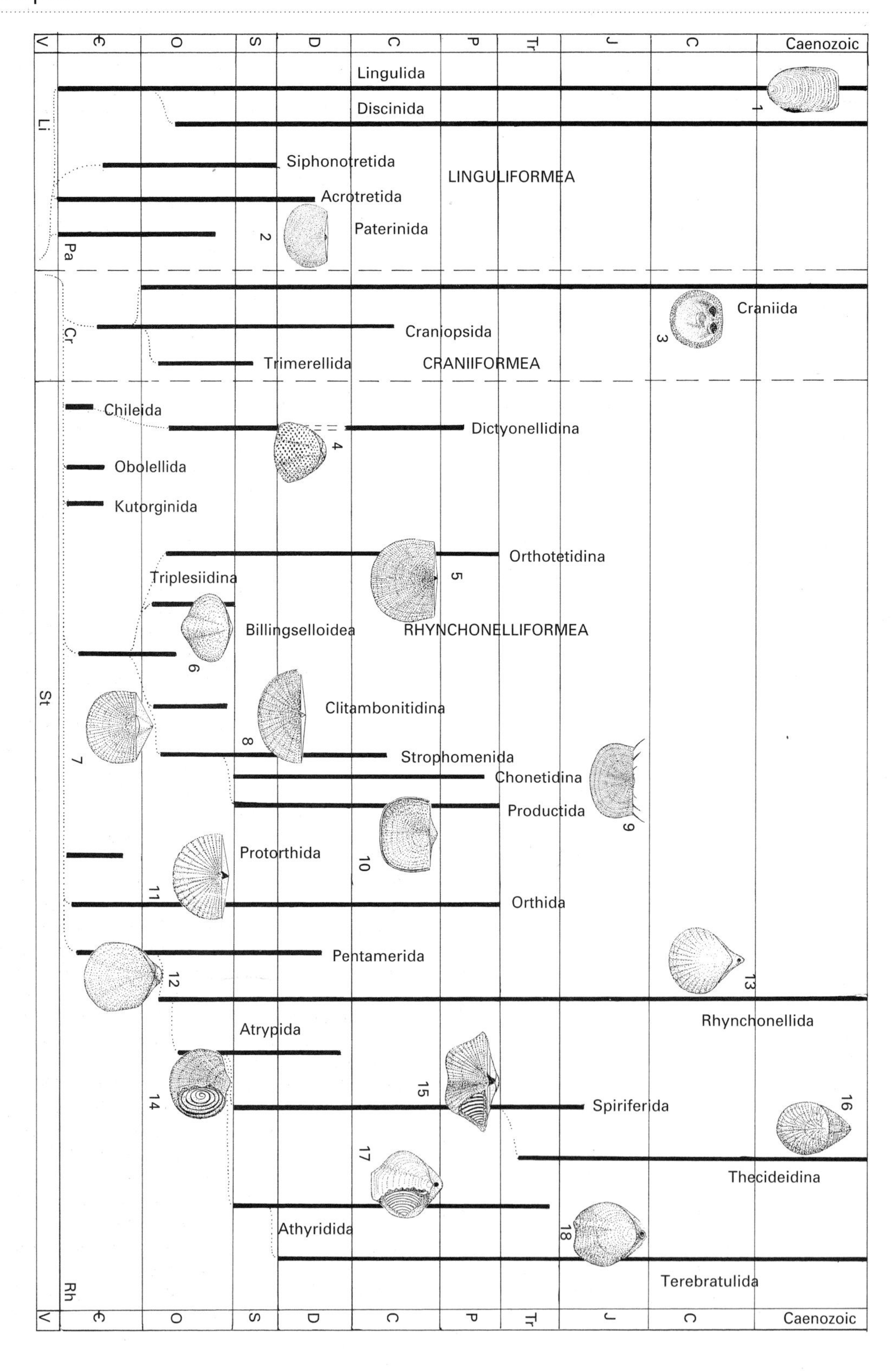
V
Є
O
S
D
C
P
Tr
J
C
Caenozoic
Li
Pa
Cr
St
Rh
Lingulida
Discinida
Siphonotretida
Acrotretida
Paterinida
LINGULIFORMEA
Craniida
Craniopsida
Trimerellida
CRANIIFORMEA
Chileida
Dictyonellidina
Obolellida
Kutorginida
Orthotetidina
Triplesiidina
Billingselloidea
RHYNCHONELLIFORMEA
Clitambonitidina
Strophomenida
Chonetidina
Productida
Protorthida
Orthida
Pentamerida
Rhynchonellida
Atrypida
Spiriferida
Thecideidina
Athyridida
Terebratulida
1
2
3
4
5
6
7
8
9
10
11
12
13
14
15
16
17
18

continuing problem remains in how to reconcile data from living and fossil brachiopods. Over 95% of all brachiopod genera are extinct and there are considerable differences in anatomy and ontogeny between the various brachiopod groups.

The main brachiopod taxa originated in the Cambro-Ordovician, and here significant new data has recently become available in the form of previously unknown faunas of very early calcareous-shelled brachiopods representing a critical stage in evolution (e.g. Class Chileata). The fact that such faunas were recovered from Kirghizistan and Antarctica explains why they had not previously been recorded. Williams *et al.* (1996) performed a substantial computer-based cladistic analysis of brachiopod relationships, incorporating this new data. This proved to be fully consistent with the stratigraphic record of all brachiopod groups. It will be adopted generally and a simplified version is given below, and illustrated in Fig. 7.16. From the student's point of view, it is perhaps unfortunate that the simple, and easily remembered classification of six major orders of 'articulated' brachiopods has now passed into history, but such is the nature of scientific progress.

PHYLUM BRACHIOPODA (Cam.–Rec.): Solitary marine bivalved coelomates, inequivalved but bilaterally symmetrical normal to the commissure plane. Shell chitinophosphatic or calcareous; pedicle attached, cemented or free. Valves formed by mantle, enclosing body and mantle cavities, the latter with a filamentous lophophore.

SUBPHYLUM LINGULIFORMEA (L. Cam.–Rec.): Valves not hinged by teeth and sockets. Shell organophosphatic. Pedicle usually present, fleshy and muscular, emerging between valves or from the apex of a valve. Planktic larva.

CLASS 1. LINGULATA (L. Cam.–Rec.): Brachiopods with organophosphatic valves lacking teeth or sockets, valves attached by muscles and body wall only. Muscular pedicle formed as outgrowth of mantle, lophophore unsupported. Gut with true anus.

ORDER 1. LINGULIDA (L. Cam.–Rec.): Shell biconvex, elongate, teardrop shaped or rounded, umbones usually located terminally, pedicle growing out posteriorly between valves; e.g. *Lingula, Obolus*.

ORDER 2. ACROTRETIDA (L. Cam.–Rec.): Shell circular or subcircular, often elongated, pedicle opening where present confined to pedicle valve; e.g. *Acrotreta, Conotreta, Numericoma*.

ORDER 3. DISCINIDA (M. Ord.–Rec.): Shell rounded, pedicle opening (at least in young shells) forms a triangular notch in posterior margin of pedicle valve; e.g. *Trematis, Discina, Orbiculoidea*.

ORDER 4. SIPHONOTRETIDA (U. Cam.–Ord.): Shell biconvex, rounded to teardrop-shaped with hollow spines, pedicle foramen circular and apical, e.g. *Siphonotreta, Nushbiella*.

CLASS 2. PATERINATA, with the single Order Paterinida (L. Cam.–U. Ord.): Shell rounded to elliptical but with a near-straight edge forming a pseudointerarea with a (often closed) delthyrium. Pedicle probably absent; e.g. *Paterina, Micromitra, Dictyonites*.

SUBPHYLUM 2. CRANIIFORMEA (L. Cam.–Rec.): Brachiopods with calcareous shells, lacking teeth and sockets, and with the pedicle reduced or absent. Pedicle valve cemented, gut with anus. Single Class CRANIATA.

ORDER 1. CRANIIDA (?M. Cam.–Rec.): Shell rounded, punctate, usually attached by cementation of the pedicle valve, brachial valve usually conical, muscle scars pronounced; e.g. *Crania, Neocrania, Valdiviathyris*.

ORDER 2. CRANIOPSIDA (M. Ord.–L. Carb.): Shell elliptical to linguloid, with marked concentric growth lines, impunctate low muscle platform present, cemented by apical region of pedicle valve or free lying; e.g. *Craniops*.

ORDER 3. TRIMERELLIDA (M. Ord.–U. Sil.): Shells large and thick with muscle scars in both valves, articulation by hinge-plate in brachial valve and cardinal socket in pedicle valve. All were free lying; e.g. *Trimerella, Dinobolus*.

SUBPHYLUM 3. RHYNCHONELLIFORMEA (L. Cam.–Rec.): Shell calcareous, endopunctate, impunctate or pseudopunctate. Teeth and sockets present, though sometimes lost secondarily. Crura usually present, sometimes prolonged as a brachidium. Valves opened by diductors, closed by adductors. Pedicle arises from larval rudiment. Gut without anus.

CLASS 1. CHILEATA (L. Cam.) Shell strophic, inarticulated, pedicle valve with cardinal interarea, ventral valve perforated. Single ORDER CHILEIDA; e.g. *Chile*.

CLASS 2. OBOLELLATA (L. Cam.–M. Cam.): Shell calcareous, biconvex, subcircular to oval. Pseudointerarea in pedicle valve, position of pedicle opening variable, beak of brachial valve marginal. Single ORDER OBOLELLIDA; e.g. *Obolella, Trematobolus*.

CLASS 3. KUTORGINIDA (L. Cam.–M. Cam.): Shell calcareous, biconvex, cardinal area in both valves, delthyrium and notothyrium present but no teeth, sockets or cardinal process, e.g. *Kutorgina*.

CLASS 4. STROPHOMENATA (M. Cam.–Trias.): Normally concavo-convex or plano-convex shells with strophic

Figure 7.16 Time ranges and relationships of major brachiopod groups, with representative genera illustrated. The nine classes include Lingulata (Li), Paterinata (Pa), Craniata (Cr), Chileata (Ch), Obolellata (Ob), Kutorginata (Ku), Strophomenata (St) and Rhynchonellata (Rh). Genera illustrated are 1, *Lingula*; 2, *Micromitra*; 3, *Crania*; 4, *Dictyonella*; 5, *Orthotetes*; 6, *Triplesia*; 7, *Vellamo*; 8, *Eoplectodonta*; 9, *Chonetes*; 10, *Dictyoclostus*; 11, *Hesperorthis*; 12, *Pentamerus*; 13, *Plicirhynchia*; 14, *Atrypa*; 15, *Cyrtospirifer*; 16, *Thecidea*; 17, *Composita*; 18, *Stiphrothyris*.

hinge, simple teeth, often lost, shell usually pseudopunctate, some umbonally cemented, tubular spines may be present , lophophore supports rarely found. Cardinal process often bilobed, pedicle foramen often closed.

ORDER 1. STROPHOMENIDA (Ord.–Trias.). Usually concavo-convex, smooth or with fine ribs, simple or reduced teeth.

SUBORDER 1. STROPHOMENIDINA (Ord.–Trias.). Plano-convex to concavo-convex, often small, with fine radial ribs, simple teeth often replaced by denticles. e.g. *Eoplectodonta, Sowerbyella, Plectambonites, Strophodonta.*

SUBORDER 2. ORTHOTETIDINA (Ord.-Trias.): Pedicle non-functional in adult, the pedicle valve is often cemented, teeth simple, a spiral lophophore is inferred for many genera; e.g. *Fardenia, Orthotetes, Meekella, Streptorhynchia..*

SUBORDER 3. TRIPLESIIDINA (L. Ord.–U. Sil.): Biconvex impunctate shells, pedicle valve interarea with arched pseudodeltidium, brachial valve interarea lacking, with long forked cardinal process; e.g. *Triplesia.*

SUBORDER 4. BILLINGSELLOIDEA (L.–M. Cam.). Biconvex, well-developed hinge, interareas and teeth, simple cardinal process, widely divergent muscle scars; e.g. *Billingsella.*

SUBORDER 5. CLITAMBONITIDINA (Ord.): Wide-hinged impunctate pseudopunctate shells, having pronounced pseudodeltidium; e.g. *Clitambonites.*

SUBORDER 6. CHONETIDINA (L. Sil.–Perm.): Small to large plano-convex shells, with prominent spines along the pedicle valve interarea; e.g. *Strophochonetes, Chonetes.*

ORDER 2. PRODUCTIDA (L. Dev.–?Trias.): Shell usually plano-convex, usually with deep mantle body cavity, occasionally conical and bizarre in form. Valves often geniculate, usually prolonged into a 'trail'; delthyrium and notothyrium frequently closed. Shell usually with tubular spines.

SUBORDER 1. PRODUCTIDINA. Usually cemented in early stages, free living as adults, cardinal process bilobate or trilobate; e.g. *Productus, Gigantoproductus, Waagenoconcha, Dictyoclostus.*

SUBORDER 2. STROPHALOSIIDINA. Early stages usually cemented, later stages anchored by root-like spines. Includes many bizarre forms; e.g. *Strophalosia, Chonosteges, Cyclacantharia, Richthofenia, Leptodus.*

CLASS 5. RHYNCHONELLATA (L. Cam.–Rec.). Rhynchonelliformes with biconvex, calcite shells, impunctate or endopunctate, a calcified brachidium may be present.

ORDER 1. ORTHIDA (L. Cam.–U. Perm.): Articulata with unequally biconvex strophic shells. Delthyrium and notothyrium usually open; cardinal process normally present. Shell impunctate, rarely punctate. Ventral muscle field small; muscle platform rare. Brachidia absent, but lophophore probably schizolophous or spirolophous. Suborders include

SUBORDER 1. PROTORTHIDINA (L. Cam.): Primitive orthides, lacking a cardinal process; e.g. *Protorthis.*

SUBORDER 2. ORTHIDINA (L. Cam.–Perm.): Normally impunctate, finely ribbed shells (other than in SUPERFAMILY ENTELETACEA in which the shells are punctate); e.g. *Orthis, Heterorthis, Valcourea, Visbyella, Schizophoria, Dalmanella.*

ORDER 2. PENTAMERIDA (M. Cam.–U. Dev.): Biconvex, impunctate, usually non-strophic shells with spondylium in pedicle valve and sometimes a pair of vertical plates (cruralium) opposite the spondylium, the whole enclosing the muscle cavity. Suborders include

SUBORDER 1. SYNTROPHIIDINA (M. Cam.–L. Dev.): Pentamerides with prominent fold and sulcus and open delthyrium; cruralium rare; e.g. *Porambonites.*

SUBORDER 2. PENTAMERIDINA (M. Ord.–U. Dev.): Large strongly biconvex shells, usually lacking fold and sulcus; cruralium well developed, e.g. *Pentamerus, Stricklandia, Sieberella.*

ORDER 3. RHYNCHONELLIDA (M. Ord.–Rec.): Shell normally non-strophic, usually with beak and functional pedicle; delthyrium partially closed; crura in brachial view. Usually with coarse ribs meeting along a zigzag commissure, and often with a pronounced fold and sulcus. Spirolophous. Muscle platforms developed only in Superfamily Stenocismatacea. Impunctate except for monogeneric Superfamily Rhynchoporacea; e.g. *Rhynchonella, Pugnax, Camarotoechia, Gibbirhynchia, Uncinulus, Tetrarhynchia, Wilsonia.*

ORDER 4. ATRYPIDA (M. Ord.–U. Dev.): Impunctate and biconvex shells with short hinge line and narrow or absent interareas. Open delthyrium or umbonal pedicle foramen. Spiralia may be dorsally or laterally directed; e.g. *Atrypa, Catazyga, Glassia, Dayia, Retzia, Davidsonia.*

ORDER 5. SPIRIFERIDA (M. Ord.-Perm.): Shells biconvex, usually strophic, often with a long hinge line, typically ribbed, interareas well developed, especially in the pedicle valve. Laterally directed spiral brachidium, with or without jugum, punctate, delthyrium open or closed. e.g. *Spirifer, Brachythyris, Mucrospirifer, Cyrtia.*

ORDER 6 SPIRIFERINIDA (L. Carb.–L. Jur.): Usually small shells, similar in most respects to Spiriferida, but mainly punctate; e.g. *Spiriferina, Liriplica, Punctospirifer.*

ORDER 7. ATHYRIDIDA (U. Ord.-Jur.): Smooth impunctate shells with narrow hinge line; beak often truncated by umbonal foramen; laterally directed spiralia joined by jugum of peculiar and complex form; e.g. *Athyris, Composita, Meristella, Tetractinella.*

ORDER 8. TEREBRATULIDA (L. Dev.–Rec.): Biconvex articulates with short non-strophic hinge, and with functional pedicle emerging through umbonal foramen. Delthyrium generally closed by delthyrial plates. Shell punctate. Brachidium loop-like, usually plectolophous. Suborders include

SUBORDER 1. CENTRONELLIDINA (L. Dev.–Perm.): Archaic terebratulides with primitive loop of oval form; e.g. *Centronella, Amphigenia, Stringocephalus*.

SUBORDER 2. TEREBRATULIDINA (L. Dev.–Rec.): Short loop; median septum usually absent; internal spicules developed; e.g. *Dielasma, Terebratula, Pygope, Plectothyris*.

SUBORDER 3. TEREBRATELLIDINA (Trias.–Rec.): Long loop with cardinalia and median septum. e.g. *Magellania, Zeilleria, Digonella, Terebratella*.

ORDER 9. THECIDEIDINA (Trias.–Rec.): Small thick-shelled articulates, usually cemented and without pedicle, capable of opening very wide. Ptycholophe recessed deep within granular interior. They have been regarded variously as strophomenides, terebratulides, or spiriferides; Baker (1990) on the basis of shell microstructure refers them to the Order Spiriferida; e.g. *Thecidea, Lacazella, Moorellina*.

Order uncertain

SUBORDER DICTYONELLIDINA (M. Ord.–Perm.): Small group of unusual brachiopods in which umbo of pedicle valve is marked with peculiar triangular umbonal plate; e.g. *Eichwaldia*.

7.5 Evolutionary history

The earliest brachiopods are of Lower Cambrian age, and they are readily differentiated into the three defined subphyla which persist until the present day. There are such evident differences between Recent linguliformes, craniiformes and rhynchonelliformes that some authorities believe the brachiopods as a whole to be polyphylectic. In this view the brachiopods are a 'clade of organization' which includes other lophophorates (phoronids and bryozoans).

The division of Brachiopoda into three subphyla (Williams *et al.*, 1996) implies monophyly, though the craniiformes and rhynchonelliformes are more closely related to each other than they are to the linguliformes. There is no record of brachiopods before the Cambrian, and although some Precambrian 'brachiopods' of teardrop shape with concentric rings were described some years ago, these turned out to be inorganic, resulting from small-scale mud flows. The earliest brachiopods do not occur until some distance above the defined base of the Cambrian. But their diversity, especially in the light of new discoveries, is already remarkable. Amongst the earliest are short-lived rhynchonelliform orders (Chileida, Obolellida and Kutorginida), the Billingselloidea which were the progenitors of the Strophomenata (Fig. 7.13), and the first orthides and pentamerides, as well as the first rhynchonelliformes. The earliest craniiform *Philhedra* is doubtfully recorded from the Middle Cambrian, but this subphylum did not really get under way until the Middle Ordovician.

Brachiopods as a whole had become quite common locally in the late Cambrian; for example, swarms of the orthide *Orusia* in Scandinavia cover endless bedding planes. It was only in the early Ordovician, however, that there was a real evolutionary burst, and it was mainly confined to the rhynchonelliformes. The linguliformes and craniiformes never thereafter played more than a subordinate role, except locally.

By early Ordovician times most of the rhynchonelliform groups which came to dominate the Palaeozoic were becoming fully established, including the early strophomenides and rhynchonellides, while the orthides and pentamerides continued to diversify. Then came the late Ordovician ice age, which had a significantly adverse effect on brachiopods, though none of the major taxa became extinct. During the Silurian there was some reduction in the diversity of inarticulated brachiopods, and of orthides and strophomenides, though rhynchonellides and spiriferides continued to expand. In the uppermost Silurian to Lower Devonian the last of the important orders, the Terebratulida, was added, probably derived from an athyridine ancestor. During the Devonian the spiriferides reached their peak, and there seems to have been a general brachiopod expansion. By the late Devonian there was a general decline and widespread extinction, and the pentamerides died out. In the later Palaeozoic there was a new burst of evolution, particularly in the Order Productida, which include many large and peculiar, though highly functional genera. These first appeared in the Silurian and extended through the Permian. Their great success was partially because they successfully colonized a habitat previously closed to brachiopods: the quasi-infaunal or almost buried (and hence protected) mode of life. Amongst these are included *Gigantoproductus*, the largest of all brachiopods. The Permian was a time of great evolutionary plasticity in productides, culminating in the origin of some very peculiar Permian groups – especially in Subfamily Strophalosiidina (e.g. *Richthofenia, Cyclacantharia, Gemellaroia*) – which colonized reef environments and which, in spite of their aberrant and quite unbrachiopod-like appearance, were highly adapted to particular environments. Some of these, interpreted functionally by Rudwick and Cowen (1968), involved very promising adaptations, such as rhythmic flow feeding supplanting ciliary pumping.

However, with the severe extinctions of late Permian time, in which all fossil groups were affected, even these new devices did not help the productides to survive. Only a very few (such as the family Thecospiridae) made the transition to the Mesozoic. The rest became extinct, as did the majority of the Palaeozoic groups. By the end of the Middle Jurassic the last spiriferides had gone, leaving only rhynchonellides and terebratulides, which together with the small suborder Thecideidina make up the groups that have survived until Recent times. Living rhynchonellides are able to extrude their spirolophes out from the shell into the water for more efficient food gathering, and possibly also for righting an overturned shell by leverage. Ager (1967) suggests that this may be one reason why rhynchonellides survived but spiriferides failed to do so.

Thecideidines (e.g. *Lacazella*) are a small group of calcareous-shelled ptycholophous articulates whose affinities have been disputed, but on the basis of shell structure are now regarded as spiriferides. They remain an interesting enigma. Detailed studies of these remarkable brachiopods have been made by Baker (1970, 1983, 1990).

Linguliformes seem to have changed little since their decline at the end of the Ordovician, though their habitats have become less diverse if the known presence of deep-water lingulides in the Silurian is a reliable guide.

Craniiformes were never more than a minor component of the brachiopod fauna as a whole, and are represented today only by three genera, of which only *Crania* is cosmopolitan.

The character of brachiopod faunas thus seems to have changed greatly throughout the Phanerozoic, and even a superficial analysis of any fauna will normally enable it to be referred to a particular system. Various theories have been proposed to account for the remarkable conservatism of post-Palaeozoic brachiopods.

Competition with bivalves, especially those with siphons, may have been a factor in limiting the post-Palaeozoic expansion of the brachiopods. It was not, however, a direct effect, for only after the Permian extinctions did the bivalves come to dominance. They seem to have 'weathered the storm' better (Chapter 1) and, being the first in the 'race' to colonize the sea floor afterwards, stayed there as a dominant component of new but balanced ecosystems, which generally excluded brachiopods. Moreover, as Rudwick (1970) has shown, the infaunal habitats so successfully exploited by the siphonate bivalves are not those in which brachiopods were ever very successful, other than the quasi-infaunal productides. Possibly it was only when the latter were extinct that it was possible for the bivalves to exploit the infaunal habitat to a higher degree than before.

Another factor may have been the rise of predatory starfish in the early Mesozoic, the common enemy of both brachiopods and epifaunal bivalves. Although the first starfish with muscular arms are known from the Carboniferous, tube feet with suckers, a flexible mouth frame and an eversible stomach did not appear until the Mesozoic. Bivalves, being primarily infaunal, in any case have better defences against predators, whereas living epifaunal, sessile brachiopods have none (other than a repellent taste and preference for cryptic habitats). Thus early Mesozoic brachiopods, already recently depleted, may have been prevented from re-radiating by predatory starfish, to which they so easily fell prey.

7.6 Ecology and distribution

Ecology of individual species

Nearly all brachiopods through time have been benthic. A few linguliformes found in graptolite shales, and rare tiny rhynchonelliformes (e.g. chonetids; Burton and Curry, 1985) may have been epiplanktonic if a convenient habitat were available. Bassett's (1984) life-strategy groupings proposed for the Silurian can be extended to cover virtually all benthic brachiopods; these categories are not mutually exclusive but often intergrade. Whereas nearly all Recent brachiopods remain attached to hard substrates, many Palaeozoic faunas were able to live on soft sediment substrates.

Epifaunal brachiopods

Fixosessile brachiopods

These remain attached throughout life. Those which have stout, single, unbranched pedicles are known as **plenipedunculate**. Most modern brachiopods are of this kind and are normally attached to rocks and dead shells on the sea floor by means of the pedicle. Fossil plenipedunculate brachiopods are sometimes found in 'nests' or clusters.

In some brachiopods the pedicle is divided into thin threads like byssus, which can penetrate like roots into deep-sea oozes and attach themselves to dead foraminiferid shells. Such **rhizopedunculate** brachiopods as the living terebratulaccan *Chlidonophora* and the rhynchonellide *Cryptopora* are thus able to anchor themselves on soft substrates and this mode of life may have been quite common in brachiopods generally. The pedicles of brachiopods have remarkable attachment strengths, and many can live abundantly in subtidal current strengths of 2 m s^{-1}. Brachiopods do not, therefore, seem to be excluded from turbulent environments by any weaknesses in the pedicle (Thayer, 1975). It has been calculated and shown experimentally (Blight and Blight, 1990) that *Austrospirifera*, anchored to the substrata by a long thin pedicle, would actually rise in the water in much the same way as a kite flies in the air. This would only happen at a current velocity of >0.5 m s^{-1}. If this takes place a region of reversed flow takes place round the anterior edge of the brachiopod. The advantage of such 'flying' would be enhanced filter feeding due to in with minimal effort, since the 'reversed flow' could be made use of. Burial would also be avoided.

Encrusting brachiopods attach to hard substrates by the pedicle valve directly the larva has settled. In the living *Crania*, and its fossil relatives, the encrusted shell follows the contours of the substratum. A few Silurian rhynchonelliformes are likewise encrusters, as are some lingulates (e.g. *Orbiculoidea*). Such an encrusting strategy was often confined to early ontogeny in **umbonally cementing** brachiopods, and is best known in productides, where a scar or cicatrix often marks the umbo of the pedicle valve indicating attachment of the juvenile shell. When the shell had grown enough to be stable independently, it broke off and became free-lying (*Waagenoconcha*; q.v.). But in such productide genera as the Permian *Falafer* (Suborder Strophalosiidina; Fig. 7.17a–c) in which, incidentally, the lophophore support is preserved, the tubular spines were used for attachment to a hard substrate.

Many such Permian productides used their spines in this way. The Lower Permian *Chonosteges* (Fig. 7.17l–n) is usually found cemented to larger brachiopods or corals. Here the cemented base of the shell is flat, the anterior part is raised up and fixed by tubular spines from the pedicle valve, and the brachial valve bears little funnels of uncertain function arranged along the edge of the shell.

Liberosessile brachiopods

Most brachiopods with a closed pedicle opening (e.g. many strophomenides, some spiriferides or orthides) must have lain upon the sea floor. Such a habit is rare in Recent brachiopods, and this may be linked with the increase in bioturbation since the Palaeozoic. Brachiopods that lie freely on the sea floor as adults are termed **ambitopic**. Many strophomenides have a dish-like recumbent habit that spreads the weight evenly. Chonetidines have spines along the posterior margin of the shell which may have supported the recumbent shell like a snowshoe. Alternatively the shell might have been held upright with the spines plunged vertically in the sediment. The Silurian spiriferid *Cyrtia* lay flat upon its expanded interarea, and the slender cone-like pedicle emerging from it could have had a 'tethering' function but would not have been supportive. Platforms and 'wings' were sometimes developed, giving maximum support on the surface of the sediment, while *Atrypa* had the shell extended as a broad flange. Co-supportive associations, with brachiopods tightly packed in an umbo-down posture, are sometimes found. In these the close crowding would give mutual support. This kind of relationship is most common in pentameraceans and other thick-shelled brachiopods of fairly high-energy, open shelf environments.

Endofaunal brachiopods

Lingula (q.v.) is a highly adapted **burrowing** brachiopod which is able to live effectively in soft substrata. A fair number of instances of fossil individuals in life position have been described. Some Silurian species of *Lingula*, however, seem to have occupied borings, made and abandoned by other organisms in large coral colonies.

Though the infaunal niche so well exploited by

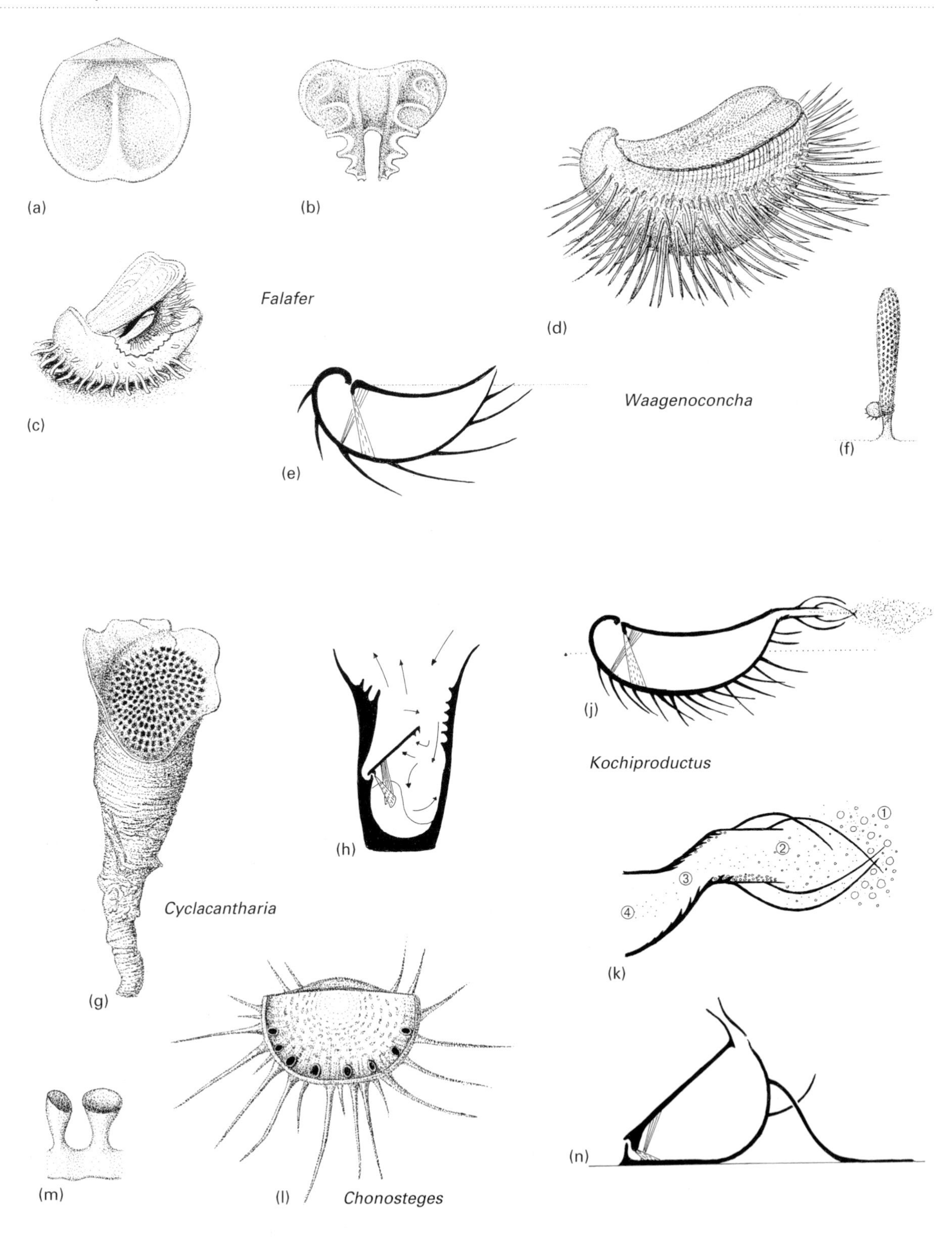
(a)
(b)
Falafer
(c)
(d)
Waagenoconcha
(e)
(f)
(g)
Cyclacantharia
(h)
(j)
Kochiproductus
(k)
①
②
③
④
(l)
Chonosteges
(m)
(n)

Lingula was never fully colonized by rhynchonelliformes, many productides and perhaps some other strophomenides seem to have spent their adult lives partially buried in sediment and with a sediment cover over the concave brachial valve. These are said to be quasi-infaunal. Adult productides (Fig. 7.17) were anchored by the strong spines that projected from the pedicle valve. These spines, formed at the margin of the valve, were originally hollow, later becoming filled with calcite as the mantle withdrew. Since the margins of both valves are sharply flexed upwards and drawn out into a narrow crescentic flange or **trail,** only this thin crescent would show above the general level of the sediment, and the brachiopod would thus be well camouflaged. *Waagenoconcha* (Fig. 7.17d–f) is a productide with delicate spines on the brachial valve which probably prevented settled sediment from being winnowed away. During the ontogeny of *Waagenoconcha* the settling larva attached itself to cylindrical bryozoans in the same environment, which it embraced with one or more pairs of curving cardinal spines and to which it remained attached for some time. When the shell became too large and heavy it broke off and, landing on the sea floor, grew long stout spines which it used for anchorage in the mud, eventually becoming buried as far as the trail (Grant, 1966a,b).

Figure 7.17 Adaptations in Permo-Carboniferous Productida: (a) *Falafer* (Perm.), interior of pedicle valve; (b) the ptycholophe lophophore; (c) reconstruction of *Falafer*, attached by spines to a hard substrate (all × 2): (d) *Waagenoconcha*, a Permian spiny productide, reconstructed; (e) median section showing disposition of valves and spines; (f) juvenile of same attached to an erect bryozoan (all × 0.8); (g) *Cyclacantharia*, a highly modified Permian reef-dwelling productide (× 2); (h) Rudwick's conception of how such a brachiopod fed by flapping the valves: a rising brachial valve would flush out the interior; (j), (k) *Kochiproductus*, a Carboniferous spiny productide of quasi-infaunal habit, shown (j) cleaning sediment from flange by clapping valves together (× 0.4); (k) cross-section through flange of *Kochiproductus*, showing various sediment baffles (1, spines excluding large particles; 2, flange acting as a settling table for intermediate particles; 3, taleolae (endospines) trapping small particles; 4, only food particles get through); (l)–(n) *Chonosteges* (Perm.), (l) in life position, dorsal view (× 5): (m) side view of funnels on brachial valve (× 20); (n) median section of *Chonosteges*, showing posterior cemented part, raised anterior part attached by tubular spines, and funnels (× 5). [Mainly based on Grant, 1966a, 1972, and illustrations in the *Treatise on Invertebrate Paleontology*; (h) based on Rudwick, 1961; (j), (k) redrawn from Sheills, 1968.]

In the case of *Kochiproductus* (Fig. 7.17j,k) the trail forms a wide, flat, horizontal flange which must have lain flat on the sea floor. Experiments on models with a flow tank (Sheills, 1968) showed that the wide flange acted as a settling table for extraneous and inedible particles, while smaller particles that passed through the first baffle were trapped by taleolae extended like a row of stakes at the entrance to the mantle cavity.

The quasi-infaunal niche does not seem to have been adopted by any Recent brachiopods.

A rather unusual adaptation of the pedicle has been described in the Recent terebratulide *Magadina cumingi*. This lives in a high-energy environment in drifting, mobile shell sand. Most shells are partially buried with only their anterior edges protruding; if their shells are buried they push themselves to the surface using the finger-like pedicle as an elevator. These observations (Richardson and Watson, 1975) extend quite unexpectedly the range of known substrate relations in brachiopods.

A very few living micromorphic brachiopods live as **interstitial** fauna in marine sands and it is possible that some tiny fossil linguliformes (e.g. acrotretides) could have done the same. Some brachiopods lived in nests or clusters whose early members anchored themselves to a chance dead shell and later their own dead shells provided a settling place for later generations. In such nests at least three annual broods have been found (Hallam, 1962), containing brachiopods of the same or different species. The nests came to a sudden end, probably by being overwhelmed by the sediment which preserved them.

Spiriferides, which sometimes grew in similar clusters with only a small basal attachment area, are sometimes found to be asymmetrical and distorted, probably through crowding by adjacent individuals (Ager and Riggs, 1964).

Recent plenipedunculate brachiopods have the ability actively to orientate themselves with respect to currents, and the positions taken up by living specimens in areas of current flow conform to hydrodynamic predictions.

If the anterior–posterior axis of the shell was parallel with the current (and especially if the excurrent region faced into it), ciliary pumping would actually be opposed by pressure distributions. If, on the other hand, the living animal had the axis perpendicular to the current (and especially if one of the incurrent regions faced into it), pumping would be assisted. This is, in fact, exactly what happens in species studied, and brachiopods may swing through a substantial arc to take up this preferred position relative to prevailing currents (La Barbera, 1977).

Functional pedicles seem to have been present in most rhynchonellides and terebratulides, most orthides and spiriferides, some pentamerides and a few strophomenides.

Brachiopod assemblages and 'community' ecology

Fossil brachiopods tend to be found in recurrent assemblages, often composed of several species. In these assemblages other invertebrates also occur, but the brachiopods are usually dominant. These assemblages have been much used in recent years in attempts to understand the structure of ancient (mainly epifaunal) palaeocommunities and the controls operating upon them.

The recurrent assemblages characterize particular environments and in life must have related to such parameters as depth, temperature, salinity and substrate. Most palaeoecologists working with recurrent assemblages have tried not only to define them in terms of their composition and the relative abundance of their faunal elements, but also to determine, as far as possible, the nature of the controls.

So much recent work has been done on this most vital of palaeoecological fields that the bulk of literature is considerable. Only a few examples are here selected for discussion.

Ordovician palaeocommunities

In recent years voluminous knowledge of Ordovician palaeocommunity structure has accumulated. Although this started later than comparable studies on the Silurian, over 50 different associations have by now been described. It is hard to summarize and integrate all this information into a coherent picture, but some key works have been successful in doing so. Lockley (1983), using cluster analysis, and working on Anglo-Welsh faunas from the type area of the Ordovician, has been able to discern some clear patterns of distribution of fossil brachiopods relative to original sediment types (Fig. 7.18).

Coarse sediments, as has long been known, are dominated by coarse-ribbed orthaceans (*Hesperorthis*, *Dinorthis*), whereas very fine sediments contain many inarticulated brachiopods and small plectambonitaceans (*Sericoidea*, *Chonetoidea*). The silty middle part of the spectrum is of wider diversity and is dominated by dalmenellids, heterorthids and large plectambonitaceans such as *Sowerbyella*, which have relatively wide facies ranges. Extrapolating further, Lockley was able to show that the range of brachiopod associations could be divided into eight major community types, which persisted throughout the whole of the Ordovician. Even though the genera changed through time, the balance, facies associations and general aspect of these eight communities remained relatively similar.

Silurian palaeocommunities

Early classic studies of brachiopod-dominated palaeocommunities were those of Ziegler (1965) and Ziegler *et al.* (1968), updated by Cocks and McKerrow (1984), which were based on recurrent assemblages of Upper Llandovery age in the Welsh Borderland (Fig. 7.19).

During the earlier Llandovery a shoreline ran eastwards from Haverfordwest and, curving northwards near the town of Llandovery, extended due northwards. On the shelf sea between the shore and the graptolite shale area to the west lived a rich fauna. By the time the shoreline had retreated to the east of the Malvern Hills in the Upper Llandovery the shelf was much broader and shelly faunas were differentiated over it. Five distinct though gradational palaeocommunities dominated by brachiopods (though with subsidiary bivalves, gastropods, corals and trilobites) lay in distinct concentric belts parallel with the shore.

Nearest the shore lay a *Lingula* palaeocommunity, with abundant small rhynchonellides (probably *Stegerhynchus* or *Rostricellula)* and *Lingula*. The next-from-shore *Eocoelia* palaeocommunity had similar rhynchonellides *(Eocoelia)* but a more diverse range of other brachiopods. In the *Pentamerus* palaeocommunity which lay farther out the eponymous genus dominated all others, often reaching a large size; specimens are sometimes found in life position. The large *Costistricklandia* was the most abundant brachiopod in the penultimate palaeocommunity, though smaller brachiopods were also important. In the outer-shelf *Clorinda* palaeocommunity there were at least a dozen genera of relatively small brachiopods, so that it was the most diverse community of all. Beyond this lay an area of graptolitic muds

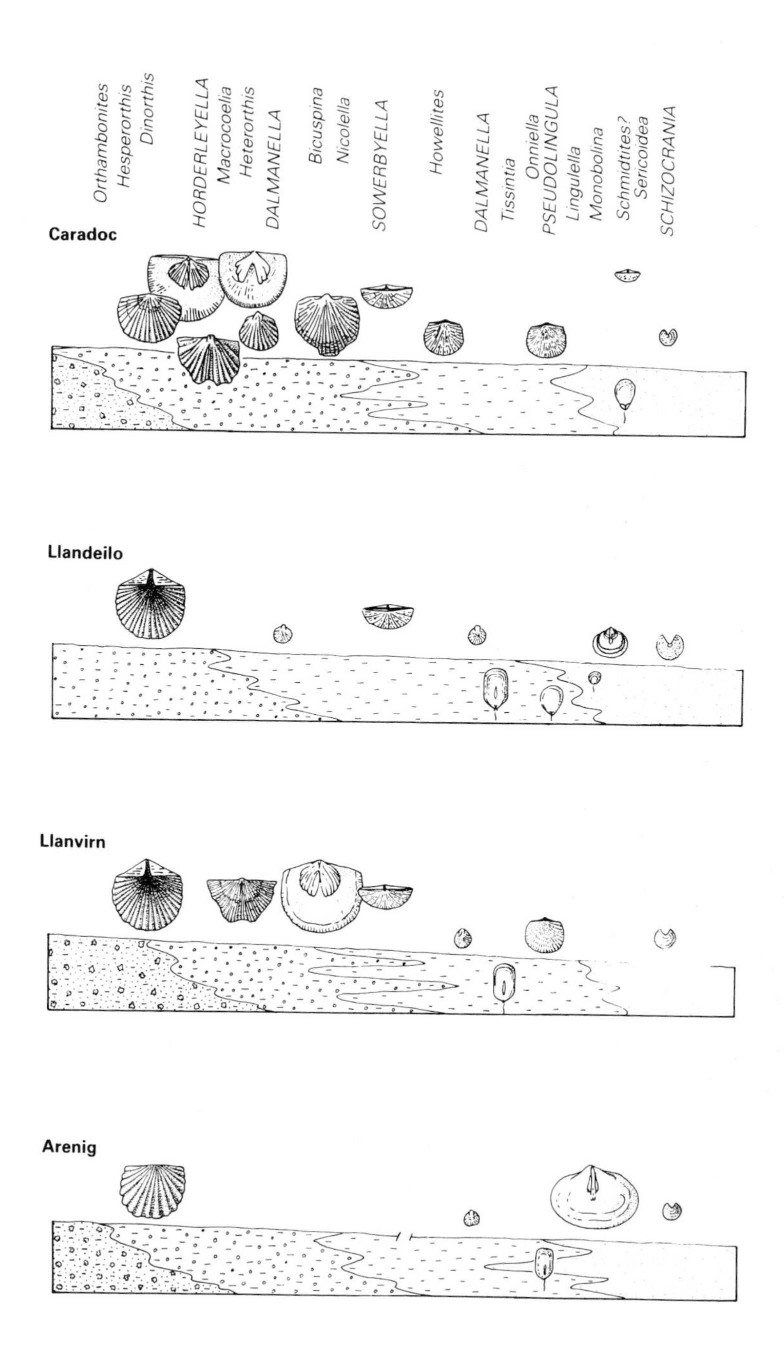

Figure 7.18 Generalized pattern of facies relationships and substrate preferences of Ordovician brachiopods through time: coarse substrates are on the left, with a predominance of ribbed brachiopods; fine substrates are on the right. (Redrawn from Lockley, 1983.)

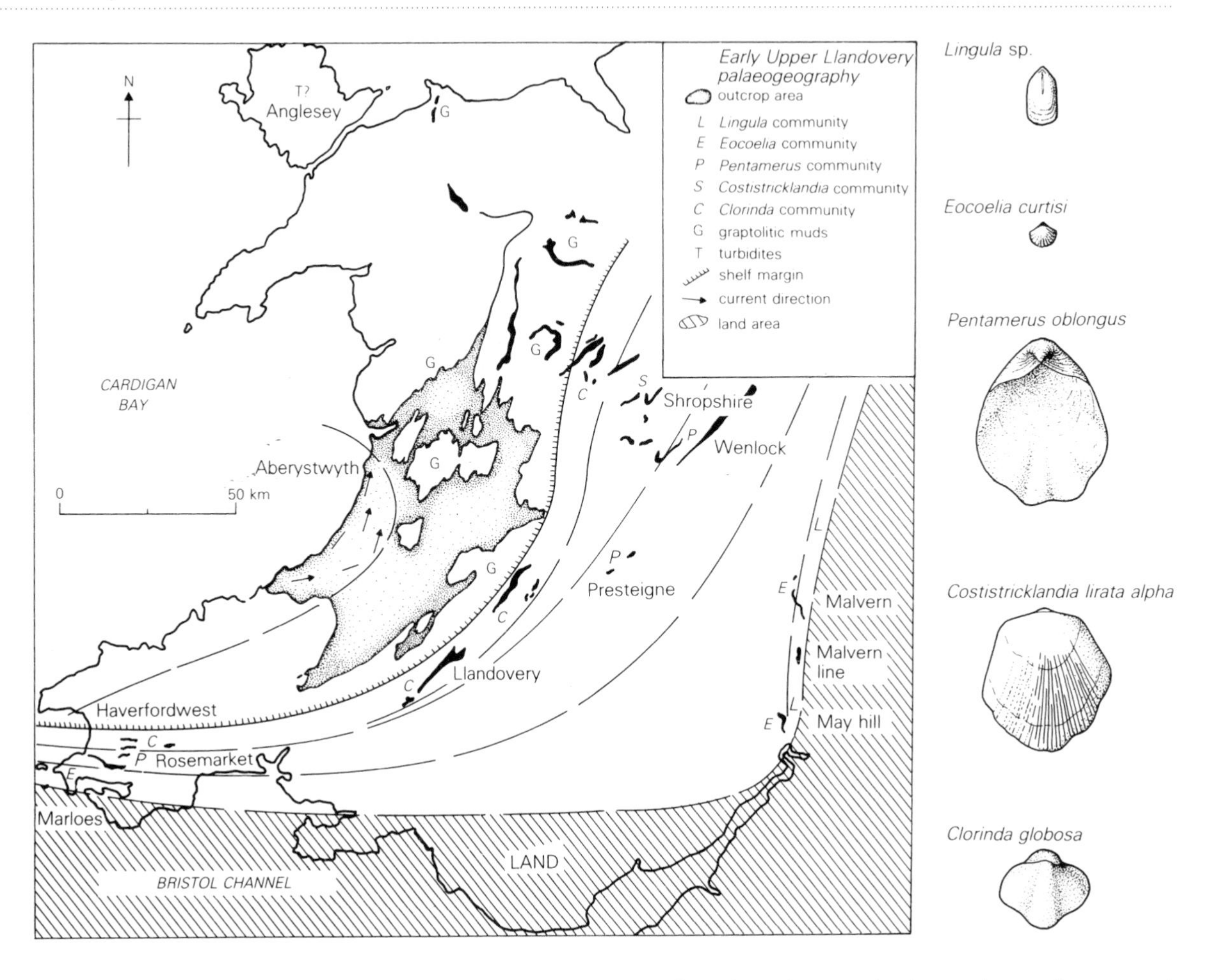

Figure 7.19 Palaeogeography and brachiopod palaeocommunities in Wales in the early Upper Llandovery, showing illustrations of the brachiopod genera defining the palaeocommunities. (Largely redrawn from Ziegler *et al.*, 1968.)

accumulating in a deeper-water environment. Similar palaeocommunities have been found in rocks of the same age in other parts of the world, and in some cases they seem to be equivalent to those defined by Ziegler *et al.* (1968).

The examples of Ziegler *et al.* (1968) show a relatively straightforward pattern of distribution in narrow bands which were largely independent of substrate and parallel with the shore. This seems to suggest that the distribution pattern is in some way related to depth of water. Depth alone, however, has been generally regarded as a rather elusive factor whose influence on sediments or animals is indirect. It is not depth as such that is a primary control, but certain important factors that vary with depth, including pressure, water turbulence, salinity, substrate and food supply. Temperature, however, which decreases in deeper water, according to Boucot (1975), may be the most important of all these variables.

Where other Silurian brachiopod palaeocommunities have been described, however, such as the Ludlovian assemblages of Lawson (1975), it is less easy to assess the controls of distribution. Lawson, suggesting that substrate may be more important than depth in some cases and noting that brachiopods belonging to different associations may have varying tolerance ranges, rejected attempts to impose the simple pattern exhibited by the Upper Llandovery palaeocommunities on other kinds of assemblage.

Devonian brachiopod assemblages

A primary task for palaeoecology is to try to integrate the existing bulk of data on Palaeozoic assemblages and to establish how they have evolved in fluctuating environments through time. This work, following that of Bretsky (1969) and Boucot (1971), promises well, but the major hazard encountered is in determining the environmental equivalence of disjunct associations which replaced each other in time but which may have few characters in common with their precursors.

The major work of Boucot (1975) has shed much light on this topic. He undertook a large-scale study of Siluro-Devonian brachiopod distribution, primarily to assess the size of breeding populations (in terms of the areas they occupied) and so to determine how rates of evolution and extinction varied in populations of different sizes. He was able to confirm the predictions of population dynamicists that evolution would be rapid in smaller populations and postulated a complex series of biological and physical controls over extinction events.

In establishing a 'standard community framework' for this study, Boucot redefined all known Ordovician to Devonian palaeocommunities in terms of **benthic assemblages** (BAs). A benthic assemblage is 'a group of communities occurring repeatedly in the same position relative to a shoreline'. According to Boucot, these were probably controlled by water temperature, decreasing with depth. Boucot thus regarded the *Lingula* through *Clorinda* 'communities' of Ziegler *et al.* (see Silurian palaeocommunities above) as BAs 1–5, zoned according to distance from shoreline. He also defined a sixth BA lying seawards of BA 5. He was able to relate all known Ordovician to Devonian assemblages to a system of six BAs, though the components might vary laterally depending on water turbulence. Separate quiet and rough water assemblages could be distinguished in most of these BAs. Absolute depths for BAs have now been investigated using several lines of evidence, and BAs 1–4 are now believed to lie within the photic zone (0–60 m), and possibly all or part of BA 5 too (Brett *et al.*, 1993)

The composition of BAs changed through time, often abruptly. Where there were reefs, with their largely endemic and rapidly evolving populations, they were usually associated with BAs 2–3. Assigning a given brachiopod association to a particular BA may be difficult where there has been an abrupt faunal replacement so that there is no continuity with former associations. However, where neighbouring associations have not changed, then the assemblage in question may be 'bracketed' between them and so referred to its correct BA. If an unknown association lies between BA 3 and BA 5 it is probably BA 4, even though it is disjunct.

Siluro-Devonian chonetaceans, as distributed within benthic assemblages, show a clear relationship between morphology and environment (Racheboeuf, 1990). Chonetaceans were not uncommon in the Silurian, but with diversification of environments in the early Devonian they underwent a major radiation. Various adaptations in shell shape and the development and distributions of hinge spines have been directly related to bathymetric gradient.

Boucot's model of benthic associations is so far only a provisional step towards the integration of 'palaeocommunity' data and the interpretation of this in environmental and evolutionary terms, but it provides a necessary framework for future developments of community studies in Palaeozoic brachiopods.

Permian reef associations

In western Texas, Southeast Asia and Sicily there are immense reefs of Permian age composed mainly of algae, bryozoans and sponges. The Capitan reef complex of western Texas is the best known of these. It marks a late stage in reef development, forming a giant barrier reef round the margins of deep-water basins which it encircled; these had only a narrow outlet to the sea on the west. Behind the reef lay large salt flats: evaporating pans in which dolomite, gypsum and anhydrite were deposited in sequence away from the reef as the water became progressively more hypersaline. These were replenished by sea water washing over the top of the reef. The reefs are not especially fossiliferous and contain only localized developments of brachiopods, in pockets of low diversity. Prior to the development of the great algal reefs, numerous smaller reefs or bioherms grew in somewhat deeper water. These patch reefs, best exposed in the Glass Mountains, grew up to 80 m high, though they were usually smaller and are abundantly fossiliferous, containing some of the most highly modified brachiopods ever to have lived (Grant, 1971).

Most of the Permian biohermal brachiopods are

of relatively normal form; the majority were pedicle-attached and able to exploit several different habitats within and peripheral to the bioherm. Other kinds were definitely 'anti-reefal' and lived in the flat sediment between the reefs. Others again were confined to the reef, often being of conical form with long tubular spines forming an interlacing meshwork which contributed to the reef fabric. These rather coral-like brachiopods were all modified strophalosiid productides in which the conical pedicle valve was cemented in the early stages, and had tubular spines, as in *Chonosteges* (q.v.), which held them up. Such spines seems to have provided an 'evolutionary springboard' for their possessors.

Some bioherms were composed largely of Richthofeniidae (e.g. *Coscinaria* and *Cyclacantharia*; Fig. 7.17g,h). These have long, pointed pedicle valves supported by tubular spines. The brachial valve is reduced to a thin, flat, horizontal plate recessed well within the pedicle valve; it could open to 80° or so and had a protective grille of spines or meshwork above the opened valve. Rudwick (1961) made a working model of one of these and showed that the brachiopod could have generated strong currents, bringing in food and flushing out the mantle cavity simply by rapidly flapping the brachial valve using the adductor and diductor muscles alternately.

Though this interpretation has been criticized by Grant (1972) – on the grounds that the known structure of the large ptycholophe in the Permian productides (Fig. 7.17a–c) would make this unlikely, and that the brachiopod could have functioned perfectly well without such a valve-flapping mechanism – the possibility still remains that the richtofeniides could have used flapping valves to supplement or replace ciliary pumping as a means of collecting food.

Some other Permian productides (*Gemellaroia* and *Leptodus* (= *Lyttonia)* are even more peculiar and although the functional morphology of some of these peculiar forms may be open to question, surely Rudwick is correct in saying that these adaptations 'bear witness to a degree of plasticity unparalleled at any other time in brachiopod evolution'.

Mesozoic brachiopod associations

In many instances there seems to be a broad and general correlation of brachiopod shell form with environments. It was recognized long ago, for example, that Lower Palaeozoic orthides with coarse ribs preferred arenaceous environments, and there is a general tendency in the Mesozoic for similar anatomical features to occur in unrelated stocks in similar environments. Ager (1965, 1967) showed how Mesozoic brachiopods can be related to seven main biofacies, and the nature of the substrate seems to have exercised a dominant control over distribution. Thus there are characteristic assemblages from (1) very shallow water, coarse sediments, (2) sand-grade shelf sediments, (3) peri-reefal sediments, (4) rocky bottoms, (5) mud-grade shelf sediments, (6) fine-grade, probably bathyal sediments and (7) epiplanktonic brachiopods. Consequently, any assemblage from very shallow waters generally includes rhynchonellides with coarse ribs and large pedicle openings supporting stout pedicles, as well as terebratulides with sharp commissure folds. A quite different assemblage was to be found in 'peri-reefal' environments, with large asymmetrical rhynchonellides and terebratulides with elongated beaks. Some peculiar brachiopods, e.g. the perforate *Pygope*, were confined to calm, deep seas, whereas small thin-shelled rhynchonellides were evidently epiplanktonic.

One well-studied section in the French Jura, for example, has four successive rhynchonellids, *Rhactorhynchia* at the base, associated with reefs, *Acanthothiris* of normal benthonic habit and found in sands, *Lacunosella*, associated with sponges, and at the top the isolated *Acanthorhynchia*, which occurs in clays and was probably epiplanktonic. Likewise, other rhynchonellids from different stocks illustrate homeomorphic convergence in adapting to reefal or peri-reefal habitats.

7.7 Faunal provinces

The global distribution of brachiopods is most fully documented in the Lower Palaeozoic, which is in any case that part of the geological column where they are of greatest stratigraphical value and where they show an instructive parallel with contemporaneous trilobites.

The most recent summaries are by Boucot (1975) and Cocks and Fortey (1990). Data on brachiopod distribution are consistent with those from trilobites and graptolites, and fit in well with modern palaeogeographical reconstructions (Scotese and

McKerrow, 1990). In the early Ordovician the large Gondwanan continent (South America, Africa, Madagascar, India, Antarctica and Australia) extended from the South Pole to some 30°N. Baltica (Scandinavia and western Russia) lay outboard, between 30 and 60°S, as did Avalonia (England, southern Ireland and eastern Newfoundland). The large Laurentian continent (North America, western Newfoundland and Scotland) sat astride the equator, as did the independent landmasses of East Asia and Siberia.

As discussed earlier (Chapter 1), the primary control of distribution of today's shelf-living faunas is temperature, though the relative isolation of the continental landmasses also has a significant role. The same controls operated in the past, and the known distribution of Ordovician brachiopod faunas is fully consistent with them. Thus there was an early Ordovician fauna of endemic lingulates and inarticulates occupying cold polar waters, later joined by dalmanellids. (This occupied the same waters as the contemporaneous calymenacean–dalmanitacean trilobite province.) Likewise the low-latitude areas of Gondwana carried their own endemic brachiopod faunas, and so did Baltica and the freely floating continents strung out along the equator. The 'subpolar' Ordovician faunas of western South America, North Africa and Europe (west Gondwana) can be distinguished from those of the more tropical 'east Gondwana', namely Australia and Southeast Asia. Specialized deep-water faunas, such as the later Ordovician *Foliomena* fauna, of small thin-shelled brachiopods colonized outer-shelf margin sites.

By the later Ordovician brachiopod faunas became somewhat less provincialized due to the closer approach of the continents, though there was still high endemicity in places.

At the end of the Ordovician an ice sheet, which seems to have existed in polar regions (west Gondwana) from the early Caradoc, expanded significantly. Associated with this is the *Hirnantia* fauna (discussed in full by Rong and Harper, 1988). Typically this consists, as in Europe, of only a very few genera, though there is a steep biogeographical gradient towards China, which was more tropical, and the diversity of *Hirnantia* faunas is appreciably higher. This late-Ordovician (Hirnantian) glaciation did much harm to brachiopods, as to the other faunas, and collectively the biotas recovered slowly.

Silurian brachiopod faunas recovering from this decimation became cosmopolitan, though two new provinces arose, one characterized by *Clarkeia* which became established in South America, and the other with *Atrypella* which arose in the former USSR, Greenland and western North America. During the Devonian there was further provincial differentiation, resulting eventually in five faunal provinces which have been fully documented by Boucot (1975).

The striking parallels between trilobite and brachiopod distribution in the Lower Palaeozoic shows how palaeobiogeography can be a powerful tool in timetabling major global events and constraining plate tectonic models.

7.8 Stratigraphical use

Many brachiopod genera and species are relatively long-ranged and thus of limited value in correlation. This is particularly so with Mesozoic brachiopods, which though sometimes useful on a local scale are of less stratigraphical applicability over wider areas.

In the Palaeozoic, however, brachiopods have proved of much greater value; together with trilobites they are the primary stratigraphically useful fossils in the shallow-water facies of the Ordovician. Brachiopod assemblages are, of course, very much controlled by facies and, on a global scale, by realms and provinces, but correlations at the stage level based upon concurrent brachiopod ranges within provinces can be very good. Thus stratigraphy and structure in the Caradoc rocks of the Girvan district of Ayrshire, Scotland, have been correlated by reference to the standard continuous brachiopod/trilobite-bearing series of Virginia, Alabama and Tennessee, both areas being part of the Scoto-Appalachian province of the American realm.

So successful has this stratigraphical work been that a case can be presented for using 'hybrid' stratigraphical tables in areas such as the British Isles, where the incursion of 'American' forms at a particular period in time allows the recognition of zones and stages originally defined in the United States.

There is now a good correlation between the stages of the Ordovician and Silurian shelly facies and concurrent graptolite zones; indeed there are seven Caradoc stages in England and Wales for only two graptolite zones, indicative of a fine refinement of correlation. The American shelly facies stages are less finely drawn but still good. However, such correlations are still more firmly based in some parts of the world than in others.

In the Carboniferous, brachiopods have been of some stratigraphical value when combined with corals in the zones erected for the Lower Carboniferous in England by Vaughan (1905) and Garwood (1913). Though still valid, these zones are now supplemented by stratigraphical work using microfossils, which gives a more precise correlation.

Bibliography

Books, treatises and symposia

Boucot, A.J. (1975) *Evolution and Extinction Rate Controls*. Elsevier, Amsterdam. (Develops concept of benthic assemblages)

Brunton, C.H.C. and Curry, G.B. (1979) *British Brachiopods*. Linnean Society Synopses of the British Fauna No. 17. Academic Press, London. (All British species)

Davidson, T. (1851) *A Monograph of the British Fossil Brachiopoda*. Palaeontographical Society, London (Standard superbly illustrated reference in five volumes: taxonomy)

McKinnon, D, Lee, D.E, and Campbell, J.D. (eds) (1991) *Brachiopods Through Time*. Balkema, Rotterdam (several up-to-date papers)

Moore, R.D. (ed.) (1965) *Treatise on Invertebrate Paleontology*, Part H, *Brachiopods*, 2 vols. Geological Society of America and University of Kansas Press, Lawrence, Kan. (Kaesler, R.L. (ed.) (1977) Vol. 1 of the revised version is now published)

Rudwick, M.J.S. (1970) *Living and Fossil Brachiopods*. Hutchinson, London. (Invaluable treatment, with a functional bias)

Individual papers and other references

Ager, D.V. (1965) The adaptation of Mesozoic brachiopods to different environments. *Palaeogeography, Palaeoclimatology, Palaeoecology* **1**, 143–72.

Ager, D.V. (1967) Brachiopod palaeoecology. *Earth Science Reviews* **3**, 157–79.

Ager, D. and Riggs, E.A. (1964) The internal anatomy, growth and asymmetry of a Devonian spiriferid. *Journal of Paleontology* **33**, 749–60. (Crowding results in asymmetry)

Baker, P.G. (1970) The growth and shell microstructure of the thecideacean brachiopod *Moorellina granulosa* from the Middle Jurassic of England. *Palaeontology* **13**, 76–99.

Baker, P.G. (1983) The diminutive thecideidine brachiod *Enallothecidea pygmaea* (Moore) from the Middle Jurassic of England. *Palaeontology* **26**, 663–70.

Baker, P.G. (1990) The classification, origin, and phylogeny of articulate brachiopods. *Palaeontology* **33**, 175–92.

Bassett, M.G. (1984) Life strategies of Silurian brachiopods, in *Autecology of Silurian Organisms* (eds M.G. Bassett. and J.D. Lawson), Special Papers in Palaeontology No. 32, Academic Press, London, pp. 237–63.

Blight, F.G. and Blight, D.F. (1990) Flying spiriferids: some thoughts on the life style of a Devonian spiriferid brachiopod. *Palaeogeography, Palaeoclimatology, Palaeoecology* **81**, 127–39.

Boucot, A.J. (1971) Practical taxonomy, zoogeography, paleoecology, paleogeography, and stratigraphy for Silurian–Devonian brachiopods. *Proceedings of the North American Paleontological Convention*, Chicago, 1969 (F), pp. 566–611. (Early study; base for Boucot, 1975)

Bretsky, P. W. (1969) Central Appalachian late Ordovician communities. *Bulletin of the Geological Society of America* **80**, 193–212. (Community structure)

Brett, C., Boucot, A.J. and Jones, R. (1993) Absolute depth ranges of Silurian benthic assemblages. *Lethaia* **26**, 25–40.

Burton, C.J. and Curry, G.B. (1985) Pelagic brachiopods from the Upper Devonian of East Cornwall. *Proceedings of the Ussher Society* **6**, 191–5.

Cherns, L. (1979) The environmental significance of *Lingula* in the Ludlow Series of the Welsh Borderland and Wales. *Lethaia* **12**, 35–46. (Deep-water lingulids)

Craig, G.Y. (1951) A comparative study of the ecology and palaeoecology of *Lingula*. *Transactions of the Edinburgh Geological Society* **15**, 110–20. (Fossil *Lingula* in life position)

Cocks, L.R.M. and Fortey, R.A. (1990) Biogeography of Ordovician and Silurian faunas, in *Palaeozoic Palaeogeography and Biogeography* (eds W.S. McKerrow and C.R. Scotese), Geological Society Memoir No. 12, pp. 97–104.

Cocks, L.R.M. and McKerrow, S. (1984) Review of the commoner animals in Lower Silurian marine benthic communities. *Palaeontology* **27**, 663–70.

Curry, G.B. (1982) Ecology and population structure of the Recent brachiopod *Terebratulina* from Scotland. *Palaeontology* **25**, 227–46.

Curry, G. (1983a) Microborings in Recent brachiopods and the functions of caeca. *Lethaia* **16**, 119–27.

Curry, G. (1983b) Ecology of the Recent deep-water rhynchonellid brachiopod *Cryptopora* from the Rockall trough. *Palaeogeography, Palaeoclimatology, Palaeoecology* **44**, 93–102. (Tethering pedicle)

Emig, C.C. (1981) Implications de donnés recentes sur les lingules actuelles dans les interpretations paleoécologiques. *Lethaia* **14**, 151–6. (Habitats of lingulids)

Emig, C., Gall, J.C., Pajaud, D. and Plaziat, J.C. (1978) Reflexions critiques sur l'écologie et la systèmatique des lingules actuelles et fossiles. *Geobios* **11**, 573–609.

Ferguson, L. (1962) The palaeoecology of a Lower Carboniferous marine transgression. *Journal of Paleontology* **36**, 1090–1107 (Classic palaeoecological study)

Garwood, J.E. (1913) The Lower Carboniferous succession in the northwest of England. *Quarterly Journal of the Geological Society of London* **68**, 449–596. (Zonation using coral and brachiopod fauna)

Grant, R.E. (1966a) A Permian productoid brachiopod: life history. *Science* **152**, 660–2. (Change in life habits during ontogeny)

Grant, R.E. (1966b) Spine arrangement and life habits of the productoid brachiopod *Waagenoconcha*. *Journal of Paleontology* **40**, 1063–9. (As above)

Grant, R.E. (1971) Brachiopods in the Permian reef environment of west Texas. *Proceedings of the North American Paleontological Convention*. Chicago, 1969 (J), pp. 1444–81. (Bioherms with unusual specialized faunas)

Grant, R.E. (1972) The lophophore and feeding mechanisms of the Productidina (Brachiopoda). *Journal of Paleontology* **46**, 213–48. (First description of productid lophophores and functional interpretation)

Hallam, A. (1962) Brachiopod life assemblages from the Marlstone Rock bed of Leicestershire. *Palaeontology* **4**, 653–9. (Successive annual broods)

Hammond, L.S. (1983) Experimental studies of salinity tolerance, burrowing behaviour and pedicle regeneration in *Lingula anatina* (Brachiopoda Inarticulata). *Journal of Paleontology* **57**, 1311–16.

Holmer, L.E. (1989) Middle Ordovician phosphatic inarticulate brachiopods from Västergötland and Dalarna, Sweden. *Fossils and Strata* **26**, 1–172.

La Barbera, M. (1977) Brachiopod orientation to water movement. 1. Theory, laboratory behaviour and field orientation. *Paleobiology* **3**, 270–87.

Lawson, J.D. (1975) Ludlow benthonic assemblages. *Palaeontology* **18**, 509–25. (Distinction of four brachiopod-dominated assemblages and their controls)

Lockley, M.G. (1983) A review of brachiopod-dominated palaeocommunities from the type Ordovician. *Palaeontology* **26**, 111–45. (Excellent treatment)

McGhee, G.R. (1980) Shell form in the biconvex articulate Brachiopoda: a geometric analysis. *Paleobiology* **6**, 57–76.

Muir-Wood, H. (1955) *A History of the Classification of the Phylum Brachiopoda*. British Museum of Natural History, pp. 1–124. (First attempt to erect a modern classification)

Percival, E. (1944) A contribution to the life-history of the brachiopod *Terebratella inconspicua* Sowerby. *Transactions of the Royal Society of New Zealand* **74**, 1–24.

Racheboeuf, P. (1990) Les brachiopodes Chonetacés dans les assemblages benthiques siluriens et devoniens. *Palaeogeography, Palaeoclimatology, Palaeoecology* **81**, 141–61.

Richardson, J.R. and Watson, J.E. (1975) Form and function in a Recent free-living brachiopod. *Paleobiology* **1**, 379–87. (Unusual life habits)

Rong, J.-Y. and Harper, D.A.T. (1988) A global synthesis of the latest Ordovician Hirnantian brachiopod faunas. *Transactions of the Royal Society of Edinburgh* **79**, 383–402.

Rudwick, M.J.S. (1959) The growth and form of brachiopod shells. *Geological Magazine* **96**, 1–24. (Growth vectors in recti-marginate and plicated shells)

Rudwick, M.J.S. (1961) The feeding mechanism of the Permian brachiopod *Prorichthofenia*. *Palaeontology* **3**, 450–71. (Working model shows operation of flapping valves)

Rudwick, M.J.S. (1964) The function of zig-zag deflections in brachiopods. *Palaeontology* **7**, 135–71. (Zigzags interpreted as protective)

Rudwick, M.J.S. (1965a) Adaptive homeomorphy in the brachiopods *Tetractinella* Bittner and *Cheirothyris* Rollier. *Paläontologische Zeitschrift* **39**, 134–46. (An extreme case of homeomorphy interpreted functionally)

Rudwick. M.J.S. (1965b) Sensory spines in the Jurassic brachiopod *Acanthothiris*. *Palaeontology* **8**, 604–17. (Spines as early warning sensors)

Rudwick, M.J.S. and Cowen, R. (1968) The functional morphology of some aberrant strophomenid brachiopods from the Permian of Sicily. *Bulletin Societa Paleontologica Italiano* **6**, 113–76. (Richthofeniids and others replace ciliary pumping by valve movement to generate currents)

Scotese, C.R. and McKerrow, W.S. (1990) Revised world maps and introduction, in *Palaeozoic Palaeogeography and Biogeography* (eds W.S. McKerrow and C.R. Scotese) Geological Society Memoirs No. 12, pp. 1–21.

Sheills, K.A.C. (1968) *Kochiproductus coronus* n. sp. from the Scottish Visean, and a possible mechanical advantage of its flange structure. *Transactions of the Royal Society of Edinburgh* **67**, 477–507. (Use of flow table in determining particle settling on flange)

Thayer, C.W. (1975) Strength of pedicle attachment in articulate brachiopods: ecologic and paleoecologic significance. *Paleobiology* **1**, 388–99.

Thayer, C.W. and Steele-Petrovic, H.M. (1975) Burrowing of the lingulid brachiopod *Clottidia pyramida:* its ecologic and paleoecologic significance. *Lethaia* **8**, 209–21. (How the pedicle is used)

Vaughan, A. (1905) The palaeontological sequence in the Carboniferous limestone of the Bristol area. *Quarterly Journal of the Geological Society of London* **61**, 181–307. (Classic study using corallbrachiopod zonation)

West, R.R. (1976) Comparison of seven lingulid communities, in *Structure and Classification of Palaeocommunities* (eds R.W. Scott and R.R. West), Dowden, Hutchinson and Ross, Stroudsburg, Penn., pp. 171–92. (Lingulids were adapted to different environments)

Williams, A. (1968) A history of skeletal secretion in brachiopods. *Lethaia* **1**, 268–87. (Excellent general review)

Williams, A. (1973) Distribution of brachiopod assemblages in relation to Ordovician palaeogeography, in *Organisms and Confinents Through Time* (ed. N.F. Hughes), Special Papers in Palaeontology No. 12, Academic Press, London, pp. 241–69. (Methodology, statistics; development and changes in faunal provinces)

Williams, A. (1976) Plate tectonics and biofacies evolution as factors in Ordovician correlation, in *The Ordovician System* (ed. M.G. Bassett), University of Wales Press, Cardiff, pp. 29–66. (Faunal provinces and stratigraphical problems raised by plate closure and faunal migration)

Williams, A. and Wright, A.D. (1961) The origin of the loop in articulate brachiopods. *Palaeontology* **4**, 149–76.

Williams, A., Carlson, S.J., Brunton, H.C. *et al.* (1996) A supra-ordinal classification of the Brachiopoda. *Philosophical Transactions of the Royal Society of London B* **351**, 1171–93.

Ziegler, A. M. (1965) Silurian marine communities and their environmental significance. *Nature* **207**, 270–2. (Assemblage distribution, with maps)

Ziegler, A.M., Cocks, L.R.M. and Bambach, R.K. (1968) The composition and structure of Lower Silurian marine communities. *Lethaia* **1**, 1–27. (Classic, well-illustrated work)

8 Molluscs

The Mollusca are one of the most diverse of all invertebrate phyla and include a whole range of animals, living and fossil, which at first sight seem to be so different as to be unrelated. The important groups include the curious plated chitons (polyplacophorans), the slugs and snails (gastropods), tooth shells (scaphopods), bivalves, their possible progenitors (the extinct rostroconchs) and finally the most complex and highly organized of all molluscs, the cephalopods, which include the modern squids, cuttlefish, octopuses and pearly nautilus, and the fossil ammonoids and belemnites. Molluscs are mainly a marine phylum, and only a few bivalves and gastropods have been successful in fresh water. Only one group of gastropods, the pulmonates, was ever able to make the transition from sea to land. The diversity of different kinds of molluscs is extreme, yet they are all united by a common ground plan, and the same basic structures are to be found in all molluscs, however much they have become differentiated.

8.1 Fundamental organization

It is probably simplest to consider the fundamental organization of molluscs with reference to a hypothetical archimollusc in which all the basic features of molluscan structure are present, though not particularly specialized in any one direction (Fig. 8.1).

While this way of presenting such concepts may be out of fashion at present, I continue to use it since most students find it helpful in understanding the complexities of molluscan anatomy. This hypothetical molluscan ancestor has close similarities to the chitons and an even closer resemblance to the modern limpet-like primitive mollusc *Neopilina* (Fig. 8.2), which was unknown until 1953 when many specimens were dredged from the deep sea during the Danish 'Galathea' Expedition.

Neopilina is a member of the molluscan Class Monoplacophora, which has fossil representatives known from the Cambrian to the Devonian. The discovery of this 'living fossil' has been of the greatest interest, especially since, contrary to expectations, it shows traces of segmentation.

The structure of the archimollusc is quite simple. It has a cap-like **shell** secreted by a layer of tissue known as the **mantle**. Below this is the body with a mouth at one end and anus at the other, the latter discharging into a posterior space, the **mantle cavity**, in which also reside the gills.

In other molluscs this structure is modified in different ways. Most molluscs have an external shell of calcium carbonate, though this has been lost in various gastropods, particularly the marine nudibranchs and the terrestrial slugs. In squids and cuttlefish, as well as in the extinct belemnites, the shell has become modified as an internal skeleton, and in the octopuses it has been entirely lost. Where the shell is present it is always secreted by the mantle, which directly underlies the shell. Throughout growth calcium carbonate is added to the edge of the shell from the mantle. In most molluscs this mantle consists of only a few cell thicknesses, but in the cephalopods it is greatly thickened and supplied with powerful muscles. The mantle cavity is a constant feature of all molluscs and (in different molluscan classes) has been put to various uses. The gills, whose primary function is respiratory exchange, are present in all molluscs except the land gastropods (pulmonates), where the internal surface of the mantle is highly vascular and is modified as a lung. In the bivalves the gills are not only respiratory but also adapted for the gathering of suspended food particles from the water. The mantle cavity has both an inlet and an outlet to the sea through which waste material from the anus discharges, as well as respired water. This outlet has also undergone a

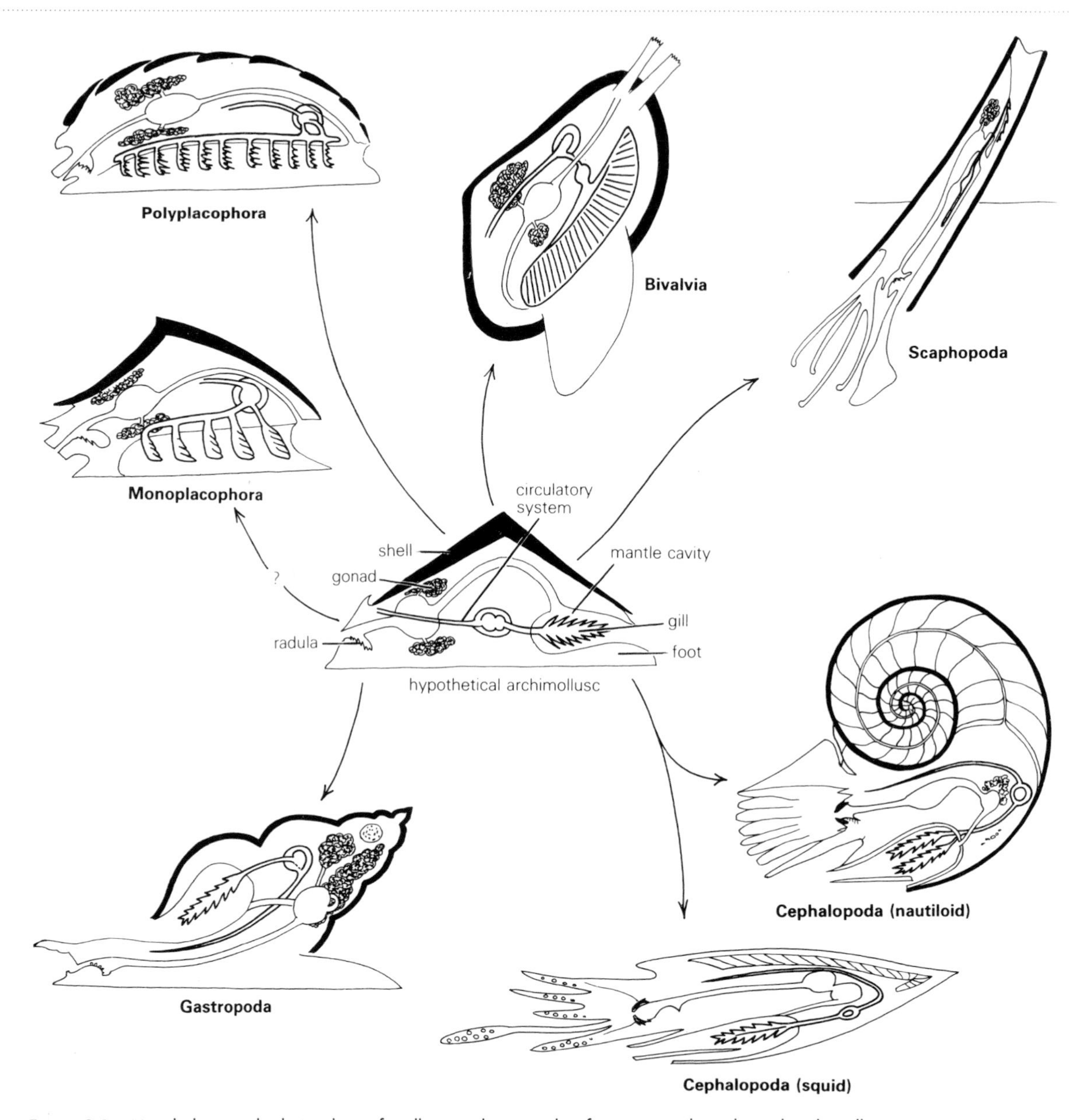

Figure 8.1 Morphology and relationships of molluscan classes with reference to a 'hypothetical archimollusc'.

differential modification, most markedly in the cephalopods where water, taken into the mantle cavity and squirted out as a jet through a funnel, is the primary means of propulsion.

Above the mantle cavity in all molluscs is the **visceral mass**, which contains the gut, digestive glands and kidneys and the nervous, circulatory and muscular systems. These too, in structure, arrangement and physiology, have most features in common.

Such then is the fundamental plan of molluscan organization, but why should molluscs have diverged from it in so many directions? It seems that much of the evolutionary differentiation of molluscs is bound up with their mode of nutrition, and it is interesting to see how their structural differentiation is closely connected with how they feed. The

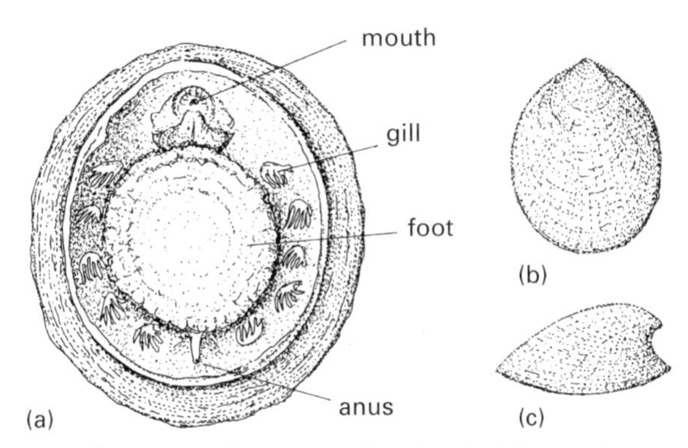

Figure 8.2 *Neopilina* (Monoplacophora): (a) ventral view, showing central foot and serially arranged gills (× 2 approx.): (b) dorsal view (× 1); (c) lateral view (× 1). [(a) based on an illustration by Lemche in *Treatise on Invertebrate Paleontology*, Part I.)

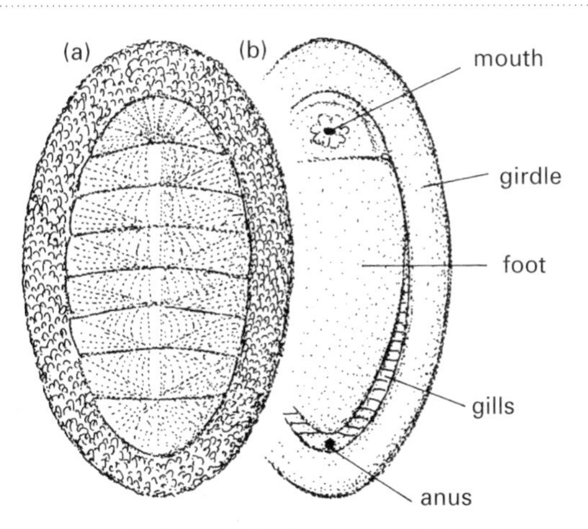

Figure 8.3 *Chiton* (Polyplacophora): (a) dorsal view; (b) ventral view (× 2).

monoplacophorans, chitons and most gastropods are all slow-moving molluscs which creep along on a muscular 'foot'. They are all provided with a radula, a belt of serially arranged teeth within the mouth, which fits them to be herbivores, carnivores or scavengers. Only the gastropods, of these three classes, have a well-developed head with sense organs which enable them actively to hunt. The bivalves are mostly suspension feeders and less commonly deposit feeders. They have no head, no jaws and normally only limited capacity for movement. The cephalopods, on the other hand, are fast-moving hunters; here the radula is used mainly in swallowing, but the cutting up of food is done by the powerful jaws. The head, with its highly developed sense organs and well-organized brain, the animal's power of movement, its buoyancy devices and the tentacular apparatus all evolved in accordance with this habit of catching active prey. It is the mode of feeding that has been at the root of molluscan differentiation; this should be remembered when considering the structural plan of the various molluscan groups.

8.2 Classification

An outline classification of the main divisions within Phylum Mollusca is as follows.

CLASS 1. MONOPLACOPHORA (Camb.–Rec.; Fig. 8.2): A group of primitive marine molluscs with univalved limpet-like shells (though unrelated to limpets proper, which are gastropods and distinguishable from them on the basis of muscle scars). They have paired, serially repeated muscles, gills, nephridia (excretory organs) as well as other internal organs. The foot is circular and central and ringed by the mantle cavity in which lie the gills. Monoplacophorans are the only molluscan class with a true internal segmentation, suggesting a zoological relationship with a segmented ancestor (but see Chapter 3). Fossils such as *Pilina* and *Tryblidium* are known only from the Palaeozoic. *Neopilina*, *Vema* and a few other genera occur today mainly in the deep sea.

CLASS 2. POLYPLACOPHORA (AMPHINEURA) (U. Camb.–Rec.; Fig. 8.3):Marine molluscs, having a bilaterally symmetrical shell with seven or eight calcareous plates, but otherwise resembling the archetypal molluscan plan in having an anterior mouth with radula, a posterior mantle cavity with anus and gills, and a ventral foot. Polyplacophorans are rare fossils but occur scattered throughout the Phanerozoic, e.g. *Chiton*.

CLASS 3. SCAPHOPODA (Ord.–Rec.; Fig. 8.4a): Marine molluscs with small tapering curving shells open at both ends. The anterior wider end with the mouth is permanently embedded in sediment; the animals feed on small organisms using specially adapted tentacles. The anus is at the upper end and the gills are much reduced. Recent species are more abundant than fossil ones and occur dominantly on the continental slope or shelf. The one detailed study so far made of the ecology of a fossil scaphopod, the Ordovician *Plagioglypta*, suggests similar habits to modern forms, e.g. *Dentalium*.

CLASS 4. BIVALVIA (LAMELLIBRANCHIA or PELECYPODA) (Cam.–Rec.): Bivalved molluscs or 'clams' with no definite head, but having the soft parts enclosed between paired but unequilateral calcareous shells united by a dorsal hinge, which may be toothed. The valves can be shut by strong internal musculature, and opened by the outward pressure of a springy ligament along the hinge when the

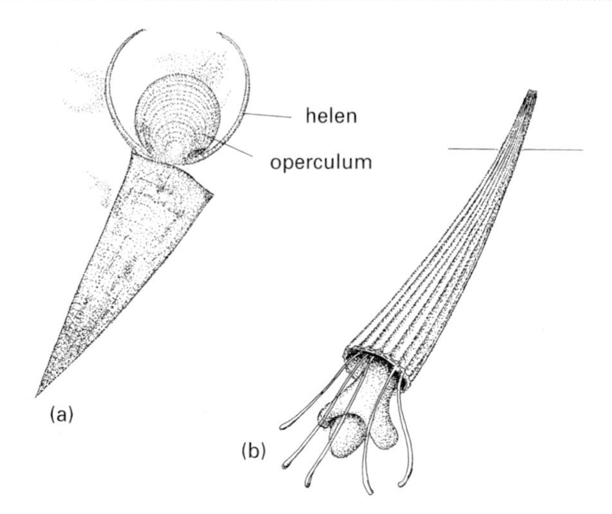

Figure 8.4 (a) a flattened hyolith, showing operculum and paired helens (M. Cam.; × 4), (b) *Dentalium*, a scaphopod, reconstructed in life position (× 3). [(a) redrawn from Butterfield and Nicholas. *Journal of Paleontology* **70**, 1996; (b) redrawn from Kershaw, *Animal Diversity*, 1983.]

muscles relax. The gills are large and modified for filter feeding, and the mantle cavity may be connected to the outer environment by **siphons**.

CLASS 5. ROSTROCONCHIA (L. Cam.–Perm.; Fig. 8.16): Extinct molluscs of bivalve-like appearance but with one or more of the shell layers continuous across the dorsal margin so that a dorsal commissure is lacking. The juvenile shells are univalved and coiled.

CLASS 6. GASTROPODA (Cam.–Rec.): Snails of all kinds – marine, land and freshwater – which usually creep along on a flattened 'foot'. A true head is present with eyes and other sense organs, and the single univalved shell is coiled, **planispirally** or more often **helically**. The internal organs are twisted by a 180° **torsion** so that the mantle cavity faces anteriorily. Some secondarily shell-less groups that have lost their torsion are known. Pteropods are marine gastropods, adapted for swimming, in which the foot is prolonged into lateral wings for life in the ocean. The tapering conical shells of these planktonic gastropods are sometimes found in fossil form.

CLASS 7. CEPHALOPODA (U. Cam.–Rec.): The most advanced of all molluscs, having external or internal chambered shells with the chambers linked by a siphuncle and giving buoyancy. They have a properly defined head with elaborate sense organs, and move by jet propulsion of water from the mantle cavity. The modern *Nautilus*, squids and octopuses are here included, as well as ammonites, belemnites and some chambered Cambrian molluscs (Fig. 8.5), extinct relatives of *Nautilus*.

The above classification is conservative and some authors would allow many more classes in the Cambrian. According to Yochelson (1979), for example, an early molluscan radiation in Cambrian times gave rise to a number of separate short-lived classes, including many kinds of simple cap-shaped forms (e.g. *Helcionella*; Fig. 8.16e). These were subsequently displaced by more effective early representatives of 'modern' groups in a second, late Cambrian–early Ordovician radiation. The alternative view

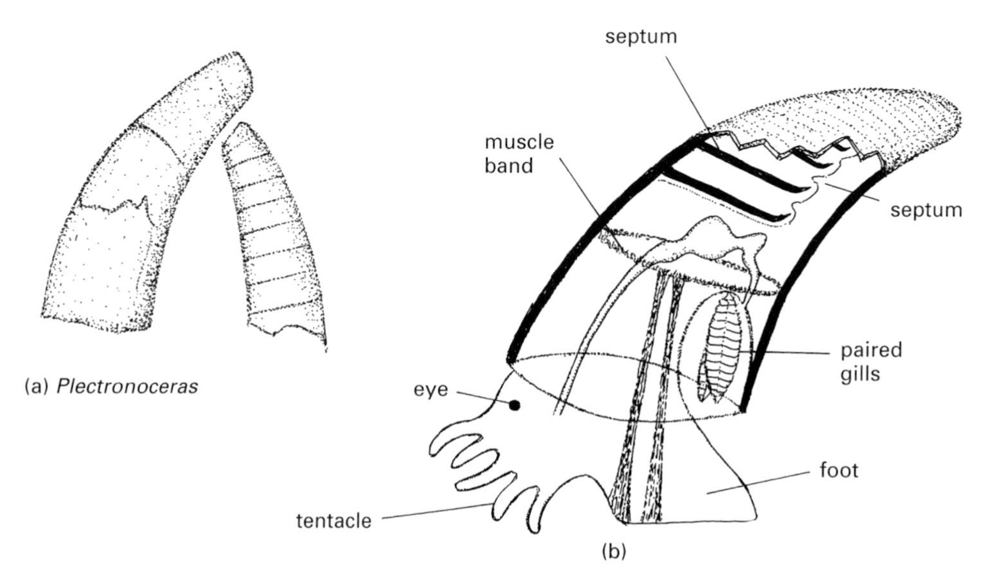

Figure 8.5 *Plectronoceras*, an early cephalopod: (a) specimens from the late Cambrian of Shantung, China (× 3); (b) inferred restoration of soft parts with shell partially removed to show mantle cavity with gills. (Based on Yochelson *et al.*, 1973.)

(Runnegar, 1996) suggests, however, that the small Cambrian molluscs do not pertain to extinct classes, but are early representatives of extant groups. Thus, according to this perspective, monoplacophorans, bivalves, gastropods and rostroconchs can be traced back to the early Cambrian while cephalopods, scaphopods and polyplacophorans, in the light of present evidence, do not seem to appear until the latest Cambrian or early Ordovician. The morphological evidence seems to favour the latter view but, since these early Cambrian molluscs are tiny, inconspicuous and of low diversity, still relatively little is yet known about them. Towards the end of the Cambrian there seems to have been a major (and if the early Cambrian origin of at least some molluscan classes is allowed) long-delayed radiation, coupled with a dramatic increase in size. It was this that led to the forceful establishment of the dominant molluscan groups which became then so important. The two models are agreed on the importance of the late Cambrian–early Ordovician radiation – the difference is in whether the molluscs of this time belong to newly established classes or are simply new, large members of ancient lineages. Of these various classes only the bivalves, gastropods and cephalopods are common and important fossils and will be treated in some detail. Some attention is also given to the Rostroconchia; however, the other three classes are rare fossils and will not be discussed further.

8.3 Some aspects of shell morphology and growth

Since the shell is the only part of the mollusc normally to be fossilized, certain aspects of shell shape should be first considered.

Coiled shell morphology (Fig. 8.6)

Although some molluscan shells are simply straight tubes or cones, very many others, whether bivalved or univalved, are coiled. This coiling is most evident in gastropods and cephalopods, but even the individual valves of a bivalve shell are coiled; the 'open' side of the valve, which in life encloses the viscera, is analogous to the **aperture** of other coiled shells. Leaving aside for the moment the highly modified shells of belemnites and squids, it is not hard to see that most coiled molluscan shells are simply hollow cones rolled up on themselves to a greater or lesser extent. In such rolled-up cones, which grow at the apertural end only, there are very interesting mathematical properties, for the coiling, represented by a line traced along the edge of the shell from the first-formed part (**protoconch**) to the aperture, invariably has the form of a **logarithmic or equiangular spiral**. d'Arcy Thompson (1917) may

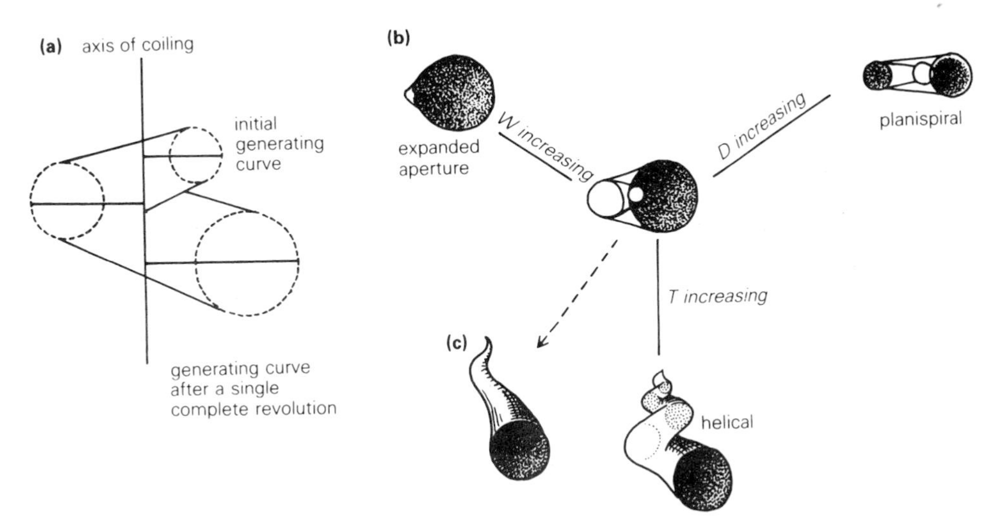

Figure 8.6 Theoretical morphology of the coiled shell: (a) schematic diagram of part of a helically coiled gastropod shell; (b) computer-simulated shell shapes derived by varying various geometrical parameters *D*, *T* and *W*; (c) a theoretical but non-functional shell shape not adopted by any molluscan group. (Based on Raup, 1966.)

be quoted here: 'it is peculiarly characteristic of the spiral shell that it does not alter as it grows; each increment is similar to its predecessor, and the whole, after each spurt of growth, is just like what it was before'. Hence the spiral shell can accrete new material at the one end only without changing its shape. In an ideal shell (such as that of *Nautilus*) of radius r, the equation of this spiral shape is $r = a\theta$ where a is a constant and θ is the whole angle through which the spiral has been traced.

The usefulness of this way of growing is undoubtedly the reason for its adoption by so many living organisms – in the shells of brachiopods and molluscs, in the foraminiferids, in the horns of mammals, and even in the eyes of trilobites – where the ability to grow at one end only without changing form has been important.

Nevertheless, in nature coiled shells often only approximate to the ideal mathematical form, and often it is the departures from the hypothetical that are of the greatest functional interest. In ammonites (fossil cephalopods) there may be several minor changes in the spiral angle throughout growth, and not all parts of the ammonites necessarily obey the rules of the logarithmic spiral form. The spacing of the ribs on the shell is nothing to do with the growth of the spiral and may alter significantly at various periods throughout growth, whereas the aperture is often peculiarly contracted.

Following d'Arcy Thompson's early studies, the theoretical morphology of the coiled shell has attracted considerable attention. The best-known recent studies are those of Raup (1966), who used a computer-based graphical method to produce various kinds of hypothetical coiled shapes and was then able to see how many of these possible types had in fact been adopted in nature. He was able to generate a large number of ideal shapes using only four parameters in the programme; the shell shapes were projected on an oscilloscope screen. Taking the example of a helically coiled gastropod (Fig. 8.6a), and allowing that its shell is no more than a hollow cone growing at the apertural end and coiling about a vertical axis as it grows, these parameters are as follows.

1. The shape of the tube in section, otherwise known as the shape of the generating curve (effectively equivalent to the shape of the aperture). In the example illustrated here it is circular, but the apertural shapes of gastropods usually depart from this form.
2. The rate of whorl expansion (W) after one revolution. In the diagram $W = 2$, since the diameter of the tube after a single revolution is twice what it was one whorl before.
3. The position and orientation of the generating curve with respect to the axis (D). In our example the circular tube is separated by a constant distance from the coiling axis, equal to half its own diameter; this is D.
4. The rate of whorl translation along the axis (T), i.e. the relative distance between successive revolutions along the axis as compared with away from the axis.

In some of Raup's models the parameters were kept constant; in others one or more of them were made to change as the model was being generated (Fig. 8.6b). Thus a gastropod with an increasing W would have a concave lateral surface. In general the coiled shells of any one molluscan or other shell-secreting group cannot readily be confused with those of others. They evolved to different functional ends. Thus gastropods are usually helically coiled and tend to have a low W, but T is very variable and may be extremely high, giving very long, thin, high-spired shells. In bivalves T is low and W very high, whereas in brachiopods W is again high but $T = 0$. Both these shells have very expanded apertures. Cephalopods are normally planispiral; W is never normally high and $T = 0$. Because of the difference in W between cephalopods and brachiopods there is no overlap in shell shape between them, though some planispiral gastropods are approximately equivalent in shape to coiled cephalopods. What makes such differences in the form of many planispiral and helical shells is, however, the shape of the generating curve, i.e. the tube in cross-section. In ammonites, for instance, the shape of **cadicone** shells (Fig. 8.25) contrasts markedly with that of a compressed **platycone**, yet it is only really this parameter which is radically different.

Of all the possible shell shapes that Raup was able to generate, only relatively few have been found to be useful biologically, and similar shell shapes have been adopted time and time again within the groups that bear them. The other types (e.g. Fig. 8.6c) have not been able to be put to useful functional purposes and so are rarely, if ever, found in any living or fossil group.

A new approach is Okamoto's moving-frame method (Savazzi, 1990). Here the direction and amount of growth are defined relative to the current position of the shell aperture, rather than a fixed reference frame as used by Raup. This method can generate various kinds of surface sculpture, and simulates more directly the biological processes involved in shell growth and morphogenesis.

Septation of the shell

There is a fundamental distinction between those shells of molluscs which are divided by internal partitions (**septa**) and those which are not. Cephalopod shells are always septate, whereas those of bivalves and most gastropods are devoid of septa, as are the shells of the Polyplacophora and Scaphopoda. But there are some gastropods, living and fossil, which do have septa in the upper part of the shell, and they are not uncommon. There is, however, a great difference between gastropod and cephalopod shells, in that cephalopods always have a tubular siphuncle running through all the septa and connecting the chambers; this structure is intimately connected with the buoyancy of cephalopods both recent and fossil. Gastropods do not have such a structure, and their viscera normally fill up the whole of the inside of the shell, except in the septate forms.

Some high-spired Cambrian monoplacophoran shells have been described which have septa (Yochelson *et al.*, 1973); these closely resemble the shells of the earliest known cephalopods of the genus *Plectronoceras* (Fig. 8.5a), which comes from the Upper Cambrian of China. *Plectronoceras*, however, is 'siphunculate', whereas the monoplacophorans are not. It is quite possible that the earliest cephalopods were derived from septate monoplacophorans, the first forms being bottom-crawling as in their ancestors; the later ones, which rapidly evolved means of buoyancy (effected by the siphuncle) and propulsion, were able to exploit the vacant nektic niche, which was hitherto closed to molluscs. Unfortunately, the fossil record gives no further evidence of how this remarkable transition could have taken place; nor is there evidence from comparative anatomy and embryology. As is commonly and frustratingly the case in attempting to unravel phylogenies, the intriguing intermediate forms were rapidly superseded and have left no trace of their existence.

8.4 Principal fossil groups

Class Bivalvia

All bivalves have a shell consisting of a pair of calcareous valves between which the soft parts of the body are enclosed. Unlike the brachiopods, which are also bivalved, they are very abundant and diverse today, and most shorelines are littered with their dead shells. The majority of bivalves are marine; most are benthic and live infaunally or epifaunally. Some genera have successfully colonized freshwater habitats. The earliest known genera, such as *Fordilla*, are Cambrian, but bivalves only became diverse in the early Ordovician. They are, however, generally of relatively limited abundance in the Palaeozoic except locally. It was not until the wake of the Permian extinctions that bivalves moved from their original nearshore habitat to the offshore shelf regions formerly inhabited by brachiopods. Thereafter, in the Mesozoic, they became much more common, as did the gastropods, and from the early Tertiary onwards they have come to dominate the hard-shelled shallow-marine fauna. The name Bivalvia was originally given by Linnaeus in 1758, adopted from the usage of Bonnani in 1681, but for rather complex historical reasons the later terms Pelecypoda and Lamellibranchia have been more commonly in use. There is much to be said, however, for suppressing the two latter names both for taxonomic correctness and to avoid confusion.

Cerastoderma (Fig. 8.7)

The modern cockle *Cerastoderma edule* (Subclass Heterodonta, Order Veneroida) is an infaunal bivalve which lives in the intertidal zone in European and other waters. It is still sometimes referred to in the literature as *Cardium edule* though, since its morphology departs too greatly from that of the type species of *Cardium* originally described by Linnaeus, the genus *Cerastoderma* has later had to be erected for it.

C. edule inhabits a burrow a few centimetres deep, in which it lives with its two valves joined dorsally by a hinge, with the line of closure (the commissure) being vertical. The two valves are normally slightly open, allowing the muscular axe-shaped foot to protrude between them anteriorly and the siphons connecting the animal with the surface to project backwards and upwards.

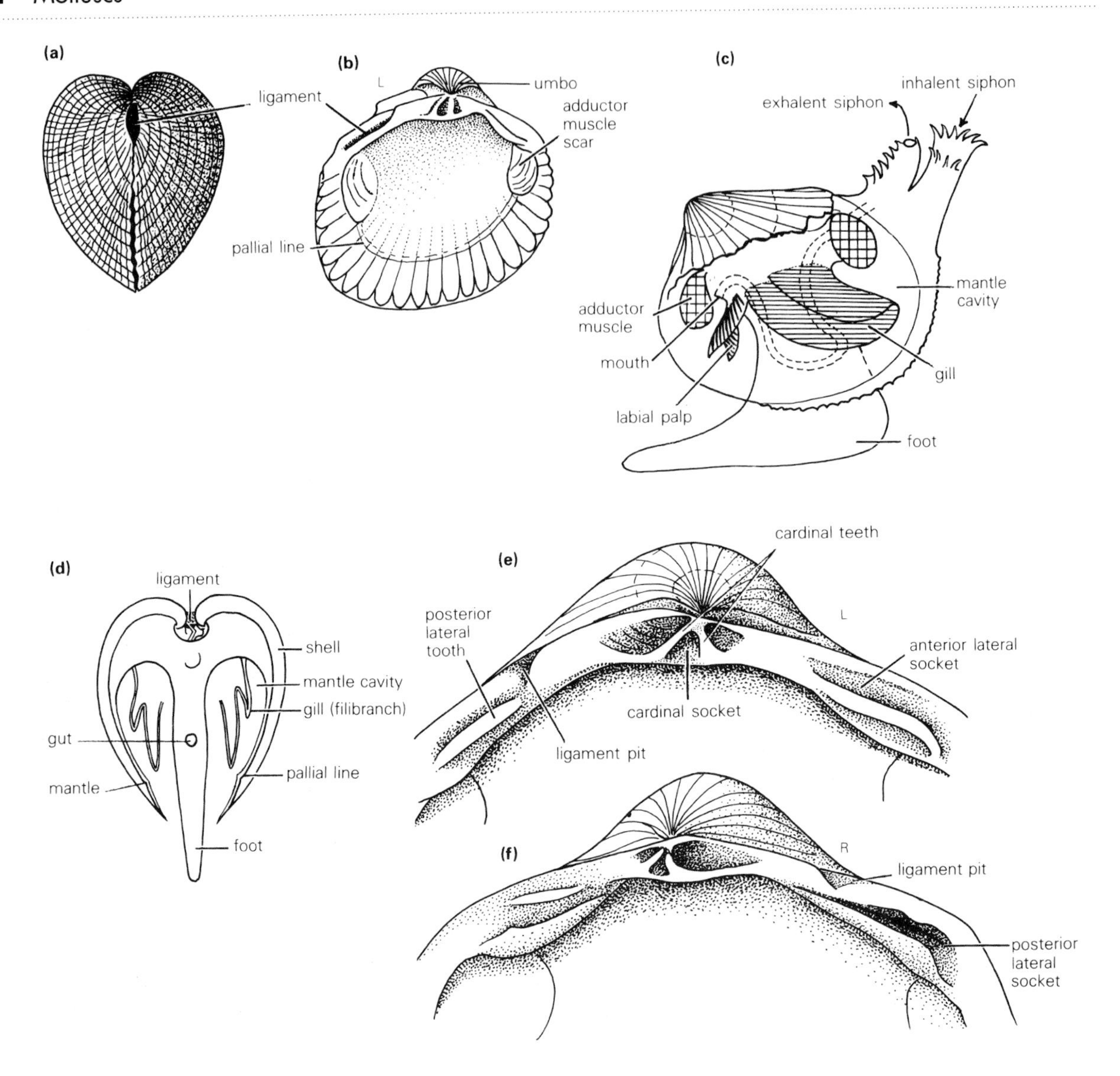

Figure 8.7 *Cerastoderma edule*, a Recent infaunal heterodont bivalve: (a) posterior view of intact shell; (b) interior of left valve; (c) right lateral view of living animal in life position with shell partially removed to show soft parts; (d) vertical section of living animal; (e), (f) cardinal region of left and right valves, respectively.

Shell morphology and orientation

The two valves on either side of the commissure are virtually mirror images of each other; they are said to be **equivalves**. An intact *Cerastoderma* when examined from either end appears heart-shaped, with the two valves symmetrical about the vertical commissure and their **umbones** close together and facing each other. These umbones, as with brachiopods, are the early-formed part of the shell. From the side the valves do not appear symmetrical but are inequilateral and somewhat lopsided. From the umbones radiate 22–28 strong ribs, crossed by concentric **growth lines** which are records of the former positions of the edge of the shell. The umbones are set slightly towards one end of the shell, and it is this which is important in determining the orientation of the shell and which valve is which. In standard orientation for *Cerastoderma* the umbones face anteriorly and are closest to the anterior end; this is typical of most though not all bivalves. If an intact *Cerastoderma* with both valves present is held in the hand with the commissure vertical, the hinge horizontal and the umbones directed away from the observer, the **right** and **left valves** are immediately distinguished. In this orientation the **ligament**, which holds the valves together and is instrumental in opening the shell, is nearest to the observer and is thus posterior.

If either valve is examined from the inside (Fig. 8.7b) the following internal features are observed.

A flattened vertical area (the **hinge plate**) bearing **teeth** between which are **sockets** corresponding to the teeth on the other valve; these act as guides ensuring that the two valves go exactly back into place when they close, making a secure and tight fit. Dentition in bivalves is of various kinds; in *Cerastoderma* the teeth are of **heterodont** type and fall into three groups. In *Cerastoderma*, directly below the umbo (Fig. 8.7e,f) are the large **cardinal teeth**, two in the left valve and two in the right; the two sets of elongated and obliquely set **lateral teeth** lie some distance away. Bivalve dentition is very important in classification and identification and is considered in detail later.

The ligament: a rubbery material connecting the two valves and holding them open. It lies in an elongated pit posterior to the umbo. The ligament is rarely preserved when the shell has been dead some time.

Two large ovoid smooth areas towards the ends of the shell. These are the **scars** or attachment sites for the **adductor muscles**, which run between the valves and when contracted keep the shell closed.

The **pallial line** which joins the two adductor scars and runs parallel with the edge of the shell and some distance within it. This marks the point of mantle attachment to the shell. In some bivalves, though not *Cerastoderma*, it also marks the transition between microstructural shell types. Though the pallial line of *Cerastoderma is* entire, many bivalves have a deep indentation (the **pallial sinus**) towards the rear of the shell. This is present only in genera with retractable siphons and allows a pocket in which these can be tucked away in case of danger (Fig. 8.12). The pallial line has nothing to do with the adductor muscles as such; it is the line along which the inner muscular part of the mantle is attached.

Internal anatomy

The shell (Fig. 8.7c,d) is formed by the mantle, which is analogous though not homologous with the mantle of brachiopods. There are three layers in the shell, all of which are secreted by different parts of the mantle. At its periphery the mantle forms three folds, only the outermost of which is secretory and shell-forming. The middle fold is sensory, the inner one muscular. The outer part of the shell is a thin layer (the **periostracum**) made of dark-coloured tanned protein. In some modern shells this is thick and distinct and protects the calcareous shell against damage and dissolution. Where it is fine and flexible it allows the manufacture of intricately sculptured shell morphology and ornamentation (Harper, 1997). In pteriomorphs it is vanishingly thin and is rarely preserved in fossils. It is formed by the outermost mantle fold at the shell edge and is the only shell layer to cross the dorsal margin. Below this are two much thicker layers of crystalline calcium carbonate, formed by deposition in a proteinaceous (**conchiolin**) matrix. The inner layer stops short at the pallial line.

Within the shell the upper region is occupied by the visceral mass and the lower by the mantle cavity. The mouth is set anteriorly and the anus at the posterior end of the mantle cavity. The large central foot is a muscular organ used in digging. The animal can alternately contract and expand it and thus can use it both to dig its way into the substratum and to move horizontally within it. On either side of the foot the long filamentous gills hang down into the mantle cavity. Near the mouth two extra gill-like structures are found: the **labial palps**. These together with the gills are used in feeding.

Since *Cerastoderma* lives as an infaunal suspension feeder protected within the sediment, it has to be able to maintain connection with the surface waters for feeding and respiratory exchange. This is facilitated by two large siphons which project upwards from the posterior end of the animal. Water is drawn into the shell through the inhalant siphon by the ciliary pumping action of the **gill filaments**. These vertical filaments form a comb-like structure lined with innumerable cilia, which beat in successive and coordinated waves of movement and collectively set up one-way pressures, thus generating quite strong inward currents. The gills and labial palps not only generate the currents but also trap the food particles, by secreting a sticky mucus. The food particles adhere to this and are conveyed to the ventral edge of the gill or palp by the cilia. They then move in a line along this edge until they come to the mouth. Waste water containing carbon dioxide, exchanged by the gills, is taken away through the exhalant siphon together with excreta from the anus, which is placed very close to the exhalant siphon.

The only other important structures within the shell are those connected with its opening and closure. *Cerastoderma*, like other bivalves, has many natural enemies, especially birds, gastropods and starfish, and its only defence against them is to withdraw the foot inside the shell and to keep the valves tightly shut for as long as possible. This closure is effected by the strong horizontal adductor muscles, which are both about the same size (the **dimyarian** condition) and attached to facing points on the internal walls of the opposing valves. When the adductors relax the valves open automatically to let the foot out, forced apart by the ligament which acts like a compressed spring and whose effect, acting alone, would keep the valves permanently open. Recently dead shells in which the muscles have decayed or been eaten may still retain the ligament, and the two valves still joined together are always found in the open position.

Bivalves thus open and close their valves by an antagonistic muscle–ligament system, which is very effective but requires continual expenditure of energy when the shell is closed. The shell closure system may be contrasted with that of brachiopods, which operates on two sets of antagonistic muscles.

Though there is no head in bivalves, there are well-developed circulatory, excretory and nervous systems, adequate for all their physiological needs. *Cerastoderma* is very well adapted for life as an infaunal suspension feeder. It has a high filtering rate and, having colonized the rather 'difficult' intertidal environment, is highly successful.

Range of form and structure in bivalves

Even though certain features of bivalve organization, such as the absence of a head, have limited their evolutionary potential, the range in their form and the adaptations that bivalves have undergone show a remarkable degree of inherent or actualized evolutionary plasticity. There are some features in the hard and soft parts that have remained fairly stable since their origin. They have not altered much or have undergone only minor evolutionary changes. Such characters are very useful in classification for defining the higher taxonomic categories; they include shell microstructure, gill morphology and dentition, though the latter has on the whole been more variable than the others. Other characters, however, including the overall form of the shell, the musculature and the presence and relative development of siphons and associated structures, are more directly related to the bivalves' specific adaptations to particular modes of life and are thus of more value at lower taxonomic levels.

Shell microstructure and mineralogy (Fig. 8.8)

The calcareous shells of bivalves are multilayered and consist of two intermixed phases (Wilbur, 1961): (1) an organic matrix and (2) crystalline calcium carbonate in the form of calcite or aragonite. Oyster shells, for example, are made of calcite, though they have a thin aragonitic pad where the muscles are attached. Other bivalve shells are entirely aragonitic, but perhaps the majority have different layers composed of calcite and aragonite. Vaterite, another form of calcium carbonate, is reported from injured shells which the mantle has been able to regenerate.

The two phases tend to be found in bivalves in a number of recurrent patterns, occurring in discrete

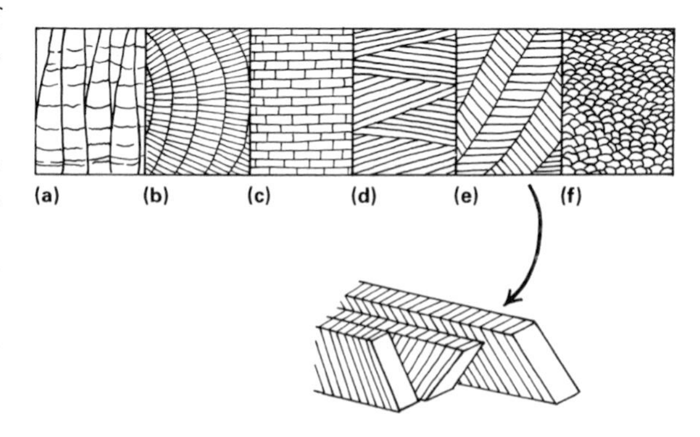

Figure 8.8 Bivalve shell layer morphology as seen in thin section: (a) simple prismatic; (b) compound prismatic; (c) sheet nacreous; (d) foliated; (e) crossed lamellar with inset showing disposition of stacked aragonite lamellae; (f) homogeneous. (Based on Taylor and Layman, 1972.)

shell layers. These have been used in the unravelling of phylogenies. Six primary types have been differentiated (Taylor *et al.*, 1969; Bathurst, 1975).

Simple prismatic structure with columnar polygonal calcite or aragonite prisms.
Composite prismatic structure with tiny radiating acicular crystals.
Nacreous structure in which tabular sheets of aragonite are found resembling a brick wall when cut in section. These are usually found in middle and inner shell layers.
Foliated structure of lath-like calcitic crystallites arranged in sheets.
Crossed-lamellar structure which is normally aragonitic. Here the shell is made of closely spaced lamellae, within each of which are found thin stacked plates of aragonite, those of adjacent lamellae being inclined in opposite directions to one another. In some cases intergrowths of blocks of crystals are found (**complex crossed-lamellar** structure) with four principal orientations.
Homogeneous structure with small granular anhedral crystals.

Of all these, nacreous structure seems to be phylogenetically the oldest, not only in bivalves but in molluscs generally; it is also the strongest, which raises the unsolved question of why the other types evolved if they are less strong? Underlying the areas of muscle attachment there is a specialized region of irregularly prismatic crystals: the **myostracum**. In *Cerastoderma* the calcareous part of the shell is two-layered, the outer layer being of crossed-lamellar form, the inner being complex crossed-lamellar. A few bivalves, such as those of Superfamily Lucinacea (Subclass Heterodonta), have three calcareous layers. There is some evidence that the type of shell structure actually present in bivalves may partially reflect their mode of life. Thus crossed-lamellar, complex crossed-lamellar and composite prismatic shell layers have the highest hardness values but do not have particularly high compressive or tensile strengths. They are commonest in burrowing bivalves, in which a good resistance to abrasion, but not necessarily resistance to bending stresses, would be useful.

The mineralogical differences in the shell may be under ecological control; for instance, in the modern Superfamily Mytilacea (Subclass Pteriomorpha) tropical species have entirely aragonitic shells whereas the percentage of calcite increases progressively towards cooler waters.

Although shell structure and mineralogy are very useful in taxonomy, they are rarely preserved unchanged in older fossils. Tertiary fossils may retain their original aragonite, but there are no known molluscs with aragonite which are older than the Carboniferous.

Gill morphology (Fig. 8.9)

The structure of the gills is seldom preserved in fossils, but it is most valuable in taxonomy. The gills hang down into the mantle cavity on either side of the foot. In Order Nuculoida (Ord.–Rec.) the gills are **protobranch**, being small and leaf-like rather like those of amphineurans and cephalopods. Their unmodified appearance and the fact that they occur in a primitive group suggest that they are not far removed from the ancestral type. **Filibranch** gills form lamellar sheets of individual filaments in a W-shape, as do **eulamellibranch** gills (occurring in *Cerastoderma*), but in the latter there are cross-partitions joining the filaments and making water-filled cavities between them. The vast majority of bivalves have gills belonging to the latter two types. **Septibranch** gills, which are confined to a single superfamily of rock borers, the Poromyacea (Subclass Anomalodesmata), run transversely across the mantle cavity, almost enclosing an inner chamber which maintains only a small connection with the outer cavity.

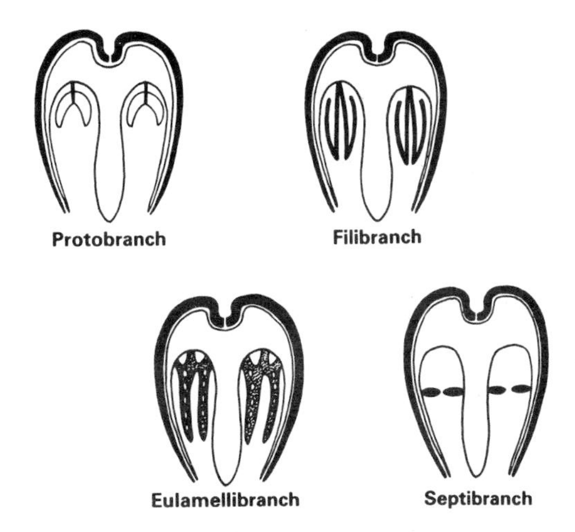

Figure 8.9 Bivalve gill morphology, four basic types shown by transverse sections: shells are shown in black with the foot projecting centrally. (Redrawn from Moore *et al.*, 1953.)

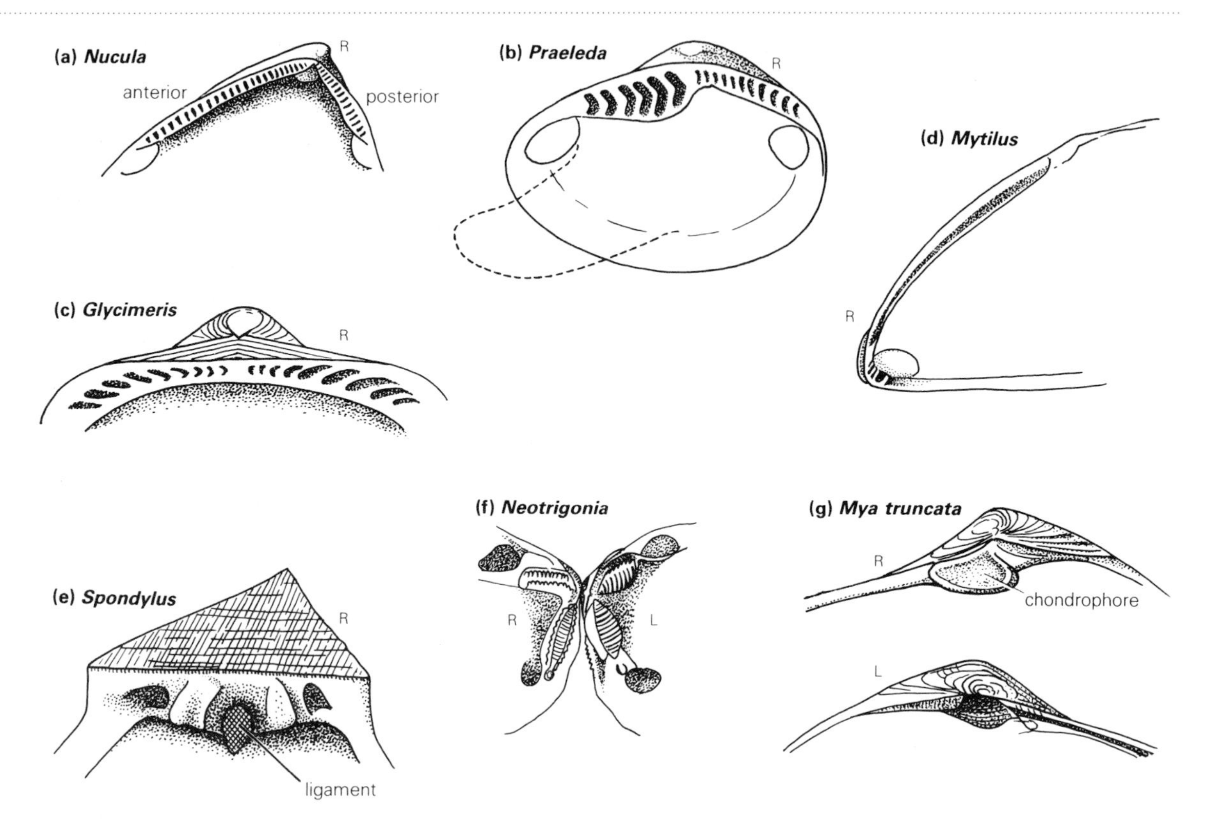

Figure 8.10 Bivalve hinge lines and dentition (not to scale): (a) *Nucula* (Tert.–Rec.), right valve (note that the umbones face posteriorly), taxodont hinge; (b) *Praeleda* (Ord.), right valve, modified taxodont hinge, foot position inferred; (c) *Glycymeris* (Tert.–Rec.), right valve, taxodont hinge; (d) *Mytilus* (Rec.), right valve, dysodont hinge; (e) *Spondylus* (Cret.), right valve, isodont hinge; (f) *Neotrigonia* (Rec.), schizodont hinge; (g) *Mya truncata* (Rec.), desmodont hinge with chondrophore. [(b) Redrawn from Bradshaw, 1970; (c) redrawn from Woods, 1946.]

Fossilized gills have been described in Upper Jurassic *Trigonia*. Unsurprisingly, they are very like the gills of *Neotrigonia*, living today.

Dentition (Fig. 8.10)

There are several kinds of tooth and hinge structure in bivalves of which the following can be clearly recognized.

In **taxodont** dentition the teeth are numerous and subparallel or radially arranged, and in modern taxodont bivalves such as *Glycymeris* and *Arca* they are all rather similar. *Nucula* and its relatives have taxodont teeth, as do many of the Ordovician bivalves, though in some of these there is quite considerable differentiation of the teeth. In the Ordovician **palaeotaxodont** genus *Praeleda*, for instance, there is a posterior section with low ridge-like teeth and an anterior part with much larger ridge-like teeth, presumably allowing the protrusion of a large foot below them which they protected from above while the shell was open.

Dysodont dentition consists of small simple teeth near the edge of the valve, as in *Mytilus*.

Isodont teeth are very large and located on either side of a central ligament pit, as in *Spondylus*. Such teeth are characteristic of one superfamily only, the Anomiacea (Subclass Pteriomorpha).

Schizodont teeth are confined to Superfamily Trigoniacea (Subclass Palaeoheterodonta, Order Trigonoida); they are very large and have many parallel grooves normal to the axis of the tooth. In *Trigonia* the left valve has three teeth, the right has two.

Most Tertiary and Recent bivalves have **heterodont** teeth, as exemplified by *Cerastoderma*. Normally there are two or three cardinal teeth below the umbo, as well as the elongated lateral teeth anterior and posterior to these. Heterodont dentition can be traced back to the Ordovician, and it has been suggested that both the palaeotaxodont and **palaeoheterodont** teeth in the bivalves of that time were derived from an ancestor with multiple-ridged teeth.

Pachydont dentition occurs only in the peculiar hippuritid (rudistid) bivalves, which cement themselves to the substratum by a very large right valve of coral-like form. The teeth are very large, heavy and blunt. Evidently this peculiar dentition is directly connected with the unusual mode of life.

Desmodont dentition is a common form of hinge structure in which the teeth are very reduced or absent, but accessory ridges lying along the hinge margin take their place, and often, as in *Mya*, a large projecting internal process (the **chondrophore**) carries the ligament. All bivalves possessing desmodont hinges are infaunal suspension feeders.

The great variety of dental structure in bivalves contrasts strongly with the situation in brachiopods, where it is almost constant.

Muscles and ligaments

The ligament is a variable structure in bivalves and may have two parts: one external and the other internal. The latter may reside in a pit between the teeth (e.g. in isodont shells) or may be supported by a chondrophore (e.g. *Mya*). Either component may be absent.

Cerastoderma is an example of an **isomyarian** bivalve, in which the two adductor muscle scars are more or less the same size. *Mytilus* (Figs 8.10d, 8.12h,j) and other byssally attached genera, however, have a greatly reduced anterior adductor (**anisomyarian** condition), and in the swimming scallop *Pecten* (Fig. 8.14a) the anterior adductor has vanished altogether and there remains only the large **monomyarian** posterior adductor.

Other shell structures

Most bivalves have the umbones anterior to the midline (**prosogyral**), but there are some **opisthogyral** genera with posterior umbones. Some bivalve shells, whether proso- or opisthogyral, have depressed areas (the **lunule** and **escutcheon**) placed in front of and behind the hinge, respectively. Their function may have something to do with burrowing, but their presence and shape depend on how the shells grow. Nevertheless the Triassic *Lima lineata* (Subclass Pteriomorpha, Superfamily Limacea) apparently found a use for its lunule, as has been shown in some interesting ecological studies. This *Lima*, like its modern species, apparently rested in life with its lunule on the sea floor, as is apparent from the orientation of epizoic pectinaceans and other small organisms on the shells and from the position of the shells themselves when fossilized. Juvenile specimens of *Lima* are often found clustered at the anterior end of the lunule. They have been interpreted (Jefferies, 1960) as photonegative young which entered the dark space between the umbones of the adult and lived there for a while. There are no large juveniles in the lunules, however, and presumably they must have moved out when they had outgrown their dark shelter. Many modern *Lima* species build nests of byssus in dark cavities, and it seems that they must have inherited this preference for the dark from their ancestors, at least as far back as the Triassic.

The orientation of bivalves and the distinction of left from right valves is facilitated if the following considerations are borne in mind.

Most bivalves are prosogyral, though *Nucula*, *Lima* and *Donax* are exceptions to this rule.

The pallial sinus is always posterior, and if the external ligament is not central, it too is usually posterior.

Though the adductors may be the same size, if one is reduced or absent it is usually the anterior one, though not, for example, in *Ensis*.

Classification

Bivalves have always been found hard to classify. The problems are not acute at lower taxonomic levels, for species, genera and families seem on the whole to fall into clearly defined natural groupings. These are based upon shell form and structure, the presence or absence of a pallial sinus, dentition and other characters. Some of the family groups have remained very conservative over long periods of time, and a number of genera have persisted with relatively little change since the Palaeozoic.

However, it is not so easy to define higher taxonomic categories of a kind that can be related in a phylogenetically meaningful way. There are too few morphological clues, and most of the more useful stable characters are in the soft parts or shell microstructure. This causes difficulties with the fossil forms, where the soft parts have vanished and where the aragonitic part of the shell is normally recrystallized or dissolved. Furthermore, parallel evolution, which has confused many lines of descent, again raises taxonomic difficulties.

A recent classification, given below and based on that in the *Treatise on Invertebrate Paleontology*, uses shell microstructure, dentition and (to some extent) hinge structure, gill type, stomach anatomy and the nature of the labial palps as the stable characters, and in this classification six subclasses are defined. Characters such as shell shape are so closely related to life habits that they are 'unstable' and useful only in classification at lower taxonomic levels.

SUBCLASS 1. PALAEOTAXODONTA (Ord.–Rec.): Includes only the one ORDER NUCULOIDA. Small, protobranch, taxodont, infaunal, labial palp feeders with aragonitic shells; e.g. *Nucula, Ctenodonta*.

SUBCLASS 2. CRYPTODONTA (Ord.–Rec.): A largely toothless (dysodont), infaunal, aragonitic-shelled, mainly Palaeozoic group with *Solemya* (ORDER SOLEMYOIDA) as the only living representative. *Cardiola* and *Eopteria* are representative of the only other order, the Palaeozoic ORDER PRAECARDIODA.

SUBCLASS 3. PTERIOMORPHIA (Ord.–Rec.): A rather heterogeneous group of normally byssate bivalves with variable musculature and dentition. Shells may be calcitic, aragonitic or both.

ORDER 1. ARCOIDA: Isomyarian filibranchs with crossed-lamellar shells, taxodont; e.g. *Arca, Glycymeris*.

ORDER 2. MYTILOIDA: Anisomyarian filibranchs and eulamellibranchs, prismatic/nacreous shells, byssate, dysodont; e.g. *Mytilus, Pinna, Lithophaga*.

ORDER 3. PTERIOIDA: Anisomyarian or monomyarian filibranchs or eulamellibranchs, shell structure varied, byssate or cemented. Includes all scallops, oysters, and pearl clams; e.g. *Pecten, Pteria, Gervillea, Inoceramus, Lima, Ostrea, Exogyra*.

SUBCLASS 4. PALAEOHETERODONTA (Ord.–Rec.): A dominantly Palaeozoic aragonitic-shelled group including the following orders.

ORDER 1. MODIOMORPHOIDA: The Palaeozoic precursors of most later bivalves with heterodont teeth (actinodonts); e.g. *Modiolopsis, Redonia*.

ORDER 2. UNIONOIDIA: Heterodont non-marine genera with a long time range; e.g. *Unio*.

ORDER 3. TRIGONOIDA: Bivalves with large trigonal shells and well developed schizodont teeth. Common in the Mesozoic when they were amongst the most numerous bivalves; now represented by *Neotrigonia*, the one living genus; e.g. *Trigonia*.

SUBCLASS 5. HETERODONTA (Ord.–Rec.): Heterodont eulamellibranchs to which most modern genera belong, nearly all with aragonitic crossed-lamellar shells, adapted to varied modes of life and especially to infaunal siphon feeding. The hinge structures may degenerate to a desmodont condition.

ORDER 1. VENEROIDA: Active heterodonts with true heterodont teeth; e.g. *Lucina, Thetis, Cardita, Crassatella, Cerastoderma, Venus, Mactra, Tellina*.

ORDER 2. MYOIDA: Thin-shelled burrowers and borers, very inequivalve, hinge degenerate, siphons well developed; e.g. *Mya, Corbula, Pholas, Teredo*.

ORDER 3. HIPPURITOIDA: Large, often coralloid, cemented extinct bivalves with pachydont dentition; e.g. *Diceras, Hippurites, Radiolites*.

SUBCLASS 6. ANOMALODESMATA (Ord.–Rec.): Burrowing or boring forms, very modified, with aragonitic shells and desmodont dentition. One order only, the Pholadomyoida; e.g. *Pholadomya, Edmondia, Pleuromya*.

Evolutionary history

Virtually all the major bivalve stocks were established by the Middle Ordovician (Pojeta, 1975, 1978), and from then on the history of bivalves is straightforward. The search for bivalve ancestors, however, has proved less simple. Such as do exist are tiny, of millimetre scale. They are variously envisaged as surface dwellers, lying on their sides, feeding with the foot and respiring with the gills, or as shallow burrowers. The best-known genus, *Fordilla*, from the Lower Cambrian, seems to be a very small bivalve, though not all would agree that it is such (Yochelson, 1981). However, most other authors claim that it has a true hinge line, a single tooth and socket in each valve, and a narrow posterior ligament trough. Another minute genus (1 mm) is *Pojetaia*, from the early Cambrian of Australia (Runnegar and Bentley, 1982). This has characteristics reminiscent of Ordovician palaeotaxodonts and it is quite likely that these early, and tiny, though distinct genera could have provided the rootstock for most known higher bivalve taxa.

Curiously, no bivalves are known from the Middle and Upper Cambrian and those formerly recorded as bivalves are now known not to be so. Early Ordovician (Tremadoc) faunas are of low diversity, and only in the Arenig did bivalves become common, when they underwent an initial great burst of adaptive radiation. This rapid evolutionary change took place in the close-inshore habitat favoured by early bivalves, but such environments are seldom preserved in the fossil record. One such occurrence from the early Arenig of Wales

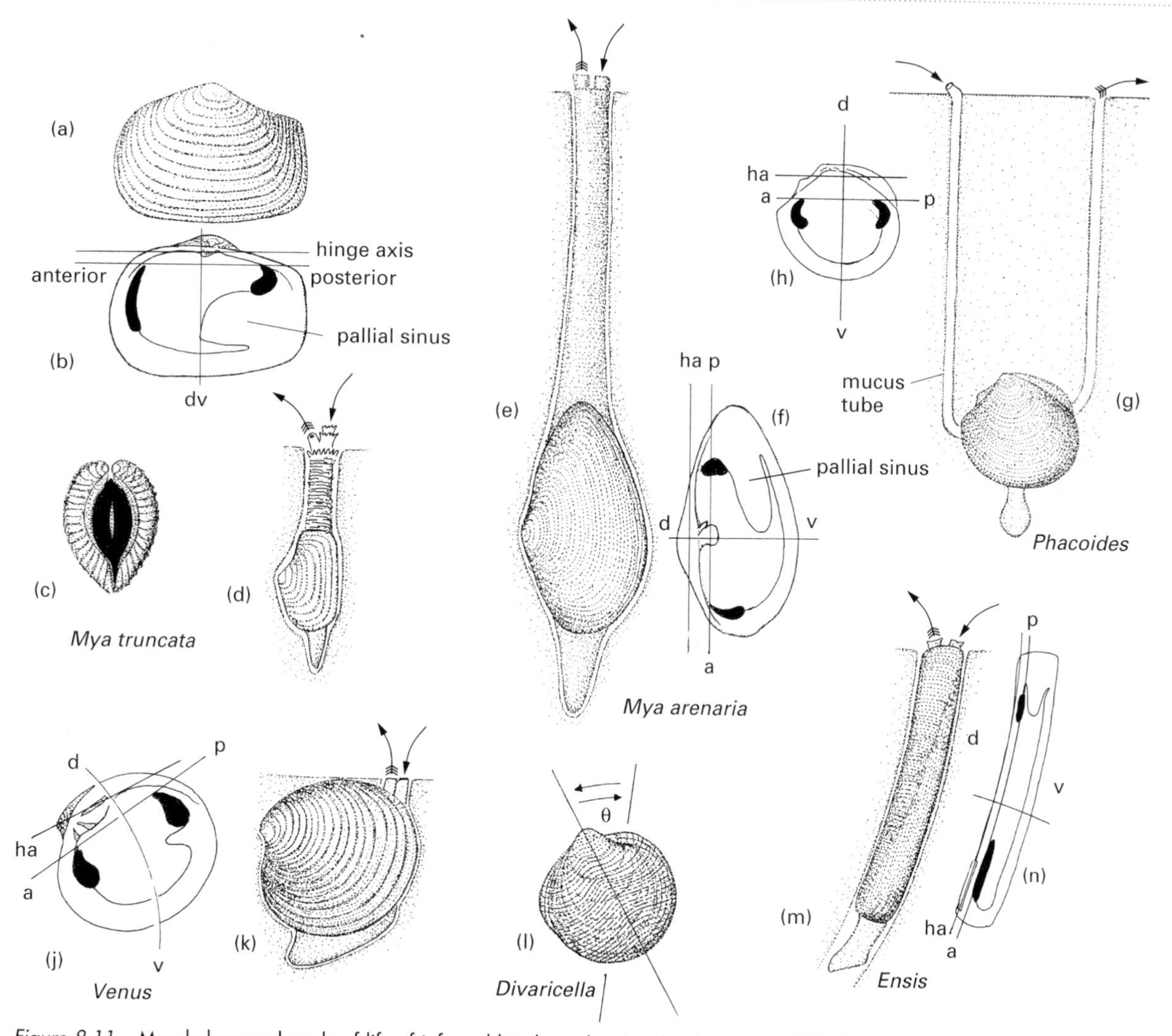

Figure 8.11 Morphology and mode of life of infaunal bivalves, showing the disposition of the hinge axis (ha), dorso-ventral axis (d–v), antero-posterior axis (a–p) and shell modification; small arrows represent incurrents (plain) and excurrents (feathered). (a)–(d) *Mya truncata*: (a) external view of left valve; (b) internal view of right valve, showing muscle scars, pallial line and pallial sinus; (c) posterior view showing gape (all × 0.5); (d) in life position (× 0.2); (e), (f) *Mya arenaria* in deep burrow with long siphons and deep pallial sinus (× 0.25); (g), (h) *Phacoides*, a lucinoid in its burrow, with mucus tube to surface (× 0.5); (j), (k) *Venus*, a venerid in its shallow burrow (× 0.5); (l) *Divaricella*, a bivalve which burrows using its surface sculpture as a saw, by rocking through the angle (× 0.75); (m), (n) *Ensis*, a razor shell in its long shallow burrow (× 0.3). (Mainly redrawn from Stanley, 1970.)

(Cope, 1996) comprises no less than 18 genera, belonging to at least seven subclasses; another is the diverse cool-water peri-Gondwana fauna, mainly of small shallow-burrowing deposit feeders described by Babin and Gutierrez-Marco (1991) from Spain.

This early Ordovician evolutionary burst gave rise to several superfamilies having taxodont, dysodont or heterodont hinges and belonging to four feeding types. First there were the earliest labial palp feeders of Order Nuculoida, which have paired flexible extensions of the palps which project from the shell and collect food particles directly from the sediment (Fig. 8.12d).

Second, there were shallow burrowing types such as the Astartidae (Order Veneroida), with no real siphons (Fig. 8.12e) Then there were epifaunal, byssally attached bivalves, the pteriomorph Order Mytiloida, amongst others (Fig. 8.12f–j), and finally there arose the infaunal mucus-tube feeders, the superfamily Lucinacea (Subclass Heterodonta),

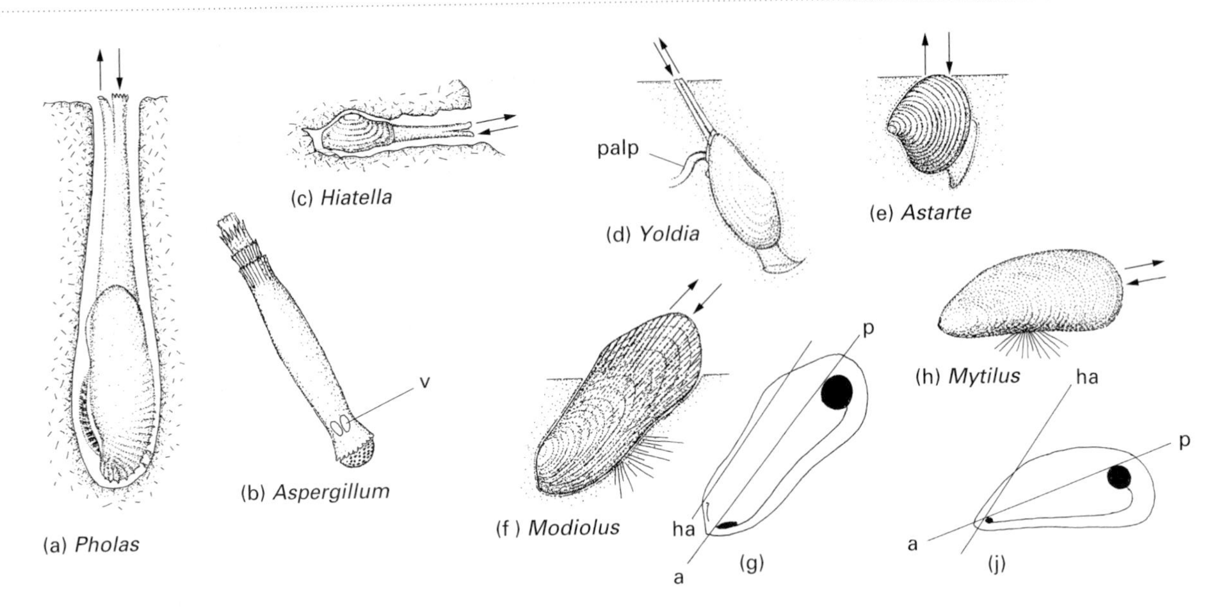

Figure 8.12 Boring, cavity-dwelling and other modified bivalves: (a) *Pholas*, a stone borer (× 0.5); (b) *Aspergillum*, a highly modified sand-dwelling bivalve with reduced valves (v) (× 0.5); (c) *Hiatella*, a 'squatter' which occupies old borings (× 0. 5); (d) *Yoldia*, a labial-palp feeder (× 0.5); (e) *Astarte*, a non-siphonate infaunal feeder (× 0.4); (f), (g) *Modiolus*, semi-infaunal and byssally attached (× 0.5); (h), (j) *Mytilus*, epifaunal and byssally attached (× 0.5). (Mainly redrawn from Stanley, 1970.)

which were the only Palaeozoic deep burrowers (Fig. 8.11g,h). All these feeding types survive to the present, as do the superfamilies in which they arose. Apparently the earliest types were infaunal, but epifaunal genera began to diversify later in the Ordovician.

In one genus of Ordovician bivalves, *Babinka*, the shell has a rather curious conformation which has attracted considerable attention (McAlester, 1962). This genus is an early lucinoid known only from the Llanvirnian of the Czech Republic. *Babinka* has an isomyarian shell of standard form, but between the large adductor scars are two chains of smaller muscle scars like strings of beads. The upper chain has eight scars, the lower very many tiny ones. These have been interpreted as foot (**pedal**) and gill (**ctenidial**) muscle scars and bear a remarkable similarity to those of the monoplacophoran *Neopilina*. *Babinka*, being a lucinoid, was probably infaunal and a mucus-tube feeder, but it does seem that it retains the muscle pattern as an inheritance from an ancestral mollusc, something quite like a monoplacophoran. The evidence is slender but is in accordance with deductions about molluscan phylogeny from other sources.

It is worth noting that the presence of the muscular foot gave an inherent advantage to bivalves over the rhynchonelliform brachiopods, which other than rostroconchs were the rival hard-shelled suspension feeders. Only by means of the foot could bivalves colonize infaunal environments successfully, which the rhynchonelliformes were never really able to do. Indeed, the second great expansion of the bivalves during the early Mesozoic and continuing throughout the Cenozoic was directly due to the fact that they could burrow. However, burrowing ability alone was not enough, for what gave the impetus for the full exploitation of the infaunal habitat was the fusion of the posterior edges of the mantle to form true siphons (Stanley, 1968). In the early Mesozoic a large number of new heterodont or desmodont superfamilies arose, of which by far the majority were siphonate, using their siphons for feeding while remaining protected deep below the sediment surface.

Many of these live in the intertidal zone where they either occupy deep (relative to body size), permanent burrows (e.g. *Mya*, *Macoma*) or, like the more shallow-burrowing *Cerastoderma*, can burrow in again quickly if washed out and thus re-establish themselves. Quite possibly the intertidal zone was not greatly colonized by bivalves during the Palaeozoic since they did not have the requisite structural potential. Thus the mantle fusion and siphon formation, originating in the Heterodonta at the very end of the Palaeozoic, appear to have given the bivalve groups that inherited this structure a great new evolutionary potential. Probably the desmodont hinge, which was derived from a heterodont predecessor, is a subsidiary modification also associated with the infaunal siphon-feeding habit. The expansion of siphon feeders into new, previously unexploited habitats (intertidal, deep burrowing, boring, etc.), dependent upon one new key character alone, has been described as an ideal model of adaptive radiation.

Some bivalve stocks have been remarkably stable since their first appearance. Thus the shallow-burrowing taxodont Family Glycymeridae has retained essentially the same form since the beginning of the Cretaceous, and its representatives have lived in the same current-swept habitats. They are morphologically unspecialized, and although they are ecologically specialized as regards substrate preference, in other respects they are remarkably tolerant of physically vigorous conditions, and associated faunas are of low diversity. The shell form is defined by rigid mechanical and geometrical constraints and, as Thomas (1975) has shown, the demands of the functional and geometrical needs of the animal in its shell did not allow the family enough evolutionary flexibility to undergo further radiations, hence its conservatism through time.

Family Trigoniidae, characterized by enormous, complex hinge teeth, is another important group which lived in coarse shifting sands. They were the dominant shallow-water burrowers of nearshore habitats during the early Mesozoic, but the extinctions of the late Cretaceous proved disastrous for them. One relict genus only, *Neotrigonia*, survives today. The large teeth maintained valve alignment at the wide gape required for extrusion of the huge T-shaped foot, for which gave the trigoniids more effective mobility than any Palaeozoic suspension-feeding bivalves (Stanley, 1977). All other features of their unusual morphology are highly adaptive, and had it not been for the disastrous events of the Late Cretaceous, they would no doubt be diverse and flourishing today.

Functional morphology and ecology

All bivalve shells have to be able to open without the umbones coming together and so the umbones have to be kept some distance apart. At the same time the elastic ligament has to be prevented from undue stretching and breakage as the bivalve grows. These two geometrical requirements are in conflict, and the shape of any bivalve shell must therefore be a compromise between them. The range of shell morphologies available is thus constrained by the geometry of the coiled shell. Many different compromise solutions have been achieved, and can be simulated by computer modelling. Among those extrapolated by Savazzi (1987) are, for example, evolving a whole range of inequivalve adaptations, very wide spacing of the umbones and having a ligament that is continually replaced as it breaks during ontogeny, cessation of growth before the shell has coiled to half a whorl, and decreasing whorl expansion rate throughout ontogeny. There are only a limited number of solutions to these common basic requirements, and these recur quite often in parallel, in unrelated lineages.

Shell form and mode of life in bivalves is illustrated in Figs 8.11–8.14. Within the bivalves the structure of the hinge, dentition, mineralogy and shell structure and composition do not appear to be characters of much adaptive significance. They are important in classification for defining major taxonomic groups but have little to say about the adaptation of the different sorts of bivalves to their environment.

On the other hand, the shape and general morphology of bivalve shells directly reflects their mode of life. Indeed our current understanding of the ways in which modern bivalves are adapted to particular modes of life enables reasonable inferences to be made as to how extinct bivalves lived.

Modern bivalves can be grouped into several morphoecological categories (Stanley, 1970):

infaunal shallow burrowing;
infaunal deep burrowing;
epifaunal, attached by **byssus** threads to the substratum;
epifaunal, cemented to the rock;
free-lying;
swimming;
boring and cavity-dwelling.

Examination of the dead shells of most bivalves and indeed also fossil ones will allow the correct category to be inferred. In the burrowing species, shell form is also related to whether or not the animal was a slow or rapid burrower and to the nature of the sediment into which it burrowed. Most bivalves, like *Cerastoderma*, are suspension feeders. There are a few genera, however, members of the Nuculacea, which are deposit feeders. Nuculaceans such as *Yoldia* (Fig. 8.12d) extrude specialized extensions of their labial palps into the sediment and so collect organic particles from it. The veneroid Superfamily Tellinacea are partly deposit and partly suspension feeders. Tellinids use their long inhalant siphons to suck up food particles from the sediment surface rather like a vacuum cleaner. Other than these, the bivalves discussed in the following sections are all suspension feeders.

Infaunal burrowing bivalves (Fig. 8.11)

Bivalves that burrow in soft substrata, such as our type example *Cerastoderma*, have a well-defined sequence of movements which enable them to penetrate the sediment. First the foot probes downwards and swells with blood from the circulatory system; then the siphons close. This is followed by a rapid adductive movement of the two valves which dilates the foot further and squeezes water out from between them. When the foot is subsequently retracted the shell sinks down into the sediment. Then the muscles relax prior to the onset of the next cycle. The anterior adductor usually contracts first, followed by the posterior one; hence a rocking movement is imparted to the shell, which can be up to 45° in some of the more discoidal shells but is normally less. While this process of 'digging in' is going on, the siphons extend to keep contact with the surface.

Not all burrowing bivalves have siphons. *Astarte*, for instance (Fig. 8.12e), which lives just below the surface has open, slit-like mantle edges, not fused into siphons. Most shallow-burrowing genera (e.g. *Lucina*, *Donax*, *Venus*), however, are siphonate. They have equivalved shells with the two adductor muscle scars of about the same size, and often have pallial sinuses. The anterior–posterior line in these (joining the dorsal tips of the adductor scars) is approximately parallel with the hinge line. The anterior and posterior sections of the commissure may be permanently parted; the shell may therefore gape at either or both ends for the foot and the siphons to come out. Pedal and siphonal gapes are more characteristic of the deeper-burrowing genera. Some shallow burrowers (e.g. *Tellina*, *Divaricella*) have a curious external sculpture of ridges on the outside of the shell. *Divaricella* has an unusual W-shaped pattern of fine ridges (Fig. 8.11l). When it burrows its rocking movement of some 45° is aided by the grip given to the shell by these ridges as it 'saws' its way down into the fine sand in which it lives. The oblique ridges of the more elongate *Tellina* assist the burrowing function in much the same way. Patterns such as these are unusual, and the majority of burrowers are smooth shelled and streamlined.

Shallow burrowers do not normally have very elongated shells. There are, however, a number of well-known, deep-burrowing genera, which have shells of very drawn-out form adapted for life in deep excavations which are virtually permanent. Representatives of many bivalve families have independently evolved to this mode of life, and their shells have become modified in very similar ways. The large *Mya arenaria*, for instance (Fig. 8.11e–f) is a sluggish bivalve which burrows in firm sand or mud. It has a long elliptical desmodont shell which is very thin, with a curiously modified ligament (chondrophore), much reduced teeth, and pronounced anterior and posterior gapes. The siphons are fused together, and though they may be collapsed as an escape reaction when the blood supporting them is drained they cannot be withdrawn inside the shell. Razor shells such as *Ensis* (Fig. 8.11m,n) and *Solen* have very long, almost tube-like shells, with reduced teeth at the anterior end alone and permanent anterior and posterior gapes. They occupy tubular burrows down which they can move for protection when threatened.

Not all deep burrowers are of such modified form. Some lucinoid bivalves such as *Phacoides* (Fig. 8.11g,h) have conventional-looking shells of near-circular form and a nearly horizontal hinge axis. There is a long posterior (exhalant) siphon, but the inhalant current is drawn in through a long mucus tube connecting with the sediment surface. This habit is probably ancient, for lucinoids are found as far back as the Ordovician.

Byssally attached bivalves (Fig. 8.12f–j)

Many bivalves secrete threads of the protein collagen, with which they attach themselves to the sea floor. The marine *Modiolus* (Fig. 8.12f,g) has a tapering cylindrical shell and lived with only the posterior part of the shell exposed. One species is adapted for life in salt marshes. From such endobyssate types originated the common mussel *Mytilus* (Fig. 8.12h,j), which lives attached to the surface, in an upright position, with the commissure vertical. The threads, known as byssus, are secreted from a gland located at the base of the foot. They form sticky secretions which harden rapidly after their formation. The shell is elongated, often with a flattened ventral surface offering support and stability to the shell. There is a **byssal notch** marking the base of the shell, from which the byssus emerges. Usually too the anterior part of the shell is much reduced, and the anterior adductor is greatly diminished in size so that the anterior–posterior line is oblique to the hinge axis, though there is no clear agreement why this should be so. Almost all juvenile bivalves of any group are byssally attached, and the retention of byssus in adults is probably paedomorphic.

Mytilus is epifaunal, but many of its modern relatives live partially or completely buried. *Tridacna* is a genus of enormous thick-shelled clams which inhabit tropical reefs and is likewise byssally attached. Its siphons are directed straight up and are greatly expanded, whereas the byssus is midventral with the hinge just posterior to it. Within the expanded siphon tissue are innumerable algae (zooxanthellae) living in a symbiotic relationship with this clam, as they do with corals.

Though many byssally attached bivalves live with the commissure plane vertical, there are others in which this is not so. *Pteria* for instance, has one of the two '**ears**' along the hinge line greatly enlarged, extending the hinge like a wing. In the specialized habit of *Pteria*, which lives attached to alcyonarian stems, the function of the wing seems to be that exhalant currents are removed as far as possible from the inhalant region and are not recycled. In benthic 'eared' bivalves the extension of the shell prevents it from being overturned by currents.

Cemented and secondarily free-lying forms (Fig. 8.13)

The best known of those bivalves which attach themselves by cementation to the substratum are the oysters (Order Pterioida, Suborder Ostreina) whose morphology and evolution have been fully documented in the *Treatise on Invertebrate Paleontology* by Stenzel (1971). These are perhaps the most successful of all bivalves, having a very efficient feeding mechanism which integrates the activities of palps and gills in a way that no other bivalves have been able to do. Oysters are abundant in ancient and modern sediments and are normally preserved in their natural life position. Many hardgrounds in the Mesozoic and Cenozoic are marked by the presence of many oysters which attached themselves to what was then a hard, recently submerged substratum.

When they settle, oyster larvae attach themselves to the sea floor by their left valve, which becomes cemented to the rock. In the process of cementation (Harper, 1991a; Fig. 8.13f), a thin sheet of periostracum forms at the mantle margin. This is secreted during cementation and through it leaks 'extrapallial fluid' from a space between the shell and periostracum and the mantle. From this fluid a calcareous material crystallizes. It is compositionally like the shell and forms a strong-bonding cement. Since the two valves must close exactly along the commissure, any irregularity in the left valve is also reflected in the right valve. Thus if the larva settles down on a dead ammonite shell, both the left and right valves will have an impression of the ammonite. Oysters have a single large adductor to close the valves and are often of somewhat arcuate form, with the gills lying horizontally and the anterior–posterior axis at some 60° to the hinge axis. Though in most genera the commissure is more or less flat, some fossil and Recent genera [e.g. the Cretaceous *Arctostrea* (Fig. 8.13b) and the modern *Ostrea frons*] have zigzag commissures like those of many brachiopods, and these presumably fulfilled a similar function (section 7.6). Oysters may build up substantial biostromes, and as they often lie in belts parallel with shorelines they have been used successfully in determining former shore positions.

Cementation has originated independently in some 20 major groups, and always from a byssate ancestor. Such cementation is thus polyphyletic, and there are even examples from fresh water. *Etheria*, for instance, forms large oyster banks in fresh water in West Africa. Many groups of cementing bivalves appeared independently in the early Mesozoic, coinciding with a great increase in population pressure. Harper's (1991b) experimental work shows why this should be so: predators (crabs and starfish) were offered a choice of byssate or cemented prey, and the latter survived far more effectively since the predators could not readily manipulate or pluck off the cemented bivalves from the substrate.

An extinct group of cemented bivalves, the hippuritoids or rudistids (Subclass Heterodonta), became very highly modified, so that they are hardly recognizable as bivalves at all. These inhabited the shallow tropical seas of the Tethyan realm during the Late Jurassic and Cretaceous. Examples are *Hippuritella* and *Radiolites* (Fig. 8.13j,k) from the Upper Cretaceous, in which the valves are very

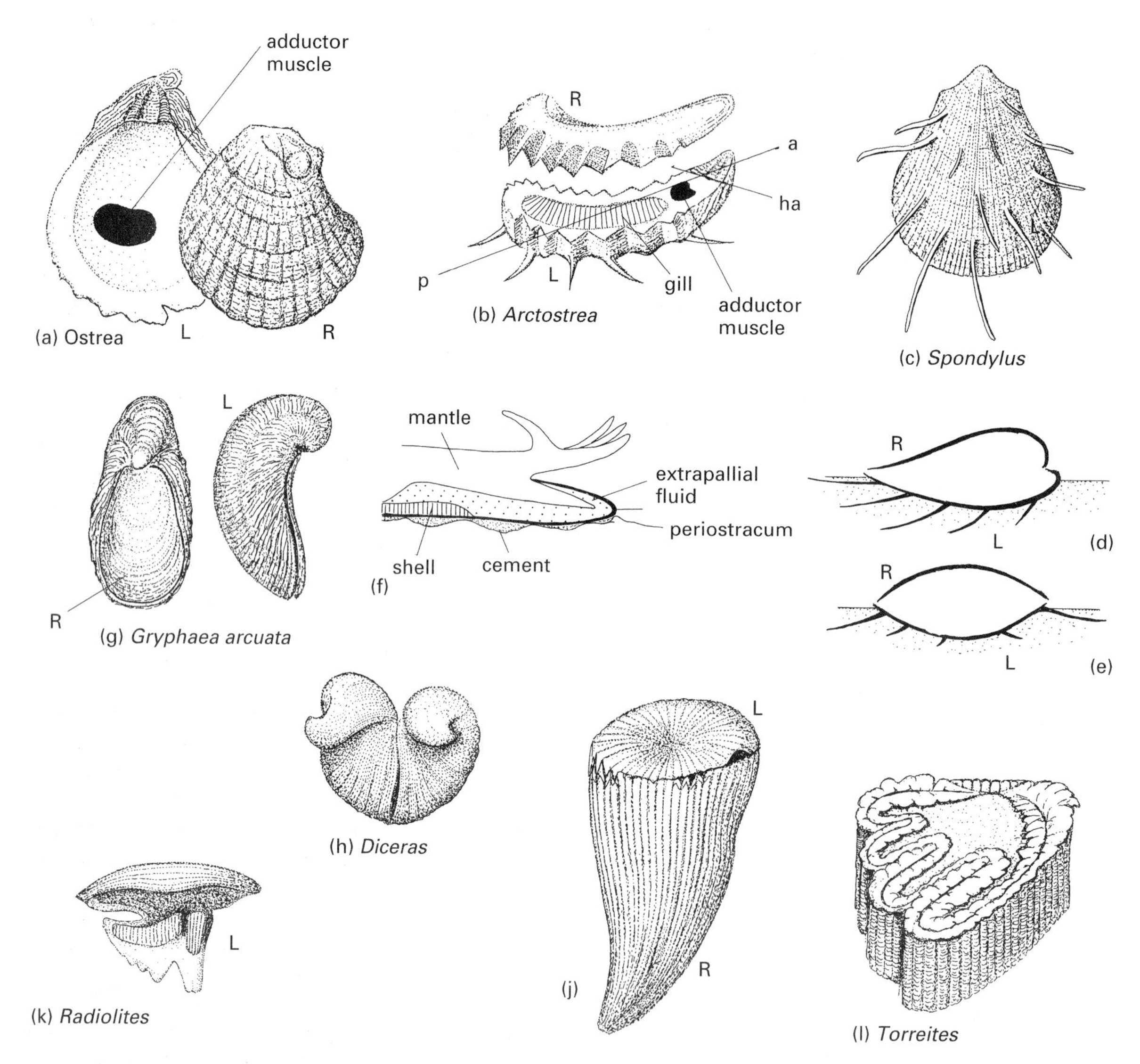

Figure 8.13 Cemented or secondarily free bivalves: (a) *Ostrea*, cemented by its left (L) valve; right valve (R) overlying (× 0.3); (b) *Arctostrea*, a Cretaceous oyster, cemented by its left (L) valve, showing gill, spines, adductor muscle scar and orientation (× 0.5); (c)–(e) *Spondylus spinosus* (Cret.), (c) left valve; (d), (e) in lateral and anterior view, showing how spines are used as snowshoes (× 0.75); (f), (g) *Gryphaea arcuata* (L. Jur.), a secondarily free-lying oyster, with the heavier left valve below (× 0.5); (g) section through the edge of an oyster shell, showing processes of cementation (see text); (h) *Diceras* (Jur.), probable ancestor of rudistids, with twisted umbones (× 0.3); (j), (k) *Radiolites* (Cret.), showing coralloid right valve (lower) and smaller left valve (upper); internal structure of latter shown in (j) (× 0.3); (l) *Torreites*, a Cretaceous rudist reconstructed as with thick extensions of mantle margin between the valve rims. [Mainly based on drawings in Zittel, *Textbook of Paleontology*, 1913; (d), (e) redrawn from Carter, 1972; (l) redrawn from Skelton and Wright, 1987.]

unequal in size. The right valve is conical; the left valve is nearly flat and sits on top of the right valve like a lid. The right valve has very thick walls of two layers, a relatively small body cavity, and a single gigantic tooth which articulates with an equally large pair of teeth hanging down from the lower surface of the left valve. In external morphology there are some similarities to the bizarre Permian strophomenide brachiopod genera *Gemellaroia* and *Cyclacantharia*, and in a remote sense even to solitary corals, but there is certainly little functional similarity to the latter. Some rudistids seem to have been able to suck in water through the perforated left valve, and detailed functional analyses have been made showing probable current directions (Skelton, 1976). Most rudistids were small, only a few centimetres high, but there are some immense species up to 50 cm high. They are found in Cretaceous limestones in France and elsewhere, sometimes forming extensive clusters which may even form small reefs, though only some species actually contribute towards reef formation.

The earliest rudists (Family Diceratidae) were encrusters, attached by the umbo of one valve. The had heavy shells with an external ligament, and the umbones were twisted in a 'spirogyral' fashion (Fig. 8.13h). This conservative construction was retained by one group until the end of the Cretaceous, but they remained as encrusters and did not invade new adaptive zones. A second group, however, the uncoiled rudists (Skelton, 1978, 1985), have an 'invaginated' ligament – it is no longer marginal but neatly tucked away within the shell. The growth form of the shell is thus not constrained by an external ligament. It was only because of this that the shells were able to uncoil, and Skelton regards the shortening and invagination of the ligament as a pre-adaptation for their radiation into hitherto unoccupied adaptive zones. Since they were now able to grow elongated tubular valves, they could invade new habitats as 'elevators' and 'recumbents'. The former are tall conical or barrel-shaped shells, often forming thickets, and feeding in calm waters well above the sediment surface and out of the zone of muddy water. Recumbents, on the other hand, are large extended shells lying freely but stably in current-swept shores on the crests of build-ups. The high diversity of such Cretaceous uncoiled rudists testified to the advantage given by the initial pre-adaptation.

The large rudist *Torreites* is reconstructed (Skelton and Wright, 1987) as having had thick extensions of mantle margin projecting out between the valve rims (Fig. 8.13l), as in the living giant clam *Tridacna*. This genus is found in both Oman and the Caribbean and was probably dispersed by larvae along shallow 'staging posts' in the Pacific and eastern Tethys.

A geometrical model for the growth and form of rudist bivalves and its implication for their classification is given by Skelton (1978), and a full treatment of their evolution, ecology and role as reef formers may be found in Kaufmann and Sohl (1981).

The very spiny isodont genus *Spondylus*, which lives today in coral reefs, is usually cemented. Some species, however, have become secondarily free. The Cretaceous *S. spinosus* (Fig. 8.13c–e) has spines arranged at right angles to the shell margins; these apparently acted as snowshoes, preventing this free-living bivalve from sinking into the soft ooze upon which it lived.

Amongst other secondarily free-lying genera is the large oyster *Gryphaea* (Fig. 8.13f,g), whose evolutionary development has been much debated in a series of papers since the early 1940s. I never used to have much enthusiasm for the unattractive-looking *Gryphaea*. I felt that it had too few morphological characters to sustain the endless flow of papers which purported to describe its evolution, and some of these works did in fact prove biometrically unsound. However, the excellent recent studies of Johnson and Lennon (1990) and Johnson (1994) have restored this genus to eminent respectability, as illustrating a complex, non-unidirectional and broadly gradual change through time. *Gryphaea* has a very thick convex left valve, necessary for stability, for if an individual was overturned the commissure would be blocked, with fatal results. In this context selection seems to have favoured a more stable, broader and flatter shell shape, and increased size as an adaptation to a stable food supply.

Amongst the strangest of all bivalves are the alatoconchids; the 'giant clams' of the Tethyan Permian. Genera such as *Shikamaia* (*Tachintongia*) reach lengths of up to 1 m and the shell was up to 3 cm thick. These shells had wide wing-like flanges on each valve, extending posteriorly and giving the appearance of a very flattened shell. The juveniles were byssally attached but the large and heavy (10 kg) adults lay freely on the sediment surface, the weight being partially borne by the flanges. This is one of the few bivalves that lay flat on the sea floor with a vertical commissure (Yancey and Boyd, 1983).

Free-swimming forms (Fig. 8.14)

The large scallop *Pecten* normally lies free on the sea floor, but it can swim by vigorous and repeated clapping of the valves together so as to expel water in successive jets on both sides of the 'ears'. Such activity is exhausting for the bivalve and cannot be sustained for very long, but it is normally used only to escape from predators.

In *Pecten* the two valves are unequal in size, the lower being the more convex, but they are nearly equilateral. On either side of the umbo the hinge is prolonged into two 'ears' of nearly equal sizes. The ligament is set centrally and internally and emplaced in a small triangular pit. A single large adductor muscle occupies much of the central space. Evidently this kind of shell was derived from a byssally attached ancestor; the two kinds of shell have many features in common, though that of *Pecten* is extended by an increased umbonal angle, assisting its capacity as a hydrofoil.

Lima is a byssally attached genus, but it can also swim as an escape reaction. Individuals can release their byssus and swim by valve-clapping with the commissure held vertically. The mantle is here prolonged into 'tentacles' around the commissure; these row like oars whilst the animal is swimming, which adds to the speed of movement.

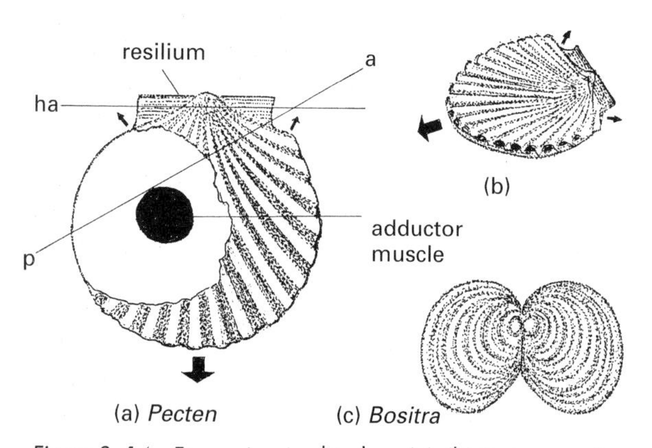

Figure 8. 14 Free-swimming bivalves: (a), (b) *Pecten*, a swimming bivalve in which the anterior adductor is reduced completely (×0.4) – broad arrow shows the direction of shell movement when water is ejected (small arrows) near the hinge; (c) *Bositra buchi*, a Jurassic bivalve, possibly nektoplanktic, with valves open in inferred life position [(a), (b) Based on Stanley, 1968; (c) based on Jefferies and Minton, 1965.]

Pecten and *Lima* probably inherited their natatory ability from a common ancestor, and there is evidence that as far back as the Carboniferous some bivalves could swim. Fossil Pectinacea are not uncommon. Some of them were like *Pecten* in morphology and habit (e.g. the Carboniferous *Pterinopecten*); others such as the Devonian to Jurassic posidonian genus *Bositra* and its relatives may have been specialized for a nektoplanktonic (entirely free-swimming) mode of life. *Posidonia* is characteristically present in black shales (ecologically equivalent to the earlier graptolitic facies) in which the only other fossils are ammonites or goniatites, but it may also occur in shallow-water limestones deposited as lime muds fine enough to preserve the shells (the parallel with graptolite preservation is again noteworthy). All posidonians are thin-shelled with only two shell layers. They have gapes on either side of the hinge as in their living pectinacean relatives, but they never had a byssal notch at any time in their ontogeny, and they probably never went through an attached phase at any time in their life history. Specimens are almost always preserved with both valves open, though in modern limaceans and other pterioids this is rarely so. Experiments by Jefferies and Minton (1965) showed that the valves would be preserved in the open position only if their normal opening angle exceeded 60°, which may indeed have been their normal angle of opening in between swimming contractions. Further experiments and calculations have indicated that such a swimming bivalve would not have sunk rapidly, especially if the drag effects were increased by a fringe of stiff tentacles around the commissure, as in the modern *Lima*.

Posidonia therefore may well have exploited an ecological niche, that of the permanently swimming nektoplankton, which no later bivalve has been able to invade since the extinction of this genus in the Cretaceous.

Boring and nestling bivalves (Fig. 8.12a–c)

Certain bivalves are adapted for life in hard substrates. The stone- and wood-boring genera *Lithophaga* (Order Mytiloida) and *Teredo* (Order Myoida, Suborder Pholadina) have elongated shells of cylindrical form, and like modified deep burrowers they live with their long axis normal to the surface and have very extended siphons. The shells of these bivalves are very thin and must be very resistant to abrasion, for the edges of the shells are used in excavating their burrows, aided by acids secreted by cells in the mantle. The insoluble periostracum prevents their own shells from being dissolved. Frequently the shell edges are provided with stout spines, used as scraping tools in excavation, which is effected by rocking movements of the shell about the long axis. Some species of *Lithophaga* bore into live coral, and have highly specialized adaptations. Not all highly attenuated bivalves are borers. Some live in firm sediments, e.g. *Aspergillum* (Order Anomalodesmata; Fig. 8.12b), and are well adapted for this mode of life; evidently this represents the primitive condition from which the rock- and wood-borers later evolved. The boring habit has evolved nine times independently since the early Palaeozoic and has become especially important from the Mesozoic onwards.

Boring bivalves are commonly found preserved in their burrows, and where the borings fit tightly round the shells it would seem probable that the occupant was the creator of the domicile. Sometimes, however, the fossil within the boring was merely occupying a vacated residence; it was, in effect, a 'squatter'. Kelly (1980) has shown that the late Jurassic *Hiatella* occurs in two habitats, first as a simple byssal nestler on hard substrates, and second, but more commonly, in borings made by other bivalves penetrating hard substrates (Fig. 8.12c). In these the contours of the *Hiatella* shell do not closely approximate the shape of the burrow.

Nestling bivalves cannot bore but are photonegative, like *Lima*, and occupy pre-existing cavities. Some are byssally attached, and their shells grow to fit the cavity even if this is of irregular shape.

Ecology and palaeoecology

While shell shape and other factors have been shown to be useful in interpreting the mode of life of extinct bivalves, other kinds of ecological study are more directly concerned with the organism–environment relationship. The following examples show how biological and geological criteria can be used together to give information about past environments with facies-controlled faunas.

A study of a Silurian bivalve fauna from Möllbos in Gotland (Sweden) combines both auto- and synecological data (Liljedahl, 1984, 1985, 1994). Most Palaeozoic bivalves possess few diagnostic features and are thus hard to classify and identify. Only when the interior is really well preserved can the anatomy of the soft parts be reconstructed with confidence and thus allow the life position and habits of the various taxa to be inferred. The Möllbos fauna is silicified (through the initial agency of endolithic microorganisms), and was preserved *in situ*. The bivalves, isolated from the muddy matrix by acid etching, are preserved in such detail that their adaptive morphology can be established.

Using muscle scars, shell shapes, size and musculature of the foot, and by comparison with living bivalves, Liljedahl was able to assess the burrowing potential of the various species. Statistical analysis of the trophic structure of the community showed that it was dominated (>90%) by deposit feeders, which unlike suspension feeders are adapted to a clay-rich mud environment. There were three different feeding levels within the sediment (Fig. 8.15): (1) surface/semi-infaunal, (2) infaunal and (3) deeper infaunal.

Only the shells of the first category were often fragmented and normally encrusted by epibionts, and likewise the infaunal species were more commonly articulated. The ecological assumptions made about this fauna of trophically tiered deposit feeders, inhabiting a muddy soft-bottomed environment, fully accord with conclusions reached through functional morphology.

There is a clear contrast between the life habits of the Möllbos assemblage and the somewhat younger Grogarnshuvud assemblage, also from the Silurian of Gotland (Liljedahl, 1991). Here the lucinoid *Ilionia prisca* is found, usually in life position in shallow subtidal lime mud. It is very similar to living lucinoids (Fig. 8.11a) and was likewise a mucus-tube feeder, living at some depth within the sediment. *Ilionia* oriented itself obliquely to the direction of wave action, thereby maximizing the intake of suspended food particles through the mucus tube and avoiding fouling by its own waste material. It may also have lived in symbiosis with sulphur-oxidizing bacteria.

The following Mesozoic example is on a larger scale but equally of interest. The Great Estuarine Series (M. Jur.) of the Inner Hebrides of Scotland consists of a shale and sandstone sequence in which bivalves, along with some gastropods and ostracodes, are the most dominant fossils (Hudson, 1963). Individual bivalves are very common but

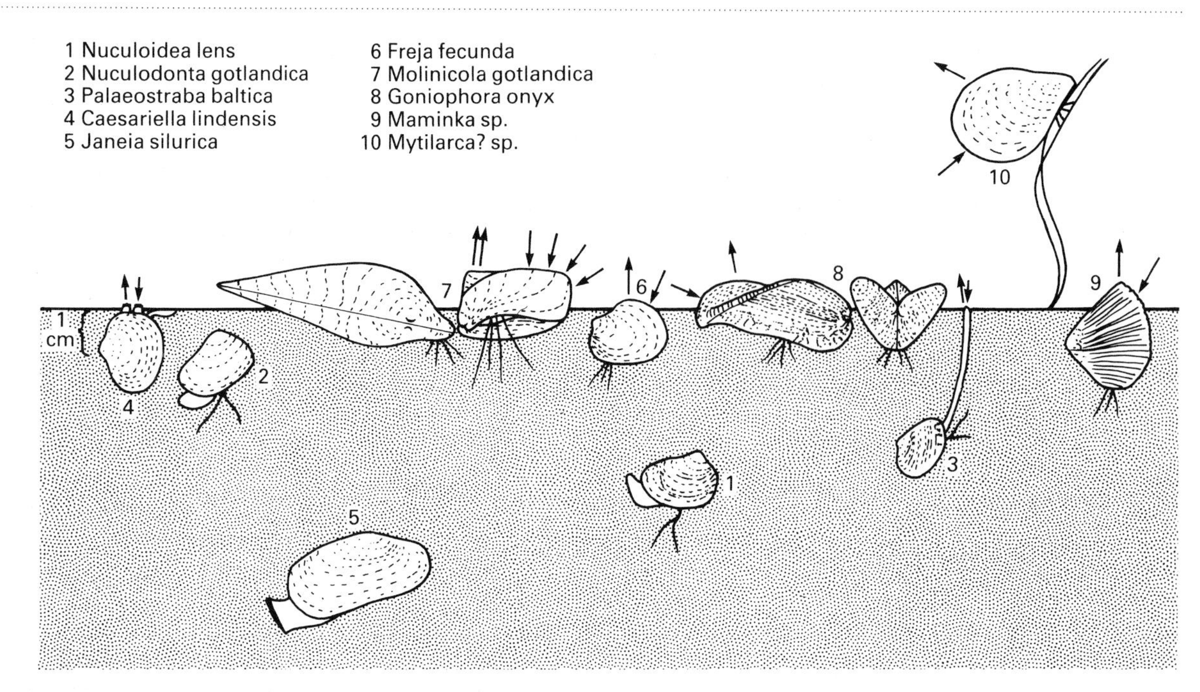

Figure 8.15 Suggested life positions of Silurian bivalves from Möllbos, Sweden (redrawn from Liljedahl, 1985.)

belong to only a few species, and within the sequence there are hardly any normal marine fossils except at the very top. Such association is normally indicative of a 'difficult' environment, which few species have been able to colonize and in which the species that have done so are successful. The beds were deposited in very shallow water as shown by mud cracks at certain horizons suggesting periodic desiccation. Successive lithologies within the sequence contain bivalve faunas. Hudson found remarkable analogies with the present shallow lagoonal bays of the Texas coast, from the point of view both of environment and of the bivalve genera living there. Several of these modern genera (*Mytilus*, *Ostrea* and *Unio*) have direct counterparts in the Great Estuarine Series; furthermore, in Texas the bivalve assemblages in faunal content, which can be closely compared with the Jurassic ones, are salinity controlled. The Texan *Crassostrea* species and mytilids live in water of rather reduced salinity, and Hudson inferred that analogous Jurassic associations likewise lived in hyposaline water.

By matching the analogues Hudson showed clearly that the overlapping assemblages were controlled by salinity variations through geological time: from fresh water (dominated by the bivalve *Unio* and the small *Viviparus*): through brackish water (the most 'difficult' environment to colonize for physiological reasons), where only the euryhaline *Neomiodon* was present; thence to brackish marine, where oysters and mytilids thrived; and finally to fully saline marine environments.

More recent work on oxygen and carbon isotopes has shown the essential correctness of this picture, though a few modifications were needed, notably that *Unio* and *Neomiodon* of the Jurassic apparently lived in normal marine conditions.

Stratigraphical use

Bivalves are on the whole far too long-ranged in time to be of much zonal value. However, they have been used in a broader stratigraphical sense, as in Lyell's division of the Tertiary, which has abundant bivalves and gastropods, into four series based upon the relative percentages of the molluscan faunas present therein now living.

The one circumstance under which bivalves have been used successfully as stratigraphical indicators is in the British Carboniferous Coal Measures, where non-marine bivalves are abundant at certain horizons. These genera (*Carbonicola*, *Naiadites* and *Anthraconaia* amongst others) are not unlike the modern freshwater *Unio*, but they differ in certain morphological characters. The species are not easy to distinguish and are rather long-ranged. But even so, six or seven zones have been defined using concurrent ranges of different species of non-marine bivalves, and the zones thus defined have been corroborated by plant and spore fossils.

Class Rostroconchia

In recent years a small group of Palaeozoic molluscs has been recognized as being of unique phylogenetic interest and has been separated out from other molluscs as Class Rostroconchia (Pojeta and Runnegar, 1976).

Rostroconchs look superficially like bivalves and were probably fairly similar to them internally, for example in the possession of a protrusible foot, signified by a marked anterior gape in the shell. Where they differ is in the morphology of the hinge line, for they do not possess a functional hinge at all. These molluscs began their growth by producing a small, limpet-like, bilaterally symmetrical protoconch. From this the adult, likewise bilaterally symmetrical shell (**dissoconch**) grew down as a pair of valves. But there is no true hinge for some or all of the shell layers are continuous across the dorsal margin. The valves must have been held rigidly together, the dorsal margin functioning at best as a poorly elastic structure.

The earliest known genus, *Heraultipegma*, is Lower Cambrian. Later rostroconchs reached their maximum development in the early Ordovician, almost rivalling bivalves at that stage, but they declined thereafter and only one order, the Conocardioida, continued until the Permian. In the early rostroconchs (e.g. *Riberoia*) all the shell layers traverse the dorsal margin, whereas in the advanced forms the outer layer does not cross it, suggestive of an independent step towards the condition already achieved by bivalves. *Conocardium* (Fig. 8.16c,d), one of these advanced forms, has a gape at one end and a very pronounced **rostrum** at the other.

It has been suggested that rostroconchs occupy a key position in molluscan phylogeny. Pojeta and Runnegar (1976) have proposed that helcionellans are extinct monoplacophorans and that these gave rise to the rostroconchs, losing their segmentation in the process. These in turn produced the bivalves on the one hand (by separation of the valves and development of a proper hinge), and possibly the Scaphopoda on the other. The cephalopods and gastropods, according to these authors, were probably

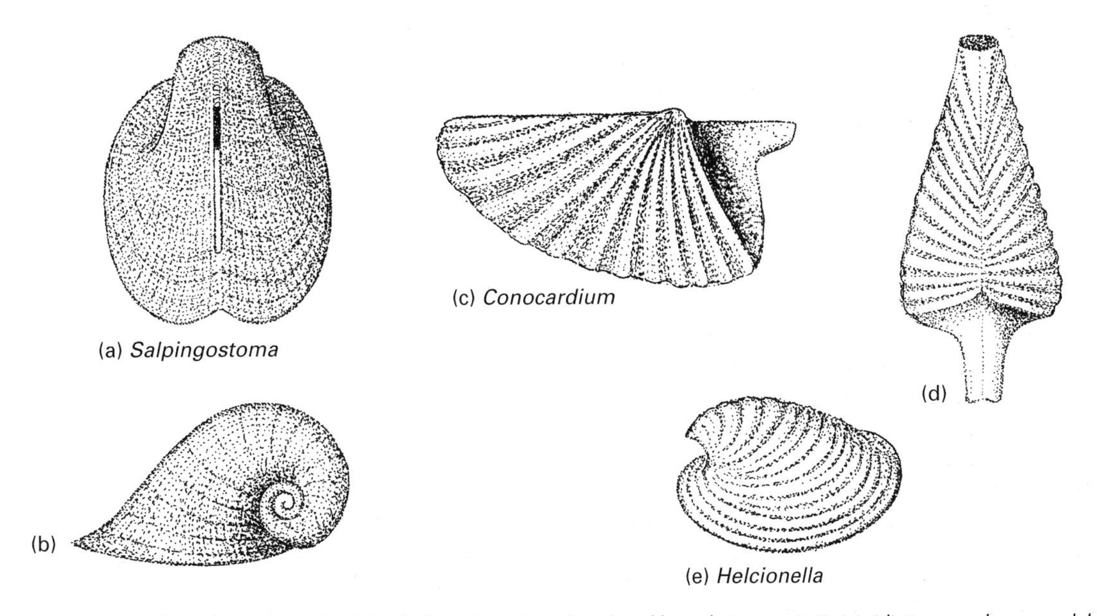

Figure 8.16 (a), (b) *Salpingostoma* (Ord.), a bellerophontid, in dorsal and lateral views (× 0.5); (c), (d) *Conocardium pseudobellum* (Dev.), a rostroconch, in lateral and dorsal view (× 1.8); (e) *Helcionella*, an early Cambrian mollusc (gastropod or monoplacophoran; × 2.5). (Redrawn from *Treatise on Invertebrate Paleontology, Mollusca 1*, 1960; Pojeta and Runnegar, 1976.)

derived independently from the monoplacophorans. When the rostroconchs had given rise to the more adaptable bivalves, competition may well have been an important factor in their demise.

These hingeless 'bivalved' molluscs have helped in an unexpected way to bridge a large morphological gap in the phylogeny of early molluscs. Perhaps more discoveries of similar kind will illuminate the relationships of molluscan classes even further, allowing a test of this evolutionary model.

Class Gastropoda

Introduction and anatomy

Gastropods include all snails and slugs living in the sea, in fresh waters and on land and also the pteropods of the marine plankton. The earliest genera are Lower Cambrian, and though they probably are more abundant now than at any other time they can be found in sedimentary rocks of all ages.

The majority of present-day gastropods, and the only ones preserved, have coiled shells. Gastropods all have a true head, usually equipped with tentacles, eyes and other sense organs, which is more or less continuous with the elongated body; this typically has a flat, sole-like lower surface upon which the animal creeps by small-scale waves of muscular contractions, lubricated by slime from **mucous glands**. The visceral part of the body largely resides inside the shell, which in such an example as *Buccinum* (Fig. 8.17a,b), the common whelk, is helically coiled.

The **head-foot**, i.e. the protrusible part of the body, can be withdrawn inside the shell by **retrac-**

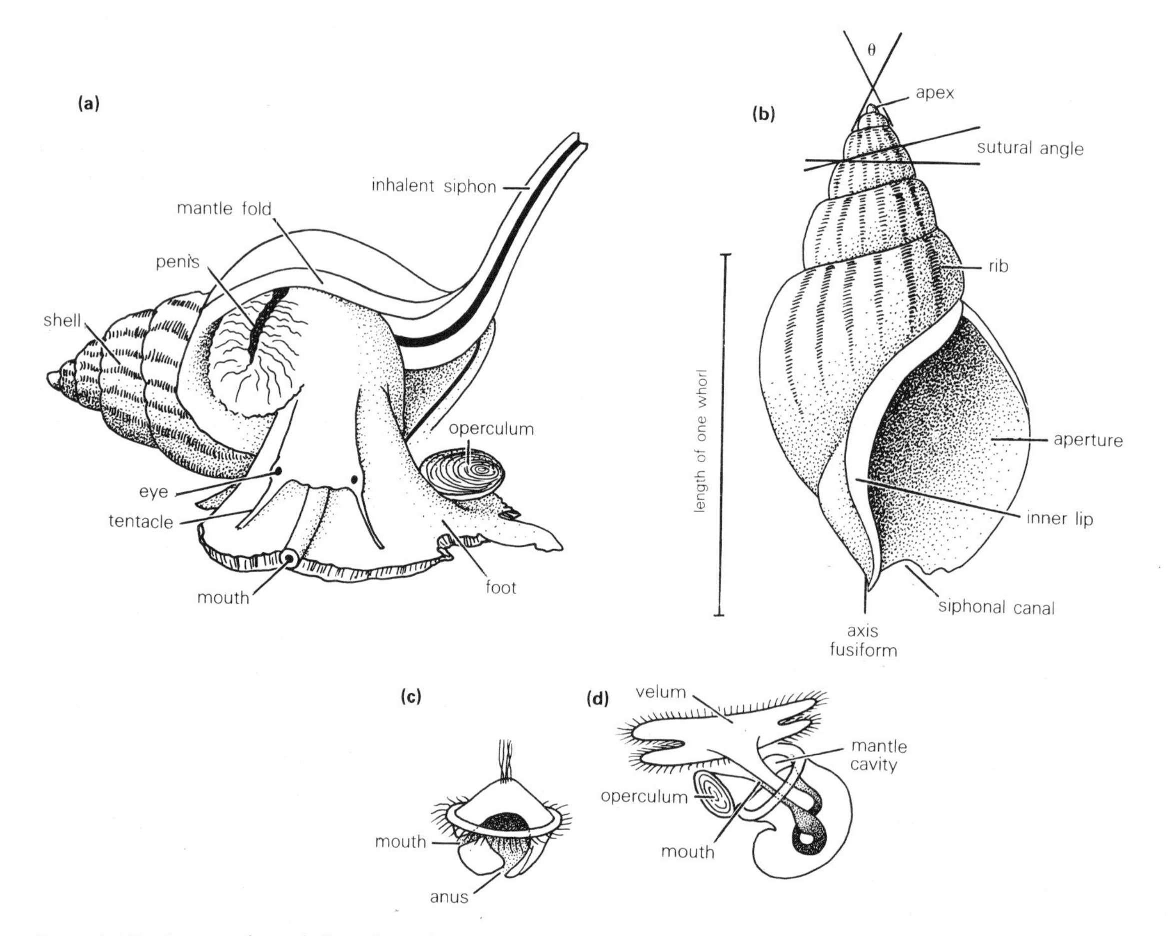

Figure 8.17 Gastropod morphology shown by *Buccinum* (Rec.): (a) living animal (× 1 approx.); (b) shell (fusiform); (c) trochophore larva; (d) veliger larva of gastropod. [(a) Redrawn from Cox in *Treatise on Invertebrate Paleontology*, Part I.)

tor muscles (the only attachment of the soft part to the shell) and closed off from the outside by a plate-like trap door (the **operculum**).

The mouth contains a rasping jaw (the radula) in its lower part, composed of a multitude of tiny teeth which scrape against a horny plate in the upper part of the mouth to shred food material. *Buccinum* has a tubular extension of the mouth (the **proboscis**) which is present in the more advanced gastropods. Like most snails, the whelk is hermaphrodite, with male and female organs in the one individual. However, individuals copulate with a complex mating pattern, involving the use of a very large male penis and the expulsion of calcareous darts, which are shot out of special apertures and embedded in the body of the other individual; this apparently acts as a stimulating procedure.

The mantle cavity lies anteriorly and in *Buccinum* communicates with the external environment by means of an inhalant siphon: a tubular organ formed from a fold in the mantle and occupying an indentation in the shell margin. Many gastropods do not, however, have such a siphon, and water is drawn in along the edge of the shell. Within the mantle cavity lie the gills, anus, mucous glands and also specialized organs (the **osphradia**) whose function seems to be to sample the water entering the cavity.

It is characteristic of gastropods that the mantle cavity faces anteriorly. Associated with this the internal organs, including the nervous and pallial systems, are peculiarly twisted. This internal asymmetry, known as torsion, is fundamental to gastropod morphology, and though it has been lost in some of the shell-less forms, this return to a more normal condition is clearly secondary.

This curious displacement of the internal organs is in no way connected with the asymmetrical coiling of the shell but has a quite different origin. Many suggestions have been proposed to account for it, the most generally accepted being that of the English zoologist Walter Garstang (1951), who has suggested that such torsion gave a singular protective advantage to the gastropod when in the larval state and was retained in the adult where it allows the animal to retract into its shell.

There are some coiled monoplacophorans, e.g. *Sylvestrosphaera* and *Sinuites*, which have reduced the number of muscle attachment sites on the shell interior so that the first muscle scars lie half a whorl back from the specture. This suggests that the visceral mass was able to retract deep into the shell. Such monoplacophorans as this would seem to be pre-adapted for torsion, and it is probable that the first gastropods arose from these (Peel, 1980). The great success of gastropods is a testament to the value of torsion in allowing full retraction within the shell. Such a facility may have been acquired independently by various groups, and thus the gastropods might be polyphyletic.

When gastropod eggs hatch they turn into planktic larvae known as **trochophores** (Fig. 8.17c), which closely resemble the larvae of certain marine worms. Trochophores are very small and more or less globular, with a fringe of cilia round their widest part and another tuft at the apex. The mouth, situated at one side just below the ciliary ring, leads to a simple gut terminating in an anus at the lower pole. The next stage in larval development is the **veliger** stage (Fig. 8.17d), during which a thin shell is secreted over the upper pole and the region under the gut becomes expanded into a large sail-like **velum**: a flat bilobed organ covered with cilia which propel the veliger through the water by their coordinated action. It is at this stage that torsion occurs: a rather sudden twisting through 180°, as a result of which the large velum can be withdrawn inside the shell in case of danger. Garstang has suggested that the advantage conferred in being able to tuck away the velum was so great that the resultant displacement and asymmetry of the internal organs was a minor price to pay for the safety rendered. (In addition to his serious papers on the subject, Garstang also wrote about his concepts in the unusual medium of comic verse, and his poem *How the Veliger got its Twist* crystallizes the arguments in a memorable fashion). The main problem that the gastropods acquired through torsion was that excreta from the anus (located in the mantle cavity) would be expelled just over the mouth. In many gastropods there are devices that cope with this by separating the inhalant and exhalant water. Exhalant siphons are present in some, and in others there is an indentation (**slit-band**) in the shell, forming a channel through which the exhaled water is carried dorsally away from the shell and head region (Fig. 8.18l).

The velum of larval gastropods eventually becomes the foot upon which the adult glides, by the loss of cilia and the development of an internal system of longitudinal muscles which allow very small rhythmic waves of contraction to pass backwards along the foot; the lubrication of the gastropod's passage is assisted by the supply of copious slime from the mucous glands.

Gastropods may feed in a variety of ways. Some, such as the muricids are actively carnivorous and can rasp away the shell of another gastropod or a bivalve to reach the flesh. They drill a neat, round hole in so doing, which may take 24–48 h. They then inject a muscle relaxant through the newly drilled hole, the shell then opens and they are able to feed. The oldest known predatorial gastropod drillholes are probably Devonian (Smith *et al.*, 1985) and drilling habits have also been recorded in the Carboniferous *Platyceras*, a specimen of which was found located directly above a conical hole in the tegmen of a crinoid. In this case, however, the gastropod was probably parasitic rather than predatorial. It was not, however, until the Lower Cretaceous that such drilling became widespread, and 'forced' evasive adaptations on the prey. Today *Murex* and related genera are the most effective of all the 'driller killers', as they have been memorably termed.

Turritelline gastropods have been extensively drilled and 'peeled' (i.e. have had their apertures broken) by other gastropods from the Late Cretaceous until the present day. Yet such predation has not apparently forced any evolutionary changes in the shells of turritellines during this time: there is no evidence of an 'arms race' and predation seems to have stayed at much the same level since the later Mesozoic (Allmon *et al.*, 1990).

Many gastropods feed on detritus or decayed material. Still others have specialized modes of feeding, such as the coprophilic (faeces-ingesting) gastropods found clustered round the exhalant spires of Carboniferous blastoids.

Classification

The classification of gastropods is largely based upon soft parts. Gill and osphradial morphology is most important, as is the structure of the nervous system, heart, kidneys and reproductive system. A condensed version of that in the *Treatise on Invertebrate Paleontology is* given here, though more recent studies (Ponder and Waten, 1988; Haszprunar, 1988) present a somewhat different arrangement.

SUBCLASS 1. PROSOBRANCHIATA (L. Cam.–Rec.): Shelled gastropods in which torsion is complete.

ORDER 1. ARCHAEOGASTROPODA (L. Cam.–Rec.): The gills here are **aspidobranch** (i.e. have their filaments arranged in a double comb on either side of the axis and are free at one end). There may be two gills, or one may be lost. The shell structure is variable; some symmetrical forms exist or are found as fossils, but the shells are normally helical spires. Nearly all marine; e.g. *Euomphalus*, *Pleurotomaria*, *Patella*, *Platyceras*, *Trochus*, *Maclurites*.

ORDER 2. MESOGASTROPODA (Ord.–Rec.): Mainly have **pectinibranch** gills, a much more elaborate and efficient system than the aspidobranch condition, which permits free flow of water through the mantle cavity. Living mesogastropods are classified on radular structure; e.g. *Strombus*, *Cypraea*, *Natica*, *Cerithium*, *Nerinea*.

ORDER 3. NEOGASTROPODA (Cret.–Rec.): The meso- and neogastropods are often combined in the single Order Caenogastropoda. They have pectinibranch gills. An inhalant siphon leads into the mantle cavity, characteristically with a short or long groove in the shell carrying a siphonal tube anteriorly from the shell; e.g. *Murex*, *Buccinum*, *Voluta*, *Conus*.

SUBCLASS 2. OPISTHOBRANCHIATA (?Carb.–Rec.): Marine gastropods which have largely or completely lost the shell. They have undergone detorsion and straightened themselves out. These include the planktonic pteropods and nudibranch sea slugs, which carry secondary gills on the dorsal surface.

SUBCLASS 3. PULMONATA (Mes.–Rec.): In these land-dwelling (and secondarily freshwater-dwelling) slugs and snails, the gills are lost and the whole surface of the mantle cavity is modified as a lung, liberally supplied with blood vessels and kept permanently moist. The pulmonates are the only molluscs that have made a really successful transition to land. Most of them have retained their shells, though the land slugs have lost them altogether.

Shell structure and morphology, which are all that is left to the palaeontologist, are not generally high on the list of criteria for determining correct systematic placement. Furthermore, the common tendency for shells of unrelated stocks to acquire similar forms through homeomorphic evolution causes additional problems.

Shell structure and morphology

The shell of a gastropod is basically an elongated cone, rarely septate, which may be coiled in a number of ways. The most characteristic shell shape is a helical spire, coiled about an axis, which is usually illustrated as vertical. There are some gastropods with planispiral and hence symmetrical shells. Usually the whorls all touch one another, but in some cases they do not embrace. Coiling is characteristically dextral but may be sinistral; in very rare cases equal numbers of individuals in a population

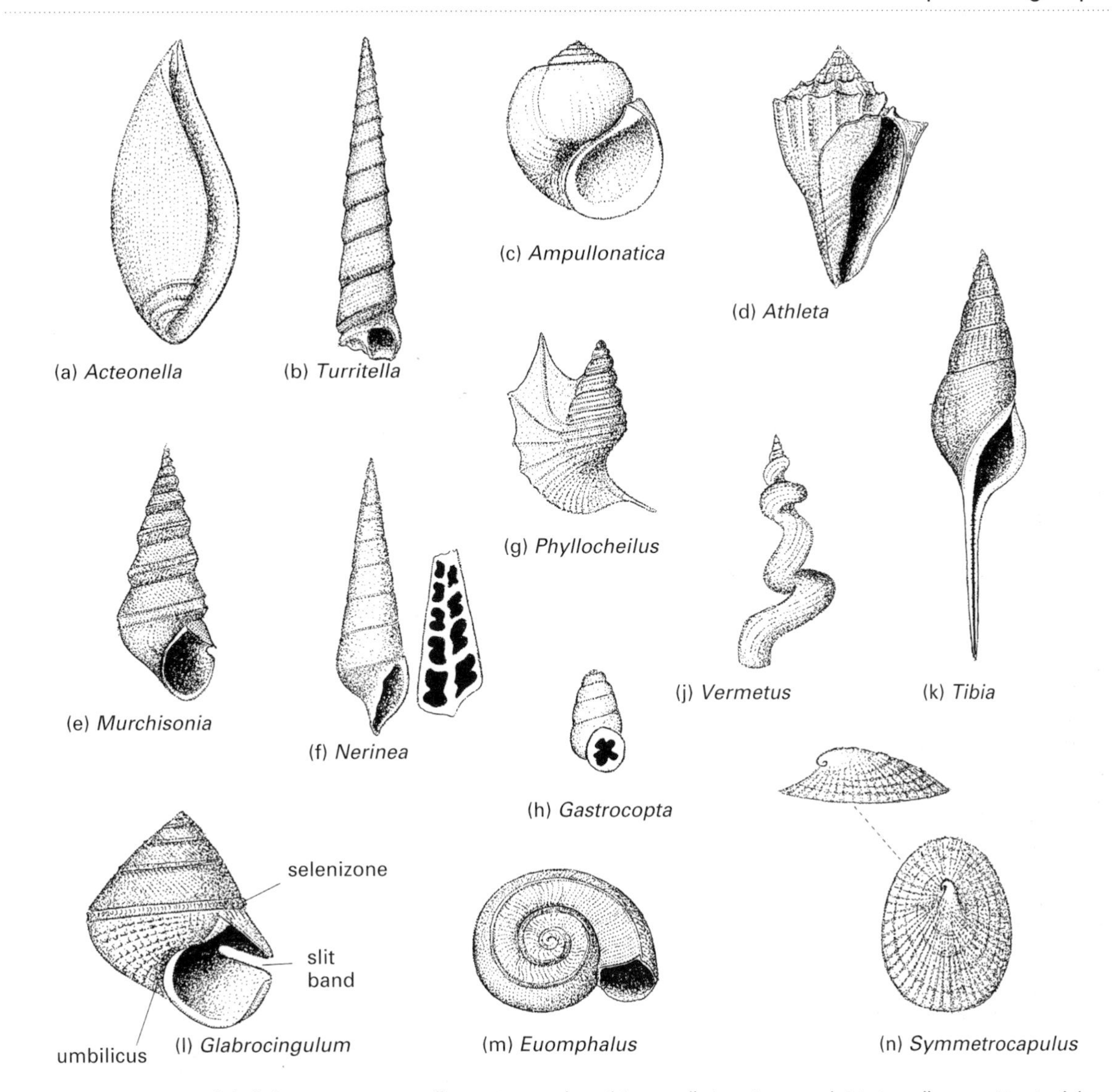

Figure 8.18 Gastropod shell shapes: (a) *Acteonella* (Rec.), convolute; (b) *Turritella* (Eoc.), turreted; (c) *Ampullonatica* (Eoc.), globular, low-spired; (d) *Athleta* (Eoc.), volutospine; (e) *Murchisonia* (Carb.), high-spired; (f) *Nerinea* (Jur.), with columella and internal folded thickenings within the whorls; (g) *Phyllocheilus* (Jur.), turbinate, with digitate aperture; (h) *Gastrocopta* (Rec.), pupiform, with constricted aperture; (j) *Vermetus* (Rec.) irregular, partially uncoiled; (k) *Tibia* (Eoc.), turbinate to high-spired, siphonate; (l) *Glabrocingulum* (Carb.), showing slit band for exhalant siphon; (m) *Euomphalus* (Carb.), discoidal, near planispiral; (n) *Symmetrocapulus* (Jur.), patellate. (Mainly redrawn from *Treatise on Invertebrate Paleontology*, Part I, and *British Cenozoic Fossils*, British Museum, 1979; not to scale.)

may be dextral and sinistral. Linsley (1977) has documented some interesting comments on the geometry of gastropod shell form. The basic terminology of gastropod shells is given in Fig. 8.17b. It is largely self-explanatory, as are some of the basic shell forms illustrated in Fig. 8.18.

It is worth noting that the earliest juvenile whorls of certain gastropods (the protoconch) are often of peculiar form and do not necessarily coil in the same axis as the rest of the shell; furthermore, they usually have a different surface ornament.

The external whorls may be ornamented with various characteristic features: ribs, spines or vertical bars (**varices**). The aperture may be entire and unmarked by any feature (**holostomatous**) or, as in neogastropod genera, may be equipped with a **groove** which is the outlet for the exhalant siphon (**siphonostomatous**). The siphon may be set in a deep slit-band which separates it far from the mouth. The calcified strip representing the 'track' of the slit-band as it is calcified is the **selenizone** (Fig. 8.18l). In such genera as the sand-dwelling *Aporrhais* and *Phyllocheilus* the aperture may be enlarged into a flange which helps to stabilize the shell while the gastropods are feeding.

There is a very clear distinction between those shells in which the later whorls do not meet centrally, and thus have an **umbilicus**, and those in which they touch. In the latter case they are welded to a central rod (the **columella**). Likewise, high-spired forms, in which the apical angle is low, are very distinct from low-spired genera, which have a higher apical angle. The shape of the whorls, any unusual coiling patterns, the presence or absence of different kinds of external ornament, apertural shape, and the presence or absence of slit-bands or of siphonal grooves are all used in classification and identification.

Shell composition

The shells of gastropods have an outer horny layer, the periostracum, below which lies the shell proper, a structure normally composed of layers of aragonite, though many gastropods, especially those living intertidally, have a calcitic outer layer. In fossil shells the aragonite either recrystallizes to calcite or may be dissolved, especially where there is leaching in the rock.

Many shells have an inner nacreous layer of very thin aragonite leaves parallel with the internal shell surface, each separated by equally thin organic layers. External to this is the crossed-lamellar layer in which thin lamellae (0.02–0.04 mm thick) are arranged normal to the shell surface. But each lamella itself consists of strips of extremely thin aragonite, arranged obliquely, the strips of adjacent lamellae being arranged at right angles to one another.

There are other kinds of shell structure in gastropods, all of systematic importance, but, as in the case of the layers described above, they almost always disappear in fossils and no trace of them remains.

Evolution

General features of evolution

Gastropods must have arisen from a bilaterally symmetrical molluscan ancestor, approximating in general morphology to the hypothetical archimollusc described earlier. The protection given by torsion and the resultant ability to withdraw inside their shells must have given gastropods a great advantage, for from their first appearance in the Lower Cambrian they seem to have been remarkably successful.

The earliest known possible gastropods (*Coreospira*, *Helcionella*) are found in the Lower Cambrian. They have coiled shells with a wide expanded aperture and are planispiral, hence bilaterally symmetrical. These very early genera are normally classified with the early archaeogastropod Suborder Bellerophontacea (L. Cam.–Trias.; Fig. 8.16a,b) named after the characteristic genus *Bellerophon* (which was the first Palaeozoic mollusc ever to be described, by the conchologist de Montfort in 1808). (The systematic position, however, of Families Sinuitidae and Bellerophontidae has been much debated. Specimens of *Bellerophon* sometimes show paired muscle scars and thus this genus and its allies are considered by some authorities more likely to be monoplacophorans than gastropods.)

Bellerophontidids have been interpreted (Harper and Rollins, 1985) as infaunal or semi-infaunal molluscs with the apertural region enwrapped by the mantle and possibly also the foot. One form, the Silurian *Pterotheca trimerelloides* (Clarkson *et al.*, 1995) shows an extraordinary range of form within the one species. Possibly this represents a survival strategy, presenting to intending predators, especially cephalopods, a mosaic of different 'search images', and thereby confusing it.

The bellerophontaceans are a most important group of fossil gastropods, and over 70 valid genera have been described. However, some of the Cambrian bellerophontiform shells have peculiarities uncharacteristic of the bellerophontaceans as a whole (Yochelson, 1967). *Helcionella* and its relatives have a simple aperture without any indentations, but all other bellerophontaceans have a pronounced notch or emargination in the plane of symmetry. There are other reasons for believing that *Helcionella is* not a normal bellerophontacean, and some authorities have suggested that it may not be a gastropod at all but rather may be a representative of the monoplacophorans or of an extinct class of molluscs.

Some other Palaeozoic bellerophontiform molluscs, the tightly coiled *Cyclocyrtonella* and the more cap-shaped *Tryblidium*, have paired muscle scars inside the shell and do not appear to have undergone torsion; their muscle scar pattern indicates that, rather than being bellerophontaceans, they are actually monoplacophorans. Hence bellerophontiform shells are not necessarily all gastropods, but such simple morphology rather seems to have been common in diverse molluscan groups during the earlier Palaeozoic.

In the Upper Cambrian are found the first asymmetrical, helically coiled shells, belonging to the important archaeogastropod Suborder Pleurotomariina (Cam.–Rec.). There is also an exclusively Palaeozoic suborder, the Macluritina; all these early forms are low-spired. Recent pleurotomariides are 'living fossils' in which the grade of organization present in some of the earliest gastropod genera can be seen by direct homology. Internally the gills of modern pleurotomariids are unmodified (aspidobranch). The reproductive organs are likewise in a relatively primitive condition, since the capacity for internal fertilization is not present and eggs and sperm are merely shed into the water. It was only the development of internal fertilization that rendered possible the later invasion of the land and fresh waters.

By Carboniferous times gastropod faunas were very rich and diverse; indeed it is becoming increasingly clear that Palaeozoic gastropods seem to have occupied a range of habitats approximating to those of the present day. In one fauna of finely preserved Carboniferous gastropods, the Visean Hotwells Limestone fauna of the Mendip Hills in Somerset, England, no less than 45 genera and upwards of 80 species occur in association with a typical Lower Carboniferous coral–brachiopod association (Batten, 1966). In this fauna there are many archaeogastropods (bellerophontids, pleurotomariids and limpet-like genera), but there are some caenogastropods as well; the ancestry of the latter can be traced back to the Ordovician. All the species of the Hotwells Limestone were apparently adapted to microniches within their environment.

There is some evidence of caenogastropods having migrated into a non-marine habitat by the Carboniferous: a first invasion of the habitat so successfully colonized by the pulmonates much later on in the Jurassic and Cretaceous.

The gastropods were affected, as were most other organisms, by the great extinction period at the end of the Permian, but they continued to evolve throughout the Mesozoic. Many characteristic groups of Mesozoic age became very important for a while, such as the Nerineidae (Fig. 8.18f). These are a family of high-spired Mesozoic mesogastropods found in carbonate sediments; in them the inside of the spire is thickened by folded calcium carbonate. Spiral calcite rods running within the spire may be the original duct system within the digestive gland (Barker, 1990). It has been argued that many nerineids were infaunal and that since they lived in organic-rich carbonate mud, on which they fed, they would not have needed the nutrient storage units that other gastropods possessed in the spire. Hence the space was taken up instead by calcite which followed the contours of the digestive gland and gonad. The calcite taken in with the food was thus conveniently disposed of. Other groups, such as the gigantic neogastropod Family Strombidae and the cowrie-shell Family Cypraeidae, appeared for the first time in the later Mesozoic. But the real acme of gastropod evolution was reached in the Tertiary, continuing until the present, with the great success of the long-siphoned neogastropods (e.g. Fig. 8.18k) which dominate today's gastropod fauna.

Gastropods seem to have been a stable and constant component of the marine fauna since early times, and may dominate some marine communities. Freshwater Tertiary limestones may be composed entirely of gastropods, and a particular long-ranged association of gastropods and ostracodes, usually also with bivalves, is often an indicator of brackish–lagoonal facies.

Though gastropods are long-ranged and evolved slowly, new structural developments allowed important advances at different times in geological history. The modern vermetids, for instance, are gastropods with peculiarly uncoiled shells which may live permanently attached to branching corals and feed by straining off food particles from water passed between the edge of the shell and the operculum. In *Vermetus* (Fig. 8.18j) and *Vermicularia*, which are typical examples, two advantages seem to be given by such uncoiling (Gould, 1969b). One is rapid upgrowth towards the source of food particles raining down as detritus from the surface. The other is the considerable flexibility rendered possible to individuals when growing round obstacles.

There are some peculiar Devonian to Triassic counterparts of these, especially abundant in the Carboniferous; the so-called 'vermiform gastropods'. They tend to form densely packed bioherms, often accompanied by calcareous algae, especially in restricted shallow-lagoonal habitats. These, however, are not actually gastropods (Weedon, 1990). They have an initial coiled bulbous protoconch and a unique three-layered shell structure. It is possible on the basis of shell microstructure that they are related to the tentaculitids (q.v.), and Weedon regards the two as a sister group of all other molluscs, having branched off early in time.

In early Tertiary times there appeared the pteropods, which are small (*c.* 2 cm) planktonic opisthobranchs. These may have thin shells which can be coiled or straight; alternatively they may have no shells at all. They spend all their lives afloat and are important components of the plankton. It is often suggested that the pteropods had a paedomorphic origin from floating veliger larvae, which seems an eminently reasonable proposition. Their primary geological importance is as one of the main components of pteropod ooze in the deep oceans.

The earliest authenticated pteropods came from the Eocene; genera described as 'pteropods' from the Cambrian and other systems are now known to be shells of hyolithids, which are extinct animals of unknown affinities, unrelated to gastropods.

Microevolution in gastropods

Gastropods are of limited value in stratigraphy but have proved to be important in microevolutionary studies. The earliest of these, by Hilgendorf (1863; see Reif, 1983) concerned populations of the freshwater planorbid snail *Gyraulus kleini* which became isolated in the 3 km diameter meteor-impact crater lake of Steinheim in southern Germany. The populations here were adapted to alkaline conditions. There are clear evolutionary changes here, especially a general change from flattened to high-spired morphology in the middle part of the succession, and later a reversal to a low-spired form again towards the top. It is not easy to establish whether or not speciation was allopatric, though even in the small compass of the isolated lake this may have been possible within different local habitats.

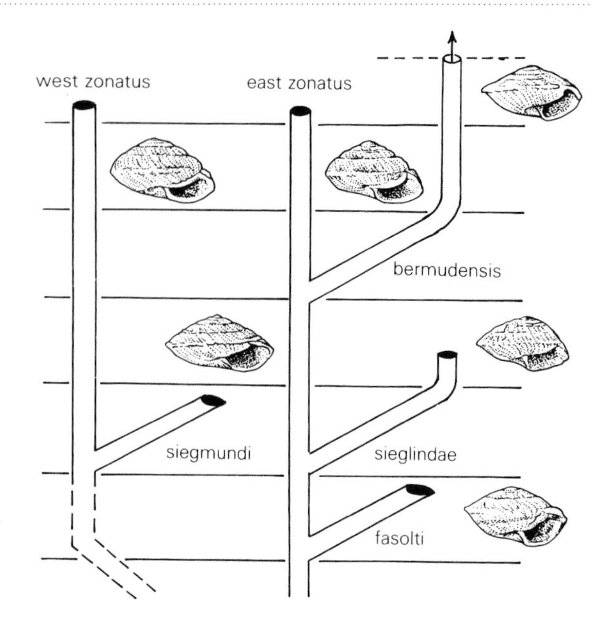

Figure 8.19 Evolution of *Poecilizonites bermudensis* by iterative development of paedomorphic subspecies in two separate areas. (Redrawn from Gould, 1969a.)

A now-classic study is that of Gould (1969a; Fig. 8.19), who elucidated the evolutionary history of the Pleistocene pulmonate gastropod genus *Poecilozonites* which lived in Bermuda during the last 300 000 years of the Pleistocene; one subspecies is living today.

Fossil shells of this gastropod occur abundantly in a sequence of alternating reddish soils and wind-blown sand, which Gould carefully documented. The shorelines oscillated through the later Pleistocene, so that islands on which the gastropods lived were alternately partially submerged and isolated, and exposed and hence linked, though often in different combinations. In his study of the changing populations throughout this time, Gould found that two populations of *P. zonatus bermudensis*, distinguished primarily by colour banding, lived in eastern and western regions and were presumably isolated throughout most or all of the time. These eastern and western *P. zonatus* populations formed the parent stocks from which other subspecies were derived. There are four derived subspecies: one originating from the western *P. zonatus* population, the other three from the eastern group. The three eastern forms appeared successively, each becoming extinct before the appearance of the next, and the last-produced subspecies is the one still living today. Gould was able to show that all four subspecies arose from isolated small groups on the periphery of the main population and thence spread to colonize other regions: a classic case of allopatric speciation in each instance. He also made it clear that the origin of each population was the result of a paedomorphic and hence instantaneous change, emphasizing its role as an important control of evolution.

Had the stratigraphical, geographical and morphological documentation been less complete, it would have been very easy and tempting to fit together the three eastern *P. zonatus* subspecies as part of a single evolving plexus rather than as three quite distinct iterative populations which started as peripheral isolates.

Therefore, in order to understand the principles whereby population differentiation takes place, it is at this level of detailed analysis that studies have to be undertaken.

Williamson (1981) studied a series of lacustrine molluscan faunas in the late Tertiary of the eastern Turkana Basin in Kenya. The 400 m of lake sediment are punctuated by mappable tuffs which proved to be useful in ordering the faunas in stratigraphic sequence. The study of bivalves and gastropods provided the first fine-scaled palaeontological resolution of speciation events so that Williamson was able to establish the nature of evolutionary phenomena in no less that 14 lineages. In his view all the speciation events confirmed the 'punctuated' equilibrium model; in the perspective of geological time, long periods of stasis alternated with short bursts of rapid speciation. Major changes seemed to occur in peripheral isolates during regressions of the lake. During these times the populations, being isolated and under stress, showed considerable phenotypic variation reflecting extreme developmental instability. From members of such populations new species arose and proliferated during subsequent lacustrine transgressions. Although the Turkana sequence has been accepted as an excellent example of precise palaeontological documentation, some doubt has been accorded (Jones, 1981) as to just how 'sudden' the speciation events actually were. The 'intermediate' unstable forms persisted in the Turkana lineages for up to 50 000 years, between very much longer periods of stability, and this means as many as 20 000 generations between stable species. To Jones we owe the comment that 'Depending upon the time scale to which the investigator is accustomed, one man's punctuated equilibrium may be another's evolutionary gradualism'. This is surely a valid comment in consideration not only of gastropod evolution specifically, but of evolutionary phenomena in general.

At this point, we should mention the tentaculitoids (Ord.–Dev.), an enigmatic group of conical, thin calcite shells, 1–3 cm long, with strikingly annulated exterior surfaces. They may or may not be molluscs. These have been extensively researched by Lardeux (1969) and Larsson (1979) and have proved of indubitable stratigraphic value. Their soft parts and life habits remain unknown.

Class Cephalopoda

The cephalopods, which are entirely marine, are the most highly evolved of all molluscs. Within this class are included the modern *Nautilus*, the argonauts, squids and octopuses, and the extinct ammonoids and belemnites. All modern cephalopods are distinguished by having a properly developed head with a good brain and elaborate sensory organs; the structural and functional parallels between the eyes of cephalopods and those of vertebrates, which are so well known, serve to illustrate the possibilities of evolutionary attainment inherent in the molluscan archetypal plan.

It may seem remarkable that the highly mobile cephalopods are constructed upon the same basic plan that is found also in the headless and mainly benthic bivalves. Yet as Fig. 8.1 shows, all the components of the archetypal mollusc are present in the cephalopods as they are in other molluscs. It is largely because the cephalopods were able, early in their evolutionary history, to develop an effective means of buoyancy using the chambered shell that they were able to free themselves from the sea floor, and colonize the nektic habitat, with its rich food resources of actively moving large-sized prey. Their evolutionary history and functional morphology shows how well they were able to exploit it, for they are nearly all active carnivores and, other than fish, the most accomplished swimmers in the sea.

There have been considerable difficulties in

classifying cephalopods, especially in erecting large natural categories.

The segregation of the cephalopods into two broad divisions, Tetrabranchiata and Dibranchiata, erected by Owen in 1832 and based on gill morphology, has now been abandoned. Whereas many authorities (e.g. Donovan, 1964; Teichert, 1967) prefer to divide cephalopods into several subclasses, others (e.g. Holland, 1987) recognize only two, the Ectocochlia (cephalopods with external shells) and Coleoidea (cephalopods with internal shells). Dzik's (1984) classification in which three subclasses are proposed (Nautiloidea, Ammonoidea and Coleoidea) is followed here.

SUBCLASS NAUTILOIDEA (U. Cam.–Rec.): Shell (phragmocone) external (ectocochliate), straight, curved or coiled, chambered with simple sutures; siphuncle central or subcentral, often of complex form. Four gills present.
SUBCLASS AMMONOIDEA (L. Dev.–U. Cret.): Shell external; coiled, often ribbed, chambered with complex sutures, siphuncle ventral or nearly dorsal, of simple form. Gill number unknown.
SUBCLASS COLEOIDEA (?Carb.–Rec.). Shell internal, straight or coiled, siphuncle may be lacking. Two gills present.

Subclass Nautiloidea

Nautilus (Figs 8.1, 8.20, 8.21a–e)

The only living cephalopod genus with a coiled external shell is *Nautilus*, of which there are six living species confined to the Indo-West Pacific faunal province between the Philippines and Samoa (Saunders and Landman, 1987; Ward, 1987). The *Nautilus* shell, some 20 cm in diameter, is planispiral and ornamented externally with a radial colour banding of irregular and bilaterally symmetrical orange–brown stripes. This shell is divided into internal gas-filled chambers (or **camerae**) by septa, concave towards the aperture; the animal resides in the last chamber (**body chamber**) and moves forwards each time a new septum is secreted. A single tube (the **siphuncle**) passes through the centre of each septum and connects the chambers. Each septum meets the inner wall of the external shell along a slightly curved line (**suture line**). Fossil *Nautilus* and its relatives are usually preserved with the shell dissolved and the chambers filled with spar or matrix. In such cases the position of each septum is marked by a suture line. Growth is very rapid in living *Nautilus*, new septa being emplaced about every 2 weeks on average. The living animal itself is separated by a fluid cushion from the last septum. The soft parts can be considered (cf. Fig. 8.21b) as two separate units: (1) the body, which is fully enclosed by the mantle and contains the viscera and the mantle cavity with its contents, and (2) the head-foot, a cartilage-supported structure with 38 tentacles surrounding the mouth with its horny, parrot-like jaws. The eyes are placed laterally on the head-foot. The dorsal part of the head-foot, above the tentacles, has the form of a **hood** which normally extends some way up the shell. This has a tough, warty outer skin, so that when the tentacles are withdrawn for protection the hood closes the aperture, presenting a largely impenetrable and uninviting surface to any predator. The full extent of the unretracted hood is marked by a black film on the shell, only exposed when the hood is closed down over the tentacles. Below the tentacles is the **hyponome** or funnel: a long tubular structure which can be turned in any direction.

Nautilus swims by jet propulsion; water enters the mantle cavity through a slit-like inhalant passage and, passing over the four gills within the mantle cavity so that respiratory exchange is effected, is squirted through the hyponome by the contraction of powerful muscles of the hyponome. (Unlike squids the mantle of *Nautilus* does not have muscular walls, and the musculature is confined to a specialized sac: the **branchial chamber**.) As in other cephalopods there is a coordinated system of rhythmic flow by means of which water is pumped through the mantle cavity with fairly gentle regular movements. *Nautilus* also has an escape reaction

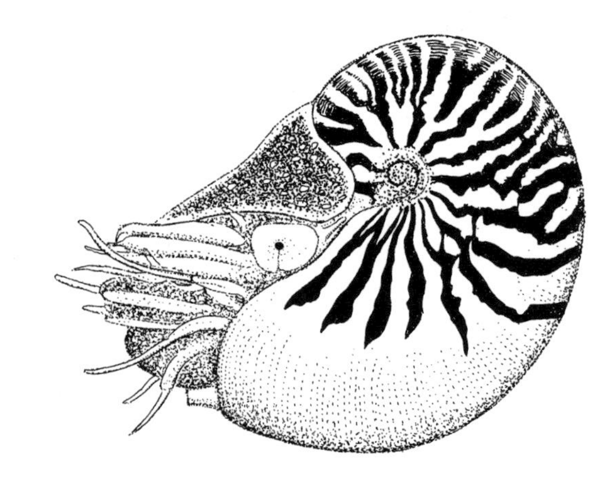

Figure 8.20 Living *Nautilus pompilius* (× 0.25), from a photograph.

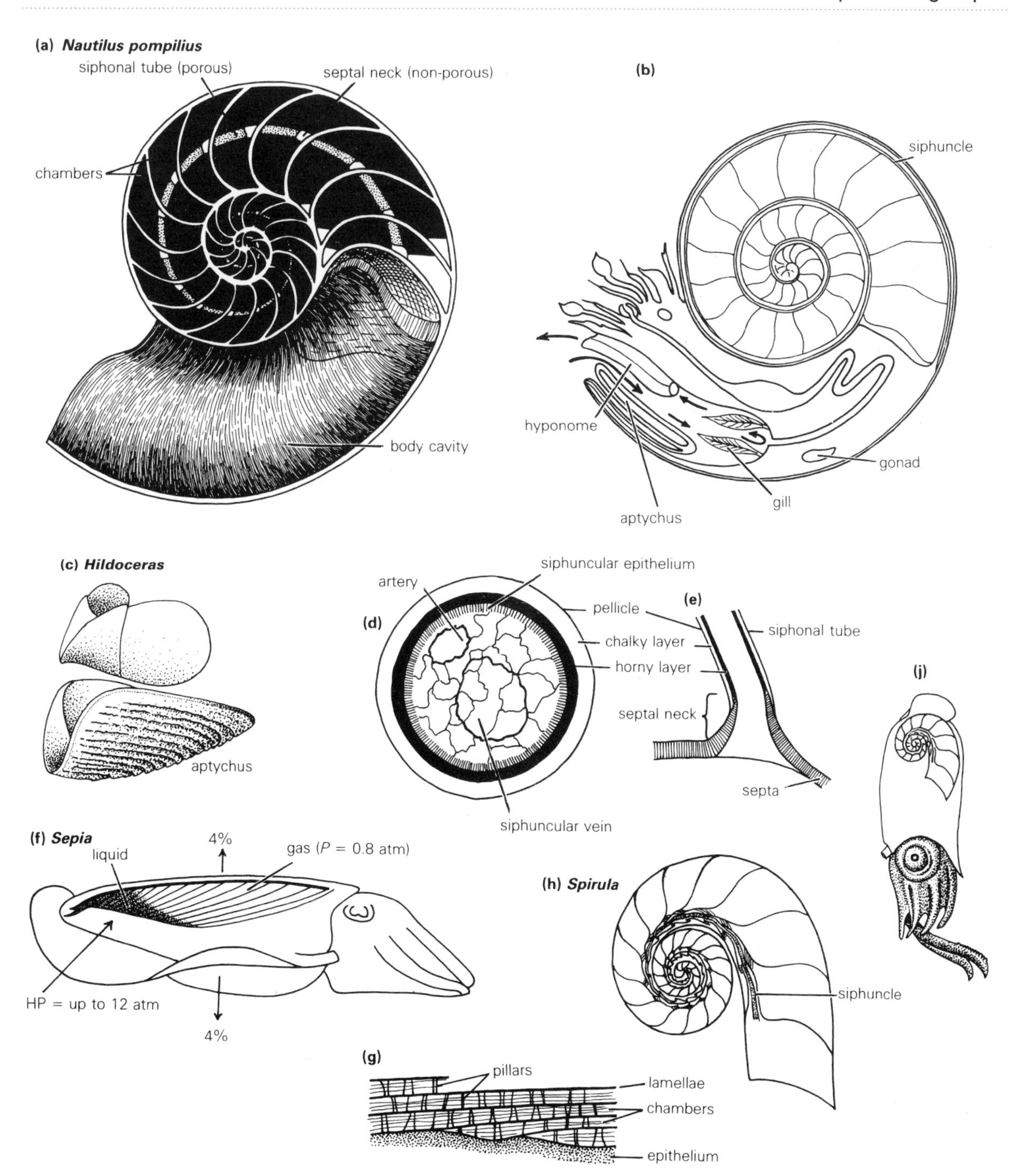

Figure 8.21 (a) *Nautilus pompilius* (Rec.) shell sectioned subcentrally (× 0.5); (b) ammonoid morphology reconstructed on the basis of the anatomy of *Nautilus* showing inferred disposition of soft parts and marginal siphuncle; (c) reconstructed jaws of *Hildoceras* (Jur.), the lower jaw having a pair of aptychi on the ventral surface; (d) anatomy of *Nautilus* siphuncle – transverse section of siphonal tube; (e) longitudinal section showing junction of septal neck with siphonal tube; (f) buoyancy mechanism in the cuttlefish *Sepia* (cuttlebone of older chambers is filled with liquid and younger chambers with gas at about 0.8 atm pressure; a lift of some 4% is imparted balancing the excess weight of the animal for hydrostatic pressures of the sea up to 12 atm; × 0.25); (g) structure of lower part of the cuttlebone with chambers supported by pillars and separated by lamellae; (h) *Spirula*, a small Recent cephalopod – median section of shell showing ventral siphuncle; (j) animal in normal life position, showing location of shell. [(a) Redrawn from Denton and Gilpin-Brown, 1966; (b) based on Trauth in *Treatise on Invertebrate Paleontology*, Part L; (c) redrawn from Lehmann, 1981; (d), (e) based on Denton and Gilpin-Brown, 1966; (f), (g) based on Denton, 1961.]

involving a violent contraction of the branchial chamber, so that the animal literally jumps out of the way of any predator; an equivalent mechanism in squids is often accompanied by the emission of ink into the water through the hyponome from an **ink sac** within the mantle cavity. This biological jet propulsion is the standard means of locomotion in cephalopods.

Nautilus, being a mobile feeder, has a highly organized brain coupled with good sense organs and a complex behaviour pattern. When hunting, *Nautilus* spreads its outer tentacles in a 'cone of search', but when it captures food it uses the inner tentacles to handle it. The food, which consists mainly of large crustaceans and fish, is cut up by the beak and stored in an expanded oesophagus prior to being passed to the stomach and digested.

Nautilus has a rather complex reproductive pattern. The sexes are separate; testes and ovaries are to be found at the posterior extremity of the body. During copulation the male transfers a ball of sperm (**spermatophore**) to the mantle cavity of the female using a specially adapted erectile group of tentacles (the **spadix**). The fertilized eggs are large and the juveniles of *Nautilus* hatch at some 25 mm in shell diameter with about seven chambers already present. Other cephalopods normally have only one modified tentacle (the **hectocotylus**), but their patterns of display, courtship and copulation are often quite complex. All species of *Nautilus* have alternate periods of rest and activity. Individuals are active at night, and during the day they sink to the sea floor where they rest with the hood pulled down and almost covering the eyes.

THE SHELL OF *NAUTILUS*

The shell of *Nautilus* is made of aragonite in a conchiolin matrix. It consists of two main layers: the outer porcellanous ostracum and the inner nacreous layer. The outer layer is formed first and grows from the mantle at the edge of the shell, beginning as aragonite seeds in a conchiolin matrix; these become larger and closely packed together as they grow into vertical prisms. The inner nacreous layer is later deposited by the mantle in a series of films; rather like the nacreous layer of bivalves it consists of a brick wall structure of hexagonal aragonite crystals with conchiolin layers between them. The septa have fundamentally the same structure. An empty *Nautilus* shell (Fig. 8.21a) shows short backward-pointing **septal necks** piercing each septum ventrally. In life these carry the siphuncle: a single strand of living tissue extending from the body to the protoconch and carrying a rich blood supply.

BUOYANCY OF THE SHELL IN *NAUTILUS* AND OTHER CEPHALOPODS

Since the whole success of the cephalopods has been so intimately bound up with their possession of buoyant shells, it is appropriate to consider how modern cephalopods of different kinds actually achieve such buoyancy.

The researches of Denton (1961, 1974), Denton and Gilpin-Brown (1966), Ward and Martin (1978) and others have shown clearly that the buoyancy of cephalopods works on a quite different principle from that of fishes. In fishes the swim bladder contains gas at an equal pressure to that of the surrounding sea; in deep-water fishes the internal pressure of the swim bladder can be enormous.

In contrast, the shells of cephalopods such as *Nautilus* and the squids *Sepia* and *Spirula* (which have internal shells; Fig. 8.21f,h) all contain gas at a pressure of less than 1 atm, and neutral buoyancy at different depths is achieved not by pressure equalization but by density control. In both *Nautilus* and *Sepia* the chambers contain gas at pressures ranging from about 0.3 atm in the more recently formed chambers to about 0.8–0.9 atm in the older chambers of the shell. The living *Nautilus* is slightly negatively buoyant, i.e. heavier than sea water. In immature specimens the chambers contain a fair amount of liquid, the cameral water, but in fully grown shells the volume of cameral water is very small. It has been removed by extraction via the siphuncle. Although it was formerly believed that the animal could rise or sink in the water by short-term secretion or extraction of cameral liquid, it has now been shown (Ward, 1979) that this operation takes place only very slowly. In the living *Nautilus* only the more recently formed chambers contain liquid, and this is present in diminishing amounts from the newest to the older chambers. Thus the chambers beyond the tenth from the body chamber are empty of liquid.

It is the siphuncle that extracts this liquid. The siphuncle (Fig. 8.21d,e) consists of two parts: the impermeable septal necks, and the permeable strands between them (**siphonal tube**). This tube has an inner core of living material with arteries and veins running the whole length of it, with a cylinder of epithelial cells. Outside this is a horny tube of conchiolin fibres, and surrounding this again is a concentric tube of irregularly arranged aragonite crystals. A very thin external pellicle of conchiolin completes the ensemble. Where the siphuncle joins with the septal necks the horny tube becomes continuous with the nacreous material of the neck. Both the aragonite and horny layers are very porous and permit the passage of liquid through them. There is also a thin layer of scattered aragonite crystals on the concave wall of the septa, which acts like blotting paper and renders the wall wettable and so able to retain the cameral liquid. In terms of its tensile strength, the siphuncle has been likened to 'a garden hose loaded internally by water pressure and internally supported against squirming instability by the septa' (Hewitt and Westermann, 1986, 1987).

The body of *Nautilus* is in contact with the septum during the time the latter is formed. Eventually the body moves away from the last septum and is separated from it by a cushion of liquid. When a new septum is formed it encloses a chamber which initially retains all this liquid. However, when the septum is fully formed and is strong enough to withstand the pressure of the sea, the siphuncle begins to pump out the liquid, leaving only a small residual amount which is reduced in salt and hypotonic to sea water. This pumping process works by osmosis, but the water actually in the siphuncle at any one time may be 'decoupled' from that in the chamber, so that the actual work done is not very great. Gas very slowly diffuses into the space left, which may explain why the gas pressures in the most recently formed chambers are low. *Nautilus* cannot adjust its buoyancy quickly. It may take weeks to make a full adjustment since the maximum rate of liquid removal is no more than 1.0 ml day^{-1}. The cameral liquid–phragmocone system may be used for long-term buoyancy changes but not as an aid to vertical movement. Cameral water is needed to support each septum as it is being formed, and it also acts as a reservoir of liquid ballast which, as the animal grows and increases in weight, is steadily extracted. In adults the little remaining cameral liquid is used only for maintaining a slight negative buoyancy. When *Nautilus* migrates upwards in the water column at night in search of food, it does so by hyponomic swimming alone.

When the *Nautilus* squirts water through its hyponome, the shell does not go into a spin, since the centre of gravity is located a few millimetres directly below the centre of buoyancy (i.e. the centre of gravity of the displaced water). This imparts a remarkable stability to the shell, and it would need a very strong couple to turn the animal on its side or through 90°.

Nautilus is unlike any other living cephalopod since it can live at remarkably low oxygen tensions (Wells *et al.*, 1992). It normally inhabits relatively deep water, in or close to the oxygen-minimum layer, growing slowly and using only a limited amount of energy. This enables it to avoid competition with fast-moving fish and squid in the upper waters of the sea; it has an optimal depth range of 150–300 m, swimming above the sea floor as a rather generalized, slow-moving scavenger or carnivore. *Nautilus* only pays brief nocturnal visits into the fish-dominated zone above, and is very rarely found above 75 m.

A final point concerns the depth to which the living *Nautilus* can sink before the shell implodes. The shell is very strong and rigid, and experiments have shown that an adult shell would not implode until a tensile strength of 131 MPa was applied, corresponding to a depth of 700–830 m. Newly hatched *Nautilus* shells, however, would implode at a depth of only 300 m (Hewitt and Westermann, 1987), the last septum being the weakest part. With age the animal is able to go down to greater depth, and although there are reports of living individuals venturing to depths approaching implosion limits, this is rare.

Evolutionary diversification

GENERAL CONSIDERATIONS

The earliest known cephalopods are the small curving shells of *Plectronoceras* from the Upper Cambrian of China (Fig. 8.5). They are endogastric, with marginal, empty siphuncles. Where these occur they are very rare, but at a slightly higher horizon there is evidence of a modest adaptive radiation. Towards the end of the Upper Cambrian, however, there was a phase of explosive radiation (which included the first ellesmeroceritids) and which Holland (1987) regards as 'a late phase of the initial metazoan radiation, in which perhaps a new combination of physiology and morphology took its niches in the animal world'. From such ancestors came the very many Ordovician genera so common in the fossil

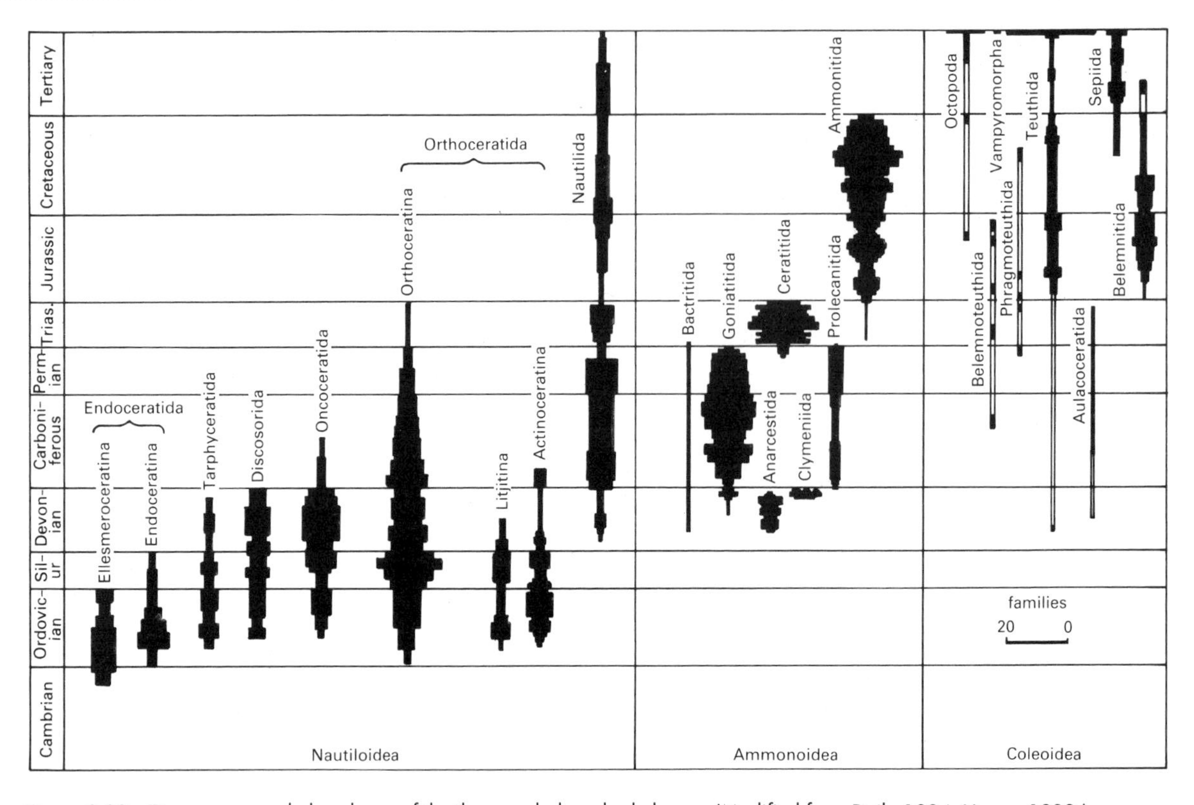

Figure 8.22 Time ranges and abundance of the three cephalopod subclasses. (Modified from Dzik, 1984; House, 1988.)

record, for the early Ordovician was marked by a further evolutionary explosion of cephalopod genera. These had mainly straight (**orthocone**) or curving (**cyrtocone**) shells and were generally of much larger size than their little Cambrian ancestors. From the one original order, the Endoceratida, there arose by the early Middle Ordovician many other orders, which persisted through all or part of the Palaeozoic (Fig. 8.22).

The latest of the orders to appear, in latest Silurian or early Devonian time, was the Nautilida, in which the living *Nautilus* is classified.

Most of these Palaeozoic cephalopods became extinct by the end of the Permian, only the coiled-shell Nautilida surviving through the Mesozoic to the present. The decline of Palaeozoic cephalopods seems to bear some relationship to the success of their progeny, namely the ammonoids, and especially to the ammonites of the Mesozoic, though as in all such cases of apparent 'takeover', many factors are involved and direct competition is not necessarily the only issue. One factor may be that the ammonites were exceptionally well adapted for resisting implosion at the depths they inhabited. Another is that they became ecological specialists invading a great many niches, unlike the comparatively unspecialized scavenger, *Nautilus*. Yet the ammonoids became extinct at the end of the Cretaceous and *Nautilus* did not. Ammonoids were very much the prey of fishes and marine reptiles and there are plenty of bite-marks on their shells to prove it (Martill, 1990), but then there is no reason why nautiloids should not have been so too. Holland's (1987) suggestion that hatching size may have been important seems eminently reasonable. The young *Nautilus* hatches as a 25 mm animal immediately able to take up the deep-water foraging life style of its parents. Ammonoids, on the other hand, had tiny hatchlings, much more vulnerable to the collapse of the planktonic ecosystem which they inhabited, at the end of the Cretaceous.

Classification

ORDER 1. ENDOCERATIDA: Primitive nautiloids with a wide, ventral, cylindrical siphuncle and a straight to endogastrically coiled shell. [Includes Suborders ELLESMEREOCERATINA (short septal necks with thick connecting rings), e.g. *Ellesmereoceras*, *Bathmoceras*, *Plectronoceras*, and ENDOCERATINA (very long septal necks which may intrude the previous septum: ventral siphuncle filled apically with 'cone-in-cone' structures – endocones), e.g. *Endoceras*, *Piloceras*.]

ORDER 2. TARPHYCERATIDA. Exogastrically coiled shell, elongated body chamber, cylindrical siphuncle, large protoconch; e.g. *Tarphyceras*, *Discoceras*, *Trocholites*.

ORDER 3. DISCOSORIDA. Endogastrically curved, compressed shell, with ventral siphuncle and inflated connecting rings; e.g. *Discosorus*, *Phragmoceras*.

ORDER 4. ONCOCERATIDA. Mostly exogastrically curved shells, ventral siphuncle with inflated connecting rings and short living chamber; e.g. *Oncoceras*, *Richardsonoceras*, *Pentameroceras*.

ORDER 5. ORTHOCERATIDA. Straight to weakly curved shell with subcentral siphuncle. In some extreme forms the shell may be short or exogastrically spirally coiled and with the siphuncle ventral to dorsal. [Includes SUBORDERS ORTHOCERATINA (long straight shells with transverse ornament ventral siphuncle and inflated, embryonic part of shell inflated, living chamber long, subcentral siphuncle, connecting ring, cylindrical to slightly inflated), e.g. *Orthoceras*, *Michelinoceras*, *Ascoceras*, *Cycloceras* (Dzik includes the BACTRITIDA here); LITUITINA (subcentral, cylindrical siphuncle with long septal necks, apical part of shell exogastrically coiled, funnel sinus narrow and deep), e.g. *Lituites*, *Sinoceras*; ACTINOCERATINA (siphuncle with considerably inflated connecting rings and with well developed cameral deposits in radial blocks, large protoconch, long straight shell, siphuncle towards ventral side), e.g. *Actinoceras*, *Rayonnoceras*.]

ORDER 6. NAUTILIDA. Exogastrically coiled, moderately elongated shell with narrow subcentral siphuncle. Larval development within egg capsule, no planktonic larval stage. [Includes SUBORDERS CENTROCERATINA (Early Palaeozoic), e.g. *Centroceras*, *Trochoceras*; TAINOCERATINA (Late Palaeozoic and Triassic), e.g. *Vestinautilus*; and NAUTILINA (post-Triassic) e.g. *Nautilus*, *Aturia*, though these cannot be satisfactorily diagnosed due to frequent homeomorphy].

Geological history (Fig. 8.21)

The post-Cambrian nautiloids are very diverse and abundant and often quite different in shell shape from the living *Nautilus*. As in their sole modern representative, the suture lines, as seen in internal moulds, are nearly always straight or slightly curving and are very rarely more complex. On this and other factors even the planispiral nautiloids can readily be distinguished from the ammonoids.

Only a few of the Palaeozoic cephalopods have planispiral shells. The majority are of either cyrtocone or orthocone shape. These may be very elongate (**longicone**) or short and rather swollen (**brevicone**). It has been generally assumed that all fossil cephalopods with straight or curved shells were **ectocochlear**, i.e. that the shell was wholly external to the body. The X-ray photographs taken by Stürmer (1970) of certain genera from the Lower Devonian Hunsrückschiefer show very clearly that in some cases there were living tissues, including lateral fins, external to the shell. Some of the orthoconic Palaeozoic cephalopods were apparently ancient squids, precursors of the later belemnites. However, at present, with only the Hunsrückschiefer 'window' to look through, it is not known how many of the Palaeozoic orthocones were really **endocochlear**, for in Stürmer's photographs both internal- and external-shelled kinds are evident.

Nautiloids are classified on various characters, but perhaps the most important are the form and arrangement of the various siphuncular structures.

Modern *Nautilus* (Order Nautilida) has only septal necks with which to support the siphuncle. In the ancient groups there are a variety of other structures, some of them of very complex form, and in particular there are often structures within the siphuncle itself (**endosiphuncular** structures).

In Order Endoceratida the shells are either orthocones or cyrtocones and the siphuncles are usually broad and normally marginal, and most genera have nested endosiphuncular conical sheaths or alternatively radial lamellae running the length of the siphuncle.

Order Orthoceratida, which persisted into the Triassic, are usually straight-shelled and have simple septal necks with only thin connecting rings, and sometimes endosiphuncular structures.

Representatives of Suborder Actinoceratina all have orthoconic shells. The septal necks are short but have inflated rings between them, within which is a delicate and complex system of radial canals.

Some orthoceratid genera have shells of decidedly odd form, such as the Ordovician *Lituites* (Fig. 8.23h) in which the early chambers are coiled and the rest of the shell is a rapidly expanding orthocone.

Even stranger are the shells of *Ascoceras* and *Glossoceras* (Fig. 8.23f). In these the shell consists of

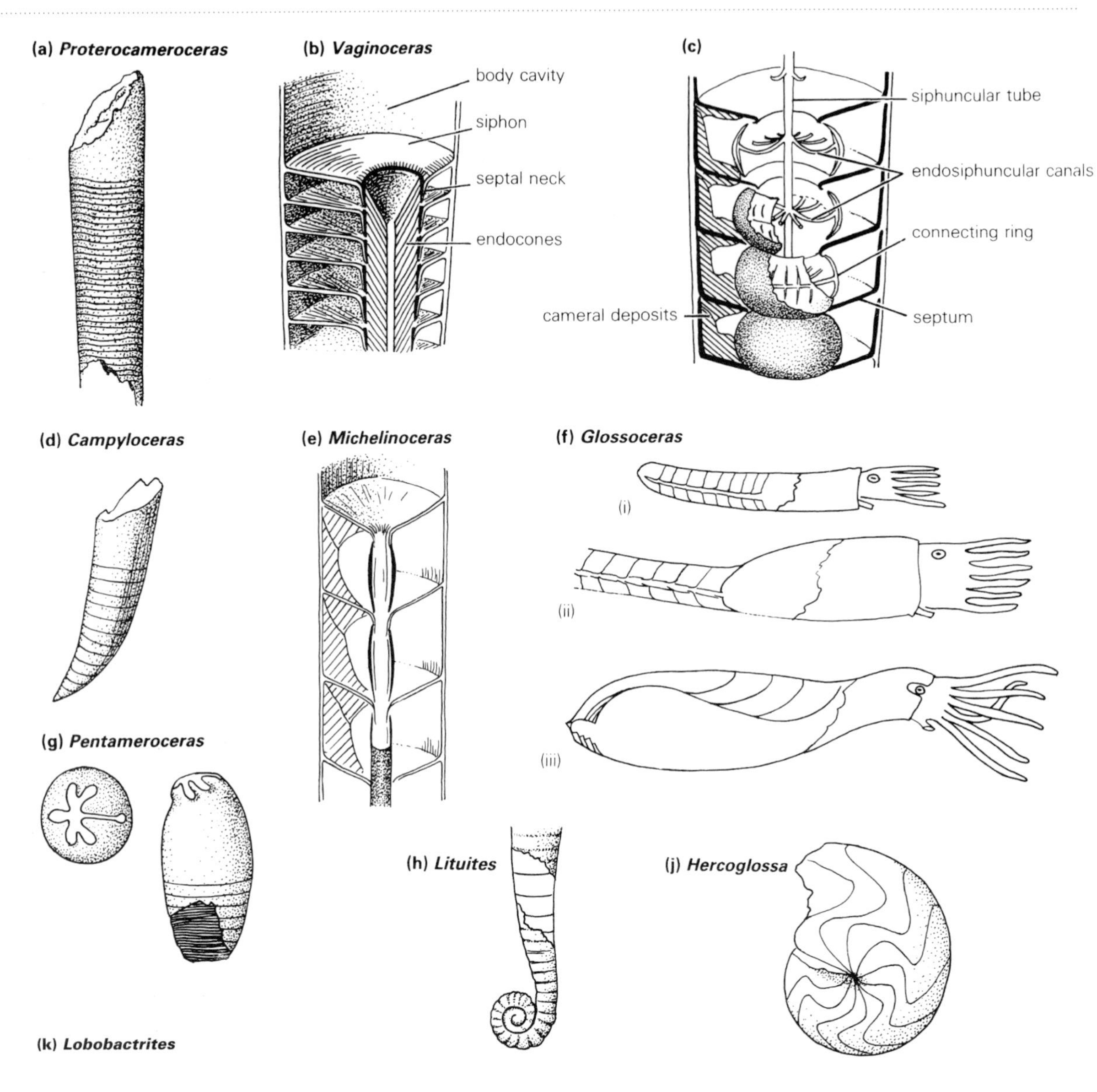
(a) Proterocameroceras
(b) Vaginoceras
body cavity
siphon
septal neck
endocones
(c)
siphuncular tube
endosiphuncular canals
connecting ring
septum
cameral deposits
(d) Campyloceras
(e) Michelinoceras
(f) Glossoceras
(i)
(ii)
(iii)
(g) Pentameroceras
(h) Lituites
(j) Hercoglossa

(k) Lobobactrites

two quite distinct parts. The first-formed part is a small, rather standard orthocone or slightly curving cyrtocone, with a thin straight siphuncle. This is sometimes found joined to a much thicker, swollen brevicone where the internal structure is highly modified. In this the siphuncle is short and confined to the apical end, with inflated connecting rings; from it the highly modified sigmoidal septa are formed to enclose large chambers located in the dorsal part of the shell only, above the body chamber. The two parts are very rarely found together, and it is generally accepted that the longiconic portion was deciduous and thrown away when the animal was mature. Probably a change in mode of life was involved; it is quite likely that the juvenile shell was a nektobenthic animal whereas the mature ascocerid, which had shed its early stage, was an active nektonic hunter.

The great diversity in shell form in the Palaeozoic cephalopods, other than in these bizarre examples, gave rise to a formerly held conception of a gradual increase in coiling from the straight to the fully coiled condition. This, however, is an incorrect view, for each shell form, even if of unusual appearance, is fully adapted to its own particular mode of life, and buoyancy adaptations in particular have been a primary evolutionary control.

Occasionally a circular structure composed of two large, symmetrical lateral plates and a triangular dorsal plate are found in association with orthocones. This **aptychopsid** has been found resting within the shell aperture in a few specimens from the Silurian of Sweden, and is surely a protective operculum though probably not a jaw structure.

Figure 8.23 Morphology of fossil 'nautiloids': (a) *Proterocameroceras* (Endoceratoidea; L. Ord.; ×0.4); (b) *Vaginoceras* (Endoceratoidea; M. Ord.), with endocones within the siphuncle, and long septal necks (×0.35); (c) actinoceratid morphology, shell and siphuncle partially dissected (×1 approx.); (d) *Campyloceras* (Orthoceratoidea; L. Carb.; ×1); (e) *Michelinoceras* (Orthoceratoidea; Ord.–Trias.), with long septal necks and cameral deposits (×1); (f) *Glossoceras* (Orthoceratoidea; Sil.; ×1.8), showing three growth stages – (i) juvenile cyrtocone, (ii) truncated cyrtocone with 'ascocerid' portion growing and (iii) mature ascocerid which has shed the cyrtocone part; (g) *Pentameroceras* (Nautiloidea; Sil.), an ovoid form with modified aperture (×0.75); (h) *Lituites* (Nautiloidea; Ord.) proximal part, partially coiled (×0.35); (j) *Hercoglossa* (Palaeocene), with 'goniatitic' sutures (×0.35); (k) *Lobobactrites* (Dev.) Wissenbacher Schiefer – an X-radiograph of an endocochlear nautiloid of total length 65 mm. [The photograph in (k) is reproduced by courtesy of W. Stürmer.]

It is generally accepted that most orthoconic cephalopods were free-swimming forms carrying their shells in a horizontal position. Some evidence of this comes from the rarely preserved colour markings that occur on the dorsal side of the shell only and appear to have been a kind of camouflage. But more importantly, the internal structures of the shell also indicate a horizontal position. Within the shell, as mentioned, there are endosiphuncular deposits. There are also regularly shaped masses of calcareous material known as **cameral deposits** (Fig. 8.23c–e). These cameral deposits were secreted progressively from the apical end as the shell grew; they are concentrated near the apex and developed less and less in the chambers nearer to the aperture. Together with the endosiphuncular deposits, they must have given extra weight to the apical end throughout the growth of the shell, thus allowing a continued equilibrium in a horizontal structure, with the centres of buoyancy and gravity staying close to one another (Flower, 1957). Only orthoconic and cyrtonic shells have cameral deposits, since in the coiled shells of the Nautilida and Ammonoidea the two centres lie in the same vertical plane so that the problem of stability is solved another way. Presumably, as in the modern *Nautilus*, all Palaeozoic cephalopods had some kind of siphuncular buoyancy control, and a slight change in density would have enabled the living animals to rise or sink. The swollen, egg-shaped shells of Order Oncoceratidae (Fig. 8.23g) are quite common in the Ordovician and Silurian. They may have floated with the long axis vertical but with the aperture downwards, searching the sea floor with their tentacles. In these the aperture is often constricted into a number of sinuses which take up their final shape only in the adult (Stridsberg, 1981, 1985). This may have been essentially a protective device.

By calculating the septal strength index for Palaeozoic nautiloids, Westermann (1985) was able to show that various types were adapted to different depths, below which they would implode. Thus in the later Silurian of Bohemia there are (1) brevicones which were probably epipelagic and generally restricted to depths of *c.* 35–150 m, (2) longicones with thin, closely spaced septa, living at less than 300 m depth and (3) longicones with thick, widely spaced septa which could have withstood depths of $<$1100 m. Likewise, the Carboniferous *Michelinoceras* has a strengthened shell which would not implode until a depth of 1125 m.

Several 'Orthoceras' limestones have been described (e.g. in Morocco, Scandinavia and China), in which there are vast concentrations of orthocone shells. These may be the results of mass mortality after mating, or salinity changes in the water. A very few orthoconic nautiloids survived into the Triassic, but otherwise only the coiled forms of the Nautilida carried on. The elaborate siphuncles of the other orders are never found here; there is only a thin, subcentrally situated siphuncular strand. Most Nautilida, except for some late Palaeozoic cyrtocones, have coiled shells which may be **involute** or **evolute**. Involute shells have the last whorl entirely covering all the former whorls; in evolute shells the former whorls are all visible. Sometimes there is external ribbing or even spines in the post-Palaeozoic genera, and occasional forms, such as the Triassic *Clymenonautilus* and the early Tertiary *Hercoglossa* (Fig. 8.23j) and *Aturia*, have sutures reminiscent of the goniatites of the Carboniferous.

Subclass Ammonoidea

Cephalopods of Subclass Ammonoidea (Dev.–Cret.) and especially the Mesozoic forms known in the vernacular as 'ammonites' are amongst the most abundant and well known of all fossils. Their beautiful planispiral shells, often strikingly ornamented with external ribbing, have been aesthetic treasures to innumerable collectors. Yet to stratigraphers their usefulness transcends their visual attraction, for by nature of their rapid evolution, abundance and widespread distribution they are the most valuable of all fossils for zoning the rocks in which they occur. They have proved of special effectiveness in the Triassic, Jurassic and Cretaceous systems, where their high turnover of species has made it possible to erect zones equivalent to time periods of less than a million years' duration.

The ammonoids may have been derived from the Bactritoidea: a straight-shelled cephalopod subclass which ranged through the Palaeozoic; the shells of these have a bulb-like protoconch and marginal siphuncle, are very similar to those of ammonoids in all respects other than coiling, and like them have no cameral deposits.

A complete morphological series from the Lower Devonian Hunsrückschiefer of the German Rhineland allows a structural sequence to be traced from straight forms to loosely coiled and then to tightly coiled ammonoids. All of these have, other than in their degree of coiling, virtually identical structure, and it is in cephalopods such as these that the ancestry of ammonoids should be sought (Erben, 1966; Chlupäc, 1976).

How ammonoid shells differ from those of other cephalopods (Fig. 8.24)

In the ammonoids, except in certain peculiar genera known as **heteromorphs**, the shell is planispirally coiled. Some Palaeozoic nautiloid shells also have this form, but these usually have a central perforation which is absent in all ammonoids except some of the early ones. The suture line normally follows a complex pattern; each suture marks the junction of a septum with the inner surface of the shell wall, and the array of suture lines is visible in nearly all ammonites since the aragonitic shell readily dissolves in diagenesis. If an individual ammonite septum or a mould of it is examined in face view (i.e. as if looking down the aperture), it appears flat or slightly curving in the centre but becomes increasingly frilled towards its point of attachment to the shell, where it becomes the suture. In the earlier ammonoids, of Devonian and Carboniferous time, the sutures are often simple zigzags, but from the Triassic to the Cretaceous the complexity of the ammonoid suture is considerable.

In most ammonoids the siphuncle is situated near the outer margin (**venter**). One order, the Upper Devonian Clymeniida, is typified by a dorsal siphuncle running along the inner margin of the shell (**dorsum**); in this group too the septal necks are retrochoanitic (backwardly pointing), as in nautiloids, and very long. This contrasts with the most advanced ammonoid condition, where the septal necks are short and **prochoanitic** (directed forwards).

Whereas the shells of coiled nautiloids are often unornamented or have only a feeble external sculpture, ammonoid shells are frequently ribbed, and the ribs may have knobs, tubercles or spines; in very compressed forms there may be a **keel**. In such genera as the Jurassic *Kosmoceras* such developments are carried to an extreme, and in addition lateral **lappets** are present on either side of a compressed aperture (Fig. 8.25). Such lappets are restricted to microconchs.

Such primary differences in morphology clearly distinguish ammonoids from those of nautiloids.

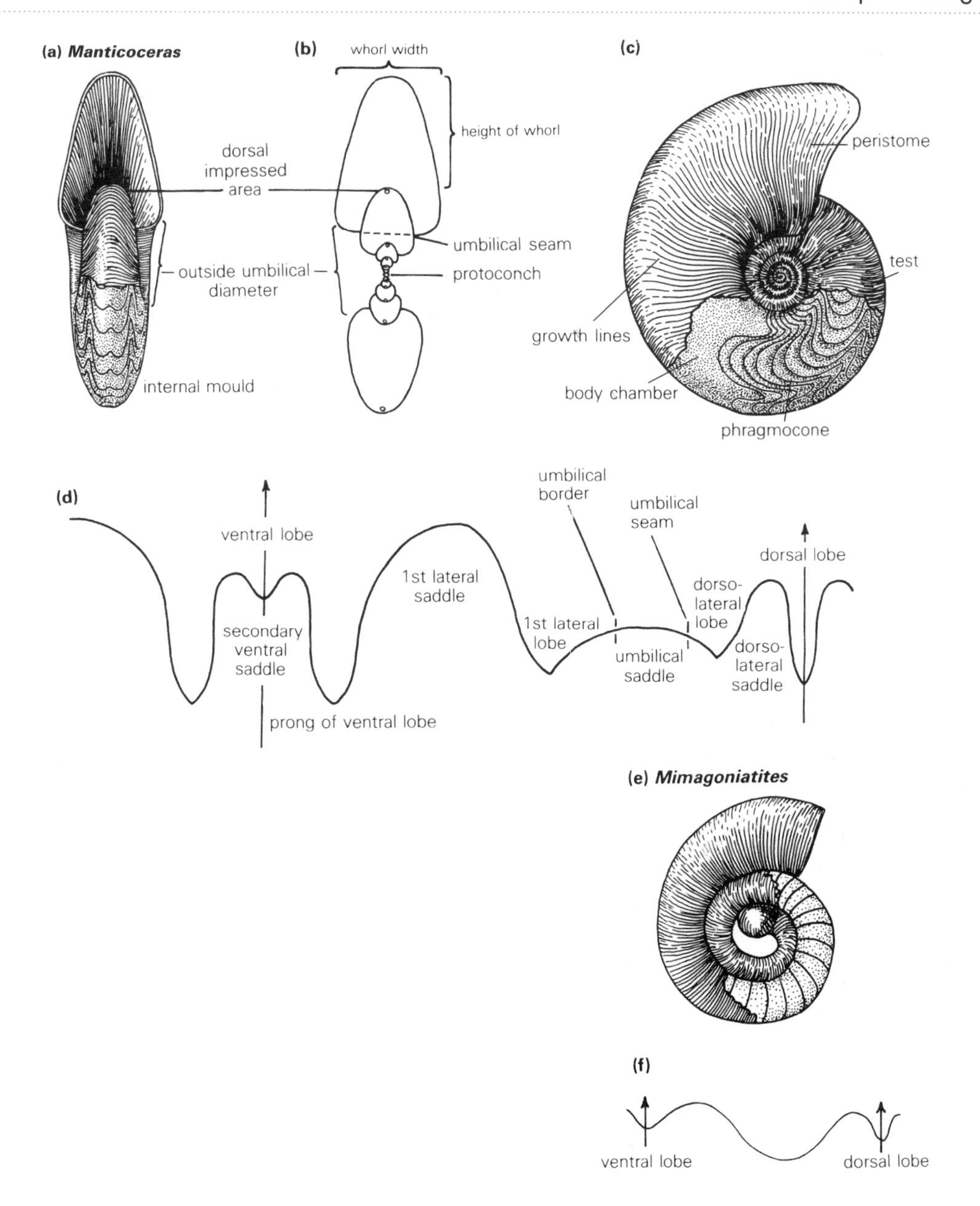

Figure 8.24 (a)–(d) *Manticoceras* (U. Dev.), a goniatite exemplifying ammonoid structure: (a) ventral view; (b) cross-section; (c) lateral view with shell partially removed, showing internal mould with sutures (all × 0.7); (d) suture line; (e), (f) *Mimagoniatites* (L. Dev.), a very early goniatite with perforated umbilicus and exposed phragmocone – the shell is partially removed to show chambers; (e) lateral view (× 5); (f) suture line. [(a)–(d) Redrawn from Miller and Furnish and (e), (f) after Schindewolf, all in *Treatise on Invertebrate Paleontology*, Part L.]

Normally the soft parts of ammonoids have been considered to be basically similar to those of nautiloids, and ammonoid reconstructions have been made very largely on this basis (e.g. Fig. 8.21b). Yet it is generally held (Engeser, 1996) that although this picture is generally correct, the biological affinities of ammonoids are closer to coleoids (dibranchiate cephalopods such as squids, cuttlefish and octopuses) than they are to *Nautilus*. Radulae have been found in a few ammonites (Nixon, 1996), and these

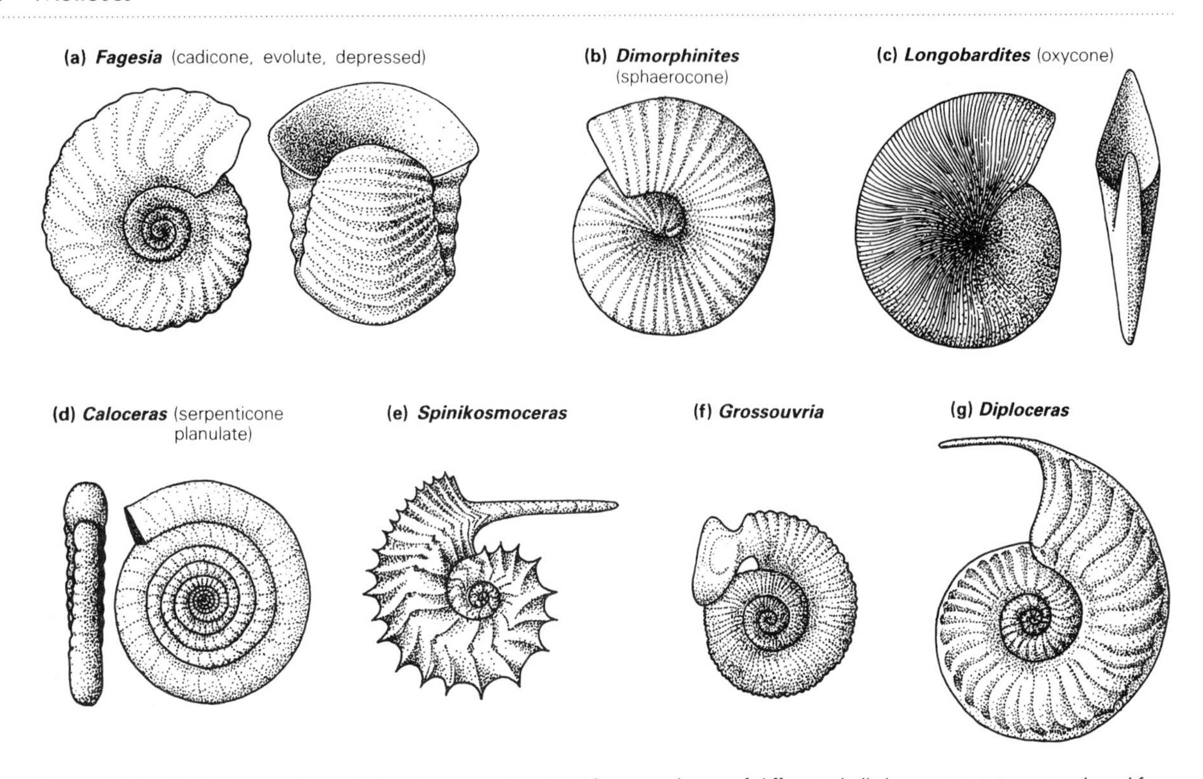

Figure 8.25 Shell shape and morphology in ammonites: (a)–(d) nomenclature of different shell shapes; (e)–(g) apertural modifications (lappets and spines) in Jurassic ammonites. (Original drawing by E. Bull.)

have seven teeth in each transverse row, as in the Recent dibranchiate cephalopods, whereas *Nautilus* has nine. In addition, ammonoid upper jaws, where known, correspond closely with those of recent octopuses; furthermore, there are known ammonoid ink sacs, which the Recent *Nautilus* does not have. Such evidence, though admittedly slender, suggests that ammonoids and coleoids stand rather close together phylogenetically, whereas *Nautilus* may be more distantly related.

Morphology and growth of the ammonoid shell

Ammonoid shells are characteristically tightly coiled in a planispiral fashion. They form rolled-up cones of which the earliest-formed part is a small bulbous ellipsoidal protoconch. This was probably initially inhabited by a planktonic larva. The protoconch is usually located in the centre of the shell, but in occasional early forms (e.g. Fig. 8.24e) the coiling is looser so that there is an umbilical perforation between the protoconch and the first whorl, like that of the early coiled nautiloids. The septate part of the shell (**phragmocone**) usually grows in a logarithmic spiral from the protoconch, but in some Palaeozoic genera the shell shape may be curiously quadrilateral or triangular. As the shell grows from the apertural end the siphuncle grows with it. The siphuncle starts as a small bulb (the **caecum**) within the protoconch; it is initially thick relative to the first-formed chambers and may wander about in its position. After the first one or two whorls, however, it settles down into its adult, normally ventral position, occupying relatively less space as the shell grows.

The septa were probably secreted in rapid growth episodes, like those of *Nautilus*, followed by resting periods, and the part of the shell where the animal actually resided (body chamber) grew progressively further from the centre during the spiral growth of the phragmocone.

There are several useful terms for describing the final form of the shell (Fig. 8.25). Those shells which are involute have the last whorl covering all the previous whorls; in evolute shells most or all of the previous whorls are exposed. The **umbilicus**, which is the concave surface on each side through which the spiral axis runs, may be almost flat or deeply excavated. The shell may be of normal rather flattened form (**planulate**), very compressed (**oxycone**), very fat and inflated (**cadicone**) or almost globular (**sphaerocone**).

Some ammonoids have pronounced ventral keels (Fig. 8.25a), which may be blunt or sharp. The keel may have a single, double or triple parallel ridge system, running along the venter. In cross-section the whorls may take many forms, which are often characteristic of particular ammonoid families and thus very useful in taxonomy.

The outer surface of an ammonoid shell is usually marked by faint growth lines, but in addition there are usually ribs, radially arranged and projecting above the surface. There are so many patterns that only a few can be illustrated here. Some ribs go over the venter; others are interrupted; some are straight whilst others curve or branch; they may be united at **nodes**; and shorter ribs may be intercalated between longer ones.

Tubercles may erupt on points on the ridge or be independent of them; they may be prolonged into spines or united by flat nodes. The external pattern of ribs and associated structures is of great value in classification and identification, but since homeomorphy is always a possibility, their use must be coupled with that of other characters, particularly the nature of the suture. [Most illustrations of fossil cephalopods, including those here, follow traditional practice and show the shell upside down in respect to its living position. It would actually be much more sensible to illustrate them as they were in life (Stridsberg, 1990); this probably will be done in future.]

The curious compression of some ammonoid apertures, and their frequent ornamentation by lappets or spines, have already been mentioned. Some of the more peculiar forms are illustrated here (Fig. 8.25e), but these are unusual and the majority of ammonoid apertures, and especially those of Palaeozoic genera, are simple like those of nautiloids.

In ammonoids of various kinds the body chamber may be long or short, as discussed in the section on buoyancy.

Ammonoid suture (Fig. 8.26)

Most ammonoids are preserved as internal moulds, and in these the junctions between septa and shell walls show up clearly as suture lines, which in ammonoids are always more complex than in nautiloids.

In the early ammonoids (Devonian and Carboniferous) the sutures were normally of a fairly simple form and lack accessory crenulations, but by Permian times some genera were showing more complex sutures of a kind that reached their full flowering in the Mesozoic, when genera with extremely complex sutures were the norm.

To clarify ammonoid suture morphology, the sutures are usually represented graphically. Neglecting surface convexity of the ammonoid shell, the sutures are drawn from the venter to the **umbilical seam** (the external suture) and thence to the **dorsum** of that septum. Involute ammonoids

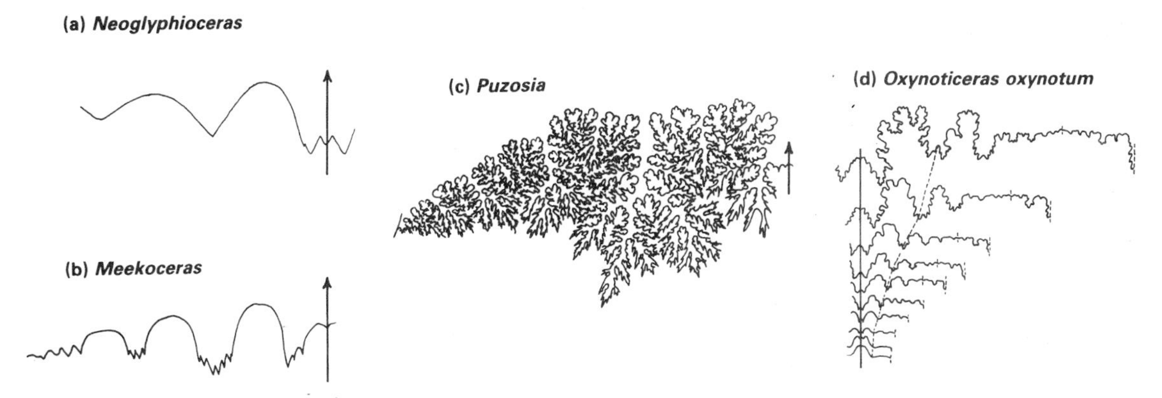

Figure 8.26 Suture morphology in ammonoids: (a) goniatitic suture (*Neoglyphioceras*; L. Carb.); (b) ceratitic suture (*Meekoceras*; Trias.); (c) ammonitic suture (*Puzosia*; L. Cret.); (d) sutural ontogeny in *Oxynoticeras oxynotum* (Jur.). (Modified from various authors in *Treatise on Invertebrate Paleontology*, Part L.)

have a distinct internal suture, but in the evolute forms the internal suture is small and insignificant and is normally not drawn. Since the full suture is symmetrical about the dorso-ventral axis only half a septum is normally drawn. Because no account is taken of the surface convexity, the suture diagram is only a projection, which can never be entirely accurate, but this is unimportant for purposes of comparison of different suture lines. Such diagrams are not only useful for comparing the sutures of different species and genera, but also essential in tracing the changes in sutural morphology throughout the growth of an individual, which has various uses.

In drawing a suture the venter is shown on the left-hand side with an upwardly directed arrow pointing towards the aperture; the umbilical seam is shown as a curving line, and another vertical arrow marks the dorsum on the right-hand side. Inflections on the suture line pointing upwards (the apertural direction) are the **saddles** (easily remembered since a horse's saddle faces forwards), whereas the backwardly pointing inflections (facing downwards on the diagram) are the **lobes**.

The suture is most important in taxonomy; particular kinds of sutures characterize distinct ammonoid families and are very useful in classification and identification. There are some broad and general terms defining ammonoid groups, based on sutural morphology, which are often used to characterize different grades of organization. Thus the **goniatites** are Palaeozoic ammonoids with sharply angular and generally zigzag sutures, without any accessory crenulations. Not all ammonoids with such sutures are goniatites; there are some unrelated Mesozoic genera with similar sutures. In the Triassic **ceratites** the lobes are frilled though the saddles are entire. Some curious Cretaceous genera (**pseudoceratites**) have similar sutures. The Mesozoic **ammonites** all have finely subdivided and complex lobes and saddles, though there are also some Permian ammonoids with sutures of this kind. Though there is thus a general stratigraphical increase in sutural complexity, the progression from goniatite through ceratite to ammonite is not, however, a direct phylogenetic line.

In many Palaeozoic and Mesozoic ammonoids the ontogeny of the suture has been worked out (Fig. 8.26d), normally by breaking off the chambers one by one as far as the protoconch and drawing the sutures as the inner whorls are exposed. The early sutures in all ammonoids are far less complicated than the later ones, and the whole history of development of a mature suture can be traced using serial diagrams of successive sutures. Such ontogenetic series have several uses. They have been used, for instance, to distinguish homeomorphs in which the mature suture in two distantly related ammonoids looks similar but has been arrived at ontogenetically in quite different ways.

Perhaps the greatest value of sutural ontogeny is in unravelling phylogenies, especially in Palaeozoic forms, for in these ammonoids the earlier stages in development of an 'advanced' suture closely resemble the mature septa of a more primitive ammonoid type. Thus ancestor–descendant relationships can be postulated, and phylogenies have been drawn up on this basis which have later been shown to hold firm in the light of more evidence.

During ontogeny, in Palaeozoic and Mesozoic genera, primary lobes and saddles appear early, and normally the first and second lateral saddles are very distinct. Though these persist into adult suture, **adventitious** lobes or saddles may appear in between the primaries and eventually grow as large as the primaries themselves. The only way to distinguish which is which is then to trace them back to the earliest ontogenetic stages.

In the ammonites, where the suture reaches its maximum complexity, the lobes and saddles are all crenulate, and the saddles all have accessory lobes. The terminology of all these, as given in Fig. 8.24d, becomes rather complex. Some rather peculiar sutures are present in a few Mesozoic genera. In certain examples each mature suture is apparently truncated against the one in front, though in fact the 'missing' parts continue below the preceding septum. Septa may be asymmetrical, and even taxonomically unstable within the one genus.

Siphuncle, suture and buoyancy

Ammonoid shells are always thinner than the shells of *Nautilus*, their siphuncles are ventral (except in Order Clymeniida), the septal necks are normally prochoanitic in ammonoids and retrochoanitic in nautiloids, and the septa in ammonoids are usually convex, not concave, towards the aperture.

It is generally held that ammonoids achieved buoyancy in much the same way as does *Nautilus* and that the fluted septa amongst other functions may have increased the strength of the shell so that it was able to resist implosion at depth. But to be more specific than this, and most particularly to be able to perform mathematical analysis, various other factors must be known, including the nature and composition of the siphuncle, to see if the shell could have functioned like that of *Nautilus*. There is good evidence from rare unaltered specimens that calcium phosphate is the primary constituent of ammonoid siphuncles, though juveniles have more calcium carbonate and probably the phosphatization was preceded by a calcitic stage. The calcitic outer tube of *Nautilus* is missing. Now in certain Mesozoic ammonites the strength of the siphuncular tube (against explosion) has been calculated. Needless to say, the tube strength is an important limitation on depth. The relative strength index $(h/r) \times 100$, where h is the wall thickness and r *is* the tube radius, has been calculated for several Mesozoic ammonoids, and it is most interesting to see that two orders, Phylloceratida and Lytoceratida, have a strength index like that of *Nautilus* ($h/r \times 100 = 10–19$), whereas their derivatives the Ammonitida have a significantly lower value (3–6.5). This information accords with the recognition that the Phylloceratida and Lytoceratida were deeper-water groups, whose derivatives repeatedly invaded shallow water by a classic process of iterative evolution. The calculations show that the Phylloceratida and Lytoceratida could withstand a water column of some 450 m, about the same as that of *Nautilus*, whereas Order Ammonitida could probably not withstand depths of more than 100 m. Other data which have been used in attempts to infer bathymetry in ammonoids, such as the thickness of the shell wall and of the septa, have not proved so useful since the relationships involved are very complex and elude simple analysis.

There has been much discussion of the function of the complex ammonite septum. Many authorities believe that the primary function of the convoluted septal periphery was to give the shell extra strength, thus preventing implosion at depth. Ammonite shells are thinner than those of nautiloids and the (adorally convex) septum would help to sustain strong pressures because of its fluting. Moreover, because of this internal buttressing shell shapes other than the standard ovate shell, such as that of *Nautilus*, are possible. The view that ammonite sutures functioned mainly as pressure resistors seems reasonable, and is the model preferred by Raup and Stanley (1971) and by Kennedy and Cobban (1976). One problem, however, is that beyond a certain level of complexity, increased septal fluting decreases the angle between septum and shell. Thus the shell of an ammonite with a very complex suture may be more prone to damage through bending stresses than one with a less complex suture.

Several alternative functions have been proposed, which involve physiological rather than structural necessity. How, for example, could the ammonites have maintained a secure attachment of the body muscles to that shell during episodic forward growth? Were these muscles attached to the septal flutings, and if so might the latter have increased the surface area and thus provided a better attachment? Or was the principal function of the septum to spread the effects of hydrostatic pressure evenly across the shell? Were the septal flutings consequent upon the form of the organic template from which it was calcified (initially forming a viscous finger pattern), itself just a by-product of geometric necessity? Or was the function of the fluted septum connected with the removal of cameral liquid? Possibly some or all of these are valid if partial explanations. A recent review of this whole issue, based on physical and engineering principles (Hewitt and Westermann, 1986, 1987) stresses the adaptation of ammonoid septa to resist hydrostatic loading, which could be applied both via the body chamber of the thin shell wall. The relative influence of each of these would depend upon (1) habitat depth, affecting hydrostatic pressure, (2) whorl profile, affecting stresses in the shell wall, and (3) the actual size of the whorl, which would affect bending stresses in the shell wall and the fluted septum. Thus a circumferential support function is primary for the ammonoid suture, and it must have operated in the earliest growth stages too, as would be necessary in swimming animals adapted for offshore breeding.

In cephalopods generally, and undoubtedly in ammonoids, the main adaptations were concerned with establishing an effective, neutrally buoyant poise of the living animal in the water. Such structures as nodes, ribs, tubercles and spines, as well as septal approximation in mature individuals, may have been concerned with fine-scale trimming of the poise. Perhaps paradoxically, external ornamentation can actually reduce drag, by introducing turbulence into the boundary layers, as with dimples on a golf ball. The hydrostatic adjustments of the ammonoid shell, in general terms, probably worked as in *Nautilus*, as did hysteresis mechanisms which allowed changes in liquid within the chambers to take place comparatively slowly. This permitted short-term excursions into different levels of the sea and automatic return to the starting level (Ward and Martin, 1978; House, 1981).

Sexual dimorphism (Fig. 8.27)

The question of whether or not ammonoids were sexually dimorphic has been debated at great length for a very long time. As early as 1869 Waagen, one of the early German palaeontologists, pointed out that in ammonites of Family Oppeliidae (Order Ammonitina) there were apparently two parallel phylogenetic lineages developing throughout the Middle and Upper Jurassic. Furthermore, he found that in some ammonites pairs of 'species' could be distinguished, differing only in the character of their outer whorls; the nuclei could not be told apart. Waagen, however, did not believe that sexual dimorphism could be invoked as an explanation, and it was nearly 100 years later that Callomon (1963) and Makowski (1963) convincingly demonstrated the criteria that could be used in establishing dimorphism unequivocally. Since then many Mesozoic dimorphic pairs have been clearly identified, and there are sporadic records in Palaeozoic ammonoids too (Davis *et al.*, 1996).

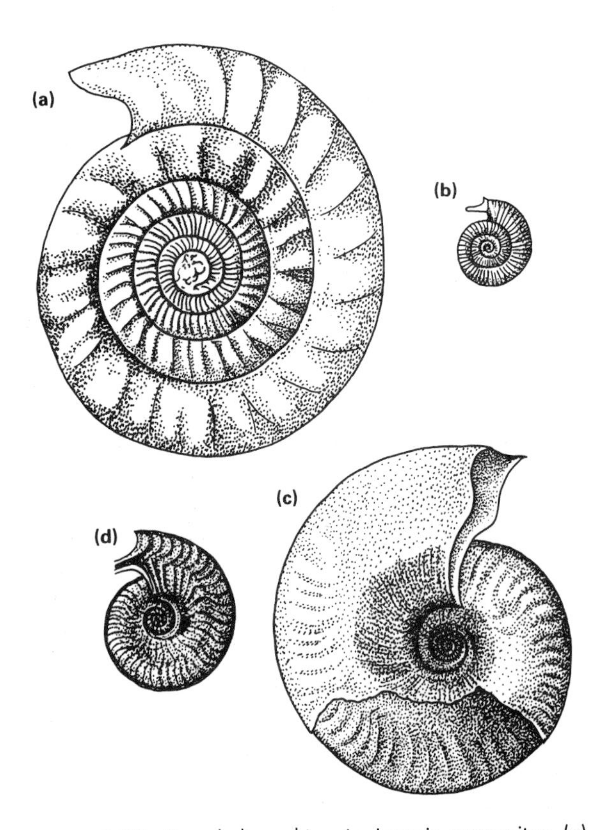

Figure 8.27 Sexual dimorphism in Jurassic ammonites: (a) *Perisphinctes (Arisphinctes) ingens* and (b) *Perisphinctes (Dichotomosphinctes) rotoides*, both from the same horizon in the Corallian (×0.15; these are probably dimorphs); (c) *Graphoceras cavatum* and (d) *Ludwigina cornu*, a Bajocian dimorphic pair (× 0.65). (Redrawn from Callomon, 1963.)

In trying to work out dimorphic pairs it is essential that all the ammonite shells studied are mature specimens, i.e. adult shells in which growth had ceased. Such maturity in ammonite shells is marked by (1) crowding (approximating) of the last few septa, due to diminishing growth rates; (2) a change in the external sculpture near the aperture – the ribbing may be different, or the aperture may be constricted or marked by lappets or horns; and (3) some slight uncoiling of the body chamber from the rest of the shell. Not all these features are found in any one ammonite, and indeed some of them are confined to one sex only; nevertheless, the presence of only one or two such characters will enable a mature shell to be distinguished. Most ammonoid shells found are in fact mature, and in specimens collected from the same bed mature shells of the same kind are normally very much alike in size.

When mature shells from the same bed and locality are separated into 'species', they often come out into two quite distinct groups, without intermediates. These are distinguished primarily by size, one group being two to four times larger than the other so that the two kinds are generally referred to as **microconchs** and **macroconchs**. In microconchs and macroconchs from the same horizon, the inner whorls are normally indistinguishable and only the outer whorls differ in number, size and morphology. If there are any peculiar features of the aperture, these are normally found only in the microconchs, whereas the macroconchs have simple apertures, and the more extreme cases of detachment of the body chamber from the preceding whorl are likewise characteristic of the microconchs.

It is theoretically possible that the apparent sexual pairs are merely closely related taxa living in the same place at the same time, but since in such lineages new characters (such as modifications of external sculpturing) appear in both groups at the same time, then sexual dimorphism appears to be the only likely possibility. However, all these criteria have to be used with caution in interpreting ammonoid populations, for they do not always, by any means, occur in pairs; the expected 50:50 ratio is only sometimes found, and in some cases the variation within a population collected from a single locality may be enormous. Hence we have to agree with Lehmann (1971b) who adjures us 'to be extremely careful when it comes to maintaining sexual relationships between hitherto blameless ammonites'.

It is hard to know which morph is male and which is female. Some authors have considered the question so academic as to be unanswerable; others have suggested positively that the microconch is the male, and the macroconch the female. *Nautilus* shows little dimorphism (the male is somewhat larger and broader) and is thus no great help, but some Mesozoic nautiloids have been described as clearly dimorphic.

Since all known dimorphic pairs were originally described as separate species, what should be done with them taxonomically when they are recognized as dimorphic? Should dimorphic 'species' be grouped together under the one name as true biological species, or should the established names continue in use as 'morphospecies'?

Undoubtedly the former practice is more technically correct, but the latter is perfectly admissible if it is clear that we are talking about 'morphospecies' and not biospecies, and it does have the advantage of greater flexibility. For the moment, palaeontologists are divided on this question, though the definition of true biospecies appears to be gaining ground.

Ammonite jaws

Commonly found in association with Mesozoic ammonites are paired calcitic plates known as **aptychi**, which superficially resemble bivalves (Fig. 8.21c). Aptychi are usually scattered on bedding planes where ammonites are abundant. Where the aragonitic shells of the ammonites are crushed, as in so many Jurassic shales, the calcite aptychi retain their original, slightly curving relief. Sometimes no ammonites are found, only aptychi, where the aragonite shells have been dissolved completely. Rarely aptychi have been found inside body chambers of ammonites, meeting like double doors and partially fitting the shape of the aperture.

There is no question but that these were part of the ammonite, and Trauth, working in the 1930s, was able to establish that different types of aptychi belonged to different ammonite groups. It was his view, at that time, that aptychi were some kind of opercula.

There are also **anaptychi**, which consist of a single plate of chitin or other organic material, normally flattened and in a butterfly shape. Anaptychi are confined to sediments of Upper Devonian to Lower Jurassic age, at which time the earliest aptychi appeared. The last aptychi are found with the last ammonites at the end of the Cretaceous. Just what the aptychi were and how they functioned remained a matter for some dispute, though most palaeontologists believed that they were opercula and that anaptychi were some kind of jaw structure. Lehmann (1967, 1971a,b, 1979) finally settled the nature of these structures by studying exceptionally well-preserved material and in particular by making serial sections. He showed conclusively that the anapytchus is a lower jaw, originally V-shaped but usually preserved flattened and spread out. In addition there is a beak-like upper jaw of the same horny material, very similar to the jaws of modern coleoids. It is usually preserved crushed on top of the anaptychus and had not previously been noticed as a separate structure. Aptychi, according to Lehmann, are calcareous deposits which have grown on the external (ventral) surface of the lower jaw, as a pair of thickened plates, and these did not develop, evidently, until the Lower Jurassic.

Lehmann's interpretations have been confirmed and extended with the discovery of complete jaws in other Jurassic and Cretaceous ammonites. In all of these the jaw structure is relatively like that of octopuses, but there is one important difference, namely that the lower jaw is very large; in relation to body size it is about four times larger in the ammonites than in any modern coleoid. To what extent this was related to the ammonites' diet (probably slow-moving live animals and carrion) is quite unknown. It now seems likely that aptychi functioned both as lower jaws and as opercula (Lehmann and Kulicki, 1990). This would certainly have been possible if the ammonite's lower jaw was relatively mobile when feeding, with most of the biting work done by the upper jaw.

Not only do ammonites possess jaws but also a radula, like that of other molluscs with serried rows of teeth. It is set in between the jaws. Such radulae were first discovered in Carboniferous goniatites from Brazil in 1967, but have since been found in several genera of Jurassic ammonites.

Phylogeny and evolution in the Ammonoidea (Fig. 8.28)

The following orders have been distinguished in the Subclass Ammonoidea.

ORDER ANARCESTIDA (L. Dev.–Perm.): Basic ammonoid stock, ventral retrochoantic siphuncle, bulb-like protonch, often with perforated umbilicus, simple sutures.

SUBORDER BACTRITINA (U. Sil.–Perm.): Ortho- or cyrtocone, probably ancestral to all other ammonoids; e.g. *Bactrites*.

SUBORDER AGONIATITINA (L.–M. Dev.) Loosely to tightly coiled, transitional between bactritids and early goniatites; e.g. *Anetoceras*.

SUBORDER ANARCESTINA (L.–U. Dev.): Serpenticone to globular–involute, simple wavy suture; e.g. *Maenioceras*, *Anarcestes*.

SUBORDER GEPHUROCERATINA (M.–U. Dev.): Serpenticone to involute, more complex suture; e.g. *Manticoceras*.

ORDER CLYMENIIDA (U. Dev.): The only ammonoid group with a dorsal, marginal siphuncle; they may have "goniatitic" sutures.

SUBORDER GONIOCLYMENIINA (U. Dev.): The shell is often triangular in lateral view; e.g. *Wocklumeria*, *Gonioclymenia*.

SUBORDER PLATYCLYMENIINA (U. Dev.): Sutures usually simple, involute or evolute; e.g. *Clymenia*, *Platyclymenia*.

ORDER GONIATITIDA (M. Dev.–U. Perm.). Ammonoids with ventral marginal siphuncle, usually prochoanitic, goniatitic suture of eight lobes.

SUBORDER TORNOCERATINA (U. Dev.–Carb.; e.g. *Tornoceras*, *Cheiloceras*, *Gattendorfia*) and SUBORDER GONIATITINA (mainly Carb.; e.g. *Goniatites*, *Eumorphoceras*, *Gastrioceras*, *Popanoceras*) are distinguished primarily on the suture.

ORDER PROLECANITIDA (L. Carb.–U. Perm.): Shell smooth with large umbilicus, ventral, marginal, retrochoanitic siphuncle, sutures goniatitic to ceratitic; e.g. *Prolecanites*, *Medlicottia*.

ORDER CERATITIDA (U. Perm.–U. Trias.): Shell often highly ornamented, ventral prochoanitic siphuncle, ceratitic suture, but some sutures attain great complexity. This order underwent a dramatic radiation in the Triassic; e.g. *Ceratites*, *Xenodiscus*, *Norites*, *Pinacoceras*.

ORDER PHYLLOCERATIDA (Trias.–Cret.) Smooth or weakly

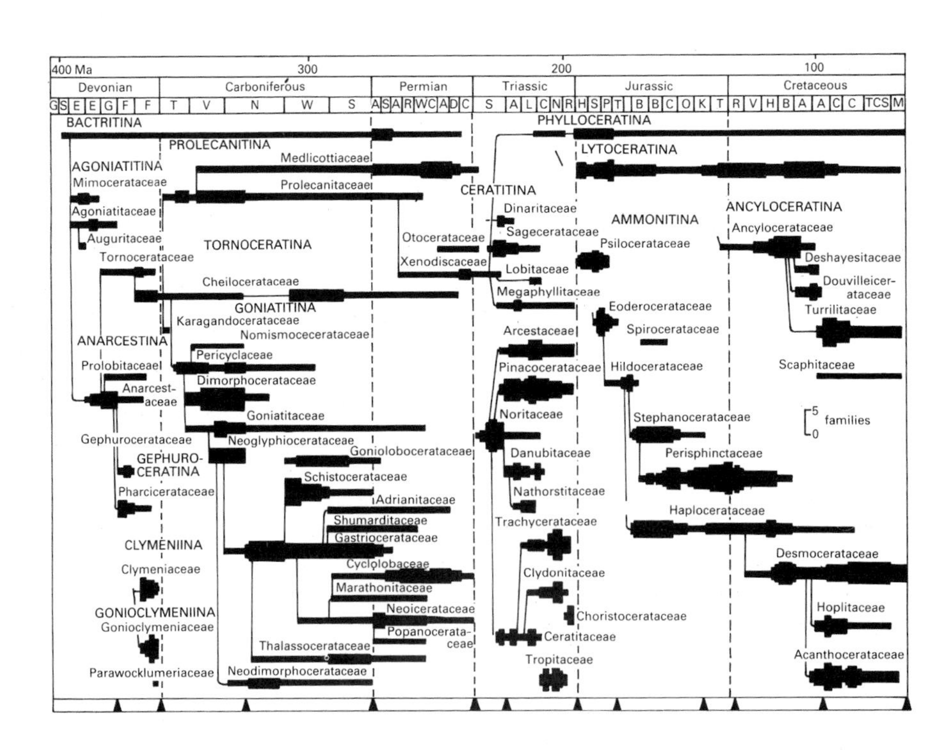

Figure 8.28 Evolutionary pattern of ammonoid orders, suborders and superfamilies. (From House, 1988.)

ornamented shells with a characteristically phylloid suture; typically involute. A very persistent stock; e.g. *Phylloceras*.
ORDER LYTOCERATIDA (Jur.–Cret.): Evolute, loosely coiled, often with 'flares'. Sutures complex, not phylloid; e.g. *Lytoceras*.
ORDER AMMONITIDA (L. Trias.–U. Cret.): The true ammonites of the Jurassic and Cretaceous, with typical ammonitic suture. Normally coiled and often dimorphic; e.g. *Asteroceras*, *Deroceras*, *Cardioceras*, *Perisphinctes*, *Schloenbachia*, *Hoplites*.
ORDER ANCYLOCERATINA (Cret.) An entirely Cretaceous order, including normally coiled forms, but also all the Cretaceous heteromorphs; e.g. *Deshayesites*, *Turrilites*, *Ancyloceras*, *Scaphites*.

The earliest ammonoids are of Lower Devonian age. They belong to an ancestral stock, Order Anarcestida, which may have been derived from the straight-shelled Suborder Bactritina (Erben, 1966). In both groups there are similar, almost straight sutures, and there is a large bulbous protoconch which is accommodated in the Anarcestida by a perforate umbilicus. Nevertheless, other early genera do not have the perforate umbilicus, and there very rapidly arose more 'advanced' features, such as involution of the shell and increased sinuosity of the sutures leading to the goniatitic condition. All the early forms have ventral retrochoanitic siphuncles, but a second Palaeozoic order is distinguished from all other ammonoids by a dorsal siphuncle. In this order, the Upper Devonian Clymeniida, the shells may be involute or evolute, and the sutures are goniatitic. The third of the Palaeozoic orders is the Goniatitida, which includes by far the majority of all Palaeozoic forms. All these have a prochoanitic siphuncle, apart from a few primitive genera, and typical goniatitic suture lines.

These Palaeozoic groups continued throughout the Devonian and Carboniferous and into the Permian, but they underwent many crises as they did so. During the latest Devonian, in particular the goniatites very nearly became extinct and only the genus *Tornoceras* survived to give rise to the great radiation of Carboniferous goniatites. By the Upper Permian the goniatites went into a final decline. Then, at the end of the Permian, as happened with so many groups, widespread extinction was the rule, and only a very few genera belonging to Order Prolecanitida crossed the Permian–Triassic boundary.

Shortly after the beginning of the Triassic the few Permian survivors gave rise to a vastly successful and 'explosive' adaptive radiation. The Triassic ammonoids nearly all belong to Order Ceratitida, which has several superfamilies. They are broadly known as 'ceratites', and indeed in most members the suture is some variant on the established ceratitic suture theme, though the external ornament became quite complex in some groups (e.g. Superfamily Tropitacea, whose members are generally ornamented with strong ribs or nodes). However, a ceratitic suture is not characteristic of all Ceratitina. Indeed, as has often been pointed out, the suture in *Pinacoceras* (Superfamily Pinacoceratacea) reaches a degree of 'ammonitic' complexity barely rivalled by that of Jurassic or Cretaceous genera. A few peculiarly coiled 'heteromorphs' are known.

In the early Triassic there also arose the ammonite stock that was to give rise to all post-Triassic ammonoids. This is Order Phylloceratida, an almost smooth-shelled group with a characteristic 'phylloid' suture (Fig. 8.29a).

The history of this order is interesting, on account of both its stratigraphical persistence (it continued until the Cretaceous) and its remarkable conservatism. Within the group there was extraordinarily little evolution, but its evolutionary offshoots became the very diverse ammonites of the Mesozoic.

From the Phylloceratida there arose another persistent order, Lytoceratida (Fig. 8.29b), which originated around the Triassic–Jurassic boundary. Though the main rootstock genera (such as *Lytoceras*) in parallel with the Phylloceratida became little differentiated, they produced a greater number of radiations than did the parent Phylloceratida. Both the Phylloceratida and Lytoceratida produced radiating lineages, but the phylogeny is so complex and the difficulties in reconstructing phylogeny are so great that in many cases it is not known for certain which of the parent stocks gave rise to which descendants. Hence the taxonomy is unclear, and ammonite students have normally retained a polyphyletic (or 'ragbag') Order Ammonitida to accommodate them, more as a matter of convenience than for scientific accuracy.

The patterns of evolution illustrated in Fig. 8.28 have been described as **iterative evolution**, in which an ancestral stock from time to time gives rise to short-lived groups which replace each other successively. These new groups expand and diversify

Figure 8.29 (a) *Phylloceras*, an Upper Triassic phylloceratid from the red limestones of Hallstatt, Austria (× 0.7). The shell has been removed and the surface polished to show the body chamber (filled with brecciated debris) and sutural patterns, picked out by crystallization within chambers. This group represents the rootstock of all post-Triassic ammonites, but itself remained very conservative; (b) *Lytoceras fimbriatus* (L. Lias.) from Lyme Regis, England (× 0.35), a representative of one of the two deep-water stocks from which many Ammonitina arose. [(a) Reproduced from a photograph of a specimen in the Royal Scottish Museum.]

for a while, but they are geologically speaking ephemeral and in due course become extinct. The niche they vacate is then occupied by descendants of the same long-lived ancestral stock. Bayer and McGhee (1984) show how in southern Germany 'similar directional changes in the physical environment are mirrored by similar morphological changes in the ammonite faunas'.

The two ancestral orders, Phylloceratida and Lytoceratida, continued throughout the Mesozoic and gave rise to superfamilial side branches (the Eoderoceratacea, Hildoceratacea, etc.). Each of these dominated the scene for a while and during its 'little hour of grace' constituted a miniature adaptive radiation, rapidly producing short-lived families so that the phylogenetic tree resembles the prongs of a toasting fork. Iterative evolution as shown by the Mesozoic ammonites has also been described under the well-chosen term 'the palaeontological relay'.

On the whole the ammonoids form a rather homogeneous group based upon a common constructional plan from which there were relatively few deviations (e.g. clymeniids and heteromorphs). Within specific lineages many evolutionary trends have been documented, in the shape of the shell, its ornament, the suture lines and even in size and ratio of sexual dimorphs (Swan, 1986; Dommergues, 1990). Though the adaptive significance of these frequently eludes analysis, there is very often a clear relationship between trends and heterochrony demonstrating here, as in so many other instances, the importance of this phenomenon as an evolutionary control. Because of the rapidity of turnover, wide distribution, abundance and ease of recognition of the ammonites, they are of outstanding zonal value.

Evolution in the Cardioceratidae

Ammonites provide some of the best examples of all evolutionary lineages. The rapidity at which they evolved, their abundance and ubiquity, and the diversity of morphological characters which they present make them especially amenable to this kind of analysis. The history of the Jurassic Family Cardioceratidae, which spanned 20 Ma, is particularly instructive (Callomon, 1985). In early Jurassic times, the northern continents clustered round an almost landlocked Boreal Sea. This sea was only sporadically colonized by ammonites in the Lower and early part of the Middle Jurassic, and then in the Upper Bajocian, descendants of the Pacific Ocean genus *Sphaeroceras* found their way into the otherwise uncolonized Boreal Sea. They were the first members of the Cardioceratidae, a family which remained broadly in the circumpolar Boreal habitat and evolved in isolation, often in the absence of other ammonites, until the abrupt extinction of this group at the top of the Lower Kimmeridgian.

The evolution of this family can be traced through 28 zones and 62 subzones, the latter approximating some 250 000 years each in duration. Most interestingly, all of the 100 or so assemblages that have been described are monospecific, which makes it easy to distinguish evolutionary lineages, and also dimorphic pairs. Within some species, such as *Quenstedtoceras lamberti*, there is an extaordinary range of intraspecific variation, embracing virtually the whole range from cadicone to oxycone, but most other species are of low variability. Throughout the evolution of this group, shapes and sizes change, and characters such as ribs and keels (corded and uncorded) may come and go; indeed virtually the whole range of sculpture and shell shapes found in the ammonites as a whole are encountered in the Cardioceratidae. Some morphs are remarkably similar to those of the earlier Liassic Family Amalthidae. Thus the evolutionary history of the Cardioceratidae is richly documented in modern terms. However, there remains a problem. Quite simply, why did these changes take place? We just do not know. There is no clear evidence of what kinds of selection pressures forced different morphologies, if indeed they did so at all. As Callomon comments 'we now know rather precisely how the Cardioceratidae evolved, but not why'.

Heteromorphs, extinction and habitat in ammonites (Fig. 8.30)

It is a well-known fact that at certain periods in geological history some groups of ammonites evolved shells of highly aberrant form. Such shells are known as heteromorphs. Some appeared during the late Triassic, some in the Jurassic, and there was a more extensive development of heteromorphs during the later Cretaceous. Heteromorphs may be loosely coiled with the whorls wholly or partially separated, e.g. *Choristoceras* (Trias.), *Spiroceras* (Jur.) and *Lytocrioceras* (Cret.). Other genera are almost straight, e.g. *Bochianites* (Cret.), resembling baculitids. Others have the body chamber hooked over the top like a walking stick, e.g. *Hamulina* (Cret.). In such bizarre genera such as the Cretaceous *Macroscaphites* and *Scaphites*, though the early whorls are of normal form, the shell is then straight and finally sharply recurved near its termination, so that the aperture faces the first-formed whorls. The Triassic *Cochloceras*, and the Cretaceous *Turrilites*, *Ostlingoceras* and other genera are helically coiled like gastropods, while the Cretaceous *Heteroceras* has its first few whorls of helical shape and the rest of the shell like *Scaphites*. *Nipponites*, from the Cretaceous of Japan, is perhaps the most extreme of all heteromorphs, having a very long tubular shell coiled in a series of U-bends into an unlikely tangle.

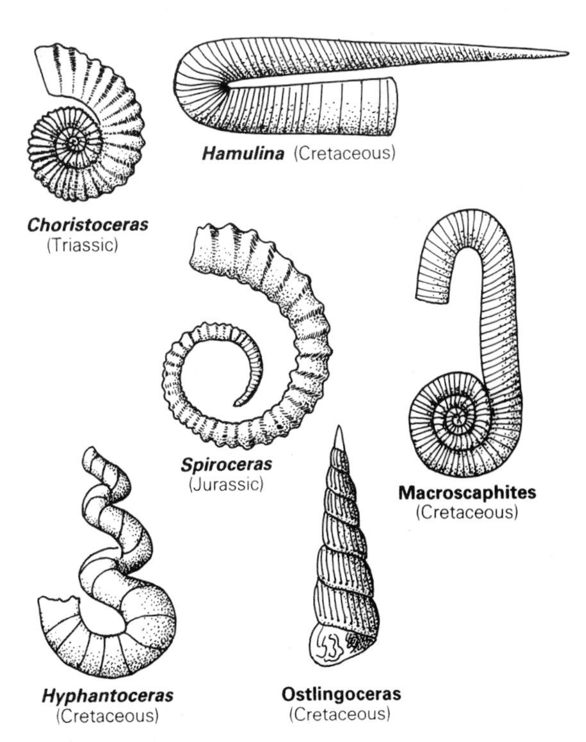

Figure 8.30 Heteromorphic ammonoids, showing variety in form. (Redrawn from *Treatise on Invertebrate Paleontology*, Part L.)

There are many other such irregularly coiled genera in the Cretaceous, and their existence has given rise to very much evolutionary speculation.

For a long time the palaeontological literature on heteromorphs was dominated by the idea that such shell forms were degenerate, retrogressive and biologically inadaptive. Furthermore, because the Triassic and Cretaceous episodes of heteromorphy took place shortly before major extinction periods for the ammonoids, it seemed to many palaeontologists that there was a definite relationship between heteromorphy and extinction. During the 1920s and 1930s the view was quite widespread that internal rhythms in evolution eventually culminated in a kind of 'racial senescence' during which bizarre and overspecialized forms were produced, as a last and final extravaganza before inevitable extinction overtook the degenerating stock.

However, the concept of heteromorphs as degenerate and inadaptive phylogenetic end-forms, plausible though it seemed to Schindewolf and its other proponents, is now no longer generally held. To begin with, it has been shown (Wiedmann, 1969) that heteromorphs appeared in certain lineages only and were not characteristic of the ammonoids as a whole at any one time. In the Triassic there were only four closely related heteromorphic genera, quite long-ranged and probably specialized for bottom living, which became extinct at about the same time as eight normally coiled superfamilies; clearly there is no likelihood of a causal connection between heteromorphy and extinction. Likewise the seven known genera of Jurassic heteromorphs were probably monophyletic and were specialized bottom dwellers, whereas Cretaceous heteromorphy, which reached its maximum development in the early Cretaceous, seems to have been a polyphyletic phenomenon. Curious shell forms appeared in several lineages quite suddenly, often associated with reduction of the primary suture. Many of the uncoiled or partially coiled heteromorphs that had been present in the Lower Cretaceous produced descendants with normal or near-normal coiling; this again falsifies the view that heteromorphs were overspecialized end-forms from which no further evolution was possible. Presumably their ability to return to normal coiling when appropriate ecological niches were available was a factor in the subsequent success of the re-coiled genera.

It has been shown, furthermore, by density–buoyancy calculations (Trueman, 1941) that heteromorphs such as *Macroscaphites* and *Lytocrioceras* were well adapted for floating in the water in particular attitudes (Fig. 8.31).

Although the apertures do not necessarily face in the same directions as those of normally coiled involute or evolute genera, there is no reason to believe that the forms were in any way unfunctional.

Helically coiled ammonite shells have the siphuncle in a dorsolateral rather than ventral position on the shell. These probably swam apex uppermost, and in this position the cameral liquid would decouple rapidly, immediately a newly formed chamber was emptied. Such decoupling increases the efficiency of vertical migration in Recent cephalopods, and the helically coiled ammonites were probably adapted to a similar vertically migrant mode of life (Ward, 1980).

Though there were some heteromorphs at the end of ammonite history, the majority of the ammonites at that time were normally coiled. In the early Upper Cretaceous the ammonites went into a slow decline over a long period of time, and towards their final end they had become restricted to certain parts of the world only. The number of genera became fewer, and finally no new characters

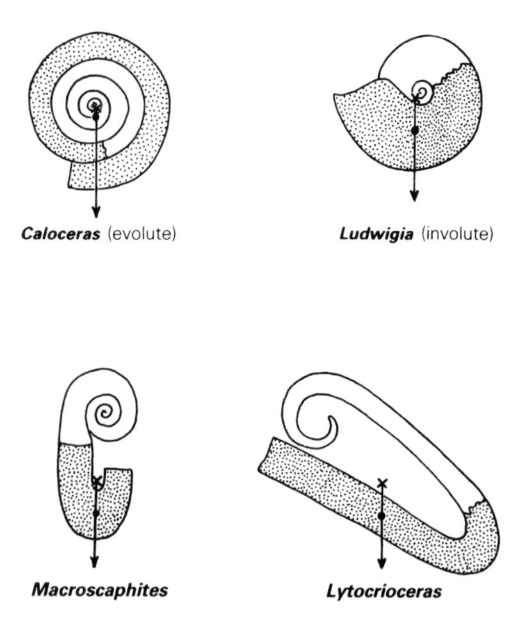

Figure 8.31 Possible life attitudes of normal and heteromorphic ammonoids as calculated by Trueman (1941). The body chamber is stippled.

appeared. There is no question of an internally weakened stock; the decline was rather the adverse effect of long-continued environmental conditions, of a kind detrimental to the ammonites and probably associated with a series of marine regressions. [There is now some doubt as to how pronounced this decline actually was; there was indeed a decline at the end of the Campanian, but intensive collection from critical horizons in the Maastrichtian led Ward (1996) to favour a more steady-state model for that time.] The last major Cretaceous regression (and the Chicxulub meteor impact), at the Maastrichtian–Danian boundary, coincided with the final and sudden demise of the last ammonites; there are cephalopods in the succeeding Danian stage, but they are coiled nautiloids of the genera *Aturia* and *Hercoglossa* (Fig. 8.23j) which have zigzag suture lines. These presumably were occupants of one of the ecological niches vacated by the ammonites, but Danian nautiloids apparently did not invade the multiplicity of microniches that their ammonite precursors had done.

We have seen that *Nautilus* lives at some depth, and in regions where the oxygen tension is low. One view of ammonite depth preferences (Wells *et al.* 1992) is that they had similar habitat requirements to *Nautilus* (even though they are actually more closely related to the coleoids). Throughout the Palaeozoic and much of the Mesozoic, oxygen levels in the deep sea and mid-waters were lower than they are today. In the Jurassic and Cretaceous there were no polar caps, and thus no way of ventilating the deep oceans by high-density melted ice-water. If ammonites had a comparable physiology to that of *Nautilus*, and a similar ability to withstand low oxygen levels, they could have successfully colonized the vast hypoxic regions of the sea between 50 and 300 m, where there was little direct competition from fish, belemnites or marine reptiles, which inhabited the highly oxygenated upper waters. Perhaps, as the oceans became increasingly oxygenated, the specialized hypoxic habitat of the ammonites began to shrink. They could not colonize the deep waters because their shells would have imploded and, chased downwards by predators from the waters above, the ammonites were caught in a trap. In the early Upper Cretaceous they were already on the decline, and the catastrophic events of the Cretaceous–Tertiary boundary only hastened their end.

A somewhat different view is that of Westermann (1990, 1996) who presents habitat diagrams for each geological period. He suggests that Jurassic and Cretaceous ammonites were mainly pelagic (drifters, swimmers and vertical migrants). While some may have colonized neritic and oceanic habitats down to a depth of 800 m, most lived between 30 and 120 m. Some of the larger streamlined oxycones may have come up close to the surface. According to this view, the neritic zone, above the continental shelf, includes (1) planktonic ammonite larvae, (2) passive pelagic drifters such as the long-bodied Jurassic serpenticone *Dactylioceras*, some sphaerocones and some Cretaceous heteromorphs, (3) pelagic vertical migrants including *Turrilites* and *Scaphites*, (4) sluggish nektobenthos, such as the highly ornate planulates (*Arietites* and *Acanthoceras*). To this habitat belonged the Jurassic *Gravesia* and the Cretaceous neoceratites, which were amongst the shallowest- water ammonites of all, and (5) mobile nektobenthos. These were streamlined forms, such as the oxycone oppeliids and some platycones (*Hildoceras* and *Grammoceras*). Many of these may have preyed actively on other ammonites.

The oceanic region had an epipelagic zone inhabited by planktonic larvae, which lived there for up to a month before moving elsewhere. In deeper offshore waters (lower epi- and upper mesopelagic) there lived mainly fragile-shelled ammonites, such as some Cretaceous heteromorphs, and importantly the deeper mesopelagic realm was inhabited in the Jurassic by *Phylloceras* (>500 m) and *Lytoceras* (>800 m), both of which were vertical migrants within this range.

Why did *Nautilus* survive when the ammonites did not? Possibly because it was an unspecialized feeder, especially well adapted to low oxygen conditions, and laid large eggs, with high survival potential, in deep waters. On the other hand, perhaps it was nothing more than good luck.

Subclass Coleoidea: dibranchiate cephalopods

Modern coleoids

The dibranchiate cephalopods or coleoids include squids, cuttlefish, octopuses, the paper-nautilus or *Argonauta*, and various extinct groups, notably the belemnites. They all have only a single pair of gills within the mantle cavity, and it is this feature that separates them from all the other cephalopods.

The modern squids and cuttlefishes range from tiny animals only 4 cm long to the gigantic

Architeuthis, which is (including the tentacles) 18 m long. Yet their geological importance is limited; only the belemnites are normally preserved as fossils, and in any case coleoids are more diverse and abundant now than they were formerly.

One kind of dibranchiate structure is shown by the cuttlefish *Sepia* (Order Sepiida), which is illustrated in Fig. 8.21f,g. It differs from *Nautilus* in having only eight arms and two tentacles, which are provided with suckers and hooks. These hooks, like the jaws, are horny and are not usually preserved as fossils, though some rare fossilized examples are known. In the mantle cavity with its two gills there is an ink sac, used for clouding the water as an escape reaction from predators. Otherwise, and apart from shell structure, the internal organization of the body is broadly comparable.

The shell of *Sepia is* wholly internal. It is a large oval body, known in the vernacular as the **cuttlebone**, located dorsally and composed of closely spaced, oblique, calcareous partitions supported by pillars. The cuttlefish controls its buoyancy by secretion and extraction of liquid from within the spaces between partitions. A shell such as this seems at first sight very dissimilar to that of a nautiloid, but it is morphologically the dorsal half of the nautiloid phragmocone which has become flattened and expanded. In the squid *Loligo* the shell (sea pen) is horny.

Since the chambers within the partitions are very narrow, the change in buoyancy through the pumping out of water can be a very slow process, at least as regards water deep within the chamber. Nevertheless, buoyancy regulation by the alteration of water balance within the chambered shell is not the only system in modern squids. Quite a large number of squids, e.g. *Cranchia*, have instead adopted the use of ammonium chloride within the tissues, which is isotonic and isosmotic with sea water and thus gives neutral buoyancy. In these cranchids the coelomic cavity is vastly distended and filled with ammoniacal liquid of relative density 1.010–1.012, which is less than that of sea water (relative density 1.026). Though such ammoniacal squids are very abundant and the use of ammonia is not confined to the cranchids, the system cannot apparently be used for rapid buoyancy changes.

Packard (1972), in discussing the evolutionary convergences and interactions between cephalopods and fish, has put the ammoniacal and other squids into an interesting evolutionary perspective. He has shown how by early Palaeozoic times cephalopods were the most advanced and mobile of all marine animals. With their chambered shells and buoyancy mechanism along the lines of *Nautilus* they were highly successful, though limited in how deep they could go due to danger of implosion. However, when fish began to diversify and invade the upper waters of the sea, in the later Palaeozoic, one effect of such competition was to put pressure on living space. The origin of the 'ammoniacal' buoyancy system was one answer to the problem, for it frees those cephalopods which possess it from the limitations of the shell, and it is perhaps not surprising that the deeper parts of the sea have been extensively colonized by ammoniacal squid. Another solution to competition from fish was that many cephalopods became more fish-like, structurally and functionally, so that the two were then competing on more equal terms; as Packard comments, 'cephalopods, functionally are fish'.

Spirula (Fig. 8.21h,j) is a little squid some 10 cm long which has an open, spiral, chambered shell enclosed entirely within the body. The shell has a ventral siphuncle which connects with the chambers only through a very small porous region in each. The lower chambers are largely filled with liquid which is normally inert within them, but this can be added to or subtracted from through the porous region so that the shell's buoyancy can be controlled in a manner similar to that of *Nautilus*. The ever-changing colour patterns of squids, in which rolling colour waves, sudden darkenings of the skin or colour flashes are usual, may mesmerize prey, camouflage the predator or facilitate communication between individuals.

The shell of the female *Argonauta* ('paper nautilus'; Fig. 8.32n) is a very thin spiral shell, but it is formed by secretion of the tentacles and carried external to the body.

It is used as a brood pouch, more rarely for carrying about a captured male! The male, incidentally, is shell-less and only about one-twentieth the size of the female. It has a very long sexual organ (the **hectocotylus**) which carries the sperm; this can detach itself from the male and swim actively towards the female – a kind of 'guided missile copulation' which seems to be unique in nature.

Despite the fragility of the argonaut shell, the fossil record can be traced back to the Oligocene

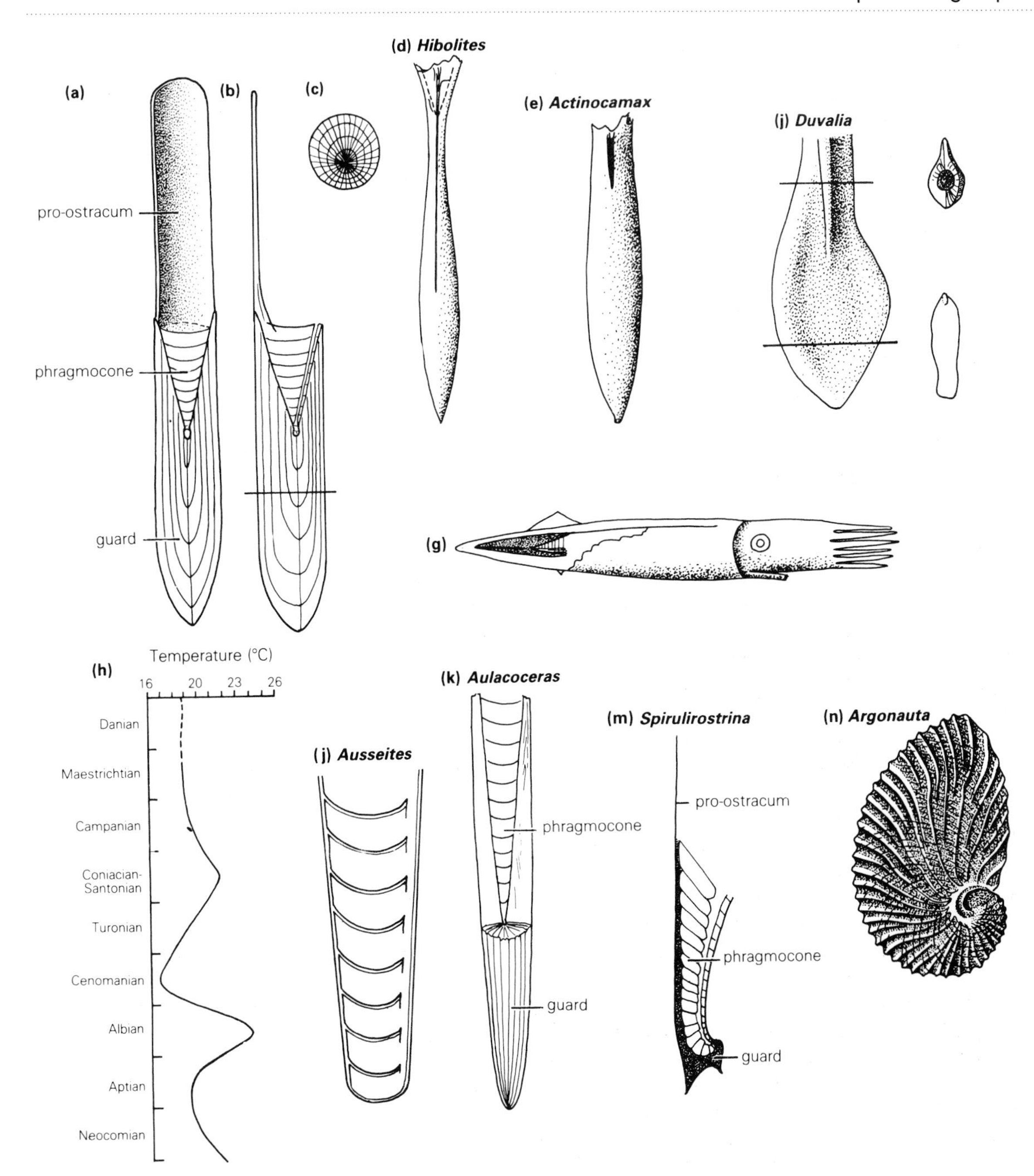

Figure 8.32 Coleoid morphology: (a) longitudinally sectioned belemnite in ventral view; (b) same, in lateral view; (c) belemnite guard cut transversely (all × 0.75); (d) *Hibolites* (Cret.; × 1 approx.); (e) *Actinocamax* (Cret.; × 0.75); (f) *Duvalia* (Cret.; × 0.75); (g) reconstruction of swimming belemnite, with the guard wholly internal; (h) climatic fluctuations in the Cretaceous, plotted from oxygen isotope ratios in belemnite guards; (j) *Ausseites* phragmocone (Trias.; × 0.75); (k) *Aulacoceras* (Trias.; × 0.75); (m) *Spirulirostrina* (Mio.) in section (× 0.75); (n) 'shell' of Recent paper-nautilus *Argonauta* (× 0.35). [Mainly based on illustrations by Naef and others in *Traité de Paléontologie*, II; (h) after Bowen, 1961.]

(Holland, 1988). The earliest argonaut shells are smooth and indeed very like those of nautiloids. It is possible that the ancestral female argonauts firstly used discarded nautiloid or ammonite shells for protection and then constructed their own on the same model.

Extinct coleoids and their evolution (Fig. 8.32)

In very many Jurassic and Cretaceous sediments there may be an abundance of fossil belemnites. These are the internal shells of fossil squid-like cephalopods, but they do not closely resemble the shells of any modern squid or cuttlefish.

In the typical genus *Belemnites* the shell has three parts. The largest and most posterior section is the **guard**, a massive bullet-shaped cylinder of solid calcite. It is parallel-sided, tapers posteriorly to a point and is indented at its anterior end by a conical cavity (the **alveolus**). If the guard is cut transversely, the structure is seen as radially oriented needles of calcite, with concentric growth rings, which are also apparent in horizontal section. The axis of growth is not central, however, but placed towards the ventral margin. Usually the surface of the guard is smooth, but it may be granular or pitted. Some genera, such as *Hibolites*, have a long ventral groove extending two-thirds of the way from alveolus to point, and the posterior part of the guard itself is somewhat swollen. *Actinocamax* is more parallel-sided but likewise has a ventral slit below the alveolus. Some species of *Duvalia* have a curious laterally flattened guard.

Within the alveolus and fitting it exactly is the **phragmocone**. This is a conical, thin-walled aragonitic structure which projects outside the alveolus. It is septate with its septa concave anteriorly and separating fair-sized camerae. A slender siphuncle threads through the septa at the ventral margin. These structures, together with a tiny bulbous protoconch, leave no doubt that the coleoid phragmocone is the direct homologue of the shell of a nautiloid, ammonoid or *Spirula*.

The guard, which is such a substantial part of the belemnite, has no direct homologue in other coleoids. It seems to have acted as a necessary counterweight to maintain the belemnite body level when swimming, i.e. fulfilling in a different way the same function as the cameral deposits of nautiloids. Since the guard consists of thick and unrecrystallized calcite, it has proved most useful in palaeotemperature analysis of the Jurassic and Cretaceous by ^{16}O:^{18}O isotope ratios (Fig. 8.32h).

The third component of the belemnite shell is the **pro-ostracum**: a long, flat, expanded tongue projecting forwards and presumably covering the anterior part of the body. It seems to be homologous to the 'pen' of the squid *Loligo*. This is rarely preserved, however, and its function is poorly understood. Sometimes specimens with eight radiating sets of hooks have been found forwards of the guard, testifying to the former presence of arms, and even fossilized ink sacs have been located in place.

Belemnites seem therefore to have been a kind of fossil squid, with a different hard-part construction to others and therefore with a unique system of buoyancy control.

Mass accumulations of belemnite guards are not uncommon in the Mesozoic, and such 'belemnite battlefields' may, as with modern squids, represent mass mortality after spawning. Similar accumulations of orthocone nautiloids are common in the Palaeozoic. However, there are other possibilities; catastrophic mass mortality, intense predation, and concentration through winnowing of sediment (Doyle and Macdonald, 1993). It is usually possible to establish which of these agencies operated in particular instances.

There is evidence of sexual dimorphism in some belemnites (e.g. *Youngibelus*; Doyle, 1985), as with ammonites. Hitherto belemnites have only been used for stratigraphical purposes in the Upper Cretaceous of northern Europe. Yet they are abundant, rapidly evolving and usually independent of facies. Though at first sight they appear to look rather similar, it is actually not hard to identify them (Doyle, 1990; Doyle and Bennett, 1995). Often they are found where the more fragile ammonites are absent, and their biostratigraphic use can but increase.

Geological history

Belemnites are grouped into two taxa: the Aulacocerida (?Dev./Carb.–Jur.), which may have retained the body chamber, and the Belemnitida proper (Jur.–Cret.), in which the latter was reduced to the pro-ostracum. Both groups have guards which are not easily distinguishable. Other than a doubtful Devonian record, the earliest aulacocerids have been reported from erratic boulders in North America believed to be of Mississippian age. It is generally agreed that these forerunners of the Mesozoic belemnites were derived from orthocone cephalopods about this time through the expansion of the thin covering of the tip of the phragmocone into the massive guard by a simple process of relative growth.

The systematics and phylogeny of coleoids is discussed by Doyle *et al.* (1994). Little is known about the earlier stages of belemnite evolution, and only in the Triassic are there encountered well-known Aulacocerida, which in some ways have characters intermediate between those of orthocones and Belemnitida. The Triassic *Ausseites*, a genus from Austria, has a small guard and a long phragmocone with a marginal siphuncle. *Aulacoceras* has a larger, externally ribbed guard, though the phragmocone otherwise resembles that of *Ausseites*. It is possible here to assume an evolutionary progression from an orthocone with external thickening, through the Triassic genera to the true belemnites. The latter expanded and diversified throughout the Jurassic and Cretaceous and then, like the ammonites, declined exceedingly, though a very few belemnites continued into the early Tertiary before becoming extinct.

A possible side branch of Mesozoic 'belemnites' is represented by the Triassic *Phragmoteuthis* and Jurassic *Belemnoteuthis*. In these the guard is present but is very thin and delicate, while the phragmocone is short and broad with a very long pro-ostracum extending from it. The first octopuses are known from the Cretaceous of Lebanon, and there are some fossil argonauts known from the Tertiary, but otherwise the fossil record of these modern types is so poor that it is not worth mentioning further. However, there are some extinct coleoids of Tertiary age whose fossil history is interesting if a little confusing. Some genera have a somewhat reduced guard and a spiral coil to the early part of the phragmocone. The Eocene *Belosepia* and Miocene *Spirulirostina* are of this kind; in both, the siphuncle is much expanded and the ventral part of the phragmocone reduced. Though this kind of structure is in some ways intermediate between that of the belemnites and *Sepia*, which has much reduced guard, the fossils are very few, and to relate the end members directly in a sequence *Belemnites–Belosepia–Sepia* is at present unjustified. It is possible, indeed, that these early Tertiary coleoids are no more than small parts of large ordinary sepiids.

Order Teuthida, to which *Loligo* belongs, seems to have developed along another line, and the pro-ostracum has become an important internal structure at the expense of the phragmocone and guard. Sepiids and teuthoids are found in the Jurassic Solnhofen Limestone, and teuthids at least are known from the Lower Devonian Hunsrückschiefer (Chapter 12). Phosphatized soft parts, resulting from very early diagenesis, are known in squids from the Oxford Clay of the English Jurassic (Allison, 1988).

The ancestry of the little squid *Spirula* with its coiled internal shell is unknown. It has no guard or pro-ostracum, but only a loosely coiled phragmocone with a large protoconch and marginal siphuncle.

Since the buoyancy mechanisms of modern coleoids are now beginning to be well understood, it may be hoped that this will give a correspondingly broader conception of the means whereby fossil cephalopods, including coleoids, were able to control their buoyancy. It is unquestioned that this has been the most important single factor in their evolution.

8.5 Predation and the evolution of molluscs

In molluscs, as in all other animals with voluminous and edible flesh, the danger of predation has had a decided impact upon the course of their evolution. Such predation has increased through time and different kinds of molluscs have responded to it in different ways. In fossils this can be assessed and its effects interpreted by examining traces of predation on the shells. Vermeij (1983, 1987) argues that through time the food-catching capacity of predators (on molluscs and other invertebrates) has increased dramatically, but has necessarily been matched, in the prey, by increasingly effective anti-predator devices. Such adaptive escalation has led to a 'biological arms race', which is most evident following periods of mass extinction.

Cephalopods depend upon escape or avoidance. Their capacity for jet propulsion, coupled with the ejection of ink when disturbed, has been an undoubted factor in their survival and success. Gastropods, especially from the Jurassic onwards, are protected and well served by their thick shell, narrow aperture, and retractability of soft tissues within the shell. The frequency of gastropod shells which have been repaired after damage testifies to the effectiveness of the gastropod shell as a fortress, for though such shells have been damaged, their inhabitants have not been killed.

Bivalve shells are much more vulnerable than are gastropods. They have an intrinsic limitation imposed upon them by their shell construction. For when the shell is broken at the margin, the seal between the valves is destroyed, body fluids leak out and predators can make short work of the shell. While this was always a problem for bivalves, it became acute in the early Mesozoic owing to the proliferation of new kinds of predators (gastropods, crustaceans and starfish) This, the Mesozoic Marine Revolution (MMR) of Vermeij, put immense pressure on the bivalves, and they responded in a variety of ways, primarily escape and avoidance, to protect themselves (Harper and Skelton, 1993). Those that remained epifaunal were especially at risk, cemented forms being less so than attached bivalves (though not against the 'driller killers', the muricid gastropods). Epibyssate forms can trap smaller assailants with byssal threads. Tightly sealed valves, with interlocking commissures (Carter, 1968), help to deter predators, and spinose bivalves, together with boring and nestling types, increased greatly after the MMR. Spines, however, can only be generated in bivalves with a flexible periostracum, not in others, and in general potential adaptations are not uniformly spread but restricted to some groups only, through structural constraint. Burrowing, which became possible through mantle fusion, is one obvious response to predation since the MMR; avoidance by active swimming, sheltering in seagrass communities, the invasion of the physiologically 'difficult' intertidal zone, colonization of deeper water and high fecundity are others. That these have been successful strategies, despite continual predation pressure, is evident from the great number of diverse bivalves living today.

Bibliography

Books, treatises and symposia

Bathurst, R.G.C. (1975) *Carbonate Sediments and their Diagenesis*. Elsevier, Amsterdam. (Fine structure of molluscan shells)

Batten, R.L. (1966) *The Lower Carboniferous Gastropod Fauna from the Hotwells Limestone of Compton Martin, Somerset*. Palaeontographical Society Monographs Nos 119 and 120. Palaeontographical Society, London. (Monographic study of a diverse fauna)

Garstang, W. (1951) *Larval Forms and Other Zoological Verses*. Blackwell, Oxford. (Gastropod torsion)

Kennedy, W.J. and Cobban, W.A. (1976) *Aspects of Ammonite Biology, Biogeography and Biostratigraphy*, Special Papers in Paleontology No. 17. (Invaluable treatment)

Landman, N.H., Tanabe, K. and Davis, R.A. (1996) *Ammonoid Paleobiology*. Topics in Geobiology No. 13. Plenum, New York and London. (Indispensable compilation, 20 papers)

Lebrun, P. (1996, 1997) *Ammonites 1 & 2. Mineraux et Fossiles 2*. Editions CEDIM, Paris. (Fine colour plates)

Lehmann, U. (1981) *The Ammonites – their Life and their World*. Cambridge University Press, Cambridge. (Clear, straightforward text)

Lehmann, U. (1990) *Ammonoideen – Leben zwischen Skylla und Charybdis* Ferdinand Enke Verlag, Stuttgart. (Second edition of above)

Moore, R.C. (ed.) (1957) *Treatise on Invertebrate Paleontology*, Part L, *Mollusca 4*. Geological Society of America and University of Kansas Press, Lawrence, Kan. (Ammonoids)

Moore, R.C. (ed.) (1964) *Treatise on Invertebrate Paleontology*, Part K, *Mollusca 3*. Geological Society of America and University of Kansas Press, Lawrence, Kan. (Nautiloids).

Moore, R.C. (ed.) (1969a) *Treatise on Invertebrate Paleontology*, Part I, *Mollusca 1*. Geological Society of America and University of Kansas Press, Lawrence, Kan. (Principles, gastropods)

Moore, R.C. (ed.) (1969b) *Treatise on Invertebrate Paleontology*, Part N, Vols 1–3. Geological Society of America and University of Kansas Press, Lawrence, Kan. (Bivalves)

Moore, R.C, Lalicker, C.G. and Fischer, A.G. (1953) *Invertebrate Fossils*. McGraw Hill, New York. (Useful discussion of molluscan anatomy)

Morton, J.E. (1967) *Molluscs*. Hutchinson, London. (Valuable summary of morphology and relationships)

Piveteau, J. (ed.) (1952) *Traité de Paléontologie, II, Mollusques*. Masson and Cie, Paris. (Shorter than the *Treatise*, but an essential compilation)

Saunders, W.B. and Landman, N.H. (1987) *Nautilus*. Plenum, New York. (14 mainly biological papers)

Thompson, d'Arcy W. (1917) *On Growth and Form*. Cambridge University Press, Cambridge. (Individual classic work on physical laws determining growth; an abridged edition of this book was edited by J.T. Bonner in 1961 and also published by Cambridge University Press)

Vermeij, G.J. (1987) *Evolution and Escalation. An Ecological History of Life*. Princeton University Press. (The evolutionary 'arms race' between predator and prey)

Ward, P.D. (1987) *The Natural History of Nautilus*. Allen and Unwin, New York, 267 pp.

Wiedmann, J. and Kullmann, J. (eds) 1988. *Cephalopods Present and Past*: Schweizerbarttösche, Stuttgart.

Woods, H. (1946) *Palaeontology; Invertebrate*, 7th edn. Cambridge University Press, Cambridge. (Student text, still useful)

Individual papers and other references

Allison, P.A. (1988) Phosphatised soft-bodied squids from the Jurassic Oxford Clay. *Lethaia* **21**, 403–10.

Allmon, W.D., Nieh, J.C. and Norris, R.D. (1990) Drilling and peeling of turritelline gastropods since the Late Cretaceous. *Palaeontology* **33**, 595–612.

Babin, C. and Guitierrez-Marco, J.C. (1991) Middle Ordovician bivalves from Spain and their phyletic and palaeogeographic significance. *Palaeontology* **34**, 109–48.

Barker, M.J. (1990) The palaeobiology of nerineacean gastropods. *Historical Biology* **3**, 249–64.

Bayer, U. (1977) Cephalopoden-septen. Teil 2. Regelmechanismen in Gehause and Septembau der Ammoniten. *Neues Jahrbuch für Geologie und Paläontologie* **155**, 165–215. (Ammonite septal functions)

Bayer, U. and McGhee, G.R. (1984) Iterative evolution of Middle Jurassic ammonite faunas. *Lethaia* **17**, 1–16.

Bowen, R. (1961) Oxygen isotope paleotemperature measurements on Cretaceous Belemnoidea from Europe, India and Japan. *Journal of Paleontology* **35**, 1077–84.

Bradshaw, M. (1970) The dentition and musculature of some Middle Ordovician (Llandeilo) bivalves from Finistère, France. *Palaeontology* **13**, 623–45.

Callomon, J.H. (1963) Sexual dimorphism in Jurassic ammonites. *Transactions of the Leicester Literary and Philosophical Society* **57**, 21–56.
(Many examples cited, with methodology)

Callomon, J.H. (1985) The evolution of the Jurassic ammonite family Cardioceratidae, *in Evolutionary Case Histories from the Fossil Record* (eds J.C.W. Cope and P.W, Skelton). Special Papers in Palaeontology No. 33, Academic Press, London, pp. 49–90.

Callomon, J.H. (1995) Time from fossils: S. S. Buckman and Jurassic high resolution stratigraphy, in *Milestones in Geology* (ed. M.J. Le Bas), Geological Society Memoir No. 16, pp. 127–50.

Carter, R.M. (1968) Functional studies on the Cretaceous oyster *Arctostrea*. *Palaeontology* **11**, 458–85.

Carter, R.M. (1972) Adaptations of British Chalk Bivalvia. *Journal of Paleontology* **46**, 325–40.

Clarkson, E.N.K., Harper, D.A.T. and Peel, J.S. 1995. Taxonomy and palaeoecology of the mollusc *Petrotheca* from the Ordovician and Silurian of Scotland. *Lethaia* **28**, 101–4.

Chlupäc, I. (1976) The oldest goniatite faunas and their stratigraphical significance. *Lethaia* **9**, 303–15.

Cope, J.C.W. (1996). Early Ordovician (Arenig) bivalves from the Llangynog Inlier, South Wales. *Palaeontology* **39**, 979–1026.

Davis, R.A., Landman, N.H., Dommmergues, J.-L. *et al.* (1996) Mature modifications and dimorphism in ammonoid cephalopods, in *Ammonoid Paleobiology* (eds N.H. Landman, K. Tanabe and R.A. Davis), Plenum Press, New York and London, pp. 464–43.

Denton, E.J. (1961) The buoyancy of fish and cephalopods. *Progress in Biophysics and Chemistry* **11**, 178–234. (Essential reference on attainment of negative buoyancy)

Denton, E.J. (1974) On buoyancy and the lives of modern and fossil cephalopods. Croonian Lecture 1973. *Proceedings of the Royal Society B* **185**, 273–99.

Denton, E.J. and Gilpin-Brown, J.B. (1966) On the buoyancy of the pearly nautilus. *Journal of the Marine Biological Association of the United Kingdom* **46**, 723–59. (Siphuncular structure and function)

Dommergues, J.-L. (1990) Ammonoids, in *Evolutionary Trends* (ed. K.J. McNamara), Belhaven, Boston, pp. 162–87.

Donovan, D.T. (1964) Cephalopod phylogeny and classification. *Biological Reviews* **39**, 259–87.

Doyle, P. (1985) Sexual dimorphism in the belemnite *Youngibelus* from the Lower Jurassic of Yorkshire. *Palaeontology* **28**, 133–46.

Doyle, P. (1990) The British Toarcian (Lower Jurassic) belemnites. Parts 1 and 2. *Palaeontographical Society Monograph*, 1–49, 50–79.

Doyle, P. and Bennett M. (1995) Belemnites in biostratigraphy. *Palaeontology*, **38**, 815–29.

Doyle, P. and Macdonald, D.I.M. (1993) Belemnite battlefields. *Lethaia* **26**, 65–80.

Doyle, P., Donovan, D.T. and Nixon, M. (1994) Phylogeny and systematics of the Coleoidea. *University of Kansas Palaeontological Contributions* **5**, 1–15.

Dzik, J. (1984) Phylogeny of the Nautiloidea. *Palaeontologica Polonica* **45**, 1–219.

Engeser, T. (1996) The position of the Ammonoidea within the Cephalopoda, in *Ammonoid Paleobiology* (eds N.H. Landman, K. Tanabe and R.A. Davis), Plenum Press, New York and London, pp. 3–19.

Erben, H.K. (1966) Uber die Ursprung der Ammonoidea. *Biological Reviews* **41**, 641–58. (Earliest ammonoids)

Fischer, A.G. and Teichert, C. (1969) Cameral deposits in cephalopod shells. *University of Kansas Paleontological Contributions* **37**, 1–30. (Cameral morphology and use as counterweights)

Flower, R.H. (1957) Nautiloids of the Palaeozoic, in *Treatise on Marine Ecology and Palaeoecology 2* (ed. H.S. Ladd), Geological Society of America Memoir No. 67, Geological Society of America, Lawrence, Kan., pp. 829–52. (Ecology and annotated bibliography)

Gould, S.J. (1969a) An evolutionary microcosm: Pleistocene and Recent history of the land snail *P. (Poecilozonites)* in Bermuda. *Bulletin of the Museum of Comparative Zoology, Harvard University* **138**, 407–532. (Allopatric speciation)

Gould, S.J. (1969b) Ecology and functional significance of uncoiling in *Vermetularia spirata*: an essay on gastropod form. *Bulletin of Marine Science* **19**, 432–45. (Uncoiling useful for rapid upward growth)

Harper, E.M. (1991a) The role of predation in the evolution of cementation in bivalves. *Palaeontology* **34**, 455–.60.

Harper, E.M. (1991b) Post-larval cemenbtation in the Ostreidae and its implications for other cementing bivalves. *Journal of Molluscan Studies* **58**, 37–47.

Harper, E.M. (1997) The molluscan periostracum, an important constraint on bivalve evolution. *Palaeontology* **40**, 71–97.

Harper, E. M. and Skelton, P.W. (1993) The Mesozoic Marine Revolution and epifaunal bivalves. *Scripta Geologica Special Issue* **2**, 127–53.

Harper, J.A. and Rollins, H.B. (1985) Infaunal or semi-infaunal bellcrophont gastropods: analysis of *Euphemites* and funcationally related taxa. *Lethaia* **18**, 21–38.

Haszprunar, G. (1988) On the origin of major gastropod groups with special reference to the Strophoneura. *Journal of Molluscan Studies* **54**, 367–441.

Hewitt, R.A. and Westermann, G.E.G. (1986, 1987) Function of complexly fluted septa in ammonoid shells. I. Mechanical Principles and functional models. II. Septal evolution and conclusions. *Neues Jahrbuch für Geologie und Paläontologie Abhandlungen* **172**, 47–69; **174**, 135–69.

Holland, C.H. (1987) The nautiloid cephalopods: a strange success. *Journal of the Geological Society of London* **144**, 1–15.

Holland, C.H. (1988) The Paper Nautilus. *New Mexico Bureau of Mines and Mining Research* **44**, 109–14.

House, M.R. (1981) On the origin, classification and evolution of the early Ammonoidea, in *The Ammonoidea. The Evolution, Classification, Mode of Life, and Geological Usefulness of a Major Fossil Group* (eds M.R. House and J.R. Senior), Systematics Association Special Volume No. 18, Academic Press, London, pp. 3–36. (Still valuable)

House, M.R. (1988) Extinction and survival in the Cephalopoda, in *Extinction and Survival in the Fossil Record* (ed. G.P. Larwood), Systematics Association Special Volume No. 34, pp. 139–94.

Hudson, J.D. (1963) The recognition of salinity-controlled mollusc assemblages in the Great Estuarine Series (Middle Jurassic) of the Inner Hebrides. *Palaeontology* **11**, 163–82. (See text)

Jefferies, R.P.S. (1960) Photonegative young in the Triassic lamellibranch *Lima lineata* (Schlotheim). *Palaeontology* **3**, 362–9. (Use of lunule as a brood chamber)

Jefferies, R.P.S. and Minton, R.P. (1965) The mode of life of two Jurassic species of *Posidonia* (Bivalvia). *Palaeontology* **8**, 156–85. (Planktonic mode of life postulated)

Johnson, A.L.A. (1994) Evolution of European Lower Jurassic *Gryphaea (Gryphaea)* and contemporaneous bivalves. *Historical Biology* **7**, 167–86.

Johnson, A.L.A. and Lennon, C.D. (1990) Evolution of gryphaeate oysters in the Mid-Jurassic of Western Europe. *Palaeontology* **33**, 453–486.

Jones, J.S. (1981) An uncensored page of fossil history. *Nature* **293**, 427–8. (Interpretation of Williamson, 1981)

Kaufmann, E.G. and Sohl, N.F. (1981) Rudists, in *The Encyclopaedia of Palaeontology* (eds R.W. Fairbridge and C. Jablonski), Dowden, Hutchinson and Ross, Stroudsburg, Penn.

Kelly, S.R.A. (1980) *Hiatella* – a Jurassic bivalve squatter? *Palaeontology* **23**, 769–81.

Kennedy, W.J., Taylor, J.D. and Hall, A. (1968) Environmental and biological controls on bivalve shell mineralogy. *Biological Reviews* **44**, 499–530. (Shell mineralogy subject to many controls)

Lardeux, H. (1969) Les tentaculites d'Europe occidentale et d'Afrique du nord. *Cahiers de Paléontologie Editions C.N.R.S.*, 1–238.

Larsson, K. (1979) Silurian tentaculitids from Gotland and Scania. *Fossils and Strata* **11**, 1–180.

Lehmann, U. (1967) Ammoniten mit Kieferapparat und Radula aus Lias Geschieben. *Paläontologisches Zeitschrift* **41**, 38–45. (First description of jaws and radula)

Lehmann, U. (1971a) Jaws, radula and crop of *Arnioceras* (Ammonoidea). *Palaeontology* **14**, 338–41. (Unusual presentation)

Lehmann, U. (1971b) New aspects in ammonite biology. *Proceedings of the North American Paleontological Convention*, Vol. 1, 1251–69. (As above, with interpretation)

Lehmann, U. (1979) The jaws and radula of the Jurassic ammonite *Dactylioceras*. *Palaeontology* **22**, 265–71.

Lehmann, U. and Kulicki, C. (1990) Double function of aptychi (Ammonoidea) as jaw elements and opercula. *Lethaia* **23**, 325–32.

Liljedahl, L. (1984) Silurian silicified bivalves from Gotland. *Sveriges Geologiska Undersokning* **78**, 1–82.

Liljedahl, L. (1985) Ecological aspects of a silicified bivalve fauna from the Silurian of Gotland. *Lethaia* **18**, 53–66.

Liljedahl, L. (1991) Contrasting feeding strategies in bivalves from the Silurian of Gotland. *Palaeontology* **34**, 219–35.

Liljedahl, L. (1994) Silurian nuculoid and modiomorphid bivalves from Sweden, *Fossils and Strata* **33**, 1–89.

Linsley, R.M. (1977) Some 'laws' of gastropod shell form. *Paleobiology* **3**, 196–206.

McAlester, A.L. (1962) Systematics, affinities and life habits of *Babinka*, a transitional Ordovician lucinoid bivalve. *Palaeontology* **8**, 231–41. (Multiple muscle scars suggest affinities with segmented ancestors)

McKinnon, D.I. (1982) *Tuarangia papura,* n. gen and n. sp. a late Middle Cambrian pelecypod from New Zealand. *Journal of Paleontology* **56**, 589–98.

Makowski, H. (1963) Problems of sexual dimorphism in ammonites. *Acta Palaeontologica Polonica* **12**, 1–92. (Comes independently to the same conclusions as Callomon, 1963)

Martill, D.M. (1990) Predation on *Kosmoceras* by semionetid fish in the Middle Jurassic Lower Oxford Clay of England. *Palaeontology* **33**, 739–42.

Mutvei, H. (1971) The siphonal tube in Jurassic Belemnitida and Aulacocerida. *Bulletin of the Geological Institute of the University of Uppsala* **3**, 27–36. (Detailed electron micrography)

Mutvei, H. (1972) Ultrastructural studies on cephalopod shells. Part I. The septa and siphonal tube in *Nautilus*. Part II. Orthoconic cephalopods from the Pennsylvanian Buckhorn Asphalt. *Bulletin of the Geological Institute of the University of Uppsala* **3**, 237–73. (Detailed micrography)

Nixon, M. (1996) Morphology of the jaws and radula in ammonoids, in *Ammonoid Paleobiology* (eds N.H. Landman, K. Tanabe and R.A. Davis), Plenum Press, New York and London, pp. 23–43.

Packard, A. (1972) Cephalopods and fish: the limits of convergence. *Biological Reviews* **47**, 241–307. (Evolutionary interactions between fish and cephalopods; a valuable work)

Peel, J.S. (1980) A new Silurian retractile monoplacophoran and the origin of the gastropods. *Proceedings of the Geological Association* **91**, 91–7.

Pojeta, J. (1975) *Fordilla troyensis* Barrande and early pelecypod phylogeny. *Bulletins of American Paleontology* **67**, 363–84.

Pojeta, J. (1978) The origin and taxonomic diversification of pelecypods. *Philosophical Transactions of the Royal Society of London B* **284**, 225–46.

Pojeta, J. and Runnegar, B. (1976) *The Palaeontology of Rostroconch Mollusks and the Early History of the Phylum Mollusca.* US Geological Survey Professional Paper No. 968. (Rostroconchs as bivalve ancestors)

Ponder, W.F. and Waten, A. (1988) Prosobranch phylogeny. *Malacological Reviews Supplement* 4.

Raup, D.M. (1966) Geometrical analysis of shell coiling: general problems. *Journal of Paleontology* **40**, 1178–90. (See text)

Raup, D.M. (1967) Geometrical analysis of shell coiling: the cephalopod shell. *Journal of Paleontology* **41**, 43–65. (See text)

Raup, D.M. and Stanley, S.M. (1971) The cephalopod suture problem, in *Principles of Paleontology,* Freeman, San Francisco, pp. 172–81. (A good treatment of this theme)

Reif, W.E. (1983) The Steinheim snails (Miocene Schwabische Alb) from a neo-Darwinian point of view: a discussion. *Paläontologische Zeitschrift* **57**,112–25 (Hilgendorf's snails – see text)

Runnegar, B. (1983) Molluscan phylogeny revisited. *Memoirs of the Association of Australian Palaeontologists* **1**, 121–44. (Most recent treatment)

Runnegar, B. (1996) Early evolution of Mollusca: the fossil record, in *Origin and Evolutionary Radiation of the Mollusca* (ed. J. Taylor), Oxford University Press, Oxford, pp. 77–87.

Runnegar, B. and Bentley, C. (1982) Anatomy, ecology and affinities of the Australian Early Cambrian bivalve *Pojetaia runnegari* Jell. *Journal of Paleontology* **57**, 73–92.

Runnegar, B. and Pojeta, J. (1974) Molluscan phylogeny: the palaeontological viewpoint. *Science* **186**, 311–17.

Runnegar, B., Pojeta, J., Morris, N.J. *et al.* (1975) Biology of the Hyolitha. *Lethaia* **8**, 181–91.

Savazzi, E. (1987) Geometric and functional constraints on bivalve shell morphology. *Lethaia* **20**, 293–306.

Savazzi, E. (1990) Biological aspects of theoretical shell morphology. *Lethaia* **23**, 195–212.

Seilacher, A. (1954) Okologie der triassischen Muschel *Lima lineata* (Schloth) und ihren Epöken. *Neues Jahrbuch für Geologie und Paläontologie Monatshefte,* 163–83.

Seilacher, A. (1960) Epizoans as a key to ammonoid ecology. *Journal of Paleontology* **34**, 189–93. (See text)

Seilacher, A. (1968) Swimming habits of belemnites – recorded by boring barnacles. *Palaeogeography, Palaeoclimatology, Palaeoecology* **4**, 279–85. (Suggestion that the guard was in life exposed)

Skelton, P.W. (1976) Functional morphology of the Hippuritidae. *Lethaia* **9**, 83–100. (Perforated 'lid' allowed ingress of water currents)

Skelton, P.W. (1978) The evolution of functional design in rudists (Hippuritacea) and its taxonomic implications. *Philosophical Transactions of the Royal Society of London B* **284**, 305–18. (Geometry of growth)

Skelton, P.W. (1985) Preadaptation and evolutionary innovation in rudist bivalves, in *Evolutionary Case Histories from the Fossil Record* (eds J.C.W. Cope and P.W. Skelton), Special Papers in Palaeontology No. 33, Academic Press, London, pp. 159–173.

Skelton, P.W. and Wright, V.P. (1987) A Caribbean rudist bivalve in Oman: island hopping across the Pacific in the late Cretaceous. *Palaeontology* **30**, 505–29.

Smith, S.A., Thayer, C.W. and Brett, C.E. (1985) Predation in the Palaeozoic: gastropod-like drillholes in Devonian brachiopods. *Science* **230**, 1033–5.

Stanley, S.M. (1968) Post-Palaeozoic adaptive radiation of infaunal bivalve molluscs: a consequence of mantle fusion and siphon formation. *Journal of Paleontology* **42**, 214–29. (Radiation dependent on one key character)

Stanley, S.M. (1970) *Relation of shell form to life habits in the Bivalvia (Mollusca).* Geological Society of America Memoir No. 125. Geological Society of America, Lawrence, Kan. (Extended treatment of Recent bivalve functional morphology)

Stanley, S.M. (1972) Functional morphology and evolution of byssally attached bivalve mollusks. *Journal of Paleontology* **46**, 165–212.

Stanley, S.M. (1977) Coadaptation in Trigoniidae; a remarkable family of burrowing bivalves. *Palaeontology* **20**, 869–99. (Well worth reading)

Stenzel, H.B. (1971) Oysters. *Treatise on Invertebrate Palaeontology,* Part N, Vol. 3. Geological Society of America and University of Kansas Press, Lawrence, Kan. (A full account of morphology, evolution and ecology: deals succinctly with evolution in *Gryphaea*)

Stridsberg, S. (1981) Apertural constrictions in some oncocerid cephalopods. *Lethaia* **14**, 269–76.

Stridsberg, S. (1984) Aptychopsid plates – jaw elements or protective operculum. *Lethaia* **17**, 93–8.

Stridsberg, S. (1985) Silurian oncoccrid cephalopods from Gotland. *Fossils and Strata* **18**, 1–65.

Stridsberg, S. (1990) Orientation of cephalopod shells in illustrations. *Palaeontology* **33**, 243–48.

Stürmer, W. (1970) Soft parts of trilobites and cephalopods. *Science* **170**, 1300–2. (Soft tissues external to shell of Devonian orthocones)

Swan, R.H. (1986) Heterochronic trends in Namurian ammonoid evolution. *Palaeontology* **31**, 1033–52.

Taylor, J.D. and Layman, M. (1972) The mechanical properties of bivalve (Mollusca) shell structures. *Palaeontology* **15**, 73–87.

Taylor, J.D., Kennedy, W.J. and Hall, A. (1969) The shell structure and mineralogy of the Bivalvia. Introduction, Nuculacea–Trigoniacea. *Bulletin of the British Museum of Zoology* Suppl. **3**, 1–125. (Standard work)

Teichert, C. (1967) Major features of cephalod evolution, in *Essays in Palaeontology and Stratigraphy*, (eds C. Teichert and E.L. Yochelson), University of Kansas Press, Lawrence, Kan., pp. 161–210. (Classification used in this chapter)

Thomas, R.D.K. (1975) Functional morphology, ecology, and evolutionary conservation in the Glycymeridae (Bivalvia). *Palaeontology* **18**, 217–54.

Trueman, A.E. (1941) The ammonite body chamber, with special reference to the buoyancy and mode of life of the living ammonite. *Quarterly Journal of the Geological Society of London* **96**, 339–83. (Classic study of functional morphology of unusual shell shapes in ammonoids)

Vermeij, C.J. (1983) Traces and trends of predation, with special reference to bivalved animals. *P alaeontology* **26**, 455–66.

Vogel, K. and Gutmann, W. (1980) The derivation of pelecypods: role of biomechanics, physiology and environment. *Lethaia* **13**, 269–75.

Ward, P.D. (1979a) Cameral liquid in *Nautilus* and ammonites. *Paleobiology* **5**, 40–9.

Ward, P.D. (1979b) Functional morphology of helically coiled ammonite shells. *Paleobiology* **5**, 415–22. (Mode of life involves vertical migration)

Ward, P.D. (1980) Comparative shell-shape distributions in Jurassic–Cretaceous ammonites and Jurassic–Tertiary nautilids. *Paleobiology* **6**, 32–43.

Ward, P.D. (1996) Ammonoid extinction, in *Ammonoid Paleobiology* (eds N.H. Landman, K. Tanabe and R.A. Davis), Plenum Press, New York and London, pp. 815–24.

Ward, P. and Martin, A.W. (1978) On the buoyancy of the pearly nautilus. *Journal of Experimental Zoology* **205**, 5–12. (Essential reading).

Weedon, M. (1990) Shell structure and affinity of vermiform 'gastropods'. *Lethaia* **23**, 297–310.

Wells, M.J., Wells, J. and O'Dor, R.K. (1992) Life at low oxygen levels; the biology of *Nautilus pompilius* and extinct forms. *Journal of the Marine Biological Association of the United Kingdom* **72**, 313–28. (Habitat of *Nautilus* and ammonites)

Westermann, G. (1985) Post-mortem descent with septal implosion in Silurian nautiloids. *Paläontologisches Zeitschrift* **59**, 79–97.

Westermann, G. (1990) New developments in ecology of Jurassic-Cretaceous ammonoids. *Atti dei Secondo Convegno Internazionale Fossili, Evolutione, Ambiente, Pergola,* 25–30 October 1987, pp. 459–78.

Westermann, G. (1996) Ammonoid life and habitat, in *Ammonoid Paleobiology* (eds N.H. Landman, K. Tanabe and R.A. Davis), Plenum Press, New York and London, pp. 608–707.

Wiedmann, J. (1969) The heteromorphs and ammonoid extinction. *Biological Reviews* **44**, 563–602. (Full documentation of heteromorphy as a functional phenomenon)

Wilbur, K.M. (1961) Shell formation and regeneration, in *Physiology of Mollusca,* Vol. 1 (eds K.M. Wilbur and C.M. Yonge), Academic Press, New York, pp. 243–81. (Secretion and development of bivalve shell)

Williamson, P.G. (1981) Palaeontological documentation of speciation in Cenozoic molluscs from the Turkana Basin. *Nature* **293**, 437–43. (Allopatric speciation)

Yancey, T.E. and Boyd, D.W. (1983) Revision of the Alatoconchidae: a remarkable family of Permian bivalves. *Palaeontology* **26**, 497–520.

Yochelson, E.L. (1967) Quo vadis *Bellerophon?* in *Essays in Palaeontology and Stratigraphy* (eds C. Teichert and E.L. Yochelson), University of Kansas Press, Lawrence, Kan., pp. 141–61. (Bellerophontids may not have been gastropods)

Yochelson, E.L. (1979) Early radiation of molluscs and mollusc-like groups, in *The Origin of Major Invertebrate Groups* (ed. M.R. House), Academic Press, London, pp. 323–58.

Yochelson, E. (1981) *Fordilla troyensis* Barrande: 'The oldest known pelecypod' may not be a pelecypod. *Journal of Paleontology* **55**, 113–25.

Yochelson, E.L., Flower, R.H. and Webers, G.F. (1973) The bearing of the new later Cambrian monoplacophoran genus *Knightoconus* upon the origin of the Cephalopoda. *Lethaia* **6**, 275–310. (Early diversification of cephalopods)

9 Echinoderms

9.1 Introduction

The familiar starfish and sea urchins which are so common in shallow waters are representative of an entirely marine phylum, the Echinodermata (Cam.–Rec.), which stands apart zoologically from nearly all other invertebrate groups. Echinoderms of all kinds have internal **mesodermal skeletons** of porous calcite plates, which are normally spiny and covered outside and in by a thin protoplasmic skin. Normally the skeletons have a five-rayed or **pentameral symmetry**, though in some fossil groups this is not so and in some modern and fossil sea urchins a bilateral symmetry is superimposed upon the radial plan. Another important feature of echinoderms is the **water–vascular system**: a complex internal apparatus of tubes and bladders containing fluid. This has extensions which emerge through the skeleton to the outside as the **tube feet** or **podia**. Tube feet have various functions, especially locomotion, respiration and feeding. They may be considered as all-purpose organs used by animals living inside a calcite box for manipulating the environment.

Because of their calcitic skeleton echinoderms are very abundant in the fossil record, and often their remains have greatly contributed to carbonate sediments. Thus crinoidal limestones, composed largely of the stem fragments of sea lilies, are very common in some rocks, notably the Carboniferous. In such rocks the porous plates have often been impregnated with diagenetic calcium carbonate but are otherwise unchanged. Echinoderms are, however, stenohaline, and their remains are only found in sediments of fully marine origin.

9.2 Classification

A modern system of echinoderm classification, as defined in the *Treatise on Invertebrate Paleontology* and modified by Sprinkle (1976) is as follows.

SUBPHYLUM 1. ECHINOZOA: Radiate echinoderms, usually globose or discoidal.
- CLASS 1. ECHINOIDEA (Ord.–Rec.): Sea urchins.
- CLASS 2. HOLOTHUROIDEA (Ord.–Rec.): Sea cucumbers.
- CLASS 3. EDRIOASTEROIDEA (L. Cam.–Carb.)

There are several other extinct classes.

SUBPHYLUM 2. ASTEROZOA.
- CLASS STELLEROIDEA (Ord.–Rec.).
  - SUBCLASS 1. ASTEROIDEIA (Ord.–Rec.): Starfish.
  - SUBCLASS 2. OPHIUROIDEA (Ord.–Rec.): Brittle stars.
  - SUBCLASS 3. SOMASTEROIDEA (Ord.–Rec.).

SUBPHYLUM 3. CRINOZOA: 'Pelmatozoans', i.e. echinoderms having a small plated body (calyx) fixed by a stem, and with pinnulate arms adapted for food gathering.
- CLASS CRINOIDEA (M. Cam.–Rec.): Sea-lilies.

SUBPHYLUM 4. BLASTOZOA: 'Pelmatozoans', often stalked, lacking free arms, but with biserial **brachioles** for food gathering, and often various respiratory structures in the cup.
- CLASSES 1 and 2. DIPLOPORITA and RHOMBIFERA (?Cam.–Dev.): 'Cystoids' – extinct groups with perforated plates in the calyx.
- CLASS 3. BLASTOIDEA (Sil.–Perm.): Extinct 'pelmatozoans' with complex respiratory structures.
- CLASS 4. EOCRINOIDEA (L. Cam.–Sil.): Primitive echinoderms with pores along the sutures.

SUBPHYLUM 5. HOMALOZOA (Ord.): Rare peculiar organisms, calcite plated but with no planes of symmetry. These have been the subject of much controversy and may be a separate chordate subphylum on their own: the CALCICHORDATA.

This is a useful and workable classification and is retained for convenience here. Smith (1984b), however, proposes an alternative one which has been derived cladistically. In his estimation, the view of the early taxonomists, that there is a natural division between fixed (pelmatozoan) and free-living (eleutherozoan) echinoderms is phylogenetically correct, and that the two are sister groups. In consequence he proposed two subphyla, Pelmatozoa (crinoids, etc.) and Eleutherozoa (starfish, sea urchins and sea cucumbers). Many details remain to be worked out, however, and here I use the same classification as in the second and third editions of this text. Some further comments are made in section 9.8. Besides the echinoderm groups noted here there are several others. In particular there was a marked proliferation of short-lived echinoderm classes in the Lower Palaeozoic. Some of these seem to combine characters typical of many groups and are hard to classify. They are often known only from a single locality and a small number of specimens. But though these are clearly echinoderms, their characters are so different from those of other known echinoderms that separate classes (e.g. the echinozoan classes Helicoplacoidea, Cyclocystoidea and Lepidocystoidea) have had to be established to accommodate them. Though there are some Cambrian echinoderms, the epoch of their maximum proliferation (at class level) was the Ordovician, as can be seen from their times of origin in the above list. At generic level, however, echinoderms were most abundant in the Carboniferous. A whole range of remarkable forms arose at that time in a great burst of adaptive radiation, but only some of these were successful; the others, to which the Cambrian classes belong, produced no new lines of descent and became extinct.

9.3 Subphylum Echinozoa

Class Echinoidea

Morphology and life habits of three genera

Echinus

The common sea urchin, *Echinus esculentus* (Order Echinoida), which lives in shallow waters around the North Atlantic, shows the fundamentals of echinoderm structure as organized in an animal well

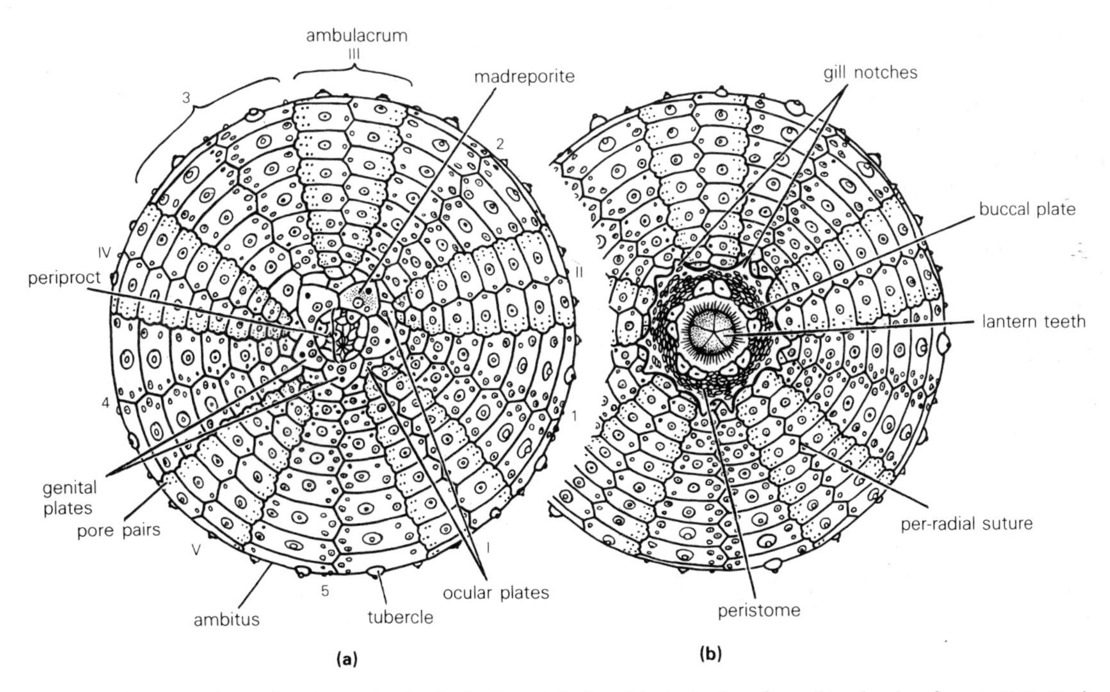

Figure 9.1 *Echinus esculentus* (Rec.), test deprived of spines: (a) aboral (adapical) surface: (b) adoral surface (× 0.8). (Redrawn from Durham in *Treatise on Invertebrate Paleontology*, Part U.)

adapted for a benthic free-living mode of life (Fig. 9.1a,b).

Echinus has a globular **test** some 10 cm in diameter which is slightly flattened at the poles. In life it is covered with short (1–2 cm) spines; if these are removed the plating structure is visible. On the upper (**aboral** or **adapical**) surface there is a central **apical disc**: a double ring of plates surrounding a central hole or **periproct**, which contains the anus. The apical disc is formed of two types of plates: the larger **genital plates**, and the smaller **ocular plates** which are usually outside the ring of genitals. Each is perforated by a pore. The **genital pores** are the outlets of the **gonads**, and the **ocular pores** are part of the water–vascular system. One genital plate (the **madreporite**) is larger than the others. It has numerous tiny perforations which lead into the water–vascular system below. The anus resides in the centre of a number of small plates attached to a flexible, rarely fossilized membrane extending across the periproct.

The test is divided into ten radial segments extending from the apical disc to the **peristome** (q.v.) which surrounds the mouth on the lower (**adoral**) surface. The five narrower segments are the **ambulacra** (ambs) which connect with the ocular plates, whereas the broader **interambulacra** (interambs) terminate against the genital plates. Both ambulacra and interambulacra consist of double columns of elongated plates which meet along a central suture in a zigzag pattern. In the ambulacrum this is the **per-radial suture**. The interambulacral plates are large and tubercular, without perforations, but the ambulacral plates each have three sets of paired pores near the outer edge of the plate. These **pore pairs** are the sites where the tube feet emerge through the test from the internal part of the water–vascular system.

The ambulacra and interambulacra are widest at the **ambitus**, which is the edge of the specimen when seen from above or below. The peristome is a large adoral area, covered in life by a flexible plated membrane, which contains the mouth centrally. In fossil specimens, however, the membrane has normally gone, leaving a large circular or pentagonal cavity. Five pairs of **gill notches** are found where the interambulacra abut the edge of the peristome, and from these project feathery bunches of gills which provide surfaces for respiratory exchange additional to those of the tube feet. Inside the peristome the test is turned back into a perforated flange which is the **perignathic girdle** (Fig. 9.2).

This girdle forms a support for the masticatory apparatus of the echinoid: the **Aristotle's lantern** (Figs 9.2, 9.25). This lantern has five strong jaws, each with a single calcitic tooth. The whole assembly is suspended by ligaments and muscles attached to the perignathic girdle. It operates as a kind of five-jawed grab and, though each jaw has only limited play, the teeth can rasp away at organic detritus or algal material on the sea floor and pass it inwards to the gut. Within the test (Fig. 9.2) most of the soft parts are related to the structures already described. Inside the test is a thin layer of protoplasm, and since the gut is only a simple tube running spirally round the inner wall from mouth to anus, the body of the test is largely empty. However, at breeding time, which is normally in the summer, the gonads swell enormously before releasing their products through the genital pores. Since echinoids often live in clumps or congregate to spawn, the chances of cross-fertilization of eggs and sperm from male and female individuals are fair. The ciliated **echinopluteus larvae** which grow from the zygotes swim actively in the plankton and undergo many transformations before finally settling down.

The **coelom** has various tubular elements. Of these the **haemal and perihaemal** systems seem to be involved in material transfer, and the **axial organ** seems to be associated with the repair of injury, but their specific functions are unclear.

The water–vascular system of the echinoderms,

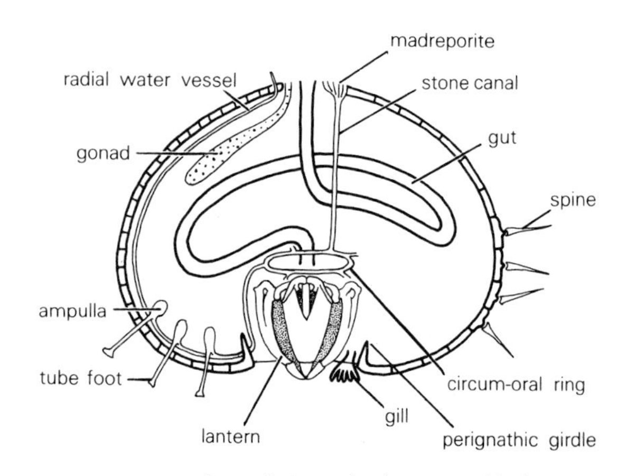

Figure 9.2 Internal morphology of *Echinus* (simplified) passing through an ambulacrum (left side) and interambulacrum (right side).

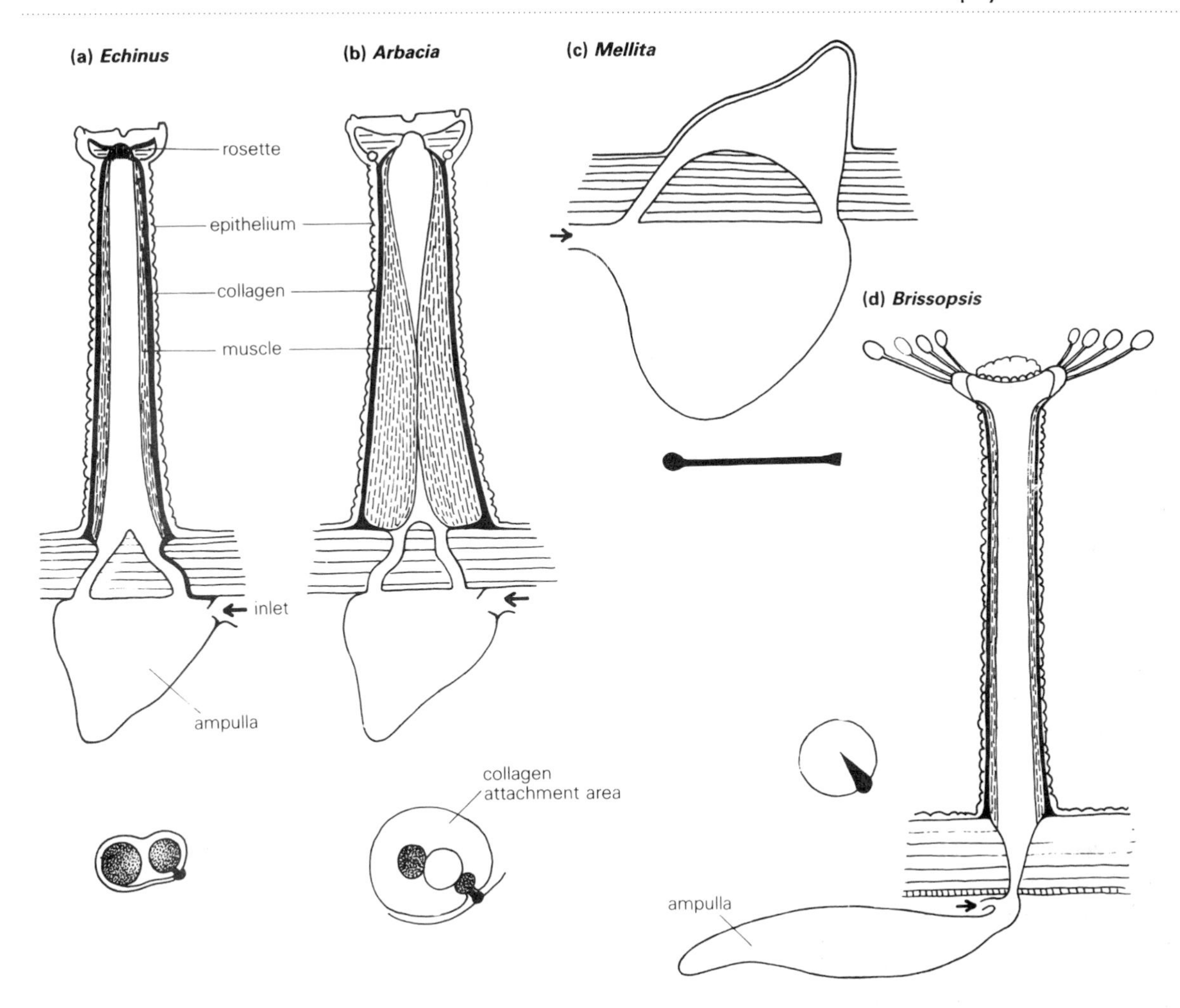

Figure 9.3 Structure of echinoid tube feet: (a) *Echinus* tube foot (with sucking disc) and pore pair; (b) *Arbacia* muscular-walled tube foot with large sucking disc and pore pair with expanded attachment area for collagen fibres: (c) adapical respiratory tube foot and slit-like pore pair: (d) sub-anal tube foot of *Brissopsis* with 'flue brush' and associated pore. (All redrawn from Smith, 1978, 1980.)

which is also coelomic, resembles nothing else in the animal kingdom. Its primary function is to operate the tube feet. In echinoids its only exit from the test is via the madreporite. From this a calcified tube (the **stone canal**) descends to near the top of the lantern. Here it joins the **circumoral ring**, from which five **radial water vessels** extend, one running up the centre of each ambulacrum. Each of these passes finally through an ocular pore, but it only forms a tiny closed tube (apparently light sensitive in some echinoids). From the radial water vessels there arise, at intervals, paired lateral tubes, each leading to a tube foot and associated apparatus (Fig. 9.3).

At the base of each tube foot is an inflatable sac (the **ampulla**), and the tube foot leads outwards from this, dividing as it passes through the pore pair and reforming on the other side. This device prevents the tube foot from being withdrawn right inside the test when retracted and, since one of the functions of the tube foot is respiration, it also separates incoming oxygen-rich water from the outgoing fluid depleted in oxygen. The tube foot possesses longitudinal muscles and has a suction cup at the end, rendering it prehensile. Within the water–vascular, haemal and perihaemal systems are many amoeboid cells (**coelomocytes**) which perform numerous functions.

Echinus moves by using its tube feet, especially those on the lower part of the body. It can extend the elastic tube feet for a considerable distance, approximately half of those in any one ambulacrum being extended at a given time, the other half retracted. A tube foot will extend when water pressure within it increases due to contraction of the ampulla, the radial water vessel itself or a neighbouring tube foot. As water comes into it the longitudinal muscles relax; when these contract water is forced back into the ampulla which correspondingly relaxes. At the time of maximum extension, the suction cups on the end of the extended tube feet adhere to an adjacent part of the sea floor. When the tube feet contract, the echinoid moves along the sea floor, supported by spines. *Echinus* moves slowly over the sea floor in this way, feeding voraciously with its jaws and defended against predators by its armament of spines. Each spine, like the individual plates of echinoids, is a single crystal of calcite. The spine base forms a socket which articulates with the ball joint of the tubercle below it; the tubercle is crystallographically continuous with the plate (Figs 9.4a–c, 9.23). Round the tubercle is a ring of muscles, attached to the spine base so that the spine can be moved in any direction.

Amongst the spines are small organs of balance (**spheridia**), on the adoral part of the per-radial suture, and **ophicephalous pedicellariae**, which are tiny spines with their heads modified as pincers (some with poison glands); these clean the surface, discourage predators and prevent larvae from settling (Fig. 9.4d). Normally pedicellariae lie recumbent on the surface, but they can be erected and will snap shut on any extraneous object. New pedicellariae are formed when any are dislodged in defence.

Echinus is a **regular** echinoid: one in which the periproct opens in the centre of the apical disc (**endocyclic**). Such regular echinoids are common today and in the fossil record, and they live either on the sea floor or, like *Strongylocentrotus*, in cavities in rocks which they may have excavated themselves.

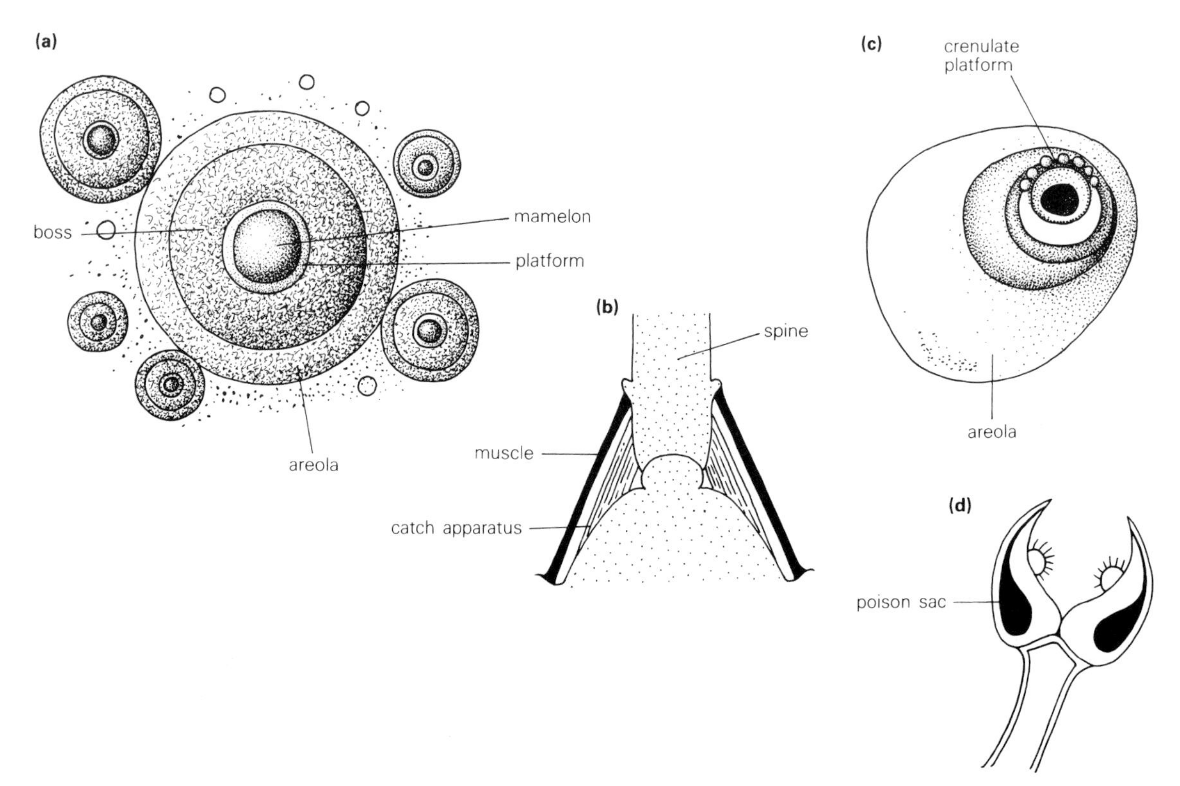

Figure 9.4 Structure of echinoid spines and tubercles: (a) surface view of standard tubercles and (b) vertical section through same type of tubercle and spine base with muscles; (c) asymmetrical tubercle with crenulate platform and wide area of muscle attachment at left; (d) opened ophicephalous pedicellaria with sensory hairs and poison sac. [(a)–(c) Based on Smith, 1980.]

Regular echinoids are normally illustrated according to a conventional orientation (Figs 9.1a, 9.19a,b). The madreporite is always shown at the right anterior and with its dependent interambulacrum is numbered 2. The numbering proceeds anticlockwise (as seen from the adapical pole) so that genital 5 is always posterior. Roman numerals designate the oculars and ambulacra, likewise numbered anticlockwise, but starting to the right of the genitals. The same system is used in numbering the plates of **irregular** echinoids: those with a dominant bilateral symmetry, marked particularly by the position of the periproct which is no longer within the apical system (**exocyclic**). *Echinocardium* and *Mellita,* described below, are two very dissimilar irregular echinoids, with different life habits.

Echinocardium (Fig. 9.5)

Echinocardium (Order Spatangoida) is abundant in shallow water, but unlike *Echinus* lives in a burrow within the sediment. Its morphology is highly modified in accordance with this burrowing mode of life. Living below the surface gives it protection against predators, but it has to maintain an adequate connection with the surface for food supply and respiratory exchange, and it also has to be able to cope with sanitation. It has, in fact, essentially the same problems of life as a subsurface-living bivalve. These problems have been solved mainly by reshaping of the test and by extreme modification of the tube feet and spines in different ways for performing different jobs.

In *Echinocardium* a bilateral symmetry is superimposed upon the radial symmetry. The test is covered with a mat of short spines, but when these are removed it is seen as heart-shaped in plan, a flattened ellipse in profile. The aboral surface possesses an elongate apical disc from which the periproct is absent; the latter is located on the nearly vertical posterior wall of the test. A single ambulacrum (III), dissimilar to the others, is located in a deep anterior groove and goes straight towards the peristome. The other ambulacra are paired (II + IV; I + V). Each of these is in two parts. The aboral parts are expanded into four recessed leaf-like '**petals**', which terminate above the ambitus. The ambulacra continue below this level but are of more normal form, flush with the surface and less pronounced. In the petals the outer pore of each pore pair is elongated, slit-like and widely separated from the round inner pore.

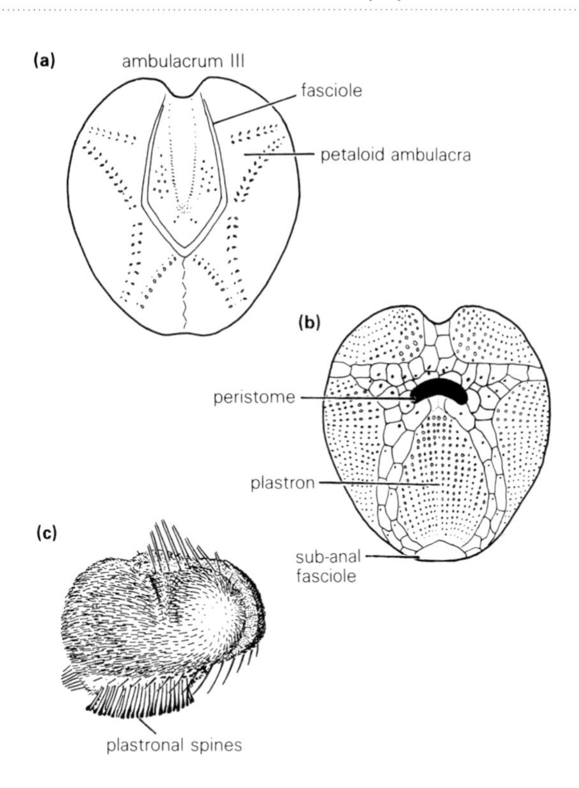

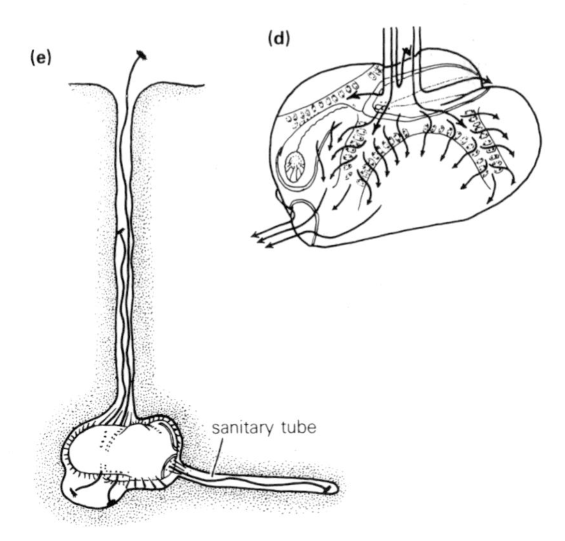

Figure 9.5 *Echinocardium cordatum* (Rec.): (a) aboral view (×0.8); (b) adoral view; (c) lateral view with all spines present; (d) oblique lateral view showing current directions; (e) in life position in burrow. (After Nichols, 1959.)

From these emerge flattened respiratory tube feet, leaf-like and rectangular. Elsewhere, including within the anterior ambulacrum, the pore pairs are more normal, though they may carry tube feet specialized for other functions. Adorally (Fig. 9.5b) the peristome is located far forwards and has a projecting lip (or **labrum**) below. There is no lantern. Adorally the plates are enlarged, and from their large pore pairs emerge sticky food-gathering tube feet. Behind the mouth is a flattened area (the **plastron**) formed from the modified posterior interambulacrum and densely covered with flat, paddle-shaped spines.

In life *Echinocardium* lives in a burrow up to 18 cm deep, with a single funnel connecting it with the surface (Fig. 9.5e). This funnel is both created and maintained by enormously long tube feet with star-shaped ends resembling flue brushes. These emerge from the adapical part of the anterior ambulacrum, their bases protected by a pyramid of spines which help in building the lower part of the funnel. A set of similar tube feet, emerging from non-petaloid regions of the two posterior ambulacra below the vertical rear wall, build a single 'sanitary tube' to receive excreta.

Two regions of the test known as **fascioles** generate currents. The larger fasciole surrounds the anterior ambulacrum as a ribbon-like strip; the smaller (**sub-anal**) fasciole is an elliptical ribbon located below the anus. In both the surface is covered with small vertical spines (the **clavulae**), each of which is covered by innumerable cilia, as in the intervening epithelium; it is the coordinated beat of these cilia that produces the currents whose direction is shown in Fig. 9.5d. A strong current goes down the anterior ambulacrum, and food particles are caught by the sticky tube feet and passed to the mouth. Other currents bathe each of the paired ambulacra, facilitating respiratory exchange, and the current from the subanal fasciole propels waste matter into the sanitary tube. When the sanitary tube is filled up *Echinocardium* moves forwards. The funnel-building spines are withdrawn, the anterior spines are erected and scrape away at the front wall of the burrow, whilst the paddle-like spines attached to the plastron move the echinoid forwards. A new funnel is created while the old funnel, burrow and sanitary tube collapse behind it.

The whole organism has the same basic elements as *Echinus*, apart from the lantern and girdle which appear only in the embryo and are soon lost through being resorbed. Other than the modified shape of the test, it is mainly a division of labour between the spines and tube feet that enables *Echinocardium* to live far below the surface while feeding, respiring and excreting effectively.

Specific bioturbation structures in the sediment, often of complex form, result from the forward movement of heart urchins such as *Echinocardium*. These are clearly recognizable and testify to the former presence of sea urchins moving within the sediment even if the animals themselves have not been fossilized.

Mellita (Fig. 9.6a,b)

Mellita quinquiesperforata (Order Clypeasteroida) is a flattened sand dollar common in littoral and sublittoral sands in southeastern North America. It lives either on the surface or buried horizontally within it. It is very flat with its ambulacra petaloid aborally and with five perforations (**lunules**) in the test: two pairs in the paired ambulacra, the other being unpaired. Adorally the peristome is central, with the anus very close behind it and just in front of the posterior lunule. On the adoral surface are five dichotomous channels (the **food grooves**) which are lined with tube feet. They run between the lunules and converge on the mouth. The whole test is covered with a dense mat of fine spines, of which the adoral ones are used for walking.

It usually lives horizontally, burrowing just below the surface and covered by a thin layer of sand.

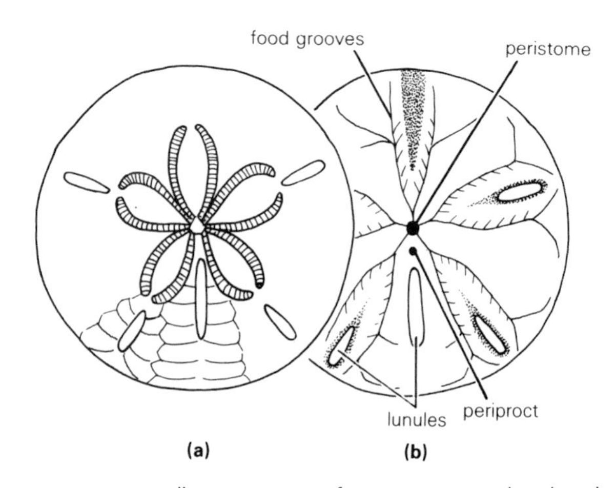

Figure 9.6 Mellita quinquiesperforata (Rec.): (a) aboral and (b) adoral views (× 1.5.)

These echinoids feed as they move through the sediment by sieving out fine detritus from the sand layer above. The food particles fall between the dense spine canopy, and are carried to the test margin. There food is picked up by the tube feet of the oral surface and carried in mucus strings to the mouth, assisted by both tube feet and tiny spines. Long oral spines form a protective mesh over the mouth; they screen out sand but help in the ingestion of organic detritus. *Mellita* normally travels forward but it can also rotate and right itself if overturned. As it moves it makes a characteristic trail through the thin veneer of sand which overlies it, and it sometimes passes sand upwards through the posterior (anal) lunule as it progresses forwards.

The ambulacral lunules, according to Smith and Ghiold (1982) have one primary function. They increase the perimeter of the test available for food gathering, thus acting as short cuts for transferring food from the aboral surface to the mouth. They may also act as hydrodynamic stabilizers limiting the tendency for sand dollars to be flipped over by sudden currents.

The anal lunule seems to serve a different function, and forms in juveniles by resorption rather than initially by indentation of the shell margin, as do the ambulacral lunules. Its role may be to act as an outlet for the excess water drawn into the mouth by ciliary feeding currents, without disrupting the flow of the feeding currents themselves. It may serve a subsidiary function as a passageway for excreta passed through to the aboral surface and left behind as *Mellita* progresses forwards.

M. sexiesperforata has six lunules but lives in a very similar manner. *Dendraster*, another sand dollar which is non-lunulate, is a filter feeder. It lives at a high angle to the substrate, with most of the posterior part showing above the surface.

Some species of *Mellita* and other sand dollars living in the surf zone ingest magnetite from the sand, and they distribute this within the body as a 'weight belt' which gives stability.

Sand dollars, like other highly modified irregular echinoids, do not grow in the same way as *Echinus* by continued secretion of plates from the apical disc. The plates are all formed in the early stages of ontogeny and become very rigidly united through interlocking sutures and internal pillar supports.

Sand dollars originated from clypeasteroids in the early Tertiary, and their success and rapid spread was a consequence of their mode of feeding. The differentiation of adapical 'sieving' spines and adoral locomotory spines, the flat-based sharp-edged test, the rigidity given by the sutural interlocking of the plates, the adoral branched food-grooves and the lunular perforations through the test make up a specialized and highly effective innovation in functional design (Seilacher, 1979, 1990; Kier, 1982). However, this seems to have originated in the Eocene at least four times over – a testament to the kind of adaptive breakthroughs possible from the echinoid 'Bauplan', but it would only work if the many component elements are successively incorporated and fully coordinated.

Classification

The taxonomic division of the Class Echinoidea into Regularia and Irregularia, originally proposed in 1925, was used until the 1950s. Nevertheless, authorities such as Mortensen (1928–1952), who continued to employ it, did not necessarily believe it to be a truly phyletic classification, but simply a convenient and provisional one. The classification erected by Durham and Melville in the *Treatise on Invertebrate Paleontology*, later amended (Durham, 1966), and used in the first edition of this book, was based upon a wide variety of stable characters defining major groupings, many of which are still valid. These characters included the overall structure and rigidity of the test, the number of ambulacral and interambulacral columns, and the morphology of the plates, lantern and girdle.

A critical and debatable issue in echinoid classification is whether irregularity had arisen once only or whether the irregular echinoids are polyphyletic. Smith's (1984b) classification, based partially on new data on lantern and tooth structure, and expanding the ideas of Markel, favours the view that irregular echinoids are monophyletic, and in consequence he revives the old group Irregularia. Smith's scheme, which is partially cladistic, is adopted here. It must be appreciated, however, that classifications generally are in a state of flux, and this may well change over the next few years.

SUBCLASS 1. PERISCHOECHINOIDEA (Ord.–Rec.): Regular endocyclic echinoids with interambulacra in many columns; ambulacra in 2–20 columns; no compound plates. Perignathic girdle simple or absent and lantern with simple grooved teeth. Includes all Palaeozoic echinoids except the cidaroids. This is a difficult group to classify: it includes five groups: (1) lepidocentrids, the earliest echinoids such as *Aulechinus*, with two columns of ambulacral plates and many imbricating ambulacral plates; (2) echinocystidids, with at least four columns of plates in each ambulacrum (*Echinocystites*, *Cravenechinus*); (3) lepidesthids, with very many plates in each ambulacrum (*Lepidesthes*), a character shared by (4) proterocidarids, but in the latter the ambulacra are much expanded on the adoral surface (*Proterocidaris*); (5) palaechinoids with polygonal ambulacral plates with small secondary tubercles (*Palaechinus*, *Melonites*).

SUBCLASS 2. CIDAROIDEA (Dev.–Rec.): Regular endocyclic echinoids with only two columns of plates in the ambulacra and in the interambulacra of later genera. Each of the interambulacral plates are ornamented with a longer single central tubercle, supporting a massive solid spine. The upright lantern has no foramen magnum, and the perignathic girdle has apophyses only. Examples include *Cidaris*, *Archaeocidaris*, *Miocidaris*.

SUBCLASS 3. EUECHINOIDEA (U. Trias.–Rec.): This group includes all post-Palaeozoic echinoids, regular or irregular, with bicolumnar ambulacra and interambulacra. The perignathic girdle has both auricles and apophyses and the lantern may be secondarily lost in some groups. This subclass is so large that it has been divided into extra divisions which come between subclass and order. There are two **infraclasses**:

INFRACLASS 1. ECHINOTHURIOIDEA: Deep-sea echinoids with a flexible test and a unique kind of pseudo-compounding of the ambulacral plates (*Echinothuria*).

INFRACLASS 2. ACROECHINOIDEA: Echinoids with an upright lantern having a deep V-shaped foramen magnum, compound ambulacra, and peristomial tube feet. This infraclass includes three **cohorts**:

COHORT 1. DIADEMATACEA: Retains many primitive features such as regularity and grooved teeth in a rather cidaroid-like lantern. Shallow-water species include coral reef inhabitants with poison spines. There are three orders. Examples include *Diadema*.

COHORT 2. ECHINACEA: All have keeled teeth, solid spines and sutured plates, all of which are advanced features. They are regular with gill slits, compound plates and a complex perignathic girdle. There are two orders (1) STIRODONTA (e.g. *Acrosalenia*, *Hemicidaris*, *Stomechinus*, *Arbacia*, *Phymosoma*) and (2) CAMARODONTA (e.g. *Echinus*, *Strongylocentrotus*), distinguished on the structure of the lantern.

COHORT 3. IRREGULARIA: According to Smith the irregular echinoids were monophyletic, and from such 'intermediate' genera as *Plesiechinus* and *Pygaster* evolved to the great diversity of nine orders grouped in three superorders. All Irregularia (other than those that have now lost the lantern) have diamond-shaped or wedge-like teeth.

SUPERORDER 1. EOGNATHOSTOMATA: Distinguished on the structure of the lantern (Fig. 9.25); have simple unspecialized spines and tube feet. In Order Pygasteroidea the periproct is keyhole-shaped and halfway out of the apical disc (*Plesiechinus*, *Pygaster*). Order Holectypoida have a fully disjunct periproct, but otherwise remain quite simple (*Holectypus*).

SUPERORDER 2. MICROSTOMATA: The lantern is often lost in the adult, but if present is wholly internal; have a fully disjunct periproct, and a small peristome. Tubercles on the adoral surface are arranged for unidirectional locomotion.

Neognathostomata have a rounded, pentagonal or elliptical outline. They retain the lantern, at least in the juveniles. They include four orders, particularly (1) ORDER CASSIDULOIDA (*Clypeus*, *Nucleolites*, *Cassidulus*), which are subglobular with phyllodes and bourrelets, and normally lose the lantern in the adult; (2) ORDER CLYPEASTEROIDA, which comprises flattened or vaulted echinoids including the sand dollars (they are the most recent order of echinoids, first appearing in the Paleocene; e.g. *Clypeaster*, *Mellita*, *Rotula*, *Scutellum*).

Atelostomata are heart-shaped urchins which have lost all trace of the lantern. There are three orders: (1) Order Disasteroida, with a split apical system (*Disaster*, *Collyrites*); (2) Order Holasteroida, with an elongated apical disc (*Holaster*, *Pourtalesia*) and many 'sub-spatangoid' features; (3) Order Spatangoida, with a compact apical system and characters exemplified by *Echinocardium* and *Micraster*.

The time ranges of all taxa are given in Fig. 9.7.

In the following discussion of morphology, function and habit the three primary echinozoan subclasses, the Perischoechinoidea (primitive), Cidaroidea and Euechinoidea (advanced), are taken separately.

Subclass Perischoechinoidea

Perischoechinoids (Ord.–Rec.) are the primitive echinoid stock which includes all extinct Palaeozoic echinoids. Only 37 genera and about 125 species of Palaeozoic echinoids are currently known. Yet since their preservation may be good, phylogeny and evolution have been studied in detail, especially in the marathon works of Jackson (1912) and Kier (1965).

Palaeozoic echinoids, including cidaroids, differ from their later counterparts in several respects.

Many of them have flexible tests, often of large size and with the plates not rigidly united, so that they are often found in a collapsed state. In some cases the plates are thick and have bevelled edges so that they can slide over one another at the margins.

Either the ambulacral or interambulacral columns

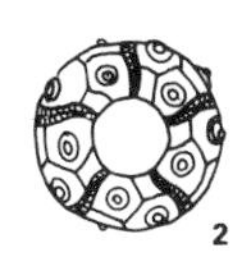

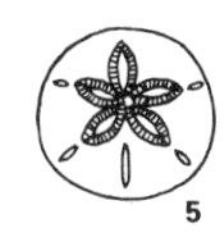

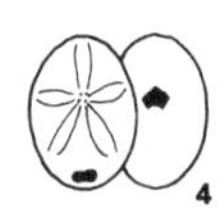

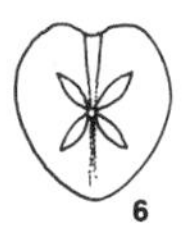

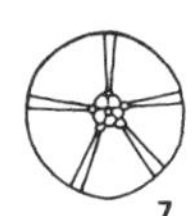

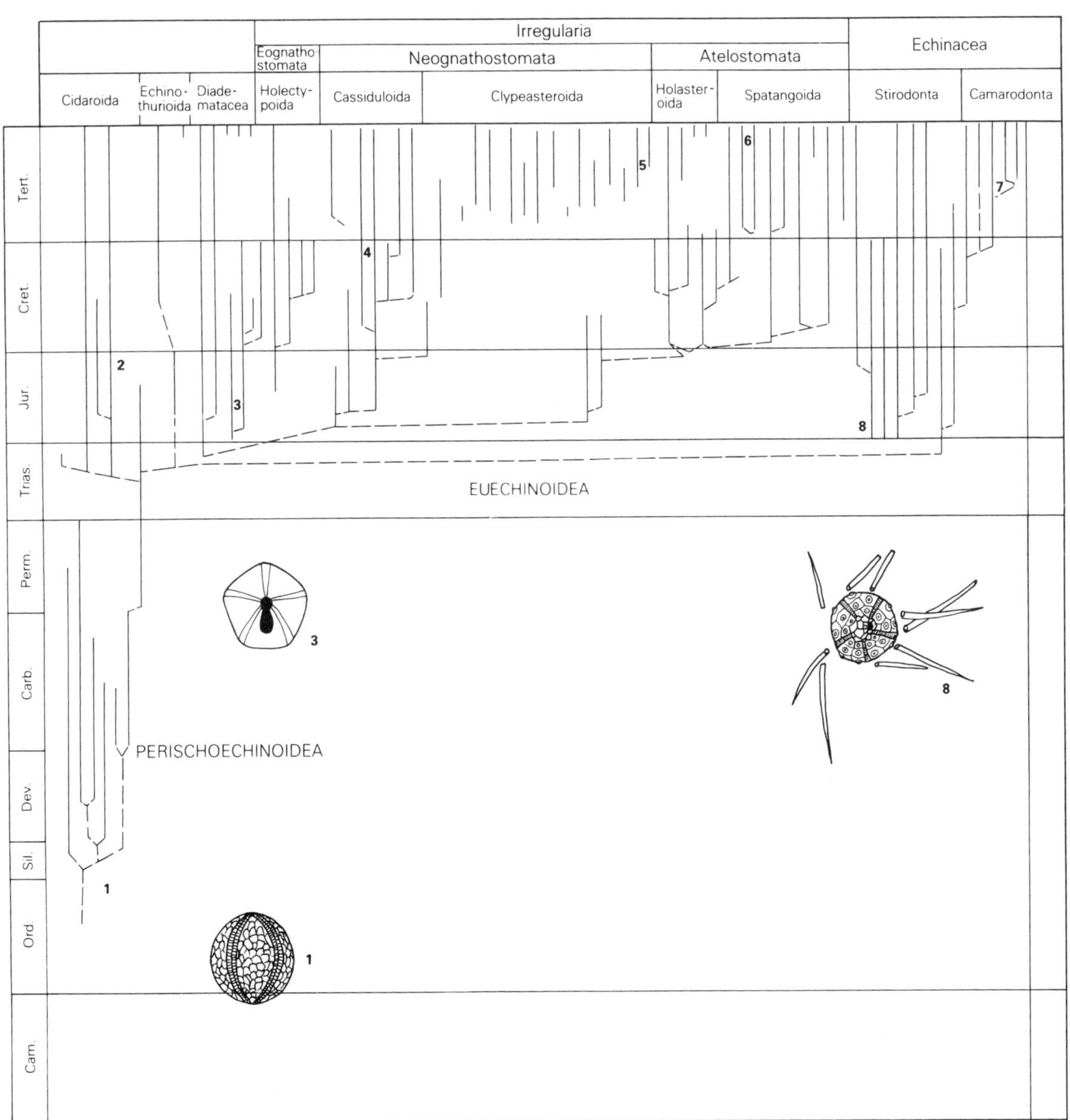

Figure 9.7 Time ranges and Smith's (1984b) classification of echinoids, illustrating some representative genera. Each vertical bar represents an order. 1, *Aulechinus*; 2, *Cidaris* (Cidaroida); 3, *Pygaster* (Eognathostomata); 4, *Catopygus* (Cassiduloida); 5 *Mellita* (Clypeasteroida); 6, *Echinocardium* (Spatangoida); 7, *Echinus* (Camarodonta); 8, *Acrosalenia* (Stirodonta).

or both may consist of many columns of plates, but compounding of the ambulacral plates is unknown; that is, there is never more than one pore pair per column.
The perignathic girdle is normally absent or of simple construction, and the lantern is flattish and less elaborate than that of later echinoids.
The test is usually globular and invariably regular, with the periproct in the centre of the apical disc (**endocyclic**). Only a few flattened genera are known.
Some of the earlier genera have the radial water vessel enclosed internally by inwardly projecting flanges from the ambulacral plates.

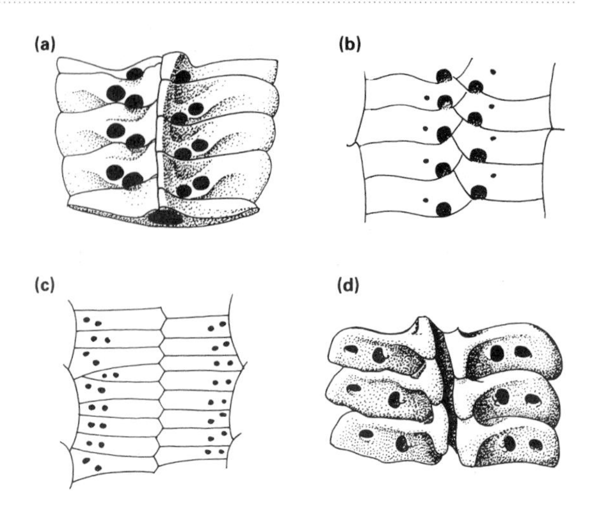

Figure 9.9 Portions of the ambulacra of perischoechinoids: (a) *Aptilechinus* (Sil.), internal view; (b) *Ectenechinus* (Ord.), with pores near the per-radial suture; (c) *Pholidechinus* (L. Carb.) with pores remote from the per-radial suture; (d) *Hyattechinus* (L. Carb.), internal view, with partial enclosure of the radial water vessel. (Redrawn from Kier, 1973, 1974; not to scale.)

Echinoids seem to have been a fairly clear-cut group from the beginning, but the presence of a lantern of rather echinoid-like appearance in Class Ophiocistitoidea, and the similarity in jaw structure between echinoids and ophiuroids, makes the distinction less complete than was formerly thought.

Palaeozoic echinoids seem to have been a relatively small and unimportant component of the Palaeozoic biota and were probably environmentally restricted. The flexible lepidocentrids include the very earliest true echinoids: the Upper Ordovician *Aulechinus* and *Ectenechinus*. *Aulechinus* (Fig. 9.8) has only two columns of plates in the ambulacra.

This is characteristic of the primitive Family Lepidocentridae to which it belongs. The ambulacral plates are curious in that the per-radial suture is situated in a deep groove with the single unpaired pore on the aboral margin of each plate close to it. Furthermore, the radial water vessel was enclosed by a tubular covering arising from the lower surface of the ambulacral plates; evidently the ampulla lay below this. The related *Ectenechinus* (Fig. 9.9b) has paired pores, though one pore is smaller than the other, and in later lepidocentrid genera, e.g. the Silurian *Aptilechinus* (Fig. 9.9a), the pore pairs are identical though still close to the per-radial suture.

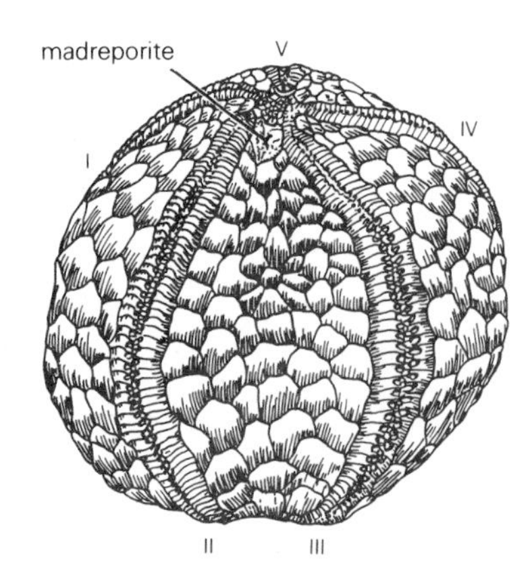

Figure 9.8 *Aulechinus* (Ord.), lateral view (×2). (Redrawn from MacBride and Spencer in *Treatise on Invertebrate Paleontology*, Part U.)

The apical system of *Aulechinus* and other lepidocentrids is peculiar in that there is only one genital plate with a single pore, though there are five oculars. Jaws and other parts of the lantern have been found, but the lantern is much less complex than in the euechinoids. In these early echinoids the interambulacra are smooth or have only small tubercles, but *Aptilechinus* has large spines on the ambulacra (Kier, 1973).

The lepidocentrids lasted until the Carboniferous but gave rise to another family, the Echinocystitidae, in which the ambulacra have more than two columns of plates and primary tubercles on the interambulacral plates. The number of interambulacral columns ranges from just one in *Cravenechinus* to 32 in *Myriastiches*. In the Lepidesthidae there are again many columns of ambulacral plates, and in the

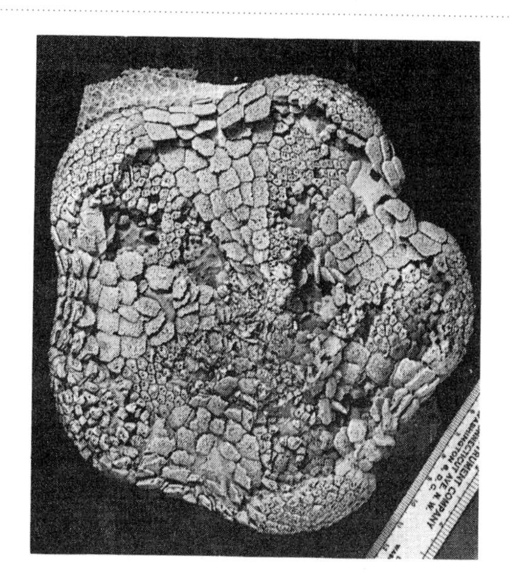

Figure 9.10 *Proterocidaris belli*, a Carboniferous echinoid, aboral view (× 0.5 approx.) – see also Fig. 9.11. (Photograph reproduced by courtesy of P.M. Kier.)

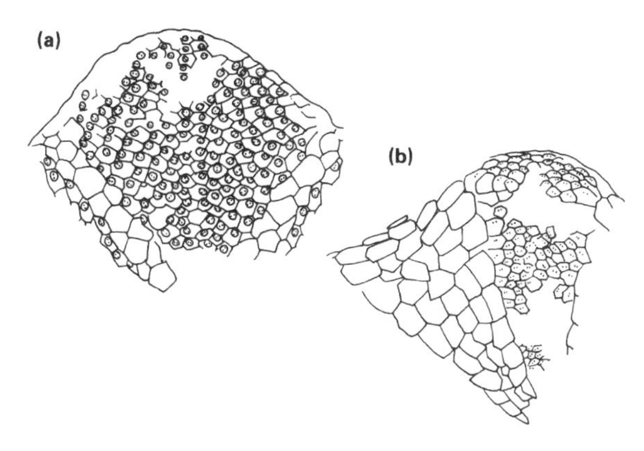

Figure 9.11 *Proterocidaris belli* (L. Carb.; Echinocystoida): (a) showing expansion of ambulacra adorally; (b) and then contraction adapically (× 0.5 approx.). (Drawings of top of specimen shown in Fig. 9.10, from Kier, 1965.)

related Proterocidaridae (Figs 9.10, 9.11) the ambulacra became greatly expanded on the adoral surface.

They consist of great numbers of small plates, each with a single pore pair in a broad, round depression. Almost certainly these were the sites of muscular suckered podia which could anchor the echinoid securely to the sea floor.

In the Palaechinidae (e.g. *Maccoya*, *Palaechinus*, *Melonechinus*), probably derived from the Echinocystitidae, there were many columns of nor-

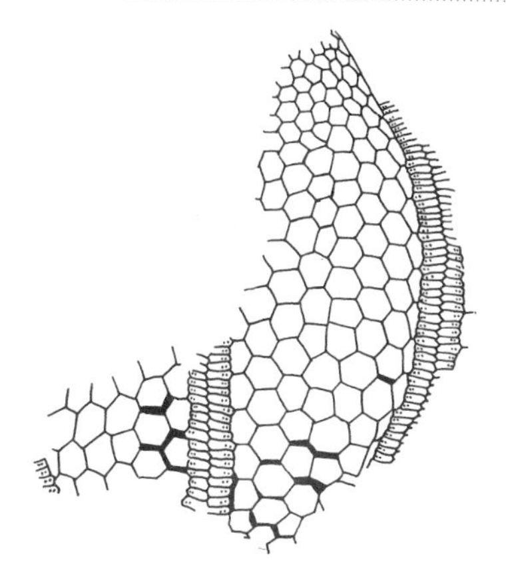

Figure 9.12 *Palaechinus merriami* (L. Carb.; Palaechinoida), preserved fragment of test in lateral view (× 2 approx.). (Redrawn from Kier, 1965.)

mally thick hexagonal plates arranged in more or less vertical columns (Fig. 9.12).

Various evolutionary trends have been noted within the echinoids of the Palaeozoic. Some, such as the flattening of the test and adoral expansion of the ambulacra, are found in Order Echinocystoida alone. Other trends are found in more advanced Palaeozoic echinoids, such as a general increase in the size and number of ambulacra; an increase in the complexity of the lantern; the development of regularity in the interambulacral plates; and the loss of the enclosure of the radial water vessel. Most changes were allometric, but some aristogenetic change is seen, i.e. in the increased complexity of the lantern and in the development of tubercles and spines for the first time.

Subclass Cidaroidea

The cidaroids are the only echinoid group to survive the Palaeozoic and they still persist today. They formed the rootstock of all post-Palaeozoic echinoids whilst themselves showing relatively little important change. The modern cidaroids, which have changed very little at least since the Cretaceous, have often been regarded as 'living fossils'. All cidaroids have relatively narrow and frequently sinuous ambulacra composed of small plates, each with a single pore pair (Figs 9.13, 9.14).

The interambulacral plates are very large, with a single large tubercle in the centre of each plate, to which a strong spine is attached (Figs 9.14a,b, 9.15).

The **mamelon**, or central boss, is surrounded by

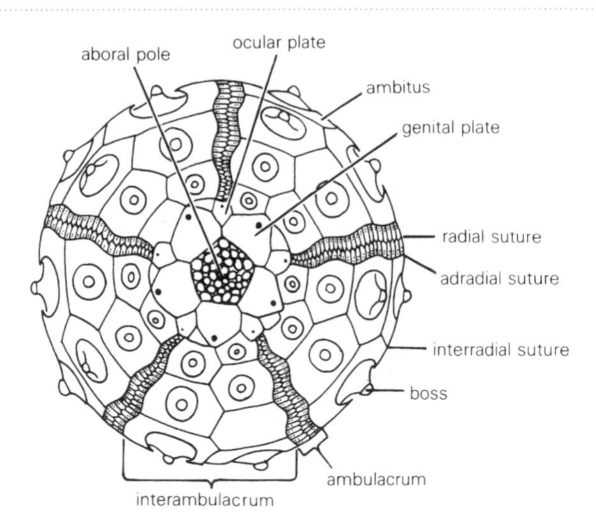

Figure 9.13 Cidaroid morphology, adapical surface, with terminology (× 1). The anus is located within the periproctal membrane at the adapical pole. (Redrawn from Fell in *Treatise on Invertebrate Paleontology*, Part U.)

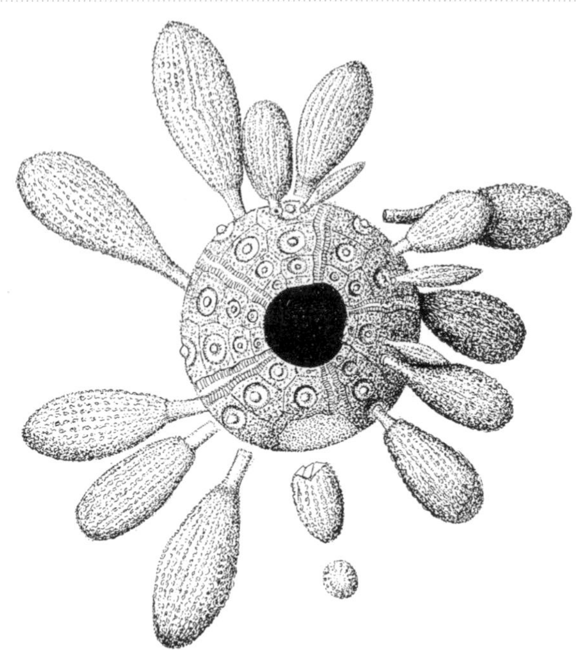

Figure 9.15 *Tylocidaris florigemma* (Cret.), showing club-shaped stabilizing spines, some of which are still in position.

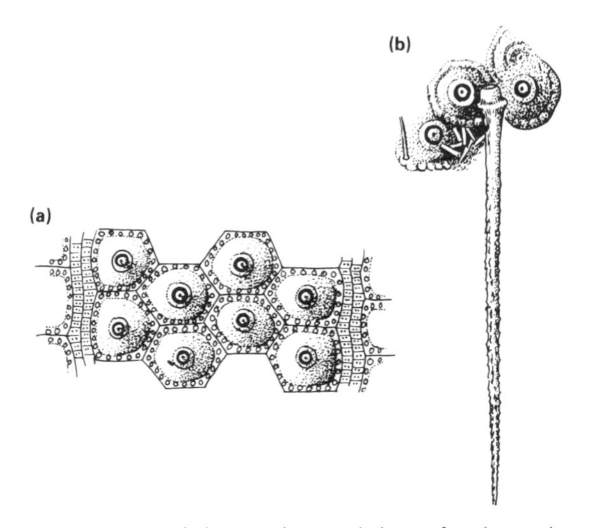

Figure 9.14 Ambulacra and interambulacra of *Archaeocidaris* (L. Carb.): (a) reconstructed and (b) with spine preserved. (Based on Jackson, 1912.)

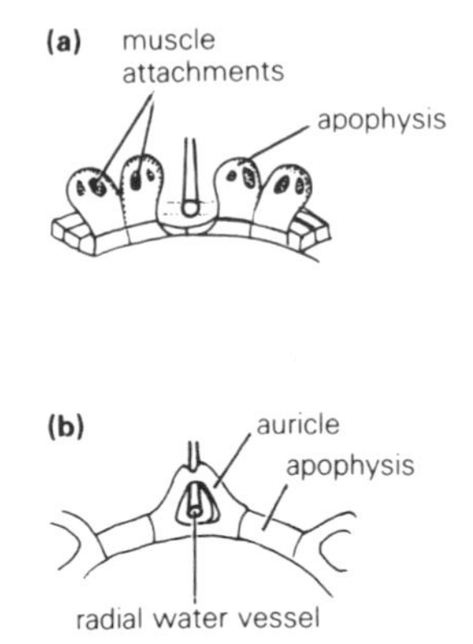

Figure 9.16 (a) Perignathic girdle of cidaroid with apophyses only showing muscle attachments (viewed from within the peristome); (b) perignathic girdle of Recent *Paracentrotus* (Echinoida) with apophyses and auricles. (Simplified from Cuenot in *Treatise on Invertebrate Paleontology*, Part U.)

a wide smooth **areola** around which is a ring of tiny **scrobicular tubercles**. Outside these a series of small secondary tubercles is irregularly dispersed.

The apical system is usually large and has five ocular and five genital plates, but in most cidaroids these are not rigidly united to the test and normally drop out on fossilization. The lantern is relatively simple; the perignathic girdle has **apophyses** only (Fig. 9.16), i.e. flanges reflected from the interambulacra.

The oldest known cidaroids, other than possibly some Silurian forms represented only by spines which have been referred to this group, belong to Family Archaeocidaridae (U. Dev.–Perm.), which like other Palaeozoic echinoids have flexible tests and multiserial interambulacra. *Nortonechinus* (Dev.) and *Lepidocidaris* (Carb.) have six to eight columns of

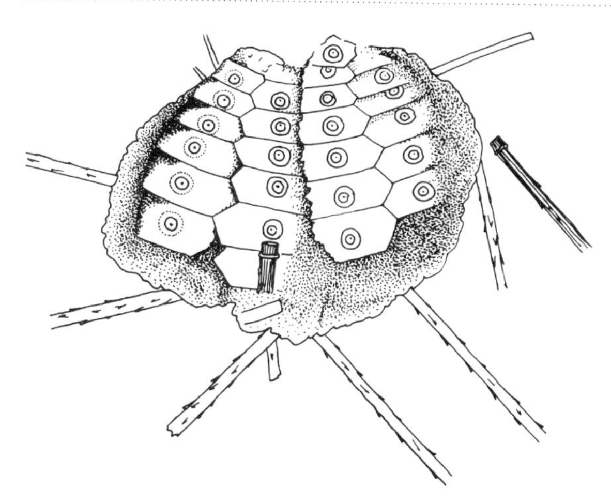

Figure 9.17 *Miocidaris* (Perm.), collapsed so as to conceal the ambulacra (× 1.75). (Based on Kier, 1965.)

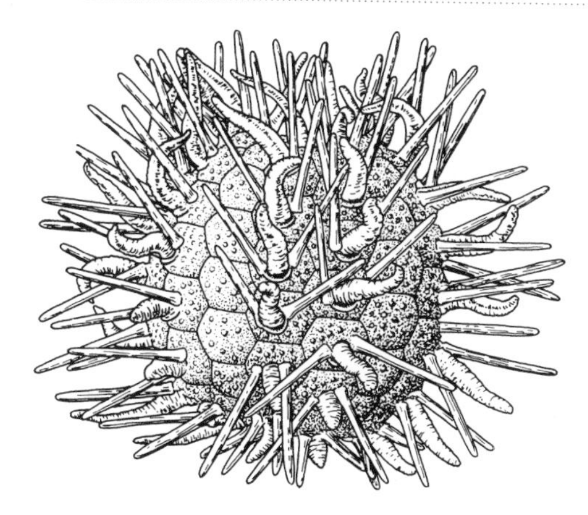

Figure 9.18 *Bothriocidaris* (Ord.), reconstructed, showing spines and tube feet. (Redrawn from Durham, 1966.)

plates; the common Carboniferous *Archaeocidaris* (Fig. 9.14a) has four. However, cidaroids of Family Miocidaridae [including *Miocidaris* (Fig. 9.17), the only known genus to survive the Permian extinction] have only two columns of plates in ambulacra and interambulacra, settling the pattern for all later echinoids, for which *Miocidaris* alone was the ancestor.

In Palaeozoic cidaroids, adaptive trends have been documented which all relate to improved locomotion on a harder substrate.

There were a few small-sized cidaroid genera in the Triassic, including the first cidaroids of modern type, but the cidaroids reached their acme in the Mesozoic, declining later, though they are still abundant today in the Indo-West Pacific province, in the North Atlantic and – as an endemic family, the Ctenocidaridae – in the Antarctic.

Cidaroids are noteworthy for the remarkable development of their spines. Fusiform, club-shaped and peculiarly shaped spines are found in the shallow-water genera. Most of these are sluggish and the heavy spines seem to be used for stabilizing them in rough water. Slender elongate spines are more usual in the deeper-water or mud-living species.

Bothriocidaris

It seems appropriate here to mention a group of echinoderm genera which may or may not be echinoids and which have been a singular puzzle for generations of echinologists. The notorious *Bothriocidaris* (Fig. 9.18), which has been one of the most controversial of all fossil echinoderms, comes from the Ordovician of Estonia, but recently specimens of a related genus have been found in Scotland (Paul, 1967) as well as a single true *Bothriocidaris*.

It has five single-columned interambulacra and five double-columned ambulacra which do not extend as far as the peristome, each ambulacral plate having two large spines and a single pore pair. The peristome has a primitive lantern but no girdle, and is surrounded by a single or double ring of plates with pore pairs. On one interpretation there are no genital plates, and the madreporite is on one of the oculars; alternatively, the apical plates traditionally regarded as ocular plates could actually be genitals (Durham, 1966).

Bothriocidaris has been variously regarded in the past as an echinoid, a diploporite cystoid, or a holothuroid. Its lantern plates are small and poorly developed, and there is some evidence that it might be more closely related to holothuroids than to echinoids. Its taxonomic position is far from clear, but it is not believed that any later echinoids were directly descended from it. *Eothuria*, another Ordovician oddity, was originally described as a plated holothurian (sea cucumber), but other than the presence of many tiny openings in each ambulacral plate it seems to have echinoid features, albeit deviating from the norm. This again was an early offshoot of the great Ordovician radiation of echinoderms and does not seem to have produced any known descendants.

Subclass Euechinoidea and the morphological characters of euechinoids

The euechinoids (Trias.–Rec.) include all Mesozoic to Recent echinoids other than cidaroids. They all have five bicolumnar ambulacra and five bicolumnar interambulacra; they may be regular or irregular; the lantern may be present but in irregular forms can be secondarily lost. The cidaroids and euechinoids are now believed to have diverged prior to the appearance of *Miocidaris* (Smith and Hollingworth, 1990). Though the fossil record is scant, at least two echinoid lineages must have passed across the Permo-Triassic boundary, one giving rise to the modern cidaroids, the other to the euechinoids which began to diversify in the later Triassic and early Jurassic, undergoing a great adaptive burst and evolving into many families. At this time there were all manner of functional innovations (Kier, 1974), which allowed ecological differentiation denied to the conservative perischoechinoids and cidaroids. The echinoids of the Palaeozoic were all regular and endocyclic. Though the regular endocyclic system has been retained by many of the Mesozoic and later taxa, numerous other groups quite independently became irregular, and all of these adapted for a wholly or partially infaunal existence. There are, however, different 'grades' of irregularity in the echinoids. Some are rather simple, with the periproct still in the adapical region, though outside the apical disc, whereas in others it has migrated to the lower surface. The extremes of functional differentiation, however, were not reached until the Cretaceous with the origin of Order Spatangoida.

The early Mesozoic adaptive radiation of euechinoids gave rise to representatives of all four superorders. In some of these the archetypal plan tended to remain rather conservative after its foundation in the early Mesozoic, and there was very little functional evolution thereafter. In others each stage, though in itself highly functional, was capable of yet further functional modification. The adaptive differentiation of the more extreme forms of burrowing echinoid, as represented by *Echinocardium*, is a far cry from the morphology of its remote, regular, Palaeozoic ancestor. Yet once the origin of irregularity had provided the initial stimulus, there were then plenty of new morphological characters for evolutionary processes to work on.

Tube feet and pore pairs

In regular and irregular echinoids alike, tube feet are often highly modified and perform various functions. The 'standard' *Echinus* tube foot (Fig. 9.3a) has a relatively thin muscular layer, and some echinoid tube feet have virtually no muscle. In the tube feet of such echinoids as live in high-energy environments, e.g. *Arbacia,* the tube feet are highly muscular and have a well-developed sucking disc for clinging to rock surfaces. Tube feet modified for respiration and for funnel building (Fig. 9.3c,d) have specific modifications and, as Smith (1978, 1980) has shown, the nature of the tube foot can readily be inferred from the structure of the pore pair by reference to modern examples. This in turn sheds light upon the life habits even of extinct echinoids.

Apical disc (Fig. 9.19)

In regular euechinoids the apical disc is entire and contains the periproct. The latter bears the anus, located centrally within a series of small plates which cover the periproctal membrane. The disc may be **exsert**, with the circlet of ocular plates outside the apical disc, or **insert**, with ocular and genital plates alternating in a single ring around the periproct. But even within the same species there may be individuals with a fully insert, exsert or intermediate condition. Genera such as *Hyposalenia* (Fig. 9.19p) and *Peltastes* have greatly expanded apical discs covering up to a third of the total surface area of the echinoid. These have an extra (**suranal**) plate by the periproct, and the ocular and genital plates are often highly ornamented.

In irregular echinoids the periproct may have wholly or partially escaped from the apical disc and be located outside it. One can see this actually happening in the development of modern irregular sea-urchin larvae. The periproct in these is initially apical, but after a short period of growth it moves externally, disrupting the embryonic plates as it does so; these then re-form, though in an asymmetrical fashion, and the genital V is usually missing. The irregular patterns are, of course, retained in the adult. Various kinds of apical discs in fossil and recent echinoids are here illustrated in a morphological series showing all stages in the migration of the periproct, though this is not intended to display the actual course of evolution: merely grades of organization. The Pygasteroida show some interesting evolutionary trends in this respect. The earliest

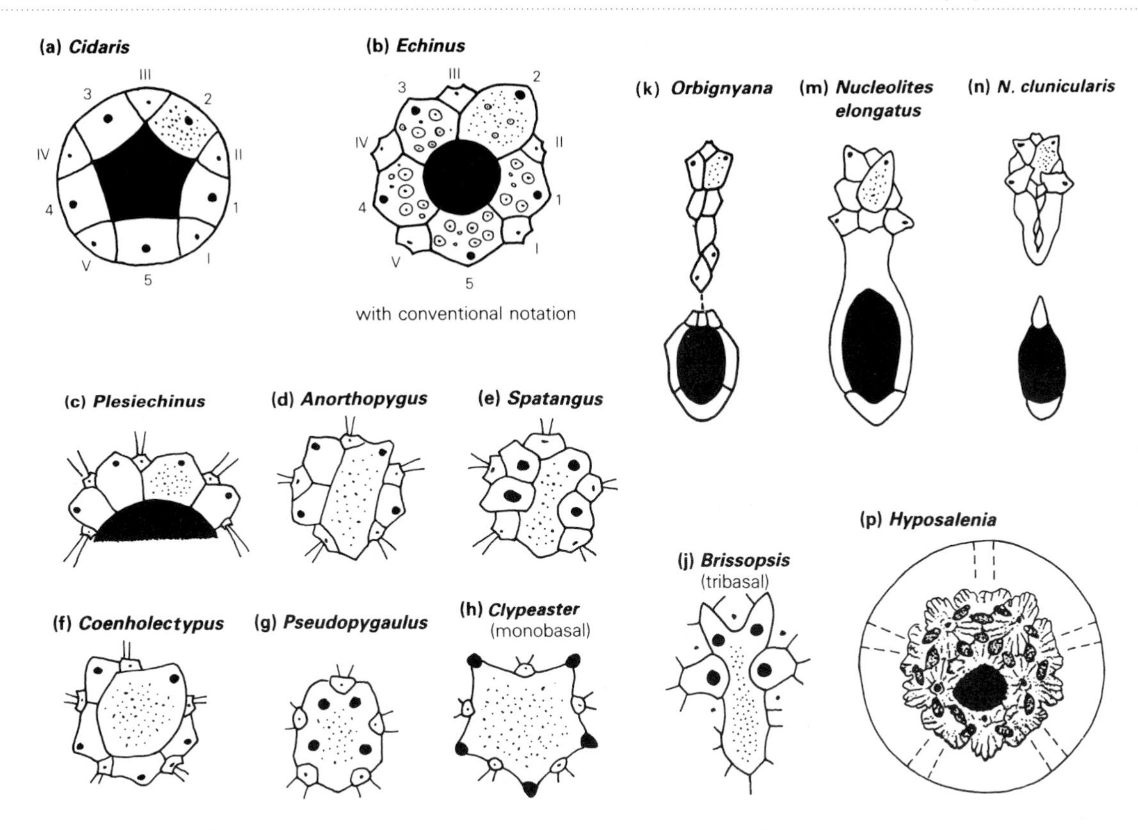

Figure 9.19 Apical discs (not to scale): periproct, where present, shown in black. (Redrawn mainly from Hawkins, 1912; Kier, 1974.)

pygasteroid, *Plesiechinus hawkinsi* from the Lias, has an apical disc like that of a regular echinoid and is purely endocyclic. In slightly later pygasteroids the periproct is displaced outside, though still enclosed by elongated plates, and the genital V is still present (an equivalent condition is seen in some early *Nucleolites* species). In more 'advanced' pygasteroids, the genital V is absent, and the periproct lies at least partially outside the apical disc, forming a keyhole-shaped depression on the adapical surface (Fig. 9.19c, 9.20).

In the majority of irregular echinoids only four genital plates remain, though occasionally a fifth is found.

Other than the breakout of the periproct, a reduction in the number of genital plates has been common in various irregular echinoid lineages. In holectypoids (Fig. 9.21) and cassiduloids (Fig. 9.24) there are defined trends from a system with four distinct genital plates (**tetrabasal**) to a system with a single large plate with four pores (**monobasal**).

A monobasal system has been adopted by all clypeasteroids. Most spatangoids are tetrabasal, some **tribasal** and a few monobasal; it seems that these too are undergoing a general reduction to a monobasal system.

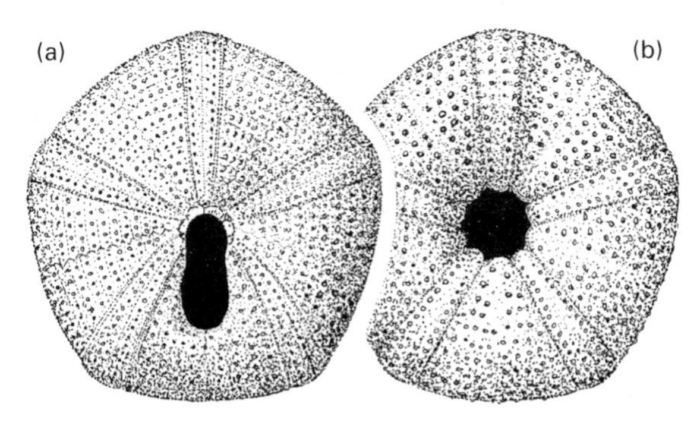

Figure 9.20 *Pygaster* (Jur.): (a) adapical and (b) adoral surfaces (× 0.75).

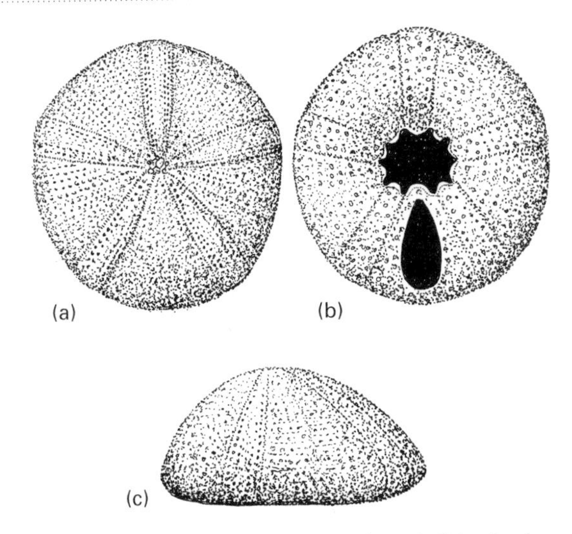

Figure 9.21 *Holectypus* (Jur.): (a) adapical; (b) adoral surfaces; (c) lateral view (× 1 approx.).

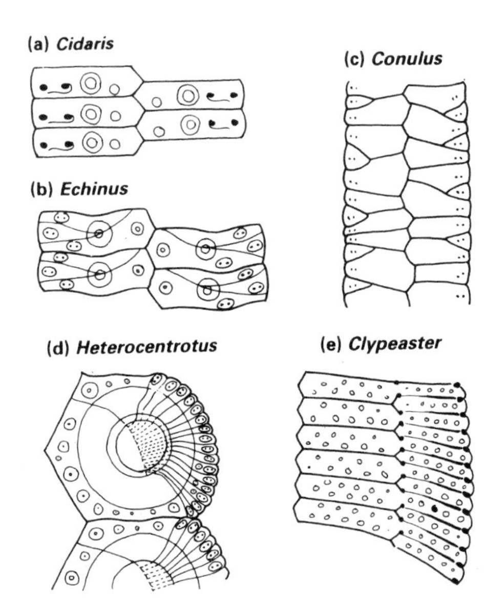

Figure 9.22 Ambulacral plating showing simple (cidaroid) and compound plates (not to scale). (Redrawn from Durham *et al.* in *Treatise on Invertebrate Paleontology*, Part U.)

Ambulacra

COMPOUND PLATES (FIG. 9.22)

In euechinoid ambulacra there are never more than two columns of plates, but there often are a very large number of pore pairs and hence tube feet. This has been achieved in different ways by various groups of echinoids. Thus in the cidaroids the plates have single pore pairs, but are very small and hence numerous. Alternatively, compounding of plates is common in many euechinoids. Compound plates have two, three or more **demiplates** within the confines of a single ambulacral plate. Each demiplate has its own pore pair. Such morphology is most pronounced in genera like *Heterocentrotus* (Fig. 9.22d) in which the large central tubercle is traversed by numerous demiplates. Compounding results from the subjugation and incorporation of embryonic plates into a smaller number of 'master plates' and is important in allowing a larger number of tube feet per unit area. It is also commonly associated with the development of a large central tubercle, and another advantage of the system must be that a larger tubercle could be supported by larger plates formed by the fusion of smaller ones. Compounding also separates the rows of tube feet within one ambulacrum, so that the test bears ten widely spaced columns of tube feet rather than five closely spaced pairs of columns as in the cidaroids.

PETALOID AND SUBPETALOID AMBULACRA

Irregular echinoids such as *Echinocardium* have ambulacra forming 'petals', i.e. with the adapical parts expanded to form a flower-like rosette. Within the petals the pore pairs are always elongated and the individual pores widely separated, one being slit-like; they are always associated with flattened respiratory tube feet. Some of the earlier and 'primitive' irregular echinoids [e.g. *Holectypus* (Fig. 9.21), *Pygaster* (Fig. 9.20)] have no petals but they do, incidentally, have gills. The next grade in organization is a slightly expanded subpetaloid condition, and a more advanced grade still is the system of some clypeasteroids and cassiduloids in which the petals are large but flush with the surface. In the early flattened echinoids, so many tube feet would have been located on the lower surface, and used for food-gathering, that respiratory exchange might have been impaired had it not been for the development of petals on the upper surface. These petaloid ambulacra were then pre-adapted for respiratory use in deeper-burrowing echinoids. Some holasteroids and spatangoids have depressed petals, especially the geologically later ones (*Echinocardium* being a prime example), and the deepening of the petals is such that they form an effective channelling structure for the respiratory currents as well as providing a basis for a protective cover of flat-lying spines. The differentiated anterior ambulacrum of holasteroids and spatangoids also arose through relatively gradual stages from an originally flush and undifferentiated type.

PHYLLODES AND BOURRELETS (FIG. 9.24A–C)

Certain irregular echinoids, especially the cassiduloids, and a few regular genera have specialized ambulacral structures in the vicinity of the mouth. These are the **phyllodes**: areas in which the ambulacrum expands into a leaf-like shape close to the peristome. The phyllodes are separated by interambulacral swellings known as **bourrelets**. In the phyllodes the plates are very crowded, a much larger number of tube feet occur than elsewhere in the ambulacra and, furthermore, the pores tend to be larger. In the irregular echinoids these extra tube feet are used primarily for feeding. Some, such as the spatangoid *Meoma,* feed on organic-rich sand, and the tube feet are used as 'sticky shovels' to deposit sand in the mouth. Regular echinoids may also have phyllodes, particularly those living in high-energy environments where the tube feet are used for clinging to the substrate. Presumably the extra adoral tube feet in *Proterocidaris* and other flexible echinoids may have served a similar purpose. A tendency to develop phyllodes has been noted in many lineages. Later members often have broad phyllodes with larger but fewer pore pairs. In many species the tube feet emerge from single pores, the respiratory function which necessitates separate channels for oxygen-rich and oxygen-poor coelomic fluid having been lost.

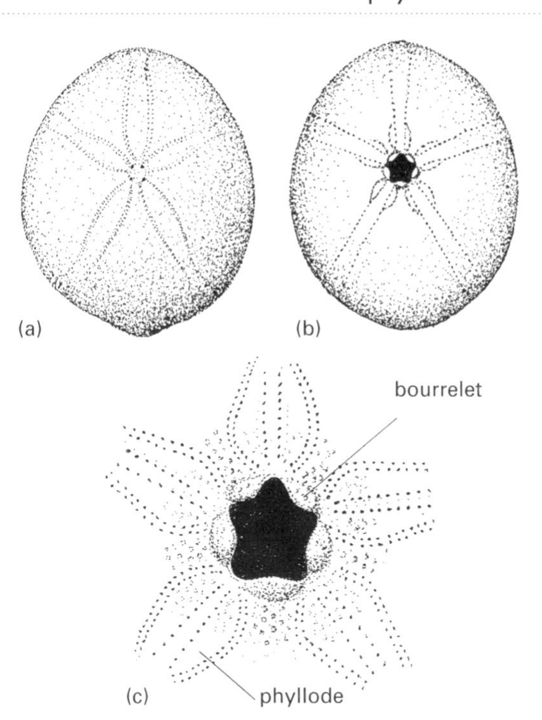

Figure 9.24 *Catopygus* (Cassiduloida; Cret.): (a) adapical and (b) adoral surface (× 2); (c) enlargement of phyllodes and bourrelets (×8).

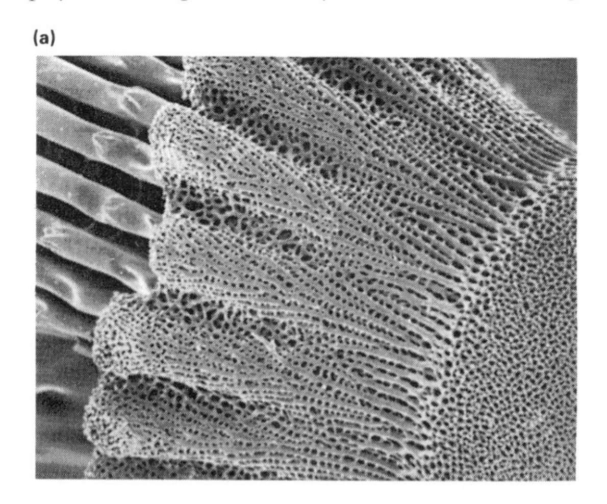

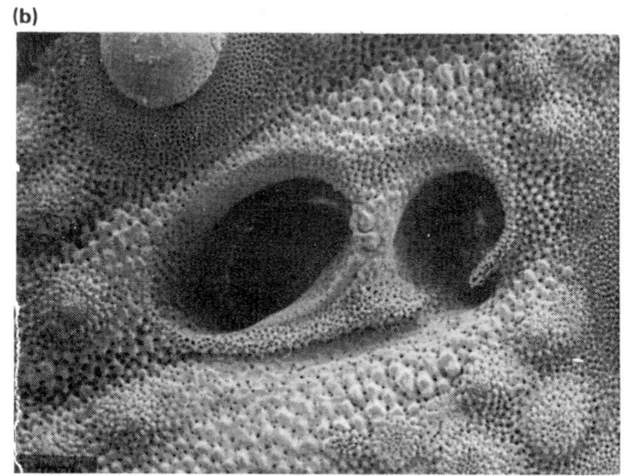

Figure 9.23 Structure of Recent echinoids: (a) *Centrostephanus nitidus* (Rec.), Indian Ocean. showing the ambital spine base – the spine muscles attach to the raised wedges on the upper part (× 100 approx.); (b) *Psammechinus miliaris* (Rec.), Torquay, England, an aboral pore pair with neural groove (× 50 approx.). (SEM photographs reproduced by courtesy of Dr A.B. Smith.)

Interambulacra and spines (Fig. 9.23a,b)

Cidaroids, as we have seen, are characterized by interambulacral plates having a single large central tubercle. This system was retained in genera such as *Hemicidaris* and *Salenia,* though the tubercles are generally smaller. However, in the majority of other euechinoids the size of the primary tubercles decreased markedly, and most particularly in the irregulars. On the interambulacral plates of sand-dollars there may be up to ten primary tubercles per square millimetre and hundreds per plate.

This tendency, which so greatly altered the whole appearance of the interambulacral plates in the evolution of euechinoids, is functionally sound, since the modern cidaroids can neither bury themselves in detritus for camouflage nor burrow. Short spines are a prerequisite for burrowing echinoids; they also provide cover and shade. Perhaps the most striking modification of all echinoids is in the modern reef-dwelling regular echinoid *Podophora* in which the adapical spines are mushroom-shaped, their tops being flat polygonal plates forming a continuous basaltiform mosaic, which gives protection in heavy surf; these echinoids can also cling to the rocks on which they live by their phyllodal tube feet.

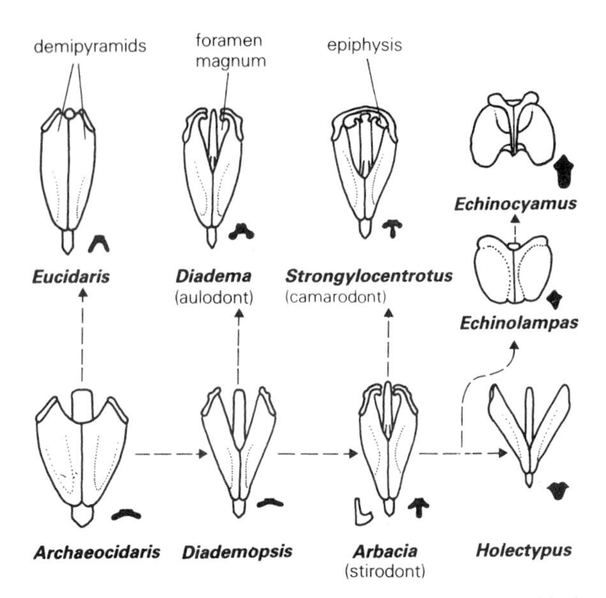

Figure 9.25 Various types of lantern pyramid with (in black) shapes of teeth, keeled or grooved, in section. The arrows represent possible relationships. (Based on Smith, 1982.)

Lantern and perignathic girdle (Fig. 9.25)

The lantern is most important in echinoid taxonomy at the super-ordinal level. Although it is rare to find it fossilized, enough specimens have now been discovered to allow a phylogeny of echinoids to be constructed using the teeth and lantern as major taxomomic criteria. The details of the lantern and its structure are beyond the scope of this work, but it is important to appreciate the major stages in its evolutionary development. The lantern consists of five identical pyramids, each composed of two demipyramids. Between these is a depressed foramen, and above are short, rod-like epiphyses. Each pyramid contains a single calcite tooth which grows down continually as it is abraded.

The classic work of Jackson (1912) distinguished four main types of lantern, cidaroid, aulodont, stirodont and camarodont, but as Smith (1981, 1982) and others have shown, the range of structure and function is rather more complex. Isolated pyramids illustrated here show some of the main lantern types in an inferred evolutionary relationship. The primitive archaeocidarid pyramid was relatively broad, with flattish grooved teeth, while later cidaroids have a more compressed pyramid with a shallow foramen magnum and deeply grooved teeth. The recently discovered lantern of the early Jurassic *Diademopsis* is quite like that of *Archaeocidaris,* though with a deeply cleft foramen magnum, and the teeth are likewise very similar. From this early rootstock genus seemingly arose various 'aulodont' types (e.g. diadematoids), in which the epiphyses grow inwards, and – by other paths – stirodont and camarodont lanterns, in which the teeth, rather than being grooved, have a pronounced strengthening keel. In camarodont lanterns the epiphyses join at the top. Some of these lantern types define major taxonomic groupings in the irregular echinoids.

From a primitive *Diademopsis*-like stock came the first irregular echinoids, in which the lantern is distinguished by the loss of the epiphyses, but in which the teeth became independently keeled. From the pygasteroid rootstock arose the holectypoids and finally the cassiduloids and clypeasteroids, the latter having remarkably expanded wedge-like teeth.

All these changes are associated with improvements in function. The lantern of cidaroids can only move up and down and open and shut like a five-jawed grab. The camarodont lantern, however, is capable of sideways scraping, which is much more effective in bottom feeding, and its whole organization seems to be related to this end. The clypeasteroid lantern with its expanded wedge-like teeth is a functional crushing apparatus, and although it cannot move sideways it is very powerful and can pulverize the bottom detritus upon which the echinoid feeds. The lantern has been dispensed with in the irregular Atelostomata, which are microphagous feeders.

Palaeozoic echinoids other than cidaroids do not have a true perignathic girdle, and the muscles and ligaments were attached directly to the inside of the test. The presence of a girdle raises the attachment of the muscles and mechanically improves their line of action. Cidaroids (Fig. 9.16) have a primitive girdle with projections (**apophyses**) in the interambulacrum. Euechinoids have both apophyses and **auricles**: supports at the base of the ambulacrum which, in more advanced echinoids, may arch over the radial water vessel. The development of auricles in addition to apophyses probably increased the spread of muscle attachment and thus allowed the possibility of lateral movement of the lantern.

Gills

Gill notches are found in echinoids from the early Jurassic onwards. They are present in nearly all regulars and in some early or primitive irregulars (e.g. *Holectypus*, *Pygaster*) but not in the later irregulars, in which petaloid ambulacra having efficient respiratory tube feet are developed.

Marsupiae (Fig. 9.26)

Sexual dimorphism is not normally evident in echinoids, and when it does occur it is in the form of minor differences such as larger genital pores in the female. However, there are some 28 species of modern echinoids in which the females have developed special brood pouches (**marsupiae**) in which to incubate the fertilized eggs. These then develop directly without the free-swimming larval stage. Nearly all of them live in cold Antarctic waters where development and growth are slow. Such brood pouches occur in recent cidaroids and spatangoids, but they are also found in some Cretaceous and Tertiary echinoids of Orders Temnopleuroidea, Spatangoidea and Clypeasteroidea from southeastern Australia (Philip and Foster, 1971), which from palaeoclimatic evidence seems to have been very cold at that time. The presence of such genera in the Australian Tertiary and their absence in modern offshore Australian fauna accord well with geological evidence on the northward drift of the Australian continent away from Antarctica in the later Tertiary.

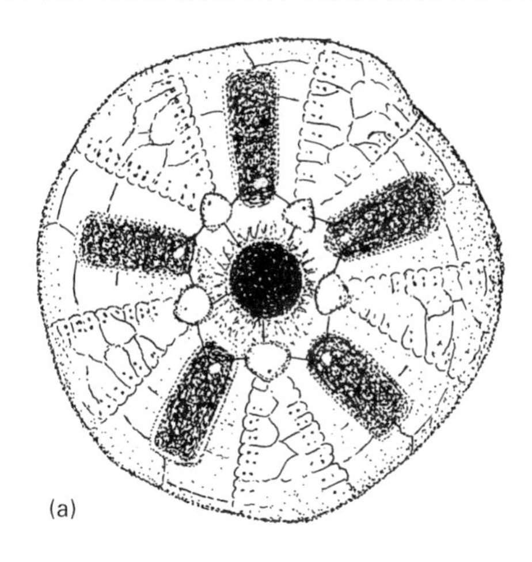

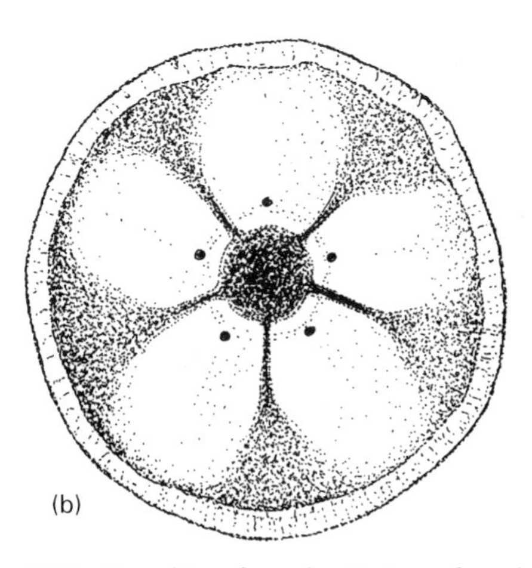

Figure 9.26 *Pentechinus* from the Tertiary of southeastern Australia, showing marsupiae (× 1) (a) adapical surface; (b) cut specimen: marsupiae seen from inside. (Based on Philip and Foster, 1971.)

The temnopleuroid *Pentechinus* has (in the female) each of the five genital pores opening into an elongated depression extending some way down the interambulacrum, whereas in *Paradoxechinus* a deep annular depression surrounds the apical disc, and the genital pores open directly into this. *Paraspatangus* has a shallow depression in which the apical disc lies. Normally in marsupiate echinoids the brood pouches are adapical, but an adoral marsupium is present in the clypeasteroid *Fossulaster*. The independent origin of marsupiae in so many cold-water forms testifies to an evident advantage of viviparity in low-temperature conditions.

Evolution in echinoids

The first recognizable echinoids are Ordovician, appearing at the same time as a whole plethora of other echinoderm types, some at least of which, in certain features of their organization, are reminiscent of echinoids. Amongst these could be numbered holothuroids and the echinozoan Class Ophiocystoidea, which have enormous plated tube feet and a lantern structure. (Apparent resemblances between echinoids and edrioasteroids or diploporites are superficial.) The features of true echinoid organization became more clearly defined a little later, and by the Carboniferous the large regular flexible-tested perischoechinoids with many columns of plates were becoming diverse, though never really abundant. The only fossil echinoid genus known to cross the Permo-Triassic boundary is *Miocidaris*, though it is now believed that the euechinoids were derived from a different Late Palaeozoic lineage. In the Triassic are found small cidaroids, and then in the early Jurassic came a great burst of euechinoid radiation. Of these taxa some are 'improved regulars' and the rest irregulars, adapted for many modes of life. Today echinoids are a vital part of the invertebrate marine realm and they are probably as abundant now as at any time in the past. In the Recent fauna, 25% are spatangoids, 16% cidaroids, 14% clypeasteroids and 14% temnopleuroids. The fortunes of various orders have varied greatly during the Tertiary; for instance, cassiduloids were very important until the Pliocene, when they declined greatly, as did the clypeasteroids, and many old groups such as the holectypoids are very nearly extinct.

Modern echinoids are very successful feeders. Some even devour other echinoids; the process of such feeding has been shown recently by time-lapse photography, which showed three large cidaroids devouring a sand dollar, eating it from the edges like a biscuit.

The precise relationship between structural and functional differentiation is perhaps more clearly seen in the echinoids than in any other invertebrate group. By contrast with the evolutionary pattern of most fossil groups, evolutionary changes and repeated trends in the Mesozoic and later echinoids have been rather slow and spread out. So often in other taxa the vitally important intermediate stages are compressed in time, and there are no clear links between one grade of organization and another. Abrupt changes of this kind are seen in trilobites, molluscs and brachiopods, and the whole process of understanding phylogeny and its functional significance is bedevilled by morphological discontinuities. Even though there are gaps in echinoid phylogeny, such as in the establishment of euechinoid orders, they are not all that abrupt, and such an important stage as the breakout of the periproct can be seen as a relatively gradual process. The evolutionary trends within both Palaeozoic and Mesozoic to Recent echinoids, which Kier (1965, 1974) has carefully documented, occur in several groups, testifying to continued selection leading to 'the gradual improvement of the animal as a living mechanism'. The remarkable functional differences between the early Mesozoic cidaroids and, say, the spatangoids or clypeasteroids has been achieved only by modification of existing organs or of elements already present in the ancestor. Even the fascioles of the spatangoids are merely tracts of modified spines, the clavulae, in which the surface is covered with a ciliated epithelium of a kind native to echinoderm organization. In fact the only new kinds of organ in the euechinoids not possessed by their cidaroid forebears are the auricles of the girdle, some small components in the lantern and the ophicephalous (pincer-like) pedicellariae.

Furthermore, the functional significance of any structural changes in extinct genera may be interpreted because of comparable functional adaptations in many modern representatives, using similar modifications of organs generally present in echinoids, e.g. spines, fascioles, etc. (for the same general purposes). In this context the record of evolution in *Micraster*, a Cretaceous spatangoid, is especially good. It has been elucidated stratigraphically, tested statistically and interpreted functionally with reference to its modern relative *Echinocardium* and others.

Evolution in Micraster (Figs 9.27, 9.28)

Micraster is an Upper Cretaceous genus widespread over much of Europe, and very common in France and England.

In 1899 Rowe, working in southern England, carefully collected some 2000 specimens of *Micraster* in rigid stratigraphical sequence through six successive zones of the Upper Chalk. He observed some clear changes in going from the lower zones to the upper, confirming previous suggestions; this was all the more striking in view of the monotonous

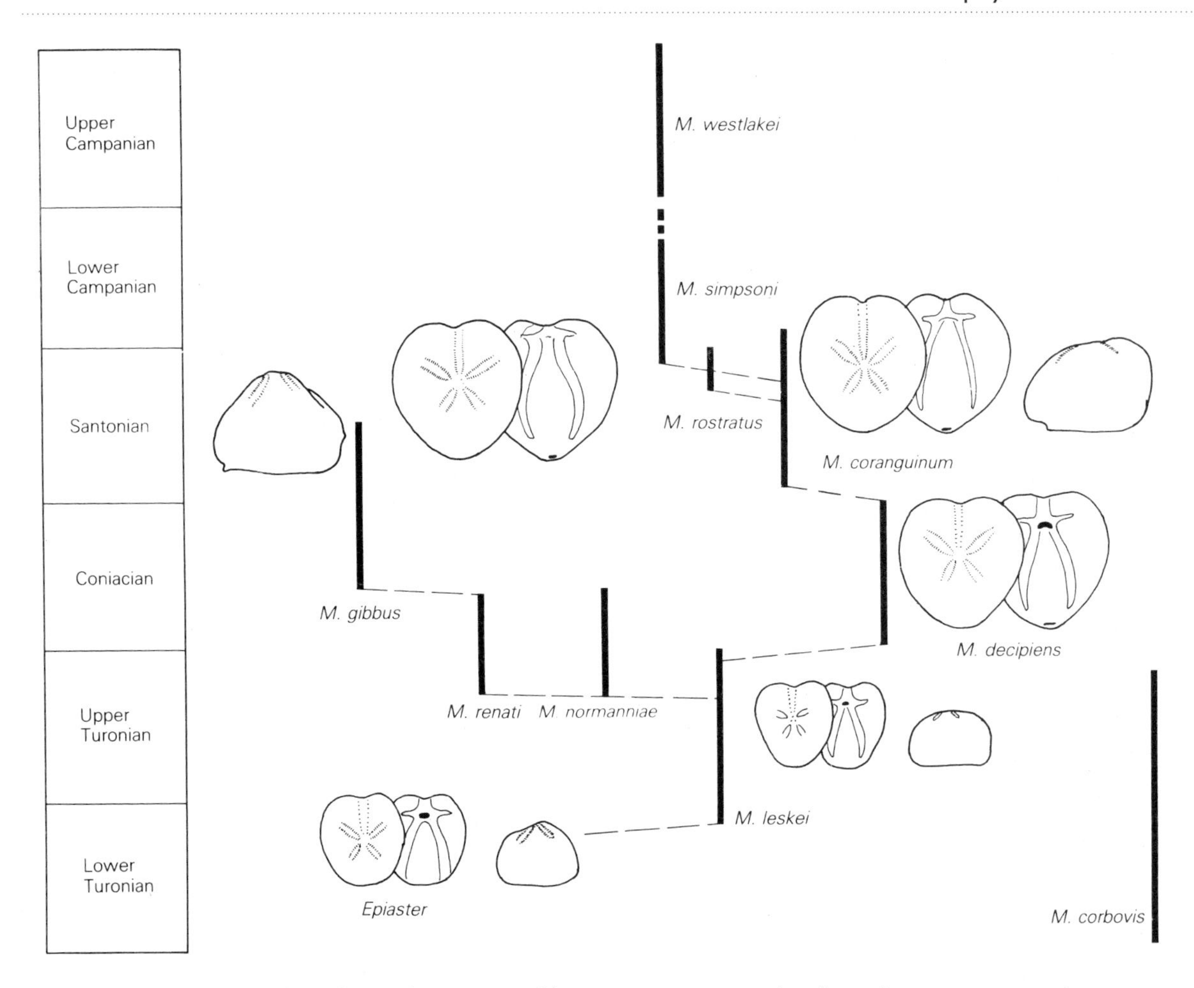

Figure 9.27 Evolutionary relationships and time ranges of the *Epiaster–Micraster* stock in the Anglo-Parisian province during Upper Cretaceous time. (Based on Ernst, 1970; Stokes, 1977; Smith, 1984b.)

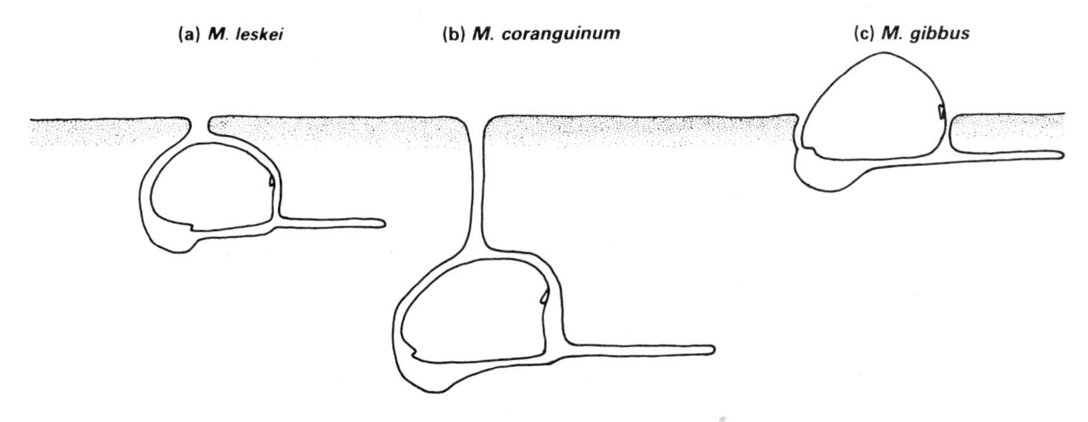

Figure 9.28 Inferred life positions and burrowing depths of Cretaceous micrasters. (Based on Nichols, 1959.)

lithology throughout. The main changes Rowe noticed on going up the sequence were as follows:

the test becomes broader, and both the tallest and broadest parts move posteriorly;
the anterior groove deepens and increases in tuberculation;
the mouth moves anteriorly, and the labrum or lip below it becomes very pronounced eventually covering the mouth;
the paired ambulacra lengthen and become straight, whereas the interporiferous areas change from being smooth through a variety of intermediates until the final pattern is one with a deep groove per-radially and inflated lateral areas.

Though Rowe imagined that the original lineage had subdivided more than once, much later work was required before the confused taxonomy could be sorted out. The early work of Kermack (1954) was greatly updated by Ernst (1970) and Stokes (1977), and it is now understood that evolution took place in two separate areas – a northern province and the Anglo-Parisian province – in the latter of which the phylogeny is best known.

Micraster leskei is a small, rather globular echinoid with a shallow anterior groove and the mouth rather far back, without a labrum. The earliest true *Micraster* originated from the earlier genus *Epiaster* found mainly in sandy facies. From this came the line of descent *M. leskei, M. decipiens* and *M. coranguinum* in which the changes noted by Rowe and confirmed by Kermack took place. The species succeed each other with very little indication of intermediates, and this established lineage is no longer considered to be gradualistic. These *Micraster* species are found only in chalk, as is *M. gibbus* (formerly known as *M. senonensis*), which descended from *M. leskei* by a different line.

Nichols (1959) shed much light on the biological meaning of these changes, by studying the precise relations between structure and function in the living *Echinocardium* and other extant spatangoids: the modern equivalents of *Micraster*. It was his view that *M. leskei* was adapted to shallow burrowing just below the surface and that the changes in the main line of descent from *M. leskei* to *M. coranguinum* give evidence of adaptation to progressively deeper burrowing conditions (Fig. 9.28).

These adaptations seem clear enough. The deepening of the anterior, food groove, the forward movement of the peristome and the development of the labrum are all advantageous in directing feeding currents to the mouth in a deep-burrowing echinoid, as are the increase in numbers of respiratory tube feet. Other than the subanal fasciole there is no other clearly developed fasciole on the adapical region, but the whole highly tuberculate surface of the animal may have acted as a 'diffuse fasciole', being covered with the bases of what were probably clavulae.

Not all are agreed that the *M. Ieskei–M. coranguinum* line necessarily has to do with adaptations for deeper burrowing. Smith (1984a) interprets *Micraster* evolution as an initial breakthrough to a new habitat (from sand-living to chalk sea-dwelling) and that the changes seen later are best interpreted to more efficient life (burrowing and feeding) within that habitat.

Micraster seems to have been initially successful in the chalk seas because it had a dense adapical canopy of tiny miliary spines which kept the sediment off the echinoid surface and allowed it to live infaunally.

A remarkably similar evolutionary line has been reported from Australian spatangids, where in the Tertiary *Linthia* led through *Paraster* to *Schizaster*. All three genera are still extant. *Linthia,* superficially like *Epiaster,* lives in coarser sediments, *Paraster* in finer material and *Schizaster* in very fine mud. These changes again are interpreted as adaptations to life in fine sediment.

Although the high zonal micrasters are interpreted as deep burrowers, *M. gibbus* was probably a shallow burrower (Fig. 9.28c). The subanal fasciole has gone, and in addition the pyramidal test would help to prevent particles from settling. There are very many respiratory feet, all of small size; this device probably ensured against predation. *M. gibbus* seems to have been derived from the mid-zonal micrasters, possibly as a peripheral isolate, and to have migrated back into the southern England area towards the end of the Upper Cretaceous, when it interbred with M. *coranguinum,* producing a number of hybrids.

In the proliferation of both spatangoids and holasteroids in the Cretaceous, numerous types arose adapted for deep or shallow burrowing, and it has been possible to interpret some of these functionally. *Echinocorys* (Fig. 9.29a–d) is a highly modified

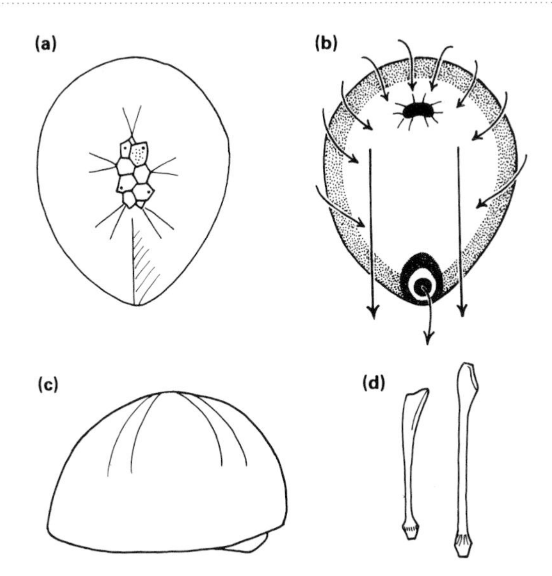

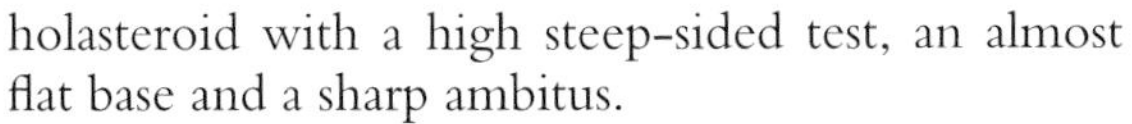

Figure 9.29 *Echinocorys scutata* (Cret.): (a)–(c) in adapical, adoral and lateral views, showing marginal diffuse and perianal fascioles (stippled) and current directions (× 0.6); (d) detached, paddle-like spines from the plastron (×15 approx.). (Redrawn from Stephenson, 1963.)

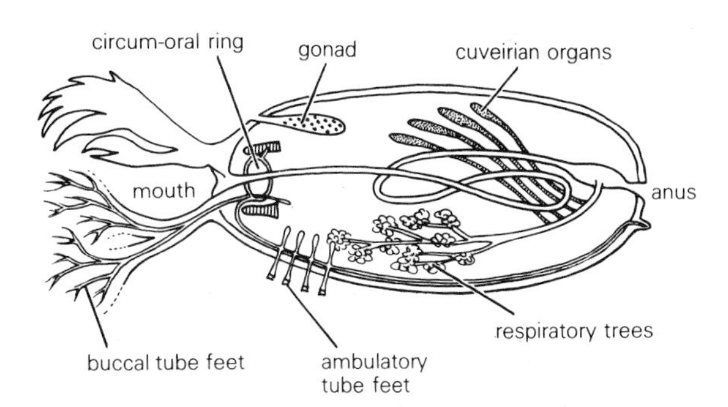

Figure 9.30 *Holothuria*, sectioned longitudinally. (Based on Nichols, 1974.)

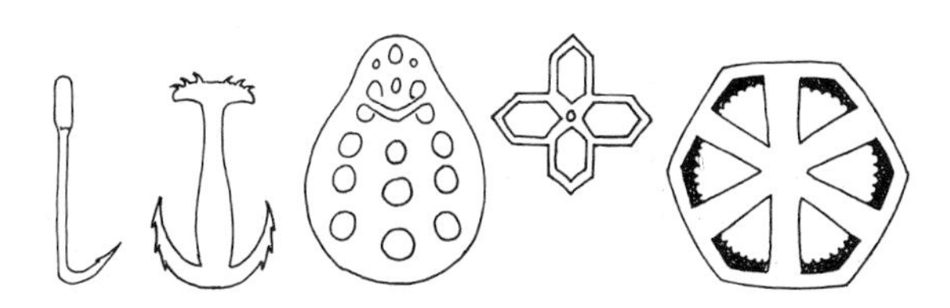

Figure 9.31 Holothurian sclerites (anchors, gratings and wheels). (Based on Frizzell *et al.* in *Treatise on Invertebrate Paleontology*, Part U.)

holasteroid with a high steep-sided test, an almost flat base and a sharp ambitus.

Short thorny spines were evidently for the protection of the adapical surface; spatulate spines attached to the tubercles of the adoral surface allowed forward progression, and ciliary currents from the marginal diffuse fasciole and the more differentiated subanal fasciole produced feeding currents. The ambulacra are only subpetaloid but are very long. *Echinocorys* was probably a shallow burrower but presumably inhabited a different niche from the high zonal micrasters (Stephenson, 1963).

In spatangoids generally, the arrangement of spines is fairly constant, but there is a wide variety of test shapes. These are adapted to different methods of burrowing (Kanazawa, 1992), which are related to habitat. Thus globular echinoids (e.g. *Echinocardium*), which live deep in sand, burrow by excavating sediment from the front and accumulating it on the back. Those with flat profiles, such as *Lovenia*, live close to the surface, though they are stable against current scouring, and when burrowing they push sediment to the posterior sides of the test. Wedge-shaped spatangoids (*Schizaster*, *Brisaster*), inhabit soft cohesive mud and burrow by repeatedly rocking the anterior end up and down and sinking forward into the sediment; the wedge-like shape is essential for this. Domed spatangoids with a flat ventral surface (e.g. *Linopneustes*) do not burrow, but live epifaunally on mud. Such modifications of test shape seem to have been of key importance in enabling spatangoids to adapt to a wide range of habitats.

Class Holothuroidea

'There were nasty green warty things, like pickled gherkins, lying on the beach' (H.G. Wells, *Aepyornis Island*, 1927). Thus has the external appearance of holothuroids (holothurians or sea cucumbers) been most graphically described. Holothuroids (Fig. 9.30) are fusiform or cylindrical echinozoans which generally lie on the sea floor (or burrow into it) with their long axis horizontal.

The mouth and anus are at opposite ends. In the type example, *Holothuria*, the mouth end has a feathery inflorescence of sticky tentacles (the **buccal tube feet**) which are used in feeding. The skin is warty and leathery, and within it are embedded the calcitic elements that are the only parts to be fossilized. These calcitic plates are normally reduced to little **sclerites**, having the form of anchors, gratings or spoked wheels (Fig. 9.31).

Very rarely they form a contiguous cover; more usually the integument is very flexible. It is from such sclerites that fossil genera can be identified and allocated to families. Only in the mouth region is there a rigid series of plates, forming a calcareous **perioral ring**, perhaps homologous with the echinoid lantern.

The five ambulacra are arranged in two sets parallel with the long axis: three ventrally and two dorsally. The tube feet of the dorsal set form small **sensory papillae**, while those of the ventral set are ambulatory and have suckers. The internal structures are much modified. There are a pair of **respiratory trees** arising from the swollen **cloaca** at the hind end of the intestine. Water coming in through the anus is pumped by rhythmic pulsations of the cloaca into these trees for respiratory exchange with the coelomic fluid. In Subclass Aspidochirota, to which *Holothuria* belongs, there are unusual defensive structures located within the body. These are the **cuvierian organs**: thread-like masses which can be shot out of the anus to entangle stickily any predator that disturbs the holothurian.

The earliest holothurian remains may be of Ordovician age; the earliest undoubted holothurian spicules (*Palaeocucumaria*) are, however, from the Devonian. *Eothuria* (Class Echinoidea, Superorder Megalopodacea) might have been somewhere near the ancestral line and indeed was first described as a plated holothurian. Reduction of the plates to sclerites, giving the body its present flexibility, seems to have been almost universal. A very peculiar Devonian ophiocystoid-like echinoderm, *Rotasaccus,* has a large lantern, but the skeleton is reduced otherwise to wheel-like sclerites. This too may be a holothurian. Since the skeletons of the majority of holothurians are reduced to microscopic sclerites only, the fossilization potential of intact specimens is very low. There are, however, rare examples of complete specimens of this kind. The best-known fauna is from the Middle Triassic of northern Spain (Smith and Gallemi, 1991), where superbly preserved plated specimens show a considerable morphological and ecological diversity. Only a few plated genera exist today, all belonging to the Dendrochirota, which with the Aspidochirota and the Apoda form the three defined subclasses of Class Holothuroidea. The first two are mainly benthic holothurians, which creep along on their ventral tube feet or on a muscular, slug-like sole in which the tube feet are reduced. However, one of the aspidochirote suborders, the Elasipoda, has swimming representatives which keep themselves afloat by pulsation, like jellyfish, and are planktivorous. Some of the benthic genera sweep the sea floor ahead of them as they move with their sticky buccal tube feet and so pick up organic detritus; others spread them out in a **horizontal collecting bowl**, as do the sluggish, burrowing Apoda.

Holothurians today are abundant in warm, shallow waters, but they are also successful in the deep sea, where undersea photography has shown dendrochirotes 'congregating in herds like grazing cows'.

Class Edrioasteroidea

The Edrioasteroidea (L. Cam.–L. Carb.) are a small extinct group of echinoderms with five distinct ambulacral and five interambulacral areas, confined to the upper (adoral) surface. These run to a central mouth. Most edrioasteroids are flattened and discoidal though some are globular or elongated.

Edrioaster (M. Ord.; Fig. 9.32a,b) has a flexible **theca** in which five biserial ambulacra are regularly plated and alternate with irregularly plated interambulacral areas.

The ambulacra are sinuous, four having an anticlockwise twist whereas the fifth turns clockwise and curls round the eccentric periproct. A small **hydropore** near the mouth presumably led to the water–vascular system. The ambulacra have two sets of plates: a lower set of **flooring plates**, arranged as in an echinoid, the margins of each pierced by a single pore; and an upper set of **cover plates**, hinged on top of the flooring plates, which could open to let out the tube feet or close down to protect them. The **peristomial plates** covering the mouth were probably fixed (as in the crinoid **tegmen**) and could not open. The whole ambulacral plating system is strikingly reminiscent of that of the Crinozoa (q.v.). As in that subphylum, the radial water vessel may have run in a median groove along the per-radial suture, though Bell (1976), in a comprehensive work, has suggested an alternative position. There is some evidence for this, for traces of both the **circumoral ring** and the radial water vessel have been preserved in *E. buchianus*. The

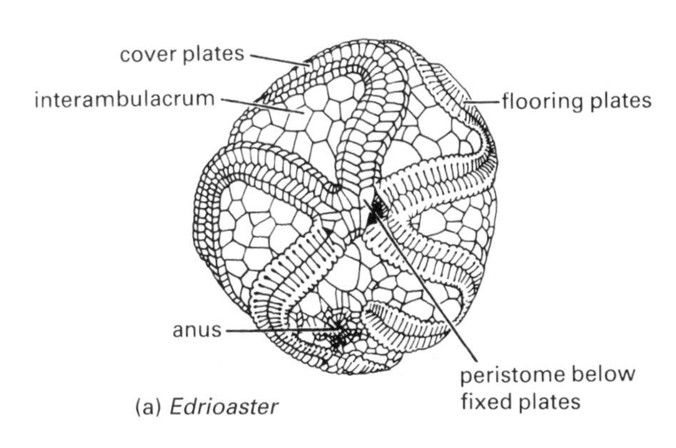

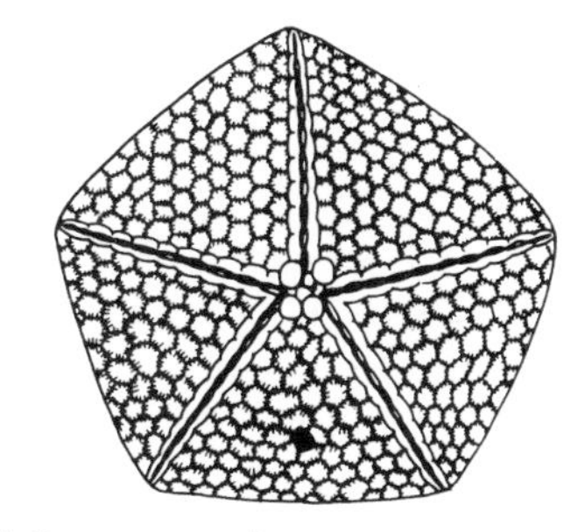

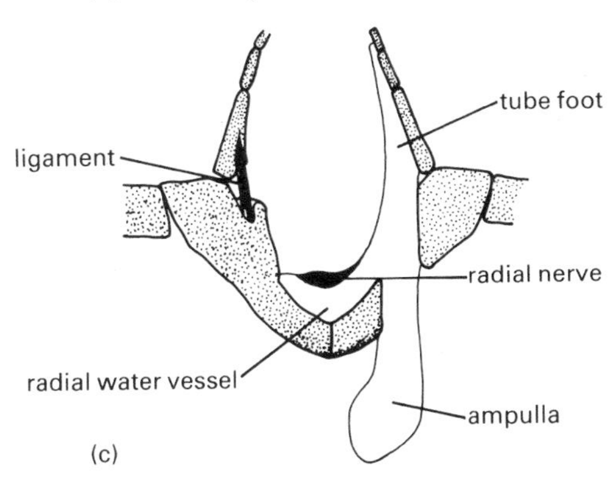

Figure 9.32 (a) *Edrioaster bigsbyi* (Ord.) with cover plates removed from three of the ambulacra (× 1.5 approx.); (b) *Stromatocystites* (L. Cam.), a stem-group echinoderm, probably a precursor to later edrioasteroids (× 1.5); (c) section through ambulacrum of edrioasteroid with inferred disposition of soft parts. [(a) Based on Kesling, (b) redrawn from Pompeckj, both in *Treatise on Invertebrate Paleontology*, Part U; (c) based on Smith, 1985.]

aboral surface consists of plates like those of the ambulacra, but the marginal plates are larger, forming a distinct and rigid ring.

In spite of the similarity of edrioasteroid ambulacra to crinoid food grooves there were no **brachioles** (short fingers extending from the test). The nature of the tube feet is unknown. *Edrioaster* may have been sessile but was not fixed to the substrate and might have moved along by pulsations of the flexible theca.

The superficial resemblance of edrioasteroids to the Precambrian *Tribrachidium* has often been noted, though this may be no more than coincidental. Though the origin of edrioasteroids is really unknown, the aboral features that they share with early crinozoans seem to suggest a common origin. There are five genera of Cambrian age, of which *Stromatocystites* (L. Cam.) is the earliest, and indeed amongst the first of all echinoderms. It is of pentagonal shape and had five straight ambulacra with cover and flooring plates. It was apparently a free-living form. It is probably best regarded as belonging to the stem group that gave rise on the one hand to the other four Cambrian genera, and all later edrioasteroids, and on the other hand to the eleutherozoans (all free-living echinoderms; Figs 9.32b, 9.51). It thus occupies a central place in echinoderm phylogeny.

Edrioasteroids reached their acme in the Middle Ordovician. They are not uncommon in the late Ordovician, but thereafter are found only in small numbers until the later Carboniferous with a slight expansion towards the end. Virtually all known species come from northern Europe and North America. Some genera were free-living, but most were permanently fixed by the marginal ring to the substrate. Sometimes living shells were apparently used as a base, and the high selectivity of particular species for one kind of shell suggests that these may have been commensal.

The majority of genera were discoidal and not unlike *Edrioaster*. In *Pyrgocystis* and *Rhenocystis*, however, there is a long tower-like stem of imbricating plates below the flat adoral surface; *Cyathocystis* is also columnar. These forms may have been permanently fixed in soft mud. In general terms edrioasteroids seem to be most abundant when encrusting hardgrounds. At one locality in the Devonian of Iowa (Koch and Strimple, 1968), many edrioasteroids (*Agelacrinites hanoveri*) along with the coral *Aulopora* and the cystoid *Adocetocystis* are present in what must have been their life habit, attached to a discontinuity surface. The sea floor at that time consisted of wave-fretted limestone, eroded into knobs. Edrioasteroids are found fixed to the sloping surfaces of these knobs; all stages in growth are found. Another edrioasteroid species lived on the sides of the cystoid thecae. The whole assemblage was overwhelmed by silt, and all specimens are preserved in place.

9.4 Subphylum Asterozoa (Fig. 9.33a–e)

Asterozoans (Class Stelleroidea) are grouped in three subclasses and include the primitive Subclass Somasteroidea, starfish (Subclass Asteroidea), and brittle stars (Subclass Ophiuroidea). They are relatively rare in the fossil record since they break up very easily after death, and the paucity of their fossil remains is no guide to their former abundance. Nevertheless they are amongst the most abundant of animals living today on rocky shores of continental shelves and in the deep sea.

In all asterozoans the central part of the body (disc) extends laterally into five or more **arms**, the mouth faces downwards and the anus, where present, is aboral. In asteroids the arms are not sharply marked off from the central disc, whereas in ophiuroids the central disc is clearly delimited and bears five flexible snake-like arms (Greek: οφιος = snake). Most of the visceral elements in asterozoans, such as the intestine and gonads, are broadly homologous with those in other echinoderms. The water–vascular system resembles that of crinoids in that the radial water vessels lie in deep grooves in the ventral (adoral) surface. This is one of the many features that suggest a relationship with crinoids rather than echinoids.

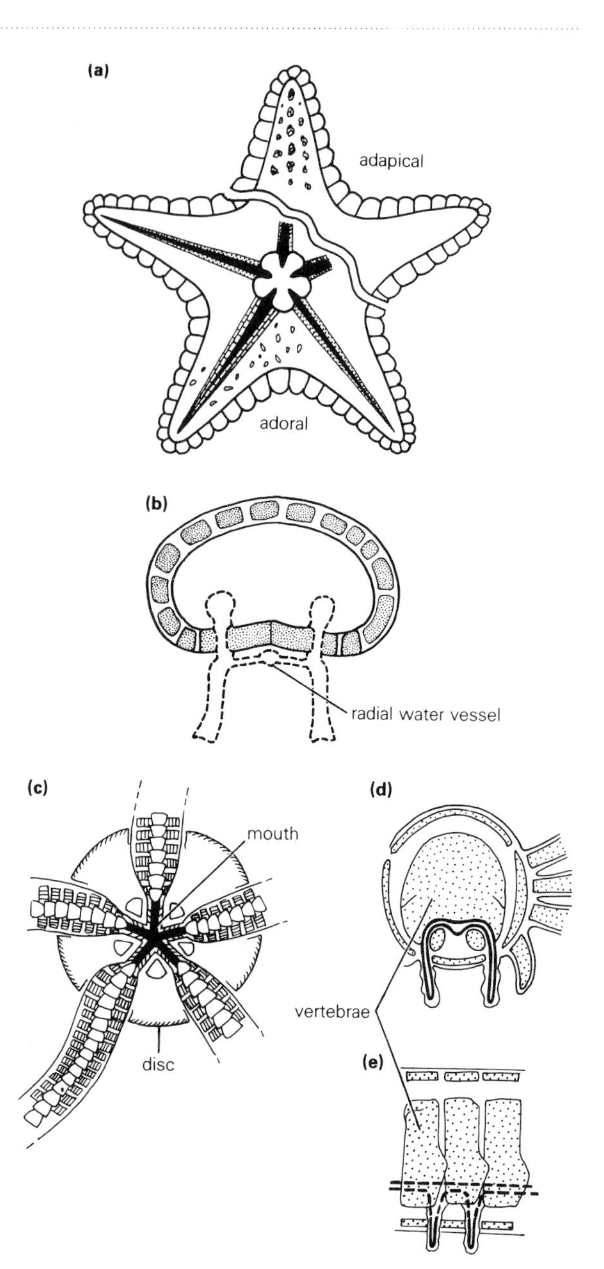

Figure 9.33 (a), (b) Asteroid anatomy: (a) external. with food grooves and florescent gills; (b) cross-section of arm, showing plating structure and external position of radial water vessel. (c)–(e) Ophiuroid anatomy: (c) adoral surface; (d) cross-section of arm with vertebrae and radial water vessel behind; (e) lateral view. [(a), (b) Modified from Sladen; (c)–(e) partially redrawn from Woods, both in *Treatise on Invertebrate Paleontology*, Part U; (c)–(e) also partially redrawn from Nichols, 1974.]

Subclass Asteroidea (Fig. 9.33a,b)

Asteroids usually have five arms, though these may be further subdivided. These arms normally extend like digits from the disc, but some starfish have a very pentagonal outline with short arms. On the upper surface there are a series of plates, which from the top down are the **carinals**, **dorso-laterals and marginals**. These may interlock so that the test is rigid, but more often they are flexible, allowing the arms to curl, which is very useful during climbing. On the lower surface the plates are larger and organized in each arm into a double column of ambulacral plates, through which the tube feet connect by pores to the ampullae even though the radial water vessel is on the outside of the test. Outside these are the **adambulacral plates** which adjoin the marginals. All the adoral plates unite around the mouth in a rigid **peristomal ring**.

Between the plates (ossicles) of the upper surface project extensions of the coelom, known as **papulae**, which are respiratory; since asteroids are very active animals, such 'extra' respiratory structures are essential. The water–vascular system of Palaeozoic asteroids has been reconstructed (Blake and Guensberg, 1988). Ampullae may or may not have been present, and if present were external in the early asteroids, by contrast with the large internal ampullae of modern types.

The first starfish probably derived from a pentameral free-living stem group, which included such echinoderms as *Stromatocystites*, the early members moving freely across the sea floor and feeding on detrital particles. The earliest true asteroids were Ordovician, derived from a somasteroid ancestor. It has been proposed (Gale, 1987) that these early asteroids were relatively inflexible, with limited capacity for arm movements, and they may have fed simply by sediment shovelling. An alternative view (Blake and Guensberg, 1988) is that the modes of life of early asteroids were diverse, and broadly parallel with those of today. Asteroid radiation is clearly related to a diversification in feeding habits. Most modern starfish are voracious predators and feed on molluscs, in particular preying actively on bivalves. When a starfish such as *Asterias* locates a bivalve it climbs on top of it, engages the two valves with its suckered tube feet, and for several minutes or longer pulls the two valves in opposite directions. Eventually the stress set up becomes too great for the bivalve's adductor muscles, and the valves open slightly. The starfish then everts its stomach out of its mouth and squeezes it bit by bit inside the opened valves. Digestion of the soft parts takes place within the bivalve shell, and when feeding is complete the stomach returns inside the starfish again. After feeding, the starfish can go without food for months without ill effects. There is some evidence from characteristic associations of disarticulated bivalve shells with fossil starfish that this extraoral feeding habit is ancient, extending as far back as the Devonian and probably further. In a well-known example (Clarke, 1912) a large bedding surface of Devonian rock revealed some 400 specimens of *Devonaster eucharis* in intimate association with the bivalves *Grammysia* and *Pterinea,* the environment having been 'invaded by starfish which congregated in vast numbers in order to feed on the clams'. There is now some doubt about this claim, and in any case not all asteroids feed in this way; some feed intraorally, as do many ophiuroids, by taking the prey, whether gastropod or bivalve, into the mouth. Even here, however, intrusion of the stomach lobes into the bivalve appears to accompany digestion. Other starfish are deposit feeders, and basket stars are suspension feeders. Since some starfish may have preyed actively on bivalves for at least 400 million years, it is highly probable (Carter, 1967) that some of the characteristic physiological and structural characteristics of bivalve shells (e.g. the interlocking commissures) can be considered as defensive structures against asteroids, the ever-present enemy. Indeed the morphological innovation of interlocking commissures in bivalves arose around the same time as the great expansion of extraorally feeding asteroids, in the later Ordovician and early Silurian.

It is possible, too, that the failure of articulate brachiopods to radiate anew following the late Permian extinctions, is directly linked to the early Mesozoic radiation of asteroids (Donovan and Gale, 1990). Against these the epifaunal brachiopods had little protection, and were especially vulnerable.

Subclass Somasteroidea

The earliest known 'starfish' (e.g. *Villebrunaster, Chinianaster*) are Tremadocian and come from southern France and Bohemia. These belong to the Somasteroidea, a short-lived group probably ancestral to both the Asteroidea and the Ophiuroidea, though primitive ophiuroids occur in the same fauna.

The Somasteroidea have a pentagonal body shape and the arms, rather than extending from the body, are just beginning to differentiate. The skeleton consists of five biserial rows of plates forming the 'arms', which have lateral rod-like 'virgalia' projecting from them, filling up the space between the arms. The skeleton has a thin rim of marginal plates.

In some respects structure of the arms resembles that of the pinnae of a biserial inverted crinoid, and some palaeontologists believed that somasteroids were derived from crinoids. This view became especially popular when Fell (1963) described a Recent deep-sea genus, *Platasterias*, as a living somasteroid, and proposed various stages in the descent of somasteroids from crinoids. Later studies suggest, however, that this may be unrealistic and *Platasterias* has been restudied in great detail by Blake (1982) who disclaims Fell's view that it is a living somasteroid. In his estimation it is an unequivocal asteroid, though primitive in some respects, such as the absence of suctorial tube feet (which it does not need since it is a microphagous feeder). It is now regarded as a subgenus of the true asteroid *Luidia*, and its morphology, rather than being a consequence of crinoid ancestry, is contingent on behaviour and habitat.

The first true asteroid is of early Arenig age, and by the Middle Ordovician the somasteroids had largely disappeared. Most Recent species of the Asteroidea, like those of the Ophiuroidea, belong to long-ranged genera which have persisted for a long period of geological time.

Subclass Ophiuroidea (Fig. 9.33c–e)

The most distinctive feature of ophiuroids, namely their thin, snake-like arms, has been an important factor in their great success, since they use sinuous movements of the arms for locomotion.

All the viscera are contained in the central disc, and all that lie within the arms are the greatly enlarged **vertebrae**, homologous with the fused ambulacral plates of asteroids. These have special articulating hinges so that the arm is very flexible. A sheath of plates surrounds the vertebrae, and the strong muscles that move the arms are found in the space between. Though the earliest ophiuroids are found in the same beds as the first somasteroids, the latter were probably their ancestors. The ambulacra enlarged to form vertebrae, and the radial water vessel became enclosed by the growth of a protective **ventral shield**. There are two orders: the Ophiurida (Ord.–Rec.), in which the arms can bend only in the horizontal plane, and the Euryalae (Carb.–Rec.), in which the arms can move in all planes; the latter are often climbing forms.

Ophiuroids are immensely successful in modern oceans, especially at bathyal and abyssal depths, where they are often found crowded together in great numbers. They are mainly suspension feeders, but there are carnivorous species which feed on small bivalves, though these do not have the extraoral feeding habit of asteroids. Ophiuroids seem to have retained much the same organization since the Ordovician, and their genera, like those of asteroids, are very long-ranged. They do not seem to have been badly affected by any major extinction periods.

Starfish beds

The remains of fossil asterozoans are all too scanty, since the plates normally disarticulate after death, not being bound together as are those of echinoids. Several well-known starfish beds are known, however, in North America, in Great Britain and in the Devonian of Germany. In the latter case starfish are almost always associated with the strange arthropods *Mimetaster* and *Vachonisia* (Stürmer and Bergström, 1976) (Fig. 12.8).

There are three or four British starfish beds which have long merited attention. Of these, one in the Lias of Dorset, is full of ophiuroids (two species), but there are no other fauna. Sedimentological criteria (Goldring and Stevenson, 1972) make it plain that these ophiuroids were smothered and rapidly buried by a thick cloud of silty sediment. Modern ophiuroids cannot escape from sediment more than 5 cm thick, and these Liassic specimens probably all died in their life position.

The famous Ashgillian starfish bed at Girvan in Scotland has an extremely rich fauna of starfish, trilobites, brachiopods, molluscs and the early echinoids *Aulechinus*, *Ectenechinus* and *Eothuria*. From the sediment inside the echinoid tests and other criteria it has been concluded that the fauna was shifted some distance and rapidly buried following turbulence in a shallow-marine environment. Other cases which have been considered confirm that only as a result of 'catastrophic' conditions are starfish likely to be preserved whole. Thus a thin lower Jurassic bed at Gmünd in Germany has a rich fauna of starfish and also of other echinoderms (Rosenkranz, 1971; Seilacher *et al.*, 1985). There is, however, very little other fauna apart from oysters killed by the same smothering event. The shelly bottom fauna was overwhelmed by an influx of sediment and, whereas the mobile pectinid and limid bivalves could swim upwards to escape, the fixed oysters could not and the echinoderms, with their ambulacral systems choked by fine sediment, had no chance of survival, but were preserved whole and as a 'selected' fauna. Such taphonomic selectivity is normal in other beds in which echinoderms are very well preserved (Brett and Eckert, 1982).

9.5 Subphylum Crinozoa

Crinozoans are primitively stalked echinoderms ('pelmatozoans') with long arms and normally lack complex respiratory structures. The comatulid crinoids have, however, lost their stalks and become secondarily free.

Class Crinoidea

Crinoids are very diverse and important in Palaeozoic faunas, and their remains have contributed substantially to Palaeozoic limestones. Complete crinoids (e.g. Fig. 9.34), however, are rarely preserved.

Crinoids are less abundant than they were, but at the present they are represented by 25 stalked genera and by some 90 genera of unstalked comatulids, which are the dominant group of modern oceans. One of these, the modern free-living *Antedon* (Fig. 9.35) is taken as a type example showing the basics of crinoid morphology.

Antedon has a stalk in its early life, formed of **columnar plates**, but it soon breaks free of it before it is fully grown and is then free to swim or crawl over the sea floor.

The body consists of a globular plated cup (the theca) from which long plated arms (**brachia**) arise. The theca has two parts: a lower region of thick rigid plates, pentamerally symmetrical, and a domed flexible roof (tegmen) with a central mouth and lateral anus. Inside the theca is a spirally twisted gut and a circumoral ring as in echinoids, but the madreporite is replaced by ciliated funnels. The gonads are borne not within the theca but on the arms.

Since there is no stem the base of the cup (or calyx) is made of a single, large (**centro-dorsal**) plate which is morphologically the top plate of the stem. Above this is a ring of five **basal plates** with another ring of five **radial plates** above it. Attached to the radial plates are the **brachial plates** of the lower part of the arms. These arms subdivide almost immediately so that there are ten arms in all, each armed with many **pinnules** constructed of **pinnular plates**. In *Antedon* and other comatulids the rigid plates of the calyx are very small and contain only a chambered part of the coelom. A number of flexible **cirri** articulate on the centro-dorsal plate and can be used for temporary fixation or locomotion of the living crinoid.

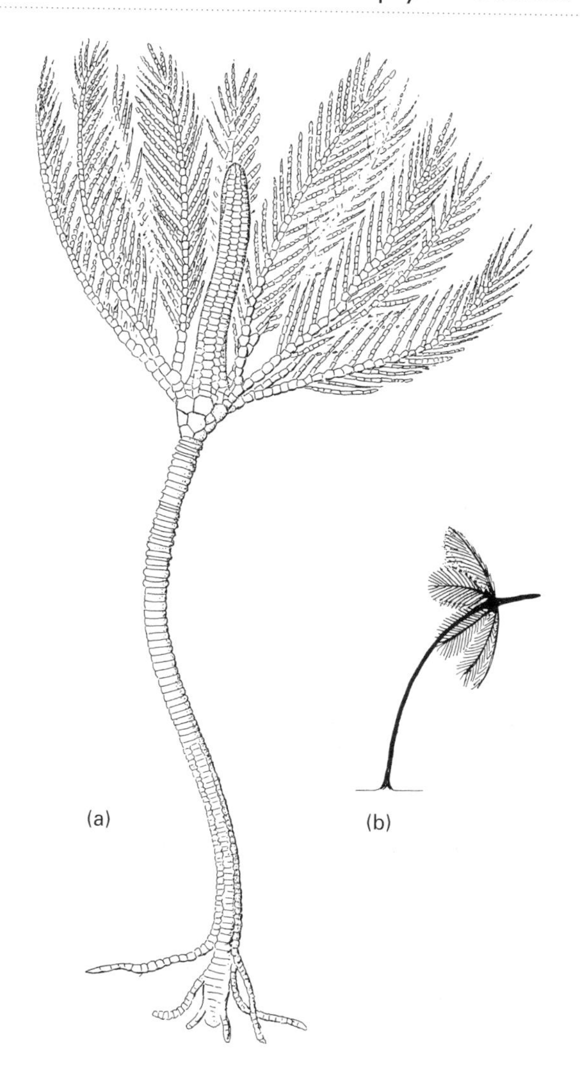

Figure 9.34 Features of crinoid morphology shown by Ordovician *Dictenocrinus*: (a) as redrawn from Bather in *Traité de Paléontologie*, 1953; (b) reconstructed as a rheophile.

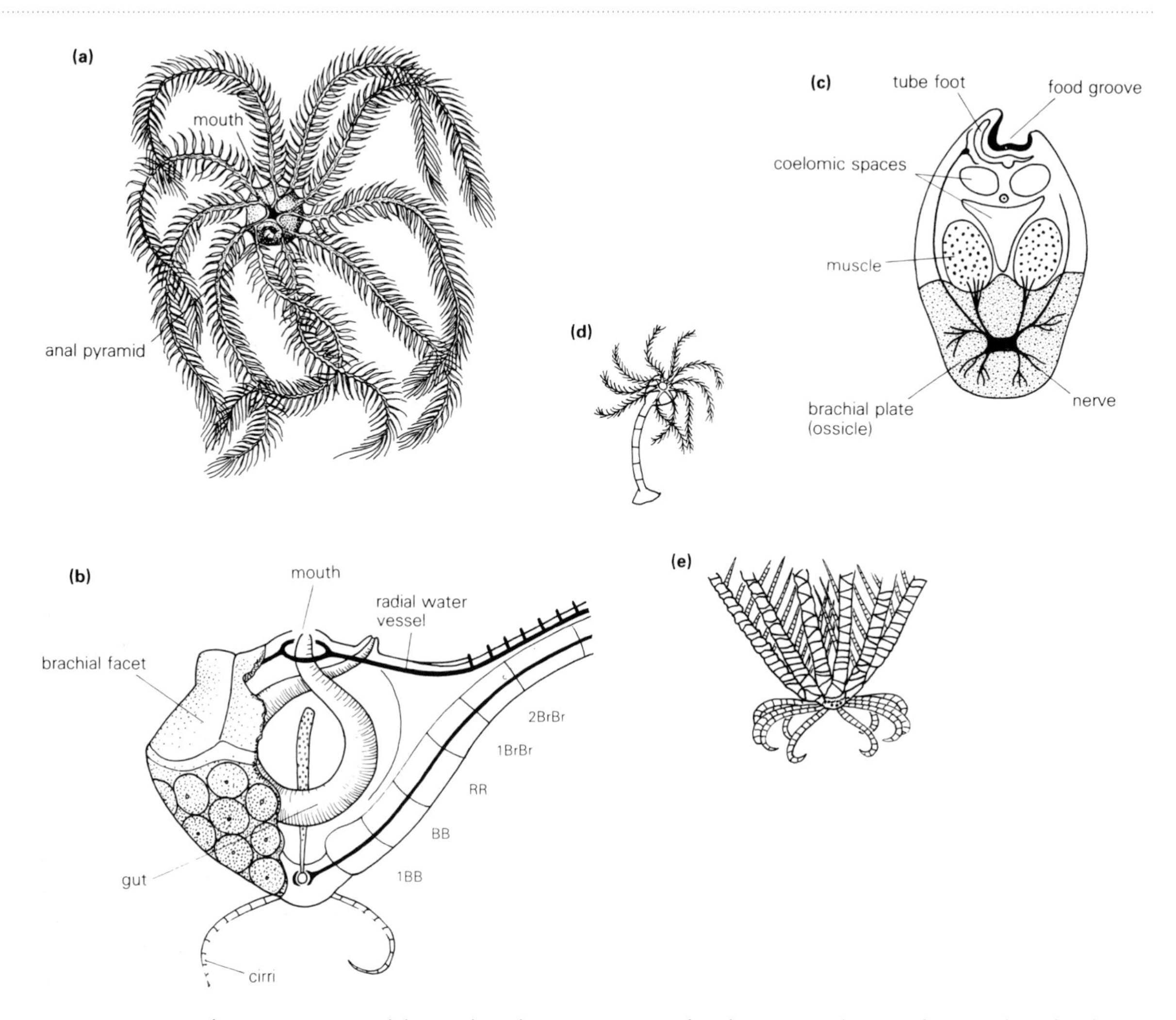

Figure 9.35 *Antedon*, a Recent comatulid crinoid: (a) living specimen in dorsal view (×1); (b) vertical section through calyx; (c) cross-section through arm; (d) larval specimen still attached to stalk; (e) lateral view of proximal part of a living specimen with flexible cirri below. [(b) Redrawn from Claus, *Textbook of Zoology*, 1884; (c) redrawn from *Treatise on Invertebrate Paleontology*, Part U.]

The arms are flexible, having articulated brachial ossicles and two segmented strands of muscle with which the arm can move in any direction (Fig. 9.35c). From the circumoral ring a radial water vessel runs down each arm, which, as in starfish but by contrast with echinoids, lies outside the plates. A double row of tube feet arises from this vessel, with a median food groove in between. When the crinoid is feeding, the pinnules with these tube feet are extended and, being sticky, catch organic detritus and small organisms which are then carried to the mouth by a **ciliary mucus tract** along the food groove. The pinnular food grooves all join with those of the brachia, which in turn unite so that five primary grooves cross the tegmen and enter the mouth. When disturbed, the tube feet can retract into the food groove, and as they do so small calcitic **cover plates** (**lappets**), normally held erect, come down like trapdoors to protect them.

Antedon has only 10 arms, but other comatulids may have as many as 200. The deeper-water species seem on the whole to have fewer arms.

Main groups of crinoids

The oldest known crinoid is *Echmatocrinus* (M. Cam.) from the Burgess Shale, but it is not very similar to later crinoids and only identified as such because of its uniserial arms and the well-preserved tube feet. Other related echinoderms, including the blastozoan Class Eocrinoidea, are also Cambrian. The relationship of *Echmatocrinus* to its Ordovician descendants of 40 Ma later is not understood, but in the early Ordovician there was a great radiation of early crinoids, contemporaneous with that of other echinoderms (Donovan, 1988a) probably concentrated in tropical seaways. It seems that the Ordovician was an 'experimental period' in crinoid morphology and Ordovician faunas reached a high level of diversity in successive adaptive radiations. Since the crinoid skeleton, however, disarticulates rapidly after death, complete specimens are rare. The study of the ring-shaped columnar plates (also known as columnars or ossicles) of which the stem is composed is helpful in appreciating diversity, since they are far more common than complete specimens or cups (Donovan, 1986, 1989). Thus in the British Ordovician crinoid fauna, columnal diversity suggests a variety of species several times greater than otherwise recognized.

Traditionally, the earliest 'true' crinoids (e.g. the Arenig *Dendrocrinus* and *Cupulocrinus*) have been held to belong to Subclass Inadunata (Fig. 9.36), which have rigid calycal plates.

There are two other Palaeozoic subclasses: the Flexibilia (Fig. 9.43c), in which calycal plates are only loosely united, and the Camerata (Fig. 9.37), where the proximal arm ossicles are incorporated into the theca.

Inadunates just crossed into the Triassic; the others became extinct in the Permian. A possible Recent survivor of the inadunates might be *Hyocrinus,* an Antarctic form, though this could merely be an aberrant articulate.

Again, traditionally, all Mesozoic to Recent crinoids are placed in a fourth subclass, the Articulata, in which the arms are very flexible. Most modern crinoids are unstalked comatulids; the acquisition of brachial flexibility and liberation from the stem were undoubtedly of great importance in determining the evolutionary potential of this crinoid group, allowing an active search for good feeding grounds. The living stalked species are unusually small, inhabiting depths greater than 100 m and having stems rarely exceeding 0.75 m.

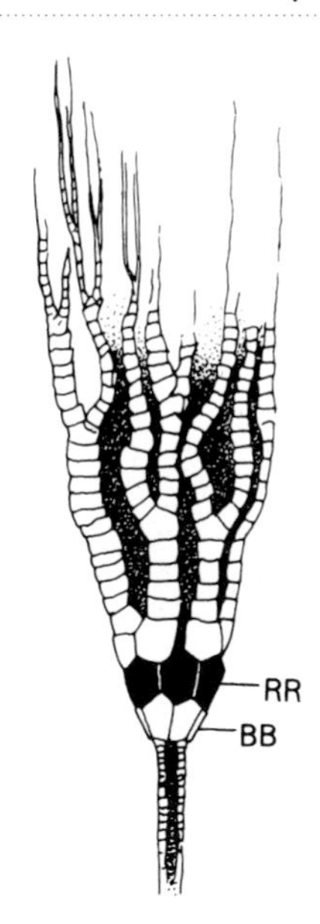

Figure 9.36 *Cupulocrinus* (Ord.), as preserved in the rock (× 2). (Drawn from a photograph by Ramsbottom, 1961.)

Crinoid classification is currently in a state of flux, and Simms and Sevastopulo (1993) ascribe many Palaeozoic groups to the Articulata; some of these likewise had flexible, muscular arms. These authors recognize the Camerata as morphologically distinct from the beginning, but they replace the Inadunata by the two subclasses Disparida (Palaeozoic) and Cladida (Ord.–Rec.) They consider the Flexibilia and the Articulata to be infraclasses of the latter, and seek the ancestry of the Articulata amongst the late Palaeozoic cladids.

Palaeozoic crinoids

Most, though not all, Palaeozoic crinoids have stems, which in the case of some Lower Carboniferous forms are of great length. They generally have larger calycal plates than *Antedon*, and a third circlet of plates (the **infrabasals**) may be present under the basals. In describing crinoids, the morphology of the calyx is usually represented by an 'exploded diagram' (Fig. 9.37c), and the plates are distinguished symbolically as brachials (BrBr), radials (RR), basals (BB) and infrabasals (IBB). Where there are only RR and BB present the crinoid is **monocyclic**; **dicyclic** crinoids have IBB as well.

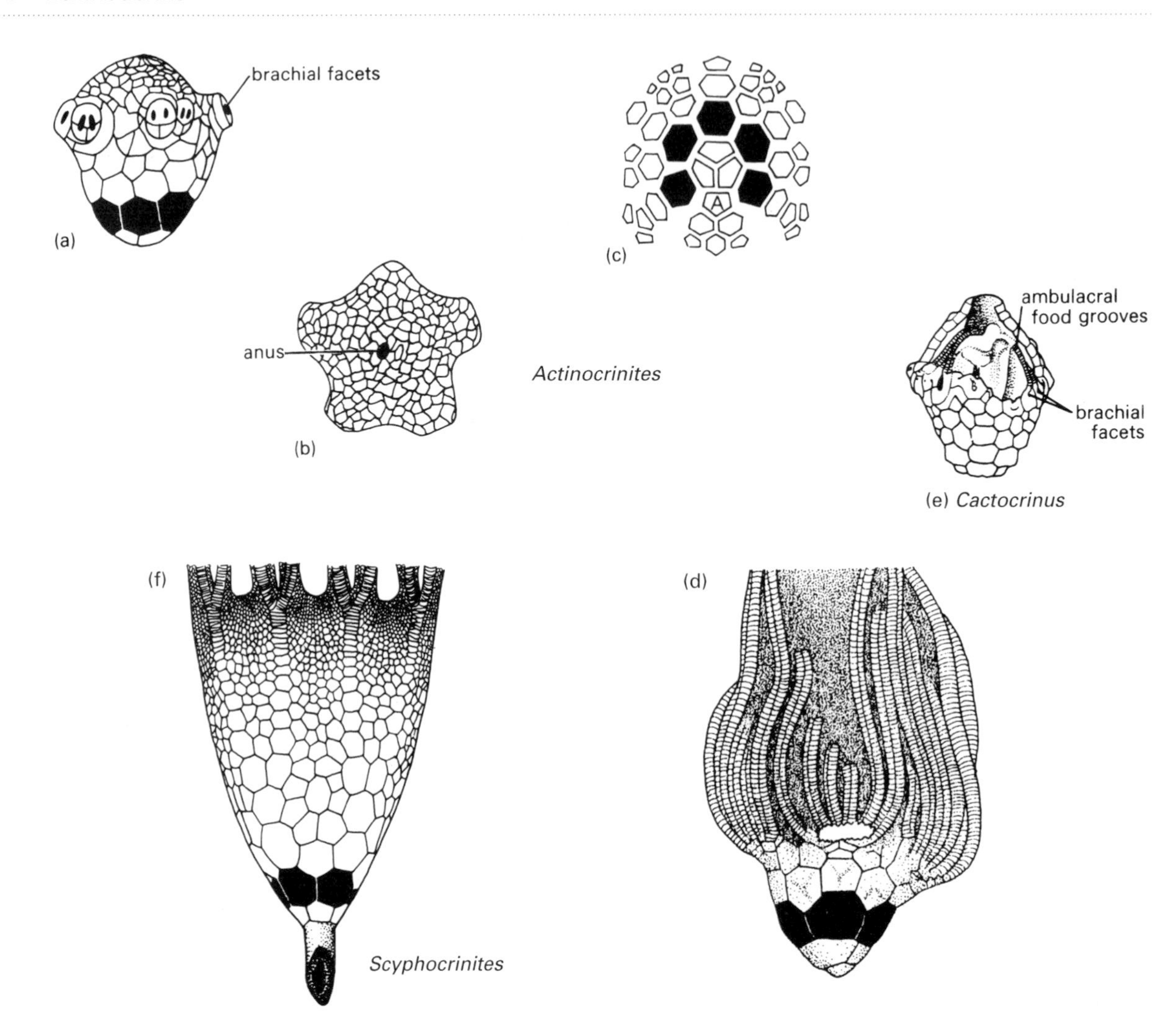

Figure 9.37 (a)–(d) *Actinocrinites* (L. Carb.), a monocyclic camerate crinoid with brachial plates incorporated into the large calyx: (a) lateral interareal view with brachial facets (× 1); (b) tegminal (oral) view; (c) exploded plate diagram; (d) specimen with brachia still attached; (e) broken specimen of the related *Cactocrinus*, showing subtegminal enclosed food grooves leading from the brachial facets to the mouth; (f) *Scyphocrinites* (Dev.), a large camerate, with pinnulars incorporated into the calyx (× 0.3). [(a)–(d) Largely based on Woods, 1946; (f) redrawn from Moore *et al.*, 1953.]

Inadunate crinoids

The Inadunata (L. Ord.–Trias.) is a very large subclass with more than 200 species. Inadunates are probably the most primitive of all crinoid groups (Figs 9.36, 9.38), having a rigid calyx with the brachials free or loosely connected above the radials.

The mouth is below the tegmen, the food grooves above it. The arms may or may not have pinnules. Commonly the pentameral symmetry of the calyx is disrupted by an extra plate, the **radianal** (RA), within the radial circlet, and there is usually an anal (X) plate in the brachial circlet (Fig. 9.38c). The vast majority of inadunates are dicyclic, a few are monocyclic. Radial plates may be 'compounded', i.e. divided transversely into **infraradials** (IRR) and **supraradials** (SSR), and there are radianal equivalents (IRA and SRA). In some an **anal tube** emerges from above the radianals; this is greatly elongated in genera such as the Ordovician *Dictenocrinus* (Fig. 9.34) and *Dendrocrinus*, where it is

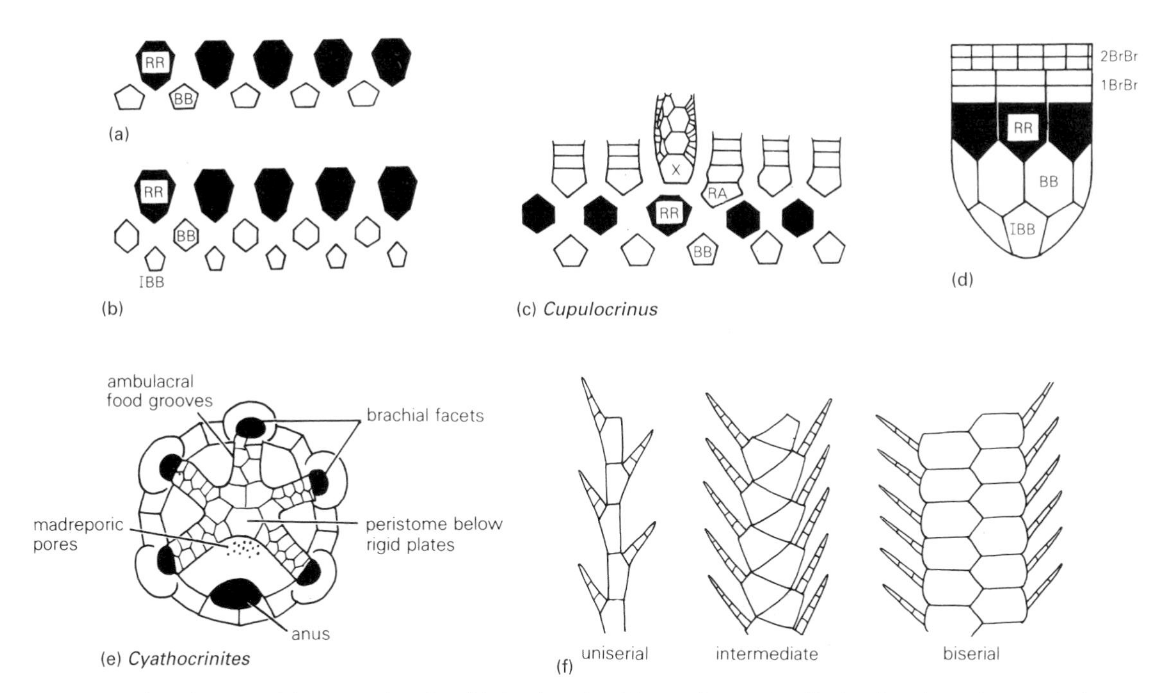

Figure 9.38 Plate diagrams: (a) monocyclic calyx; (b) dicyclic calyx; (c) *Cupulocrinus* (Ord.), dicyclic calyx with an anal tube (as in Fig. 9.36); (d) lateral view of a dicyclic calyx (IBB, infrabasals; BB, basals; RR, radials, in black; RA, radianal; X, X-plate; 1BrBr, primibrachs; 2BrBr, secundibrachs); (e) tegmen of crinoid *Cyathocrinites*; (f) evolution from a uniserial through an intermediate to a biserial condition in the brachia. [(c) Redrawn from Ramsbottom, 1961; (d) redrawn from Bather in *Treatise on Invertebrate Paleontology*, Part U; (f) modified from Moore *et al.*, 1953.]

made of thin, folded plates. The earliest inadunates had an elongated straight cup, but in many later genera the calyx became flattened and expanded laterally. Furthermore, the almost vertical direction of the dichotomous arms in the early inadunates was generally replaced by a more flower-like crown.

Numerous other evolutionary trends have been distinguished, and many specialized forms are known. *Petalocrinus* (Fig. 9.39) is a tiny Silurian inadunate resembling a minute palm tree.

Here the stem is short, the calyx small, and each of the five brachia fused into a leaf-like divided plate, with a series of dichotomous food grooves impressed upon its surface. *Hybocystis* has only three free arms, the other two being represented only by food grooves extending down the sides of the cup. The Ordovician Porocrinidae (e.g. *Porocrinus* and *Triboloporus*; Fig. 9.40a,b) are a strange family which have been placed in the inadunates (Kesling and Paul, 1968).

In these the calyx is strengthened by thickened ridges linking the plate centres. Within this framework, where the plates join there are thin circular areas of the calyx bearing folded calcitic membranes (**goniospires**). Each goniospire consists of three sets of folds, one for each of the adjacent plates. The

Figure 9.39 *Petalocrinus* (Sil.), a small inadunate, having the brachia welded into flat plates with food grooves. The cross-section shows the cover plates. (Redrawn from Tasch, *Paleobiology of the Invertebrates*, 1973.)

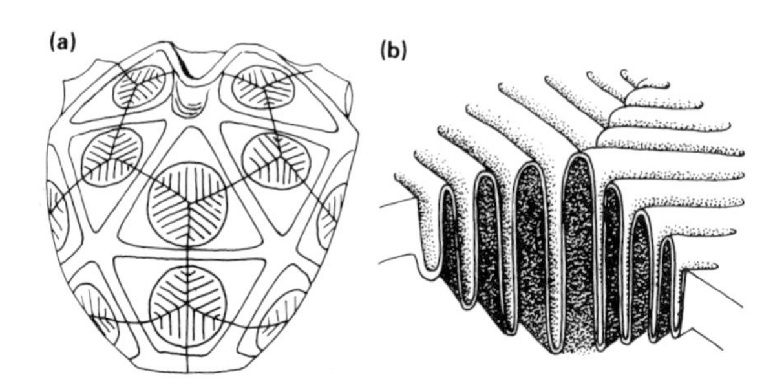

Figure 9.40 *Porocrinus* (Ord.), with folded membrane structures at the plate junctions: (a) lateral view of calyx (× 4 approx.); (b) section through folded membrane. (Redrawn from Kesling and Paul, 1968.)

goniospires probably had a respiratory function, but of greatest interest is their similarity to other types of folded membrane structure in both cystoids and blastoids, which arose quite independently. Some functional and other considerations on such structures are given later. The last known inadunates are Triassic. The well-known *Encrinus lilliformis* which occurs in the German Muschelkalk is one of these.

Flexible crinoids

The Flexibilia (M. Ord.–U. Perm.; e.g. *Protaxocrinus*; Fig. 9.43c) are less numerous and diverse than inadunates. They are all dicyclic with three infrabasals, one smaller than the other two. The lower brachials are incorporated into the cup but are not rigidly joined so that the whole cup is flexible. The mouth and food grooves are on the upper surface of the tegmen. There is a radianal and an anal plate. Some evolutionary trends, such as the relative widening of the calyx and the tendency for the lower part of the arms to droop over the cup, are parallel with those in the inadunates.

Camerate crinoids (Fig. 9.37)

Subclass Camerata (Ord.–M. Perm.) is the largest of all the Palaeozoic groups. In this group, which appeared some 4 Ma after the first inadunates, the cup is rigid, the radianal plate is absent and some of the lower brachial plates are incorporated into the calyx. The mouth and food grooves lie below the tegmen, and in some cases (e.g. *Cactocrinus*) the thin subtegminal tubes linking the brachia with the mouth have been preserved and can be dissected out. The arms always have pinnules. There are both monocyclic and dicyclic genera classified in the suborders Monobathrida and Diplobathrida, respectively. The earliest camerates, such as the Lower Ordovician *Reteocrinus*, have many features reminiscent of inadunates, implying a relationship. A crinoid such as *Actinocrinites* exhibits the basics of camerate structure clearly, in the exploded plate diagram and in the lateral and oral views. The tegmen is large and dome-shaped and has a subcentral anus. Arms arise from the circular **brachial facets**, which are lateral projections from the calyx. The many **interbrachial plates** (IBrBr) and brachial plates which are incorporated into the calyx make a large rigid theca.

In both monobathral and diplobathral crinoids there are evolutionary trends parallel with those of the Flexibilia. Thus in the early genera such as the diplobathral *Reteocrinus*, the interbrachial plates are small and irregular, lying in depressed zones between the arms. In the descendants of *Reteocrinus* they have become large, regular polygonal and less depressed. Furthermore, the anal and interbrachial plates tend in the later genera to move upwards and out of the cup, permitting a perfect radial symmetry with the radials all in contact. Many other apparent evolutionary trends have been reviewed (Simms, 1990, 1991), but have to be interpreted with caution. An evident trend common to all crinoids is the increasing prevalence of forms with pinnulate arms. This may, however, represent only an increase in 'variance' (i.e. the number of types with pinnulate arms) followed by the extinction of all groups with non-pinnulate arms, rather than a true trend impelled by functional necessity.

There are many unusual camerates. The Devonian *Scyphocrinites is* a rather extreme form in which not only the lower brachials but also the lower pinnular plates have become incorporated into a very large theca. Many genera, such as *Barrandeocrinus*, developed curiously downturned arms, drooping over the cup. The arms here are very broad, with the pinnules bent over to meet medially so that a trough was formed, presumably acting as some kind of food groove. The melocrinitids are a group of camerates in which the crown is rigid and bears five arms with five sets of parallel pinnules between, leading towards the centre. Cowen (1981) drew the remarkably close analogy between this arrangement and the ideal layout of a banana plantation! The problems are the same, harvesting an evenly distributed particulate resource, and its delivery to a central processing point with minimal expenditure of energy. As Cowen put it, 'cost–benefit analysis' should play a part in functional studies.

Mesozoic to recent crinoids: articulates

The earliest articulate crinoids are Triassic, and the early genera bear a considerable resemblance to inadunates, from which they probably evolved. Some articulates, however, could possibly have come from the Flexibilia; the order would thus be polyphyletic.

In articulates the calyx is relatively reduced: five basals, five infrabasals and five radials usually being present. The mouth and food grooves are exposed on the surface of the flexible tegmen, but brachials are never incorporated in the cup. Articulates are further and most importantly distinguished by the flexibility of their long arms, which can move up and down, coil up or twist about. It is this character that has given them a great evolutionary advantage, since the many genera of Order Comatulida which break free from their stalks in the early stages are able actively to move around on the sea floor rather than being fixed. Of modern crinoids only 25 genera have stalks, and these are rarely more than 0.75 m long, contrasting with some of the giant 3 m stalked forms of earlier times. The stems frequently have elongate cirri erupting from nodes and these cirri are muscular, enabling the crinoid to 'walk about' on them.

Stalked isocrinids such as *Pentacrinites* and *Seirocrinus* are quite abundant in the Jurassic, and there are many mid-Cretaceous forms. Mesozoic and Tertiary genera are generally long-ranged and include a number of micromorphic forms, e.g. *Eugeniacrinus*. The unstalked comatulids are very abundant today, there being 90 genera. In certain regions of the world, Antarctica in particular, there are endemic comatulid faunas which have evolved in isolation since the beginning of the Pleistocene. The Antarctic genera now seem to be spreading northwards along the many submarine ridges radiating from the Antarctic continent. Comatulids first became common in the Jurassic; evidently the potential for movement inherent in the unstalked crinoids' arms was realized early.

Successive comatulid genera continued through the Mesozoic and Tertiary. The Jurassic *Saccocoma* (Fig. 9.41b) from the Solnhofen Limestone has very long feathery arms and curious flattened lappets, being extensions of the proximal arm plates.

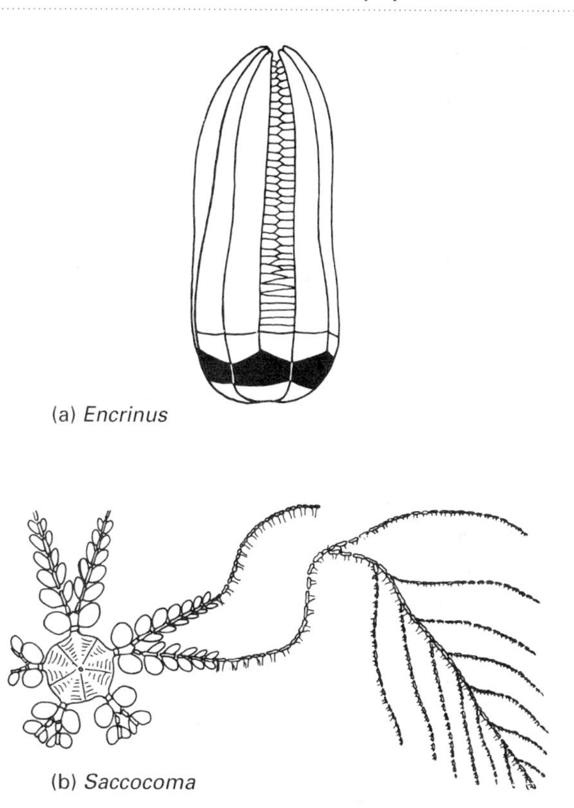

Figure 9.41 (a) *Encrinus* (Trias.), an advanced inadunate (× 0.75); (b) *Saccocoma* (Jur.), a possibly planktonic articulate (× 5). (Redrawn from Moore *et al.*, 1953.)

It is the most numerous macrofossil in the said lithographic limestone (Chapter 12), and because of its peculiar morphology has traditionally been considered as a free-swimming form, floating mouth downwards with the lappets increasing the surface area and thus retarding sinking. Milsom (1994), however, has reinterpreted *Saccocoma* as a benthic form, lying passively on the sea floor with the arms arching upwards from their midpoint and forming a horizontal collecting bowl for feeding, or used to intercept currents. *Saccocoma* was probably able to crawl and, since there was little other benthos, it seems to have been an opportunist, flourishing in vast numbers during periods when the sea floor was favourable to life. Large, plated, stemless genera (e.g. *Marsupites* and *Uintacrinus*) are well-known Cretaceous fossils. Most living genera are unknown as fossils.

Ecology of crinoids

Modern unstalked crinoids live at all depths from sublittoral to abyssal; stalked forms are normally found only below 100 m. Crinoids are uncommon in shallow water and tend to be most abundant in rather inaccessible regions, so that until recently little was known about their life and feeding habits. First-hand diving observations have illuminated many unsuspected adaptations and stimulated functional considerations on ancient forms. However, knowledge is still very limited, and most observations so far have been made largely on comatulids.

Present-day comatulids divide into two quite distinct feeding types: **rheophilic** (current seeking) and **rheophobic** (current avoiding). Rheophilic crinoids (Figs 9.34b, 9.42, 9.43a) live in relatively shallow current-swept waters and make use of the currents in feeding.

During the day individuals may hide in cavities, but come out at night to feed, climbing with their cirri onto a high position on a rock or sea grass and unfurling their arms. The arms spread out in a **vertical filtration fan**, with the polar axis horizontal, the aboral side facing the current and the pinnules spread out to form a grating. The tube feet are extended from these to form a fine filtering net. Even with a current speed of only 2 cm s^{-1} a surface of 500 cm^2 can filter 1 l s^{-1}. Such a system is remarkably efficient, making use of existing currents.

Figure 9.42 *Cenocrinus asterius* Linné; a Recent rheophilic crinoid living at depth of 200–300 m of the north coast of Jamaica. The arms are of radius 20–30 cm and form a vertical filtration fan. (Photograph reproduced by courtesy of Dr D.B. Macurda.)

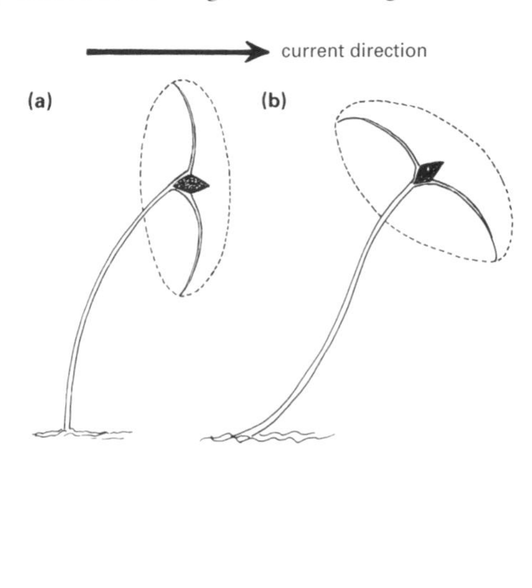

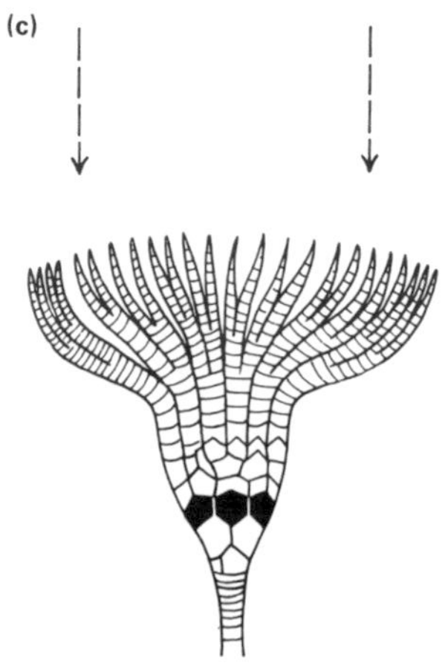

Figure 9.43 (a), (b) Stalked rheophilic crinoids with calyces arranged to act as vertical filtration fans: (a) with no lift from the current (b) with lift; (c) rheophobic crinoid (based on *Protaxocrinus*) with head acting as a horizontal bowl. [(a), (b) modified from Breimer and Webster, 1975.]

Rheophobic crinoids live in current-free waters (though most deep-water crinoids are rheophilic). They rely entirely, as far as is known, on gravity feeding; only through a continuous rain of detrital particles can they obtain enough to eat. In the laboratory *Antedon* behaves as a rheophobe, though it is naturally a rheophile. Rheophobes (Fig. 9.43b) are much less common. They lie on the bottom, the arms outspread to form a **horizontal collecting bowl**, capturing organic particles falling onto it. Some rheophobes live in deeper waters, so that a fair amount of detrital material has accumulated by the time it reaches them, but they do not consume anything like the quantity that rheophiles do. They are simply exploiting a different source of food, and in compensation their growth and metabolic rates are probably less. Abyssal crinoids (and others) may filter horizontal currents.

Most fossil crinoids, stalked or unstalked, were probably rheophiles or rheophobes; hence their mode of life may be interpreted from their morphology (Breimer, 1969; Breimer and Webster, 1975). Camerate crinoids have long and very flexible stems, as do some inadunates. Sometimes crinoid faunas consisting largely of camerates occur in high-energy carbonate sequences. In the Lower Carboniferous Burlington Limestone of Missouri, the many camerates (Families Actinocrinidae, Batocrinidae, etc.) are interpreted largely as rheophiles. The stems are strong but flexible, and their flexibility increases towards the cup. The crown could thus assume the attitude of a vertical filtration fan, with the aboral surface facing the current and the long anal tube facing away from it (Figs 9.34b, 9.43a). Since the arms of most camerates are biserial with very many pinnules, these crinoids would seem to have been highly efficient filterers, straining off food particles borne by the currents which, as is clear from the sediments, are known to have swept the Burlington Limestone sea.

Some camerate stalks have cirri on one side only and were probably fixed to the sea floor in a recumbent attitude, with only the crown and upper stalk rising above the sea floor, again being held as a vertical filtering fan.

Most flexible crinoids have short rigid stalks, and the crown could not have bent into a vertical position. Furthermore, the arms of the Flexibilia do not have pinnules, so all the evidence is against a rheophilic mode of life. They are best interpreted as rheophobes with the crown acting as a horizontal collecting bowl (Fig. 9.43c), though some may have been macrophagous like modern basket stars.

Despite the flexibility of their stems, no living crinoids have any muscles running down the column and it is most unlikely that any fossil ones did either (Donovan, 1988b, 1989). The only fibres extending through the column are non-contractile ligaments of collagen. These, however, are mutable and connected to the nervous system, and they are involved in controlling the posture of the column. This is also aided by the form of the articulation between adjacent columnals (Fig. 9.44).

Synostosial articulations, with their plane surfaces, allow almost no flexibility of the column. They are rare and confined to early Palaeozoic crinoids. Symplexial articulations, with their radial ridges, would prevent twisting of the column and hence snapping, but would allow 'rocking' between adjacent columnals to allow the stem to bend. They would also act as guides to allow the columnals to return to their original position after such flexure. Synarthrial articulations, with their central fulcral ridges, allow substantial rocking between each columnal and hence a very flexible stem.

A number of very diverse crinoid faunas are known, mainly from the USA, and can be interpreted ecologically. The Crawfordsville Limestone of Indiana is full of the stalks and crowns of inadunate and flexible crinoids and was deposited in quiet and poorly aerated waters. On morphological grounds most of the crinoids were probably rheophobes (though they might possibly have been

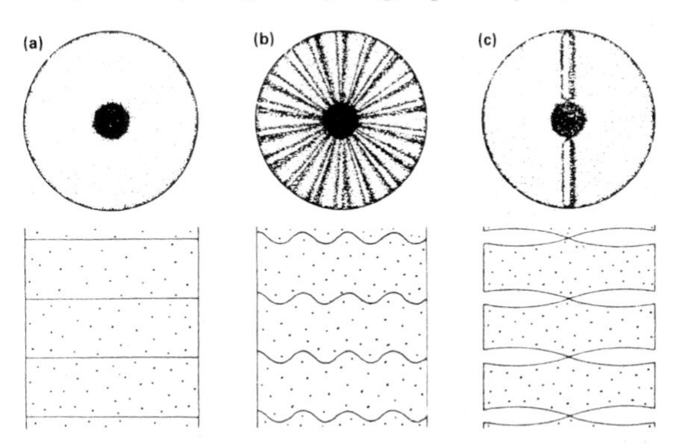

Figure 9.44 Different kinds of articulating geometries in crinoid columnals (ossicles), with longitudinal sections through columns below: (a) synostosial articulation; (b) symplexial articulation; (c) synarthrial articulation. (Redrawn from Donovan, 1988b.)

able to create their own currents by undulatory movements of the arms). The very different aspects of the Burlington and Crawfordsville crinoid fauna definitely point to a different ecology and feeding type.

Perhaps the largest and most diverse fauna of echinoderms known is in the Middle Ordovician Bromide Formation of Oklahoma. In this extraordinarily rich deposit some 61 genera belonging to 13 classes have been described (Sprinkle, 1982). They lived in a variety of warm-water shelf deposits with depths ranging from 3 to 75 m and were commonest in environments with a substrate of lime mud and skeletal debris, especially where these were small bryozoan mounds providing suitable attachment sites. Although the fauna is crinoid-dominated, so many other echinoderm groups are present that this is surely the richest of all echinoderm faunas known from any place or time.

Palaeozoic crinoid gardens may have been stratified like a tropical jungle, partitioning resources by feeding on different food at different levels.

Fossil articulates can likewise be interpreted as rheophiles or rheophobes. However, in the case of the Mesozoic crinoid family Pentacrinitidae (*Pentacrinites* and *Seirocrinus*) different interpretative criteria have been used. These are very large, often spectacular crinoids, often found completely preserved in black shales (e.g. Simms, 1989). There are two possible suggestions as to their mode of life. One is that they were benthic, but their long stems gave the calyx and arms the advantage of elevation, so that the feeding crown was able to filter from a region of the water not exploited by other organisms. A long-stalked crinoid could have gained lift from the current, provided that it was able to orientate the parabolic fan at the correct angle (Fig. 9.43b). The whole system can be likened to a kite tethered by a cord. But unlike a kite the crinoid could probably regulate its own degree of lift by altering the curvature of the arms, the angle of the fan or the attitude of the pinnules. Given certain conditions, the crown could rise or sink in the water to exploit food-rich levels.

An alternative, currently more popular view (Seilacher *et al.*, 1968; Simms, 1989; Wignall and Simms, 1990) is that these giant crinoids lived a pseudoplanktonic life, hanging downwards from a floating log and thereby feeding from the surface plankton. Overcrowding is a problem for such pseudoplankton, hence the pendent life of such obligate forms as these. They are not uncommonly found with fossilized driftwood, and further evidence for this life habit comes from a single bedding plane, now excavated and displayed in the Tübingen Museum, in which lie some 50 crinoid specimens, each with a crown about 80 cm across and a stem up to 15 m in length. They are preserved with the crowns face down in black euxinic shales. One interpretation of these is that they lived a pseudoplanktonic life, hanging downwards from a floating driftwood log which slowly became waterlogged and sank, so that the crowns touched down first, followed by the stems which settled in loops as the log drifted slightly before coming to rest. In *Seirocrinus* the basal part of the stem is the more flexible, by contrast with the norm. This too is in accordance with a pseudoplanktonic mode of life, for the stem needs to be flexible at the base to withstand storms yet rigid near the calyx to maintain control of the inverted posture. Species of *Pentacrinites* and *Seirocrinus* show little morphological change through time, and indeed have remarkable longevity as compared with contemporaneous benthic crinoids. This conforms well to an interpretation as pseudoplanktonic – a case of stabilizing selection for a very specialized habitat. One group of *Pentacrinus* species, incidentally with a shorter, densely cirrate stem, may have departed from the pseudoplanktonic habit to have become benthic.

Formation of crinoidal limestones

Many Palaeozoic crinoids seem to have lived in 'crinoid gardens', i.e. in clumps or patches isolated from other such areas and in which diversity may be high or low. Other crinoids inhabited muddier water conditions.

After death, crinoids usually break up through decay, disarticulation or the action of scavengers. The specific gravity of the various components first increases, owing to the removal of investing tissue, then decreases whilst the stereoplasm (tissue penetrating the stereom cavities) is replaced by sea water. Where isolated columnals are found they have been transported by rolling and may have come from some distance. Commonly great thicknesses of limestone are found, formed of short or long lengths of stem, isolated ossicles or calycal plates. In the Carboniferous limestone of England crinoidal debris of this kind can often form the bulk of the rock. These rocks are more or less autochthonous and represent the debris of crinoid gardens where the material was buried prior to complete disintegration.

When orientated, short stem lengths are found in abundance; they were rolled by the current into an area of accumulation, and the direction of the current that moved them can normally be told from their orientation. Unorientated debris found on the same bedding planes apparently became stuck and was not moved by the current. The rare cases when crown and stem are preserved represent almost immediate burial before decomposition or, as in the case of *Seirocrinus*, deposition in quiet and possibly anoxic waters.

9.6 Subphylum Blastozoa

Classes Diploporita and Rhombifera: cystoids

'Cystoids' are a heterogeneous group of Palaeozoic blastozoans, all of which have respiratory 'pore structures' traversing the plates of the theca.

Haplosphaeronis (Class Diploporita; Fig. 9.45) is an Ordovician cystoid known mainly from northern Europe.

It has an ovoid theca less than 30 mm in height with a flat base by which it was attached to the substratum. The theca has two circlets of polygonal plates, the lower circlet of seven plates (LL) and the upper of seven **perioral plates** (POO). Above these lies the peristome, constructed of five perioral plates and having at its centre the slit-like mouth. From the mouth radiate five food grooves or ambulacra, each of which has several branches terminating in small studs (**ambulacral facets**) which were the points of attachment of the food-gathering arms (brachioles). Brachioles are rarely preserved in cystoids but are very similar in structure and function to the arms of crinoids, though small and simple. They were probably spread out in a kind of horizontal collecting bowl, extracting food from the water and passing it to the mouth via the ambulacra.

A well-defined 'anal pyramid' of five or more triangular **anal plates** is set at the junction of two of the POO plates. The anal plates could hinge open along their bases and close again after evacuation. A small slit known as the **hydropore**, probably the entrance to the water–vascular system and characteristic of cystoids, is set near the peristome. All plates except these of the peristome are densely covered with twin perforations, surrounded by a raised rim and reminiscent of the pore pairs of echinoids. These are known as **diplopores**. They probably led to an external 'papula', uncalcified and resembling an elongated tube foot extended into the surrounding water for respiration. This genus has been studied in detail by Bockelie (1984).

Structural characteristics

Some cystoids like *Haplosphaeronis* were fixed by their bases to the sea floor or to a shell; others had stems which have been occasionally preserved with the holdfast intact, but these stems were always short and never reached more than a few centimetres in length. Many of the Ordovician cystoids had thecae composed of numerous irregular plates (e.g. the diploporite *Aristocystites* and the rhombiferan *Echinosphaerites*).

The food grooves in cystoids were protected by cover plates. These food grooves may be short as in *Sinocystis* (Fig. 9.45) or long, but only in Family Cheirocrinidae (Class Rhombifera) is there any documented trend to lengthening the ambulacra with time.

The curious Ordovician genus *Asteroblastus* has a star-shaped system of food grooves, very similar in appearance to the ambulacra of blastoids. Brachioles arose laterally from these, and the diplopores were restricted to the interambulacral regions alone. It has been suggested that blastoids descended from cystoids through *Asteroblastus*. This is unlikely, however; it is more probably a case of convergent evolution. Brachioles are rarely found in cystoids, and where they are known they are always short and slender. The **gonopore** (for the shedding of genital products) and the hydropore are usually distinct but sometimes combined.

Pore structures (Fig. 9.46)

The pore structures of cystoids, which are of very many kinds, provide a firm basis for classification (Paul, 1968, 1972). Diplopores, as in *Haplosphaeronis*, are pairs of perpendicular canals through the theca, each opening in an external depression (the **peripore**). Other kinds of pore structure are based on U-shaped tubes of various

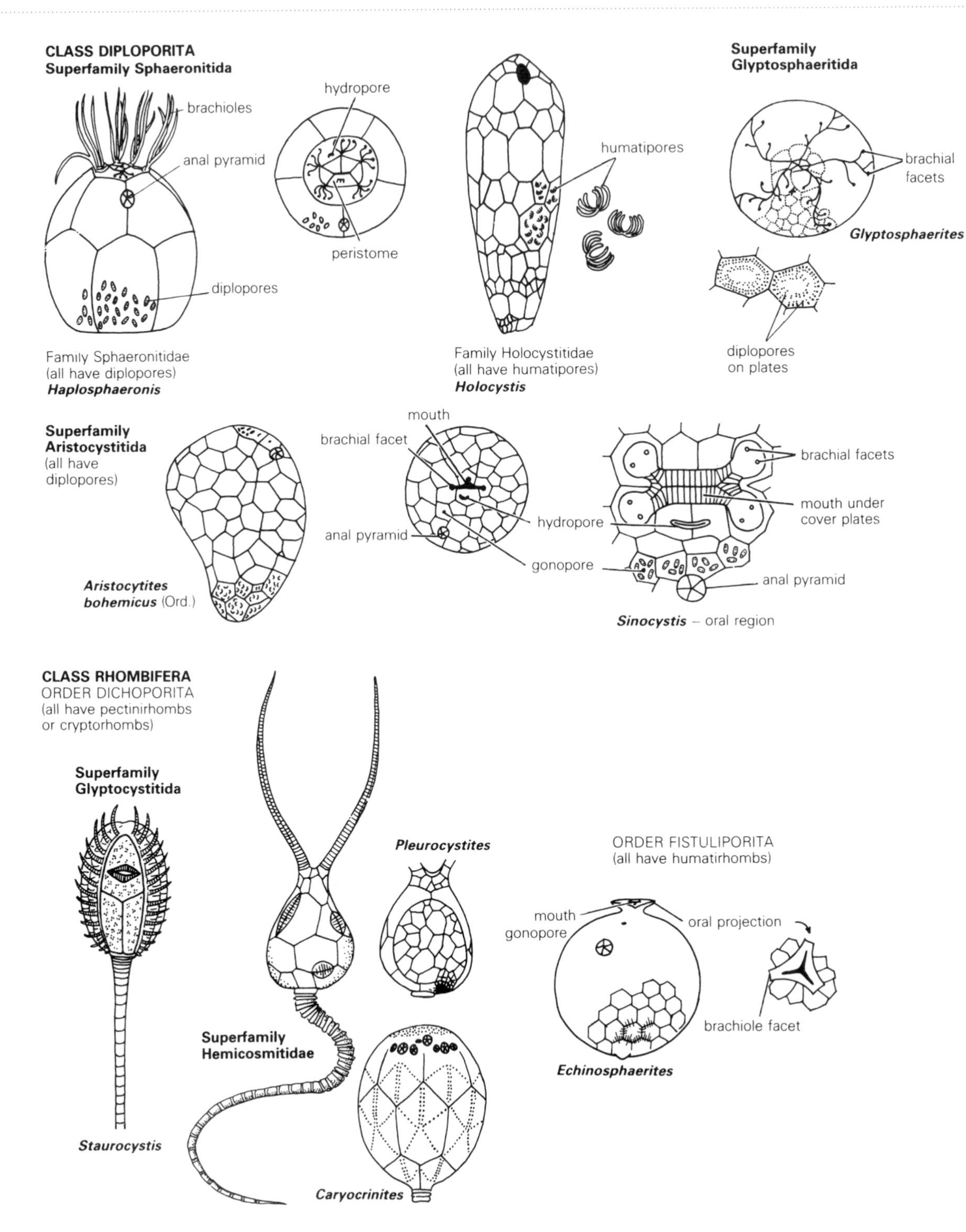

Figure 9.45 Cystoid morphology and classification according to Paul's system, showing the high taxonomic value of pore structures (Diagrams based mainly on Kesling *et al.* in *Treatise on Invertebrate Paleontology*, Part S.)

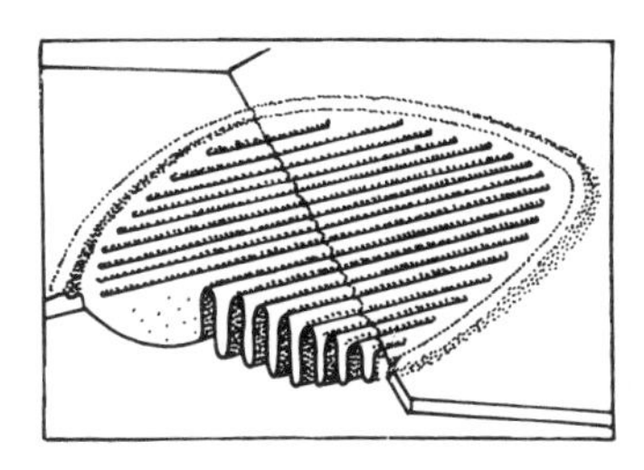

Figure 9.46 A pectinirhomb, crossing plate boundaries and showing the folded membrane. (Based on Paul, 1968.)

kinds known as **thecal canals** which open to the external or internal surface through **thecal pores**. External pores are round and simple, sieve-like or slit-shaped; internal pores are always simple. Thecal canals are always tubular and may traverse the theca perpendicularly or tangentially (diplopores are paired thecal canals forming U-pairs in the soft parts).

Two quite distinct functional types of pore structure can be distinguished. **Endothecal** pore structures (**dichopores**) had canals that ran below the external surface of the theca and communicated with the surrounding sea water through external pores. If sea water was pumped through the canal, presumably by cilia, respiratory exchange could have been effected by gaseous diffusion through the thin calcified wall of the thecal canal which bathed the internal body fluids. Respiration was thus internal. By contrast, **exothecal** pore structures (fistulipores) had internal pores and an external canal running outside the theca. Thus the body fluids would have to be brought through the canal and back down again within the theca, so that respiratory exchange was outside the theca. These two alternative methods of facilitating respiratory exchange have their advantages and disadvantages. Endothecal structures were liable to choking by foreign particles, but they may have slits, sieve plates or other devices to keep such foreign bodies out. A sieve usually protects the inhalant pore alone, enabling it to be identified. Exothecal structures did not have this problem but were liable to abrasion or breakage, and in any case they were much less efficient in terms of relative areas of surface exchange.

Diploporite cystoids such as *Haplosphaeronis* have numerous exothecal diplopores, as already described. These are simple paired tubes crossing the plate perpendicularly and leading to uncalcified external papulae. An alternative kind of structure is the **humatipore**, in which the paired perpendicular tubes are linked by a complex of exothecal tubes. These are either flush with the surface or occupy raised humps. Diploporites like *Holocystis* (Fig. 9.45) have their plates replete with such humatipores.

Rhombiferans may have exo- or endothecal pore structures which are always arranged in parallel sets with a rhomb-shaped contour. Each rhomb crosses a plate boundary. **Pectinirhombs** are highly organized units constructed of parallel dichopores. There are normally only two to four in any one cystoid. Each dichopore may open in a single slit (**conjunct**) (Fig. 9.46) or have the central part roofed over (**disjunct**) to form a separate entrance and exit. Dichopores may be joined laterally (**confluent**) forming a single folded membrane. Such pectinirhombs have a raised external rim and may be conjunct or disjunct. If disjunct, they have a thickened bar running along the line of the suture carrying a raised ridge on one side only. This has been explained as a mechanism for preventing the recycling of currents. The ciliary action of the cellular membrane lining the dichopores would produce a one-way current. Small particles dropping through to the base of the dichopore would be light enough to be disposed of by the ciliary current. Endothecal **cryptorhombs** are made of parallel dichopores without the raised external structure of pectinirhombs. They have sieve-like incurrent and simple excurrent pores. **Humatirhombs** are made of exothecal fistulipores likewise parallel and arranged in rhombs.

Classification

Many taxonomic schemes have been proposed for this complex and difficult group. The scheme of Paul (1972) adopted here (Fig. 9.45) differs somewhat from that of Kesling in the *Treatise on Invertebrate Paleontology*. Paul has given high-level taxonomic value to the pore structures. In Class Diploporita there are only diplopores scattered over the thecal surface, many to a plate, other than in Family Holocystitidae where there are humatipores instead. In Class Rhombifera the members of Order Dichoporita all have pectinirhombs or cryptorhombs, which are used to divide the order further into superfamilies. Humatirhombs characterize the other order, the Fistuliporata.

Ecology

Most cystoids were sessile and fed on small organic particles using their brachioles. These brachioles are seldom fossilized but do not normally seem to have been very large. Judging by the size of the ambulacral facets in the Holocystitidae, however, the brachioles may in this family have been of fair dimensions. Possibly the closely spaced brachioles of such genera as *Haplosphaeronis* formed a food trap like that of some modern Phoronidea.

Cystoids generally lived in shallow waters, and evidently diploporites were amongst the first colonizers of areas in which sediment deposition began again after a break. Cystoids had some tolerance of suspended sediment; conceivably exothecal pore structures could have been adapted to sediment-laden water. Where conditions were particularly favourable, cystoids lived together in clumps, and occasionally these assemblages were preserved in life position.

Some cystoid genera are strikingly modified. The rhombiferan *Pleurocystites* (Fig. 9.45) has an asymmetrical, flattened theca which has large irregular plates with pectinirhombs on one side and small flexible plates on the other. The two brachioles are large and forward projecting; the anus is located by the side of the stem, which is very flexible and tapering and lacks a holdfast. *Pleurocystites* probably lay on the sea floor with the flexible part of the theca down (Paul, 1967). The stem morphology would allow undulating movements which would permit forward propulsion, and thus the whole morphology of *Pleurocystites* seems to be specialized for a mobile, benthic life. As such it bears a remarkable superficial resemblance to similarly specialized 'homalozoans' (e.g. *Cothurnocystis*; Fig. 9.49c), which quite independently acquired similar characters through convergent evolution.

Why did cystoids need such elaborate pore structures for respiration when the crinoids, which outlasted them, got on perfectly well without them? Paul (1976) has presented a reasonable hypothesis about this based on the known palaeogeographical distribution of cystoids. One of the earliest known cystoids, the Tremadocian *Macrocystella* (which was originally thought to be an eocrinoid), lacks any pore structures, though its plates are thin and have radial ridges organized in rhombs. It lives in cold circumpolar waters, which would have had a fairly high oxygen concentration. Pore structures occurring in its descendants, in other cystoids, and indeed in all other echinoderms with elaborate respiratory structures, are found predominantly in shallow, tropical seas. Nearly all Ordovician occurrences of pore bearers are between 30°N and 30°S of the palaeo-equator. In such warm, shallow, tropical seas oxygen tensions are very low during the night, and it is suggested that the pore structures were well adapted for coping with this. Furthermore, from evidence available at present it seems that the more efficient pore structures (pectinirhombs) occur in genera living within 15°N and 15°S of the equator; the less efficient ones have a wider distribution.

Class Blastoidea (Figs 9.47, 9.48)

Blastoids (Sil.–Perm.) are an extinct class of small, pentamerally symmetrical 'pelmatozoans' characterized by internal respiratory structures known as **hydrospires**. An individual blastoid had a short, fixed stem surmounted by a calyx from which arose a crown of brachioles. The stem and brachioles are rarely preserved, and the majority of species are known from the calyx alone. The plating differs from that of crinozoans. In some genera there are extra and complex plates, but only the standard plating structure is given here. Some 80 genera are known.

Pentremites comes from the Mississippian of North America. Nearly 70 species belong to this genus. *P. symmetricus* (Fig. 9.47) from the Lower Carboniferous of Illinois has a bud-shaped calyx with five petaloid 'ambulacra' running from the summit to two-thirds of the way to the base. Each ambulacrum is indented by a vertical median groove. There are three basal plates (BB). Above them are five radial plates (RR), deeply indented by the ambulacra, and surmounting the radials are five small rhomb-shaped interambulacral **deltoid plates** (Δ). Below each ambulacrum lies a long, spear-shaped **lancet plate**, largely covered by the small plates of the food grooves.

At the summit there is a central, star-shaped mouth, and surrounding it are five large **spiracles**, each set at the summit of a deltoid plate. One of these (the **anispiracle**) is larger than the others

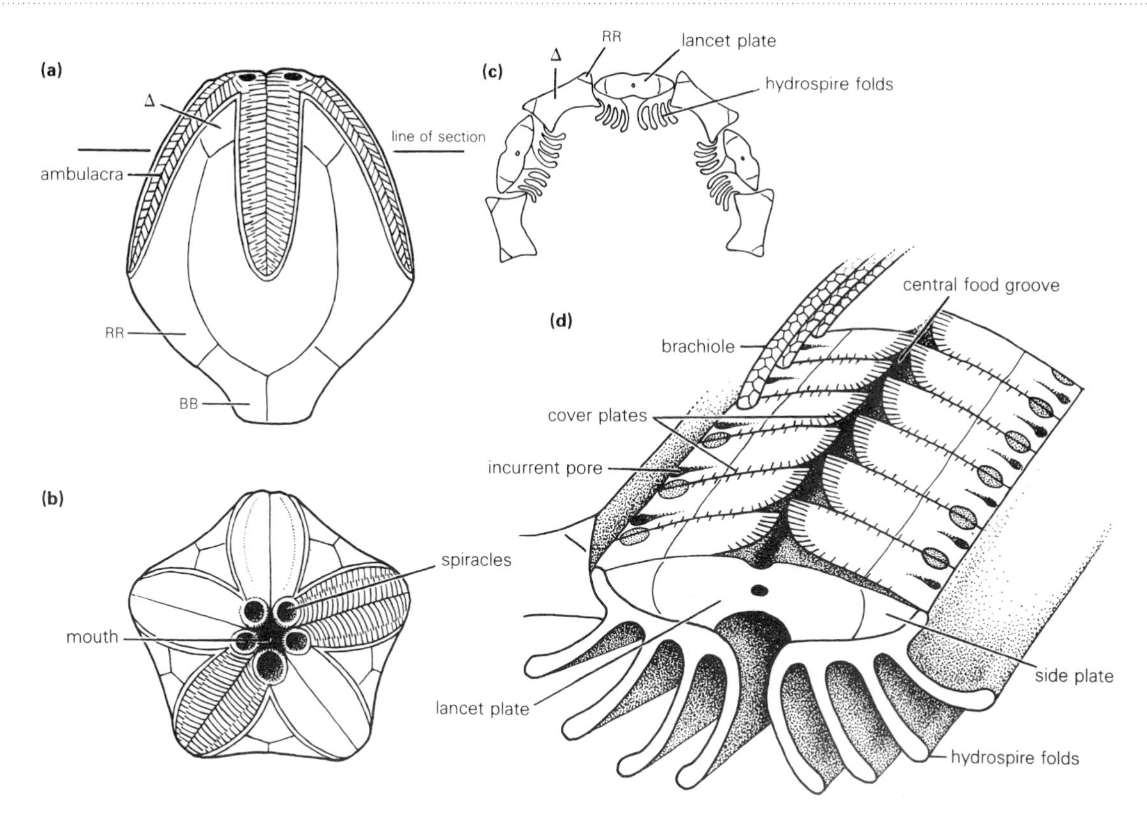

Figure 9.47 *Pentremites,* a Lower Carboniferous spiraculate blastoid: (a) lateral view (× 2); (b) adoral view (× 2); (c) transverse section with hydrospires (× 2); (d) oblique, three-dimensional section showing brachioles and hydrospire system (× 10).

because it contained the internal anus, suggesting that the spiracles were some kind of outlet system.

The ambulacral and subambulacral structure is complex and is best studied by making thin sections. These show that the lancet plate bears two lateral **side plates** associated with the elements of both the feeding and respiratory systems. Along each edge of an ambulacrum ran a single line of brachioles which, when broken off, left distinct facets. These brachioles led to food grooves with biserial cover plates crossing the ambulacrum and joining with the median food grooves which ran vertically to the mouth. Feeding was probably like that of crinoids, traces of the water–vascular system having recently been discovered. The hydropore was probably internal. Between adjacent brachioles is a narrow slit from which a pore leads down to the hydrospire system. Each of the paired hydrospires is simply a thin-walled, rigid, calcified tube, with a convoluted inner surface, lying below the ambulacrum and connected to the many **hydrospire pores** by individual thin tubes. A pair of hydrospires from adjacent ambulacra connects with a single spiracle. The subdivision of the inner face of the hydrospire into four smaller parallel tubes results from the infolding of the calcified membrane of which it is constructed.

Diversity and function of hydrospires

Hydrospires may, as in *Orbitremites* (Carb.; Fig. 9.48e), consist merely of straight undivided tubes. Usually, however, as in *Pentremites,* the hydrospires are convoluted (Fig. 9.47c,d), forming a **hydrospiraculum**, which may have from two to seven folds. The hydrospire pores were probably lined with cilia, and their beat presumably created a unidirectional current, drawing water in through the pores, passing it along the hydrospiraculum and exhaling it through the spiracles. The folding would undoubtedly allow a greater surface for respiratory exchange.

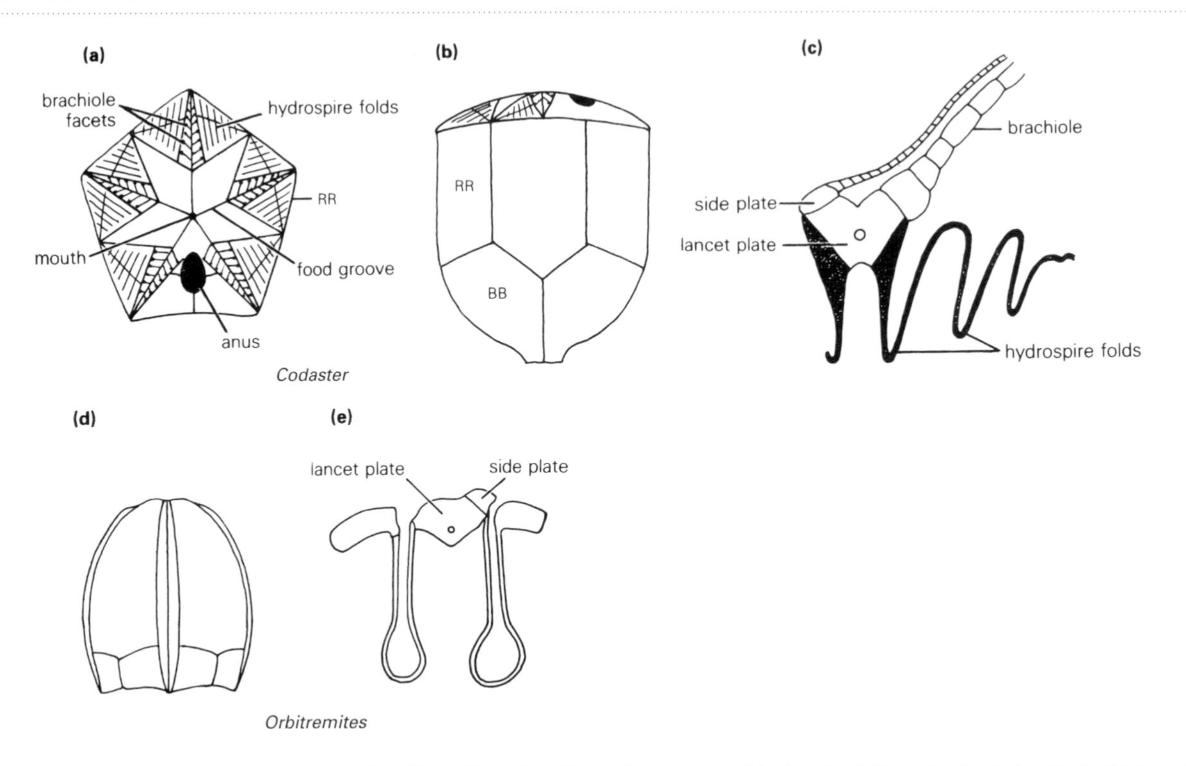

Figure 9.48 (a)–(c) *Codaster*, a Carboniferous fissiculate blastoid with exposed hydrospire folds: (a) adoral view (× 3); (b) lateral view; (c) ambulacra with hydrospires in section; (d), (e) *Orbitremites*, a Lower Carboniferous spiraculate blastoid; (d) in lateral view (× 3); (e) hydrospires in section.

How efficient were these structures? It is a fairly simple matter to calculate input and output flow rates by measuring cross-sectional areas of the hydrospire pores and the spiracles. In *Globoblastus*, where a complete study through ontogeny has been made (Macurda, 1965b), the potential outflow rate in an adult could have been six times the inflow rate. Thus at an inflow velocity of 0.1 mm s^{-1}, the water volume could be changed completely in 100 s in an adult and 40 s in a juvenile. In *Globoblastus* differential growth gradients seem to relate to hydrodynamic and feeding efficiency by controlling its external form.

Classification and evolution of blastoids

Blastoids arose in the Silurian from an undetermined ancestor. Two orders are distinguished: the Fissiculata and Spiraculata. The Fissiculata have open hydrospire folds, whereas the Spiraculata (e.g. *Pentremites*, *Globoblastus*, *Orbitremites*) have spiracles.

Fissiculate blastoids are apparently more primitive. *Codaster* (Fig. 9.48a–c), for example, has a subconical theca with a flattened upper surface. The upper edges of the radial plates form a ring around the horizontal deltoid, radial and lancet plates. The anus is off-centre, at the junction of two radials and a deltoid. Parallel with the edges of the lancet plates are the upper edges of the hydrospire folds, graded in size away from the lancet so that the hydrospire system has a half-rhomb-shaped outline. There are only a few reduced hydrospire folds between the anus and the adjacent lancets. The side plates are large and almost cover the lancet plates, as shown in Fig. 9.48c. It is presumed that the hydrospires functioned in a manner broadly similar to that of the spiraculate blastoids, but the system was probably less efficient.

The Fissiculata are widespread and lasted until the Permian. Such genera as *Orophocrinus* (e.g. Macurda, 1965a) are well-known components of the Mississippian reef knoll and carbonate shelf fauna. Bizarre genera include *Astrocrinus* (L. Carb.) which has one aberrant ambulacrum and a displaced anus. Amongst several odd genera from the Permian of

Timor is *Thaumatoblastus*, which has immensely extended lancets and a rounded base.

Fissiculates could have given rise to spiraculates by the lateral growth of the lancet and side plates and the inward migration of the hydrospires, though there is no unequivocal evidence. All spiraculates have true hydrospire pores and spiracles. They generally have the characters of *Pentremites* but may be long and thin (e.g. *Troostocrinus*) or near-globular with the ambulacra extending right down to the concave base. Genera are distinguished by the plating structure and the number of hydrospire folds.

Ecology and distribution of blastoids

Blastoids were never an abundant component of any fauna except locally, where in thin bands they are sometimes so numerous as to make up the bulk of the rock. There are, for instance, well-known bands in the Visean of northern England, where *Orbitremites* is so abundant as to permit mathematical study of variation and relative growth (Joysey, 1959). These individuals were especially abundant in crinoid bank deposits capping reef knolls.

The few Silurian blastoid genera, both fissiculate and spiraculate, are entirely restricted to the North American continent, but by the Devonian blastoids had become worldwide. They reached a maximum in abundance and distribution in the Lower Carboniferous with about 45 genera. Though they are rare in the higher Carboniferous, blastoids, often of peculiar form, are found in profusion in some very localized places in the Permian of the eastern hemisphere such as Timor. By the end of the Permian they were extinct.

9.7 Subphylum Homalozoa, otherwise calcichordates (Fig. 9.49)

The calcite-plated asymmetrical fossils known variously as homalozoans, calcichordates, or by the older and more non-committal name 'carpoids', are perhaps the most bizarre and controversial of all known extinct invertebrates. They have traditionally been regarded as echinoderms, but following Jefferies (1968) many modern scholars have ascribed them to Phylum Chordata (which includes the vertebrates). Jefferies and others use the term 'calcichordates'; they believe them to be calcite-plated chordates with echinoderm affinities which gave rise to other chordates, three times independently, by loss of the calcite skeleton (Jefferies, 1986, 1997; Jefferies *et al.*, 1996). So involved is this controversy that I can do no more here than sketch out a few details and direct the student to recent literature. None of the diverse organisms included in the carpoids has radial symmetry and most show no planes of symmetry at all (except in the bilaterally symmetrical 'tail'). They do, however, have projecting plated arm and tail-like structures, internal cavities and pores, and the problem has always been how to interpret them. Here terms from both the traditional classification of the *Treatise on Invertebrate Paleontology* and that of Jefferies are both referred to.

The Ordovician *Dendrocystoides* (Class Homoiostelea; or Soluta of Jefferies, 1990) has a lobed, asymmetrical, flattened theca of small irregular plates, a single 'arm' , and an elongated 'tail' with a stout polyserial proximal part and a long, tapering distal portion. A hydropore (which indicates that a water–vascular system was present) and a gonopore lie close to the base of the arm, while the anus is near, but to the right of, the insertion of the tail. There is a single gill slit on the left-hand posterior side. *Ceratocystis* (Class Stylophora, Order Cornuta) bears some resemblance to *Dendrocystoides* despite having larger plates, but there are also many differences, including the presence of seven gill slits on the left-hand side of the head. The most extreme version of cornute organization is found in *Cothurnocystis*. This is a laterally flattened form with a boot-shaped theca and a 'tail' not unlike that of *Dendrocystoides*. A marginal frame of stout plates surrounds the theca, extending into a 'tongue', 'heel strap' and 'toe' for the boot. On one of the flat surfaces there are projecting studs and a strengthening strut crossing the theca. Small rounded plates forming a flexible cover occupy the surfaces of the theca within the marginal plates. These are slightly different on the two sides. On the opposite side to the strut are some 15 slits arranged in a curving arc, each with a hanging flap of small plates hinged above and fixed externally (Fig. 9.49c). Behind the 'tongue' is a pyramid of plates closing an orifice.

It is agreed by all authors that *Cothurnocystis* probably lay flat on the sea floor supported by the marginal spines and the studs, with the surface with the strut lying in proximity to the sea floor, the slits uppermost and the aulacophore fixed in the mud

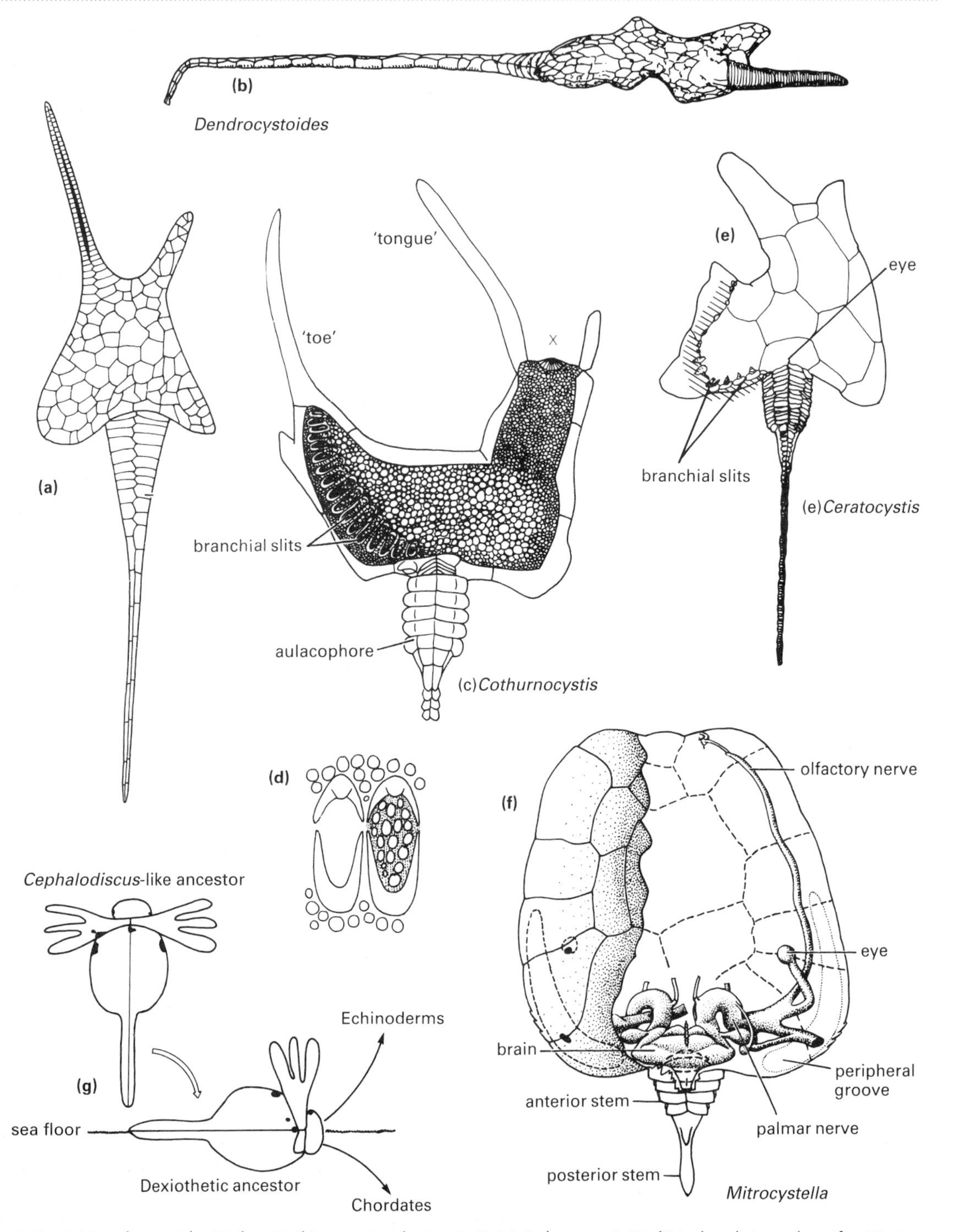

Figure 9.49 (a) *Dendrocystoides* (Ord.; × 1); (b) same, in side view (× 1); (c) *Cothurnocystis* (Ord.) in dorsal view – the orifice X is interpreted by Jefferies as the mouth and by Ubaghs as the anus (× 1.5); (d) branchial slits seen externally, showing the 'hanging curtain' of covering plates; (e) *Ceratocystis* (Ord.), in dorsal view (× 1.5); (f) *Mitrocystella* (Ord.) in dorsal view with right-hand plates removed showing nervous system as interpreted by Jefferies (× 5); (g) Jefferies' view of a *Cephalodiscus*-like ancestor giving rise to a dexiothetic progenitor of both chordates and echinoderms. [(a), (b), (e), (g) Redrawn from Jefferies, 1990; (c), (d), (f) based on Jefferies, 1968.]

or free. Experiments with models indicate that the tail was used to pull the head rearwards across the sea floor (Woods and Jefferies, 1992), and the recent discovery of a trace fossil with the allied *Rhenanocystis* lying at the end of its trail confirm that this was very likely the case. There is much disagreement as to how these structures are to be interpreted. The more traditional view of Ubaghs (1971) as to the homologies of the various structures vies exceedingly with that of Jefferies and his coworkers. Ubaghs has considered the slit-like pores to be the inhalant openings of respiratory organs, the large orifice behind the 'tongue' to be the anus, and the aulacophore to be a feeding organ. Jefferies, on the other hand, has proposed vertebrate homologies for the various organs. He has suggested that the theca is broadly homologous with the chordate head, the aulacophore is equivalent to the chordate tail, that the slit-like pores are exhalant and homologous with vertebrate gill slits, and that the large orifice is the mouth and entrance to the respiratory system. The presence of the hanging curtain of small plates outside the pores does in fact suggest that they were exhalant structures. The anus and gonopore, according to Jefferies, are a small perforation near the origin of the stem. Gill-slits on the left-hand side of the head recall the structures of the larval amphioxus, which likewise has left gill-slits only. According to Jefferies both *Dendrocystoides* and *Cothurnocystis* belong to the stem group of the Chordata.

Even more of a problem is the Ordovician genus *Mitrocystella* (Class Stylophora, Order Mitrata) which has a large, flattened, though asymmetrical theca of irregular plates and a short, stumpy, down-curved stem or tail. Jefferies (1968) has described this form as resembling a 'large calcite-plated tunicate tadpole'. The theca has a marginal rim of large plates, while the slightly smaller plates of both sides of the theca are irregular. An opening at the opposite end to the stem leads to a central cavity internally divided by ridges. In the inner dorsal surface of the theca are impressed series of stellate and elongate grooves; these have been reconstructed by Jefferies into a complex series of structures reminiscent of the brain and cranial nerves of a fish. These and other structures (e.g. the left-lateral position of the anus, as in a tunicate tadpole) led Jefferies to homologize the slit-like anterior orifice with a mouth, the stem with a vertebrate tail and the ramifying canal system with the higher nervous complex within the vertebrate skull. In support of this view, the asymmetrical pharynx of a tunicate, for example, would fit neatly into the head of a *Mitrocystella*. Ubaghs (1971), on the other hand, has interpreted the fringed opening as a periproct and located the mouth where Jefferies has suggested the anus is. His argument that the 'carpoid' skeleton is identical in structure with that of echinoderms is countered by the 'calcichordate school' who argue that there is no reason for it not to have been; such a skeleton has been retained in the echinoderms, and lost in the later chordates. Whereas the whole issue remains controversial and unresolved, it retains its fascination.

It has been known for a long time on embryological and anatomical grounds that the echinoderms and chordates are indeed related, and some authorities would group them together as Dexiothetica. Jefferies (1986, 1990) and Jefferies *et al.* (1996) view the latest common ancestor of both as something like the modern hemichordate *Cephalodiscus*, but lying on its right side in the mud (the dexiothecate condition) and thereby losing the openings and arms of the primitive right side. The various remaining organs can be homologized with those in the 'calcichordates', the stalk of the dexiothetic ancestor being the equivalent of the tail of *Dendrocystoides*, *Cothurnocystis* and by inference, modern chordates. *Dendrocystoides*, which had developed a calcite-plated skeleton, is seen to occupy a critical position in evolution of the chordate/echinoderm plexus since it retained both a tail and a water–vascular system. The former was lost in the echinoderms, while the latter disappeared in the later chordates along with the primitive calcite-plated skeleton. Here again the mitrates, according to the 'calcichordate school' are of critical importance, since they seem to have given rise respectively to the tunicates, cephalochordates and craniates, by independent loss of the calcite-plated skeleton, on three separate occasions. Hence the mitrates, from the cladistic point of view, should be assigned to the stem groups of these important taxa.

It is of course equally possible that the lines of descent leading to modern chordates took place amongst soft-bodied animals which left no fossil record, and that the 'carpoids', rather than being in the direct line of descent, were merely a rather peculiar sideline whose evolution was in some ways parallel. The debate triggered off by Jefferies' original work has continued for 30 years, and has proved

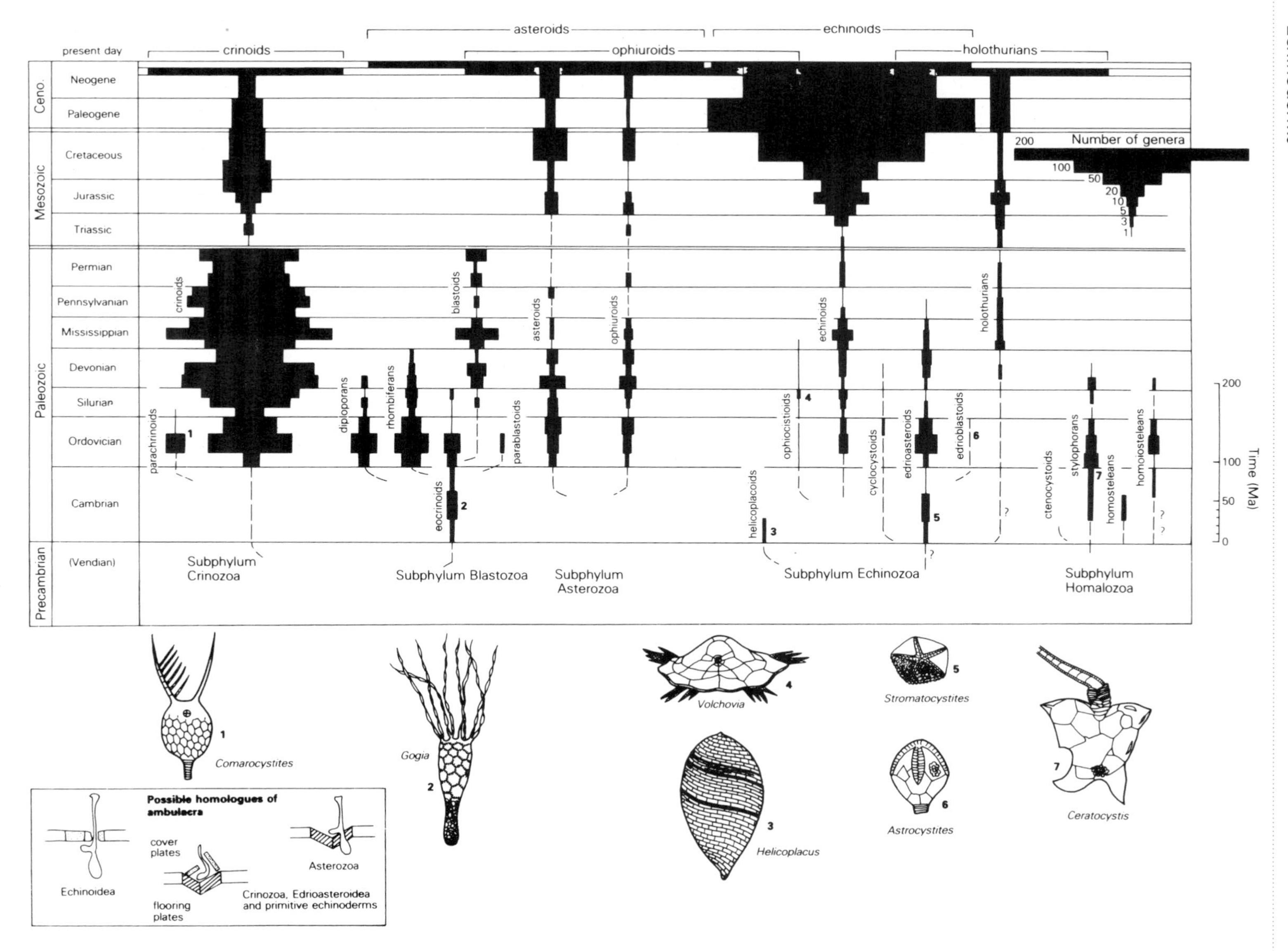

Figure 9.50 Time ranges and abundance of the various echinoderm groups. (Based on Sprinkle, 1983.)

to be perhaps the liveliest and certainly the most partisan controversy in the whole of palaeontology. Gee's (1996) work is the most objective recent synthesis, and treats the subject in great detail. He comments 'my sympathy for the calcichordate theory lies in the wealth and interest of questions it engenders . . . not because I advocate that it is "right" or "better"'. Admirably fair comment.

9.8 Evolution

There is no geological evidence as to the ancestry of echinoderms, since the first representatives appear already differentiated in the Lower Cambrian. On biological grounds, it seems probable that the chordate–echinoderm group (Dexiothetica) on the one hand, and the lophophorate (brachiopod–bryozoan) group on the other, were independently derived from a common worm-like ancestor, allied perhaps to the modern Sipunculida.

We have already referred to the model of Jefferies (1986) and Jefferies *et al.* (1996), which would derive chordates from calcichordates by loss of the calcite-plated skeleton. There are, of course, other views. An alternative proposal is that the ancestral adult of both echinoderms and chordates closely resembled the '**dipleurula**' larval stage of modern echinoderms. This small planktic larva is cylindrical, with a ventral mouth and anus and symmetrical coelomic pouches on either side. There is also a ciliated band forming a loop and running down either side. According to the 'dipleurula theory', the ciliated band was present in the ancestral adult, together with its coordinating nerve plexus. In the branch that led to chordates this band fused in the midline to provide the rudiments of the dorsal nerve cord. In the ancestral blastozoan–crinozoan stock, on the other hand, torsion led to the twisting of the gut into its present position with a central, upwardly directed mouth and a lateral anus, while one of the coelomic pouches was lost.

Earliest echinoderms and their radiations (Figs 9.50, 9.51)

The earliest echinoderms were already highly diversified on their first appearance in the Lower Cambrian (Paul and Smith, 1984). Perhaps the most primitive belong to the Cincta, disc-shaped forms such as *Trochocystites*, which has small plates surrounded by a marginal ring of stouter plates, and a short tail. Amongst the earliest echinoderms is the monogeneric Class Helicoplacoidea, specimens of which were first discovered in 1963 and are known from a few localities only in western North America. The body is fusiform, lacking pentameral symmetry and with the plates (and internal muscles) spirally arranged. In three respects it shares characters (autapomorphies) with living echinoderms; the plates have typical stereom structure, the ambulacra (three in number) have biserial cover plates, and they are radially arranged round the mouth. The mouth is laterally positioned and the anus is probably apical.

Three other, slightly later Lower Cambrian echinoderms have been described. *Lepidocystis* has a stalk, a cup, and arms with ambulacra running down them, characters unique to crinoids. *Stromatocystites* has a free-living disc-shaped body with five ambulacra on the first upper surface. It looks much like an edrioasteroid. *Camptostroma* likewise has a flat upper surface with a kind of pentamery, and a conical body below, but it retained a spiral internal musculature. Even in the Lower Cambrian these early echinoderms are widely separated in morphology and their relationships can only be understood cladistically. Smith (1988), using this approach, sees *Helicoplacus* as being so generalized that it belongs to the stem group of all later echinoderms, and from which evolution took place in two main directions, of which it was close to the latest common ancestor. The first derived group were the 'pelmatozoans', crinoids, blastozoans and other 'fixed' echinoderms of which *Lepidocystis* was an early representative. The Class Eocrinoidea (Subphylum Blastozoa) which belongs here became very diverse by the Middle Cambrian and persisted into the Silurian. In many respects these fossils resemble the Crinoidea but their plates close right up to the peristome and they do not have a tegmen. Subsequent developments included the later Blastozoa proper, of which the Eocrinoidea may have been the ancestral stock. There arose the cystoids and later the blastoids; short-lived forms of bizarre appearance, the paracrinoids and parablastoids, also originated and soon became extinct. Meanwhile, in the early Ordovician true crinoids had originated to become the dominant pelmatozoan group.

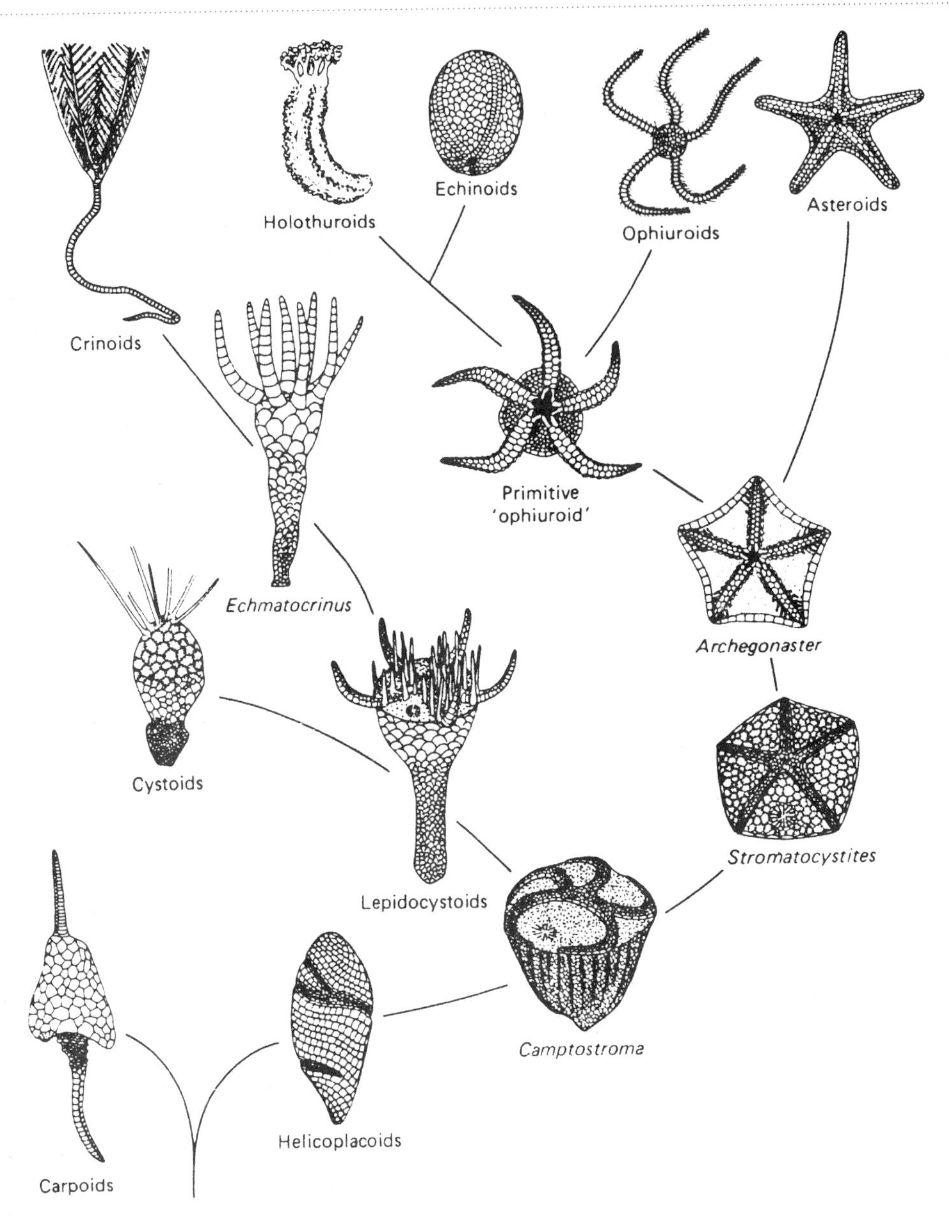

Figure 9.51 A possible scheme for the relationships of early echinoderms. (Redrawn from Paul and Smith, 1984.)

The second line of descent led to all eleutherozoans (asteroids, ophiuroids, echinoids and holothurians). Smith (1988) identifies the Middle Cambrian *Cambraster* as a member of the stem group which gave rise to the eleutherozoans. It is a small, disc-shaped echinoderm with five pentameral ambulacral arms leading to a central mouth, near which lies the anus, and a prominent marginal ring of plates. This is generally considered as an edrioasteroid (and undoubtedly a line of descent to true edrioasteroids led off from it through *Stromatocystites*). Smith regards it as belonging to a stem group which also gave rise to all later eleutherozoans. The asterozoans derived from this morphology, by extension of the arms and reduction of the marginal frame of plates, as seen in the early Ordovician *Villebrunaster* and the contemporaneous ophiuroid *Praedesura*. There are certain synapomorphic characters which the primitive ophiuroids share with the earliest echinoids: loss of the marginal frame, enclosure of the radial water vessels, the presence of articulated spines, and the modification of the oral ambulacral plates into a jaw apparatus. It has been argued, therefore, on cladistic grounds that the earliest echinoids (the late Ordovician *Aulechinus* and *Ectenectinus*) together with *Bothriocidaris* form the sister group of the ophiuroids. These first echinoids had a 'good Bauplan' but its potential was not fully to be realized until the Mesozoic. A possible sister group of the echinoids is the Ophiocystoida which have bun-shaped inflated bodies and enormous plated tube feet, but which also possess Aristotle's lanterns. The holothurians (of which the earliest is the Devonian *Palaeocucumaria*) probably arose from ophiocystoids by plate reduction; the Silurian *Rotasaccus* is in many ways intermediate between ophiocystoids and holothurians. Ophiuroids, echinozoans and holothurians are thus seen to be isolated and Smith (1984a,b) places them together as Superclass Cyrtosyringida.

The above analysis (Fig. 9.51), based on Paul and Smith (1984) and Smith (1988) is not exhaustive, but it testifies to the value of cladistic analysis in such a case as this, where it is sought to understand the interrelationships of such diverse and widely different organisms as the early radiation of the echinoderms produced.

In summary, we see an initial Cambrian radiation of very diverse forms but living only in small, localized and scattered pockets. A further radiation in the Ordovician led to a great and rapid radiation, especially of the pentameral groups. During subsequent periods the great range of higher taxa established in the Ordovician was drastically pruned. The non-radial forms were extinct by the Lower Devonian, and many of the pentameral groups became extinct in the Permian if not before. Evolution in the later Palaeozoic and afterwards occurred only in well-tried and adaptable taxa, which survive with great success today. The pattern of evolution within these groups varied greatly. In asterozoans, for instance, the peak of organization was reached early, and there seem to have been no major innovations since the Palaeozoic. In echinoids, by contrast, many new types emerged throughout successive geological periods, and in certain lines of descent each evolutionary step acted as a springboard for the next.

Evolution of the tube feet

The efficient use and differentiation of the tube feet has clearly been very important to echinoderm evolution (Nichols, 1974). Crinoids have the simplest and apparently most primitive system. In crinozoans there are no ampullae and tube feet are extruded only by the contraction of the radial water vessel, which is the only pressure generator. Echinoids and dendrochirote holothurians have ampullae, which are the main pressure generators, but contractions of the radial water vessel are still important. Aspidochirotes, on the other hand, can extend a relaxing tube foot only when a neighbouring one contracts, sending water into the extending one.

Asteroids and ophiuroids have ampullae, but to a varying extent inflate the tube feet by contracting the lateral and radial water vessel. The tube foot extensor system in extinct echinoderms has been inferred by analogy with that of modern forms.

Why pentamery?

Not all echinoderms have pentameral symmetry. Early cystoids are radially symmetrical but may have five symmetrical plates in the anal pyramid, whereas later cystoids acquire pentameral thecae. But other than in 'carpoids' and other early groups, pentameral symmetry is so characteristic that there must surely be a good reason for it. Nichols (1974) has suggested that the sutures binding echinoderm plates must have acted as lines of weakness. Only if there had been an odd number of plates would 'opposite' sutures have been offset so as to give greater strength against breakage. A circlet of three plates is too few, whereas five seems to be an ideal number for maximum resistance. A similar angular offset is needed to guard against accidental rupture along lines of pores, as in an echinoid, which has been called 'postage stamp weakness'. The offsetting of pore rows in multiples of five has been an ideal compromise between strength of test and maximum number of tube feet.

On the other hand, damaged echinoid tests rarely break along the sutures, since the latter are reinforced with strong collagen fibres; hence the sutures are not necessarily the weakest part of the test. Another suggestion is that a pattern of five combines a nearly constant width presented in almost any direction in the horizontal plane with a small number of rays.

Convergent evolution and intermediate forms

A striking characteristic in the early echinoderms is for the same kind of structure to develop over and over again in distinct though unrelated stocks. An example is the folded membrane structure found in blastoids, cystoids and the crinoid family Porocrinoidea, in all of which it seems to have evolved independently. Paired pores have arisen independently in diploporite cystoids and in echinoids. Lanceolate ambulacra and a stem are present in blastoids and in the peculiar genus *Astrocystites*, which in other respects is more like an edrioasteroid. It has been proposed that a new class, the Edrioblastoidea, be established to accommodate this genus, but it may simply be better to regard this genus as an edrioasteroid convergent on blastoids.

Very many of the Lower Palaeozoic echinoderms are thus of a composite type which, as Regnell (1960) has pointed out, lie in the direct lineage of none of the later echinoderm groups but present features from several of them. Perhaps the most extreme case may lie in the convergence between the 'carpoid' *Cothurnocystis* and the cystoid *Pleurocystites*. Indeed the controversy over the relationships of these same 'carpoids' illuminates the fact that within the early echinoderm–chordate line only a few kinds of structures seem to have been able to evolve and carry out particular functions.

The 'intermediate forms' that have evolved, often with a puzzling combination of characters, have been the bane of echinoderm taxonomy, though this is being resolved to a considerable extent using cladistic methodology. It is not easy to try to produce an objective phylogenetic scheme where such extreme genera are involved. Quite apart from the taxonomic question there are deeper evolutionary issues. Why, from the genetic point of view, should early echinoderms have produced similar structures, independently but often and so consistently? It is of course true that early members of a group tend to be rather generalized, whereas later ones are frequently specialized, having suppressed or eliminated more generalized traits. Perhaps the genetic potential for, say, folded membrane structures was present in many Palaeozoic echinoderms, though suppressed in the majority and only realized in a few isolated groups. Whatever the ultimate answer, these perplexing early invertebrates, more than those of any other groups, seem to spotlight some complex and little-known aspects of evolutionary theory that have so far received but scant attention.

Bibliography

Books, treatises and symposia

Gee, H. (1996) *Before the Backbone: Views on the Origin of the Vertebrates.* Chapman & Hall, London. (Critique of echinoderm–chordate relationships)

Jackson, R.T. (1912) *Phylogeny of the Echini with a Revision of Paleozoic Species.* Memoirs of the Boston Society for Natural History No. 7. (Large monograph of Palaeozoic echinoids, profusely illustrated)

Jefferies, R.P.S. (1986) *The Ancestry of the Vertebrates.* British Museum (Natural History), London. (Vertebrate affinities of homalozoans)

Moore, R.C. (ed.) (1966) *Treatise on Invertebrate Paleontology,* Part U, *Echinodermata 3,* Vols 1 and 2. Geological Society of America and University of Kansas Press, Lawrence, Kan.

Moore, R.C. (ed.) (1967) *Treatise on Invertebrate Paleontology* Part S, *Echinodermata 1,* Vols 1 and 2. Geological Society of America and University of Kansas Press, Lawrence, Kan.

Moore, R.E. and Teichert, C. (eds) (1978) *Treatise on Invertebrate Paleontology,* Part T, *Echinodermata 2.* Geological Society of America and University of Kansas Press, Lawrence, Kan. (Crinoids and related groups)

Moore, R.C., Lalicker, C.G. and Fisher, A.G. (1953) *Invertebrate Paleontology.* McGraw-Hill, New York.

Mortensen, T.H. (1928–1952) *A Monograph of the Echinoidea,* 5 vols. Reitzel and Oxford University Press, Copenhagen and Oxford. (The largest work ever published on echinoids, with full descriptions and photographic plates of all known Recent species)

Nichols, D. (1974) *Echinoderms.* Hutchinson, London. (Invaluable text on morphology and evolution of living and fossil forms)

Nichols, D. (1975) *The Uniqueness of Echinoderms.* Oxford University Press, Oxford. (Short, descriptive, functional approach)

Paul, C.R.C. and Smith, A.B. (1988) (eds) *Echinoderm Phylogeny and Evolutionary Biology*. Clarendon Press, Oxford. (Several important papers)

Smith, A.B. (1984a) *Echinoid Palaeobiology*. George Allen and Unwin, London. (Excellent text)

Wright, J. (1954) *A Monograph of the British Carboniferous Crinoidea*. London: Palaeontographical Society. (Fully illustrated monograph)

Individual papers and other references

Bell, B.M. (1976) *A Study of North American Edrioasteroidea*. New York State Museum Memoir No. 21. (Monographic interpretation of ambulacral structure)

Blake, D.B. (1982) Somasteroidea, Asteroidea and the affinities of *Luidia (Platasterias) latiradiata*. *Palaeontology* **25**, 167–91. (*Platasterias* not a somasteroid)

Blake, D.B. (1987) A classification and phylogeny of post-Palaeozoic sea-stars (Asteroidea, Echinodermata). *Journal of Natural History* **21**, 481–528.

Blake, D.B. and Guensberg, T.E. (1988) The water vascular system and functional morphology of Palaeozoic asteroids. *Lethaia* **21**, 189–206.

Bockelie, F. (1984) The Diploporita of the Oslo Region Norway. *Palaeontology* **27**, 1–68.

Breimer, A. (1969) A contribution to the palaeoecology of Palaeozoic stalked crinoids. *Proceedings Koninklig Nederlands Akademic Wetterschappen B* **72**, 139–50. (Application of knowledge of recent crinoid ecology to fossils)

Breimer, A. and Webster, C.D. (1975) A further contribution to the palaeoecology of fossil stalked crinoids. *Proceedings Koninklig Nederlands Akademic Wetterschappen B* **78**, 149–-67. (As above)

Brett, C.E. and Eckert, J.D. (1982) Palaeoecology of a well-preserved crinoid colony from the Silurian Rochester Shale in Ontario. *Life Sciences Contributions of the Royal Ontario Museum* **131**, 1–20. (Excellent illustrations)

Carter, R.M. (1967) On the biology and palaeontology of some predators of bivalved molluscs. *Palaeogeography, Palaeoclimatology, Palaeoecology* **4**, 29–65. (Starfish habits and anti-starfish devices in bivalves)

Clarke, J.M. (1912) A remarkable occurrence of Devonic starfish. *Bulletin of the New York State Museum* **15**, 44–5. (Bivalve–starfish association)

Cowen, R. (1981) Crinoid arms and banana plantations: an economic harvesting analogy. *Paleobiology* **7**, 332–43. (Inventive and interesting)

Donovan, S.K. (1986, 1989) Pelmatozoan columnals from the Ordovician of the British Isles, parts 1 and 2. *Palaeontographical Society Monograph*.

Donovan, S.K. (1988a) The early evolution of the Crinoidea, in *Echinoderm Phylogeny and Evolutionary Biology*, (eds C.R.C. Paul and A.B. Smith), Clarendon Press, Oxford, pp. 235–44.

Donovan, S.K. (1988b) The improbability of a muscular crinoid coelom. *Lethaia* **22**, 307–15.

Donovan, S.K. (1989) The significance of the British Ordovician crinoid fauna. *Modern Geology* **13**, 243–55.

Donovan, S.K. and Gale, A.S. (1990) Predatory asteroids and the decline of the articulate brachiopods. *Lethaia* **23**, 77–86.

Durham, J.W. (1966) Evolution among the Echinoidea. *Biological Reviews* **41**, 368–91. (Classic study; illustrations of early echinoids)

Ernst, G. (1970) Zur Stammgeschichte und stratigraphischen Bedeutung der Echiniden-Gattung *Micraster* in der nordwest deutschen Oberkreide. *Mitteilungen der Geologischer-Paläontologische Institut der Universität Hamburg* **39**, 117–35. (Distribution of *Micraster*)

Fay, R.O. (1961) Blastoid studies. *University of Kansas Paleontological Contributions* **3**, 1–147. (Monographic treatment)

Fell, H.B. (1963) Phylogeny of sea-stars. *Philosophical Transactions of the Royal Society of London B* **246**, 381–435. (Possible crinoid origins of somasteroids)

Gale, A.S. (1987) Phylogeny and classification of the Asteroidea (Echinodermata). *Zoological Journal of the Linnean Society* **89**, 107–32.

Goldring, R. and Stevenson, D.G. (1972) The depositional environment of three starfish beds. *Neues Jahrbuch für Geologie und Paläontologie Monatshefte* **10**, 611–24. (Rapid burial essential for starfish preservation)

Hawkins, H.L. (1912) Classification, morphology and evolution of the Echinoidea, Holectypoidea. *Proceedings of the Zoological Society of London* **1912**, 440–97.

Hawkins, H.L. (1943) Evolution and habit among the Echinoidea: some facts and theories. *Quarterly Journal of the Geological Society of London* **99**, 1–75. (Stimulating paper on echinoid phylogeny and adaptations)

Jefferies, R.P.S. (1968) The subphylum Calcichordata, primitive fossil chordates with echinoderm affinities. *Bulletin of the British Museum of Natural History (Geology)* **16**, 243–339. (Proposes that 'carpoids' are chordates)

Jefferies, R.P.S. (1979) The origin of chordates – a methodological essay, in *The Origin of Major Invertebrate Groups* (ed. M.R. House), Systematics Association Special Volume No. 12, Academic Press, London, pp. 443–77. (What are the homalozoans, otherwise calcichordates?)

Jefferies, R.P.S. (1990) The solute *Dendrocystites scoticus* from the Upper Ordovician of Scotland and the ancestry of chordates and echinoderms. *Palaeontology* **33**, 631–80.

Jefferies, R.P.S. (1997) A defence of the calcichordates. *Lethaia* **30**, 1–10. (Cladistic approach)

Jefferies, R.P.S., Brown, N.A. and Daley, P.E. (1996) The early phylogeny of chordates and echinoderms and the origin of chordate left–right asymmetry and bilateral symmetry. *Acta Zoologica* **77**, 101–22. (Summary of calcichordate theory)

Joysey, K.A. (1959) A study of variation and relative growth in the blastoid *Orbitremites*. *Philosophical Transactions of the Royal Society of London B* **242**, 99–125. (Classic study of a large population from blastoid-rich horizons)

Kanazawa, K. (1992) Adaptations of test shape for burrowing and locomotion in spatangoid echinoids. *Palaeontology* **35**, 733–50.

Kermack, K.A. (1954) A biometrical study of *Micraster coranguinum* and *M. (Isomicraster) senonensis*. *Philosophical Transactions of the Royal Society of London B* **237**, 375–428. (Statistical evaluation of *Micraster* evolution)

Kesling, R. and Paul, C.R.C. (1968) New species of Porocrinidae and brief remarks upon these unusually crinoids. *Contributions in Paleontology of the University of Michigan* **22**, 1–32. (Folded-membrane structures)

Kier, P.M. (1965) Evolutionary trends in Palaeozoic echinoids. *Journal of Paleontology* **39**, 436–65. (Very important paper; numerous illustrations)

Kier, P.M. (1973) A new Silurian echinoid genus from Scotland. *Palaeontology* **16**, 651–63. (Description of *Aptilechinus*, with reconstruction)

Kier, P.M. (1974) Evolutionary trends and their functional significance in the post-Palaeozoic echinoids. (Paleontology Society Memoir No. 5.) *Journal of Paleontology* **48** (Suppl). (Major reference work)

Kier, P.M. (1982) Rapid evolution in echinoids. *Palaeontology* **25**, 1–9. (Sand dollars)

Kier, P.M. and Grant, R.E. (1965) Echinoid distribution and habits. Key Largo Coral Reef Preserve, Florida. *Smithsonian Miscellaneous Collections* **149** (6), 1–69. (Life habits of several modern species)

Koch, D.L. and Strimple, H.L. (1968) A new Upper Devonian cystoid attached to a discontinuity surface. *Iowa Geological Survey*, Reports of Investigations, No. 5. (Edrioasteroids and cystoids in life position)

Macurda, D.B. (1965a) The functional morphology and stratigraphic distribution of the Mississippian blastoid genus *Orophocrinus*. *Journal of Paleontology* **39**, 1045–96. (Morphology and dynamics)

Macurda, D.B. (1965b) Hydrodynamics of the Mississippian blastoid genus *Globoblastus*. *Journal of Paleontology* **39**, 1209–17. (Inhalant–exhalant flow through spiracles)

Macurda, D.B. and Meyer, D.L. (1974) Feeding posture of modern stalked crinoids. *Nature* **247**, 394–6. (Life habits of rheophilic crinoids)

Milsom, C.V. (1994) *Saccocoma*, a benthic crinoid from the Jurassic Solnhofen Limestone, Germany. *Palaeontology* **37**, 121–9.

Moore, R.C. and Laudon, L.R. (1943) *Evolution and Classification of Palaeozoic Crinoids*, Geological Society of America Special Paper No. 46. (Major monograph on crinoid evolution)

Nichols, D. (1959) Changes in the chalk heart-urchin *Micraster* interpreted in relation to living forms. *Philosophical Transactions of the Royal Society of London B* **242**, 347–437. (Classic validation of evolution in *Micraster*)

Nichols, D. (1974) The water-vascular system in living and fossil echinoderms. *Palaeontology* **15**, 519–38. (Structure and evolution)

Paul, C.R.C. (1967) The functional morphology and mode of life of the cystoid *Pleurocystites* E. Billings 1854. *Symposia of the Zoological Society of London* **20**, 105–23. (Benthic crawling habit postulated)

Paul, C.R.C. (1968) Morphology and function of dichoporite pore structures in cystoids. *Palaeontology* **11**, 697–730. (Functional morphology and its bearing on classification)

Paul, C.R.C. (1972) Morphology and function of exothecal pore structures in cystoids. *Palaeontology* **15**, 1–28. (Functional morphology and its bearing on classification)

Paul, C.R.C. (1976) Palaeogeography of primitive echinoderms in the Ordovician, in *The Ordovician System* (ed. M.G. Bassett), University of Wales Press, Cardiff, pp. 553–74. (Temperature control of distribution of echinoderms with folded-membrane structures)

Paul, C.R.C. and Smith, A.B. (1984) The early radiation and phylogeny of echinoderms. *Biological Reviews* **59**, 443–81.

Philip, G.M. and Foster, R.J. (1971) Marsupiate Tertiary echinoids from south-eastern Australia and their zoogeographic significance. *Palaeontology* **14**, 666–95. (Description of marsupiate echinoids illustrated by stereo-pairs, and controls of distribution)

Ramsbottom, W.H.C. (1961) *British Ordovician Crinoidea*. Palaeontographical Society, London, pp. 1–37.

Regnell, G. (1960) 'Intermediate' forms in early Palaeozoic echinoderms. *Proceedings of the 21st International Geological Congress, Copenhagen* **22**, 71–80. (Taxonomic problems raised by some peculiar echinoderm groups)

Rosenkranz, D. (1971) Zur Sedimentologie und Geologie von Echinodermen–Lagerstätten. *Neues Jahrbuch für Geologie und Paläontologie* **138**, 221–58. (Various kinds of echinoderm beds)

Rowe, A.W. (1899) An analysis of the genus *Micraster*, as determined by rigid zonal collecting from the zone of *Rhynchonella cuvieri* to that of *Micraster coranguinum*. *Quarterly Journal of the Geological Society of London* **55**, 494–547. (Classic evolutionary study)

Seilacher, A. (1979) Constructional morphology of sand dollars. *Paleobiology* **5**, 191–222.

Seilacher, A. (1990) The sand-dollar syndrome; a polyphyletic constructional breakthrough, in *Evolutionary Innovations* (ed. M. Nitecki), University of Chicago Press, Chicago, pp. 231–52.

Seilacher, A., Drozozewski, G. and Haude, R. (1968) Form and function of the stem in a pseudoplanktonic crinoid (*Seirocrinus*). *Palaeontology* **11**, 275–82. (How did perfectly preserved crinoids come to lie in position on the sea floor?)

Seilacher, A., Reif, W.-E. and Westphal, F. (1985) Sedimentological, ecological and temporal patterns of fossil Lagerstätten, in *Extraordinary Fossil Biotas and their Evolutionary Significance* (eds H.B. Whittington and S.C. Conway Morris). *Philosophical Transactions of the Royal Society of London B*, **311**, 5–23.

Simms, M.J. (1989) *British Lower Jurassic Crinoids*. Palaeontographical Society Monograph.

Simms, M.J. (1990) Crinoids, in *Evolutionary Trends* (ed. K.J. McNamara), Belhaven, Boston, pp. 188–204.

Simms, M.J. (1991) Patterns of evolution among Lower Jurassic crinoids. *Historical Biology* **1**, 17–44.

Simms, M.J. and Sevastopulo, G.D. (1993) The origin of articulate crinoids *Palaeontology* **36**, 91–109.

Smith, A.B. (1978) A functional classification of the coronal pores of regular echinoids. *Palaeontology* **21**, 759–89. (Illustrated classic)

Smith, A.B. (1980) The structure, function and evolution of tube-feet and ambulacral pores in irregular echinoids. *Palaeontology* **23**, 39–83. (See text)

Smith, A.B. (1981) Implications of lantern morphology for the phylogeny of post-Palaeozoic echinoids. *Palaeontology* **24**, 779–801. (Basis for revised classification)

Smith, A.B. (1982) Tooth structure of the pygasteroid sea-urchin *Plesiechinus*. *Palaeontology* **25**, 891–6.

Smith, A.B. (1984b) Classification of the Echinodermata. *Palaeontology* **27**, 431–59.

Smith, A.B. (1985) Cambrian eleutherozoan echinoderms and the early diversification of edrioasteroids. *Palaeontology* **28**, 715–56.

Smith, A.B. (1988) Fossil evidence for the relationships of extant echinoderm classes and their times of divergence, in *Echinoderm Phylogeny and Evolutionary Biology* (eds C.R.C. Paul and A.B. Smith), Clarendon, Oxford, pp. 98–104.

Smith, A.B. and Gallemi, I. (1991) Middle Triassic holothurians from northern Spain. *Palaeontology* **34**, 49–76.

Smith, A.B. and Ghiold, J. (1982) Roles for holes in sand-dollars (Echinoidea): a review of lunule function and evolution. *Paleobiology* **8**, 242–53.

Smith, A.B. and Hollingworth, N.T.J. (1990) Tooth structure and phlogeny of the Upper Permian echinoid *Miocidaris keyserlingi*. *Proceedings of the Yorkshire Geological Society* **48**, 47–60.

Smith, A.B. and Paul, C.R.C. (1982) Revision of the class Cyclocystoidea (Echinodermata). *Philosophical Transactions of the Royal Society of London B* **296**, 577–84. (Exquisite illustrations)

Sprinkle, H.J. (1976) Classification and phylogeny of pelmatozoan echinoderms. *Systematic Zoology* **25**, 83–91. (Basis of classification used here)

Sprinkle, J. (1980) An overview of the fossil record, in *Echinoderms, Notes for a Short Course* (eds J.T. Dutro and R.S. Boardman), University of Tennessee Studies in Geology No. 4, Knoxville, pp. 15–26.

Sprinkle, J. (1982) Echinoderm Faunas from the Bromide Formation (Middle Ordovician) of Oklahoma. *University of Kansas Paleontological Contributions Monograph* **1**, 1–369. (Very detailed compilation)

Sprinkle, J. (1983) Patterns and problems in echinoderm evolution, in *Echinoderm Studies* (eds M. Jangoux and J.M. Lawrence), Balkema, Rotterdam, pp. 1–18.

Stephenson, D.G. (1963) The spines and diffuse fascioles of the Cretaceous echinoid *Echinocorys scutata* Leske. *Palaeontology* **6**, 458–70. (Functional morphology of a shallow burrower)

Stokes, R.B. (1977) The echinoids *Micraster* and *Epiaster* from the Turonian and Senonian of Southern England. *Palaeontology* **20**, 805–21.

Stürmer, W. and Bergström, J. (1976) The arthropods *Mimetaster* and *Vachonisia* from the Devonian Hunsrück Shale. *Paläontologisches Zeitung* **50**, 78–111.

Ubaghs, G. (1953) Crinoides, in *Traité de Paleontologie 111. Onychophores, Arthropodes, Echinodermes, Stomochordes* (ed. J. Piveteau), Masson, Paris, pp. 658–773.

Ubaghs, G. (1971) Diversité et specialisation des plus anciens echinodermes que l'on connaisse. *Biological Reviews* **46**, 157–200. (Early radiations of echinoderms; disclaims that carpoids are chordates)

Wignall, P.B. and Simms, M.J. (1990) Pseudoplankton. *Palaeontology* **33**, 359–78.

Woods, I.S., and Jefferies, R.P.S. (1992) A new stem-group chordate from the Lower Ordovician of South Wales, and the problem of locomotion in boot-shaped cornutes. *Palaeontology* **35**, 1–25.

10 Graptolites

Abundant stick-like fossils are sometimes found in Lower Palaeozoic black shales or more rarely in other argillaceous rocks of the same age. These are the graptolites, so called from their resemblance to written marks on the shale (Greek: λιθοξ = to write; γραφειν = stone). They may be straight or curved, sometimes spiral in form, single-branched, bifid or many-branched. When preserved in the most common way, i.e. flattened in shale, one or both edges appear serrated like a tiny saw blade, but otherwise not much structural detail may be visible.

These fossils are of singular importance in establishing a stratigraphical timescale for the Lower Palaeozoic and have been of zonal value since the mid-nineteenth century. Though much of the stratigraphical work that has been done is based upon specimens preserved as flattened carbonized or kaolinitized films (even so, they are still identifiable), the anatomy of graptolites can only be properly understood with reference to the relatively rare specimens preserved in three dimensions.

10.1 Structure

'Graptolites' is a vernacular term for members of Class Graptolithina (Phylum Hemichordata). They are colonial marine invertebrates now believed to bear a distant affinity to vertebrates, though for a long time they were regarded as being allied to corals, which have a much simpler grade of biological organization. Within Class Graptolithina there are several orders, of which only two are of any real importance: the Graptoloidea (Arenig–Pragian) and the Dendroidea (Tremadoc–Carboniferous). Examples of really well-preserved genera of these two main orders will be described below, but it has to be made clear that graptolites are rarely preserved like this; in the normal course of preservation very much fine detail is lost. Even so, it is surprising how much remaining structure can sometimes be found even in crushed and carbonized specimens.

Order Graptoloidea

Saetograptus chimaera (Fig. 10.1)

This Ludlovian species (Urbanek, 1958) is found only in glacial erratic boulders, originally derived from rocks under the Baltic Sea and now scattered over the North German and Polish plain; specimens can be isolated from the rock with acid.

As in all graptolites the 'skeleton' of *Saetograptus* consists of a series of hollow interlinked tubes, constructed of a thin sheet-like material known as **periderm**.

The first-formed part of the graptoloid is the **sicula**: a conical tube with its aperture pointing downwards and terminating at its apex in a long, hollow, rod-like **nema**, extending well beyond the upper limit of growth of the sicula. The sicula is divided into two parts: an upper **prosicula** and a lower **metasicula**. The prosicula has a thin proximal part, the **cauda**, and a broader distal **conus** has an ornament of longitudinal and spiral **striae**; the metasicula, like all other parts of the graptoloid, is ornamented with well-marked rings representing periodic, perhaps daily, growth increments and known as **fusellae**. **Fusellar tissue** consists of thin half-rings or complete rings of skeletal material stacked one above the other and uniting along zigzag **sutures**. In the sicula the suture on one side has the characteristic zigzag form shown in all graptolites; on the other side the fusellar half-rings are joined to a stout rod (the **virgella**) which projects below the aperture and may curve slightly under it.

From the sicula there grow up a number of cup-like **thecae** (singular: theca), the first-formed of

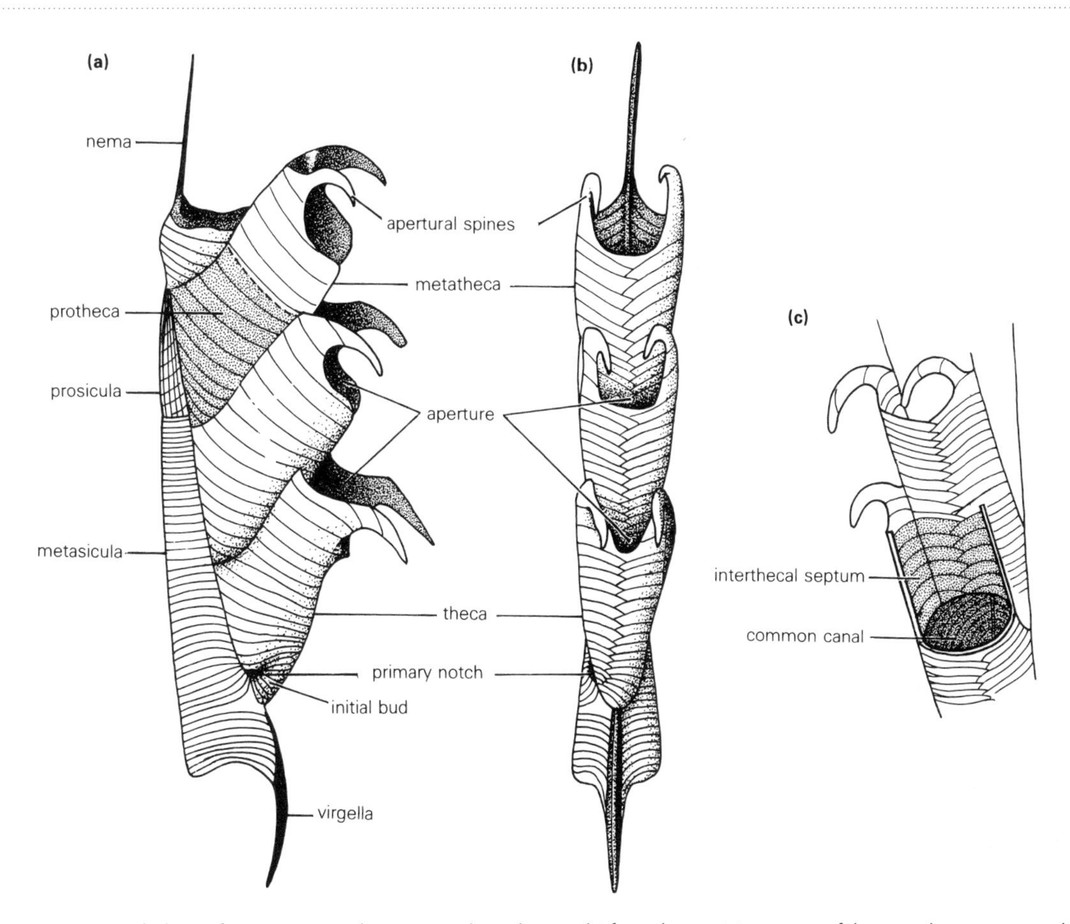

Figure 10.1 Morphology of *Saetograptus chimaera*: (a) lateral view; (b) frontal view; (c) structure of theca and common canal with part removed (× 40 approx.). (Modified from Urbanek, 1958.)

which begins at a **primary notch** in the growing edge of the sicula and on its virgellar side, before the development of the latter is complete. The **initial bud** grows up from the primary notch to form the first of the thecae, which in turn gives rise to successive thecae increasing in size away from the sicula until about the fifth or sixth theca, after which thecal dimensions remain constant. The thecal apertures point upwards but are obliquely inclined and provided with paired **apertural** spines.

The cavity of the first-formed theca connects with that of the sicula through a **foramen**, and all other thecae likewise link up with one another through a **common canal** close to the nema, since the partition walls of the thecae do not extend from the distal part of the nema. Where the nema is embedded in the wall of the graptoloid it is known as the **virgula**. The partition walls themselves are double-layered structures, the wall being secreted from both sides (Fig. 10.1c). Each theca has two parts: the **protheca**, proximal to the nema, through which the common canal runs; and a **metatheca**, which is the external part divided by partition walls and possessing apertural spines. The growth lines of the thecae tend to be more widely spaced toward the aperture, presumably signifying a period of faster growth towards the end of thecal formation.

Since graptolites were colonial animals, with the inhabitants of the thecae and the sicula (whatever they may have been like) linked to one another through the common canal, it is presumed that food caught by one individual would be ingested and shared by the whole colony. This colony is known as a **rhabdosome** (an older and less correct name is

polypary), and the branches are **stipes**; *Saetograptus* has only one stipe.

There is no direct evidence about what sort of animal was found in each theca. It is assumed that in each theca there was a **zooid** – a simple animal with some sort of food-gathering apparatus, possibly a lophophore – and that all the zooids were connected internally by tissue. The pyritized remains of retracted graptolite zooids, and tubes connecting them, have been found in the Australian anisograptid *Psigraptus* (Rickards and Stait, 1984). These, however, reveal no details of tentacles or other fine structure. The nature of the zooids will be considered after the affinities of the graptolites are discussed in more detail (section 10.3).

Diplograptus leptotheca (Fig. 10.2)

Diplograptus is a Caradocian graptoloid with two stipes in the rhabdosome. Both of these are united back to back with the nema in between; the rhabdosome is said to be **scandent** and **biserial**. This graptoloid is structurally more complex than *Saetograptus*, especially in the proximal end, i.e. the sicula and the first-formed thecae which surround it. Though the structure of the proximal end is difficult to work out in mature specimens, individuals in all growth stages have been found, and these can be arranged in a developmental series showing exactly how the different thecae originate and form (Bulman, 1945–1947).

The sicula is broadly similar to that of *Saetograptus* but has two apertural spines, and the initial bud arises, not by the formation of a notch in the growing edge of the sicula, but by the resorption of a foramen high up on the metasicula before budding. This system is actually more normal in graptolites than the specialized mode of development in *Saetograptus*.

The thecae are numbered in pairs according to a conventional notation, so that the first-formed theca is th. 1^1, the second which originates from it is th. 1^2, the third 2^1, the fourth 2^2, etc. In a biserial graptoloid such as *Diplograptus* the thecae on one side are numbered 1^1, 2^1, 3^1, 4^1, . . . and on the other are numbered 1^2, 2^2, 3^2, 4^2, ...

In *Diplograptus* th. 1^1 grows down almost vertically to about the level of the sicular aperture and is then sharply hooked upwards to terminate with an apertural spine at its lip. Th. 1^2 arises from th. 1^1 near the sicular foramen, before the growth of th. 1^1 is complete.

Th. 1^2 is first directed upwards, then is twisted into a recumbent S-shape, and finally grows across the front of the sicula, becoming welded to it and forming a **crossing canal**. Curiously enough, the fusellae that form the crossing canal are laid down not only at the growing aperture of th. 1^2, but also as a flange along the already-formed tubular surface of th. 1^1. The two parts of th. 1^2 unite to form a tube open at both ends, so that when fully formed th. 1^2 terminates in an upward curve, while th. 2^1 arises on the opposite side from th. 1^2. Then th. 2^2 grows up from th. 2^1, forming another crossing canal. Likewise the later-formed thecae all pass in front of the sicula or the nema, each time forming a crossing canal.

The mode of growth is marked by the fusellae (cf. Fig. 10.4) so that the developmental history is clear from them, but it is more useful to be able to see all the stages in growth from the developmental series here illustrated.

Order Dendroidea

The dendroids are the most ancient of all the graptolites and were apparently ancestral to the graptoloids. However, their morphology is complex, and they have various organs which are not possessed by the descendent graptoloids. Normally dendroids are preserved, as are most graptoloids, as compressions in shale, but rare three-dimensional specimens can be isolated from the rock or studied by means of thin section, so their detailed morphology is quite well known.

The basic structure of dendroids was worked out in the late nineteenth century by the Swedish palaeontologist Carl Wiman, mainly from serial sections. It was later elaborated by Bulman (1945–1947) and most particularly by Koszlowski (1948), who has described in great detail a rich and varied fauna of dendroids and other graptolites from the Tremadoc of Poland.

Dendrograptus (Fig. 10.3)

Dendrograptus, known from the three-dimensional material studied by Koszlowski, exhibits the representative dendroid structure admirably. The inverted conical sicula stood upright upon the sea floor with its apical base expanded into a holdfast. At a point about halfway up the sicula arises the

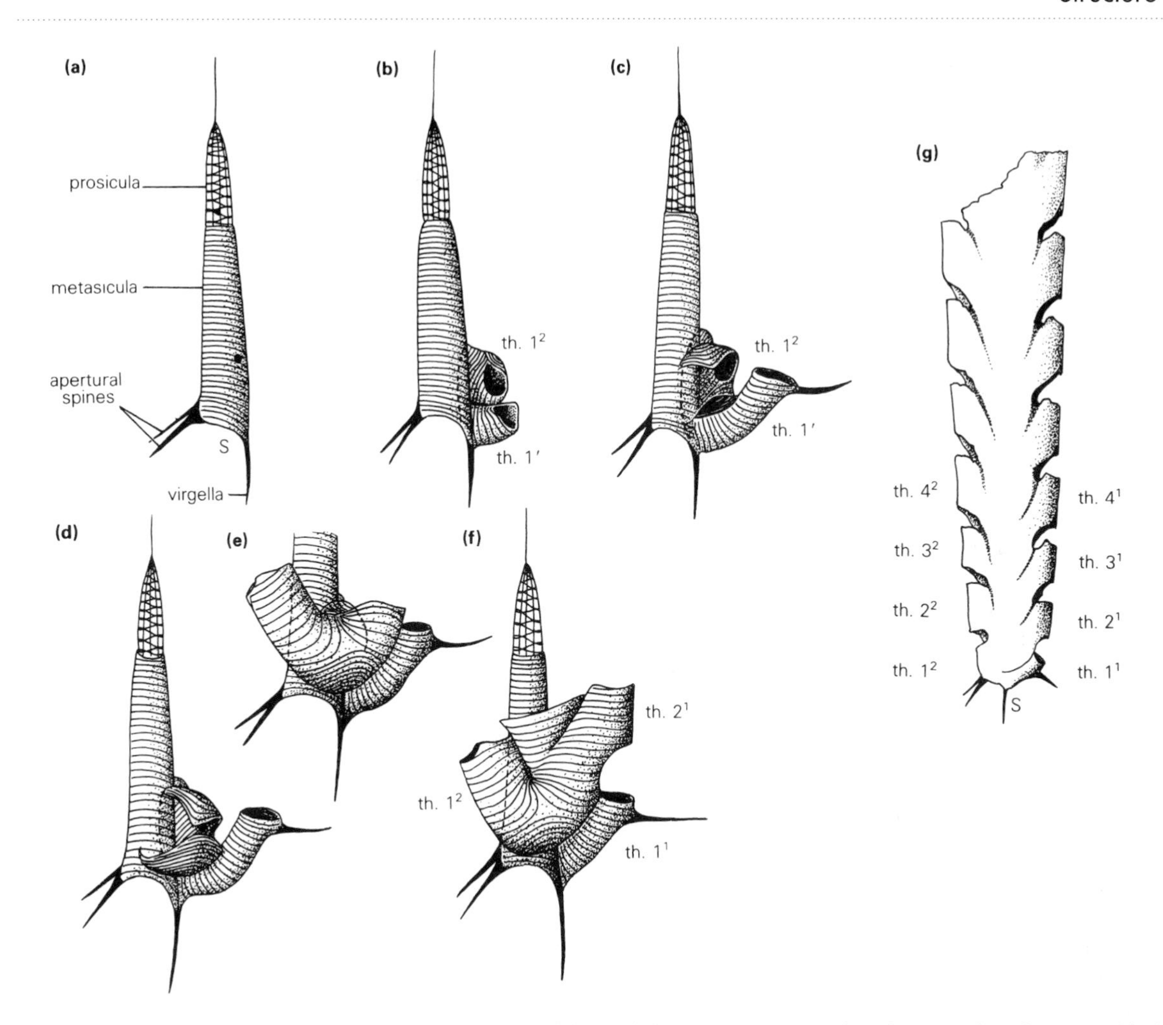

Figure 10.2 (a)–(f) Development (× 30 approx.) and (g) adult morphology (×12 approx.) of *Diplograptus leptotheca*. (Modified from Bulman, 1944–1947.)

stolotheca, which is equivalent to the prothecal series surrounding the common canal in *Saetograptus* and, with daughter stolothecae, forms a continuous closed chain all the way up the rhabdosome. From this there arise two kinds of thecae at successive and equally spaced nodal points. These are the large **autothecae** and the smaller narrower **bithecae**, which always come off the stolotheca at the same level. The bithecae in *Dendrograptus* maintain constant width, but each is normally looped over between the associated autotheca and the stolotheca, so that the aperture is on the opposite side of the stolotheca to the origin. In other dendroids the bithecae may be simple straight cylinders or be variously looped or coiled. The autothecae are very much larger and expanded upwards, each terminating in an outwardly inclined but often elaborately sculptured aperture with a median tongue.

The arrangement of the autothecae and bithecae follows the 'Wiman rule' of alternating triads; bithecae usually arise at alternate nodal points on opposite sides of the main stem and are carried distal to the autothecae. When the rhabdosome branches, the stolotheca splits and the Wiman rule alternation continues on each daughter stolotheca. In some species the auto- and bithecae arise in sequence at angles of 90° rather than 180° from the one below.

Within each rhabdosome there is an internal

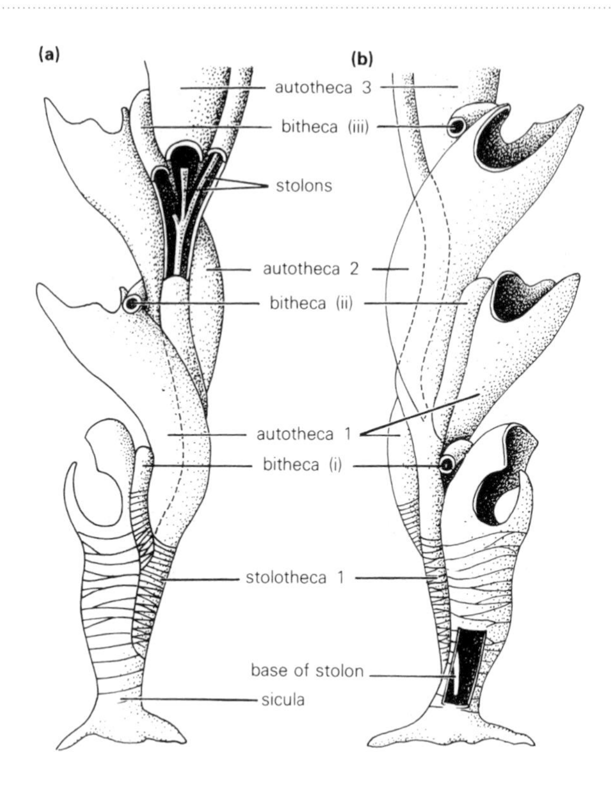

Figure 10.3 Morphology of the proximal end of a dendroid, *Dendrograptus communis*: (a) frontal view; (b) reverse view – the cut-away panels show the stolons (× 40 approx.). (Modified from Kozlowski, 1948.)

tubular system running from the sicula all the way along the stolotheca and the base of each autotheca and bitheca. This **stolon system** is only found in well-preserved specimens, but Wiman detected it in his serial sections and believed it to have carried an internal apparatus such as a nervous system. Koszlowski confirmed this suggestion when studying modern pterobranchs, which are small, tube-dwelling, marine animals, for they are apparently the nearest living relatives of graptolites and have an equivalent stolon enclosing just such a nervous system as Wiman postulated. No stolon system is found in the Graptoloidea, but it is developed in some of the other orders, especially the Stolonoidea, where there are very many peculiar and irregularly branching stolons.

The fusellar tissue of dendroids is broadly similar to that of graptoloids. However, there is also another kind of peridermal material: the **cortical tissue**, which is laid down in thin, flat bandages covering and enveloping the earlier thecae. (It is usually present in graptoloids as well but is very thin and only clearly seen with the electron microscope. Dendroid cortical tissue is much thicker and is clearly visible.) Such cortical sheets form the holdfast which affixes the colony to the sea floor.

Preservation and study of graptolites

The three genera described above are of some of the best-preserved graptolites known. They all come from fine argillaceous limestones, and though specimens may be very abundant locally such occurrences are rare. Three-dimensional specimens of such a kind must be freed from the rock so that they can be studied in the round. The rocks are initially broken into fragments, treated first with hydrochloric acid and then with hydrofluoric acid, and after washing may be picked out and individually transferred to a concentrated nitric acid and potassium chlorate solution which renders them translucent. They are then dehydrated in a concentrating alcohol series, cleaned in xylol and finally mounted on slides in Canada Balsam. Such specimens are then of clear brown hue and virtually transparent.

With the light microscope, at relatively low magnifications, individual thecae and the sicula may be very clearly marked with fusellar rings. Excellent detail may be seen with scanning electron micrography (Fig. 10.4).

Such fusellae are occasionally seen even in flattened specimens, though in typical specimens of *Saetograptus chimaera* the carbonization of the periderm has often gone too far to have conserved the microstructure, and growth rings may be lost, yet the three-dimensional structure of the rhabdosome is still preserved. When carbonization is complete the graptolite may be isolated from the rock, as with *S. chimaera*. Sometimes, however, carbonization is incomplete, and attempts to free the graptolite cause it to crumble when extracted. Then the only way to elucidate the structure is to make closely spaced serial sections through the specimen after embedding it in plaster or resin, cutting the sections normal to the long axis of the rhabdosome (Fig. 10.6). If the structure is very complex, especially in the region of the sicula and the first few thecae (the proximal end), it is often useful to build up models in wax from these sections, by superimposing one layer after another.

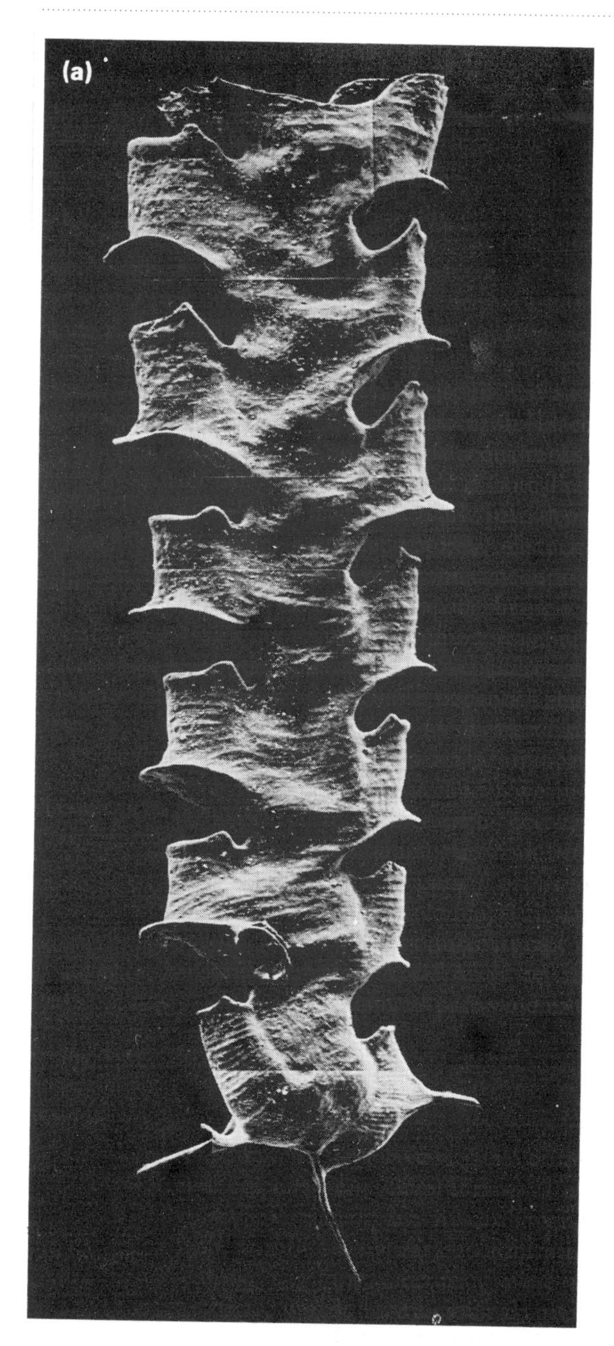

Figure 10.4 *Climacograptus inuiti*: (a) a montage built up from SEM photomicrographs showing banded fusellar tissue (total width at top approx. 1 mm); (b) cortical bandages secreted by individual zooids (× 500). (Photographs reproduced by courtesy of Dr Peter Crowther.)

In practice the wax is cut to show the internal structure of the graptolite preserved as a core, for in this way relationships of the thecae are more clearly visible (Fig. 10.5).

Not infrequently the periderm is completely destroyed, but an internal mould of the whole rhabdosome may remain, retaining the three-dimensional organization as a pyrite core. Sometimes these cores retain carbonized periderm on the outer surface, but this tends to flake off. Such preservation is common with specimens preserved in black shale, the pyrite being derived from organic matter decaying in an anoxic environment. As with the wax model the periderm has gone, but the shapes and convolutions of the thecae remain in their original relationships, and internal walls or septa may be clearly visible. However, the most common, and unfortunately the least revealing, of the various kinds of preservation is that where the rhabdosome has been completely flattened and preserved only as a compression: a carbon or aluminosilicate film on the rock. Even here, however, thecal shapes are usually retained so that the fossils can be readily identified, and it is from such specimens that most graptolite species and faunal sequences are known.

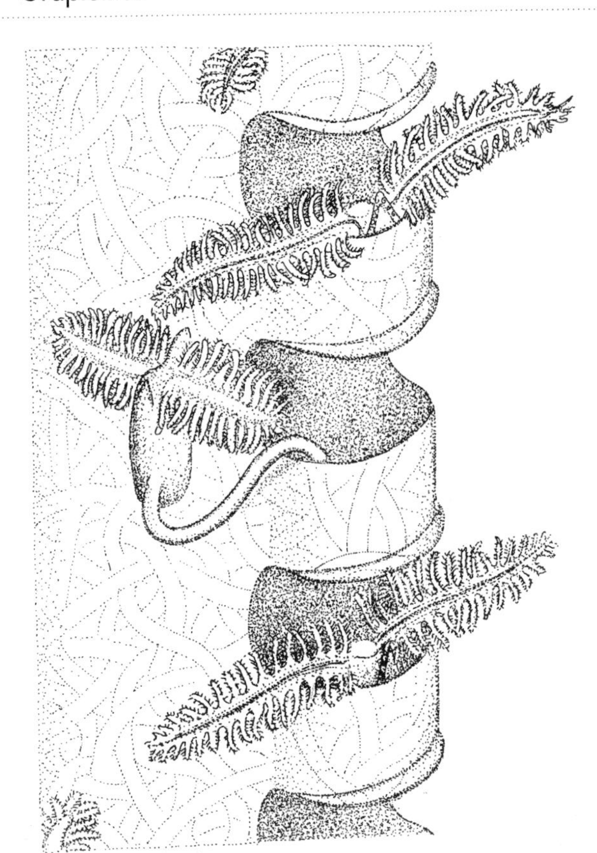

Figure 10.5 Reconstruction of part of a *Climacograptus* stipe. The upper and lower zooids are withdrawn inside the thecae but with extended lophophores, the central one is extended and is 'painting' cortical bandages on the outside of the rhabdosome. (Redrawn from the complete model of Crowther and Rickards, 1977.)

Ultrastructure and chemistry of graptolite periderm

For a long time it was believed that the graptolite periderm was constructed of chitin or a related hydrocarbon. The translucent brown appearance of isolated graptoloids, the widespread use of chitin throughout the animal kingdom as a structural component, and the reported presence of glucosamine which is a known chitin breakdown product all seemed to suggest chitin. However, more recent work has failed to detect glucosamine, and it is now agreed that, whatever the original material of the graptolite periderm may have been, it was not chitin.

The ultrastructure of the graptolite periderm, which has now been intensively studied (Urbanek and Towe, 1974, 1975), sheds light upon this complex topic. Transmission electron micrography of exceptionally well-preserved graptoloid and dendroid material shows that the fusellar and cortical tissues are constructed of several kinds of fabric (Fig. 10.7).

Furthermore, the existence of cortical tissue in graptoloids as well as dendroids has been confirmed, though in the latter it may be very thin. **Cortical fabric**, which makes up most of the cortical tissue, consists of closely packed parallel **fibrils**. In most graptolites these fibrils are arranged in short, ribbon-like **cortical bandages** deposited one after the other in successive layers. This gives the impression, in thin section, of stacking in successive layers with alternating orientation. The manner of packing and the size of the cortical fibrils is strikingly reminiscent of the appearance of the protein collagen, which is an important structural component of many animal tissues, and this close resemblance indeed indicates that the original periderm material was collagen.

Fusellar fabric, though probably also collagenous in nature, is made of fibrils in a more open meshwork. The bulk of the fusellar tissue is made of this fabric. There are also at least two kinds of non-fibrillar sheet-like material. Such **sheet fabrics** are found delimiting particular layers within the cortical tissue, forming a thin external coat on the outside of the fusellae, deposited as a thin secondary sheet inside the thecae, and occurring inside the stolon sheath.

All these kinds of fabric may occur in either cortical or fusellar tissue, but one kind is usually predominant. Thus cortical tissue is largely cortical fabric, though sheet-like fabrics form layers within it as well as outer and inner linings. Fusellar tissue may also be sporadically present in patches.

Fusellar tissue is mainly fusellar fabric and may be entirely so, but in some dendroids laminae of cortical fabric may be present, and sheet fabrics are normally present.

Graptoloids may have, in addition to all the other fabrics, a peculiar **virgular fabric** which has been found, for instance, in the monograptid nema, and also in retiolitid graptoloids in which the periderm is reduced to a meshwork of girders. This fabric has long fibrils set parallel with one another in an electron-dense matrix, and the fibrils themselves have a

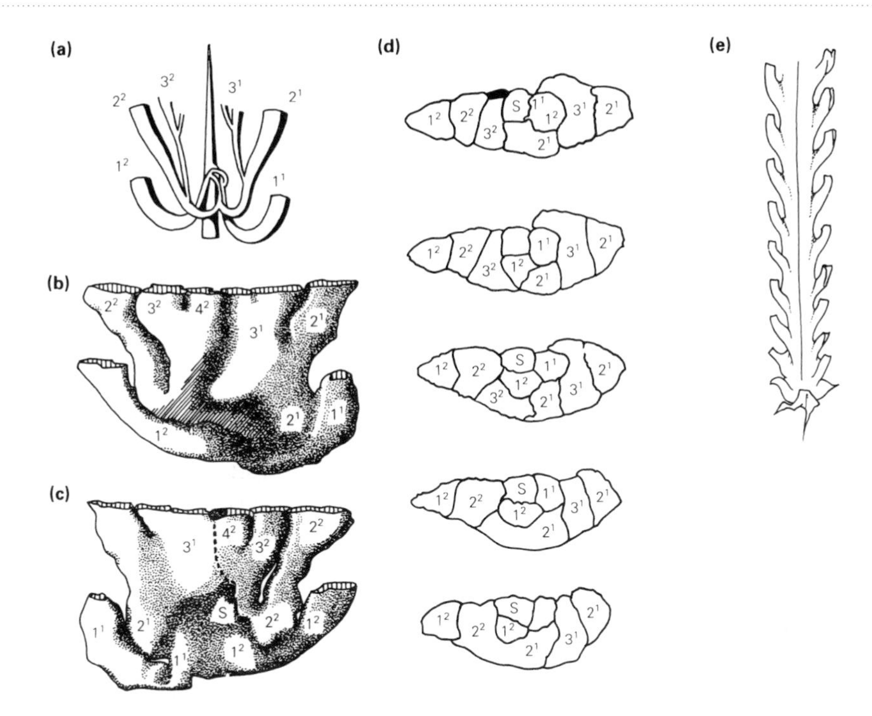

Figure 10.6 Morphology of *Glyptograptus austrodentatus americanus*, an early (Llanvirn) diplograptid: (a) simplified diagram showing relationships of thecae and sicula at the proximal end; (b), (c) frontal and reverse views of a wax model (× 30 approx.) built up from serial sections, some of which are illustrated in (d) (× 25 approx.); (e) adult rhabdosome (× 6). (Redrawn from Bulman, 1963.)

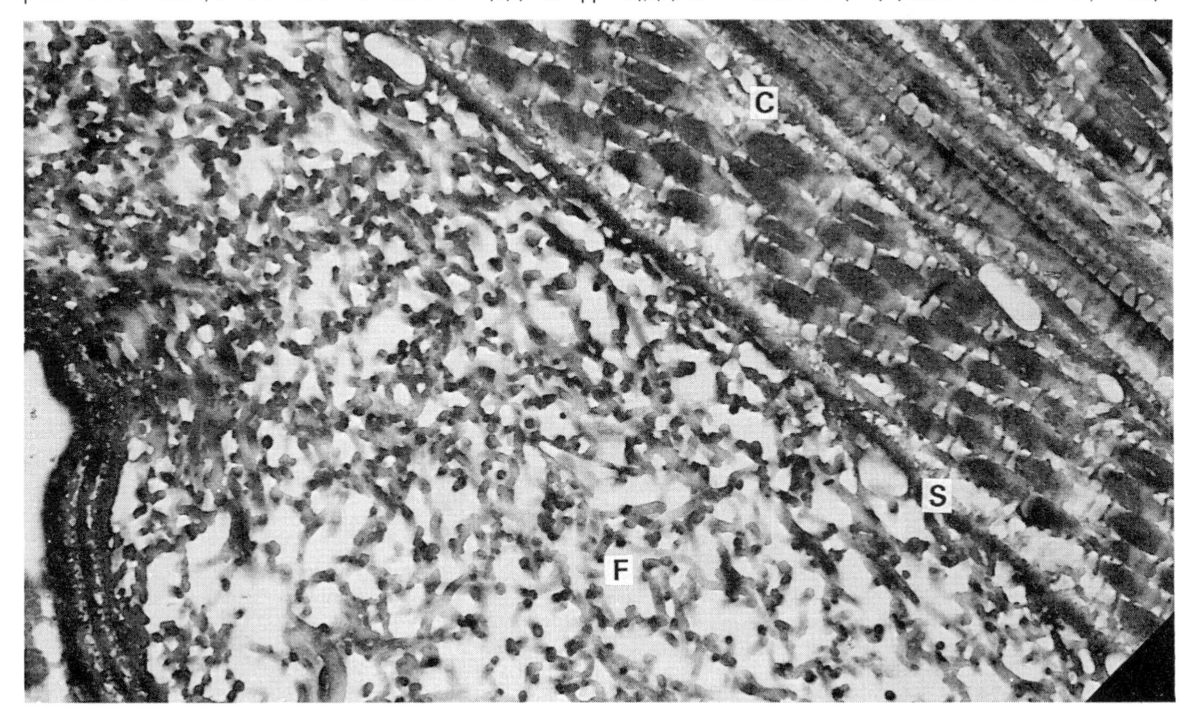

Figure 10.7 Ultrastructure of *Dictyonema* with fusellar tissue (F), cortical tissue (C) and sheet fabric (S) (× 14 000). (Photograph reproduced by courtesy of K.M. Towe.)

radial internal structure which has not been found elsewhere.

Ultrastructural study of numerous graptolites has shown that the cortical bandages swathe the whole outer and inner surface of the rhabdosome as with a mummy, thereby thickening and strengthening the tube walls. In most graptoloids and in dendroids the bandage pattern is remarkably constant despite an often sinuous course, and there is a close relationship between the width of the cortical bandages and the size of the thecae. Moreover, the bandages are often seen to radiate from the thecal aperture, and all these facts suggest that the zooids inhabiting the thecae were directly responsible for secreting the bandages. Some authorities (Crowther and Rickards, 1977; Crowther, 1981) believe that the only way in which such bandages could have been secreted is that the zooids crept out of the thecae, each tethered by a long and flexible stalk, and slowly moved over the surface, plastering on new cortical material in a continuous ribbon as they did so, as if by a paintbrush (Fig. 10.5). The oral disc was probably the secretory site. Bandages inside the thecae were secreted in the same way by the zooid within the thecal tube.

Novel though this model seems, it is at least partially supported by a study of pterobranchs, the nearest living relative to graptolites. In the pterobranch genus *Rhabdopleura* the tubes have a superficial resemblance to graptolite fusellar tissue, though ultrastructurally and biochemically the two are rather different. There is no extrathecal tissue in the pterobranchs and, in *Cephalodiscus* at least, the zooids can wander about over the surface of the colony, moving on their cephalic shields. Furthermore, although no living pterobranch has actually been observed in the act of secreting its skeleton, the secondary tissue at least could only have been produced by the zooids during these wanderings. In graptoloids it is probable that both fusellar and cortical tissue were produced by the zooids' cephalic disc.

In the opinion of some, the form and arrangement of cortical bandages makes it unnecessary to postulate the existence of extrathecal tissue in graptolites, but the construction of nemata, nemal veins and homologous structures in dendroids is not easily explained in this way.

The nema has been shown to be a hollow rod, open at its distal end. It must have been constructed by secretory cells within the rod, and these may have migrated over the tip to lay down cortical material on the outside. This could, in fact, have extended into a cover of extrathecal tissue over the whole of the sicula and even over the rest of the rhabdosome.

The study of the nature and distribution of these structural fabrics, which is actively being investigated, suggests a wide variety of distributional and compositional patterns amongst the graptolites, especially in the Graptoloidea. This will probably have considerable value in taxonomy.

10.2 Classification

Class Graptolithina (colloquially graptolites) is divided into the two important orders: the Dendroidea and Graptoloidea (Fig. 10.11). However, there are also some short-lived orders (Tuboidea, Stolonoidea, Camaroidea and Crustoidea; Fig. 10.9) largely confined to the Tremadocian which will be referred to only in passing.

ORDER 1. DENDROIDEA (M. Camb.–Carb.): Many-branched graptolites with large numbers of small thecae, sometimes with connecting links (**dissepiments**) between stipes. Stipes with two kinds of thecae (autothecae and bithecae) opening off a continuous closed chain: the stolotheca. Lower part of rhabdosome has concentric sheets of periderm (cortical tissue) covering the standard fusellar tissue. Most genera were shrubby, upright and fixed to the sea floor, like *Dendrograptus*. Dendroids are classified into three families, largely based upon their external appearances, for their detailed structure is known only in a few genera:

FAMILY 1. DENDROGRAPTIDAE (M. Camb.–Carb.), e.g. *Dendrograptus*.

FAMILY 2. PTILOGRAPTIDAE (L. Ord.–U. Sil.), e.g *Ptilograptus*.

FAMILY 3. ACANTHOGRAPTIDAE (U. Camb.–M. Dev.), e.g. *Acanthograptus*.

ORDER 2. GRAPTOLOIDEA (L. Ord.–L. Dev.): Rhabdosomes have few stipes, up to eight in the early forms, but reducing in later genera to two and finally to one. Thecae are of one kind only, equivalent to the autothecae of dendroids, and may be arranged on one side of the stipe or on both. Bilaterally symmetrical colonies are primitive; the loss of this condition in the Monograptidae is secondary. A sicula bearing a nema is always present in the adult and is usually prominent. The terminology of different thecal shapes is given in Fig. 10.8d.

Within the graptoloids there is much variation in thecal morphology, by contrast with that of dendroids. Graptolites are classified in three suborders.

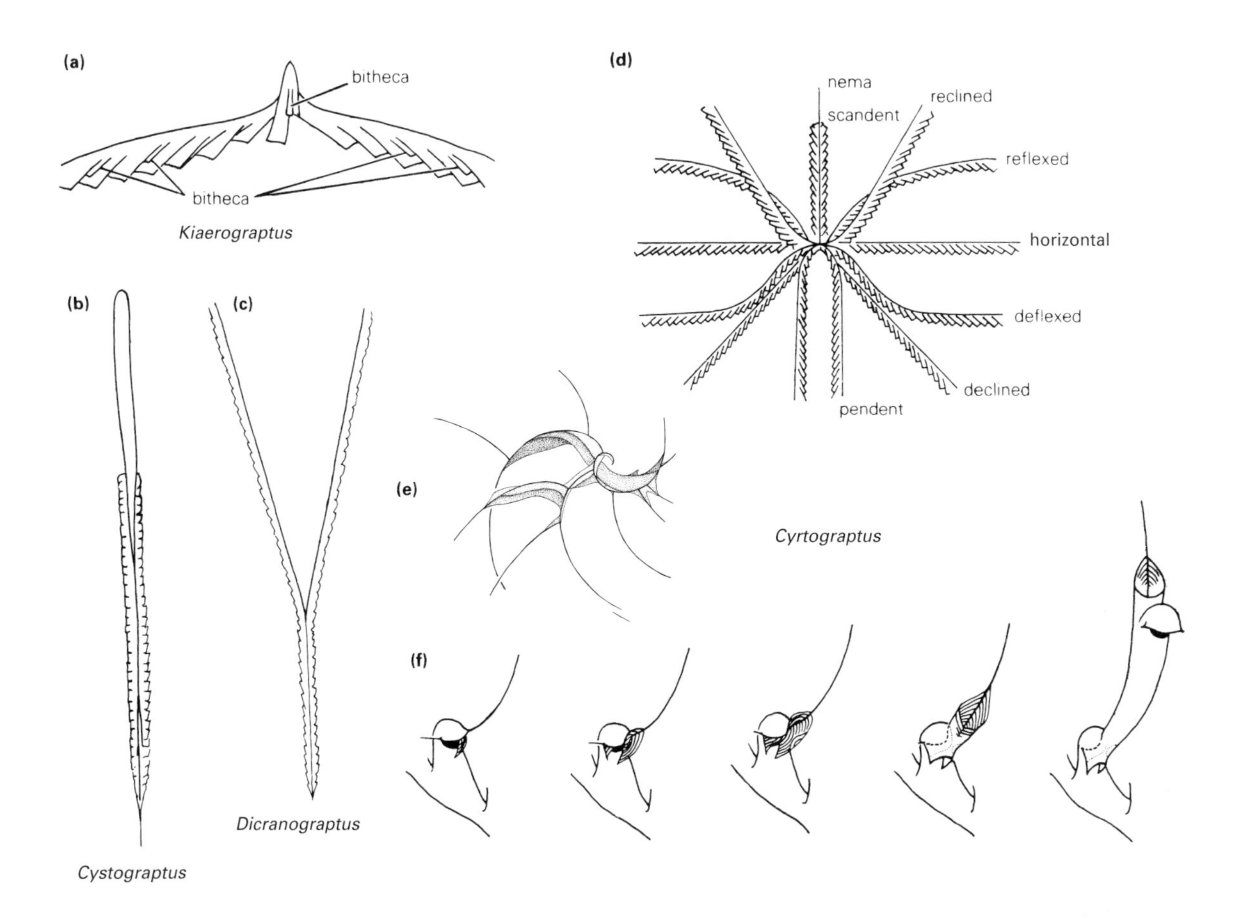

Figure 10.8 Various dendroid and graptoloid rhabdosomes: (a) *Kiaerograptus* (Tremadoc) with rare bithecae (× 6 approx.); (b) *Cystograptus vesiculosus* (Lland.) with a trifid nemal vane (× 1.5); (c) *Dicranograptus ramosus* (Caradoc; × 1); (d) terminology applied to the different shapes of rhabdosomes in graptoloids; (e) *Cyrtograptus*, with webs connecting the stipes (× 1); (f) *Cyrtograptus* (M. Sil.), stages in early cladial development (× 15). [(a) Redrawn from Spjeldnaes, 1963; (e) based on Underwood, 1995; (f) redrawn from Bulman in *Treatise on Invertebrate Paleontology*, Part V.]

Graptoloids, following Fortey and Cooper (1986) and Mitchell (1987), may be classified in three defined suborders, but the taxonomic position of some groups (anisograptids, pendent dichograptids and retiolitids remains uncertain).

SUBORDER 1. (Unassigned) contains the paraphyletic Family Anisograptidae (Trem.) which has characters in many ways intermediate between dendroids and graptoloids, but the presence of a nema groups them with the latter. Most anisograptids retain bithecae. The anisograptid rhabdosomes may form large spreading colonies as in *Radiograptus*, *Rhabdinopora* and others. Smaller rhabdosomes, with fewer branching points may be **bilateral** (i.e. with two primary branches) as in *Clonograptus*, **triradiate** as in *Bryograptus*, or **quadriradiate** as in *Staurograptus*. The structure of the thecae is known only in a few anisograptids, since their preservation is normally poor. *Kiaerograptus* (Fig. 10.8a) and some early species of *Bryograptus* have both autothecae and bithecae, but other species of *Bryograptus* only have autothecae.

SUBORDER 2. DICHOGRAPTINA (Ord.): Early graptoloids lacking bithecae, and without a virgella. SUPERFAMILY DICHOGRAPTACEA (FAMILIES DICHOGRAPTIDAE (e.g. *Tetragraptus*, *Dichograptus*, *Azygograptus*), SINOGRAPTIDAE (e.g. *Sinograptus*), SIGMAGRAPTIDAE (e.g. *Sigmagraptus*, *Goniograptus*) consists mainly of early graptoloids with only a few dichotomous branching points. They may have eight, four or two stipes and have a simple proximal end structure. In SUPERFAMILY GLOSSOGRAPTACEA the graptoloids always show 'isograptid symmetry', where the sicula and th. 1[1] form a symmetrical pair so that the axis of symmetry of the rhabdosome passes between them. [FAMILIES ISOGRAPTIDAE (bilaterally symmetrical, e.g. *Isograptus*, *Oncograptus*, *Skiagraptus*); PSEUDOTRIGONOGRAPTIDAE (scandent triserial or quadriserial *Pseudotrigonograptus*); GLOSSOGRAPTIDAE (scandent biserial forms in which the thecae may curve in order to enclose the sicula in front and behind, so that the stipes are aligned side by side but face in opposite directions, often spiny; e.g. *Glossograptus*, *Corynoides*, *Pseudisograptus*). SUBORDER 3. VIRGELLINA (Ord.–Dev.): A virgella is always present in the sicula and in the single SUPERFAMILY DIPLOGRAPTACEA the margin of the sicula bears apertural spines. FAMILY MONOGRAPTIDAE (Sil.–Dev.), which includes the Cyrtograptids, is uniserial and scandent (e.g. *Monograptus*, *Monoclimacis*, *Cyrtograptus*). FAMILY GLYPTOGRAPTIDAE (Ord.) comprises scandent biserial graptoloids which have a characteristic 'glyptograptid' proximal end pattern (e.g. *Glyptograptus*, some *Climacograptus*). FAMILY DIPLOGRAPTIDAE (Ord.–Sil.) comprises scandent biserial forms, usually with a median septum and 'diplograptid' proximal end pattern (e.g. *Diptograptus*, *Orthograptus*, *Climacograptus*). In FAMILY DICRANOGRAPTIDAE (Ord.) rhabdosomes may be extensiform, reclined (uniserial) or partially scandent (uni-biserial) (*Dicranograptus*, *Dicellograptus*, *Nemagraptus*, *Leptograptus*). FAMILY PSEUDOCLIMACOGRAPTIDAE (Ord.) comprises scandent biserial forms with a straight or zigzag median septum and a characteristic complex proximal end. Stem group PHYLLOGRAPTIDAE (Ord.) comprises quadriserial scandent graptoloids with simple thecae, and 'isograptid' development of the proximal end (e.g. *Phyllograptus*).

The taxonomic position of the pendent dictograptids (e.g. *Didymograptus*) and the retiolitids (q.v.; e.g. *Retiolites*, *Phormograptus*, *Gothograptus*) in which the periderm is reduced to a meshwork remains to be assessed.

ORDER 3. TUBOIDEA (L. Ord.–Sil.; Fig. 10.9a): Dendroid-like, with auto- and bithecae, but with irregular branching structure and a reduced stolotheca; e.g. *Reticulograptus*.

ORDER 4. CAMAROIDEA (Ord.; Fig. 10.9b): Encrusting forms, with autothecae of peculiar shape, each having an inflated balloon-like base, with a vertical **collum** (chimney). Stolotheca and bithecae present, e.g. *Bithecocamara*.

ORDER 5. CRUSTOIDEA (L. Ord.–U. Ord.; Fig. 10.9c): Encrusting forms, with dendroid-like morphology, but the autothecae have modified apertures, e.g. *Bulmanicrusta*.

ORDER 6. STOLONOIDEA (Ord.–M. Sil.; Fig. 10.9d): Encrusting forms of irregular morphology; stolothecae have many irregular ramifying stolons; autothecae present but no bithecae, e.g. *Stolonodendrum*.

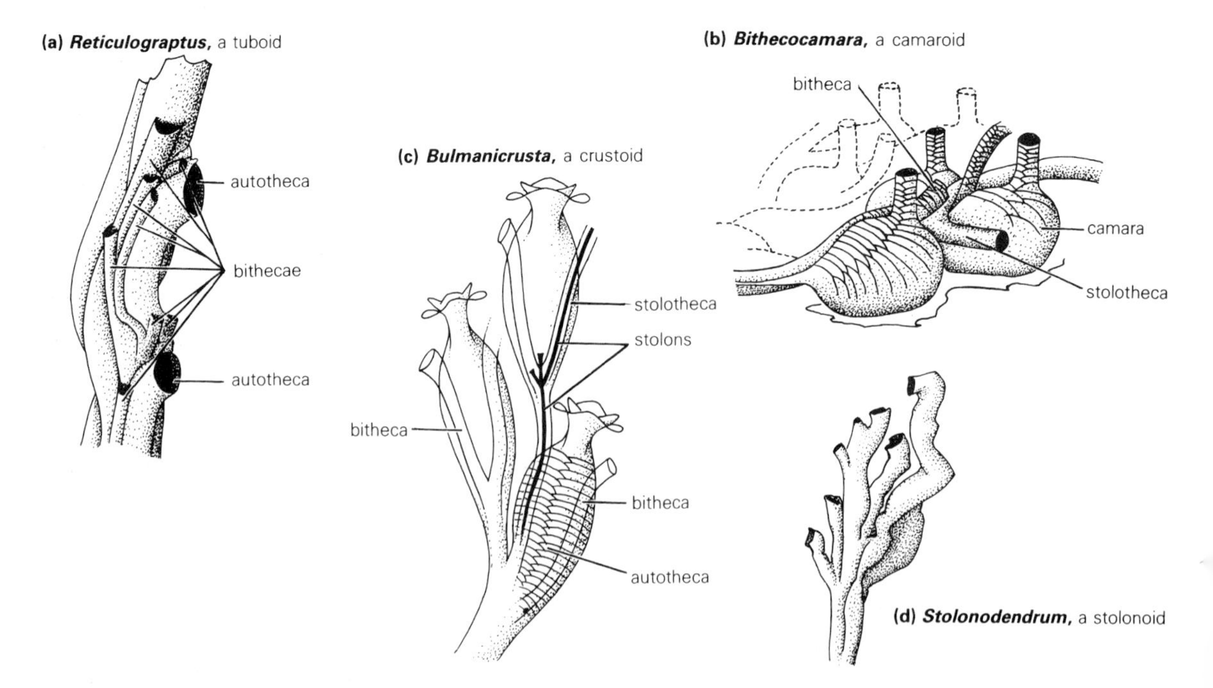

Figure 10.9 Minor graptolite orders. In *Stolonodendrum* the irregular tubes are stolons; the enclosing stolothecae are rarely preserved [(a), (b) × 50; (c) × 30; (d) × 15]. (Redrawn from Bulman in *Treatise on Invertebrate Paleontology*, Part V.)

10.3 Biological affinities

For a long time it was accepted that the affinities of the Graptolithina lay with Phylum Cnidaria and most probably with the hydrozoans. The publication of Koszlowski's work in 1948, however, made it clear that this view was no longer acceptable. Koszlowski noted pronounced resemblances in structure between the graptolites and the small, modern, *Rhabdopleura*: a colonial organism with a wide depth range (usually 100–300 m), belonging to Phylum Hemichordata and with relatives known from scanty material in the Cretaceous and the Lower Palaeozoic. The hemichordates are an 'advanced' phylum allied to the vertebrates; if Koszlowski's suggestions are correct, then the Graptolithina must be of a much higher grade of organization than was originally imagined.

The Hemichordata comprise two classes: the modern Enteropneusta, which are the 'acorn worms', such as *Balanoglossus*, and the Pterobranchia, to which *Rhabdopleura* belongs. In both cases there are only a few Recent genera, and in the pterobranchs the only important ones are *Rhabdopleura* and *Cephalodiscus*, which is similar to *Rhabdopleura* in some ways but does not form true colonies.

Rhabdopleura (Fig. 10.10) is of very small size and is colonial with individual exoskeletal tubes, each housing a zooid and arranged in a creeping habit.

Several tubes arise from a horizontal basal tube, the last one having a closed end with a terminal bud inside. The zooid which secretes and lives in

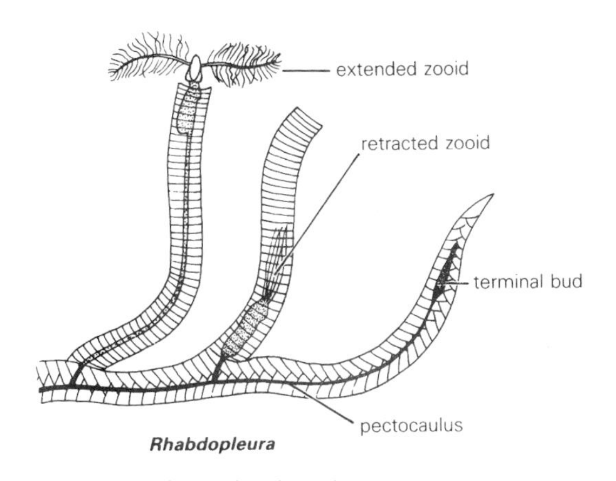

Figure 10.10 The modern hemichordate *Rhabdopleura* (×18). (Modified from Kozlowski, 1948.)

each tube is small and has a pair of tentacular food-gathering arms known as the lophophore, though this is not necessarily equivalent to the lophophore of other invertebrates. Each individual zooid is supported by a contractile stalk which links it with a **pectocaulus** running through the colony as a stolon system does in a dendroid. Initially the pectocaulus is soft, but it later becomes rigid. The pectocaulus with its similarity to the dendroid stolon gives some indication of affinity, but perhaps the relationship of the two groups is most clearly suggested by the presence of collagenous fusellar tissue which forms the periderm of both graptolites and *Rhabdopleura*. Several genera of rhabdopleurans have been found in the Lower Palaeozoic (Urbanek and Bengtson, 1986), e.g. *Rhabdotubus* (M. Cam.), *Graptovermis* (Tremadoc) and some authorities suggest that Stolonoidea are actually rhabdopleurans. The earliest recorded *Rhabdopleura* is Middle Cambrian in age (Durman and Sennikov, 1993).

The study of living pterobranchs serves as a model for many aspects of graptolite biology and skeletal growth. Significantly, in *Rhabdopleura* (Rigby, 1994) the zooids often die off but are replaced by others, budded off from the stolon, which inhabit the same tube and continue to add to it. Such multiple occupancy is marked by clear growth discontinuities, and the same are found in graptoloids. By analogy, therefore, graptolites might well have survived adverse conditions if the old zooids died off, but with improved conditions new buds arose; moreover, old or damaged zooids could have been continually replaced throughout the life of the colony. To build any kind of robust and long-enduring residence is expensive and, as with the human sphere, it makes sense to build a house for serial occupancy.

10.4 Evolution

Shape of graptolite rhabdosomes (Fig. 10.11)

The following summary of evolution in graptolites is concerned with gross morphology: the kind of details that can be seen in flattened specimens. Later some particular details will be added that can only

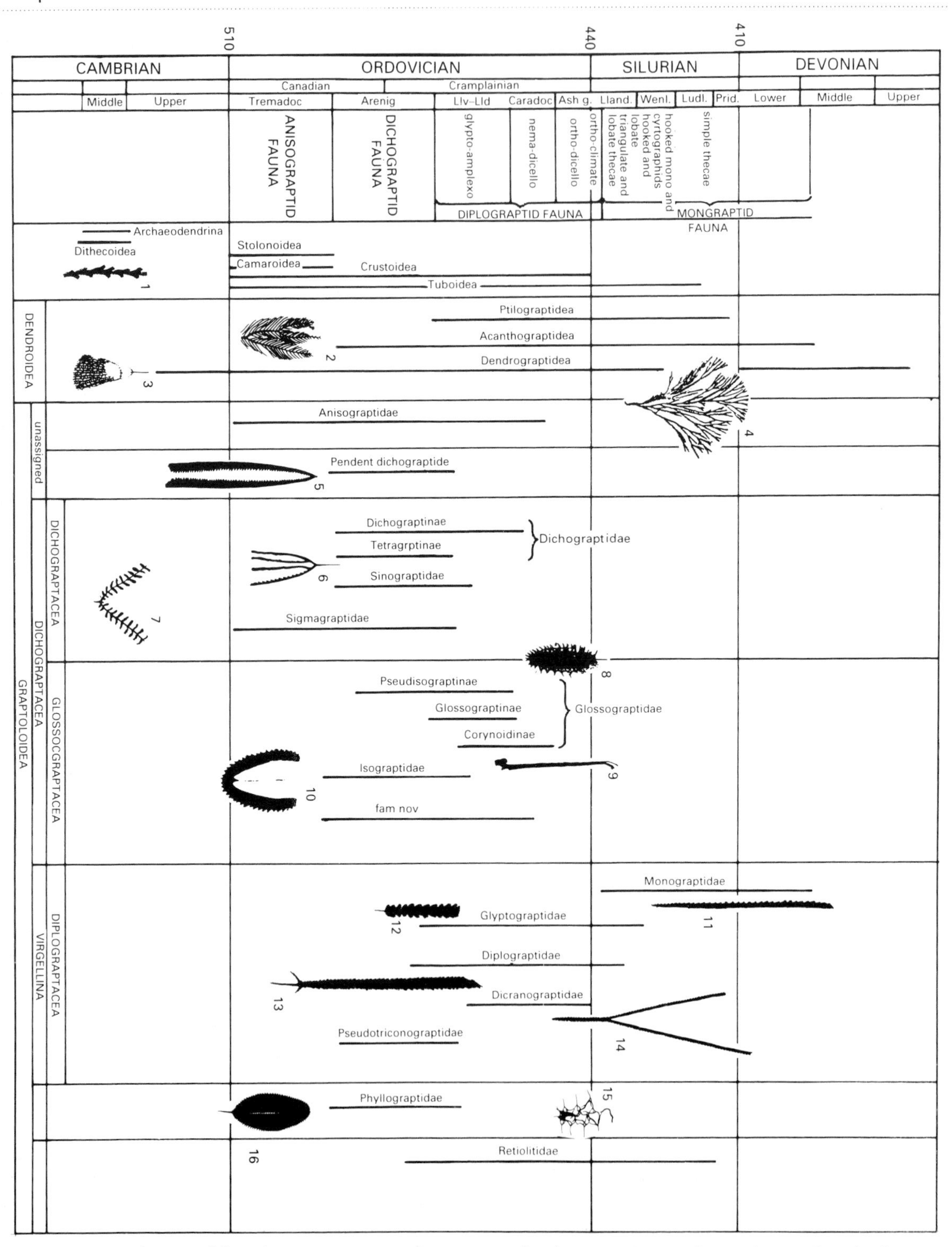

Figure 10.11 Evolution and faunal sequence in graptolites: a generalized picture. Genera illustrated: 1, *Dithecodendrum* (M. Cam.); 2, *Ptilograptus* (Ord.–Sil.); 3, *Rhabdinopora* (Trem.); 4, *Dendrograptus* (U. Cam.–L. Cam.); 5, *Didymograptus* (Arenig); 6, *Tetragraptus* (Arenig); 7, *Sinograptus* (Arenig); 8, *Paraglossograptus* (M. Ord.); 9, *Corynoides* (Caradoc); 10, *Isograptus* (Arenig); 11, *Monograptus* (Sil.); 12, *Glyptograptus* (Llandeilo); 13, *Orthograptus* (Caradoc); 14, *Dicranograptus* (Caradoc); 15, *Orthoretiolites* (M. Ord.); 16, *Phyllograptus* (Arenig).

be established through the examination of three-dimensional material.

The earliest graptolites known are Middle Cambrian and come from the Siberian Platform; they are small twig-like genera discovered and classified by Obut (1974) in two new orders, the Dithecoidea and Archaeodendrida, though more material needs to be forthcoming if this taxonomy is to be sustained. In the Tremadocian and later the evolution of graptolites is better known. The Upper Cambrian graptolites are all dendroids, along with representatives of the short-lived orders.

Dendroids were quite diverse in Upper Cambrian times, and though they were thereafter never a very abundant component of any fauna, they continued until the Carboniferous, various genera having different time ranges. **Dendritic** or twig-like forms were perhaps the most common, though there were also some **pinnate** genera (e.g. *Ptilograptus*). The peak of dendroid complexity is reached in the Ordovician to Devonian genus *Koremagraptus*, an acanthograptid with a complex and anastomosing branch system. Each branch has many subparallel stolothecae all united together in columns, with their attendant autothecae and bithecae irregularly twisted up together in a thick stem. The vast majority of dendroids grew upright on the sea floor like small shrubs, fixed by holdfasts which, like their 'stems', were strengthened by cortical tissue. However, one or two species of *Rhabdinopora* (formerly ascribed to *Dictyonema*), amongst the earliest of the anisograptids (a more rigidly engineered and cone-shaped genus), apparently departed from this ancestral habit and lived in an inverted position, probably suspended by the thin nema from floating seaweed. Small **attachment discs** have been found attached to the end of the nema; hence a suspended mode of life seems likely, though it was certainly not the norm for most graptolites.

The sessile true *Dictyonema* species probably lived attached to the sea floor, apex down like a conical bryozoan (Chapter 6), feeding in the same general way by extracting food particles from currents flowing in through the mesh, and expelling exhalant water through the open end of the cone (Rickards, 1975). In a Scottish Silurian fauna described by Bull (1996) several *Dictyonema* specimens show evidence of damage by predation or disease, but were able to repair many, though not all, kinds of injury and to return to a normal growth pattern as soon as possible. The earliest planktic descendants of *Dictyonema* floated apex upwards, and dispensed with the attachment; in its own way quite a major evolutionary step. The transition from benthos to plankton by graptolites is paralleled by virtually all other planktic organisms; no less than 18 out of 21 planktic groups had their origins in the benthos (Rigby and Milsom, 1996). Such invasions, which took place successively through time seem to have been random, and independent of any global events. The first planktic graptolites probably arose in deeper, cooler waters and thence radiated into open pelagic and neritic environments in the early Tremadoc (Erdtmann, 1982).

The earliest planktic graptolites have traditionally been regarded as arising from the floating *Rhabdinopora flabelliforme*, presumably because of the widespread occurrence of this fossil just below the anisograptid fauna. *R. flabelliforme*, however, is of conical, bell-like form, rather than being flattened. A better candidate for the *Rhabdinopora* ancestor would seem to be the flat, radially organized *Radiograptus*, the earliest siculate graptolite, which has three or four primary stipes, and in which, as in the anisograptids, the dissepiments are but weakly developed. Where this genus occurs, in Newfoundland, it underlies the zone of *R. flabelliforme*. It is suggested (Cooper and Fortey, 1982), that *Radiograptus* was the ancestor both of the (very similar) four-stiped anisograptids (e.g. *Staurograptus*) and the three-stiped *Anisograptus* as well as *Rhabdinopora* itself, from which in turn arose *Bryograptus*.

Young stages of rhabdinoporids in which dissepiments are weak or absent, however, are often found 'discoidally' preserved, i.e. flattened radially, at all stratigraphic levels, and these are sometimes dominant on particular bedding planes. Erdtmann therefore believes that the ancestors of the planktic graptolites may be sought in paedomorphic *Rhabdinopora* of *flabelliforme* type rather than necessarily in *Radiograptus*.

Most anisograptids are Tremadocian in age; there are records of a possible extension into the later Ordovician, but this is not certain. They may have two, three or four primary branches as in *Clonograptus*, *Bryograptus* and *Staurograptus*, respectively; they are rarely preserved other than as compressions. However, some Scandinavian genera are pyritized, and these are of particular interest for they show characters intermediate between those of dendroids and graptoloids at the microscopic level.

The Upper Tremadocian genera *Kiaerograptus* (Fig. 10.8a) and *Bryograptus* apparently have bithecae, but these are fewer in number than those of standard dendroids, and there is only one bitheca for every two autothecae; furthermore, they are reduced in size. Why the bithecae were dispensed with during the later evolution of graptolites is not clear.

Koszlowski believed that the bithecae and autothecae housed male and female zooids, respectively, and that graptoloids were hermaphrodite. Alternatively, autothecae may have been sites of feeding zooids only, while bithecae were solely reproductive. Whatever the function of the bitheca was, it must have been taken over by autothecae; this might have happened through an intermediate kind of graptolite such as the Llandeilian *Calyxdendrum*, in which the bithecae, rather than having external apertures, open directly into the autothecae. The anisograptids became extinct by the end of Llandeilian time and their progeny, the 'real graptoloids', became dominant. Continuous sections in the Tremadoc are few and far between; the best is in the Yukon Territory and has recently been described to show a continuous sequence of graptolite faunas. The fauna here is mainly of anisograptids and dendroids, but there are also some early 'real' graptoloids (dichograptids), the forerunners of the main stock that dominated the lowermost Ordovician.

Dichograptids have no bithecae. They are usually symmetrical, branch dichotomously and may have many or only a few branches. A burst of adaptive radiation in the early Ordovician resulted in a profusion of types; eight-, four- or two-branched, with rhabdosomes pendent or declined, or even scandent, like *Phyllograptus*, in which the four stipes are arranged back to back and are in contact all the way up from the basal sicula, or *Tristichograptus*, which has three scandent stipes. Other genera are *Tetragraptus*, with four declined or pendent stipes, and the common, zonally useful *Didymograptus*, in which the twin stipes may be pendent like a tuning fork or extended in line. Several of the early **multiramous** (many-branched) dichograptids had a weblike structure between the stipes at the proximal end, which increased the surface area and probably gave assistance in flotation.

From an unspecified two-stiped dichograptid ancestor there arose at some time during the Lower Ordovician the various stocks that came to dominate the Upper Ordovician (Fig. 10.14a–e). These never have more than two stipes. First there are the diplograptids, which are scandent forms. These have their origin well back in Llanvirnian time, their first representative being *Glyptograptus dentatus*. Thence they can be traced to the end of the Llandovery. The several diplograptid genera are distinguished one from another mainly on the characters of their thecae. Thus *Diplograptus* has simple, straight thecae, *Climacograptus* has thecae of a square-cut appearance (Fig. 10.4), *Glyptograptus* has gently curving thecae, etc. All these graptoloids are more or less elliptical in cross-section and are easy enough to identify when flattened, provided that they are preserved with the 'long axis' of the ellipse parallel with the bedding plane, for then the thecae are clearly seen in lateral view. If, however, the long axis of the elliptical cross section is perpendicular to the bedding, then only the apertures show, and since they all look similar in this **scalariform view** they are not easy to identify.

Other Upper Ordovician genera include the V-shaped, stoutly built *Dicellograptus*, the more slender but otherwise similar *Leptograptus*, and Y-shaped *Dicranograptus* [including *D. ramosus* (Fig. 10.8c) which can be more than 30 cm long]. In the last-mentioned genus the first-formed part of the rhabdosome is scandent; in subsequent growth the branches diverge. The occurrence of diplo-, dicello- and dicranograptids in the Upper Ordovician gave rise to a belief that the diplograptids evolved from a dicellograptid ancestor by 'zipping themselves up' and that the condition in *Dicranograptus* represented an incomplete zipping process. Attractive though this suggestion appears, it is unrealistic, for the first diplograptids appear well before the earliest dicello- and dicranograptids, and in any case the detailed structures of the 'proximal end' – the few first-formed thecae and the sicula, which are all taxonomically diagnostic – are quite different.

The dicello-, dicrano- and leptograptids became extinct before the Silurian, though the diplograptids continued until the end of the Llandovery as important members of the total graptolite fauna.

The Silurian is dominated by monograptids: scandent forms with the thecae arranged along only one side of the stipe. They first appear just above the base of the Silurian. Monograptids come in various

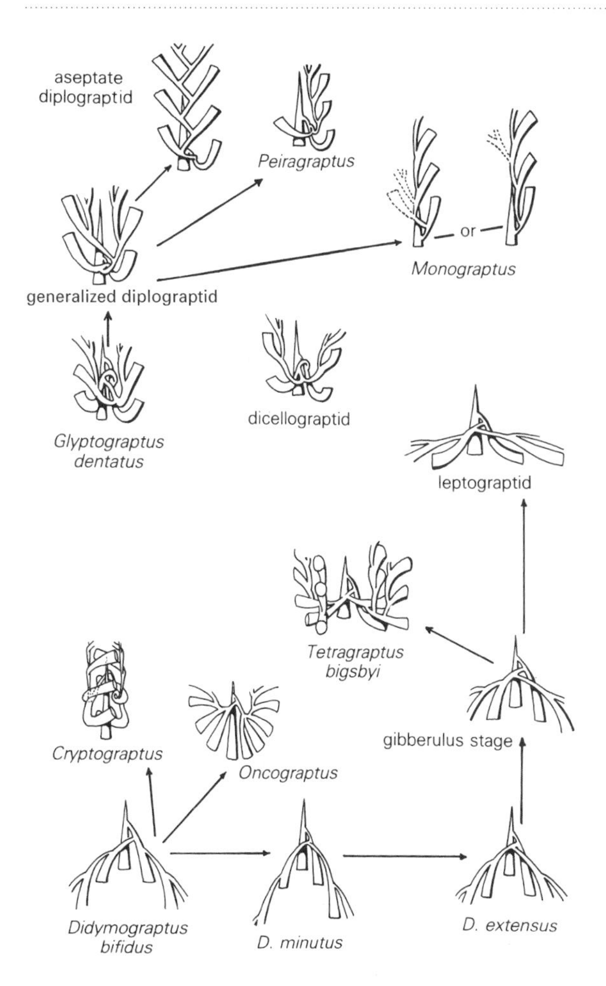

Figure 10.12 Evolution in graptolite proximal end structures. (Redrawn from Bulman in *Treatise on Invertebrate Paleontology*, Part V.)

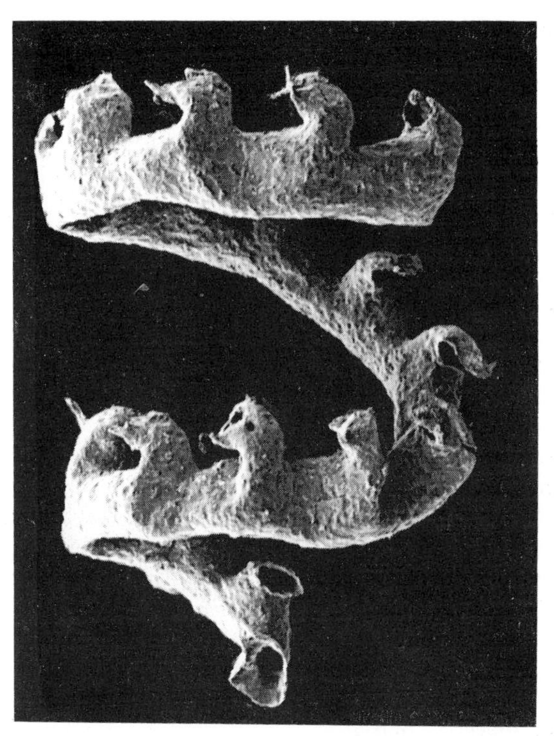

Figure 10.13 *Spirograptus guerichi*: three-dimensional specimen showing helical coiling, from the Silurian, Osmundsberg, Sweden. (SEM photograph reproduced by courtesy of Dr Dennis Bates.)

shapes and sizes; some are very long and straight, others short and stumpy, and there are various highly modified forms including curved or spirally coiled genera (Fig. 10.13), some of which, like *Cyrtograptus* (Fig. 10.8e,f), may have lateral arms (cladia).

Various changes in thecal morphology, as will be explained in detail later, appeared in succession during the Silurian. The last monograptids flourished in the early Devonian but were restricted to certain parts of the world only.

The picture of graptolite evolution presented here has been known for a long time and is well established. However, it is based mainly upon compressions, and reference to three-dimensional material expands and illuminates it to a remarkable degree.

Proximal end in graptoloids

The Graptoloidea differ from the Dendroidea in the absence of bithecae, the reduction of cortical tissue, the form of the sicula, the presence of a nema, the mode of rhabdosome branching and the structure of the thecae. They are also unique in the way in which the first few thecae develop from the sicula. In general terms, the early graptoloids have a rather simple kind of proximal end development; the later Ordovician ones are much more complex, such as that of *Diplograptus* which may now be seen in evolutionary perspective. Various types of proximal end development are shown in Fig. 10.12, illustrating the evolution of structure, but it must be remembered that the models represented here are greatly simplified from the actual structures encountered in real graptoloids and that the pattern of proximal end evolution is much more complex than this simplified picture would suggest. Most dichograptid proximal ends are of unmodified form. In these (as in *Didymograptus bifidus*) th. 1^2 grows out from th. 1^1 across the sicula and is welded to it; it forms a single crossing canal. But in later dichograptids (e.g. *Didymograptus extensus*) th. 2^1 is also welded to the face of the sicula, and so there are two crossing canals. The *'D. bifidus'* type would seem to be primitive and to represent the rootstock type of proximal ends in graptoloids. Cooper and Fortey (1982), however, have shown that it is the *'D. extensus'* type which is actually primitive for graptoloids, and *D. bifidus* morphology is secondarily derived. In species such as *Tetragraptus bigsbyi*, for all its apparent complexity, there are still only two crossing canals, since th. 2^2 is not welded to the front of the sicula.

Leptograptus usually has two crossing canals, and though it may have three its development is hardly modified from that of the later dichograptids. The dicello- and diplograptids never have less than three crossing canals. The structure of the proximal end in the scandent diplograptids differs markedly from that in their precursors, and the thecae are often twisted up together in a knot. Diplograptids may be **aseptate** (e.g. *Diplograptus*), or they may have a **median septum** separating the later thecae so that there are normally only three crossing canals rather than an indefinite number. Thus the later-formed thecae (usually those succeeding th. 2^1) open on the same side of the median line as they arise. A close analysis of proximal end structure in the Diplograptacea reveals no less than nine separate 'astogenetic' patterns (Mitchell, 1987). These separate structural patterns, once established, were conserved during evolution and thus provide a reliable guide to phylogeny. In general terms, while new diplograptacean proximal-end patterns appeared successively, they became less complex through time. The classification adopted here is based partially upon such proximal end structures.

In *Climacograptus*, for instance, th. 3^1 arises from th. 2^1, whereas in *Diplograptus* th. 3^1 arises from th. 2^2 and th. 3^2 from th. 3^1.

Similar in development to diplograptids in many ways are the dicellograptids, which likewise have three crossing canals with curiously twisted early thecae. Normally in *Dicellograptus* and its relatives the sicula reclines against one of the stipes.

There are some very modified modes of development in certain scandent graptoloid genera. *Cryptograptus*, for instance, has a thecal system quite different from that of the diplograptids; its most peculiar feature is the torsion of the thecae which results in the series th. 1^1, 2^1, 3^1, . . . opening on opposite sides (frontal and reverse) of the sicula to th. 1^2, 2^2, 3^2, ... *Skiagraptus* has a modified version of this structure which is even more complex.

The origin of the scandent biserial graptoloids has been much disputed, but some evidence suggests a common origin of both diplograptid and glossograptid types from a broad leaf-like scandent dichograptid, such as *Apiograptus*, though the problem of how the crossing canals arose is far from clear.

Monograptids presumably arose from a diplograptid ancestor, but there are no direct links between them. As already shown, the monograptids' th. 1^1 arises from a notch low on the metasicula, and the serially arranged th. 2^1, 3^1, 4^1, . . . grow upwards at an angle. But the change in proximal end structures from the complex and twisted system of the diplograptid precursors to the simple construction of the monograptid proximal end is very sudden, and there are no known intermediates.

The earliest known monograptids have been identified in the lowermost Llandovery monograptid genus *Atavograptus*, where the thecae are of simple construction and the rhabdosomes are straight or gently curved, but the proximal end is of typical monograptid construction. This gives no help with the problem of monograptid origins, but in the English Lake District in the lowermost zone of the Silurian there are some bedding planes that are covered with specimens of *Atavograptus ceryx* in which the population appears to be **dithyrial**, i.e. consisting of both uniserial and biserial individuals, the latter being similar in many ways to *Glyptograptus* (Rickards and Hutt, 1970). Perhaps it is in such populations that monograptid origins are to be sought (Rickards, 1974). Presumably selective action on such populations resulted in a changed ecological balance, with the uniserial forms eventually becoming dominant.

It is worth noting that there were some 'false monograptids' in the early Llandovery as well as species truly of monograptid type. *Peiragraptus*,

thecal form reaches a peak. Many forms have reverted to a simple thecal type, but they exist side by side with genera in which the apertures are very modified (as in *Cucullograptus*) or positively asymmetrical. The graptoloids of the Lower Devonian are quite diverse in thecal form, with hooked or straight thecae having cowl-like apertural overhangs. Apparently graptoloids went on evolving to the last, and there is no clear reason why they should have become extinct.

In Silurian graptoloids especially, it has been possible to make detailed microevolutionary studies of the changes in thecal shape through time (Fig. 10.15).

Many detailed lineages have been established for Siluro-Devonian graptoloids (Sudbury, 1958; Rickards *et al.*, 1977). In these a number of evolutionary trends have been identified, which curiously enough affected several evolving lineages, both uniserial and biserial, at around the same time or for similar time spaces. Such trends are, of course, evident in the generalized stratigraphic succession of thecal types in Monograptidae (hooked, lobate, isolate, etc), but are also evident, for example, in the development of thecal elongation in the unrelated Llandovery genera *Cephalograptus* and *Rastrites*, and of thecal isolation in the Llandoverian rastritids, dimorphograptids and non-rastritid monograptids. Other trends affect the form of the rhabdosome, e.g. curvature which was dominant in the Llandovery but also found independently until the Devonian, and the separate and common tendency in monograptid stocks to produce very slender rhabdosomes at an early stage in evolution, and more robust rhabdosomes at a later stage. These and other trends are fully documented by Rickards *et al.* (1977) and, as has been known for some time, new thecal types are normally introduced at the proximal end, later spreading along the rhabdosome in descendants of the original type. Thus from *Monograptus triangulatus* there is a trend towards isolation of the thecae, with *Rastrites maximus* as an end point and *R. peregrinus* (which has the more isolate thecae confined to the proximal region) as an intermediate; in any population collected from a single bedding plane, individuals may vary greatly in their state of 'advancement'. Perhaps the best documented of these Silurian series is the trend from *M. argenteus*, where only the early thecae are hooked, towards *M. priodon*, in which all the thecae have the form of open hooks. In all of these the new type of theca came in at the proximal end and spread from it, but there are some cases where distal introduction was the rule. Urbanek (1960) has postulated that certain growth-stimulating substances like the auxins of plants diffused from the sicula as the graptoloid grew, promoting differential curvature in the proximal thecae but not affecting the latter thecae, since their quantity decreased during ontogeny. In the descendants of these early forms the growth stimulators continued to be produced from the sicula until a later stage of development. Where new types were introduced distally, Urbanek assumed that a growth inhibitor, rather than a stimulator, diffused from the sicula. Although the operation of this system cannot be proved, it remains a valuable and interesting suggestion.

Many lines of descent are now known in graptolites and are accurately tied in to stratigraphy. Sometimes, however, certain forms may appear, very similar to those in the main line of evolution, but out of phase with these by several million years (Rickards, 1988). Such anachronistic evolution may be 'heraldic', i.e. antedating the main group, and having 'advanced' characters too early. On the other hand it may be 'echoic', the graptolites being 'pale echoes' of former successful types. The best examples here are the partially biserial stipes of *Dimorphograptus* and similar genera, which appeared in the early Silurian, up to 3 Ma after the origin of the monograptids from which they were derived, but looking something like their remoter diplograptid ancestors. The timescale of a heraldic–main-line–echoic relationship may extend over 20 Ma. The heraldic and echoic types are normally not very successful. Perhaps, as Rickards indicates, the full genetic potential for a particular morphological type was always present in a particular line but could only be realized when time and environment were right for it.

Cladia

Cyrtograptus, amongst other genera, produces lateral branches (cladia) at intervals. The development of these can be followed in well-preserved material (Fig. 10.8e,f). Each theca has a hood-like **lappet** with a pair of short lateral whiskers projecting from it. When a cladium develops from such a thecae, one of the two whiskers elongates and becomes a nema. New fusellar tissue grows along this as if from an initial bud. The aperture of the basal theca is constricted by the growth of the fusellar tissue but remains open. By the time the first theca of the new cladium has developed, the primary stipe has advanced by some seven or eight thecae. Since thecal form in *Cyrtograptus* changes on going along the rhabdosome, the characters of any one theca on the primary stipe can be matched exactly with one on the cladium; always there is a lag so that the first cladial theca is paralleled by a theca on the primary stipe seven or eight thecae ahead. Both ends of the rhabdosome, on the primary stipe and cladium, have been growing in the same way at the same time.

Though best known in *Cyrtograptus*, cladia are also present in certain other genera. *Diversograptus* has a sicular cladium, so that the straight main stipe grows in the opposite direction to the cladium. There may be accessory lateral cladia growing either from the main stipe or from the primary cladium. The related genus *Abiesgraptus* has a number of sicular cladia.

Structure of retiolitids

The retiolitids are a group of scandent biserial graptolites which ranged from the Llandeilian through to the end of the Silurian. They differ from other graptolites in that the skeleton largely consists of a meshwork of girders (lists) which may have a reticulate net covering it (Fig. 10.16).

In some retiolitids it seems as if the periderm is simply reduced to a set of girders, which outline the sicula and thecae. In others, however (Bates and Kirk, 1984), the sicula and first few thecae develop in a generally diptograptid mode. Then spines grow from the sicular aperture and link to form a curious thin ring, the **ancora**, lying normal to the axis of the rhabdosome. From this there grows up a skeleton of lists and meshwork which mantles the thecal skeleton, but lies well outside it. The primary thecal skeleton is thus a quite separate structure from this, consisting of only the sicula and a few thecae. The outer meshwork which has elaborate canals running through it is a secondary structure, and as yet its function is incompletely understood. Possibly the light, springy skeleton may have added in buoyancy, or enabled the retiolitids to colonize more turbulent waters. Judging by their persistence through time, they must have been a highly successful group.

10.5 How did graptolites live?

In general terms tuboids, camaroids and crustoids were encrusting colonies resembling in habit the rhabdopleuran hemichordates. Dendroids, on the other hand, were generally upright shrubby benthos and the graptoloids to which they gave rise and which they outlasted were the only wholly plank-

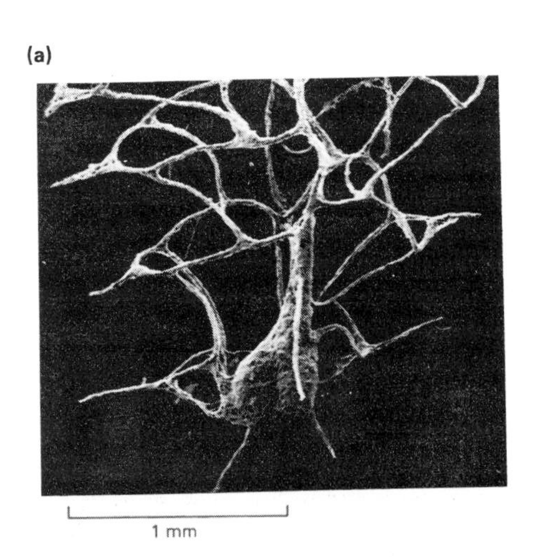

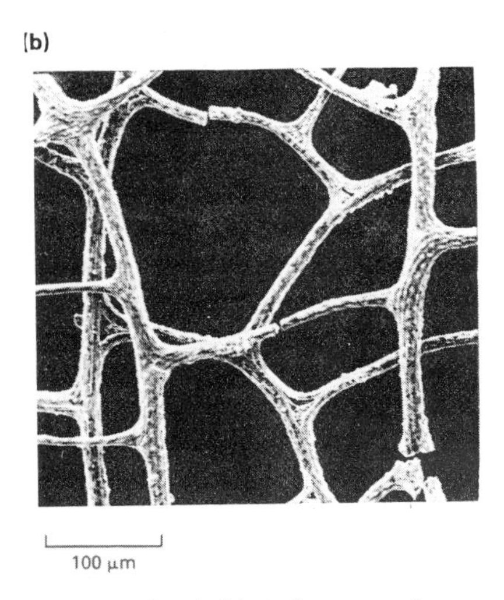

Figure 10.16 Retiolitid ultrastructure: (a) *Orthoretiolites hami* (Ord.), Oklahoma, proximal end; (b) *Gothograptus* cf. *intermedius* (Ord.), clathrial network, from an erratic boulder, northern Germany. (Photographs courtesy of Dr Dennis Bates.)

tonic order. The bottom-dwelling dendroids and others are found mainly in silts, sands and limestones of shallow-water inshore origin. They have no flotational structures and the cortical tissue of dendroids such as *Dendrograptus* appears to have been used primarily for strengthening the proximal end of the rhabdosome into a root-like holdfast. Thus the dendroids can be envisaged as having grown upright on the sea floor with the sicular end down and the stipes outstretched as in a small shrub. With the continued addition of cortical tissue the stem would have been greatly strengthened, but the early thecae were occluded and the zooids presumably atrophied. However, *Radiograptus* and *Rhabdinopora* are conical rather than shrubby and had little cortical tissue. The thread-like nema could not have supported a rhabdosome growing on the sea floor. Furthermore, small tufts of thin fibres are sometimes found terminating the nema, which could have acted as fixing devices for the attachment of the dendroid to floating seaweed. This may be suggested by the presence of much carbon in the black shales in which many graptolites are found. Presumably *Rhabdinopora* evolved from the standard sessile dendroids by inversion of the rhabdosome, sicular end up, though an 'upright' orientation seems to have been restored in the scandent graptoloids of the later Ordovician and Silurian.

A nemally attached mode of life is likely for *Rhabdinopora* and probably some other graptoloids as well, but the majority were probably free-floating (holoplanktonic), microphagous feeders, forming the main preserved part of the Lower Palaeozoic plankton. In this there were undoubtedly other organisms, some of which, such as the small epiplanktonic brachiopods and occasional small trilobites, are sometimes found preserved in graptolitic shales. Two opposing views have been put forward as to how the planktonic graptolites actually lived. The more traditional ides (Bulman, 1964; Rickards, 1975) has been that graptoloids passively floated in the ocean, perhaps at different levels, and at the mercy of currents. The alternative view (Kirk, 1969, 1972) is that graptoloids, after a larval benthic stage, became planktonic as young adults, and were actively mobile. Rather than passively drifting, they are envisaged as actively swimming, with the sicula upwards while feeding currents were drawn by the thecae towards the ciliary ornamented zooids.

Passive drifting

If graptoloids were drifters they would need some mechanism to prevent them sinking. Possible adaptations for floating are numerous; modern organisms achieve flotation either by having gas or fat bubbles within their tissues or by having the tissues isotonic with the surrounding sea water. Bulman (1964), Rickards (1975) and others believe that gas bubbles within the nema, or vacuolated extrathecal tissue containing tiny gas pockets, would give enough buoyancy. Such extrathecal tissue may have been secreted by and associated with the open end of the nema; supplementary flotation mechanisms of various kinds may have acted as a support for such vacuolated tissue. Vanes, once thought to be flattened bladders, are not uncommon in the scandent compact diptograptids. They form lateral extensions to the nema; some graptoloids have two vanes at 180°, others have three at 120°. If these were the sites of vacuolated extrathecal tissue, they would have imparted buoyancy and stability to the graptoloid. Proximal end vanes or plates occur rarely and their function is unclear.

Some dichograptids have expanded webs surrounding the proximal end which increase the surface area and may have prevented sinking. Certain spirally coiled graptoloids (*Cyrtograptus* and *Monograptus turriculatus* amongst others) were considered by Bulman (1964) as floating forms, which would be rotated round and round, spiralling upwards in the water under the influence of currents and thus buoyed up. The presence of a flat proximal membrane in *Cyrtograptus* may have been an extra flotational aid, and *M. turriculatus* is now known to have the tightly coiled stipes connected by dissepiments; such would strengthen the colony as it rotated.

No graptolites have ever been found actually attached to algae and some authorities (Kozlowski, 1971; Rickards, 1975) believe that nemal attachment by extrathecal tissue exuding from the tip of the nema was unlikely for any graptolite. On the other hand, radially arranged associations of graptoloid rhabdosomes, with their nemas directed towards the centre, have been reported. These are known as synrhabdosomes, and although they are rare (see p. 343) some may be genuine life associations in which presumably the rhabdosomes were connected by a mass of extrathecal tissue from the nemal tips. Bubble-like flotation structures, originally described at the centres of synrhabdosomes, have proved to be illusory.

The final change to the monograptid pattern produced long, slender, less dense colonies, which seemed able to manage without such extra flotation structures, as had the diptograptids, and probably made do with nemal vacuolated tissue. At the same time, since the nature of the extrathecal tissue in graptoloids is as yet not understood, this vital gap in our knowledge may well prevent any further development of our understanding of graptoloid ecology.

Automobility

The basic concept here is that feeding currents produced by zooids would have been powerful enough to propel the colony. (Since cortical bandages were put on from the outside there is no likelihood of ciliated extrathecal tissue which could have aided.) Very many members of today's plankton are automobile and there is no reason why graptoloids could not have been. Vanes, webs and other such structures could have been equally functional in mobile graptoloids as in drifters. The reduction in stipe number, the increasing symmetry and the change in inclination of the stipes are all seen as part of an evolutionary response to this new mode of life, and though the whole matter is very controversial and has been much debated it remains an interesting hypothesis.

An argument against automobility is that many species of graptoloid after suffering breakage could regenerate and grow in the opposite direction.

Use of models in interpreting the mode of life of graptoloids

There have to be functional reasons why dendroid and graptoloid rhabdosomes are the shapes they are. Such morphologies result, as in other organisms, from the interplay of adaptation by Darwinian selection and constraints imposed by the fundamental design of the organisms and by the materials of which they are constructed. The ultimate form of any such organism is thus a compromise between inherited constraints and adaptive possibilities. The analysis of the anatomical structure of the organisms and how it changed through time are classic pointers which help in understanding functional adaptations, but in the case of the graptolites other recently developed indicators are also a help.

The branching structure of multiramous graptolites, for example, has proved particularly amenable to computer analysis and simulation (Fig. 10.17).

Multiramous graptolites occur typically in the Lower Ordovician (dichograptids) but also at higher levels (e.g. nemagraptids and cyrtograptids) where some forms become secondarily convergent on dichograptids. Computer simulations have shown that various branching patterns just like those in 'real' multiramous graptoloids can be generated by permutations of very simple rules (Fortey and Bell, 1987). In most dichograptids branching is dichotomous, i.e. each growing stipe splits into two equal daughter stipes at the same time, and as the colony grows there will come a further zone of dichotomous branching.

The simplest computer-generated models were based on just such standard dichotomies. Instructions were given (1) for branches always to split dichotomously at nodal points, but (2) to alter the angle of dichotomy in some forms and/or (3) to delay the initiation of dichotomy. The 'graptoloids' thus generated bear a singular resemblance to the well-known quadiradiate (*Staurograptus*), triradiate (*Bryograptus*) and biradiate (*Clonograptus*) genera of the Lower Ordovician. Another permutation produced a 'graptoloid' with an S-shaped main stipe with fan-shaped daughter stipes (approximating *Nemagraptus*). Again, a computer model with instructions to generate curving stipes produced a close match with *Cyrtograptus*, a monograptid (Fig. 10.8e) which has become secondarily multiramous by the regular generation of cladia, some of which may themselves grow secondary cladia.

But what may be the significance of such patterns? From the analysis both of real and computer-drawn graptolites it is evident that branches do not interfere with each other. Whereas 'possible' models can be generated where the branches do overlap, actual graptoloids of this kind never occur in nature. Such multiramous graptoloids must therefore have been planar, either flat or slightly conical. Moreover the stipes within this flat plane are regularly arranged with spaces between them, and in some instances potential dichotomies within a rhabdosome have been suppressed where otherwise the stipes would

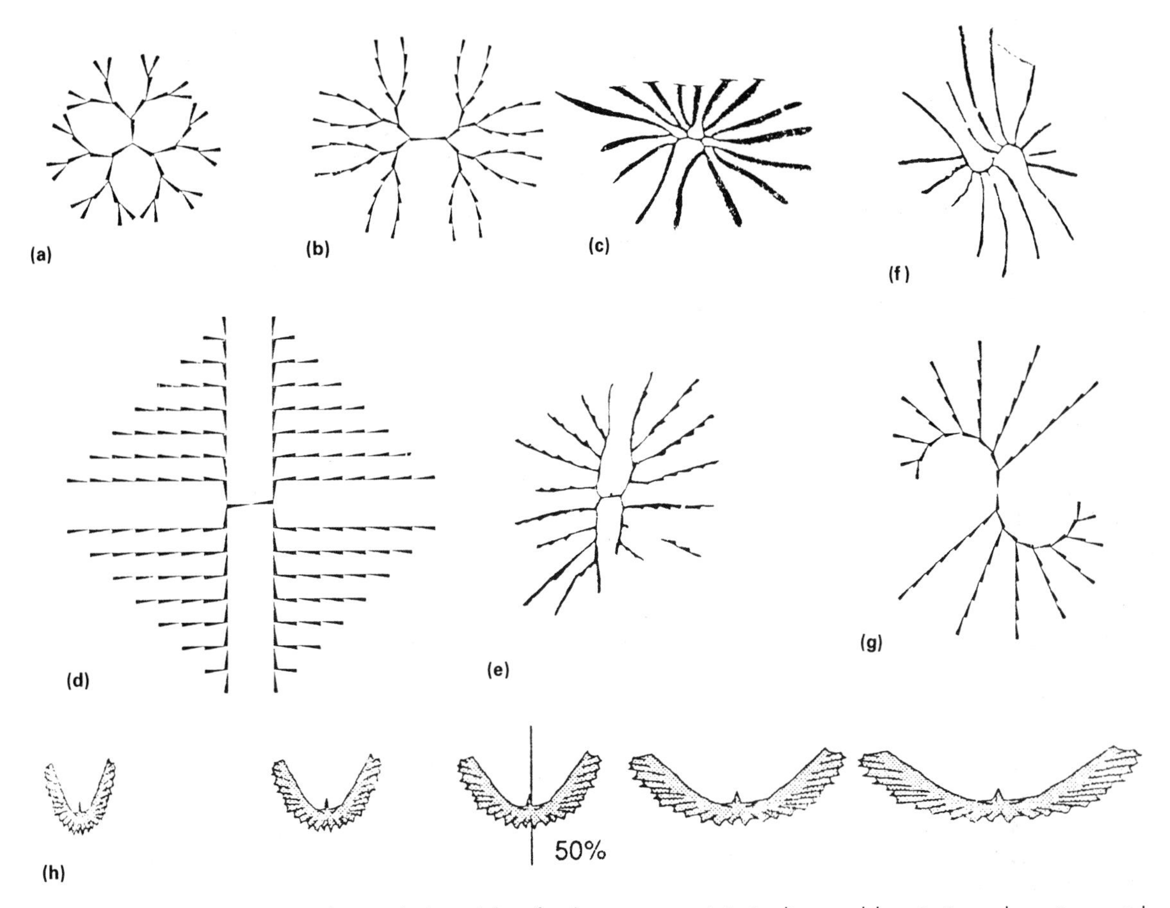

Figure 10.17 Computer-generated graptoloids and their fossil counterparts. (a) triradiate model, as in Tremadoc anisograptids, reducing angle between dichotomies, thereby avoiding overlap; (b) biradiate model, as in later Tremadoc anisograptids, with distal dichotomies suppressed and slight curvature of distal stipes, thereby avoiding overlap, compared with (c) *Loganograptus logani;* (d) in this model stipes are distributed evenly through the area enclosed by the rhabdosome, paralleled by (e) *Brachiograptus etaformis;* (f) the actual *Nemagraptus gracilis* which is directly comparable with (g), the computer-generated yin-yang model. (All redrawn from Fortey and Bell, 1987). (h) Computer-generated deformed images of *Isograptus* (left) which simulate actual graptolites in deformed rocks, useful in stress–strain measurements (after Williams, 1990).

have grown too closely. If we envisage the zooids of adjacent stipes covering the feeding area in between, then a multiramous rhabdosome can be interpreted as an effective 'harvesting machine'. It could either feed by rising or sinking vertically through the water, while the zooids strained off the nutrients, or by remaining stationary while currents passed through it.

Recently, our conception of how such multiramous (and other) rhabdosomes actually behaved has been refined by the use of experimental model graptolites (Rigby and Rickards, 1989; Rigby, 1991; Fig. 10.18).

These were accurately constructed mainly at life size, of aluminium tubing covered with 'clingfilm' to simulate the density of a living graptolite. The models were allowed to fall freely through a column of still water, in horizontal orientation, and their behaviour was observed. Most of the multiramous forms and *Nemagraptus* models rotated slowly as they sank, spiralling downwards through the water. In life such forms would have been able to 'sweep up' or harvest nutrients as they fell, aided by the spiral rotation (which allows a longer path than straight fall, without overlap between the paths of the descending zooids). The models also clarified how

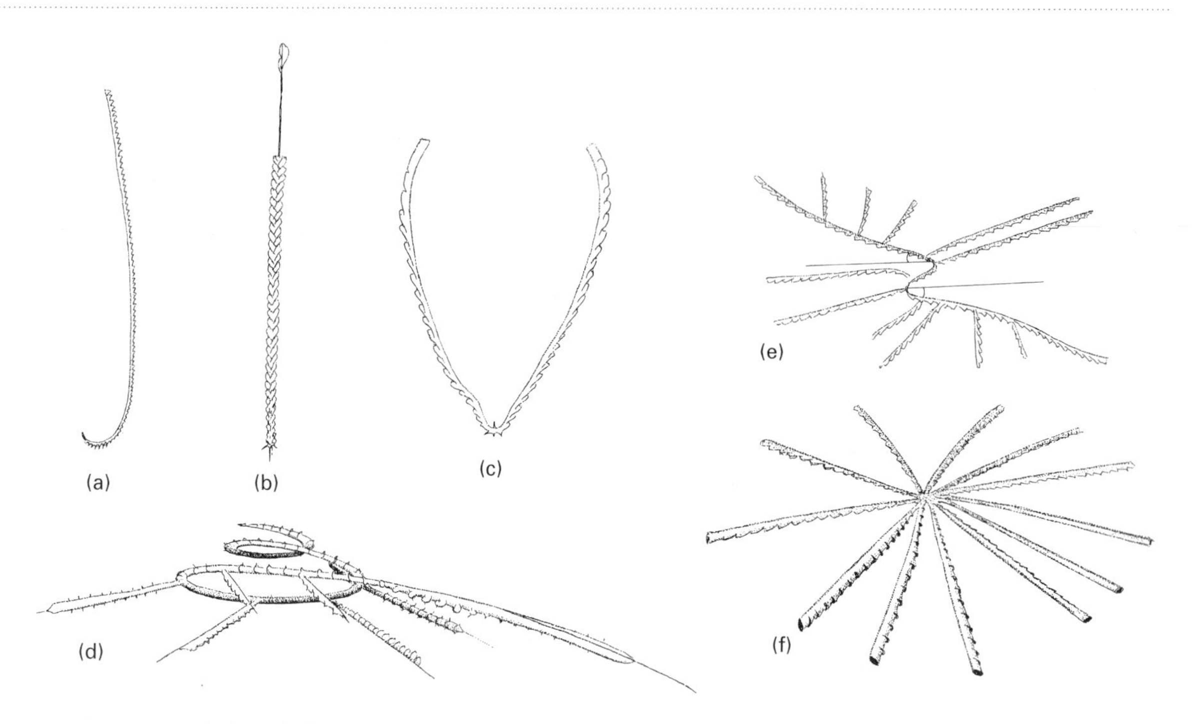

Figure 10.18 Morphological adaptations which cause graptoloid rhabdosomes to rotate, as tested by physical modelling: (a) proximal hook on monograptid rhabdosome; (b) vane on biserial rhabdosome; (c) thecal offset on a dicellograptid; (d) *Cyrtograptus*, with spiral form, cladia and thecal offset; (e) *Nemagraptus* would rotate only if the rhabdosome had this twisted S-shape, rather than being flat; (f) *Loganograptus* would rotate if stipes were angled relative to one another, as in a propeller. (Redrawn from Rigby and Rickards, 1989.)

the stipes were probably arranged in life. *Nemagraptus* models, for example, constructed as planar forms, did not rotate. If, however, the main S-shaped stipe was bent symmetrically above and below the plane, with the secondary stipes forming a 'spiral staircase', then the model rotated effectively. In other multiramous genera (*Loganograptus*) twisting of the stipes at an angle to one another as in a propeller likewise facilitated rotation, and in *Cyrtograptus* further aid was given by secondary cladia. In this genus (Underwood, 1995) there is a complex suite of overlapping webs, connecting the cladia with the main stipe (Fig. 10.8e). These are composed of thin peridermal sheets and give the rhabdosome a screw-like form. They may have acted to direct water into separate channels as *Cyrtograptus* rotated.

Many model pauciramous graptolites also rotated, *Dicellograptus* by thecal offset, biserial rhabdosomes by vanes (whether bladed or twisted ribbon type), and curving monograptids by the curve itself. Straight monograptids, on the other hand, fell without rotation.

The advantage of spiral fall is evident both in large multiramous forms, sweeping out a broad harvesting path, and in thin, scandent forms exploiting a narrow path. No rhabdosome, however, could continue falling indefinitely: it would eventually have to rise, which presupposes some degree of automobility, or at least the capacity to change its buoyancy. The nature of such processes in the absence of direct evidence is still uncertain; but the use of both computer and experimental models has at least partially reconciled opposing hypotheses, and added a new dimension to our understanding of how graptolites lived.

Graptoloids seem to have been a primary component of the plankton of the Palaeozoic, and generally tropical and temperate in distribution. Their common association with black shale (graptolitic facies) with its high carbon content and presence of syngenetic pyrite suggests that these planktonic organisms were preserved in conditions where no benthos flourished; then they would be able to be preserved undisturbed by bioturbating animals.

Since they are found in other kinds of sediment they must have been widespread, but their association with a particular argillaceous facies is largely a matter of preservation failure elsewhere.

Rigby's (1993) studies of graptoloid orientation on bedding surfaces in the Middle Ordovician of Quebec showed that the two species present (an *Amplexograptus* and an *Orthograptus*) reached the sea floor at different times. This suggests that the separate populations lived as monospecific shoals, and possibly at different depths in the water column. This work also gave evidence of seasonal growth of graptoloid rhabdosomes, with different sizes of rhabdosome at various times of the year. The nature of 'synrhabdosomes', i.e. radial aggregates of monospecific graptoloids present here and in other beds, is still disputed. Some authorities believe that these are real associations, possibly indicating a method of rapid asexual cloning, but Rigby favours the suggestion that they result from 'marine snow'. i.e. organic detritus stuck together by mucus webs of zooplankton, falling though the water and collecting graptoloids as it sank. Graptoloids were evidently eaten by predators (Underwood, 1993), and there is evidence for crunching or absorption of whole rhabdosomes, or delicate plucking of individual zooids; graptoloids may also have hosted small parasites.

10.6 Faunal provinces

The distribution of graptolites on a global scale, at least in the Ordovician, was not always uniform (Skevington, 1974; Finney and Chen Xu, 1990; Berry and Wilde, 1990; Rickards *et al.*, 1990). In the early Arenig it seems that most graptolite genera had a cosmopolitan distribution, but from the later Arenig onwards until the end of the Llandeilo there were two well-defined graptolite faunal provinces – the 'Atlantic' or 'European' province and the 'Pacific' province – whose faunas are not closely similar. The area of the Atlantic province includes England and Wales, southeastern Ireland, most of Europe, North Africa and, rather oddly, Peru and Bolivia. On the other hand, the graptolites of the Pacific province are found in North America, Argentina, all of Australia and Australasia and (at first sight paradoxically) in Scotland, northwestern Ireland and western Norway. The graptolites collected from any part of one of these provinces bear a fairly close resemblance to each other at the generic level.

Thus amongst other distinctive factors, pendent (tuning fork) *Didymograptus* species are very common in the Atlantic but not in the Pacific province, which has a more diverse fauna with iso-, cardio-, onco- and sinograptids. Biserial scandent graptoloids common to both provide a good basis for stratigraphical correlation.

The zenith of provincialism was reached in the Arenig–Llanvirn, and in the later Ordovician the two faunal provinces were replaced by a single one.

What could have been the controls of provincialism? With free-floating organisms, geographical barriers are likely to be less important than they are with shelf-living benthos, unless they cross temperature zones. Such more or less latitudinal zones defining regions of differing temperature (tropical, subtropical, warm-temperature, cold-temperature, boreal, polar) are the most important single control of the global distribution of organisms today, and presumably they were of the Lower Palaeozoic plankton also. The latitudinal thermal gradient is also responsible for the differentiation of oceanic waters into separate water masses. It is very probable (Finney, 1984; Finney and Chen Xu, 1990) that specific graptolite faunas were confined to particular water masses; indeed water-mass specificity has been cited as the prime control of graptolite biogeography. [Berry *et al.* (1987) have suggested that some graptolites inhabited the oceanic oxygen minimum layer.]

The Pacific faunas of the Lower Ordovician were apparently circumtropical and were confined entirely between the latitudes 30°N and 30°S. Atlantic faunas, on the other hand, lay south of 30°S, except for the European region, where Atlantic faunas are found nearer to the equator. It has been proposed that the isotherms more or less followed palaeolatitudes, except where a tongue of colder water projected northwards in the European region (as with today's Peru Current), carrying with it its own indigenous fauna.

Such a pattern could well account for Lower Ordovician graptolite distribution, with water-mass specificity (ultimately dependent on temperature) as the primary control, but why then did the faunas of the later Ordovician become more uniform? One reason is that Families Dichograptidae and Sinograptidae, which had been important in defining the two provinces, had by this time become extinct, but it is most significant that almost no later Ordovician graptolites are found outside the palaeolatitudes 30°N and 30°S; they are confined within the region formerly occupied by the circum-tropical Pacific province. This reduction of the geographical distribution of graptolites in the later Ordovician may well have been under climatic control. High latitudinal realms no longer supported graptolites, because the waters were too cold, and the graptolites of the tropical zones were unable to adapt to cold conditions.

By the latest Ordovician (late Ashgill) only some five or six species of graptolites remained: a decimated stock living at a time of widespread glacial conditions, from which came the earliest monograptids. During the milder climates of the Silurian their descendants were able to expand and diversify and to spread uniformly and widely across the world once more, to the regions that their ancestors had occupied in the earlier Ordovician. For a time, a tropical Pacific province and a cooler Atlantic province were re-established. From the Ludlow, however, until their final extinction in the Pragian, graptolites appear to have been confined to tropical seas alone.

10.7 Stratigraphical use

To geologists, graptolites are primarily of interest as stratigraphical indicators. The sequence of graptolite faunas has been used for the subdivision of Ordovician and Silurian rocks since the time of Lapworth, whose work in southern Scotland in unravelling the complexities of geological structure is classic. Their value for long-range correlation is because (1) they were planktonic and widely dispersed; (2) most species were eurythermic and thus not confined to latitudinal belts; (3) the majority of graptoloids were epipelagic, and thus may be preserved both in deep-water and shallow-water facies, subject to preservability; and (4) the stratigraphic range of many species is short. The main problems in the stratigraphical use of graptolites are as follows: (1) graptolites are usually preserved as compressions, and there may be problems in precise identification; (2) the time range of some graptoloid species is rather long, and it is not possible with such species to give the kind of stratigraphical precision that is possible with (for instance) ammonites; (3) graptolites are normally confined to black or grey shaly facies by preservational factors, but often great thickness of Ordovician and Silurian strata, which might be expected to contain graptolites, do not in fact have any fossils at all. Furthermore, it is relatively rare to find graptolites in coarser-grained rocks away from graptolitic facies, and there are consequently singular problems in correlating between areas of different facies.

Usually geologists look for areas where, because of shoreline oscillations, there are exposed sequences of alternating graptolitic and shelly facies, so that a particular graptolite assemblage is time-bracketed within a trilobite–brachiopod assemblage and hence the two may be directly correlated. Alternatively, the discovery of mixed graptolitic and shelly faunas in the same argillaceous sediment is most helpful. Even so, there are successions where the shelly faunas are indigenous to the region or have very long ranges and where in the absence of graptolites precise correlations are not possible.

Individual graptolite zones are based and defined on the time ranges of particular short-ranged species, though the named species is normally only one of a number in the total fauna. In the European succession 13 Ordovician, 32 Silurian and three Devonian zones have been erected and are in common use. Some widespread species, such as *Glyptograptus teretiusculus*, *Nemagraptus gracilis* and *Dicellograptus complanatus*, are normally considered as representing time-equivalent horizons, but there may be some diachroneity. The problem of diachroneity is a real one and correlations have to be based upon such events that can be shown to be the least diachronous. The only way to do this is to use 'as many closely spaced events (first and last appearances of species in time), rather than a few selected "correlation fossils"' (Cooper and Lindholm, 1990). This high-resolution approach to stratigraphy using 130 graptolite species, has enabled these authors to correlate Tremadoc–Llandeilo sequences on a global basis, with great precision.

Though some graptolites may be very widespread, there are many others which are geographically restricted and useful for strictly local correlation alone. Because of this difficulty several palaeontologists have tried to define faunal units which are larger than zones but of more or less worldwide application. These successive **graptolite**

faunas (Fig. 10.11) are easily identified and in any case epitomize, as Bulman (1958) has shown, the geological history of the group.

The following sequence of graptolite faunas through time is from Bulman (1958). The various faunas and subfaunas here noted are defined on the first appearance of new graptoloid types.

The Tremadocian is characterized by an **anisograptid fauna** consisting mainly of *Rhabdinopora* species and other anisograptids. Since provinciality was well marked at this time there are some problems in intercontinental correlation. The time equivalence of different *Rhabdinopora* species in various parts of the world is now becoming reasonably clear, even though these species may be mutually exclusive.

Very few graptolites of Upper Tremadocian age are known, other than one fauna from the Yukon Territory and another from Norway in which *Bryograptus*, *Kiaerograptus* and possibly a *Didymograptus* have been described. But other than this doubtful record there are no true graptoloids known from Tremadocian rocks.

The succeeding **dichograptid fauna** of the Arenig is marked by the incoming of the earliest dictograptids. *Tetragraptus* is found right at the base, and is usually associated with didymograptids as well as with remaining anisograptids. The extensiform didymograptids of the Arenig were replaced in the Llanvirn by tuning fork species of *Didymograptus* (though by the Llanvirn the succeeding diplograptid fauna became established), and the genus continued into the Caradoc before becoming extinct. Provinciality is well marked in the Lower Ordovician, and the endemic Pacific genera *Oncograptus*, *Cardiograptus* and *Isograptus* have been found stratigraphically useful in beds of equivalent age in western North America and Australia.

A few biserial graptoloids are found at the very top of the Arenig, but the real flowering of the biserial genera is from the Llanvirn until well into the Silurian. The **diplograptid fauna** spans the period from the Llanvirn to the lowermost Silurian, before the incoming of the **monograptid fauna**. The diplograptid fauna is divided into four subfaunas. The *Glyptograptus–Amplexograptus* subfauna of the Llanvirn and Llandeilo contains many tuning fork graptoloids in addition to the genera from which the fauna takes its name; evidently the main period of differentiation of the biserial graptoloids took place at this time, even though they are not numerically abundant. Lower Caradocian beds contain a *Nemagraptus–Dicellograptus* subfauna in which *Dicranograptus* is also found and which contains the final dichograptids. The *Orthograptus–Dicellograptus* subfauna replaces it in the Upper Caradoc and Ashgill. In the later Ashgill this subfauna is somewhat impoverished as is the *Orthograptus–Climacograptus* subfauna of the lowermost Silurian, which immediately predates the arrival of the monograptid fauna.

This fauna extends throughout the whole of the Silurian (other than in the two lowermost zones) and into the Pragian. Though it has not been divided into subfaunas, different monograptid types, characterized mainly by their thecal construction, are found successively and can be used stratigraphically. These assemblages allow the Silurian to be divided into zonal units of less than a million years duration.

Bibliography

Books, treatises and symposia

Bulman, O.M.B. (1945–1947) *A Monograph of the Caradoc (Belclatchie) Graptolites from Limestones in Laggan Burn, Ayrshire*. Palaeontographical Society Monograph (i) 1945, 1–42 (ii) 1946, 43–58 (iii) 1947, 59–78. (Detailed morphology and ontogeny; profusely illustrated)

Crowther, P.R. (1981) *The Fine Structure of Graptolite Periderm*. Special Papers in Palaeontology No. 26.

Elles, G.L. and Wood, E.M.R. (1922) *A Monograph of British Graptolites* (with synoptic supplement by I. Strachan, 1971). Palaeontographical Society, London. (Descriptions and stratigraphical use of all British species then known; standard work)

Palmer, D. and Rickards, R.B. (eds) (1991) *Graptolites: Writing in the Rocks. Fossils illustrated: 1.* Boydell Press, Woodbridge, 182 pp. (Professional, though popular account, 138 photos – unequivocally recommended)

Rickards, R.B., Jackson, R.E. and Hughes, C.P. (eds) (1974) *Graptolite Studies in Honour of O.M.B. Bulman*. Special Papers in Palaeontology No. 13. (20 original papers and bibliography of Bulman's works)

Teichert, C. (ed.) (1970) *Treatise on Invertebrate Paleontology*, Part V, *Graptolithina* (revised). Geological Society of America and University of Kansas Press, Lawrence, Kan. (Updates Bulman's original treatise of 1955)

Individual papers and other references

Bates, D.E.B. and Kirk, N.H. (1984) Autecology of Silurian graptoloids, in *Graptolite Studies in Honour of O.M.B. Bulman*, Special Papers in Palaeontology No. 32, Academic Press, London, pp. 121–39.

Berry, W.B.N. and Wilde, P. (1990) Graptolite biogeography: implications for palaeogeography and palaeo-oceanography, in *Palaeozoic Palaeogeography and Biogeography* (eds W.S. McKerrow and C.R. Scotese). Geological Society Memoir No. 12, pp. 1–435.

Berry, W.B.N., Wilde, P. and Quinby-Hunt, M.S. (1987) The oceanic non-sulfidic oxygen minimum zone – a habitat for graptolites? *Bulletin of the Geological Society of Denmark* **35**, 103–14.

Bulman, O.M.B. (1958) The sequence of graptolite faunas. *Palaeontology* **1**, 159–73. (Stratigraphical use)

Bulman, O.M.B. (1963) On *Glyptograptus dentatus* and some allied genera. *Palaeontology* **6**, 665–89.

Bulman, O.M.B. (1964) Lower Palaeozoic plankton. *Quarterly Journal of the Geological Society of London* **119**, 401–18. (The graptolite biocoenosis as the preservable plankton of its time)

Bull, E.E. (1996) Implications of normal and abnormal growth structures in a Scottish Silurian dendroid fauna. *Palaeontology* **39**, 219–46.

Cooper, R.A. and Fortey, R.A. (1982) The Ordovician graptolites of Spitzbergen. *Bulletin of the British Museum of Natural History (Geology)* **36**, 157–302. (Proximal end structures)

Cooper, R.A., Fortey, R.A. and Lindholm, K. (1991) Latitudinal and depth zonation of early Ordovician graptolites. *Lethaia* **24**, 199–218.

Cooper, R.A. and Lindholm, K. (1990) A precise worldwide correlation of early Ordovician graptolite sequences. *Geological Magazine* **127**, 497–525.

Crowther, P. and Rickards, R.B. (1977) Cortical bandages and the graptolite zooid. *Geologica et Palaeontologica* **11**, 9–46.

Dilly, P.N. (1993) *Cephalodiscus graptolitoides* sp. nov., a probable extant graptolite. *Journal of Zoology* **229**, 69–78.

Durman, P.N. and Sennikov, N.K. (1993) A new rhabdopleurid hemichordate from the Middle Cambrian of Siberia. *Palaeontology* **36**, 283–96.

Erdtmann, B.-D. (1982) Palaeobiogeography and environments of planktic dictyonemid graptolites during the earliest Ordovician, in *The Cambrian–Ordovician Boundary: Sections, Fossil Distributions and Correlations* (eds M.G. Bassett and W. Dean), National Museum of Wales, Cardiff, pp. 9–28.

Finney, S.C. (1979) Mode of life of planktonic graptolites: flotation structure in Ordovician *Dicellograptus* sp. *Paleobiology* **5**, 31–9.

Finney, S.C. (1984) Biogeography of Ordovician graptolites in the Southern Appalachians. *Palaeontological Contributions of the University of Oslo* **295**, 167–76.

Finney, S. and Xu Chen (1990) The relationship of Ordovician graptolite provincialism to palaeogeography, in *Palaeozoic Palaeogeography and Biogeography* (eds W.S. McKerrow and C.R. Scotese). Geological Society Memoir No. 12, pp. 1–435.

Fortey, R.A. and Bell, A. (1987) Branching geometry and function of multiramous graptoloids. *Paleobiology* **13**, 1–19.

Fortey, R.A. and Cooper, R.A. (1986) A phylogenetic classification of the graptoloids. *Palaeontology* **29**, 631–54.

Kirk, N. (1969) Some thoughts on the ecology, mode of life, and evolution of the Graptolithina. *Proceedings of the Geological Society of London* **1659**, 273–92.

Kirk, N. (1972) More thoughts on the automobility of the graptolites. *Quarterly Journal of the Geological Society of London* **128**, 127–33. (A different interpretation from that of Bulman)

Koszlowski, R. (1948) Les graptolithes et quelques nouveaux groupes d'animaux du Tremadoc de la Pologne. *Palaeontologia Polonica* **3**, 1–235. (Classic, profusely illustrated study of three-dimensional dendroids and other groups)

Koszlowski, R. (1971) Early development stages and the mode of life of graptolites. *Acta Palaeontologica Polonica* **16**, 313–43.

Mitchell, C.E. (1987) Evolution and phylogenetic classification of the Diplograptacea. *Palaeontology* **30**, 353–406.

Obut, A.M. (1974) New graptolites from the Middle Cambrian of the Siberian Platform, in *Graptolite Studies in Honour of O.M.B. Bulman* (eds R.B. Rickards, D.E. Jackson and C.P. Hughes), Special Papers in Palaeontology No. 13, Academic Press, London, pp. 9–13. (Oldest known graptolites)

Rickards, R.B. (1974) A new monograptid genus and the origins of the main monograptid genera, in *Graptolite Studies in Honour of O.M.B. Bulman* (eds R.B. Rickards, D.E. Jackson and C.P. Hughes), Special Papers in Palaeontology No. 13, Academic Press, London, pp. 141–7. (Monograptids may have originated from dithyrial population)

Rickards, R.B. (1975) Palaeoecology of the Graptolithina, an extinct class of the Phylum Hemichordata. *Biological Reviews* **50**, 397–436.

Rickards, R.B. (1977) Patterns of evolution in the graptolites, in *Patterns of Evolution* (ed. A. Hallam), Elsevier, London, pp. 333–58.

Rickards, R.B. (1988) Anachronistic, heraldic, and echoic evolution: new patterns revealed by extinct, planktonic hemichordates, in *Extinction and Survival in the Fossil Record* (ed. G.P. Larwood), Systematics Association Special Volume 34, Clarendon Press, Oxford, pp. 211–30.

Rickards, R.B. and Hutt, J.E. (1970) The earliest monograptid. *Proceedings of the Geological Society of London* **1063**, 115–19. (Monograptid origins)

Rickards, R.B. and Stait, R.B. (1984) *Psigraptus*, its classification, evolution and zooid. *Alcheringa* **8**, 101–11.

Rickards, R.B., Crowther, P.R. and Chapman, A.J. (1982) Ultrastructural studies of graptolites – a review. *Geological Magazine* **119**, 355–70.

Rickards, R.B., Hutt, J.E. and Berry, W.B.N. (1977). Evolution of the Silurian and Devonian Graptoloids. *Bulletin of the British Museum of Natural History (Geology)* **28**, 1–120.

Rickards, R.B., Rigby, S. and Harris, J.H. (1990) Graptoloid biogeography: recent progress, future hopes, in *Palaeozoic Palaeogeography and Biogeography* (eds W.S. McKerrow and C.R. Scotese). Geological Society Memoir No. 12, pp. 139–46.

Rigby, S. (1991) Feeding strategies in graptoloids. *Palaeontology* **34**, 797–814.

Rigby, S. (1993) Population analysis and orientation studies of graptoloids from the Middle Ordovician Utica Shale, Quebec. *Palaeontology* **36**, 267–82.

Rigby, S. (1994) Hemichordate skeletal growth, shared patterns in *Rhabdopleura* and graptoloids. *Lethaia* **27,** 317–24.

Rigby, S. and Milsom, C. (1996) Benthic origins of zooplankton: an environmentally determined macroevolutionary effect. *Geology* **24**, 52–4.

Rigby, S. and Rickards, R.B. (1989) New evidence for the life habit of graptoloids from physical modelling. *Paleobiology* **15**, 402–13.

Skevington, D. (1974) Controls influencing the composition and distribution of Ordovician graptolite faunal provinces, in *Graptolite Studies in Honour of O.M.B. Bulman* (eds R.B. Rickards, D.E. Jackson and C.P. Hughes), Special Papers in Palaeontology No. 13, Academic Press, London, pp. 59–73.

Spjeldnaes, N. (1963) Some upper Tremadocian graptolites from Norway. *Palaeontology* **6**, 121–31. (Bithecae in anisograptids)

Sudbury, M. (1958) Triangulate monograptids from the *Monograptus gregarius* zone (Lower Llandovery) of the Rheidol Gorge (Cardiganshire). *Philosophical Transactions of the Royal Society B* **241**, 485–555. (Evolutionary lineages)

Underwood, C. (1993) The position of graptolites within the Lower Palaeozoic planktic ecosystem. *Lethaia* **26**, 189–202.

Underwood, C. (1995) Interstipe webbing in the Silurian graptolite *Cyrtograptus*. *Palaeontology* **38**, 619–26.

Urbanek, A. (1958) Monograptidae from erratic boulders in Poland. *Palaeontologia Polonica* **9**, 1–105. (Fine structural detail; many illustrations)

Urbanek, A. (1960) An attempt at biological interpretation of evolutionary changes in graptolite colonies. *Acta Palaeontologia Polonica* **5**, 127–234. (Develops theory of chemical growth regulation in graptolites)

Urbanek, A. and Bengtson, S. (1986) *Rhabdotubus*, a Middle Cambrian rhabdopleurid hemichordate. *Lethaia* **19**, 293–308.

Urbanek, A. and Mierzesewski, P. (1991) The fine structure of a camaroid graptolite. *Lethaia* **24**, 129–38.

Urbanek, A. and Towe, K.M. (1974) Ultrastructural studies on graptolites. 1: The periderm and its derivatives in the Dendroidea and in *Mastigograptus*. *Smithsonian Contributions to Paleobiology* **20**, 1–48.

Urbanek, A. and Towe, K.M. (1975) Ultrastructural studies on graptolites. 2: The periderm and its derivatives in the Graptoloidea. *Smithsonian Contributions to Paleobiology* **22**, 1–24. (Transmission electron microscope studies of well-preserved cuticle, e.g. Fig. 10.7)

Williams, S.H. (1990) Computer-assisted graptolite studies, in *Microcomputers in Palaeontology*, (eds D.L. Bruton and D.A.T. Harper). *Palaeontological Contributions of the University of Oslo* **370**, 46–55.

Williams, S.H. and Rickards, R.B. (1984) Palaeoecology of graptolitic black shales. *Palaeontological Contributions of the University of Oslo* **295**, 159–66.

11 Arthropods

11.1 Introduction

Arthropods have, as their most distinctive characteristics, a hard outer coat or **exoskeleton** and **jointed appendages** which they use for movement and feeding. Those living today fall into three recognizable groupings, the Crustacea, (shrimps, crabs and lobsters), the Chelicerata (spiders, mites and scorpions), and the Uniramia (insects and other forms). Of the extinct forms, trilobites and eurypterids (the giant Palaeozoic water scorpions) are the best known. Arthropods are perhaps the most successful and diverse of all invertebrates, and since their tough exoskeleton offers considerable potential for fossilization (especially if mineralized) their geological record is good.

Arthropods are segmented coelomate metazoans, as are the annelid worms with which they possibly share a common ancestry. In annelids the only skeleton is a hydrostatic one provided by the coelomic fluid, which does not give a rigid anchorage for muscles. The arthropods, however, because of their firm exoskeleton, and also since many also have an internal skeleton (**endoskeleton**), possess a rigid base for attachment of the internal muscles that move the limbs; hence they have the potential for rapid locomotion. Many, furthermore, have hard jaw structures (**mandibles**) which can grind, crush or bite. Thus arthropods, in terms of movement and feeding, have an overriding superiority over the annelid worms and have been able to invade many different environments which remain closed to the latter.

11.2 Classification and general morphology

Diversity of arthropod types

Arthropods take their name from the jointed appendages that are a constant feature of their organisation (Greek: αρθοφς = joint, ποδος = foot). These same appendages, which include the jaw structures, have become very differentiated, and their number, arrangement and morphology are often of critical importance in taxonomy. Many different taxonomic schemes have been erected for the arthropods, and there is still much controversy over whether they arose from single or from multiple ancestors. One view is that of Manton (1973, 1977) that the arthropods are polyphyletic. She considered that 'arthropodization' occurred at least three times, leading to three or more distinct phyla. According to Manton, therefore, the 'Phylum Arthropoda' is an unreal entity consisting of heterogeneous elements. Other authorities (e.g. Briggs and Fortey, 1989, 1992) while recognizing the three main living groups, prefer the concept of a small number of major body plans with a common arthropodan ancestor. The present debate, fuelled by cladistic analysis and the discovery of very early crustaceans, is likely to continue for a long time yet, and so for convenience Manton's system is followed here. This covers both recent and fossil arthropods, though the taxonomic status of trilobites, which are the most abundant fossil arthropods, is not yet fully resolved.

PHYLUM UNIRAMIA (Cam.–Rec.): Jaws bite with the tip of a whole limb. Trunk lacks **biramous** (two-branched) appendages; the limbs are **uniramous**.

CLASS 1. ONYCHOPHORA (Cam.–Rec.): These have a non-rigid segmented body, with unjointed legs that can be inflated with blood for rhythmic walking. Jaws short, ventrally directed, with blade-like terminal claws. Examples include *Peripatus* (Recent terrestrial), *Aysheia* (Cambrian marine).

CLASS 2. MYRIAPODA (Carb.–Rec.): Myriapods and centipedes; entirely terrestrial. Have jointed mandibles biting in the transverse plane. Some giant fossil representatives, such as the terrestrial Upper Carboniferous *Arthropleura*.

CLASS 3. HEXAPODA (Dev.–Rec.): Insects, aerial and terrestrial. The most diverse and numerous of all terrestrial animal species. Mandibles primitively roll and grind, but in some cases strong secondary transverse biting is possible.

PHYLUM CRUSTACEA (Cam.–Rec.): Dominantly marine arthropods with many fossil representatives. A very varied phylum with highly differentiated limb morphology. The less advanced aquatic crustaceans pass food forwards from behind towards the mouth along a median food groove using the limbs, or by feeding currents generated by the limbs. In advanced aquatic crustaceans the food is lifted up from the substratum by specialized appendages. The biting jaws (mandibles) are formed from **gnathobases**: internal extensions of the appendages which meet in the median plane and roll or grind together as the limb moves backwards and forwards. The outer part of the leg may disappear.

The earliest crustaceans were SUBCLASS PHYLLOCARIDA (Cam.–Rec.), which have a large bivalved **carapace** almost covering the body (e.g. Jones and Woodward, 1888–1899; Rolfe, 1962). These, abundant in certain facies in the earlier Palaeozoic, may have been ancestral to the more advanced shrimp-like or lobster-like forms, the earliest of which are latest Devonian in age and which have become extremely important since then.

PHYLUM CHELICERATA (Cam.–Rec.): A large arthropod group in which all representatives have the body divided into two parts: the **prosoma** (fused head and thorax) and the **opisthosoma** (abdomen). The jaws can primitively bite together in the transverse plane, using a biting movement quite unlike the secondary transverse biting movement of the crustaceans. The presence of a pair of **chelicerae** (pincers) in front of the mouth is characteristic.

CLASS 1. MEROSTOMATA (Cam.–Rec.): Aquatic chelicerates, with two important subclasses well represented as fossils:

SUBCLASS 1. XIPHOSURA (Cam.–Rec.): Includes the modern horseshoe crab *Limulus* and its fossil representatives (Order Xiphosurida) and the Cambrian Order Aglaspida.

SUBCLASS 2. EURYPTERIDA (M. Ord.–Perm.): A group of large freshwater and marine water scorpions (similar in appearance to terrestrial scorpions, but not very closely related).

CLASS 2. ARACHNIDA (Sil.–Rec.): All are eight-legged; some are marine, but most are terrestrial. Includes spiders, harvestmen and scorpions. Fossil spiders are occasionally found in Upper Carboniferous coal-bearing sequences, and many examples are preserved in late Tertiary amber, which was exuded as a resin from the bark of conifers and trapped spiders and insects. Pycnogonida (sea spiders) have small bodies with four pairs of stout or spindly walking legs. They are all marine and especially common in modern Antarctic waters where there are over 100 species, including giant forms. They have been regarded as an independent subphylum, but Manton's studies show that they are more likely to have derived from an early group of marine arachnids.

Although the taxonomic status of trilobites and similar forms is yet uncertain, they have been considered as a possible phylum (as in the third edition of this text) but at present most authorities would prefer to regard them as a class, a scheme provisionally, and with some uncertainty, adopted here.

CLASS TRILOBITA (Cam.–Perm.): Marine arthropods known only as fossils; the earliest known arthropods, with hard, three-lobed exoskeletons tagmatized anteriorly as a cephalon, with free thoracic segments and a plate-like posterior pygidium. They have a pair of antennae and serially repeated biramous limbs all the way down the body.

Considering the great diversity of arthropods, living and fossil, and in view of their full treatment by Manton (1977), discussion of arthropods here is limited to trilobites, chelicerates and crustaceans.

Features of arthropod organization

The exoskeletal **cuticle** has undoubtedly been one of the principal factors in the success of the arthropods. It gives a physical and chemical barrier between the animal and its environment, yet allows a degree of osmotic and temperature regulation. Furthermore, it supplies good protection against predators and a firm base for the attachment of the internal muscles that move the limbs. It also provides a satisfactory location for various kinds of sense organs, being linked to the nervous system through fine **tubular canals** in the cuticle and strategically positioned for environmental monitoring.

Although the advantages of having a cuticle are clear, there are also problems which the inhabitant of a hard outer casing has to contend with. Of these, articulation of the joints, respiration and growth are perhaps the most acute.

All arthropods have segmented bodies. The exoskeleton of most modern ones consists of hard

sclerites, each usually consisting of a dorsal **tergite** and a ventral **sternite**, making a ring for each segment. These may form a rigid cylinder or, alternatively, the exoskeletal rings may be able to move against one another by means of internal muscles running from one segment to the next. The joint between each segmental ring is protected by flexible, unmineralized cuticle attached to the junction between segments. Similarly, each 'leg' (Fig. 11.1) is a chain of hard cylinders connected by short links of soft and flexible material and powered by internal muscles.

These muscles may be attached to internal knobs (**apodemes**), often formed by simple infolding of the exoskeleton, or to an endoskeleton. The muscles operate according to the normal antagonistic system, so that when one set of muscles contracts to move the limb in a particular direction, the opposite set will simultaneously relax (as with the biceps and triceps muscles of the human arm). When the opposing set contracts, the limb moves the other way.

Respiration in aquatic arthropods normally takes place through **gills**. These are usually lamellate

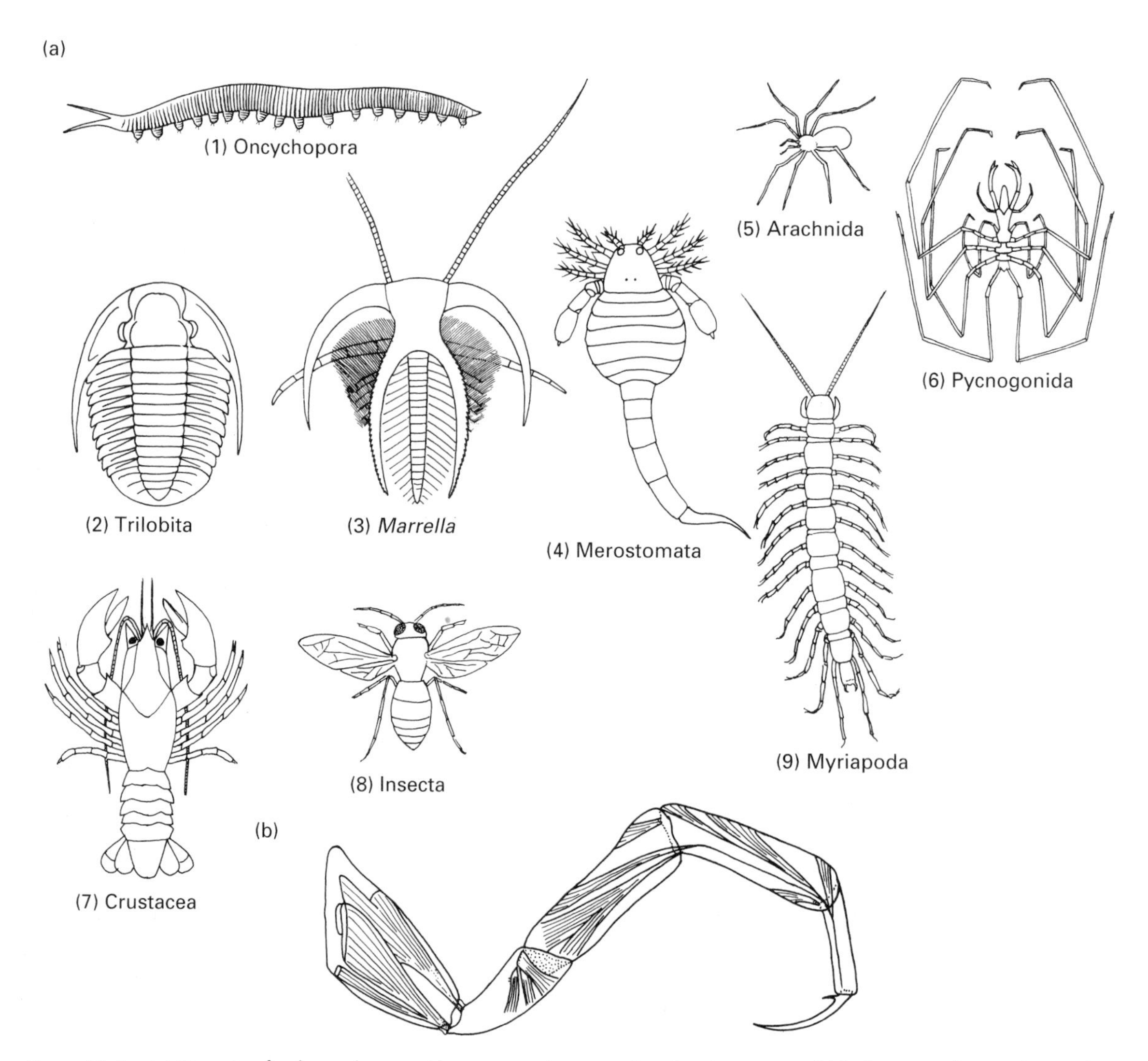

Figure 11.1 (a) Diversity of arthropod types with representative examples of major groups; (b) limb joint and internal musculature in an arthropod leg. [(b) Redrawn from Wigglesworth, *Principles of Insect Physiology*, 1965.]

organs with a very thin cuticle, extending from the bases of the appendages into the surrounding water. In very small arthropods diffusion over the whole body surface suffices, and there are no gills, but in larger ones there are gills which can sometimes be preserved fossil. Sometimes gills are located in specialized internal chambers. Modified gills in internal chambers are also found in some terrestrial arthropods, such as the **lung books** of arachnids. Many terrestrial arthropods breathe by **tracheae**, which are branched, spirally thickened tubes bringing air direct from the outside to the tissues. Such tracheal respiration is possible only in animals of relatively small dimensions, which is perhaps a critical factor in controlling the upper size limit of insects.

Growth is perhaps the most difficult physiological problem for arthropods, since the rigid exoskeleton encasing them cannot enlarge once it is formed. It has to be shed or moulted at intervals while a new and larger exoskeleton forms. This moulting process is known as **ecdysis**. It is a limiting system in arthropods and not an entirely perfect one, for some 80–90% of arthropod mortality occurs during moulting. Before the exoskeleton is cast, a new cuticle, somewhat larger than the existing hard shell, is formed below the old one. At this stage it is elastic, soft and wrinkled. While this is being formed, the lower part of the old cuticle is partially dissolved from below by corrosive fluid poured out from cutaneous glands below the new cuticle. Just before the old exoskeleton is cast, the animal stops feeding but takes up much oxygen and water. As it swells it makes spasmodic movements of the body to shake off the hard shell, and finally it is able to withdraw itself completely. The old skeleton may have special lines of weakness which facilitate its splitting; this was the case in trilobites, where the lines of weakness are known as **sutures**. The final swelling of the body to its full size takes place through the uptake of water after the casting process is complete. The animal is now soft-shelled and cannot move much; cuticle hardening and further secretion take place when this stage is over. The initial cuticle is paper-thin, and the acquisition of the full cuticle thickness may take some time.

During moulting the animal is very vulnerable, both to predators and to the possibility of tearing. Ecdysis is, furthermore, a wasteful process as much organic material is lost each time. But because of ecdysis we have a good record of the moult stages of many fossil arthropods, and by arranging the cast shells in a size series it is possible to elucidate the various transformations that the fossil arthropod went through from the larval to the adult stage. In trilobites especially this has proved to be of the greatest interest.

11.3 Trilobita

Trilobites are the earliest of all known arthropods. Their first representatives are found in rocks of Lower Cambrian age just above the earliest non-trilobite (Tommotian) Cambrian fauna. The last died out in the late Permian. Throughout their 350 million years of geological history, they preserved a remarkable constancy of form, though with many variations. All trilobites were marine. Well over 1500 genera are known, and there are several thousand species of which many, especially those of Cambrian and Ordovician rocks, have great stratigraphical value.

General morphology

Certain characteristics distinguish trilobites from other arthropods. In all trilobites the head (**cephalon**) is a single plate, made up of several fused segments. There are usually sense organs on the head, and there are also certain lines of weakness, known as **cephalic sutures**, which look like cracks on the surface but apparently facilitated ecdysis. The body (**thorax**) consists of a number of **thoracic segments** hinged to one another and allowing some capacity for enrollment of the body. The tail (**pygidium**), also segmented, is fused into a single plate like the head. These three primary divisions of the body are set at right angles to the unique, three-lobed, longitudinal division from which the class takes its name.

The limbs (or appendages) which were attached to the lower (ventral) surface are rarely preserved. When they are found they are seen to have a surprising structural uniformity within an individual trilobite. Except for the flexible uniramous **antennae** they are all two-branched (biramous), and their structure is virtually identical all the way down the body.

These special features are not shared by other

arthropods, and trilobites evidently form a group of their own which apparently left no descendants.

Acaste downingiae (Fig. 11.2)

The Silurian *Acaste downingiae*, which is common in the carbonate facies of the borders of Wales and England, shows most of the characteristic features of an 'advanced' trilobite. The body has the shape of a laterally flattened ellipse. The head (cephalon) is somewhat larger than the tail (pygidium), and the central part of the body (thorax) has 11 articulated segments.

The cephalon is quite strongly vaulted with a pentagonal to semicircular shape and transverse posterior edges. In the centre of the cephalon is a raised central hump (the **glabella**), bounded at the sides by diverging **axial furrows** and reaching to the **anterior border**. The glabella is indented by three short and more or less transverse pairs of furrows (the **glabellar furrows**) and is closed off at the back of the cephalon by an arched **occipital ring**. Stretching out sideways from the occipital ring are the **posterior border furrows**, which delimit a thin strip of the cephalon at the posterior edge; this is the **posterior border**. Yet another indentation (the **lateral border furrow**) runs parallel with the semicircular lateral border, and in this trilobite it joins the narrower **anterior border furrow** round the front of the glabella.

Placed laterally to the glabella are the eyes, which here are large and crescentic. The lenses are borne on a visual surface and are arranged in a system of hexagonal close packing. There are about 100 lenses forming a compound eye like those of insects and crustaceans, but they are unusually large and separate from one another. This is characteristic of **schizochroal** eyes, which are confined to the particular suborder (Phacopina) to which *Acaste* belongs. Above the visual surface lies the flat **palpebral lobe**, separated by a crescentic **palpebral furrow** from the small raised **palpebral area**. The anterior edge of the eye is very close to the glabella, the posterior one farther from it.

A thin though very distinct lineation (the **facial suture**) runs between the palpebral lobe and the visual surface; this suture extends forwards around the front of the glabella, being continuous with the facial suture on the other side of the head. Posteriorly, the facial suture turns out transversely on either side to terminate well in front of the **genal angle**, which is the most postero-lateral point of the cephalon. This sort of suture is said to be **proparian**. The region lateral to the glabella, though within the suture, is the **fixigena**; outside it is the **librigena**. In many trilobites the anterior branch of the facial suture cuts across the antero-lateral border, so that there are two distinct sutures rather than a single continuous one. In such cases, which include the majority of Cambrian species, the cephalon may distintegrate before burial into three components: the two librigenae and the central **cranidium**, which is the glabella plus fixigenae. This disintegration does not usually happen in *Acaste* because the suture is continuous.

The lower surface of the cephalon shows that the antero-lateral borders are continued ventrally as a narrow flange (**doublure**), whose inner edge is concentric with the edge of the cephalon and directly below the antero-lateral border furrow. It does not extend, however, all the way to the occipital ring, being phased out along the posterior border. Two pairs of pronounced knobs (the **apodemes**) project ventrally from the posterior two glabellar furrows. These seem to be associated with the attachment and articulation of the ventral appendages. Identical apodemes are present on the thoracic segments, and evidently there were legs under the head as well as under the thorax, though such appendages are not preserved in *Acaste*.

A large plate with a central 'blister' is attached to the rear edge of the anterior cephalic doublure. This is the **hypostome**, which lies below the glabella. The mouth apparently lay directly behind the hypostome, and from it the oesophagus ran forward to the stomach which lay in the space between glabella and hypostome. The gut evidently ran posteriorly from the stomach along the thoracic axis to end under the pygidium. The shape and position of the hypostome is very variable amongst trilobites, and many genera can be identified on the basis of detached hypostomes found in the rock, even if the rest of the exoskeleton is missing. A couple of small swellings (the **maculae**) lie towards the rear of the hypostome. It has been suggested (Lindström, 1901) that these are ventral eyes such as are possessed by various modern arthropods, but their fine structure is normally indistinct, and this may not have been so.

In *Acaste* there are 11 thoracic segments. They are

all identical in form, though the posterior ones are slightly smaller. In each there is an arched **axial ring** identical in form to the occipital ring of the cephalon and defined laterally by paired **axial furrows**. The axial ring is bounded anteriorly by a groove (the **articulating furrow**), in front of which projects a semicircular **articulating half-ring** which in life fits neatly under the axial ring in front. The paired **pleura** (singular pleuron) project horizontally from the axial ring, though their outer extremities are

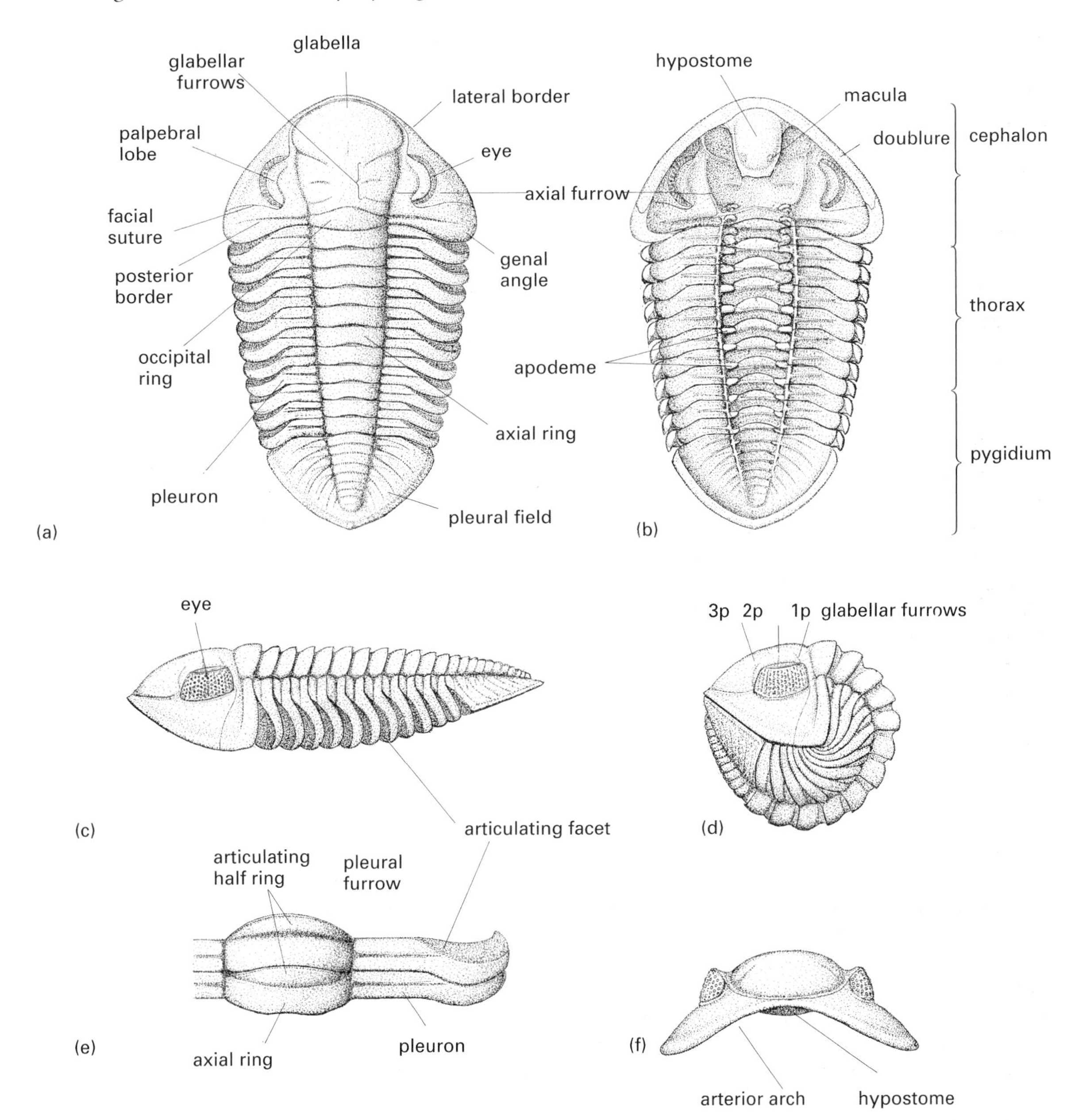

Figure 11.2 Morphology of *Acaste downingiae*. Silurian, England: (a) dorsal surface; (b) ventral surface; (c) side view in natural life attitude; (d) enrolled specimen; (e) dorsal view of two articulated thoracic segments in an enrolled specimen, with the articulating half-ring exposed; (f) frontal view showing anterior arch. [All × 2 except (e)]. For additional terminology see text.

sharply turned down. Each pleuron is indented by an oblique **pleural furrow**. There is a short doublure on the outer edge of each pleuron. The anterior edge of the downturned distal part of each pleuron is a flat and truncated **articulating facet**. When the trilobite rolled up in a ball for protection the paired facets slid below the rear edge of the preceding pleura, whilst the axial region, now expanded into a hoop, was protected by the series of articulating half-rings which were then exposed. The pleura are separated by **interpleural furrows**.

The pygidium is a flat plate of fused segments resembling those of the thorax. It has a similar half-ring at the front and is articulated with the rear thoracic segments in the same way as the thoracic segments are all linked. The pygidial axis has a series of furrows equivalent to articulating furrows, becoming more closely spaced and fainter towards the rear. The lateral parts of the pygidium (**pleural fields**) are sculpted by two kinds of indentation: one series equivalent to the edges of the thoracic segments (interpleural furrows), the other to the thoracic pleural furrows. Often the latter are more strongly pronounced in the trilobite pygidium. The pygidial doublure has about the same width as the cephalon.

Since the trilobite cuticle or 'shell material' is fairly thin, each furrow (indentation) on the dorsal surface is marked as a ridge on the ventral side and vice versa. The exception is the fine **granulation** on the outer surface of *Acaste* which is not reflected ventrally. The small granules are remains of sense organs, probably the sites of small hairs susceptible to vibrations in the water.

Acaste probably spent most of its time near the sea floor in an outstretched attitude. The reconstruction given here is based upon several assumptions; for example, only when the cephalon is in the orientation illustrated, with the anterior border raised to form an **anterior arch**, will the visual field of the eye be horizontal. However, it seems to be a reasonable reconstruction, since the axis and the tips of the anterior pleurae are parallel; the trilobite could easily rest or crawl upon the sea floor in such an orientation. The uplifted front of the cephalon (anterior arch) below which the hypostome projected, and the **posterior arch** formed by the progressively shortened posterior thoracic segments and tail uplifted from the sea floor, would allow the free passage of water below, aerating the feathery gills.

Acaste specimens, like certain other trilobites, are often found enrolled with the cephalic and pygidial doublures in contact. The outline of the two borders is identical, and the opposing surfaces when in contact are mirror images of one another. So how was such enrollment achieved? It is probable that two longitudinal sets of paired internal muscles with antagonistic action were responsible for holding the body in an extended posture or rolling it up. One set (the **flexors**) ran all the way along the body, joining up the apodemes. There was thus a continuous pair of parallel lines of longitudinal muscle. The other sets of muscles were the **extensors**. Muscle scars on the ventral surface show that they joined the underside of each articulating half-ring to the lower surface of the preceding axial ring. When the flexors contracted the extensors simultaneously relaxed, in the normal pattern of antagonistic musculature found in invertebrates and vertebrates alike. Contraction of the flexor muscles would shorten the distance between the apodemes, and since the pleural edges all formed serial parallel hinges, the thorax and pygidium could only move downwards; thus the tail was swung into the position under the head, the legs presumably having been lifted out of the way first. Contraction of the extensor muscles would bring the body back into the extended position. Hence the whole body of *Acaste* is finely adapted both for perfect spheroidal enrollment and active life in a functional, mobile, outstretched attitude.

Detailed morphology of trilobites

Trilobita are characterized on the one hand by a confining evolutionary conservatism and on the other by a remarkable plasticity within the limits dictated by the defined pattern of organization. The overall range in form is well shown in the photographic atlases of Whittington (1992), Levi-Setti (1993) and Lebrun (1995). The morphology of the earliest and latest trilobites does not depart very radically from that of *Acaste*, and much of the observed range in morphology can be related directly to the biological functions of its various components. Hence it is appropriate to consider the morphological range in functional terms before proceeding with the classification.

Cuticle (Fig. 11.3)

Arthropodan cuticles normally consist of several layers, and the cuticles of trilobites are no exception. Many arthropod cuticles are constructed of chitin (a hydrocarbon allied to cellulose) and sometimes reinforced by mineral. In trilobites, however, the cuticle consists largely of low-magnesian calcite (Wilmot and Fallick, 1989) arranged in microcrystalline needles orientated normal or near-normal to the outer surface and set in an organic base whose nature has yet to be determined (Dalingwater, 1973b; Teigler and Towe, 1975). Chitin has not yet been detected. This cuticle consists of two layers: a relatively thin outer layer with large calcite (or possibly calcium phosphate) crystallites having their *c* axes normal to the surface, and a much thicker inner or principal layer of microcrystalline calcite. The inner layer is laminated, the individual laminae being concentrated in three zones; the outer and inner zones have closely spaced laminae, whereas in the central zone they are much more widely spaced. From the biomechanical point of view (Wilmot, 1990) trilobite cuticles are best regarded as ceramics, having a low percentage of organic matter and behaving in a linearly elastic manner. Whereas the thin prismatic outer layer had good compressive strength, it had

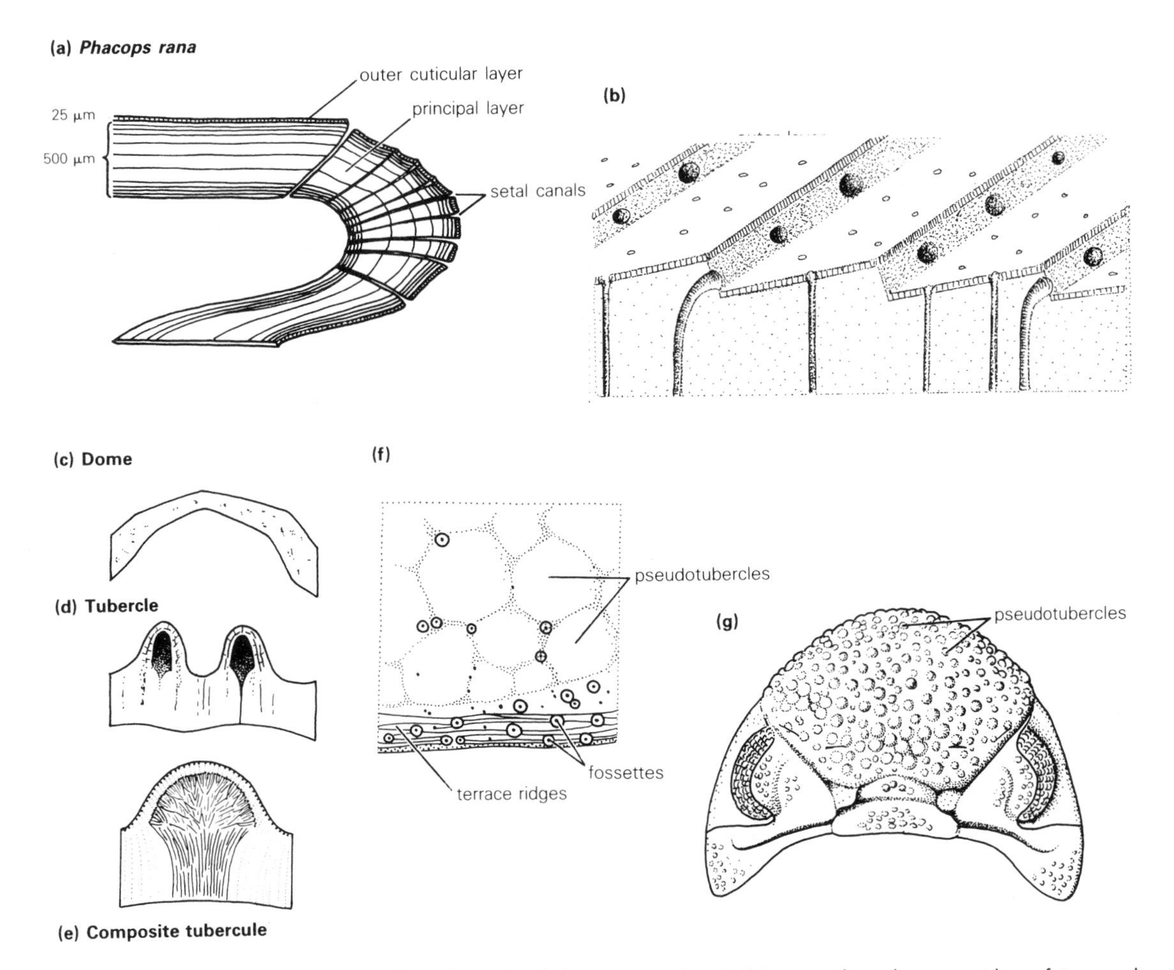

Figure 11.3 Trilobite cuticle: (a) section through cuticle of *Phacops rana* (Dev.); (b) section through terrace ridges of *P. rana* showing outer cuticular layer absent over the 'scarp slope' and the two kinds of pore canals opening on 'scarp' and 'dip' slopes, respectively; (c) section through dome; (d) section through tubercle; (e) section through composite tubercle; (f) section through anterior border and part of cephalon of *P. rana*; (g) cuticular sculpturing of *P. rana* (×1.5). [(c), (d), (f) Redrawn from Miller, 1976; (e) redrawn from Störmer, 1980.]

limited resistance to cracking. The underlying microcrystalline principal layer, however, was highly resistant to crack formation. Accordingly trilobite cuticles could resist compressive and tensile forces. Further resistance to breakage is given by the architectural properties of the exoskeleton, for structurally the exoskeleton is effectively a monocoque shell: a strong 'thin shell' behaving as a 'stressed skin'. Its construction as a series of modified domes strengthens the shell, as do doublures, cell polygons and terrace ridges. The cuticle is well supplied with **sensillae**: small structures interpreted as of sensory function. Thus many kinds of tubular canals traverse the cuticle, leading from the inside to the surface; the majority of these are straight or helically coiled **pore canals**, often with trumpet-shaped outer ends and in the Cambrian *Ellipsocephalus* with disc-like swellings within the principal layer (Dalingwater *et al.*, 1991). These are most closely packed where the cuticle is highly curved. They were probably mainly sensory, carrying small hairs (**setae**) externally, each connected to the central nervous system by a nerve running up the canal. By analogy with modern arthropods, most of these would have been sensitive to vibrations or chemical change in the water.

Sometimes the pore canals are associated with a system of parallel ridges, closely spaced and forming a regular 'fingerprint' ornament over all or part of the surface. When examined, this system resolves itself as a series of **terrace ridges** (Fig. 11.3b) giving a serially repeated dip-and-scarp topography with the sensory pore canals opening at the base of the scarps (Miller, 1975). Flume experiments have shown that this could have functioned as a current-monitoring system, sensitive to change in water current direction. Terrace ridges have also been interpreted, however, as devices for facilitating friction in burrowing trilobites (Schmalfuss, 1981). Such terrace ridges are normally concentrated on the doublure, but they may also occur on the upper surface of the trilobite. The ventral terraces may have been used to consolidate the walls of a filter chamber underneath the resting trilobite. In certain orders (e.g. the Proetida) terrace ridges are frequently developed, in others hardly at all.

Many trilobites have cuticular **tubercles** (Fig. 11.3c–g) as their dominant surface sculpture. In thin section these appear to be of various kinds (Dalingwater, 1973a,b; Miller, 1976). Some are merely **domes** with a space below; others are true tubercles enclosing an internal space connected by pore canals to the outer and inner surfaces of the cuticles. Some kinds of true tubercles seem to have been the sites of large numbers of grouped pore canals or sensory organs of other function. There are also **pseudotubercles** which do not have the discrete appearance of tubercles proper. They tend to be concentrated in particular areas of the exoskeleton, especially on the glabella and in areas where the doublure is likely to come in contact with the sea floor.

The most complex of these, also known as **composite tubercles**, are found in phacopids (Fig. 11.3e). These have a series of tiny spherical cavities directly below the surface, from which fine canals run radially inwards to a club-shaped central mass of fine branches arranged in a fan (Störmer, 1980). Their function has been debated but is unresolved.

In the example illustrated here, *Phacops rana* (Fig. 11.3f), different kinds of structures – terrace lines, domes, tubercles and pore canal openings of different sizes – have been mapped. Different parts of the cuticle were supplied with different combinations of sensillae acting as environmental sensors. Most cuticular structures seem to have been sensory, though specific functions can only ever be postulated for a given organ by analogy with sensors in modern arthropods; since exact analogies seldom occur, this procedure can be hazardous.

Some other kinds of cuticular 'ornamentation' were not apparently sensory, such as the **caecae** commonly encountered in Cambrian trilobites. The cuticle of many Cambrian trilobite genera was thin and relatively flat, and it is frequently impressed with a series of radial and ramifying ridges forming an elaborate and symmetrical pattern over the cephalon and sometimes the rest of the body. These caecae seem to be moulded to the form of a tubular system originally lying below; either a circulatory or respiratory apparatus (Jell, 1978), or the diverticulae of a complex subcephalic set of digestive organs (Öpik, 1960). These ridges have been referred to as **alimentary prosopon** and seem to have been connected with the oesophagus (Fig. 11.4a,b).

Nutrients digested in the gut could presumably be passed directly to the other parts of the body by means of this system; alternatively, and more probably, they may have functioned as auxiliary respiratory structures. However, it is very rare to find

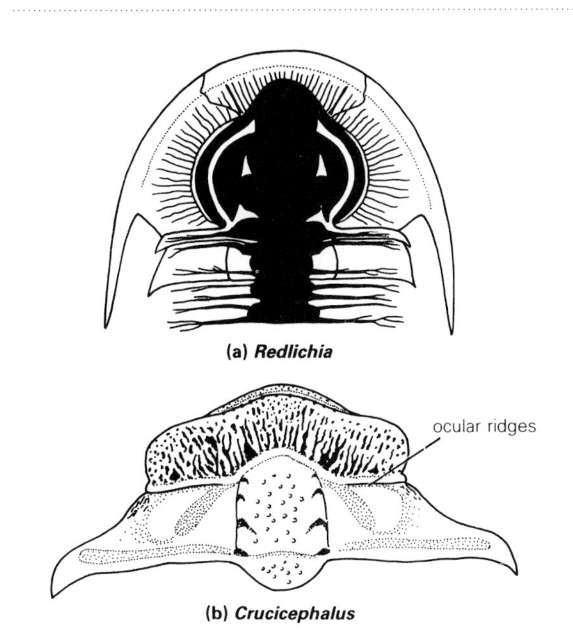

Figure 11.4 Cuticular sculpturing: (a) *Redlichia* (L. Cam.), alimentary prosopon (in black); (b) *Crucicephalus* (U. Cam.) showing peculiar cuticular sculpturing and ocular ridges (× 6). [(a) Redrawn from Öpik, 1960; (b) redrawn from Shergold, 1971.]

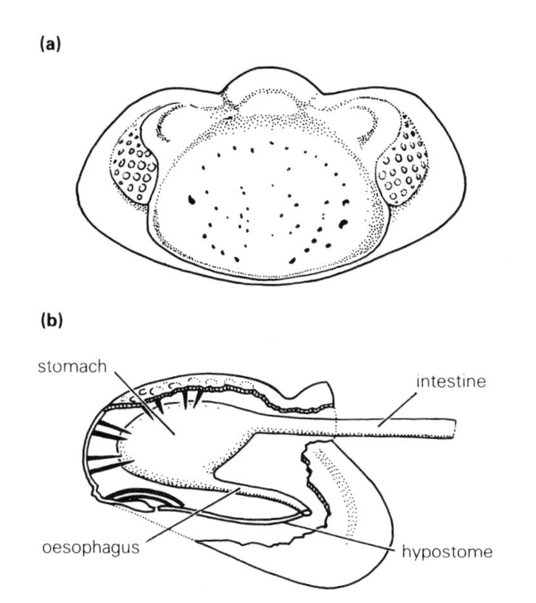

Figure 11.5 (a) Frontal view of *Phacops rana* (Dev.) with cuticle removed, showing impressions, possibly muscular, on the underlying matrix (×1.5); (b) reconstruction of cephalon in lateral view, showing possible arrangement of internal organs with stomach supported by muscles (black) and mouth opening behind hypostome. [(a) Redrawn from Eldredge, 1972a; (b) modified from Eldredge, 1972a.]

alimentary prosopon in post-Cambrian trilobites, in which the shell is generally much thicker. Presumably the internal organs, of which the prosopon is the external impression, are still there but located farther below the cuticle and no longer partially within it. Likewise, the transverse **ocular ridges** that are normal in Cambrian trilobites, linking the eye with the glabella (and incidentally usually carrying two prosopon caecae), are rarely present in post-Cambrian trilobites. The presence of such caecae usually suggests that an unknown trilobite is of Cambrian age, and the sculptural patterns have been found useful in taxonomy.

Cephalon

Most trilobite cephala have the form of semicircular plates with well-defined structures, as represented in *Acaste*. There are some, however, in which the furrows limiting the different structures are all but effaced, so that the various parts are very indistinctly defined, e.g. *Trimerus*. Particular organs – the glabella, facial sutures, eyes, hypostome and certain specialized characters – merit further attention.

Glabella

The shape, size and structure of the glabella are widely variable. Glabellae may or may not reach the anterior border; in some cases they may be greatly swollen, have lateral lobes, be entire or indented with up to five pairs of glabellar furrows, as in the examples shown in Fig. 11.6. The stomach probably lay below the glabella and above the hypostome, and its size, and hence the size of the glabella, was probably related to the trilobite's diet. Some symmetrical indentations on the glabella of certain trilobites [e.g. *Chasmops*, where they are arranged in a V-shaped pattern, and *Phacops* (Fig. 11.5a), in which they form a pair of subcircular patches] have been said to be the scars of **suspensory ligaments** or muscles holding the stomach in place and allowing it to expand and contract.

A reconstruction by Eldredge (1972a; Fig. 11.5b) has shown how these might have been arranged.

Cephalic sutures (Fig. 11.6)

The cephalic sutures, which are unique amongst arthropods, include the facial sutures and **ventral cephalic sutures** which are sometimes present. The facial sutures are of three main kinds: **proparian** (Fig. 11.6h,j), where the posterior branch passes

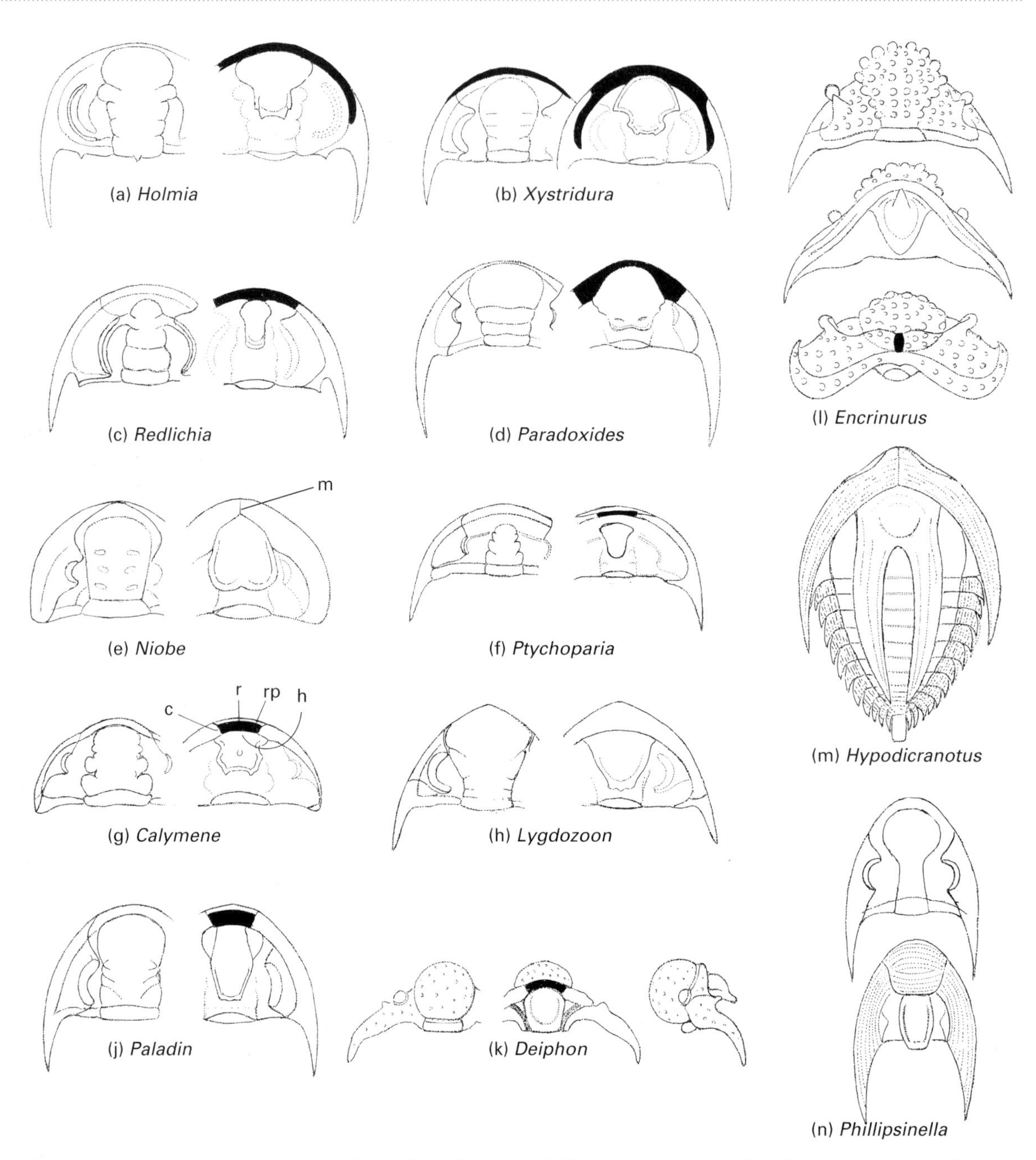

Figure 11.6 Dorsal and ventral views of the cephala of various trilobites, showing relationship of hypostome to cephalon, and course of sutures. The maximum number of ventral sutures is shown in (g) (*Calymene*), where connective sutures (c), hypostomal suture (h) and rostral suture (r) enclose the rostral plate (rp). (a) *Holmia* (L. Cam.); (b) *Xystridura* (U. Cam.); (c) *Redlichia* (L. Cam.); (d) *Paradoxides* (M. Cam.); (e) *Niobe* (L. Ord.); (f) *Ptychoparia* (U. Cam.); (g) *Calymene* (Ord.–Sil.); (h) *Lygdozoon* (Sil.); (j) *Paladin* (Carb.); (k) *Deiphon* (Sil.); (m) *Encrinurus* (Sil.); (n) *Hypodicranotus* (Ord.); (o) *Phillipsinella* (Ord.). Rostral plate in black. (Mainly redrawn from Whittington, 1988a,b, and in *Treatise on Invertebrate Paleontology*, Part O, 1959.)

in front of the genal angle or spine; **opisthoparian** (Fig. 11.6b–d,k), in which it cuts the posterior border anterior to the genal angle; and **marginal** (Fig. 11.10a–c,g), where it runs along the edge and is not visible on the dorsal surface. In one family, the Calymenidae, the facial suture is **gonatoparian** (Fig. 11.6) and runs directly through the genal angle.

These main sutural systems were believed to define natural groupings within the trilobites thoughout much of the nineteenth century and the first part of the twentieth century. They were originally used as a primary basis of classification, erecting orders: the Proparia (proparian and gonatoparian genera), the Opisthoparia (opisthoparian genera) and the Hypoparia (marginal-sutured genera). This classification eventually came under severe criticism and was finally abandoned when it was appreciated that the Proparia and Hypoparia were both composed of very heterogeneous elements which had no close natural relationships. Stubblefield (1936), for instance, has pointed out that Hypoparia could not be a natural grouping, since marginal sutures had evidently been derived independently in several groups of trilobites, the sutures becoming more marginal as the eyes were lost. There may be no primitively eyeless trilobites, and all of the blind trilobites (with the possible exception of the Cambrian agnostids) were derived from sighted ancestors.

Recent classifications have used the whole complex of trilobite characters to divide the trilobites into natural groupings, especially those of the axial region, and not merely a single character, however important it may appear to be.

On the ventral side of the cephalon there may be several other sutures, especially where the facial sutures cross over the anterior border and continue across the doublure. *Calymene* (Fig. 11.6g) is an example showing the maximum number of possible sutures. Here the facial sutures are continued and join with the lateral **connective sutures** (**c**) with which they may be homologous. An elongated **rostral plate** (**rp**) is isolated by these and by the anterior **rostral suture** (**rs**) and posterior **hypostomal suture** (**h**). This rostral plate is not present in all trilobites, and its absence or reduction has given rise to much phylogenetic speculation. The morphological series here illustrated (based on Stubblefield, 1936) suggests that it was originally present as an important ventral structural component in early trilobites such as the olenellids and that in many later lines of descent it was reduced or lost by different evolutionary pathways. In *Encrinurus* (Fig. 11.6l) the rostral plate is very small, and the two connective sutures are close together, whereas in *Niobe* (Fig. 11.6e) the connective sutures are represented by a single **median suture** (**m**). In *Acaste* and its relatives there is no trace of a median suture, and the two lateral facial sutures unite around the front of the glabella so that the librigenae are fused.

Hypostome

The hypostome may be large or small, short or long. It has proved of great value in classification. Usually it lies directly below the most convex part of the glabella. The primitive condition for trilobite hypostomes seems to have been **conterminant** (Fortey, 1990). Here, the hypostome is attached to the cephalic doublure and in shape closely corresponds with the outline of the front of the glabella. In the **natant** condition the hypostome was not attached to the cephalic doublure. Instead it lay on the ventral membrane, behind the cephalic doublure and separated from it by a gap. It may have been anchored by ligaments. Usually the anterior margin is the same shape as the front of the glabella. A third condition is an **impendent** hypostome. Here the hypostome is attached to the doublure but its shape bears little correspondence to that of the glabella.

The natant condition is derived and is usually conservative. Some trilobites originating from a natant stock become secondarily conterminant or impendent. Rarely the hypostome projects backwards beyond the cephalon. Hypostomes usually have a pair of lateral wings which may turn up inside the cephalon to rest close against its inner wall, so that the hypostome was firmly braced against the dorsal exoskeleton. A limited degree of movement may have been possible with natant hypostomes (Whittington, 1988a,b, 1989). In some cases the hypostome is rigidly fixed, as is particularly evident in *Encrinurus* (Fig. 11.6l), in which the lateral wings are pronounced though slender and delicately formed, and which has a large, three-lobed impendent hypostome projecting forwards below a pronounced anterior arch. It is connected to the doublure by a V-shaped hypostomal suture, and is presumably immovable. Likewise in the case of

Niobe (Fig. 11.6e) and its relatives no movement was possible, since the anterior border of the hypostome is strongly curved and is let into the rear of the doublure. Thus the Ordovician *Hypodicranotus* (Fig. 11.6m) has a very long hypostome extending almost to the pygidium and prolonged into a pair of long blades with a median space between.

Eyes (Figs 11.7, 11.8)

The eyes of trilobites are the most ancient visual system known, and indeed they are the earliest of all well-developed sensory systems. Their evolution can be followed through some 350 million years of geological time (Clarkson, 1975, 1979). Trilobite eyes are compound, and like the lateral eyes of modern crustaceans and insects they were composed of radially arranged visual units pointing in different directions and often encompassing a wide-angled visual field.

Though compound eyes are typical of arthropods, they have evolved from separate beginnings in a number of arthropod stocks; and though the eyes of trilobites are analogous to those of modern arthropods, they are not necessarily homologous. In most modern arthropods the visual units are the **ommatidia** (Fig. 11.8f), each being a cylinder of cells with the photosensitive elements (**rhabdom**) located deep within it. Each ommatidium is capped by a **corneal lens**, underlying which is a subsidiary dioptric apparatus (the **crystalline cone**). The lens and cone together focus light on the rhabdom. The rhabdom consists of a cylinder of stacked plates, each made of parallel **microvilli**. Alternate plates have their blocks of tubules arranged at right angles to one another. These tubules are the site of the photoreceptive pigments whose chemical alteration by light triggers an electrical discharge in the **ommatidial nerves**. The nerve impulses are processed in a complex **optic ganglion** deep below the ommatidia, and some kind of integrated image is produced from the mosaic effect of light coming down individual and separate ommatidia. How 'good' the arthropod eye is in contrast with its vertebrate counterpart is very hard to assess, for the two are different kinds of eyes performing basically different functions.

The eyes of trilobites may also have had sublensar ommatidia, though this may not have been the case in all. Little evidence of internal structure survives. The lenses alone are preserved because they, like the cuticle, were constructed of calcite.

Most trilobite eyes are **holochroal** (Fig. 11.7a), having many round or polygonal lenses whose edges are all in contact and which are covered by a single **corneal membrane**: the equivalent of the outer cuticular layer with which it is laterally continuous. Holochroal eyes are the most ancient kind of eye in trilobites; they are found in the earliest Cambrian trilobites and persist until the final extinction of trilobites in the late Permian.

The oldest well-preserved eyes are known in Lower Cambrian eodiscids from China (Zhang and Clarkson, 1990). These are highly organized, though a somewhat irregular lens-packing system suggests

(a)

(b)

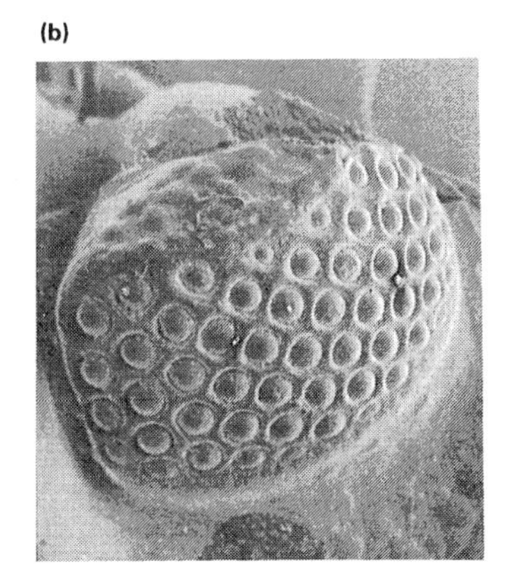

(c)

Figure 11.7 (a) Holochroal eye of *Paralejurus brongniarti* (Dev.), Bohemia, Dvorce-Prokop Limestone (× 8); (b) schizochroal eye of *Phacops rana* (Dev.), Silica Shale, Ohio (× 18); (c) *Acernaspis (Eskaspis) sufferta* (Sil.), Pentland Hills, near Edinburgh, Scotland (× 2.5).

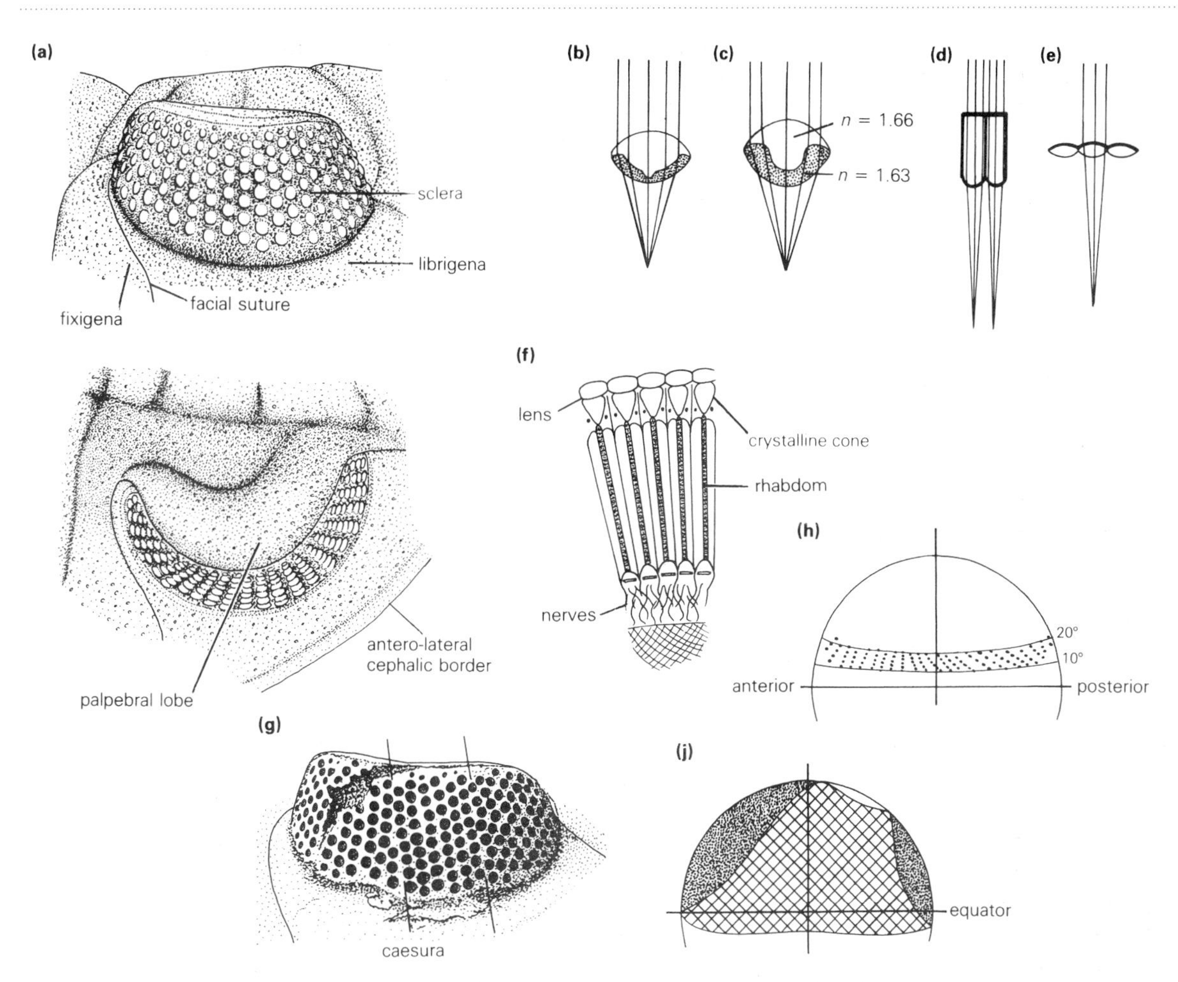

Figure 11.8 Trilobite eyes: (a) *Acaste downingiae*, lateral and dorsal views of schizochroal eye (×7.5); (b), (c) lenses of *Dalmanitina socialis* (Ord.) and *Crozonaspis struvei* with (shaded) intralensar bowl, conforming to ideal correcting lenses of Descartes and Huygens, respectively – models with a small refractive index difference between bowl and upper unit focus light sharply; (d) passage of light rays through prismatic lenses of *Asaphus* (L. Ord.); (e) *Sphaerophthalmus* (U. Cam.), an olenid; (f) internal organization of a modern arthropod eye; (g) *Ormathops* eye (L. Ord.) preserved as an internal mould, and showing irregular disposition of identical lenses (×7.5); (h) visual field of *Acaste downingiae* eye with individual lens-axis bearings plotted on a Lambert equal area net; (j) panoramic visual field of *Bojoscutellum*, only bearings of peripheral lenses plotted. [(a) Redrawn from Clarkson, 1966; (b), (c) based on Clarkson and Levi-Setti, 1975; (f) redrawn from Snodgrass in *Treatise on Invertebrate Paleontology*, Part O; (g) redrawn from Clarkson, 1971; (h) redrawn from Clarkson, 1966; (j) redrawn from Clarkson, 1975.]

that the separate developmental programmes for growth of the eye and emplacement of the lenses were not fully coordinated and sometimes got out of phase. In general, the eyes of Cambrian trilobites are poorly known, because in the majority the whole visual surface is encircled by an **ocular suture**, so that the visual surface dropped out after death or during moulting. This system was abandoned in most post-Cambrian groups, probably through paedomorphosis, and the visual surface was thereafter welded to the librigena, the only suture then running along the top of the eye as the **palpebral suture**, which forms part of the dorsal facial suture. The post-Cambrian radiation of holochroal trilobite eyes was substantial. The visual surface may be small or large, even hypertrophied in some groups, and encompasses a variable angular range.

Lenses in Cambrian trilobites, where found, are all thin and biconvex (Fig. 11.8e), but in some of the Ordovician groups, especially those with thick cuticles (Fig. 11.8d), the 'lenses' have the form of long prisms with a flattish outer surface and a hemispherical inner end; such lenses have, however, a similar focal length to those that are biconvex. Each lens is a single crystal of calcite, which being highly birefringent does not at first sight seem to be a very suitable material for a dioptric apparatus. However, all the lenses are arranged with their *c* axes normal to the visual surface, so light travelling parallel with the axis is not broken into two rays but continues unaltered to the rhabdom. Oblique rays may have been screened out by pigment cylinders below the lens, as in modern arthropod eyes, to which holochroal eyes seem to bear a fair structural and functional resemblance.

On the other hand, **schizochroal** eyes (Figs 11.7b, 11.8a–c,g), which are confined to Suborder Phacopina (Ord.–Dev.), have no known modern counterparts; they are a unique visual system unlike anything else in the animal kingdom (Horvath *et al.*, 1997). In schizochroal eyes, such as those of *Acaste*, the lenses are large and separated from each other by an interstitial material (**sclera**) of the same structure as the rest of the cuticle. Each lens has its own corneal covering, which plunges through the sclera and internally probably terminates in a sublensar conical structure, which is rarely preserved.

The structure of the lenses is most interesting. Whereas holochroal eyes have lenses ranging in diameter from 30 to 200 mm, though most are less than 100 mm, schizochroal lenses are usually much larger, being in the range 120–750 mm. They are usually steeply biconvex and of compound structure. Within each lens is a bowl-like unit, separated from the upper part of the lens by a wavy surface which has been shown in various genera to be similar in shape to the surface of aplanatic correcting lenses, as designed by Huygens (Fig. 11.8c) and Descartes (Fig. 11.8b). Experimental models (Clarkson and Levi-Setti, 1975) have shown that a slight difference in refractive index between the oriented calcite of the upper unit and the bowl would operate together with the correcting surface to produce a sharp anastigmatic focus. In addition, the bowl appears to increase transmissivity of light to the photoreceptors; indeed the whole system is fully bio-optimized for maximal efficiency, even during the stages after moulting (Horvath and Clarkson, 1993)

The lenses of *Phacops* are ultrastructurally very complex (Clarkson, 1979; Miller and Clarkson, 1980). Electron micrography shows that the bowl consists of dense calcite whereas the upper unit is built of calcite sheets arranged radially round the *c* axis of the lens. Each of these sheets in turn consists of a palisade of calcite fibres parallel with the axis in the lower part but turning outwards, fanwise, near the outer surface of the lens. This may have minimized the effects of birefringence. A central core found in *Phacops* may, as Campbell (1975a, 1977) noted, also have some correcting function. After ecdysis the lens is re-formed in stages, the bowl being added last of all.

Schizochroal lenses, in spite of being made of calcite, were corrected against astigmatism and possibly also against undue birefringent effects from oblique rays. But why were they so large and of such high optical quality? It is generally believed (Campbell, 1975a; Fordyce and Cronin, 1989) that rather than an ommatidium lying below each lens, there was a relatively short 'ocellar capsule', floored by a flat layer of narrow cells. Schizochroal eyes are thus considered as an assemblage of simple eyes of high resolving power and with overlapping visual fields. Such eyes (Stockton and Cowen, 1976) may have been capable of using adjacent lenses for stereoscopic vision through 360°. They would have been particularly useful in nocturnal animals whose large lenses enabled them to see in the dim light. These highly specialized visual organs are quite different from those of any modern compound eyes, and their design may have offered unique potential for the analysis of colour, form and depth.

Schizochroal eyes originated from holochroal ones, probably by paedomorphosis, since larval eyes of holochroal-eyed trilobites are in many ways like tiny schizochroal eyes. The earliest schizochroal genus (*Ormathops*; Fig. 11.8g) had a rather haphazard and irregular system of lens packing, which arose from the geometrical constraints of packing lenses of identical size on a curving visual surface (Clarkson, 1975). In later derivatives of *Ormathops* regular packing was achieved by graduating the lens size. Despite the elegance of the eyes of trilobites, secondary blindness is quite common (Fortey and Owens, 1990a) and characteristic of major taxa such as the Agnostida and Trinucleida. In other trilobite groups the loss of the eye is probably related to the adoption of a dark environment or infaunal habitat. Progressive eye reduction has been shown in two successive lineages of Upper Devonian tropidocoryphine trilobites in southern France (Feist and Clarkson, 1989), and contemporaneously in phacopids (Feist, 1991). Such unidirectional evolution, involving the adoption of an endobenthic habit, is an adaptive response to constant and persistent environmental influences.

Cephalic fringes (Figs 11.9, 11.10)

In Family Trinucleidae (Ord.) and Family Harpidae (Suborder Harpina; Ord.–Dev.) the antero-lateral cephalic border is developed into an extensive pitted **fringe**, the suture has become marginal and the eyes are very reduced or absent. Two kinds of fringe were independently evolved and are dissimilar in appearance. In a few other trilobite groups the anterior border may be pitted, though never in the same way nor so extensively; yet other trilobites possess expanded anterior flanges though without the pits.

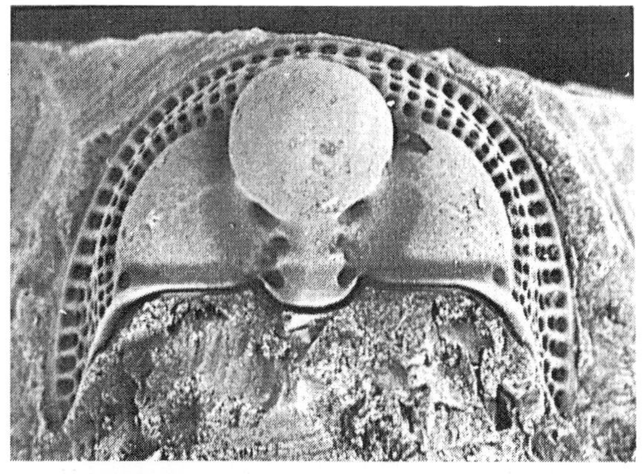

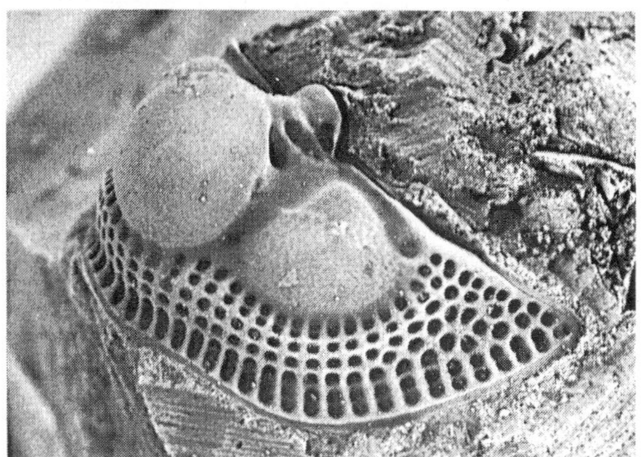

Figure 11.9 *Tretaspis ceriodes angelina* (Ord.), Norway (× 5): upper lamella of cephalon in dorsal and frontal lateral views. (Photographs reproduced by courtesy of Dr Alan Owen.)

Cephalic morphology and fringe of trinucleids (Fig. 11.10a–f)

The cephalic features of the Ordovician Family Trinucleidae are unique. In trinucleids the glabella is usually highly convex, has shallow glabellar furrows and is bounded by broad axial furrows. Lateral to the glabella are the quadrant-shaped **genal lobes**, which may range from gently swollen to greatly inflated and are devoid of eyes and sutures. In *Tretaspis* and *Reedolithus* there are small nodes set on the genal lobes. Each of these is a dome-shaped thin area of the test, though lens-like structures have been reported at the summit of the dome (Störmer, 1930). Rarely, ocular ridges may link the eyes with the glabella. In all trinucleids the **genal spines** are very long and terminate well behind the short body, which has only six thoracic segments and a short triangular pygidium.

The fringe is a bilamellar structure in which the two lamellar counterparts are separated by a marginal suture, which becomes dorsal only where the genal spines join the cephalon. The upper lamella is thus contiguous with the genal lobes, whereas the lower has the genal spines attached. A series of funnel-shaped **fringe pits** indent both lamellae, each dorsal pit located directly above an inverted ventral counterpart, so that the whole structure is really a hollow pillar. The floor of each pit (the ceiling in the case of the ventral ones) is formed by a terminal disc or 'nozzle' with a minute central perforation, and at the suture which runs between them the two counterpart terminal discs are juxtaposed. The space between the upper and lower lamellae opens to the lower side of the cephalon by a pair or series of ventral perforations. There must have been communication between the fringe cavity and the body cavity; probably part of the digestive or circulatory system was housed therein.

Some of the early trinucleids have an almost flat fringe with poorly ordered and irregular fringe pits. In later stocks, however, the fringe pits became ordered into a symmetrical pattern of ordered arcs, concentric with the antero-lateral border, intersecting with radial rows. Some small irregular 'F-pits' are found in small patches lateral to the genal lobes in the genera *Marrolithus* and *Cryptolithus*. The homologies of the arcs in various trinucleid genera, which is essential in taxonomy and identification, can be assessed with reference to a thickened ventral ridge – the **girder** (Fig. 11.10c) – concentric with

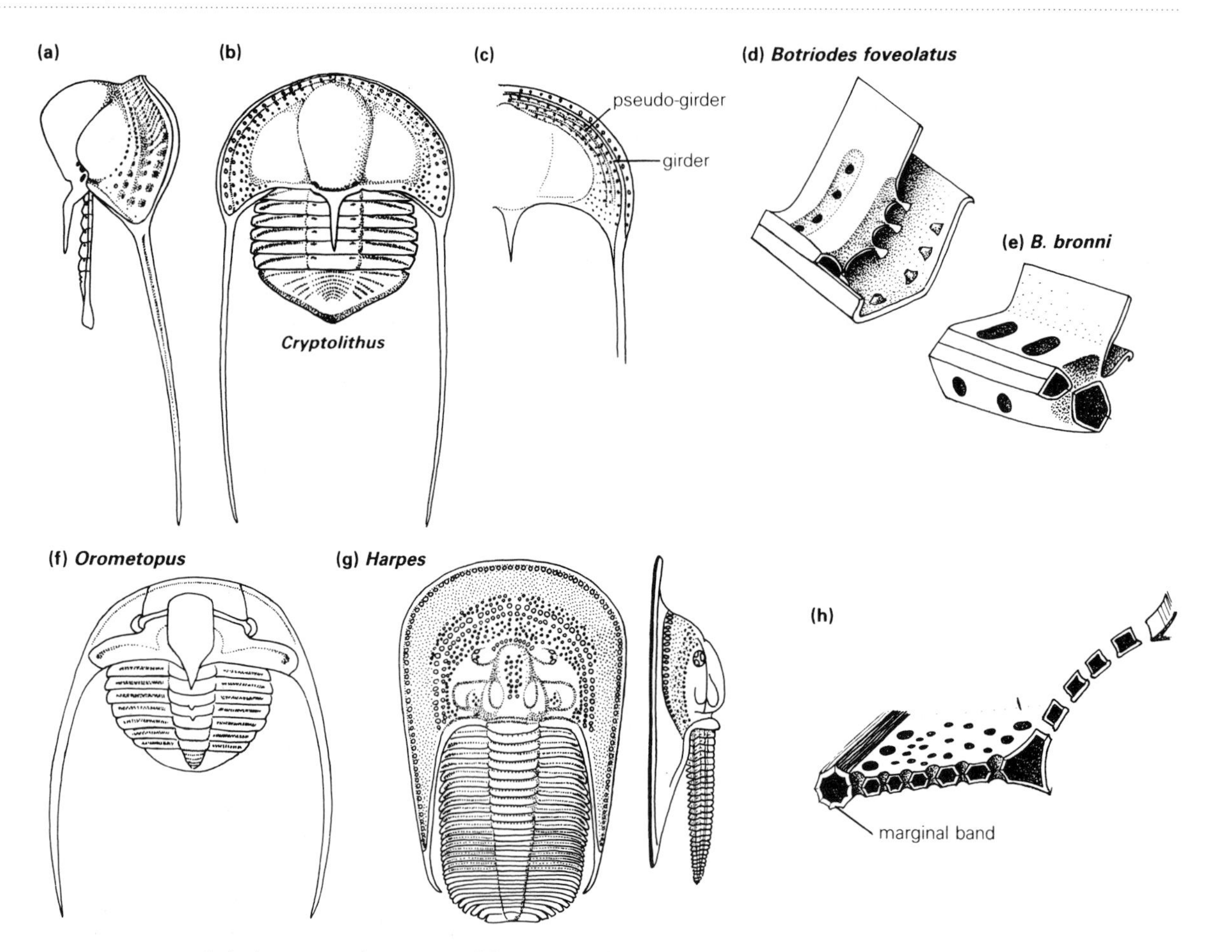

Figure 11.10 Cephalic fringes: (a), (b), (c) *Cryptolithus* (Ord.) in lateral, dorsal and ventral views (× 2); (d), (e) reconstruction of the brim in *Botroides foveolatus* and *B. bronni*; (f) *Orometopus* (Tremadocian), a possible trinucleid ancestor (× 4.5); (g) *Harpes* (Dev.) in dorsal and lateral view (× 1); (h) reconstructed section through brim and cheek showing perforations and thickened marginal band. [(a), (b) Redrawn from Whittington in *Treatise on Invertebrate Paleontology*, Part O; (c) redrawn from Hughes *et al.*, 1975; (d) modified from Störmer, 1930; (f), (g) redrawn from Whittington, as above.]

the border and now known to be homologous in all trinucleids. Arcs external to the girder are numbered E1, E2, . . ., while internal arcs are I1, I2, ... The development of **pseudogirders** concentric with the main girder in some genera has confused taxonomy in the past, but as the homologies are now clear the identification of species in this most stratigraphically important group, though time consuming, no longer presents the problems it once held (Hughes *et al.*, 1975).

How did trinucleids live, and what was the fringe for? Ventral appendages known in *Cryptolithus* have very long filamentous gills and stout walking legs. Trace fossils made by trinucleids show that these trilobites could use their legs to excavate shallow burrows, in which imprints of the fringe, genal spines and appendage scratch marks are often visible. Evidently trinucleids sat in their burrows with the head down (Campbell, 1975b). Some series of superposed burrows made by the one animal while shifting position show that during such movement the cephalon always faced in the same general direction. It was probably rheotactically oriented to the current in the only stable position; lateral currents would have overturned it. Hence the fringe may be interpreted as a sensory organ, each pit having sensory hairs at its base which were responsive to changes in current direction and so enabled the animal to keep its head to the current. This may not, however, have been the sole function. Other suggestions made include the possibility that the fringe was some kind of filter, but this is unlikely because the pronounced anterior arch would let in currents below and because the tiny holes in the terminal discs (or nozzles) are not likely to have acted as filters. Our understanding of the trinucleid fringe and its functions must proceed, as in all cases of this kind, by extremely detailed morphological analysis coupled with a stratigraphical perspective of the range and development through time of the various modifications of the basic structural theme.

Harpid fringe (Fig. 11.10g,h)

The fringe of trilobites of Suborder Harpina, unlike that of trinucleids, is flat and prolonged backwards into a pair of curving horns rather than genal spines. Furthermore, the distribution of pits on its surface is highly irregular, and the pits themselves are rather small and are found on the genal regions as well as on the fringe. Some genera have a small triangular anterior arch, but generally the fringe is flat all the way round. The fringe is bilaminar, with two closely juxtaposed laminae, and the pits, which in section have the form of opposed vertical funnels, perforate right through it. There is a pronounced marginal band, supplied with many sensillae, round the external periphery of the fringe. *Harpes* and its allies could enroll, having a pronounced flexure of the thorax in the region of the first few thoracic segments (discoidal enrollment). It may have been a sedentary animal, using the brim to spread the weight of the body like a snowshoe. The pits undoubtedly lightened the body, but such an elaborate structure may have had more than one function.

Enrollment and coaptative structures

The enrollment system exhibited by *Acaste* is very common in post-Cambrian trilobites and is known as **spheroidal enrollment**. More rarely, trilobites may roll up by tucking the pygidium and last few thoracic segments under the cephalon, this being **double enrollment**; and in harpids and trinucleids there is **discoidal enrollment** in which only the first few thoracic segments bend, the rest being held as a flat laminar plate.

Cambrian trilobites, with the exception of the isopygous agnostids, had poor facilities for enrollment. Some, e.g. the olenellids, were prevented from doing more than curl up in a half-sphere because the distal free edges of the pleurae then came in contact. Such arching of the body, however, was an essential prerequisite for moulting, for only in this position, and after opening of the cephalic sutures, could the newly moulted trilobite crawl out anteriorly (Whittington, 1990). While such partial coiling was necessary for exuviation, the structures which allowed this were usefully preadapted for the development of true enrollment. Some Cambrian trilobites were able to coil in a loose spiral, while in others from the latest Cambrian (Stitt, 1983) proper enrollment was fully achieved. Many Ordovician and later groups, however, developed an elaborate enrollment system, articulating facets being particularly well developed in Order Phacopida, to which *Acaste* belongs. In addition, within this order there are many different kinds of 'tooth-and-socket' structures, seemingly of quite independent origin, which are found on the cephalic and pygidial doublure and help to 'lock' the rolled-up sphere together. These are known as **coaptative structures** (Clarkson and Henry, 1973; Henry, 1980; Hammann, 1983), and though particularly well developed in the Phacopina they are present also in trinucleids, calymenids and asaphids.

Among phacopids *Acaste* has no real tooth-and-socket structures, though the cephalon and pygidium are the same shape and the opposing doublures fit together neatly. In the related Upper Ordovician *Kloucekia*, a single median projection in the cephalic doublure interlocks with a corresponding excavation in the pygidial doublure. *Morgatia* (Fig. 11.11g) has a series of lateral sockets for the reception of the pleural tips of the thoracic segments and the first pygidial segment.

In *Phacops* a deep marginal groove (the **vincular furrow**) is often present on the cephalic doublure. The edge of the pygidium fits into this, and the thoracic segments come to rest in extensions of the vincular furrow divided into individual lateral pockets. Such lateral pockets may still be present even when the median part of the vincular furrow is absent, as in *Eskaspis* (Fig. 11.11h). Morphological differences such as these are of some taxonomic value. Techniques of enrollment have been fully explored by Bruton and Haas (1997) in Devonian phacopids by serial sectioning. Other examples of quite different mechanisms are shown by *Encrinurus*, and in the cheirurid *Placoparia* another stratigraphically docu-

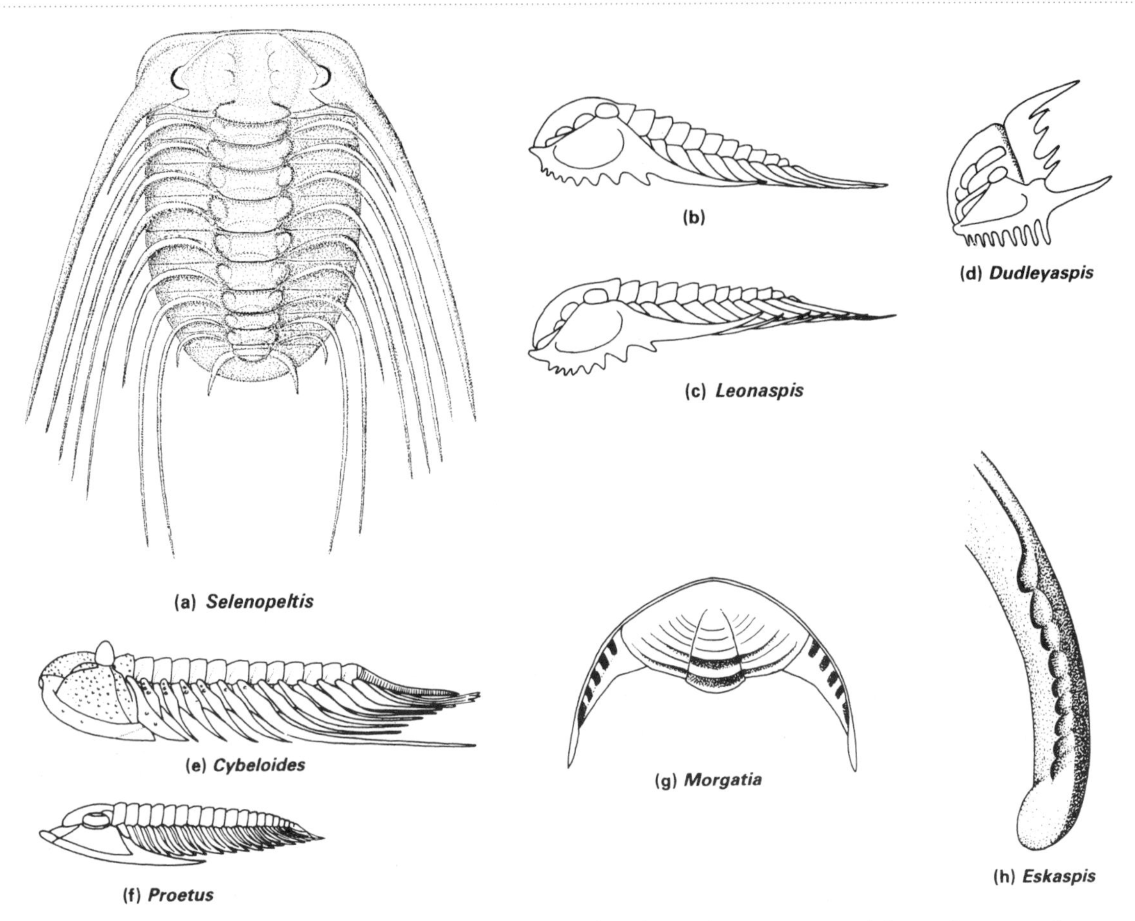

Figure 11.11 (a) *Selenopeltis macrophtalma* (Ord.), an odontopleurid (×1); (b), (c) *Leonaspis deflexa* (Sil.), in two alternative life attitudes for resting and swimming or browsing; (d) *Dudleyaspis* (Sil.) cephalon in side view, attitude equivalent to (c); (e) *Cybeloides* (Ord.) in life attitude, lateral view, showing the body supported on long macropleural spines on the sixth thoracic segment; (f) *Proetus* in life attitude, lateral view, showing high carriage of the body; (g) *Morgatia* cephalon from below, showing press-studs (vincular notches) for the reception of thoracic pleurae and with the pygidium in place (× 4); (h) *Acernaspis (Eskaspis)* cephalic doublure with vincular furrow and vincular notches (× 7.5). [(a) Redrawn from Hammann and Rabano, 1987; (b), (c) redrawn from Clarkson, 1969; (g) redrawn from Henry and Nion, 1970; (h) redrawn from Clarkson *et al.*, 1977.]

mented series shows a progressive increase in the number of depressions on the dorsal surface of the cephalic anterior border. When enrolled, the pygidial spines rested in these like clutching fingers (Henry and Clarkson, 1975).

Some other trilobite groups (e.g. the Proetida and Agnostida) appear to have been well adapted for enrollment, but only the Phacopida and Calymenida reached the summit of enrollment ability.

Enrollment structures in some dalmanitid trilobites have been portrayed by Campbell (1977). In these the pygidium does not form a tight seal with the cephalon but projects forward of it, leaving a triangular gap between the anterior margin of the cephalon and the pygidial doublure. In some genera forwardly projecting spines or rows of denticles on the cephalon partially cover this open space but leave symmetrical gaps, which presumably allowed some degree of water circulation in the enrolled animal. Enrollment structures may well be very important in classification.

Thorax

Most trilobites are 'polymerids', that is, they have several thoracic segments. In the agnostids, however, there are only two, and in the eodiscids two or three thoracic segments: this is the 'miomerid' condition.

The axis of the thorax in most trilobites is about as broad as the occipital ring. These axial rings with their half-rings and apodemes are of fairly standard construction throughout the trilobites, though there are differences in the length of the apodemes, the half-rings are sometimes absent, and surface sculpture and spinosity are variable. Certain phacopid genera from the Devonian of South America – a region long isolated so that a rich endemic fauna of trilobites developed from *Acaste*-like ancestors – repeatedly developed dorsal spines of a particular erect construction, sometimes singly, sometimes one for each segment of the thorax and on the occipital ring and pygidium as well.

The pleura tend to be rather flat in some Cambrian genera, though not in all, but they are often arched or bent down distally in the later groups and are normally marked with a pleural furrow or less often with a ridge. Usually the proximal parts of the pleura have parallel edges against which the edges of adjacent pleura can hinge; the distal parts do not touch and are free. The junction between the free and hinged parts is the **fulcrum**, at which the pleura are often sharply downturned. Cambrian genera have a simple articulating hinge structure in which contiguous pleural edges alone are used for articulation. Specialized hinge structures and articulating pleural facets did not appear until the Upper Cambrian; many of these seem to be associated with enrollment ability. The pleural doublure may be wide or narrow and is often ornamented with terrace ridges. Also to be found in post-Cambrian stocks, notably Family Asaphidae, are small protuberances on the inner faces of the thoracic doublures, which presumably acted as stops preventing overgliding of the pleura as they came to rest during enrollment. These are the **organs of Pander**; they are not to be confused with **panderian openings** which, though similarly located, probably were the orifices of some kind of segmented and possibly excretory organs.

The most highly modified pleura of all are to be found in Family Cheiruridae (Ord.–Dev.) and Superfamily Odontopleuroides (Ord.–Dev.; Figs 11.11a, 11.16). In the former the distal ends may be curiously pointed, but the contiguous parallel edges are equipped with interlocking ledges which presumably facilitated highly efficient hinge articulation. The Odontopleurida are a very spiny group of trilobites, and each pleuron terminates in two spines. The anterior pleural spine is normally vertical; the posterior one is horizontal and usually short. However, in modified odontopleurids (e.g. *Ceratocephala*) the vertical spines are very long and, being equipped with many secondary spines, virtually close off a subthoracic 'box' from the external environment: a unique feature whose function is quite unknown.

Pygidium

The pygidium is a fused plate which in the Lower Cambrian olenellids may consist of just one segment, but in later genera it consists of as many as 30 segments. It articulates with the last thoracic segment by means of an articulating half-ring, and if the thoracic segments have pleural facets, so does the pygidium. With the exception of the Agnostida most Cambrian trilobites have small (**micropygous**) pygidia, whilst post-Cambrian genera tend to have **heteropygous** (smaller than the cephalon) or **isopygous** (equal-sized) pygidia. Rarely pygidia can be **macropygous** (larger than the cephalon), as in the Lichida. In *Encrinurus* the number of axial segments is supernumerary, probably through the division of existing axial rings.

In all pygidia the pleural furrows are homologous with the axial furrows, whilst the interpleural furrows are equivalent to the edges of the pleura; the ontogenetic reasons for this are discussed later.

Appendages

The limbs are known only in a few species of trilobites as these delicate structures did not preserve well. Appendages were first described in 1876 and have been reported, often in a very fragmentary state, in about 20 species. They are well known and have been fully described in *Olenoides* (Whittington, 1975) and *Kootenia* from the Middle Cambrian Burgess Shale of British Columbia, and also in the pyritized Ordovician *Cryptolithus* by Raymond (1920) and from enrolled specimens of the Ordovician *Ceraurus* by Störmer (1939). The pyritized Lower Ordovician *Triarthrus* and Devonian *Phacops* (Fig. 11.14) have yielded much detail through dissection and X-radiography. In all these the structure of trilobite appendages shows remarkable constancy in that there is always first a pair of uniramous antennae located on either side of the hypostome, followed by paired biramous (two-branched) appendages, under the cephalon and one pair for each segment all the way down the body and also in the pygidium. These biramous appendages are almost identical to each other in all but size. The constancy of morphology here contrasts with that of the various crustacean groups, where there is wide variation even within the one genus. Yet within the trilobites whose appendages are known there is indeed some diversity, as will be shown.

About 15 specimens of *Olenoides serratus* (Figs 11.12, 11.13a–c) with limbs are known from the Burgess Shale, and the appendages are preserved as a fine dark film, usually in place but sometimes detached.

They are flattened and have a reflective surface so that many details can be captured on film, given light of suitable inclination. The antennae are composed of many jointed rings, as are the **cerci**, which are posterior equivalents of the antennae. Like the antennae they almost certainly had a sensory function; they are provided with minute setal hairs. Behind the antennae there are three pairs of biramous appendages under the cephalon, seven on the thorax and between four and six on the pygidium. Each biramous appendage was apparently attached to an apodeme, and even where the appendages are not preserved the number of cephalic apodemes will indicate the relative number of cephalic appendages. The basal joint of the appendage (coxa) gives rise to two branches: one the **walking leg** and the other the **gill branch**. Such terminology presupposes definite functions, and probably the gill was used for functions other than purely for respiration. However, other terminology presupposes homologies, and though the term **telopodite** is often used for the walking leg, the correct terminology for the gill is not agreed; for the moment the term 'gill branch' is retained, though **outer ramus** is a good alternative.

The coxa is about half the length of the axial ring with which it is associated, and its inner margin is supplied with long sharp spines, straight and curved, between which are numerous shorter spines. A precoxa, or initial segment other than the coxa, has been described but recent work does not seem to substantiate its presence. The walking leg is continued in the direct line from the coxa and consists of six jointed segments (**podomeres**). The proximal podomeres are very spiny on their inner surfaces, the more distal ones less so but with fine setae, and

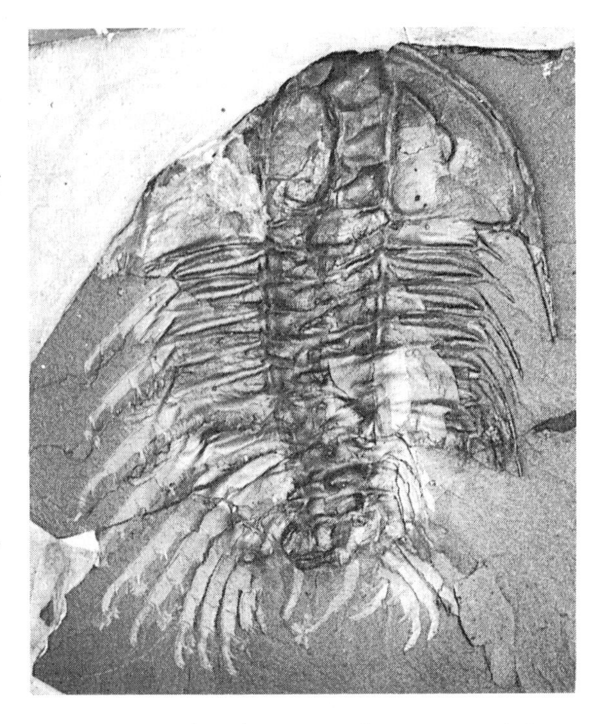

Figure 11.12 *Olenoides serratus* from Burgess Shale (M. Cam.), British Columbia: US National Museum no. 58589. (Photograph reproduced by courtesy of Professor H.B. Whittington.)

the last podomere has three terminal spines. From the coxa there also projects the gill branch, lying above the leg branch and directed posteriorly. The individual gill branched overlap markedly. Each has two lobes: a long **proximal lobe** (three times as long as it is broad) and a shorter ovoid **distal lobe**. The former has a posterior fringe of long parallel filaments diminishing in size towards the distal lobe, whereas the latter has no filaments and only a fringe of setae. An anterior marginal rim originally described on the gill branch is now shown to have been a preservational artefact.

Precisely how the coxa fitted on to the apodeme is uncertain; presumably the whole appendage could rotate in a horizontal plane, but it is not known whether it was capable of any dorso-ventral or transverse movement.

Before considering how the appendages functioned, it seems appropriate to cover the range in form of known appendages. *Kootenia* appendages are fairly similar to those of *Olenoides* but are known only from one specimen. *Triarthrus is* a Lower Ordovician

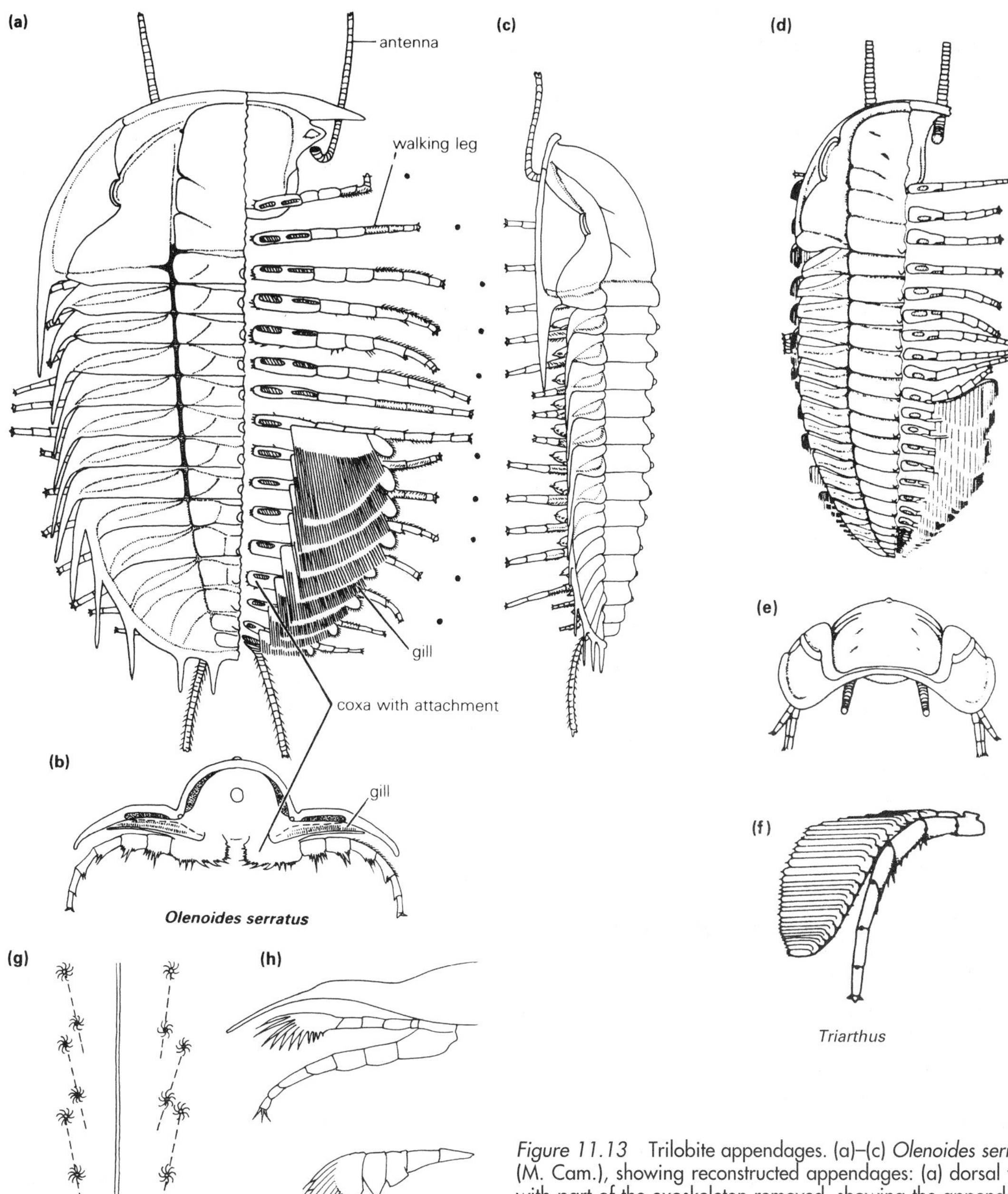

Figure 11.13 Trilobite appendages. (a)–(c) *Olenoides serratus* (M. Cam.), showing reconstructed appendages: (a) dorsal view with part of the exoskeleton removed, showing the appendages in place and the gill branch lying above the leg; (b) section; (c) lateral view; (d)–(f) *Triarthrus eatoni* (L. Ord.); (d) reconstruction with right side removed to show appendages; (e) frontal view of walking individual; (f) a detached appendage; (g) *Phacops* (Dev.) leg with a terminal brush believed to have made stellate markings on either side of a central groove; (h) *Ceraurus* (Ord.): (h) limb in frontal view (above) and gill branch in dorsal view (below). [(a)–(c) Redrawn from Whittington, 1975; (d)–(f) redrawn from Whittington and Almond, 1987; (g) based on Seilacher, 1962; (h), (j) based on Störmer in *Treatise on Invertebrate Paleontology*, Part O.]

olenid. Where found in the 'Utica Slate' (actually a black shale) of New York State the appendages are pyritized. First discovered a century ago (e.g. Raymond, 1920), they have been the subject also of recent investigations (Cisné, 1975, who used X-radiography; Whittington and Almond, 1987). The latter authors' reconstruction (Fig. 11.13d–f) shows antennae and three pairs of biramous cephalic appendages uniform with those of the thorax. The walking legs are spiny, with a coxa and six podomeres. Each gill consists of filaments joined to a flexible, segmented, anterior rod, and the distal lobe is reduced and tiny. The appendages were steeply inclined below the exoskeleton and did not project much outside it.

The use of soft X-rays has also been useful in interpreting the appendage structure of Devonian *Phacops* and *Asteropyge* specimens from the North German Hunsrückschiefer (Stürmer and Bergström, 1973). These specimens of *Phacops* (Figs 11.13g, 11.14) have stout walking legs with the usual six podomeres.

There are three pairs of biramous appendages below the cephalon, the last two having strongly developed coxae prolonged internally and armed with spines. These could have functioned as gnathobases equivalent to those in some modern arthropods, and by rolling movements in a horizontal plane they may have been able to work as crushing jaws. There is one pair of appendages for each segment of the thorax and there are also several small sets in the pygidium. The main difference between the appendages of *Phacops* and those of other trilobites is that the gill branch filaments emerge directly from the coxa and are transverse and subparallel with each other. The whole array of gill branch filaments forms a kind of slatted curtain lying dorsal to the walking legs, which may have had both a respiratory and a filtering function. Legs in the phacopid trilobite *Rhenops* have recently been reported (Bergström and Brassel, 1984).

The Ordovician cheirurid *Ceraurus* (Fig. 11.13h) has peculiar appendages, as shown by Störmer (1939), who made serial sections through enrolled specimens collected from a limestone of Trenton (U. Ord.) age. The branch structures were worked out from a wax model built up layer by layer from these sections, as has been done with the graptolites (q.v.). The walking leg is relatively unmodified, but the gill branch has five segments increasing in size distally, the last of which forms a flat paddle-like lobe fringed with filaments. These filaments are of

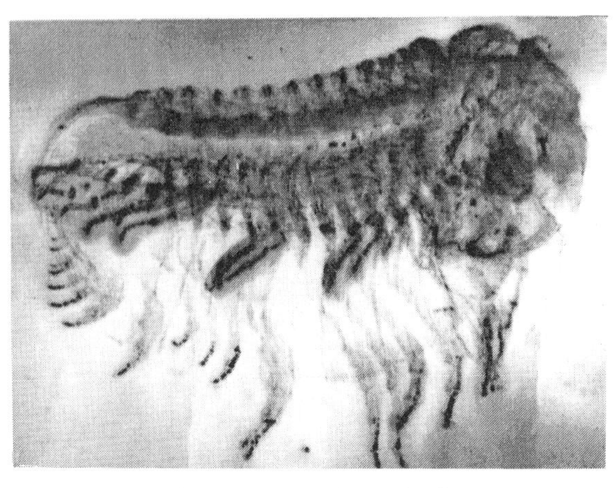

Figure 11.14 *Phacops*, an X-radiograph of a slightly crushed specimen from the Lower Devonian Hunsrückschiefer, Germany, showing lightly pyritized gill and leg branches. (Photograph reproduced by courtesy of Professor Wilhelm Stürmer.)

normal construction, i.e. elongated thin plates arranged subparallel with one another and having each a terminal bristle.

The diversity of form in the gill branch implies some differences in function. It is generally accepted that the gill branch did in fact function for respiratory exchange, though Bergström (1973) has pointed out that since they are relatively hard structures they could have functioned merely as aerators, to agitate the water while the animal was moving and so to feed more oxygen to the gills, which he imagined as soft structures attached to the ventral membrane. The paddle-like gill branches of *Ceraurus* and *Olenoides* could probably have facilitated swimming, but there is no direct evidence as to how well trilobites could swim, and the question remains open. There is, however, good evidence as to how the walking legs operated, and the detailed study of these has yielded a surprising amount of information about trilobite behaviour.

Trilobite tracks and trails

Early in the history of palaeontology elongated trail-like markings were described from many different sedimentary rocks of Palaeozoic age. The commonest kind, first called *Cruziana* (Fig. 11.15a) and later referred to by several authors as 'bilobites', are ribbon-like markings with paired oblique chevron structures.

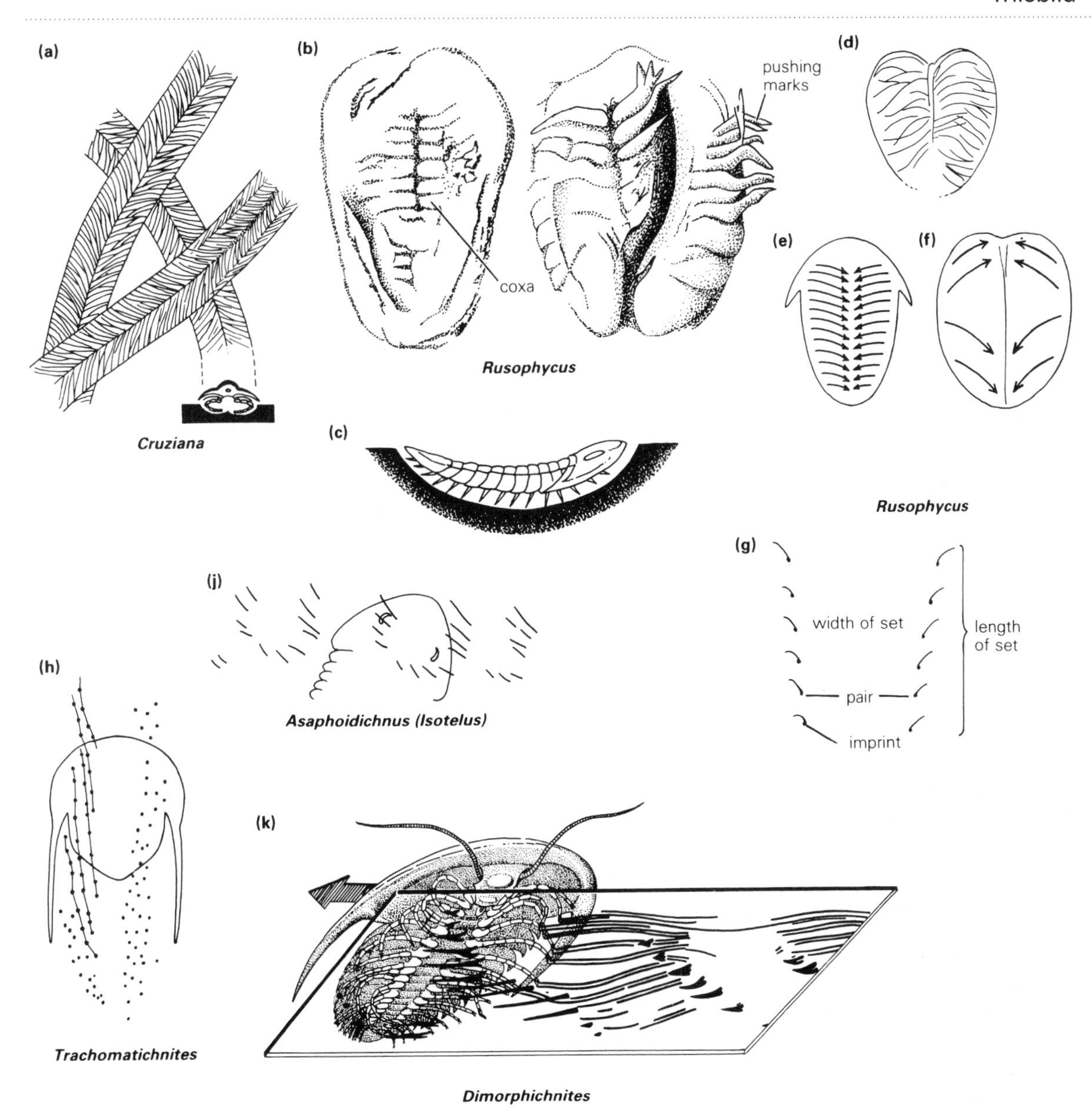

Figure 11.15 Trilobites as trail makers: (a) intersecting *Cruziana* trails showing how a 'ploughing' trilobite could have made them by pushing backwards with its legs – the lateral striations on one trail could result from the gill branches moving across the surfaces (see also Fig. 12.12a); (b) *Rusophycus* from the Cambrian of Poland, showing resting marks of coxa, and pushing marks made by appendages as the trilobite pulled itself sideways out of its burrow; (c) trilobite in a shallow burrow; (d) leg marks preserved in a *Rusophycus*; (e) directions of movement of the legs; (f) a burrow made by anterior and posterior legs moving in different directions; (g) terminology of a set of walking marks; (h) *Trachomatichnites*, a trilobite (trinucleid?) walking trail in which individual sets can be distinguished; (j) *Asaphoidichnites*, sideways movement marks made by a large (asaphid?) trilobite, possibly *Isotelus*; (k) reconstructed ventral view showing how *Dimorphichnites* was made by a sideways-crawling trilobite. [(c)–(f) Redrawn from Seilacher, 1962; (g)–(j) redrawn from Osgood, 1970; (k) redrawn from Seilacher, 1955.]

They were first believed to be of algal origin. However, Nathorst (1881) called attention to the fact that 'bilobites' always occurred at the interfaces of sandy and shaly layers in the rock. This, he presumed, was more likely to have been the result of a moving animal crawling over or rather ploughing through a firm muddy sea floor, leaving marks which had been filled with sand soon after, the difference in sediment type being responsible for their preservation. Not infrequently *Cruziana* markings are cut across by other and presumably later trails made by the same kind of animals, confirming the mobile animal hypothesis. Other kinds of markings existing on shale–sandstone interfaces could also have been made by moving animals.

It is generally believed that *Cruziana* markings were made by trilobites, though Whittington (1980) has criticized previous interpretations, and suggests that they were more likely to have been made by some other organism.

There is little direct evidence that these and other kinds of markings were actually made by any one type of organism, since trace fossils of any kind are rarely found with body fossils. Calcified shells are not normally retained by the permeable sandstones in which the trace fossils occur. However, a few specimens of intact trilobites have been found in the Ordovician of Cincinnati (Osgood, 1970), where *Calymene* specimens with intact cuticles have been found in their resting excavations (called *Rusophycus*). Other criteria for attributing particular trace fossils to triobite movements are indirect, though compelling. Thus Cambrian and Ordovician *Cruziana* are numerous and diverse, whereas the generally poorer Siluro-Devonian *Cruziana* fauna reflects the decline of the trilobites. Although trilobites are not found in *Cruziana*-bearing sandstones, they may occur in intercalated beds of the same sedimentary sequence, as in the Ordovician of Poland and the Devonian of Germany. Most *Cruziana* and other similar markings were clearly made by 'legs' of uniform and undifferentiated structure, such as are possessed by trilobites, and sometimes the 'prod marks' made by walking legs can be matched with the actual known structure of the distal part of the leg. Furthermore, the common bilobed and normally ovoid trace fossils known as *Rusophycus* are just the right size to have been made by trilobites, and sometimes indentations caused by the impression of a cephalon, genal spines or lateral parts of the body resting in the sediment are preserved in association with the *Rusophycus* marks (Fig. 11.15b).

Walking movement in arthropods

In all crawling arthropods the movement of the limbs relative to one another takes place in the same general way. Leg movements on each side of the body are synchronized in regular waves of forward movement, known as **metachronal rhythm**. Each leg comes forwards in turn, is placed on the ground and pushes backwards, helping to support and move the body in the process behind. At any one time a leg will be slightly the one in front of it in its forward movement, so that in a long-bodied arthropod (e.g. a centipede) successive waves of motion may be seen travelling up the body from the tail to the head. As the waves of movement sweep forwards, every tenth leg or so is in approximately the same position and at the same angle. Even in short-bodied arthropods (e.g. woodlice) the pattern of movement is clear, and the waves of movement always travel forwards along the body. This was evidently the case in trilobites, and the walking marks especially may be interpreted on this basis.

Different kinds of trilobite trails

The various sorts of trace fossils believed to have been made by trilobites are classified according to the binomial system of nomenclature. The retention of Linnaean practice for such **form genera** and **form species** is unavoidable, since it is very rare that a particular trace fossil can be directly related to a known producer, and even when it can, both the trail and the body fossil retain their own names. Of the various form genera that have been attributed to the life activities of trilobites some of the most important are

Protichnites, *Trachomatichnites:* walking or striding trails with individual leg impressions (Fig. 11.15h);
Diplichnites, *Petalichnites*, *Asaphoidichnus:* as above but usually oblique (Fig. 11.15j);
Cruziana: bilobed chevron-marked trails which are the traces of crawling, loughing, shovelling or burrowing movements (Fig. 11.15a);
Rusophycus: bilobed ovoid traces which are probably resting nests, burrows or surface excavations (Fig. 11.15b –f).

These movement trails would be generally classified as Repichnia (crawling traces) and *Rusophycus* as Cubichnia (resting traces), according to the taxonomic scheme of Seilacher (Chapter 12). A single trilobite could well have made different kinds of marking, depending on exactly what it was doing.

Though trilobites often walked straight ahead, they frequently progressed obliquely or even sideways like a crab. In the case of forward movement, as represented by the tracks of *Protichnites* and *Trachomatichnites*, the pygidial appendages were the first to touch the ground, followed by those of the last thoracic segment, and so on until the wave of movement reached the head; meanwhile new waves of movement were progressing forwards from the hinder end. The result of such successive waves of movement is a series of sets of paired markings impressed like open-ended (truncated) V-shapes (Fig. 11.15g; Seilacher, 1955, 1964; Osgood, 1970). Since the trilobite widens anteriorly from the pygidium, the V is directed forwards (though some xiphosurids moving forwards had a series of reversed Vs). Where the sides of the V do not curve in again at the front it seems that the head appendages cannot have been used in walking.

Trilobites moving directly forwards made a track consisting of a series of superimposed sets of V-markings, so that it is not easy to sort out the number of imprints per set, though a good example is shown in Fig. 11.15h. When trilobites walked with their movement oblique to the body axis (as in *Diplichnites*), the imprints are closely crowded and interfering on one side but quite separate on the other, and so they can more easily be counted. There is then some chance that the trail maker can be identified, as there should be some kind of parity between imprint number and the number of walking legs. *Diplichnites* might have been made by a trilobite walking obliquely as its normal mode of progression, or it could have resulted from the animal trying to keep steady on a slippery mud surface while a lateral current was flowing.

In these various walking trails the marks of the walking leg often retain the detailed imprint of the terminal digit which made them. In the Devonian Hunsrückschiefer of Germany many specimens of *Phacops* are found, some with their appendages intact (Fig. 11.13g; Seilacher, 1962). These terminate in a distal tuft of bristles, encircling the terminal claw, which if extended like an umbrella would fit ideally with the stellate impressions arranged in paired series on either side of a median groove, such as have been found at the silt–sandstone interfaces of beds in the same sequence. This is one of the few cases of unequivocal matching. In the Ordovician of Cincinnati the large asaphid *Isotelus* is present in a sequence that also contains the ichnofossil *Asaphoidichnus*, which consists of paired sets of trifid imprints (Fig. 11.15j). This trace fossil is large enough to have been made by *Isotelus*, and indeed it may give a clue as to the structure of the appendages of this trilobite, which are otherwise unknown.

Evidence of strongly oblique movement is known only in rare cases. The well-known *Dimorphichnites* (Fig. 11.15k) from the Cambrian of the Salt Range of Pakistan is one of these. In each 'unit' of *Dimorphichnites* there are two elements: on one side a set of round and deeply impressed imprints, on the other a set of lightly impressed and slightly sigmoidal raking marks. Seilacher's (1955) interpretation of these is that a trilobite (probably an olenellid) moving crabwise from left to right would press its right-hand legs into the substrate and, using these for support, would heave its body sideways by contracting its leg muscles. Then it would drag its left-hand legs across the mud, only using them for minimal support, and by withdrawing its right-hand legs one after the other and replacing them some distance to the right would then be in a position to repeat the procedure. A number of complete movements of this kind can be clearly seen on the lower surface of a sandstone unit. In one case a Lower Cambrian olenellid, moving from the left by such a mode of progression, encountered the track of another olenellid, turned round again and moved back again parallel with its own set of tracks. Then it jumped over its own tracks and moved away at right angles to its first track by slightly oblique forward movements. Seilacher interpreted these as grazing tracks since the trilobite obviously avoided previously made tracks. Similar tracks have been found in the Ordovician, apparently made by trinucleids.

Cruziana is a very widely distributed trace fossil, and up to 30 different form species have been described, though these can often be attributed to somewhat different burrowing techniques by the same animal or to the inclination of the body of the animal on different gradients. However, the basic *Cruziana* is simple: an elongated bilobed trail, each lobe being convex downwards and usually striated with oblique, closely spaced ridges and grooves forming a herringbone pattern (Figs 11.15a, 12.12a). The trilobite must have ploughed its way along the upper surface of the sediment, incising these markings as it went; in the best-preserved specimens there was probably a thin layer of sand already overlying the plastic mud that took the impressions, protecting it from current erosion. More sand later filled the markings. The two lobes, with their oblique structures, were probably made by the walking legs moving inwards towards the median line and backwards. The oblique scratches seem to have been made by divided terminal claws, though as they are generally superimposed one upon another, the marks of individual legs are often difficult to distinguish.

Variants of the basic structure are often found. Sometimes lateral furrowed borders are present, probably the scratch marks of gills or pleurae. Certain *Cruziana* species have an extra pair of lobes within the outer furrows, perhaps the brush marks of the gill branches as the trilobite ploughed forwards. *Cruziana* has also been found with apparent pygidial scratch marks as well. One species shows a herringbone pattern clearly subdivided into sets with between three and nine parallel scratches; each successive set overlaps and truncates the preceding one in a manner suggestive of alternate stronger and weaker pushes. On rare occasions a particular *Cruziana* species can be matched with a particular trilobite, where there is only one kind of *Cruziana* and one species of associated trilobite (Fortey and Seilacher, 1997), but this is uncommon.

Rusophycus (Fig. 11.15b–f), interpreted as trilobite burrows or temporary resting traces, is the name given to short ovoid markings which are normally bilobed and deepest in the centre. These may be smooth (coffee bean marks) or have herringbone ridges as in *Cruziana*. They usually occur singly, implying that a swimming trilobite had landed, excavated a temporary 'nest' and then swum away. They may also terminate *Cruziana* trails, as if the trilobite had crawled, rested and then swum off, and they have been found in association with striding trails. The smooth *Rusophycus* was probably formed beneath the body of a resting trilobite rheotactically directed upstream and hollowed out and smoothed by current scour. Some deeply excavated hollows are best interpreted as nests in which the trilobites lay for quite long periods, and it is these that have been found in the Ordovician of Cincinnati still with trilobites in them, though only a very few examples have actually been discovered. Some interesting examples described by Seilacher were clearly produced by the anterior appendages shovelling forwards and the posterior ones working backwards; hence *Rusophycus* has a bidirectional structure. Other examples were apparently excavated by using the anterior border itself as a shovel and scrabbling with the posterior appendages. Such modifications of normal practice allowed the formation of deeper burrows than the normal shallow resting traces. The trilobite could easily get out of its nest by pushing with the walking legs, but one intriguing example from Poland (Orlowski *et al.*, 1970; Fig. 11.15b) shows imprints of the legs outside the burrow, though only on one side as evidently the animal had levered itself out sideways before swimming away. This and other examples show the grooves where the coxae had rested.

Some 'hunting burrows' or *Rusophycus* from the Lower Cambrian of Sweden give evidence of predation by trilobites, probably olenellids, or infaunal 'worms' (Bergström, 1973; Jensen, 1990). Twenty-three known examples of *R. dispar* are dug down to U-shaped spreite burrows, which were probably made by a priapulid. This prey must have been located, chemically or visually, at the apertures of the burrows. The *Rusophycus* traces are always located in contact with but slightly to one side of the worm burrow. This suggests that the prey was captured by the trilobite digging down deep, then hooking one set of legs round the tubular body of the worm and finally impaling and shredding it with the spinose coxae. Where the spreite burrow is short, the trilobite was able to align itself prior to digging. Longer worm burrows required several successive diggings by the trilobite in order to achieve the correct alignment.

Life attitudes, habits and ecology

Acaste (Fig. 11.2c) shows a standard pattern of construction in side view. This attitude is very common, and even in such a spiny trilobite as *Cybeloides* (Fig. 11.11e) the spatial and angular relationships of the various parts of the body are similar. *Cybeloides*, however, has both genal spines and macropleural spines on the sixth thoracic segment. These would hold the body in an outstretched attitude on the sea floor with the posterior thoracic segments and the pygidium rising away from it, though some relaxation of the muscles would lower the 'tailgate'. Besides being used as props, the spines could have been used in burying the trilobite by a series of alternate flexion and relaxation movements.

Spinosity in trilobites was once regarded as an adaptation for a planktonic mode of life, the spines inhibiting the animal's sinking. However, most trilobites are far too large for the spines to have been of any real value in this context, and it is more likely that they were, as well as being protective, used for spreading the weight of the trilobite when resting on the sea bottom. Thus long genal spines, as in *Proetus* (Fig. 11.11f), bore the weight of the whole exoskeleton, being the most ventral part of the animal; the thorax and pygidium were carried slightly higher than these and not in contact with the sea floor. This is a common posture, especially in the Cambrian, and is probably primitive for trilobites.

Perhaps the most striking case where spinosity and life attitudes are correlated is in the Odontopleuroidae (Figs 11.11a, 11.16).

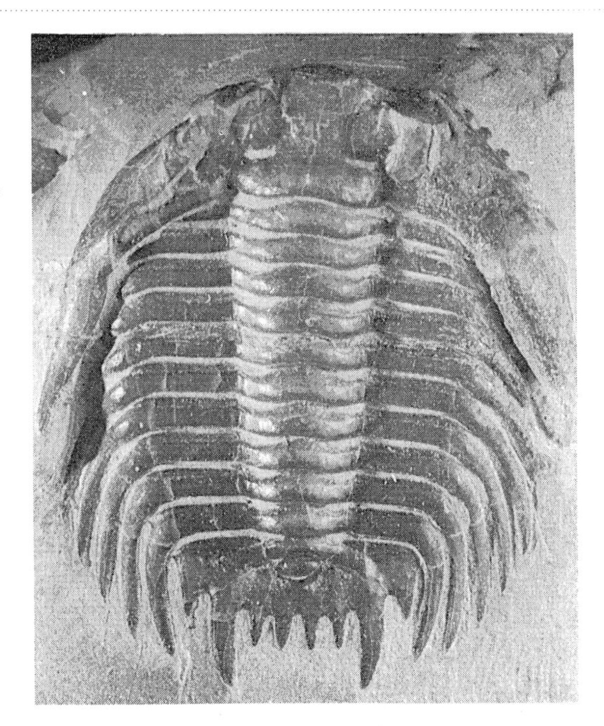

Figure 11.16 *Leonaspis coronata* (Sil.), Wenlock Limestone, Dudley, England: a complete specimen in dorsal view cephalon slightly crushed (× 4).

Odontopleurids are always spiny and have not only lateral but also ventrally directed spines. Genera such as *Acidaspis* and *Dudleyaspis* (Fig. 11.11d) have a fringe of modified **denticles** extending along the antero-lateral border on either side of the glabella. Since they terminate in a flat plane they seem to have been adapted for supporting the cephalon in a particular attitude, presumably while feeding, for the short hypostome would then be in close proximity to the sea floor and so would the mouth which lay just behind it. There is no anterior arch in these modified genera nor in *Ceratocephala*, which could support itself in a similar attitude even though it does not have the denticles (Whittington, 1956).

The rootstock genera of the odontopleurids, *Leonaspis* (Figs 11.11b,c, 11.16) and *Primaspis*, have the anterior denticles lying in a curve rather than in a plane, and it has been postulated (Clarkson, 1969) that these trilobites could take up two alternative life attitudes, both equally functional but for different purposes. In one the trilobite lies on the sea floor or crawls over it with an open anterior arch, with the cephalon supported by the genal spines and posterior denticles and with the thorax and pygidium, which slope posteriorly, resting on their spines (Fig. 11.11b). In the other attitude the cephalon is tilted forwards like that of *Dudleyaspis*, resting on the anterior denticles, while the body is held out horizontally (Fig. 11.11d). The first attitude seems to have been a resting position, the second an active or 'browsing' stance.

Dudleyaspis was a presumed derivative of the *Leonaspis* stock, permanently specialized for life in the active attitude, since the posterior thoracic spines and the pygidial spines are bent down so as to terminate in the same plane as the anterior denticles. In *Ceratocephala* it is the anterior thoracic spines that are modified; they are large and thick and terminate in the same plane as the anterior border of the cephalon, showing how specialization for the active attitude has been independently evolved.

The relatively unspecialized *Leonaspis* stock, with its dual mode of life, was very versatile, and the stratigraphical range of this one genus has been estimated as 170 Ma. However, its derivatives (*Dudleyaspis*, *Acidaspis* and others) were of relatively short duration, illustrating the general rule that the highly specialized taxa did not usually last very long.

The specialized odontopleurid *Selenopeltis* (Fig.

11.11a) (which was capable of discoidal enrollment) has been interpreted as an active swimmer but capable of resting for short periods of time on the sea floor (Hammann and Rabano, 1987).

Throughout the history of trilobites certain specialized 'morphotypes' have occurred again and again, in different stocks. This convergent evolution suggests similar ecological niches. Fortey and Owens (1990a,b) have documented eight such morphotypes resulting from iterative evolutionary trends. Thus 'phacomorphs' are tuberculate, convex, nearly isopygous trilobites with well-developed eyes. Such morphology is characteristic of many unrelated Ordovician to Permian forms. Likewise some kinds of 'illaenimorphs' have a highly convex exoskeleton in which the head in side view is hemispherical and the furrows are effaced. The body cannot articulate with the head in the normal way but declines backwards very sharply from a highly inclined occipital ring (e.g. the illaenine *Bumastoides*). Such a distinctive morphology is poorly suited to epifaunal crawling or swimming, and it is highly probable that these unusual trilobites lived infaunally in finer substrates and were largely sedentary suspension feeders (Stitt, 1976; Bergström, 1973; Westrop, 1983). Other trends include loss of eyes, development of marginal spines or pitted fringes, miniaturization, and adaptation to a dysaerobic habitat (olenimorphs), or to a pelagic mode of life. The interpretation of some trilobites as pelagic is based upon three independent criteria: functional morphology, analogy with living arthropods, and geological evidence (Fortey, 1975, 1985).

Trilobites such as the equatorial *Opipeuter* have enormous hyperhophied eyes, capable of all-round vision, and long fusiform bodies, axially vaulted but with reduced thoracic pleurae giving longitudinal flexibility (Fig. 11.17).

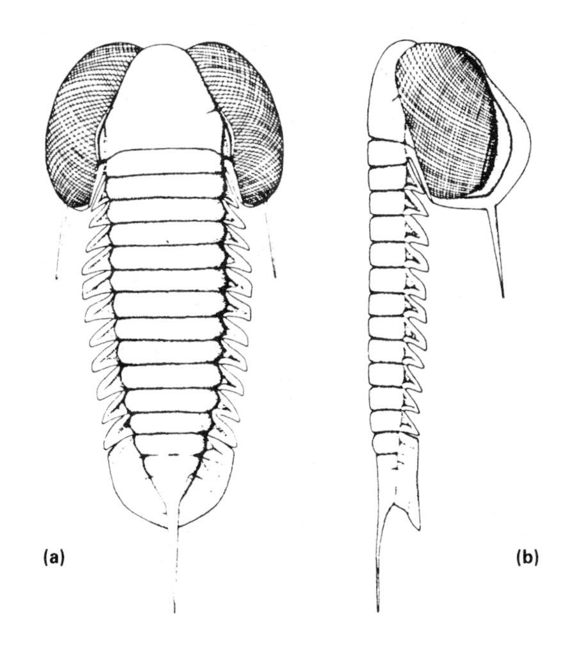

Figure 11.17 *Opipeuter*, an Ordovician pelagic trilobite in (a) dorsal and (b) side view. (Redrawn from Fortey, 1974a.)

The lateral profile shows that there is no way they could have rested on the sea floor like other trilobites. These details suggest that the lifestyle was pelagic, and analogies may be made with large-eyed pelagic crustaceans of the present day belonging to normally benthic groups (e.g. the amphipod *Hyperia*) whose eyes and bodies are modified in similar ways. Moreover, unlike the normal, largely facies-controlled benthic assemblages, *Opipeuter* seems to be independent of facies, as is *Carolinites*, another large-eyed Ordovician trilobite, which to a degree is homeomorphic on *Opipeuter*. This gives a clear confirmation of a pelagic life habit for genera such as *Opipeuter*, *Carolinites* and *Cyclopyge* which are vaulted, often spinose and poorly streamlined. Such trilobites were probably slow swimmers like the living amphipod *Hyperia*; their morphology contrasts with the remarkably streamlined form of cyclopygids like *Novakella*, which is elliptic in form, lacks spines, and has the eyes flush with the surface of the long-snouted head. Such shapes have been experimentally tested for streamlining in flume experiments and support the idea of adaptation to rapid swimming. Trilobites like *Novakella* were thus adapted for fast movement in the pelagic zone and are likely to have been predators. Cyclopygids were the high-latitude equivalents of *Opipeuter* and *Carolinites*.

A pelagic mode of life has also been proposed for many agnostid trilobites (Robison, 1972). These small, blind, isopygous Cambro-Ordovician forms have only two thoracic segments (Fig. 11.18) and their limb morphology (Chapter 12) differs from that of other trilobites.

The global distribution of these agnostids (where best known in the Middle Cambrian) is very wide; the same species occur in Scandinavia and western North America. This contrasts dramatically with

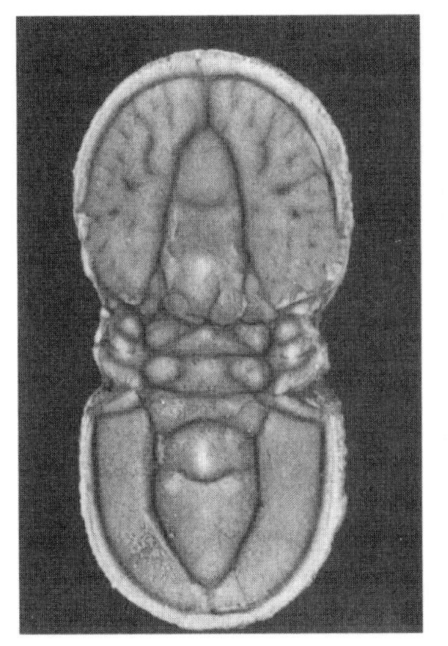
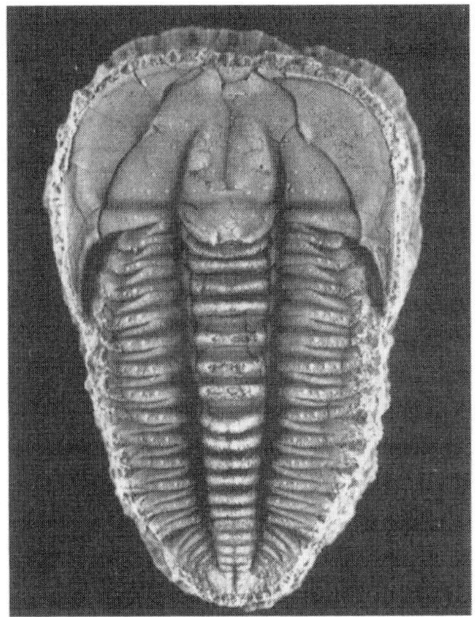
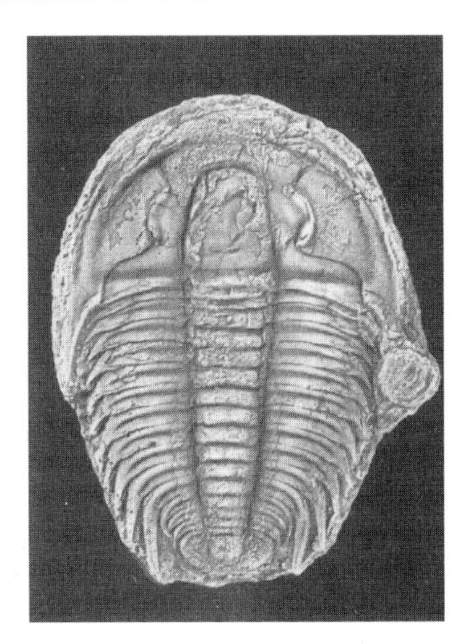

Figure 11.18 Agnostid (miomerid) and polymerid trilobites from the Middle Cambrian of Utah: (a) *Ptychagnostus atavus* (× 7); (b) *Brachyaspidion microps* (× 5); (c) *Modocia typicalis* (× 2). Agnostids can be used for intercontinental correlation, and may have been pelagic; this can be tied in with a stratigraphic scale for the Laurentian continent, based upon the benthic polymerids. (Photographs by courtesy of Dr R.A. Robison.)

that of contemporaneous polymerid trilobites, which remain endemic to their own continental masses. Thus the polymerids of the Laurentian continent (at that time firmly anchored on the equator) is quite different from that of Baltica (isolated in southern latitudes), yet the agnostids are the same. It seems reasonable to conclude that while the polymerids were benthic, the agnostids were pelagic, but the latter do not appear adapted for a planktic life. Possibly they had very extended larval stages, but some indeed may have swum in a partially enrolled condition in the upper waters of the sea; others may have adopted different modes of life.

Trilobites were adapted to specialized ecological niches (Thomas and Lane, 1984) and are usually found in characteristic associations, confined to particular sedimentary environments or bathymetries. Of the many synecological studies made recently, those of Henry (1989; Ordovician), Thomas (1980), Chlupac (1987) (Silurian) and Chlupac (1983) (Devonian) are recommended for further reading.

It is worth noting that 'opportunistic' trilobites have been recorded: these are species able to invade and flourish in resource-rich but otherwise inhospitable environments. They are usually recognized (Brezinski, 1985) by unusual abundance within a single bed.

Ecdysis and ontogeny (Fig. 11.19)

Like all other arthropods, trilobites had to moult periodically in order to grow. The cast shells (**exuviae**) were thrown off as the the animal ecdysed, being commonly disarticulated and broken along the cephalic sutures during the moulting process. Meanwhile, the soft shell which had formed below the old exoskeleton was inflated to a larger size and hardened. Occasionally the moulted exuviae remain intact or only slightly displaced into a particular and fairly constant relationship. Many different kinds of such moulting configurations have been described and different trilobites had particular techniques for escaping from the old exoskeleton (Henningsmoen, 1975; McNamara and Rudkin, 1983; Whittington, 1990). Sometimes, especially in very fine argillaceous sediments, a gradational size series can be collected which gives a partial or complete record of ontogenetic development from the earliest larval stages (Whittington, 1957). The first ontogenetic series known were described by Barrande in 1852; one of these, the Cambrian *Sao hirsuta*, is illustrated here (Fig. 11.19a) and shows the general system of growth.

The earliest stage is a **protaspis**, usually some 0.75 mm in diameter. This is usually a cambered disc, open ventrally, which carries a segmented

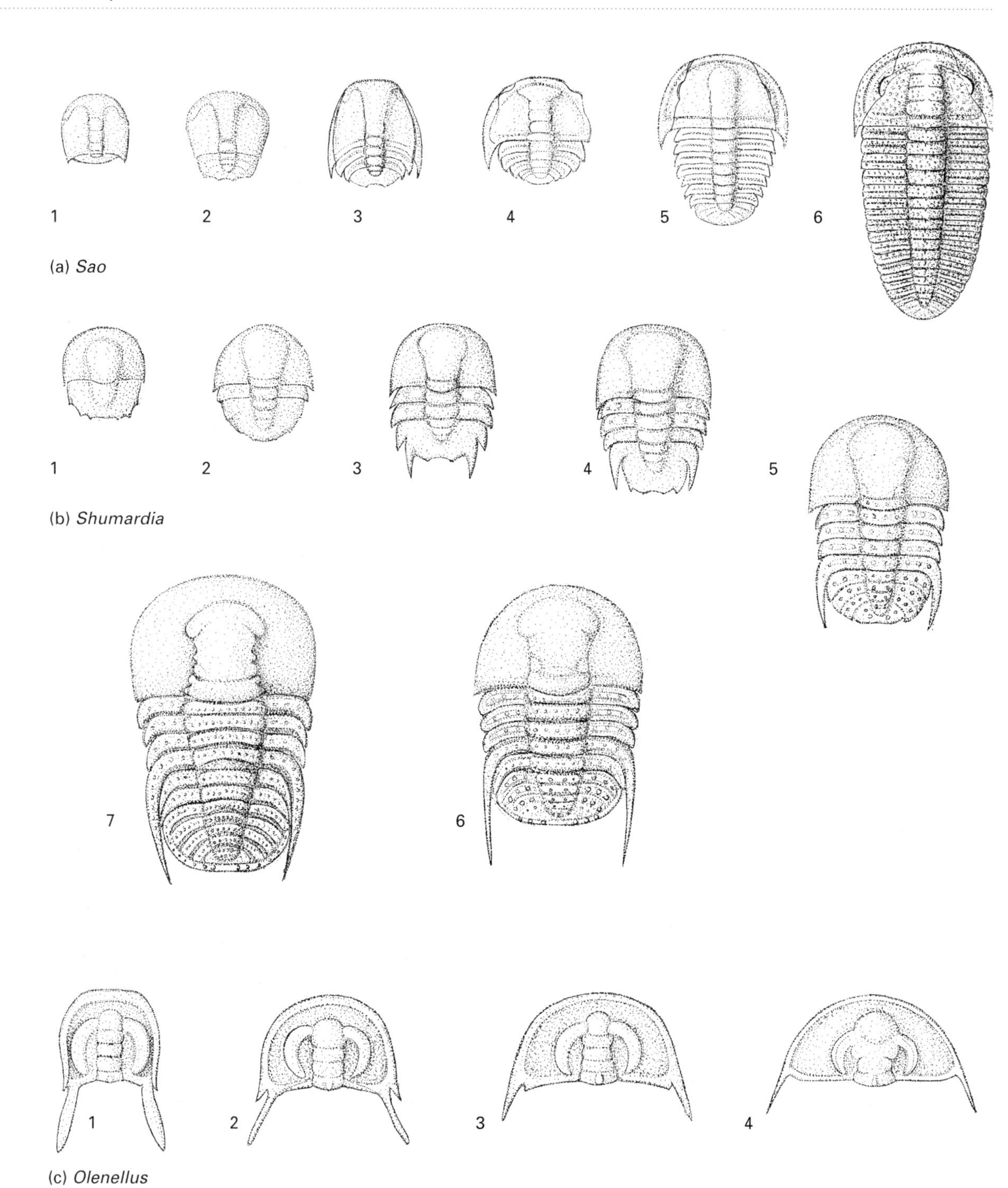
1
2
3
4
5
6
(a) *Sao*
1
2
3
4
5
(b) *Shumardia*
7
6
1
2
3
4
(c) *Olenellus*

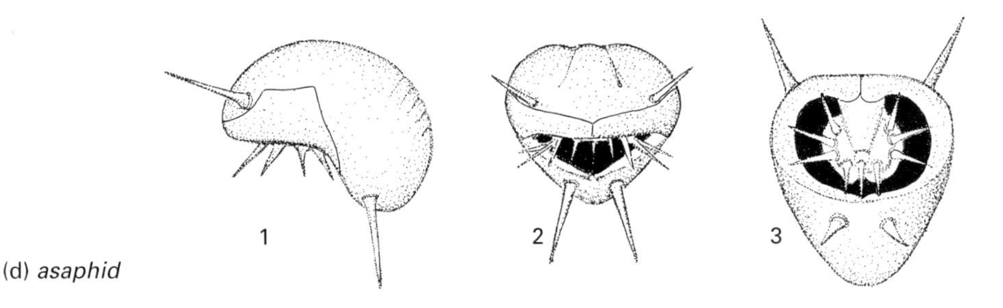

Figure 11.19 Trilobite ontogeny: (a) *Sao hirsuta* (M. Cam.), Bohemia, showing protaspides (1–3; × 14); meraspides (4, 5; × 10) of degrees 0 and 6; and holaspis (6; × 1.5); (b) *Shumardia (Conophrys) salopiensis* (L. Ord), Shropshire, England: (1–6) meraspides (degrees 0–5; × 30), showing the development of the macropleural spine; (7) holaspis (× 15); (c) *Olenellus gilberti*, (1) early meraspis (2, 3; × 8); intermediate stages (× 6, × 3); (4) adult cephalon (× 0.5). The larval (intergenal) spines are almost completely reduced in the adult and are not homolgous with the genal spines (5). (d) Lateral, anterior and ventral views of the late protaspis of an Ordovician asaphid, showing the spiny hypostome (× 40). [(a) After Barrande in *Treatise on Invertebrate Paleontology*; (b) redrawn from Fortey and Owens, 1991; (c) redrawn from Palmer, 1957; (d) redrawn from Evitt, 1961.]

central lobe, later to become the glabella. The eyes at this stage are tiny and are located on the anterior margin; later they migrate inwards bringing the facial suture with them. The hypostome is not known in *Sao*, but in genera such as *Asaphus* (Fig. 11.19d) and *Gravicalymene*, in which it has been described, it is extremely spiny, suggesting a functional change as the hypostomal spinosity diminishes. Many protaspides are generally spiny, and the spines may later disappear totally in the adult. Olenellid trilobites, for instance (Palmer, 1957; Fig. 11.19c), have protaspides with a pair of long blade-like spines, whose adult equivalent is merely a pair of tiny knobs lying within the genal spines, which are quite separate organs developing later.

As the protaspides grow by further moults a transverse furrow develops, separating the larval cephalon from the presumptive pygidium; the two can freely articulate against one another. The **meraspid** stage begins when the pygidium becomes free. The thoracic segments then form in a zone of growth along the front of the pygidium. They are actually part of the pygidium for a while, and then they are released in turn and liberated from its anterior part. The pygidium is thus known as a **transitory pygidium** as it does not acquire its permanent identity until the last thoracic segment has been released. Meraspides are numbered in degrees – 0, 1, 2, 3, . . . – according to how many thoracic segments have been freed from the anterior edge of the transitory pygidium. The process is perhaps best seen in the development of *Shumardia (Conophrys) salopiensis* (Fig. 11.19b) from the Tremadocian of Shropshire, England, since the macropleural spines on the fourth thoracic segment are a good marker. These are liberated at meraspid degree four while the fifth and sixth segments are still being formed at the front of the transitory pygidium. Meanwhile the cephalon is acquiring adult proportions.

When the adult number of thoracic segments has been reached the trilobite is now a **holapsis**, though it may have to pass through many more moults before it is of fully adult proportions. Rarely, as in *Aulacopleura* which had up to 25 thoracic segments, new segments are added until a very late stage in development.

Some curious phosphatized objects found in association with cybelinind trilobites and smaller than the earliest protaspides have been interpreted as pre-protaspid larval stages (Fortey and Morris, 1978). These **phaselus** larvae are ovoid, vaulted bodies with a ventral free edge and doublure.

Exquisitely preserved details of structure in trilobite larval stages are now known from silicified specimens which has greatly added to our knowledge (e.g. Chatterton, 1980). Two basic kinds of protaspides have been distinguished (Speyer and Chatterton, 1989), highly inflated 'non-trilobite-like' forms such as those of *Asaphus* (Fig. 11.19d), and the more typical 'trilobite-like' cambered discs

(Fig. 11.19a) as in *Sao*. The former are believed to be planktic, the latter benthic.

Quite apart from their intrinsic interest, ontogenies are potentially useful in suggesting phylogenetic relationships between established taxa. If two distinct stocks have fairly similar protaspides, differing in characters from those of other stocks, they may share a common ancestry even if the adults look quite different. Thus when the ontogeny of the corynexochid trilobite *Bathyuriscus fimbriatus* was elucidated (Robison, 1967), it became clear that there were similarities between its protaspides and the protaspides of many ptychopariids. Furthermore, adult corynexochids show many characters that are found in the protaspid and meraspid stages of ptychopariids, and so it has been possible to establish a real phyletic relationship between the two groups, as well as showing that major differences in the holaspid morphology of the two groups have resulted mainly from differential growth rates.

Ontogenies in trilobites also point to the importance of paedomorphosis as an important factor in evolution, especially where new kinds of organization seem to have arisen very rapidly and without preserved intermediates (McNamara, 1978, 1983; Whittington, 1981). Thus the suture of trilobites can be traced through ontogeny as initially proparian, though it may later become opisthoparian. Since there is good evidence that phacopid schizochroal eyes arose paedomorphically, it is not unlikely that the proparian suture of the Phacopina is likewise paedomorphic.

How trilobites reproduced is unclear. The occasional discovery, however, of monospecific, age-sorted clustered assemblages (Speyer and Brett, 1987) suggests, as with some modern arthropods, that trilobites aggregated together prior to moulting and mass copulation.

Size–frequency analysis of trilobite populations have revealed different reproductive strategies. Thus *r*-strategists have high fecundity and substantial juvenile mortality; the population curve is left-skewed. *K*-strategists on the other hand, such as *Ogygiocarella*, an inhabitant of fairly deep water, have broad, flat distribution curves (Sheldon, 1988). Here, fecundity, and hence recruitment, is low, small individuals never formed more than a small part of the total fauna, and the population was maintained in a steady state. Such strategies were probably characteristic of particular environments.

Classification (Fig. 11.20)

The classification given below is a modified version of that in the forthcoming second edition of the *Treatise on Invertebrate Paleontology*. It is based upon the whole complex of axial and other characters, without any one or more being singled out as being of overriding importance. Virtually all the characters used by the earlier systematists are herein incorporated, but in a more realistic perspective. In addition, characters such as resemblances at early ontogenetic stages become increasingly important. The character and location the hypostome are now seen being as of particular importance for higher level classification (Fortey, 1990).

CLASS TRILOBITA (Cam.–Perm.): Marine arthropods with largely calcitic exoskeleton divided into three longitudinal lobes and with a distinct cephalon, articulated thorax and pygidium. Cephalon with primitively furrowed glabella, facial sutures, compound eyes (sometimes lost) and doublure with free or attached hypostome. From 2 to 40 thoracic segments, each with axis and pleura. Pygidium variable in size and shape. Spinosity variable. Ventral appendages, apart from uniramous antennae, are one pair per segment, biramous and all of the same kind, decreasing in size posteriorly.

ORDER 1. REDLICHIIDA (L.–M. Cam.): An early group of trilobites with a large semicircular cephalon having strong genal spines, numerous and usually spiny thoracic segments and a tiny pygidium. Eyes large, hypostome conterminant. Includes SUBORDERS 1. OLENELLOIDEA (no dorsal sutures), e.g. *Olenellus*; 2. EMUELLOIDEA, e.g. *Emuella*; 3. REDLICHOIDEA, e.g. *Redlichia*; 4. PARADOXIDOIDEA, e.g. *Paradoxides* – dorsal sutures present.

ORDER 2. AGNOSTIDA (L. Cam.–U. Ord.): Small trilobites with subequal cephalon and pygidium, usually blind, sutureless, enrollment typical. Thoracic segments number only two (SUBORDER 1, AGNOSTINA [two subfamilies]) or three (SUBORDER 2, EODISCINA) [three subfamilies including PAGETIIDAE; the only group to have eyes and sutures] Hypostome natant, with ribbon-like wings; e.g. *Agnostus*, *Eodiscus*, *Pagetia*.

ORDER 3. NARAOIIDA (M. Cam.): Uncalcified trilobites with no thoracic segments; e.g. *Naraoia*, *Tegopelte*.

ORDER 4. CORYNEXOCHIDA (L. Cam.–M. Dev.): A rather heterogeneous group. Glabella of varied form but usually parallel sided or expanding anteriorly; sutures opisthoparian with subparallel anterior branches, thorax usually with seven or eight segments, often isopygous. Hypostome conterminant to impendent, may be fused with rostral plate. Includes SUBORDERS 1. CORYNEXOCHIINA, with conterminant hypostome (e.g. *Olenoides*, *Zacanthoides*, *Bathyuriscus*); 2. SCUTELLUINA with impendent hypostome (e.g. *Scutellum*, *Illaenus*, *Bumastus*, *Phillipsinella*); 3. LEIOSTEGIINA (e.g. *Chuangia*).

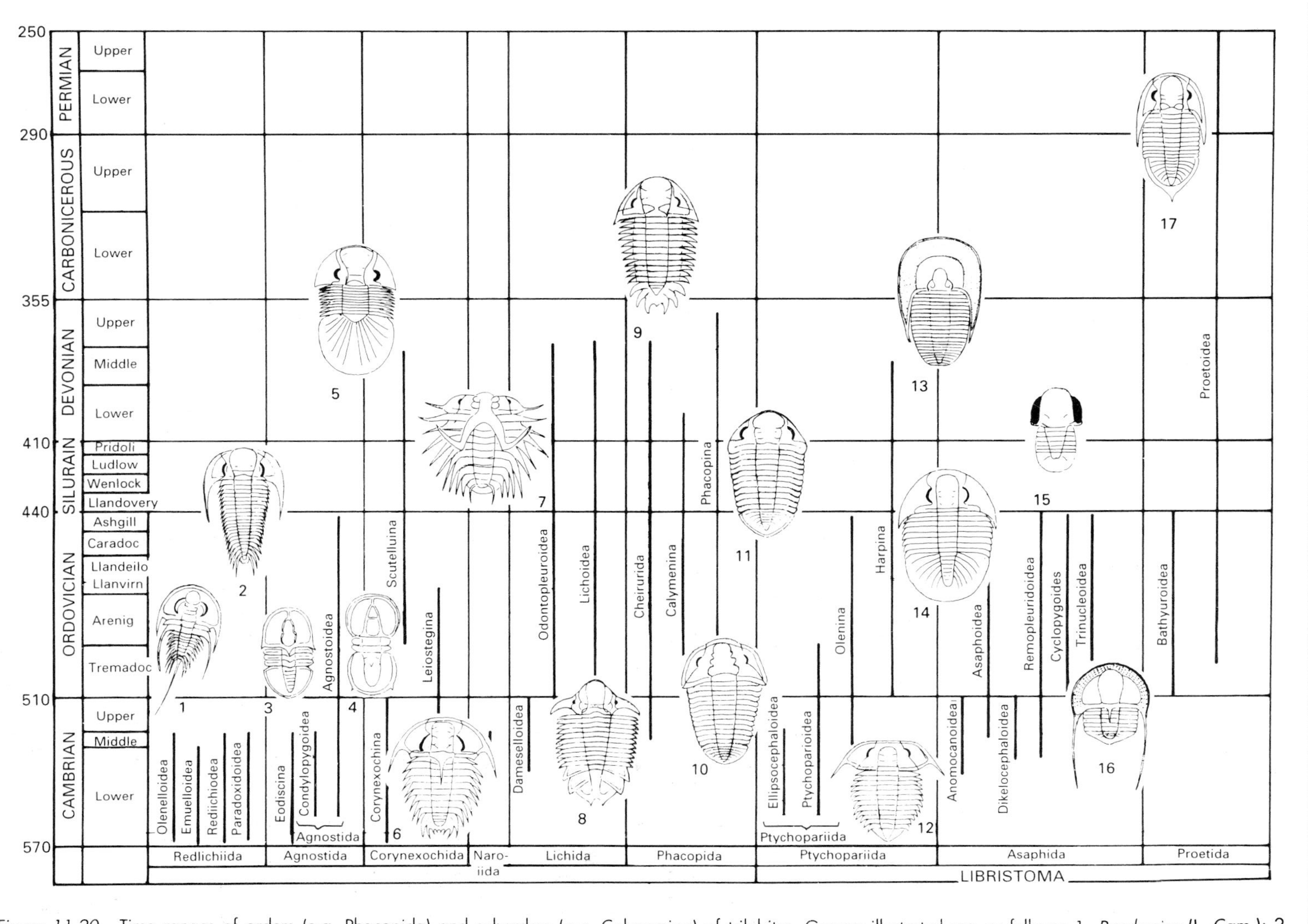

Figure 11.20 Time ranges of orders (e.g. Phacopida) and suborders (e.g. Calymenina) of trilobites. Genera illustrated are as follows: 1, *Paedumias* (L. Cam.); 2, *Paradoxides* (M. Cam.); 3, *Eodiscus* (L. Cam.); 4, *Agnostus* (U. Cam.); 5, *Scutellum* (L. Dev.); 6, *Kootenia* (M. Cam.); 7, *Dicranurus* (Dev.); 8, *Trochurus* (U. Ord.–M. Sil.); 9, *Cheirurus* (U. Ord.–M. Sil.); 10, *Calymene* (Sil.–Dev.); 11, *Acaste* (Sil.); 12, *Olenus* (U. Cam.); 13, *Harpes* (M. Dev.); 14, *Ogygiocaris* (M. Ord.); 15, *Cyclopyge* (Ord.); 16, *Trinucleus* (Ord.); 17, *Paladin* (L. Carb.).

ORDER 5. LICHIDA (M. Cam.–M. Dev.): Usually spiny trilobites, with conterminant hypostomes. Superfamilies 1. LICHOIDEA (L. Ord.–U. Dev.), medium-sized to very large trilobites, with unmistakably distinctive cephala and pygidia. Cephalon with broad glabella, often with fused lateral and glabellar lobes, opisthoparian sutures. Pygidium often larger than cephalon; tuberculate exoskeleton (e.g. *Lichas*, *Hemiarges*, *Terataspis*); 2. ODONTOPLEUROIDEA (U. Cam.–M. Dev.). Very spiny trilobites. Glabella with three pairs of lateral lobes, outside which on the librigenae lie the ocular ridges, sutures opisthoparian, hypostome small. Short spiny pygidium, tuberculate or spiny exoskeleton (e.g. *Acidaspis*, *Leonaspis*, *Ceratocephala*); 3. DAMESELLOIDEA (M.–U. Cam.): East Asian, usually spiny trilobites; probably sister group to odontopleuroids.

ORDER 6. PHACOPIDA (L. Ord.–U. Dev.): Large order of dominantly proparian trilobites, divided into three clearly defined suborders:

SUBORDER 1. CHEIRURINA (L. Ord.–M. Dev.): A very variable, normally proparian group. Glabella with up to four pairs of glabellar furrows and usually expanding forwards, small holochroal eyes, rostral plate present, hypostome free. Thorax with eight to 19 segments. Pygidium usually lobed or spiny; e.g. *Cheirurus*, *Deiphon*, *Ceraurus*, *Sphaerexochus*.

SUBORDER 2. CALYMENINA (L. Ord.–M. Dev.): A fairly homogeneous, usually gonatoparian group characterized by an anteriorly tapering glabella with four or five pairs of lateral lobes, diminishing in size forwards. Eyes small, holochroal. Thorax with 11–13 segments. Pygidium rounded or subtriangular. In FAMILY HOMALONOTIDA the axis is very broad and the furrows are largely effaced or very shallow; e.g. *Calymene*, *Trimerus*.

SUBORDER 3. PHACOPINA (L. Ord.–U. Dev.): Cephalon with forwardly expanding glabella, schizochroal eyes, proparian sutures and no rostral plate. Thorax of 11 segments. SUPERFAMILY PHACOPACEA has a small pygidium and well developed enrollment mechanisms. Members of SUPERFAMILY DALMANITIDAE are larger, isopygous and large-eyed and have enrollment structures poorly developed; e.g. *Phacops*, *Dalmanites*, *Acaste*, *Chasmops*.

SUBCLASS LIBRISTOMATA (Fortey, 1990): Includes all trilobites with natant hypostomes, together with secondarily conterminant or impendent trilobites with natant ancestors.

ORDER 7. PTYCHOPARIIDA (L. Cam.–U. Dev.). A large, paraphyletic trilobite order, all with natant hypostomes and including many highly modified groups. 'Typical' ptychoparoids such as SUBORDER 1. OLENIDA have a simple, forwardly tapering glabella with normally straight glabellar furrows, large thorax and small pygidium (e.g. *Olenus*, *Leptoplastus*, *Triarthrus*). [SUPERFAMILIES 1. ELLIPSOCEPHALOIDEA and 2. PTYCHOPARIOIDEA also belong here (e.g. *Ellipsocephalus*, *Ptychoparia*, *Conocoryphe*, *Shumardia*)]. SUBORDER 2. HARPINA have pronounced cephalic fringes; e.g. *Harpes*.

ORDER 8. ASAPHIDA (U. Cam.–Sil.): Median ventral suture present, only primitive forms retain natant conditions, most being conterminant or impendent. Higher asaphids have a distinctive inflated protaspis (Fortey and Chatterton, 1988). SUPERFAMILIES 1. ANOMOCAROIDEA (e.g. *Anomocaris*); 2. ASAPHOIDEA (e.g. *Asaphus*, *Ceratopyge*); 3. DIKELOCEPHALOIDEA (e.g. *Ptychaspis*); 4. REMOPLEURIDOIDEA (e.g. *Remopleurides*); 5. CYCLOPYGOIDEA (e.g. *Cyclopyge*, *Nileus*); 6. TRINUCLEOIDEA (e.g. *Trinucleus*, *Dionide*, *Ampyx*, *Orometopus*).

ORDER 9. PROETIDA (Ord.–Perm.): Glabella large and vaulted, well defined, usually with genal spines, narrow and backwardly tapering rostral plate, opisthoparian, eyes holochroal and usually large, long hypostome. Thorax with eight to 10 segments; isopygous; pygidium usually furrowed and not spiny (Fortey and Owens, 1975). Distinctive proetoid protaspis. Two SUPERFAMILIES, 1. PROETOIDEA; 2. BATHYUROIDEA (e.g. *Proetus*, *Phillipsia*, *Bathyurus*, *Aulacopleura*).

Evolution

General pattern of evolution

Almost nothing is known about the ancestors of trilobites. In common with other arthropods it may be presumed that trilobites were possibly derived from the same ancestors as the annelids, and the Ediacara 'soft-bodied trilobites', recently described (p. 60, Fig. 3.1j), may have been somewhere in the ancestral line, but this is only speculation. The earliest trilobite group to appear is Suborder Olenellina, which is micropygous. These, together with the secondarily blind Agnostida, which appeared a little later, diversified through the Lower Cambrian and were joined by the Corynexochida and Ptychopariida, which continued through after the demise of the Olenellina at the end of the Lower Cambrian. The Upper Cambrian fauna is dominated mainly by the Ptychopariina which show, at least superficially, relatively limited variation, by contrast with that exhibited by many Ordovician taxa. This could point either to a period of evolutionary stagnation in which there was both limited genetic potential and low selection pressure, or alternatively to a rather rigorous selection weeding out of all but the morphologically defined forms. The former alternative is favoured, since in the few groups in which evolution has been studied in detail (e.g. the Olenidae) there is, in spite of the retention of a standard model of organization, a fair degree of evolutionary plasticity within the group.

At the end of the Cambrian there was a major crisis in trilobite history. Most of the comparatively unspecialized Upper Cambrian stocks died out, few reaching the Ordovician (Whittington, 1966). The reasons for such extinction are unknown; perhaps the marine regressions of late Cambrian time and the rise of predatory cephalopods may be invoked as being amongst the causes. Some short-lived, rapidly evolving Upper Tremadocian groups then originated, and after their extinction came the first representatives of the important and dominant Ordovician groups: the Illaenina, Phacopida, Trinucleina and others, all highly differentiated and diverse, most of them of cryptogenetic origin, and surviving for various periods thereafter. It is an interesting feature of trilobite evolution that after this great burst of new constructional themes in the early Ordovician very few entirely new patterns of organisation arose; afterwards evolution in trilobites was very largely a matter of experimentation and development of the new genetic material that first came into being in the early Ordovician.

It seems clear that these strong and dominant new Ordovician stocks were able to colonize and exploit new environments in a way that had not been possible for their Cambrian progenitors. For the first time, for example, there are many genera colonizing reef environments, some persisting for a long time, others being replaced by different genera of broadly similar morphology. These reef faunas are often difficult to correlate stratigraphically with contemporaneous faunas living in other environments. The pure limestone faunas of early Ordovician age have a preponderance of large, rather flattened and rather lightly furrowed asaphids, for instance; when these died out, scutelluids of similar morphology occupied their ecological niche. Despite the success of the Ordovician trilobites as a whole, however, many distinct groups (e.g. Family Trinucleidae) became extinct by the end of the Ordovician. The survivors were long-ranged and presumably successful and versatile groups. Yet these 'well-tried' groups were not devoid of diversification potential, for there is evidence of several adaptive radiations in particular isolated areas of the world, where an original ancestor invading a fairly empty environment had paved the way for a burst of evolutionary diversification in its descendants.

The trilobite faunas of the Silurian and Devonian contain many of the same elements: phacopids, dalmanitids, cheirurids, calymenids, harpids, etc. Consequently a Silurian assemblage can easily be confused, at least superficially, with a Devonian one. Encrinuridae (Suborder Cheirurina), however, though so characteristic of many Silurian faunas, do not extend far into the Devonian.

The extinctions of the Middle and Upper Devonian disposed of the majority of trilobite groups, yet the remaining Proetida went all the way through to the Upper Permian. Most of these, especially those living in shallow waters, were large-eyed, robust genera capable of enrollment, but there were also specialized, thin-shelled, lightly constructed and often eyeless forms, especially in the deep water of the Variscan fold belt in northwestern Europe. These are found from the Upper Devonian, where reduced-eyed proetids and phacopids are present, and through the Carboniferous in the Culm facies.

Extinction of the last Permian trilobites was probably related to the lowering of sea level that contemporaneously affected so many other invertebrates.

In general, some fairly clear evolutionary trends can be distinguished (Fortey and Owens, 1990a,b), such as the origin of new eye types, the improvement of superior enrollment and articulating mechanisms, the change from micropygy to isopygy, the development of extreme spinosity in certain groups and the reduction in the rostral plate. Yet trilobites as a whole remained constructed on the same archetypal plan defined in the earliest Cambrian and, especially after the early Ordovician, the number of changes of real significance remained surprisingly low.

Microevolution

There have been many studies on trilobite evolution (summarized in Eldredge, 1977), especially on Upper Cambrian, Ordovician, Devonian and Lower Carboniferous genera; of these only three are selected for further discussion. Other case histories are documented in Clarkson (1988).

Olenidae

Trilobites of Family Olenidae occur in vast numbers in the Upper Cambrian of Scandinavia and also in Britain, North America and Argentina. They have been studied in great detail and are of great stratigraphical value. Their evolution spans over 40 million years, extending into the Ordovician.

The 'olenid sea' reached from Eastern Canada to Scandinavia. It appears to have been fairly shallow, but may have been largely stagnant near the bottom or just below the sediment surface and the olenids seem to have been adapted to dysaerobic environments. Some of them may have been adapted to bottom-dwelling and feeding off the nutrient-rich flocculent layer on the sea floor.

The overall pattern of olenid evolution in Scandinavia has been splendidly documented by Westergård (1922) and Henningsmoen (1957), who deal with Upper Cambrian Swedish and Norwegian faunas, respectively. Fortey (1974b) has described a rich Lower Ordovician olenid fauna from Spitzbergen and Clarkson (1973) dealt with eye morphology. A number of olenid subfamilies have been distinguished, of which the most important five are illustrated here (Fig. 11.21).

Subfamily Oleninae is the earliest, and its type genus *Olenus* dominates the lowest two (of six) Upper Cambrian zones in Scandinavia. *Olenus* gave rise to Leptoplastinae and Pelturinae which proliferated and became very diverse in the succeeding zones, along with some derived Oleninae. Some genera, such as the leptoplastine *Ctenopyge*, became very spiny. There was a relatively rapid turnover in olenid genera and species, most of which are of great stratigraphical value. At the end of Upper Cambrian time the Olenid Sea contracted, and then expanded in the Tremadocian. The olenids of Tremadocian time were again quite variable, and it was at this time that Subfamily Triarthrinae arose, which became dominant in the early Ordovician in North America. Subfamily Balnibarbiidae, on the other hand, which are known only from Spitzbergen, evolved rapidly in virtual isolation in the early Ordovician, a product of a remarkable burst of evolutionary activity. Only a few pelturines and triarthrines and members of smaller olenid subfamilies still remained at this time, subordinate to the Balnibarbiidae. Many studies have been made of the overall pattern of evolution in olenids, and a

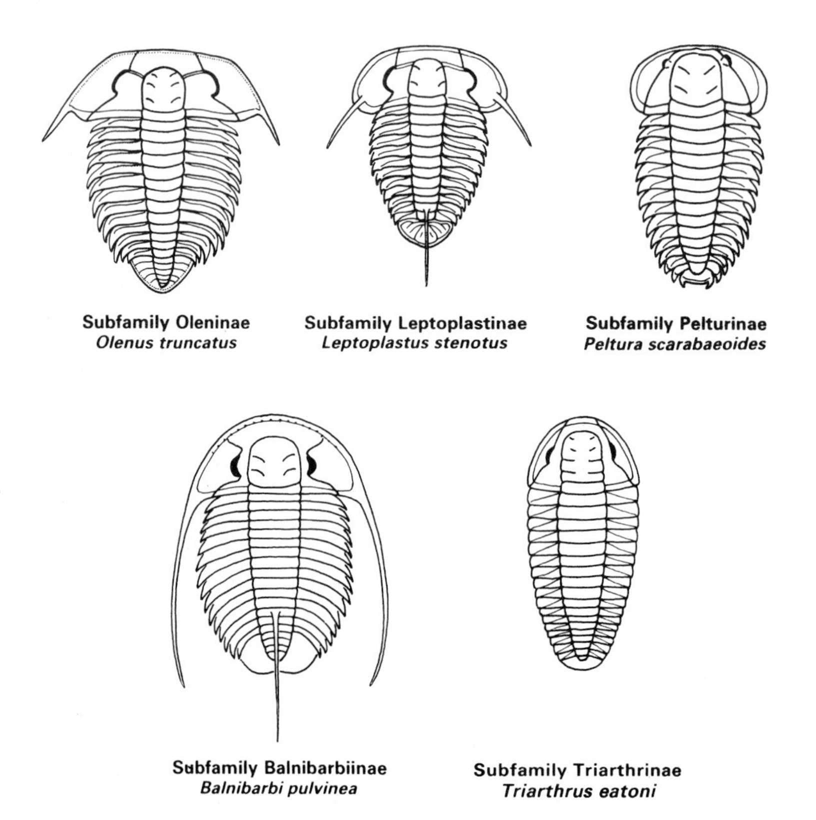

Figure 11.21 The trilobite family Olenidae: representative genera and species of each of the five main subfamilies.

very important contribution to our knowledge of speciation in fossils came from Kaufmann (1933). He made a statistical study of *Olenus* in a 2.5 m part of a condensed sequence of one of the lower zones of the Upper Cambrian. The pygidia were found to be particularly valuable, and he was able to elucidate changes in pygidial shape in no less than four distinct lineages. Where these trends were apparent within individual species he distinguished the different stages anteriosus, medius and posteriosus, going from conservative to progressive, respectively. In all of these the character mean in the population shifts gradually, and in each case there is a general progression in pygidial shape from broad and short to long and narrow. Lineages 3 and 4 overlap; the others are distinct. It was postulated that the parent lineage of *Olenus* was at this time living elsewhere, and that each of the four distinctive lineages actually seen in the section began with an invasion of an evolutionary offshoot of the parent population into the area. The subsequent development of *Olenus* through time generally followed an iterative pattern of evolution, where similar trends arise in successive stocks from a more or less unchanging ancestor. The morphological characters of the ancestral stock are not known with certainty but can be inferred from the earliest members of lineages 1, 2 and 4. Lineage 3, which partially overlaps with lineage 4, has somewhat different characters and seems to have come in from a related but distinct ancestor, probably living elsewhere. This suggestion of the origin of new groups from peripheral isolates of a main population is amply confirmed by other trilobite evolutionary studies, though only in lineage 3 did the observed trends lead to the development of a new species.

Phacops rana complex

Eldredge's (1972b) study of evolution in the Middle Devonian *Phacops rana* of North America describes a classic case of allopatric speciation (Chapter 3). *P. rana* was widespread over North America during Middle Devonian times, having probably been derived from a European ancestor such as *P. latifrons*. To the west of the area in which *P. rana* lived, *P. iowensis* resided, the two species being more or less mutually exclusive. Whereas *P. iowensis* is never very abundant and showed little evolutionary change throughout the Middle Devonian, there was much greater variability among the subspecies of *P. rana*, of which individuals are often very common. These subspecies are not very easy to distinguish, their definition being mainly based on eye morphology, and an extensive multivariate analysis was used in elucidation of this history. Eldredge's study spanned a large area of eastern North America, in which the sediments both of the shallow sea overlying the continent and of the marginal 'exogeosyncline' were represented. The region is well correlated stratigraphically, three stages being present. In the oldest (Cazenovian) stage *P. rana crassituberculata* and *P. rana milleri* were both present (conceivably they could be sexual dimorphs of a single species) in the epicontinental sea to the west, while *P. rana rana* lived contemporaneously in the eastern area. In the extreme west *P. iowensis* was resident. When the Cazenovian trilobites had become extinct *P. rana rana* spread from the east into the epeiric sea, and when this subspecies died out at the end of the Tioughniogan stage *P. rana norwoodensis*, a new invader from the eastern exogeosyncline, replaced it. Thus the trilobites of the epicontinental sea, though they appear to form a successive evolutionary sequence in which the eye is reduced, are actually derived from an ancestral population living in the east, as a series of invaders, originally peripheral isolates of a parental population which as in the case of the olenids changed relatively little throughout time. This study also illustrates Eldredge and Gould's model of evolution (e.g. Eldredge and Gould, 1972) as a series of 'punctuated equilibria'. The individual populations of *P. rana* once established were relatively stable and underwent little evolutionary change. It was only when a parent population became extinct that the descendants of small, rapidly evolving populations living as gamodemes peripheral to the main ancestral population could opportunistically invade the now-vacant territory. These then became the dominant widespread species: a successful balanced population with new and stable character states.

Sheldon's Welsh trilobites

Ordovician (Llandeilian) trilobites in Central Wales occur in great numbers in a virtually continuous series of black shales. Sheldon (1987) studied a sequence spanning some 3 Ma, in which there are eight common trilobite lineages (Fig. 11.22).

In all of the eight genera, measured from 15 000 specimens, there was a net increase in the number of pygidial ribs, a character used in species diagnosis. It is a striking example of gradual evolution occurring in parallel in the various genera. Equally, Sheldon demonstrated that there are character reversals from time to time, such as temporary decrease in rib number. There is no reason why character reversals such as this should not take place, and here they have been clearly demonstrated. This is one of the best examples of gradualistic genetic change known from the fossil record, though the selection pressures that caused it remain uncertain.

Faunal provinces

Trilobites, as much as any other faunal group, show provincial differences throughout the Palaeozoic which can be useful indicators of the biogeography of that time. There seems to have been fairly well-marked provincialism in the Cambrian and especially in the earlier Ordovician. This provincialism decreased throughout the later Ordovician until, by the late Ashgillian, trilobite faunas were more or less cosmopolitan, a condition that persisted through the Silurian. In the early Devonian separate provinces began once more to differentiate, most particularly a southern Malvino-Kaffric province, but following the extinctions of late Devonian time provinciality becomes less easy to trace.

The distribution of trilobites in the Cambrian is complex (Cowie, 1971). In a broad and general sense it appears that there were two Lower Cambrian main faunal realms, characterized by olenellid and redlichiid trilobites, respectively. By the Middle and Upper Cambrian the (possibly) pelagic agnostids had a very broad distribution while endemic benthic polymerids remained confined to their isolated continental blocks.

Trilobite distribution in the Ordovician has been fully explored, and attempts to work out the controls of distribution have on the whole been quite successful (Whittington and Hughes, 1972; Cocks and Fortey, 1990). It is known that the Lower Ordovician trilobites are clearly distributed in provinces, apart from a very few genera of cosmopolitan distribution. Some of these widespread trilobites (e.g. *Geragnostus*, *Telephina*, *Seleneceme*) could well have been planktonic genera. Otherwise no less than four provinces have been defined, named after characteristic groups or genera. Each province has endemic families and endemic genera of more widely distributed families. These provinces, when plotted on new palaeogeographical maps (McKerrow and Scotese, 1990), accord very well with Early Ordovician continental patterns,

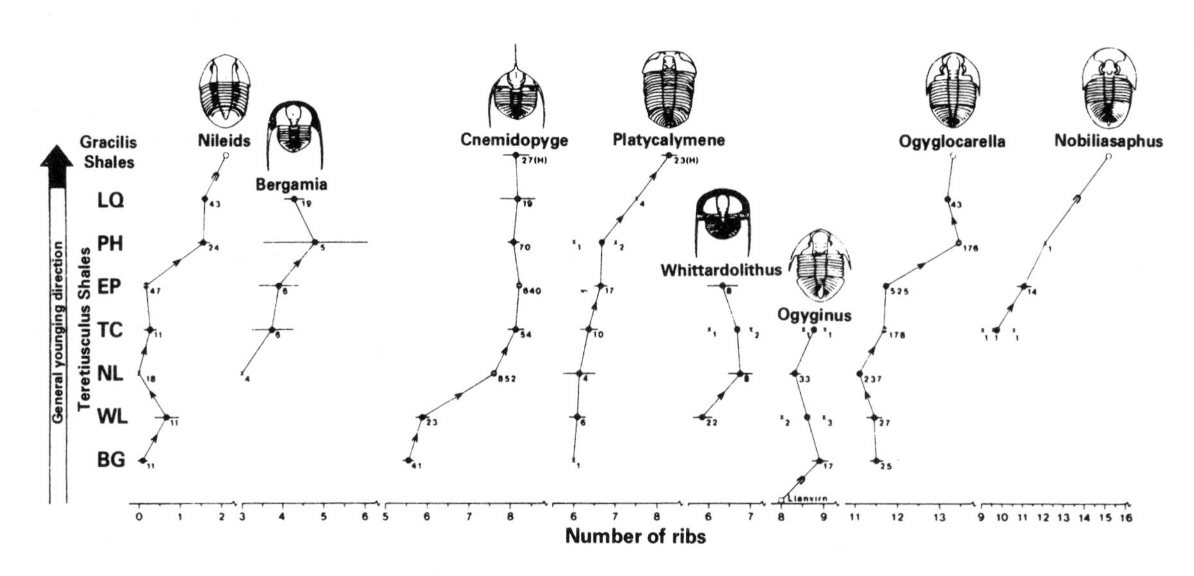

Figure 11.22 Gradual change in rib number in eight contemporaneous trilobite lineages from the Ordovician of Wales. (Redrawn from Sheldon, 1987.)

and confirm the importance of palaeolatitude as a primary control. The vast continent of Gondwanaland spread from the South Pole to the equator and displays a temperature-controlled latitudinal cline. First, the cold-water high-latitude shelves of this continent carried an indigenous '**calymenacean–dalmanitacean**' fauna (e.g. France, Spain, Central Europe and Turkey). Second, the low-latitude shelves of Gondwana (now south China and Australia) carried a '**dikelocephalinid**' fauna. Third, the tropical platforms of Laurentia (North America, Siberia and north China), carried a fauna of **bathyurid** trilobites. Fourth, the isolated Baltic–Russian platform, lying at intermediate latitudes, bears an endemic fauna of **asaphid** trilobites. Other fossils, e.g. brachiopods and molluscs, are endemic to these same four provinces.

In the later Ordovician the separate provinces tended to mingle as a result of the closer approach of the Gondwanan and Baltic continents. By the Caradoc only the 'calymenacean–dalmanitacean' province retained its integrity. The fate of some of the microcontinents can be traced by means of their faunas. The small continental fragment of Avalonia (mainly southern Britain), for example, is known to have broken off Gondwana, migrated northwards and united with Laurentia.

Towards the end of the Ordovician a polar ice sheet formed and greatly expanded during the Hirnantian (the final stage). This caused widespread extinction and is considered to be the primary cause of the breakdown of provinciality around the Ordovician–Silurian boundary.

As well as shelf faunas during the Ordovician there were deeper-water benthics marginal to the palaeocontinents (e.g. the olenid biofacies in the early Ordovician); such deeper-water faunas are less evidently controlled by palaeolatitude and continental configurations.

Following the Hirnantian glaciation, faunas did not recover very quickly. Because the continents were close together by this time tropical/temperate faunas became widely distributed; hence, apart from a cold-water *Clarkeia* fauna, Silurian faunas seem to have been more or less cosmopolitan. The differentiation of the Malvino-Kaffric province in Siluro-Devonian times (South America, Falkland Islands and southernmost Africa), which is so well elucidated in the brachiopods, appears also to be substantiated in trilobites, though this is less well known. Derivatives of an acastid stock migrated into an almost empty region, and diversified exceedingly, constituting one of the most remarkably distinct of all faunal provinces (Eldredge and Ormiston, 1979). Provinciality in the later rocks is poorly known but does not seem to have been great.

In Arctic North America cosmopolitanism began to decrease after Gedinnian time, so the Devonian as a whole seems to have been a time of increasing provinciality.

Stratigraphical use

Trilobites are of considerable stratigraphical value in the Cambrian and Ordovician but much less so in later rocks, except perhaps locally. The Cambrian system is zoned almost entirely on trilobites, other than the basal (Tommotian) zone. Cambrian trilobites obey most of the requirements of good zone fossils, being abundant and easily recognizable and often having short time ranges and wide horizontal distribution. They are limited by being distributed in the faunal provinces described above and by being rather facies controlled. Though good intraprovincial sequences are established, it is only possible to correlate between them where faunas are mixed or by using pelagic trilobites.

The Ordovician system is zoned by graptolites (q.v.), established for the offshore sequences. However, the stratigraphical subdivision of the nearshore shelly facies is based upon trilobites and brachiopods, which give a refinement of correlation of special merit in the Caradoc and Ashgill. In the British type area no stages have yet been defined in the Arenig, Llanvirn and Llandeilo, but only Upper and Lower divisions (the Llandeilo has a middle division). In the Caradocian, however, eight stages have been erected, based mainly on brachiopods though with trilobites (especially trinucleids) giving confirmatory evidence. In the Ashgillian many kinds of trilobites have been used to define several zones within the four known stages, so effectively that selected areas of lithological monotony have been mapped using the trilobite faunas as markers. There are still problems with correlation using Ordovician trilobites, mainly because of facies control and provinciality; there is still, for example, no close time correlation between the faunas of various

Upper Ordovician reefs – Dalarna (Sweden), Keisley (northern England) and Kildare (Ireland) – and those of contemporaneous bedded carbonate or mudstone facies.

In Silurian and later rocks trilobites are not of great stratigraphical value, for other fossils have generally shorter time ranges or may otherwise be used more effectively, such as the Devonian ammonoids and Carboniferous microfossils. On the other hand, for local correlation they are certainly of some use, and their value as stratigraphical indicators may increase (Thomas *et al.*, 1984).

11.4 Phylum Chelicerata

Chelicerates (Cam.–Rec.) include spiders, mites and scorpions, as well as the horseshoe crab *Limulus* and its living and fossil allies, and the extinct eurypterids. This diverse group of animals which appear so heterogeneous are united by possessing the following:

an anterior **prosoma** of six segments which is equivalent to the fused head and thorax of other arthropods;
a posterior **opisthosoma** (abdomen) with 12 or less segments;
a pair of jointed pincers (the **chelicerae**) which give the subphylum its name and which are always present in the first (**pre-oral**) segment of the prosoma.

Two main groups within the chelicerates are of singular palaeontological and evolutionary interest.

CLASS 1. MEROSTOMATA (Cam.–Rec.): Aquatic chelicerates, often of large size. There are two subclasses: SUBCLASS XIPHOSURA which with its ORDER XIPHOSURIDA and ORDER AGLASPIDA includes the *Limulus* group and their relatives; and SUBCLASS EURYPTERIDA which includes the water scorpions of Palaeozoic time.
CLASS 2. ARACHNIDA (Sil.–Rec.): Dominantly terrestrial forms, the spiders, mites and scorpions; the latter bear a superficial resemblance to eurypterids but are not closely related. Though scorpions are known from the Silurian and mites and spiders from the Devonian, all are rare as fossils, other than perhaps the Tertiary spiders preserved in amber. Fossil scorpions are known in considerable morphological detail, but their study is beyond the scope of this book, and arachnids will not be considered further.

Class Merostomata

Subclass Xiphosura
Limulus

The characters of xiphosures are well displayed by the modern *Limulus:* one of three closely related genera which are the only living representatives of this subclass.

Limulus polyphemus (Fig. 11.23a–e), the type species, live in shallow waters along the northwestern Atlantic shores.

In dorsal view it has a vaulted semicircular prosoma, which as in all chelicerates is a fused cephalothorax articulating with a less vaulted plate-like abdomen, there being a narrow channel incised obliquely between the two. From the rear of the abdomen springs a strong terminal spine, the **telson**. The axial part of the prosoma (**cardiac lobe**) is defined by axial furrows so that it has a superficial similarity to a trilobite glabella. Outside the axial furrows and parallel with them are the **ophthalmic ridges** which curve anteriorly to join medially, at which point there are a pair of small **ocelli**. There are also **compound eyes** situated in the middle of the ophthalmic ridges. These eyes have many lenses, but their structure is quite unlike that of trilobites, the dioptric parts being composed of invaginations of the cuticle which form long, inwardly projecting cones. These are of parabolic constructions optimized for maximum light collection. Below these are apposition-type ommatidia, each with a marked excentric cell.

The abdomen articulates to the prosoma along a hinge which cuts across the primary segmentation. Axial ridges are present, and traces of segmentation are evident in the fringe of six movable spines. The strong styliform telson, ridged along its length, is nearly as long as the rest of the body and can turn in almost any direction.

The segmentation of the body is readily worked out from embryology, but is also clear in the ventral morphology of *Limulus*. The pre-oral segment carries the short, paired, three-jointed chelicerae, and the five pairs of post-oral appendages are arranged radially around the mouth within the indented ring formed by the anterior doublure. These five pairs of post-oral appendages are morphologically similar to one another and function as walking legs. Each has a basal spinose coxa and several joints. The first four

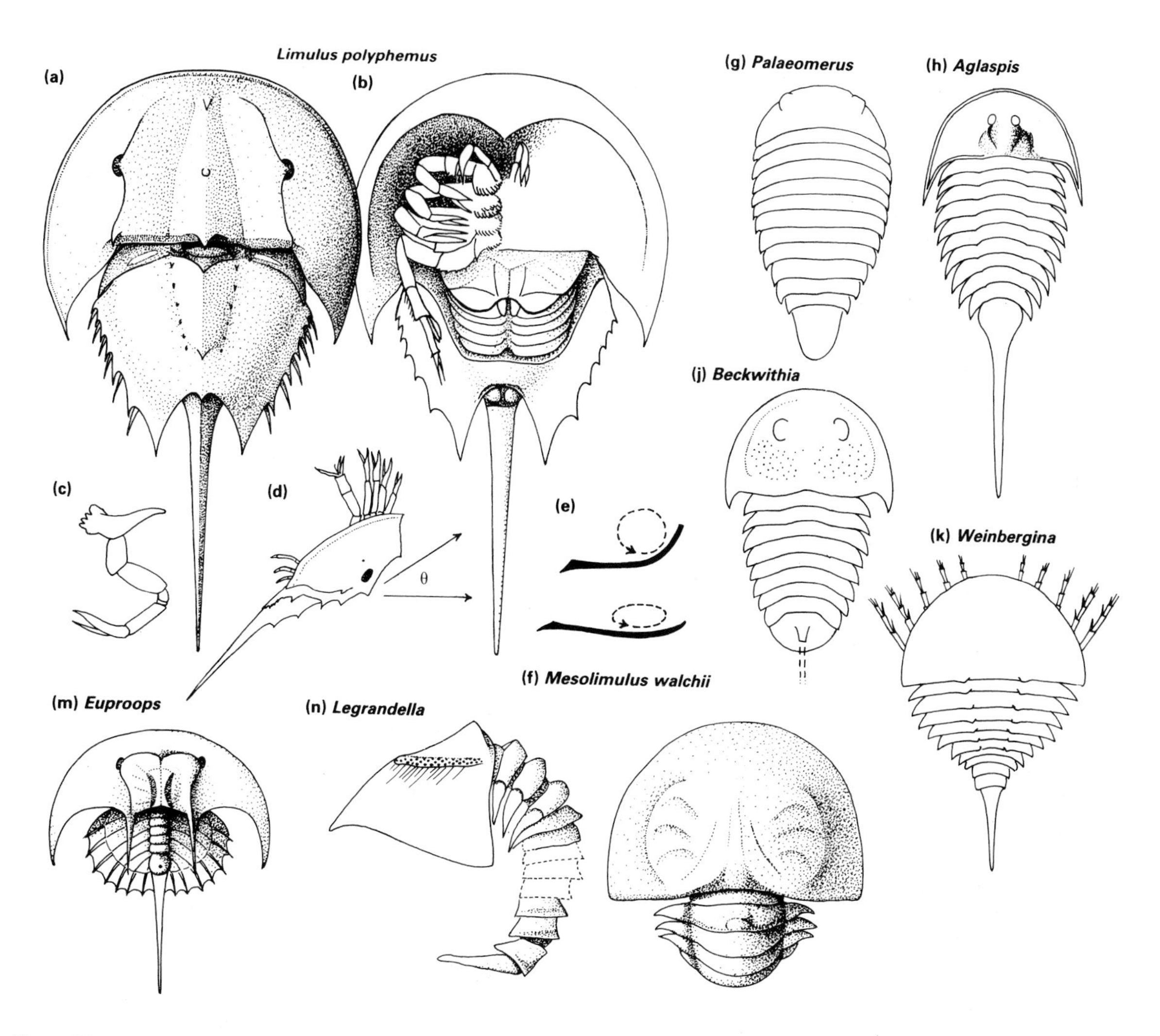

Figure 11.23 Merostomata. (a)–(e) *Limulus polyphemus* (Rec.): (a) dorsal view (× 1); (b) ventral view (× 1); (c) fourth prosomal appendage (× 1); (d) swimming upside down with the body inclined at an angle to the horizontal; (e) section through prosoma of swimming *Limulus* showing recirculating vortex; (f) *Mesolimulus walchii* (Jur.), section through prosoma (more flattened than that of *Limulus*) showing vortex; (g) *Palaeomerus* (L. Cam.); (h) *Aglaspis* (reconstructed; L. Cam.; ×1); (j) *Beckwithia* (L. Cam.; × 0.75); (k) *Weinbergina* (L. Dev.), Germany (× 0.75); (m) *Euproops* (U. Cam.; ×1); (n), (p) *Legrandella* (Dev.) in lateral and dorsal views (× 0.75). [(d)–(f) Redrawn from Fisher, 1975; (g) redrawn from Bergström, 1971; (h), (j) redrawn after Raasch, 1939; (k)–(m) redrawn from Störmer in *Treatise on Invertebrate Paleontology*, Part P; (n), (p) redrawn from Eldredge, 1974.]

pairs terminate in pincer-like **chelae**; the last pair are equipped instead with numerous spines and have a peg-like outgrowth (the **flabellum**) attached to the coxa.

Within the broad ventral doublure of the abdomen are six overlapping plates which are modified appendages. The first of these is the **operculum**, morphologically the eighth segment which, as in all chelicerates, bears the **genital openings**. The seventh segment is rudimentary and is apparently represented only by a pair of reduced plates. Behind the operculum are the succeeding five gill appendages which bear the respiratory **gills** on their inner (dorsal) side, where they are protected against desiccation.

Life habits of Limulus

Limulus is a shoreline inhabitant tolerant of a wide range of salinities, living in shallow waters and capable of crawling out for short distances on land. It can walk on the sea floor on the prosomal appendages, but it can also swim and right itself with the telson if overturned. It is an exceedingly versatile animal whose life habits have been used with some success in interpreting those of trilobites, to which it is the closest living relative and to which it has some morphological resemblances. However, there must be particular functional reasons for these similarities, and the specific uses of the various organs must be fully understood before such analyses are drawn.

SWIMMING *LIMULUS*

Limulus normally swims upside down (Fig. 11.23d), inclined at about 30° to the horizontal and moving at about 10–15 cm s^{-1}. It is propelled by both the prosomal and opisthosomal appendages. The opisthosomal and sixth pair of prosomal appendages move in anteriorly advancing waves of metachronal rhythm, while the first four pairs of walking legs move in phase, extending as they push backwards, withdrawing into the prosomal cavity or vault as they recover their position. This stroking cycle of the first appendages begins directly after the sixth prosomal pair have completed their stroke.

The hydrodynamics of flow in the swimming *Limulus* have been analysed using a model *Limulus* in a flow chamber (Fisher, 1975). In the normal inclined swimming attitude a strong recirculating vortex forms within the prosomal vault. This appears to break down or be shed at intervals into the wake of the animal prior to re-forming. The vortex is a normal consequence of flow dynamics for an object of such a shape and inclination, but it seems to be exploited by the animal in aiding the forward movement of the anterior prosomal appendages in the recovery stroke, which would otherwise have to push against the current. At greater or lesser inclinations the vortex would be less effective or absent (Fig. 11.23e,f).

Similar models of the Jurassic *Mesolimulus walchii*, which has a much more flattened morphology, show that an equivalent vortex would operate in a similar way provided that the animal's angle of inclination was less. Hence there is a direct relationship between the shape of the prosoma and swimming ability, the different vaultings of *Mesolimulus* and *Limulus* being different answers to the same problem. However, *Limulus* has more room in which to retract its legs and is more effectively adapted for burrowing.

BURROWING *LIMULUS*

Limulus is active at night, but during the day it burrows in below the sediment surface for as much as 12 h at a time (Eldredge, 1970). Adult specimens have an anterior arch, as do trilobites. At the start of the burial procedure the arch is first lowered so that the anterior edge of the prosoma comes in contact with the sea floor; it then digs into the sand as the prosomal legs perform normal walking movements. Sand is pushed backwards, covering the prosoma and the anterior third of the opisthosoma. The telson is buried and only makes horizontal movements thereafter. The walking legs excavate a burrow, pushing the sand up between the prosoma and opisthosoma. One of these channels remains open when the animal is completely buried. When this first stage of burial is effected, the animal is then quiescent and lies for a while oxygenating the gills. There is then a third stage, when the prosomal legs walk forwards once more while the opisthosoma is flapped vigorously, using the telson to some extent, and stirs up a cloud of sand which settles to cover the animal completely. The final stage is one of absolute quiescence, only a single respiratory channel being left open on one side.

It has been shown experimentally that the dorsal setae are used as mechanoreceptors which indicate to the animal when it is fully buried. The telson is covered from start to finish and is only used to a limited extent in the third stage. It functions mainly as a stabilizer or rudder in the walking or swimming individual or in righting one that is overturned. The comparative morphology of *Limulus* and trilobites in such details as the presence of dorsal sensors and the vaulted shape of the cephalon suggests that trilobites might have spent some of their time burrowed in sand, and this is amply borne out by trace fossil evidence.

Other xiphosures and their geological history

Xiphosures are rare in the fossil record and, other than *Limulus* and its modern relatives, are generally confined to non-marine sequences.

The oldest well-known xiphosures belong to the Cambrian Order Aglaspida, which are known very largely from the Upper Cambrian of southwestern Wisconsin (Raasch, 1939). These have a phosphatic exoskeleton in which the prosoma is typically xiphosuran, though with very prominent compound eyes, in which unfortunately no structure remains. In *Aglaspis* (Fig. 11.23h) the 11 or 12 thoracic segments are free as in most aglaspids, but in *Beckwithia* (Fig. 11.23j) the posterior three or four are fused into a pygidium-like plate. In some specimens appendages are preserved. Evidently the first pair bore chelae while the rest, including the abdominal ones, were single walking legs. It would be possible to infer that evolution in the xiphosures proceeded towards fusion of the abdominal segments and modification of their appendages. However, xiphosure phylogeny is obscure, and it is not certain whether aglaspids were ancestral to any later xiphosures or whether, as is more likely, they were simply an early and sterile 'experimental' group. The oldest arthropod which has been referred to Subclass Xiphosura is the Lower Cambrian *Palaeomerus* (Fig. 11.23g), but it is known from only three specimens and could be a relative of some of the Burgess Shale arthropods ('merostomoids').

The Xiphosurida normally have the distinctive cardiac lobe and ophthalmic ridges well developed, and there are 10 or fewer abdominal segments. There are two suborders.

ORDER XIPHOSURIDA (Cam.–Rec.)

SUBORDER 1. SYNZIPHOSURINA (Ord.–Dev.): Most or all of the abdominal segments are free, and the appendages do not appear to be chelate (other than the chelicerae, which are not known with certainty).

SUBORDER 2. LIMULINA (L. Dev.–Rec.): The abdominal segments may be fused, though not always, and the appendages are normally chelate.

Much confusion is associated with attempts to classify these animals, since the fossil record is so scanty and the preservation is often poor. In such cases the discovery of a new specimen, especially of a new taxon, often necessitates the taxonomic revision of the whole group. A recent classification of the Synziphosurina (Eldredge, 1974) includes only four genera: *Weinbergina* (Fig. 11.23k), *Legrandella* (which has a large and very Limulus-like eye; Fig. 11.23n), *Bunodes* and *Limuloides*. The Limulina, other than the peculiar Family Pseudoniscidae, are mainly rather similar in construction to *Limulus*. The oldest, the Lower Cambrian *Eolimulus*, is known from two partial prosomas only. It was marine, as are the much better-preserved Devonian genera, but some of the Carboniferous and later groups invaded and colonized brackish water habitats. The Carboniferous *Belinurus* and *Euproops* (Fig. 11.23m) are small forms, often found in nodules deposited in coal swamps. In the Carboniferous of Illinois (Chapter 12), which has been very thoroughly investigated, *Euproops* almost always seems to be associated with plant remains rather than with other invertebrates of ponds and lagoons. Fisher (1979), adducing much other evidence, believes that it may have been at least partially subaerial, crawling up forest trees, and disguising itself to resemble the surface of the trees.

Both *Belinurus* and *Euproops* have well-developed cardiac lobes and ophthalmic ridges. *Belinurus*, like the Devonian *Neobelinuropsis*, has some free abdominal segments, but in *Euproops* and *Limulus* the whole of the abdomen is fused, though surface traces of segmentation are very pronounced and incidentally very trilobite-like. It is of some interest that the first free larva of *Limulus*, when hatched from the egg, has a similar pronounced segmentation in the fused abdomen to the adult *Euproops*. It is colloquially known as a 'trilobite larva' without any implication of a close relationship. The prosomal appendages of all Limulina are very similar to those of *Limulus*. The earliest true limulid (*Rolfeia*) arose from a belinurid ancestor in the early Carboniferous. One of the largest known fossil limuloids is *Xaniopyramis*, described from northern England (Siveter and Selden, 1987). It is found in shallow-marine deltaic sediments and the single specimen is preserved with a current-scour mark in front of the prosoma in the enclosing sediment. This large individual, 14 cm across, was probably a burrower with poor swimming abilities.

From the known fossil Limulina it may be inferred that there was a general evolutionary trend towards a fused short abdomen and larger size. The Permian *Palaeolimulus*, the Triassic *Limulitella* and the Jurassic *Mesolimulus* differ from modern genera mainly in their smaller dimensions and their prosomal lobation. They were all marine forms. Amongst fossil limulids, several specimens exist of the Upper Jurassic *Mesolimulus* from the Solnhofen Limestone, preserved at the end of a trail in which the imprints of the appendages can be clearly seen.

Subclass Eurypterida

The eurypterids are relatively rare but always spectacular fossils. They ranged from Ordovician to Permian and are found in marine, brackish, hypersaline and freshwater lithofacies.

Eurypterid morphology is well displayed by the Silurian *Baltoeurypterus tetragonophthalmus* (Fig. 11.24a–g), formerly known as *Eurypterus fischeri*.

The morphology of this species was one of the first to be really well understood, largely from the work of Holm (1898; and later by Wills, 1965), who isolated fragments of the exoskeleton from rock, as he did with graptolites, so that they could be studied as transparencies. In dorsal view the **prosoma** is large and trapezoidal, with prominent compound eyes between which are small paired organs, probably ocelli. Behind this is the **opisthosoma**, consisting of a broad flattened **pre-abdomen** of seven segments, and a narrower and more cylindrical **post-abdomen** of only five segments, terminating in a stout, pointed telson. Each of the abdominal segments is composed of a dorsal tergite and a ventral sternite. The ventral appendages can be seen from the dorsal surface projecting well beyond the body. The ventral morphology is complex. The prosoma has a broad inflected doublure with a marginal suture, used during ecdysis, and often with paired connective sutures isolating a median doublural plate (the **epistoma**). Within the doublure is a softer integument surrounding the mouth, to which the appendages are attached. The first (preoral) pair of appendages are small chelicerae. Behind these post-orally are four pairs of stout, jointed walking legs, increasing in size towards the rear legs, which are cylindrical and spiny. The sixth pair of prosomal appendages are very large and have the terminal parts flattened like paddles. These last appendages were capable of being protracted and retracted and were probably used as in swimming, in a breaststroke manner. The coxa of each 'leg' is large and armoured internally with spines projecting as gnathobases. A small U-shaped plate (the **endostoma**) borders the mouth; this is normally covered by a much larger plate (the **metastoma**) which is actually part of the abdomen and is all that remains of the reduced seventh segment. It also covers the proximal parts of the coxa.

The pre- and post-abdomen of seven and five segments, respectively, have been defined on the position of the waist, but the abdomen can also be separated into a **mesosoma** and **metasoma** of six segments each. The former bears appendages; the latter does not. These appendages look like sternites but are in fact true appendages. The seventh segment is reduced as the metastoma, while the eighth, as in all merostomes, has the genital aperture. This segment is known as the operculum and, together with the four succeeding appendages, is plate-like and attached along the anterior border so that they all overlap. In the centre of the operculum and directed posteriorly is the **genital appendage**, an elongated and elaborately sculptured rod of which two kinds may be present in a single eurypterid population, suggesting sexual dimorphism. In *Baltoeurypterus* one kind (type A) is elongated, and the other kind (type B), though tubular, is much shorter; other genera have genital appendages that are of approximately equal length, though differently shaped (Fig. 11.24h,j). In *Pterygotus* and some other genera type A is consistently associated with 'clasping organs' on the prosomal appendages and seems to have been a male copulatory organ; the other type is best interpreted as a female ovipositor. Traces of internal ducts have been distinguished in both kinds.

The operculum and the succeeding four appendages of the pre-abdomen covered chambers in which the gills were situated. The structure of these is known in *Slimonia* and better still in *Tarsopterella*, where the different patterns of internal and external surface sculpture, even in flattened specimens (Waterston, 1975), suggested that the gills were specialized vascular tracts of the ventral body wall protected by the appendages (Fig. 11.24k). More recently (Manning and Dunlop, 1995) it has been shown that eurypterids have a dual respiratory system. The branchial chambers housed 'book lungs' in stacked thin sheets with ribbed surfaces providing extra potential for respiratory exchange. They are very similar to the book lungs of scorpions, but are delicate and seldom preserved. In addition, the gill-tracts, or more properly 'Kiemenplatten' on the roof of the branchial chamber, have downwardly facing patches of little spines covered with small projections in hexagonal rosettes. These are interpreted as water-trapping devices, linked to the vascular system, and providing an additional surface for gaseous exchange. Whereas eurypterids were primarily water dwellers, they may well have been able to make excursions on land for periods of time. If so the 'Kiemenplatten', with their

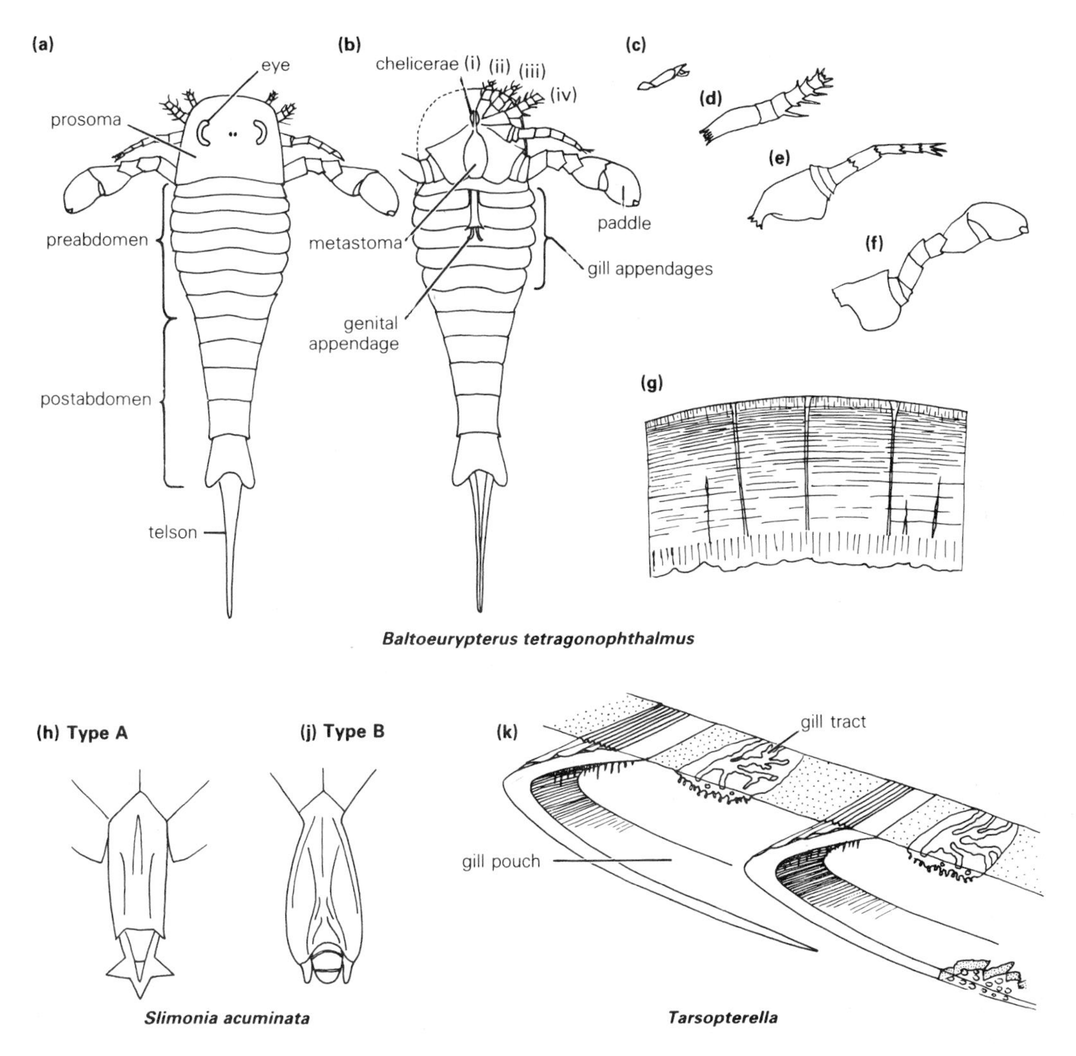

Figure 11.24 Eurypterida. (a)–(e) *Baltoeurypterus tetragonophthalmus* (Sil.): (a) dorsal view (× 0.35); (b) ventral view (× 0.35); (c) chelicera (i) (× 0.35); (d) walking leg [prosomal appendages (ii)–(iv); × 0.35]; (e) fifth prosomal appendage (× 0.35); (f) sixth, paddle-like appendage (× 0.35); (g) section (diagrammatic) through cuticle of *B. tetragonophthalmus*, showing inner, non-laminar layer with fine vertical canals, middle (laminar) layer with laminae more closely spaced externally, and outer layer with indications of vertical elements; (h) *Slimonia acuminata* (Sil.), type A (?male) genital appendage; (j) *S. acuminata* type B (?female) genital appendages; (k) *Tarsopterella* (Dev.), simplified reconstruction of a gill chamber. [(a)–(e) Redrawn from Störmer in *Treatise on Invertebrate Paleontology*, Part P; (g) based on Dalingwater, 1975; (j) redrawn from Waterston, 1960; (k) redrawn from Waterston, 1975.]

permaently moist water-covered surfaces, would function as terrestrial lungs.

Cuticle

The cuticle of eurypterids is very thin, and specimens are usually crushed. It may bear different kinds of external sculpture, notably terrace lines round the border and sometimes on the lower surfaces, and a characteristic scale-like ornament from which fragmentary remains in sedimentary rock can be immediately distinguished as being eurypterid in origin.

Ultramicrographic work (Dalingwater, 1973a, 1975) has shown that eurypterid cuticles are preserved as silica, though undetermined organic material was originally present too. The outer cuticular zone (Fig. 11.24g). The laminations resemble those of *Limulus* in nature, which could be of phylogenetic importance.

Range in form, evolution and ecology (Fig. 11.25)

Eurypterids are relatively rare fossils, and though the morphology of some genera is well known, others are represented only by very poorly preserved material, and there are acute taxonomic problems in classifying such isolated remains. Rather than describing the taxonomic problems in detail, the following discussion is confined to a few selected types which show notable and presumably adaptive differences in size, prosomal shape, the location of the eyes, and the morphology of the body, the appendages and the telson. A current classification (Störmer, 1970–1974) divides eurypterids into three suborders, each with associated families and superfamilies:

SUBCLASS EURYPTERIDA.
ORDER EURYPTERIDA.
SUBORDER 1. EURYPTERINA (Ord.–Perm.): Genera with small untoothed chelicerae; e.g. *Drepanopterus*, *Baltoeurypterus*, *Stylonurus*, *Slimonia*, *Mixopterus*, *Hughmilleria*.
SUBORDER 2. PTERYGOTINA (Ord.–Dev.): Eurypterids with enormous denticulate chelicerae; e.g. *Pterygotus*, *Jackelopterus*.
SUBORDER 3. HIBBERTOPTERINA (Carb.): Genera in which the posterior prosomal legs have a basal extension; e.g. *Hibbertopterus*, *Campylocephalus*.

The Eurypterina, by far the largest suborder, have the greatest diversity. *Hughmilleria* is relatively small and unmodified. *Slimonia* has a quadrate prosoma, antero-lateral eyes and a laterally expanded telson. *Carcinosoma* has a large discoidal pre-abdomen with a marked waist and a cylindrical post-abdomen and, like the fiercely armoured *Mixopterus*, it has a telson, like that of a scorpion, apparently modified as a poison spine. The prosomal appendages of *Dolichopterus* are of typical form though enormously extended, whereas in *Stylonurus* and its relatives all the legs are slender and elongate, even the sixth prosomal appendage, and presumably modified for lightly walking over muddy surfaces.

Most eurypterids seem to have been active benthos, though actively able to swim. *Baltoeurypterus* has recently been the subject of a detailed functional study by Selden (1981), who used exquisitely preserved, isolated pieces of eurypterid, including whole legs, from localities in the Baltic Silurian. First, it was shown that while the radially arranged coxae could masticate food by adduction and abduction, i.e. by coming together and moving apart, they could not be swung to and fro. Second, the actual mode of operation of the legs was worked out from the position and nature of the joints between the podomeres. In eurypterids, as in arthropods generally, joints are either **hinges** (with a single articulation) or **pivots** (with two articulations on opposite sides of the joint). The legs of *Baltoeurypterus* had pivots proximally and hinges distally, and each podomere could rock in one plane only relative to the next one but, except with a few specialized joints, could not rotate. Successive joints, however, have their articulations set in different planes (Fig. 11.26a–c), that of the coxa or first podomere being subvertical.

When the leg moved forwards the set of the joints ensured that it was lifted up and set down in a forward position for the succeeding backward stroke. It is probable that only the last three appendages were actually used for walking, including the paddle-like sixth appendage.

The sixth appendage, however, could also be used for rowing the animal through the water, as an analysis of joint patterns shows. If the paddle were to function as an oar it would need to present a flat vertical surface to the water when pushing backwards, but be held horizontally during the recovery stroke for minimal resistance; this, in fact, it can do, for the joints are so arranged that an actual rotation of the blade occurs both at the proximal and the distal joints of podomere 6. The blade is thus outstretched and vertical during the backward propulsive stroke and collapsed and horizontal during the forward recovery stroke.

This rowing technique has been shown to be functionally equivalent to the sculling action of modern swimming crabs (portunids; Plotnick, 1985), and it has been calculated (Selden, 1984) that *Baltoeurypterus* could have reached a maximum velocity of 2.5 times its body length per second. Pterygotids, on the other hand, being more hydrofoil-shaped probably cruised at low speeds with occasional bursts of power, using the telson for steering.

Some eurypterids at least may have been able to crawl out onto land for short periods. A large trail from the Silurian of Ringerike, Norway, with in-phase gait patterns, seems to have been made in this way. It is believed to have been made by a large *Mixopterus* (Hanken and Störmer, 1975; Fig. 11.25f–h). The trail is 520 mm long and about 160

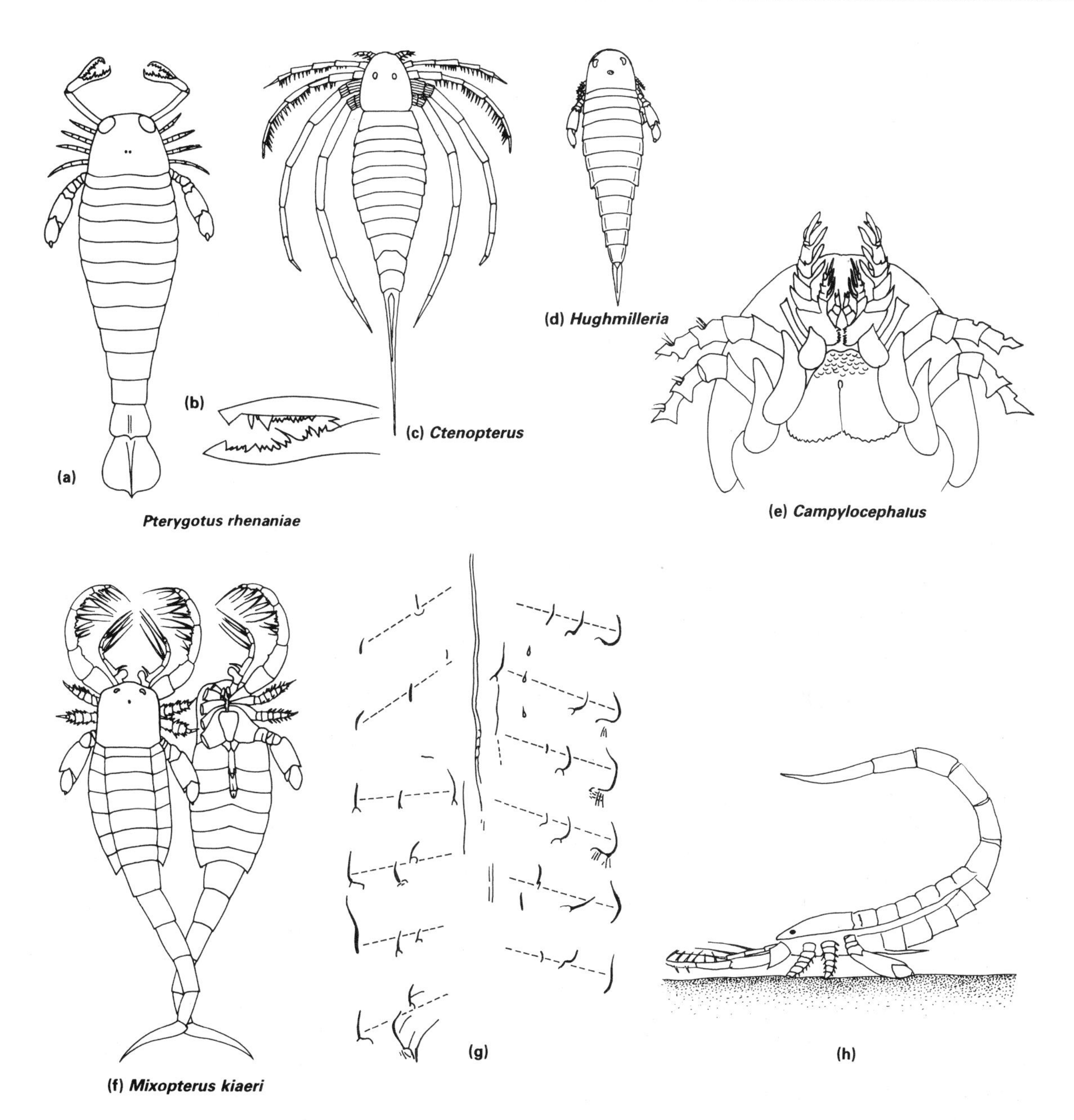

Figure 11.25 Eurypterida. (a), (b) *Pterygotus rhenaniae* (Dev.): (a) dorsal view; (b) detail of chelicera (× 0.1); (c)*Ctenopterus* (Sil.), a stylonurid (× 0.5); (d) *Hughmilleria* (Sil.; × 0.35); (e) *Campylocephalus* (Carb.), ventral view of prosoma, showing large metastoma and basal extensions on posterior legs typical of Hibbertopteridae (× 0.2); (f) *Mixopterus kiaeri* (Sil.), dorsal and ventral views (× 0.1); (g) a trail in Silurian red-beds at Ringerike, Norway, believed to have been made by *M. kiaeri* (see text) while walking in the position illustrated in (h) (× 0.25). [(a)–(d), (f) Redrawn from Störmer in *Treatise on Invertebrate Paleontology*, Part P; (e) redrawn from Waterston, 1957; (g) redrawn from Hanken and Störmer, 1975.]

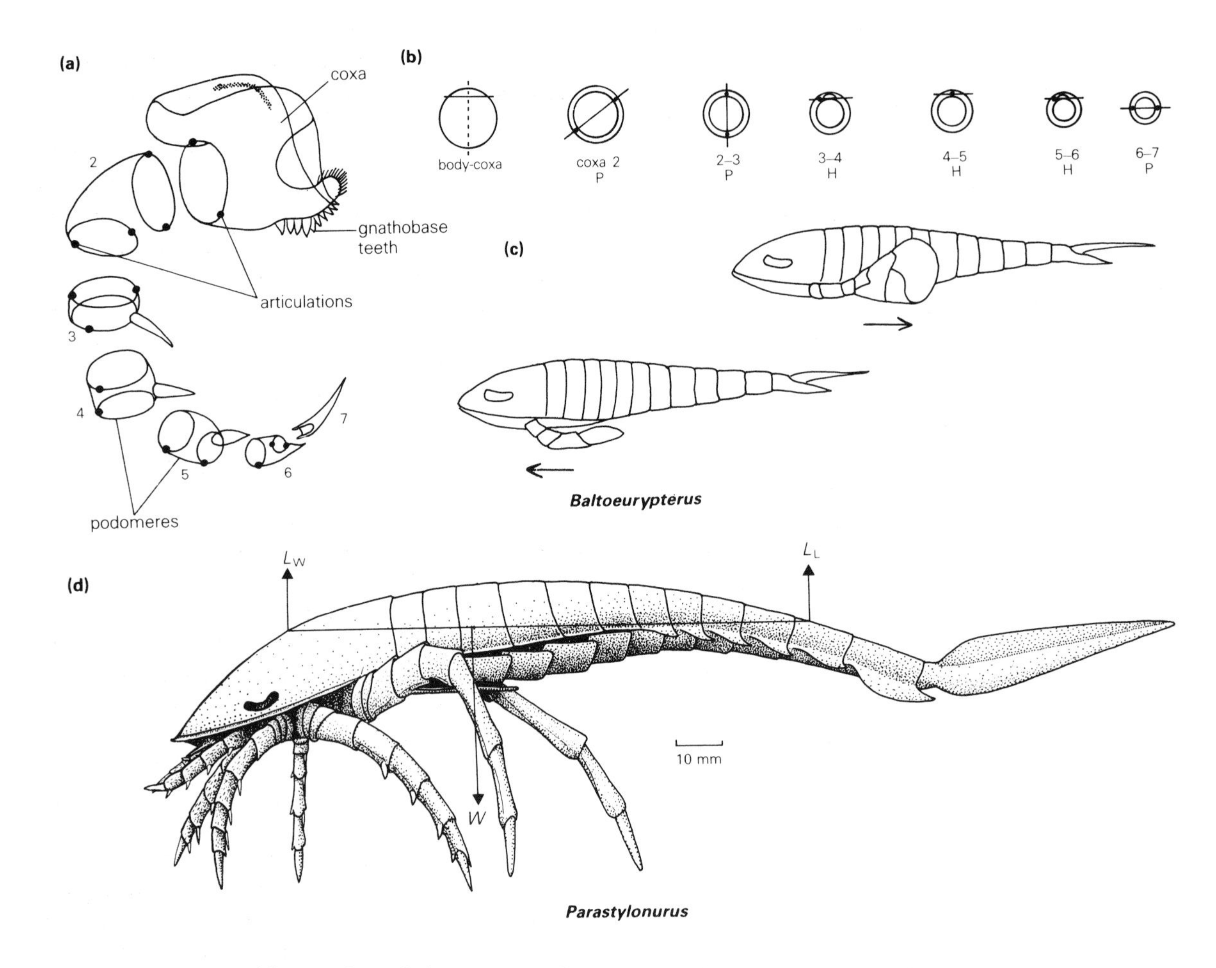

Figure 11.26 Eurypterid functional morphology. (a)–(c) *Baltoeurypterus* (Sil.): (a) exploded view of second prosomal appendage, showing hinge and pivots; (b) arrangement of hinges and pivots between podomeres which facilitates (c), for the sixth prosomal appendage, an oar-like thrust during backward stroke, and minimal resistance to the water during forward recovery stroke; (d) *Parastylonurus* (Sil.) in walking posture. [(a)–(c) Redrawn from Selden, 1981; (d) redrawn from Waterston, 1979.]

mm broad and consists of three sets of paired parallel tracks made up of separate successive imprints, decreasing in size inwards. The outermost 'A-tracks' are hook-shaped and would fit with the imprint of the paddle-shaped end of the sixth prosomal appendage. The intermediate 'B-tracks' and the innermost 'C-tracks' seem, however, to have been made by appendages with flattened spinose ends. A long median groove seems to have been made by the tip of the genital appendage scraping along the ground. These various facts fit with the concept of the trail being made by a *Mixopterus* walking on the fourth, fifth and sixth pairs of appendages, while keeping the grasping second and third pairs of appendages held out in front of it as a kind of cage. *Mixopterus* was contemporaneous and was the right size to have made such a trail. Presumably the eurypterid could also swim, using the retractable sixth appendage as a paddle.

Special adaptations for stability in long-legged walking eurypterids have been described by Waterston (1979), especially in the Silurian stylonurid *Parastylonurus* (Fig. 11.26d). This genus has a long pre-abdomen, and it is estimated that the centre of gravity would have lain near the widest point of the abdomen, at about the third pre-abdominal segment. The broad, scimitar-shaped flanges (**epimera**) of the post-abdominal segments, and the broad, keeled telson would have acted as a hydrofoil, giving lift to the posterior part of the animal, and thereby increasing stability in light currents with minimal muscular effort.

Such generalized types as *Baltoeurypterus* were probably unspecialized feeders. Most eurypterids, however, were probably predators, their prey being other eurypterids and fishes. Two different feeding types were derived from the unspecialized condition, the enlarged spinose limbs II and III of *Mixopterus*, and the greatly enlarged chelicerae of *Pterygotus*. The latter were probably kept folded as the animal cruised around looking for prey, and then rapidly extended for capture.

In some cases, characteristic eurypterid associations can be distinguished and related to environmental conditions. In the Silurian of the Welsh Borderland, for example (Kjellesvig-Waering, 1961), there seem to be three such associations: (1) carcinosomatids and pterygotids, associated with a rich, normal marine fauna; (2) Eurypterinae found with rare marine fossils, in sediments probably deposited in an inshore environment, probably lagoonal or estuarine; (3) brackish to freshwater assemblages, dominated by stylonurids and hughmilleriids. Most representatives are less than 20 cm long, but there are several giant forms including the Devonian *Pterygotus* (*Erettopterus*), the largest arthropod of all time, which reached nearly 2 m in length.

11.5 Phylum Crustacea

The arthropod Phylum Crustacea (Fig. 11.27) is singularly diverse and highly successful.

Crustaceans are mainly marine but there are some freshwater and a few terrestrial groups; the phylum includes not only the familiar crabs, lobsters and shrimps, but also ostracods, copepods, barnacles and many other taxa. Although crustaceans normally have a chitinous exoskeleton, their range and variety is such that some groups (e.g. adult barnacles with their pyramidal external skeleton of calcite plates and some of the parasitic taxa) are not immediately recognizable as crustaceans at all. Fossil barnacles are well known, especially from Tertiary deposits (Buckeridge, 1983); indeed they were the one group of which Charles Darwin made a taxonomic study before embarking upon his *Origin of Species*.

In spite of this diversity, however, the homogeneity of the phylum has usually been agreed, for in all cases the embryo, developing by spiral cleavage, subsequently becomes a **nauplius larva**, common to all except the most specialized groups. The nauplius is a small ovoid larva, invariably with three pairs of appendages. These are the uniramous **antennules**, the biramous **antennae** and the **mandibles**. In later developmental stages these mandibles become differentiated as true jaws, but they are used in the nauplius along with the other two pairs of appendages for swimming and feeding. The shape and movements of the nauplius were described in memorable terms by Garstang (1951): 'The nauplius is a wobbly thing, a head without a body. He flops about with foolish jerks, a regular Tom-Noddy . . .', for the nauplius is invariably a swimming entity, and only after several other developmental stages, which differ in the various main groups, is the adult form attained. When the adult stage is reached, the crustacean retains the antennules and the antennae, and the mandibles are joined by two other pairs of food-processing appendages, the maxillules and maxillae. These five pairs of differentiated appendages belong to the head, behind which there is usually a thorax and an abdomen. These three tagmata (head, thorax and abdomen) are possessed by most crustaceans, though not by ostracodes. Likewise provided with appendages, the head may be fused with the thorax and commonly the two are covered by a saddle-like dorsal shield, the carapace, which may be free posteriorly, as in many shrimp-like forms, or, as in crabs and lobsters, may be fused to the body. Despite these similarities there is some possibility that crustaceans are not a natural but a paraphyletic group (Briggs and Fortey, 1989), and the position of Cambrian crustacean-like arthropods both from the Burgess Shale and from Swedish 'Orsten' is still much debated.

The classification of crustaceans is based on many characters, which include the form of the body, number of segments, presence or absence of a carapace and in particular the nature of the appendages.

Eight classes are recognized by McLaughlin (1980), ranging from the tiny primitive Cephalocarida to the largest and most important group, the Malacostraca, which is the only class (other than the Ostracoda, usually treated as microfossils) with a good fossil record. Only four classes, however, (Remipedia, Malacostraca, Phyllopoda

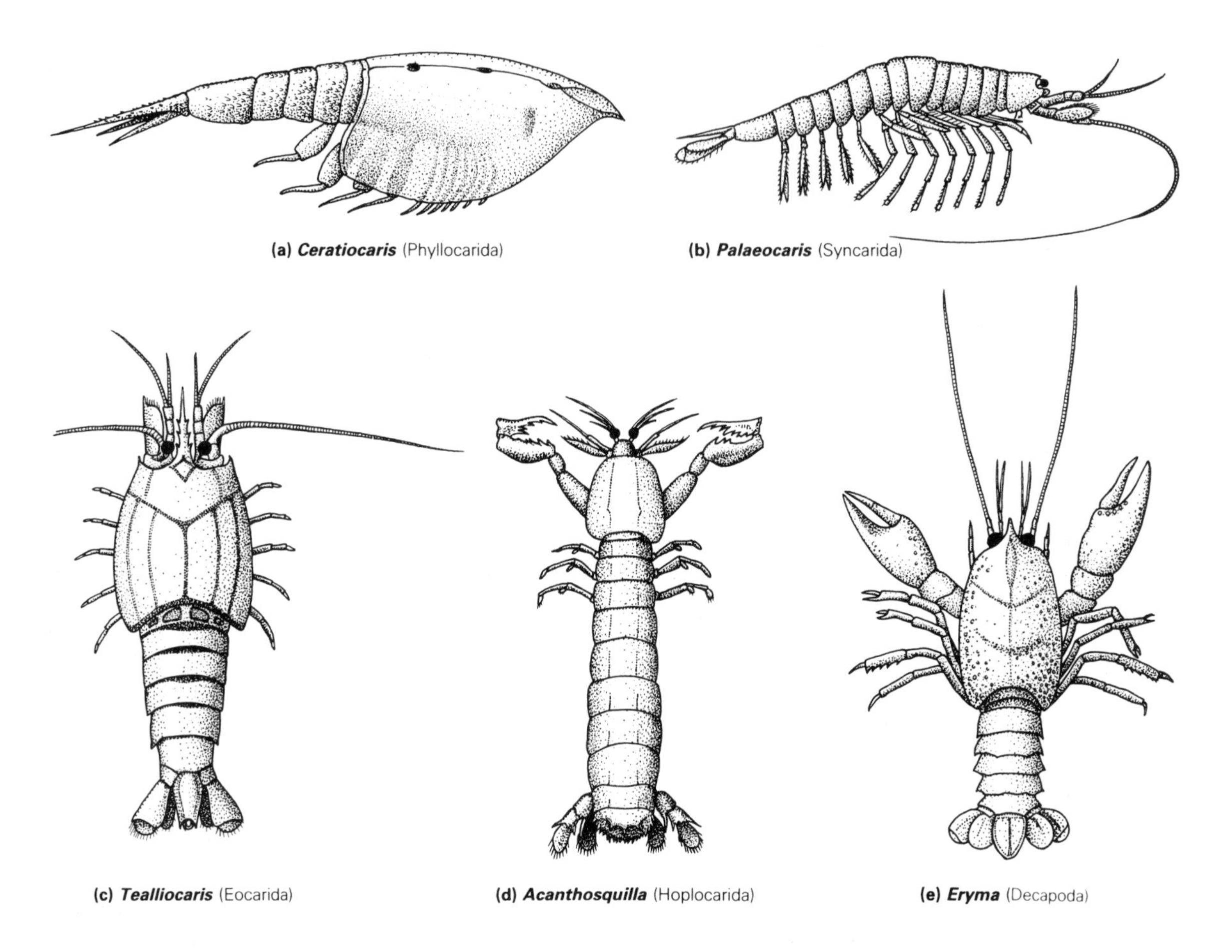

Figure 11.27 Malacostracan diversity: (a) *Ceratiocaris* (Phyllocarida; Sil.; × 0.5); (b) *Palaeocaris* (Syncarida; Carb.; × 2); (c) *Tealliocaris* (Eocarida; Carb.; × 1); (d) *Acanthosquilla* (Hoplocarida; Rec.; × 0.25); (e) *Eryma* (Decopoda; Jur.; × 0.5.)

and Maxillopoda), are recognized by Schram (1986), who used computer-based cladistic methodology. In view of the current flux in crustacean classification, I here use the more conservative classification of McLaughlin.

In the Malacostraca there are generally paired compound eyes, and the antennules and antennae are biramous, the latter having a shield-like antennal scale or scaphocerite. The biting jaws (mandibles) are formed from **gnathobases**, internal extensions of the mandibular appendage that meet in the median plane of the body and roll or grind together as the limb moves backwards and forwards. The outer part of the mandibular limb may disappear. In the Malacostraca there are usually eight pairs of biramous thoracic limbs, uniform and unspecialized in the more primitive groups, but often becoming greatly differentiated in the more advanced malacostracans. Many of the more successful predators have chelate or pincer-like appendages, and indeed much of the success of the malacostracans is linked to the differentiation of the limbs for various purposes: feeding, swimming and respiration. The malacostracan abdomen consists of six segments bearing five biramous pleopods which are often well developed and used in swimming (though they are lost in some groups). The last segment has a triangular central telson flanked by paired biramous uropods.

According to McLaughlin (1980) there are three malacostracan subclasses (followed here). Schram

(1986), on the other hand, removes the Phyllocarids to a separate class.

SUBCLASS 1. PHYLLOCARIDA (Cam.–Rec.): In this group a large, bivalved carapace, with an articulating rostrum, covers the anterior part of the body but is not fused to it. The antennae and antennules are short, while the mandibles form powerful biting jaws. Behind these are the eight biramous and foliaceous thoracopods used in filter feeding. The abdomen projects posteriorly from the carapace; it is provided with up to six appendages and terminates in a pair of leaf-like plates (**furca**). The earliest known phyllocarid is possibly Lower Cambrian in age. Exquisitely preserved phyllocarid-like crustaceans, *Canadaspis* and *Perspicaris*, with preserved appendages are known from the Middle Cambrian Burgess Shale (Briggs, 1978) and genera such as *Hymenocaris* are quite common in the late Cambrian. Phyllocarids may be abundant in certain facies in the early Palaeozoic, and especially in marginal marine and possibly brackish-water habitats. Some of these grew very large; the Silurian *Ceratiocaris* (Fig. 11.27a), for example, had some representatives reaching 75 cm in length, whereas the common recent *Nebalia* is only about 1 cm long, and the largest modern phyllocarid *Nebaliopsis* is only 4 cm in length.

SUBCLASS 2. HOPLOCARIDA (Carb.–Rec.): Represented by today's *Squilla*, the mantis-shrimp, which like all hoplocarids is a marine predator with a distended abdomen and modified raptorial thoncopods (cf. Fig. 11.27d). Some generally similar, though smaller forms from the Carboniferous have also been referred to this subclass.

SUBCLASS 3. EUMALACOSTRACA (Dev.–Rec.): The vast array of crabs, lobsters, shrimps, mysids, isopods, amphipods and others which make up this great group are too numerous to detail here – McLaughlin (1980) provides a good modern summary and illustrated descriptions of representatives of various groups. There are four eumalacostracan superorders, as defined by her, but eumalacostracan taxonomy is still much debated.

SUPERORDER 1. SYNCARIDA (M. Carb.–Rec.): Small, rather elongated crustaceans, primarily lacking a carapace, and with biramous thoracic appendages. They have seemingly evolved by a different path to other eumalacostracans, in which the carapace is primary and, if not present, has been secondarily lost. Syncarids flourished in the late Palaeozoic (Fig. 11.27b), and were only known as fossils for many years. In 1892, however, a living genus, *Anaspides*, was discovered living in Tasmanian lakes and rivers, and a few other genera have since been reported from the same area.

SUPERORDER 2. EOCARIDA (M. Dev.–Perm.): A relatively diverse group of small, shrimp-like or **caridoid** forms, swimming crustaceans in which there is a carapace, an elongated body with the abdomen often partly curled under the body, and a tail fan comprised of the telson and flattened uropods which can be rapidly flicked under the body so that the animal darted out of the way of predators. The eocarids are highly organized and of very modern appearance (Fig. 11.27c), though they have comparatively unspecialized appendages which are not provided with chelae. The eocarids were undoubtedly ancestral to all the modern crustacean groups. These latter include the following two superorders.

SUPERORDER 3. PERACARIDA (?Carb./Perm.–Rec.): Comprises, among other groups, the mysids (caridoid), isopods (dorso-ventrally flattened) and amphipods (laterally flattened). They are generally rather small crustaceans with unmineralized cuticles, and hence with a rather poor fossil record. The females have a brood pouch or marsupium in which the eggs develop and whence the young are released as miniature adults.

SUPERORDER 4. EUCARIDA: In this category come the great majority of common modern crustaceans, divided into two main orders, ORDER EUPHAUSIACEA (Rec.) and ORDER DECAPODA (Trias.–Rec.). The euphausiids are oceanic swimming caridoids of cosmopolitan distribution which are very important in today's oceans, but with a negligible fossil record. The decapods, which all have five pairs of thoracic appendages, include swimming forms (shrimps and prawns) as well as benthic representatives (the crabs and lobsters, and the freshwater crayfish).

Shrimp-like crustaceans of distinctly modern type ('eocarids') first appeared in the late Devonian and underwent a major adaptive radiation in the Carboniferous. They are usually found in sediments of marginal-marine and brackish-water origin (coastal plain and low-lying swamp environments), since it is only these environments which provided the exceptional circumstances required for the preservation of their unmineralized chitinous exoskeletons. It is actually quite likely that 'shrimps' were common in contemporaneous seas, but were not preserved (Briggs and Clarkson, 1989, 1990). These Palaeozoic forms were largely extinct by the late Permian. The appearance of calcified cuticles in Mesozoic marine eucarids allows much greater preservation potential.

All eucarids have the carapace fused with the thoracic somites. The compound eyes are stalked, and the eggs are carried attached to the abdominal appendages. The young, primitively hatched as nauplii, may undergo various free-swimming transformations, and are often of bizarre appearance before becoming adults. Lobsters and crabs have appendages and are highly successful marine predators. It is probable that the possession of such grasping pincers (absent in their Palaeozoic precursors; Fig. 11.27e) has been the most important single factor in their success. They are the largest marine crustaceans living today and, since their skeletons are reinforced with calcium carbonate, they have a good fossil record. Although some Triassic genera are known, the first real burst of adaptive radiation was delayed until the early Jurassic. Most of these were of rather lobster-like form, though some, such as the Eryonidae (e.g. Fig. 12.10), had dorso-ventrally compressed carapaces. Most of the Jurassic groups have survived to the present day but some have declined in numbers. The eryonids, for example, which formerly inhabited shallow waters, are known today only as blind species from abyssal depths. Although modern lobsters are widespread and successful, they are represented by many fewer genera than the crabs (Brachyura), in which the abdomen is very reduced and permanently turned under the thorax. These originated in the Lower Jurassic and have expanded and diversified ever since, having many kinds of adaptively specialized characteristics fitting them for life in innumerable habitats, and these can be studied from the Mesozoic onwards and especially through the Tertiary. From their Devono-Carboniferous origins the eumalacostracans have expanded enormously into diverse environments: a striking example of the radiation of a group in which 'Bauplan' and physiology combined the required degree of evolutionary plasticity to allow them to become one of the most successful life forms in the sea today.

Bibliography

Books, treatises and symposia

Bergström, J. (1973) Organisation, life and systematics of trilobites. *Fossils and Strata* **2**. (Large-scale work; presents a different classification from that in the *'Treatise'*)

Bowes, D.R. and Waterston, C.D. (eds) (1985) *Fossil Arthropods as Living Animals. Transactions of the Royal Society of Edinburgh* **76**, 101–399. (24 papers)

Briggs, D.E.G. and Lane, P.D. (eds) (1983) *Trilobites and Other Arthropods*. Special Papers in Palaeontology No. 30. Academic Press, London. (16 papers).

Clarke, K.U. (1973) *The Biology of Arthropods*. Edward Arnold, London. (Short simple text)

Garstang, W. (1951) *Larval Forms, and Other Zoological Verses*. Blackwell, Oxford.

Lebrun, P. (1995) *Trilobites. Mineraux et Fossiles 2*. Editions CEDIM, Paris. (excellent colour plates)

Levi-Setti, R. (1993) *Trilobites: a Photographic Atlas*, 2nd edn. University of Chicago Press, Chicago. (Large numbers of full page plates, with discussion of eyes)

McLaughlin, P. (1980) *Comparative Morphology of Recent Crustacea*. Freeman, San Fransisco. (Illustrations and classifications)

Manton, S.M. (1977) *The Arthropods: Habits, Functional Morphology and Evolution*. Oxford University Press, Oxford. (Definitive work on arthropod morphology and evolution)

Moore, R.C. (ed.) (1955) *Treatise on Invertebrate Paleontology* Part P, *Arthropoda* 2. Geological Society of America and University of Kansas Press, Lawrence, Kan. (Chelicerates, pycnogonids and spiders)

Moore, R.C. (ed.) (1959) *Treatise on Invertebrate Paleontology* Part O, *Arthropoda* 1.Geological Society of America and University of Kansas Press, Lawrence, Kan. (Trilobites and related groups) [Kaesler, R.L. (ed.) (1997) Vol. 1 of the revised version is now published.]

Moore, R.C. (ed.) (1961) *Treatise on Invertebrate Paleontology* Part Q *Arthropoda* 3. Geological Society of America and University of Kansas Press, Lawrence, Kan.

Moore, R.C. (ed.) (1969) *Treatise on Invertebrate Paleontology* Part R, *Arthropods,* Vols 1 and 2. Geological Society of America and University of Kansas Press, Lawrence, Kan.

Schram, F.R. (1986) *Crustacea*. Oxford University Press, Oxford.

Whittington, H.B. (1992) *Trilobites*. Fossils Illustrated: 2. Boydell Press, Woodbridge. (120 photographic plates)

Individual papers and other references

Bergström, J. (1971) *Palaeomerus* – merostome or merostomoid? *Lethaia* **4**, 393–401.

Bergström, J. and Brassel, G. (1984) Legs in the trilobite *Rhenops* from the Lower Devonian Hunsrck Slate. *Lethaia* **17**, 67–72.

Brezinski, D.K. (1985) An opportunistic Upper Ordovician trilobite assemblage from Missouri. *Lethaia* **19**, 315–25.

Briggs, D.E.G. (1978) The morphology, mode of life and affinities of *Canadaspis perfecta* (Crustacea: Phyllocarida), Middle Cambrian Burgess Shale, British Columbia. *Philosophical Transactions of the Royal Society of London B* **281**, 439–87.

Briggs, D.E.G. and Clarkson, E.N.K. (1989) Environmental controls on the taphonomy and distribution of Carboniferous malaeostracan crustaceans. *Transactions of the Royal Society of Edinburgh: Earth Sciences* **80**, 293–301.

Briggs, D.E.G. and Clarkson, E.N.K. (1990) The late Palaeozoic radiation of malaeostracan crustaceans, in

Major Evolutionary Radiations (eds P.D. Taylor and G.P. Larwood), Systematics Association Special Vol. 42, Clarendon Press, Oxford, pp. 165–86.

Briggs, D.E.G. and Fortey, R.A. (1989) The early radiation and relationships of the major arthropod groups. *Science* **246**, 241–3.

Briggs, D.E.G. and Fortey, R.A. (1992) The early Cambrian radiation of arthropods, in *Origin and Early Evolution of the Metazoa* (eds J.H. Lipps, and P.W. Signor), Plenum Press, New York and London, pp. 336–74.

Bruton, D. and Haas, W. (1997) Functional morphology of Phacopinae (Trilobita) and the mechanics of enrollment. *Palaeontographica Abteilungen A* **245**, 1–43.

Buckeridge, J.S. (1983) Fossil barnacles (Cippipedia: Thoracica) of New Zealand and Australia. *New Zealand Geological Survey Palaeontological Bulletin* **50**, 1–151.

Campbell K.S.W. (1975a) The functional anatomy of phacopid trilobites: musculature and eyes. *Journal and Proceedings of the Royal Society of New South Wales* **108**, 168–88. (Lens structure)

Campbell, K.S.W. (1975b) The functional morphology of *Cryptolithus*. *Fossils and Strata* **4**, 65–86. (Trinucleid fringe function, and ichnology)

Campbell, K.S.W. (1977) Trilobites of the Haragan, Bois d'Arc and Frisco Formations (Early Devonian), Arbuckle Mountains Region, Oklahoma. *Bulletin of the Oklahoma Geological Survey* **123**, 1–227. (Eyes and enrollment structures)

Chatterton, B.D.E. (1971) Taxonomy and ontogeny of Siluro-Devonian trilobites from near Yass, New South Wales. *Palaeontographica* **137**, 1–108.

Chatterton, B.D.E. (1980) Ontogenetic studies of Middle Ordovician trilobites from the Esbattaotine Formation, Mackenzie Mountains, Canada. *Palaeontographica A* **171**, 1–74. (Silicified fauna illustrated)

Chlupac, I. (1983) Trilobite assemblages in the Devonian of the Barrandian area and their relations to palaeoenvironment. *Geologica et Palaeontologica* **17**, 45–73.

Chlupac, I. (1987) Ecostratigraphy of Silurian trilobite assemblages of the Barrandian area, Czechoslovakia. *Newsletter in Stratigraphy* **17**, 169–86.

Cisné, S. (1975) Anatomy of *Triarthrus* and the relationships of the trilobites. *Fossils and Strata* **4**, 45–64. (X-radiography, with limb morphology)

Clarkson, E.N.K. (1966) Schizochroal eyes and vision of some Silurian acastid trilobites. *Palaeontology* **9**, 1–29.

Clarkson, E.N.K. (1969) A functional study of the Silurian trilobite *Leonaspis deflexa* (Lake). *Lethaia* **2**, 329–44. (Lateral view and functional evolution of odontopleurids)

Clarkson, E.N.K. (1971) On the early schizochroal eyes of *Ormathops* (Trilobita, Zellszkellinae). *Memoires du Bureau des Recherches Géologiques et Miniéres* **73**, 51–63.

Clarkson, E.N.K. (1973) Morphology and evolution of the eye in Upper Cambrian Olenidae (Trilobita). *Palaeontology* **16**, 735–63.

Clarkson, E.N.K. (1975) The evolution of the eye in trilobites. *Fossils and Strata* **4**, 7–31. (Summary and bibliography on eye function)

Clarkson, E.N.K. (1979) The visual system of trilobites. *Palaeontology* **22**, 1–22. (Most recent summary)

Clarkson, E.N.K. (1988) The origin of marine invertebrate species: a critical review of microevolutionary transformations. *Proceedings of the Geologists' Association* **99**, 153–71.

Clarkson, E.N.K. and Henry, J.L. (1973) Structures coaptatives et enroulement chez quelques trilobites ordoviciens et siluriens. *Lethaia* **6**, 105–32. (Enrollment and interlocking mechanisms)

Clarkson, E.N.K. and Levi-Setti, R. (1975) Trilobite eyes and the optics of Des Cartes and Huygens. *Nature* **254**, 663–7. (Lens structure and function)

Clarkson, E.N.K., Eldredge, N. and Henry, J.L. (1977) Some Phacopina (Trilobita) from the Silurian of Scotland. *Palaeontology* **20**, 119–42.

Cocks, L.R.M. and Fortey, R.A. (1988) Lower Palaeozoic facies and faunas around Gondwana, in *Gondwana and Tethys* (eds M.G. Audley-Charles and A. Hallam), Geological Society of London Special Publication No. 37, pp.183–200.

Cocks, L.R.M. and Fortey, R.A. (1988) Arenig to Llandovery faunal distributions in the Caledonides, in *The Caledonian-Appalachian Orogen* (eds A.L. Harris and D.J. Fettes), Geological Society of London Special Publication No. 38, pp. 223–46.

Cocks, L.R.M. and Fortey, R.A. (1990) Biogeography of Ordovician and Silurian faunas, in *Palaeozoic Palaeogeography and Biogeography* (eds W.S. McKerrow and C.R. Scotese), Geological Society Memoir No. 12, pp. 97–104.

Cowie, J.W. (1971) Lower Cambrian faunal provinces, in *Faunal Provinces in Space and Time, Geological Journal* Special Issue No. 4, pp. 21–46. (Recent study of trilobite provinces)

Dalingwater, J.E. (1973a) The cuticle of a eurypterid. *Lethaia* **6**, 179–86. (Morphology)

Dalingwater, J.E. (1973b) Trilobite cuticle microstructure and composition. *Palaeontology* **16**, 827–39. (Fine structure)

Dalingwater, J.E. (1975) Further observations of eurypterid cuticles. *Fossils and Strata* **4**, 271–80. (Eurypterid cuticles were siliceous)

Dalingwater, J.E., Hutchinson, S.J., Mutvei, H. and Siveter, D.J. (1991) Cuticular ultrastructure of the trilobite *Ellipsocephalus polytomus* from the Middle Cambrian of Öland, Sweden. *Palaeontology* **34**, 205–17.

Eldredge, N. (1970) Observations on burrowing behaviour in *Limulus polyphemus* (Chelicerata, Merostomata) with implications on the functional anatomy of trilobites. *Novitiates of the American Museum of Natural History* **2436**,1–17. (Valuable study of living animals)

Eldredge, N. (1972a) Patterns of cephalic musculature in the Phacopina (Trilobita) and their phylogenetic significance. *Journal of Paleontology* **45**, 52–67. (Reconstruction of internal organs)

Eldredge, N. (1972b) Systematics and evolution of *Phacops rana* (Green 1832), and *Phacops iowensis* (Delo 1935) (Trilobita) from the Middle Devonian of North America. *Bulletin of the American Museum of Natural History* **147**, 49–113. (Allopatric evolution)

Eldredge, N. (1974) Revision of the suborder Synziphosurina (Chelicerata, Merostomata) with remarks on merostome phylogeny. *Novitiates of the American Museum of Natural History* **2543**, 1–41. (Taxonomy and morphology)

Eldredge, N. (1977) Trilobites and evolutionary patterns, in *Patterns of Evolution* (ed. A. Hallam), Elsevier, Amsterdam, pp. 305–32. (What do trilobites have to tell us about evolutionary processes?)

Eldredge, N. and Gould, S.J. (1972) Punctuated equilibria: an alternative to phyletic gradualism, in *Models in Paleobiology* (ed. T.J.M. Schopf), Freeman, Cooper and Co., San Francisco, pp. 82–115. (Allopatric evolution)

Eldredge, N. and Ormiston, R. (1979) Biogeography of Silurian and Devonian trilobites of the Malvinokaffric Realm, in *Historical Biogeography, Plate Tectonics and the Changing Environment* (eds J. Gray and A. Boucot), Oregon State University Press, Portland, Ore., pp. 147–67.

Evitt, W.R. (1961) Early ontogeny in the trilobite family Asaphidae. *Journal of Paleontology* **35**, 986–95.

Feist, R. (1991) The Late Devonian trilobite crises. *Historical Biology* **5**, 197–214.

Feist, R. and Clarkson, E.N.K. (1989) Environmentally controlled phyletic evolution, blindness and extinction in Late Devonian tropidocoryphine trilobites. *Lethaia* **22**, 359–73.

Fisher, D. (1975) Swimming and burrowing in *Limulus* and *Mesolimulus*. *Fossils and Strata* **4**, 281–90. (Important functional study)

Fisher, D. (1979) Evidence for subaerial activity of *Euproops danae,* in *Mazon Creek Fossils* (ed. M. Nitecki), Academic Press, New York, pp. 379–447.

Fordyce, D. and Cronin, T.W. (1989) A comparison of the fossilised compound eyes of phacopid trilobites with the eyes of modern marine arthropods. *Journal of Crustacean Biology* **9**, 554–69.

Fortey, R.A. (1974a) A new pelagic trilobite from the Ordovician of Spitzbergen, Ireland, and Utah. *Palaeontology* **17**, 111–24.

Fortey, R.A. (1974b) The Ordovician trilobites of Spitzbergen, 1. Olenidae. *Norsk Polarinstitut Skrifter* **160**, 1–128.

Fortey, R.A. (1975) Early Ordovician trilobite communities. *Fossils and Strata* **4**, 331–52. (Three onshore–offshore benthic and one planktonic assemblage)

Fortey, R.A. (1985) Pelagic trilobites as an example of deducing the life habits of extinct arthropods. *Transactions of the Royal Society of Edinburgh: Earth Sciences* **76**, 219–30.

Fortey, R.A. (1990) Ontogeny, hypostome attachment and trilobite classification. *Palaeontology* **33**, 529–76.

Fortey, R.A. and Chatterton, B.D.E. (1988) Classification of the triilobite suborder Asaphina. *Palaeontology*, **31**, 165–222.

Fortey, R.A. and Morris, S.F. (1978) Discovery of nauplius-like trilobite larvae. *Palaeontology* **21**, 823–33. (Pre-protaspis stages)

Fortey, R.A. and Owens, R.M. (1975) Proetida: a new order of trilobites. *Fossils and Strata* **4**, 227–40. (Taxonomy; used in the classification presented here)

Fortey, R.A. and Owens, R.M. (1990a) Trilobites, in *Evolutionary Trends* (ed. K.J. McNamara), Belhaven, Boston, pp. 121–42.

Fortey, R.A. and Owens, R.M. (1990b) Evolutionary radiations in the Trilobita, in *Major Evolutionary Radiations* (eds P.D. Taylor and G.P. Larwood), Systematics Association Special Vol. 42, Clarendon Press, Oxford, pp. 139–64.

Fortey, R.A. and Owens, R.M. (1991) A trilobite fauna from the highest Shineton Shales in Shropshire, and the correlation of the Tremadoc. *Geological Magazine* **128**, 437–64.

Fortey, R.A. and Seilacher, A. (1997) The trace fossil *Cruziana semiplicata*, and the trilobite that made it. *Lethaia*, **30**, 89–168.

Fortey, R.A. and Whittington, H.B. (1989) The Trilobita as a natural group. *Historical Biology* **2**, 125–38.

Hammann, W. (1983) Calymenacea (Trilobita) aus dem Ordovizium von Spanien: ihre Biostratigraphie, Ökologie und Systematik. *Abhandlungen der Senckenbergische Naturforschende Gesellschaft* **542**, 1–177. (Enrollment structures)

Hammann, W. and Rabano, I. (1987) Morphologie und Lebensweise der Gattung *Selenopeltis* (Trilobita) und ihre Vorkommen im Ordovizium von Spanien. *Senckenbergiana Lethaea* **68**, 91–137.

Hanken, N.M. and Störmer, L. (1975) The trail of a large Silurian eurypterid. *Fossils and Strata* **4**, 255–70. (How eurypterids used their legs)

Henningsmoen, G. (1957) The trilobite family Olenidae. *Norske Videnskaps Academie Mathematism-Naturvidenskaps Klasse Skrifte* **1**, 1–303. (Morphology, taxonomy and evolution)

Henningsmoen, G. (1975) Moulting in trilobites. *Fossils and Strata* **4**, 179–200. (Exuvial assemblages and various ways in which trilobites moulted)

Henry, J.-L. (1980) Trilobites ordoviciens du Massif armoricain. *Mémoires de la Societé de Géologie et Minéralogie de Bretagne* **22**, 1–250. (Enrollment structures: exquisite photos)

Henry, J.-L. (1989) Paléoenvironnements et dynamique de faunes de trilobites dans l'Ordovicien (Lanvirn supérieur–Caradoc basal) du Massif Armoricain (France). *Palaeogeography, Palaeoclimatology, Palaeoecology* **73**, 139–53.

Henry, J.-L. and Clarkson, E.N.K. (1975) Enrollment and coaptations in some species of the Ordovician trilobite genus *Placoparia*. *Fossils and Strata* **4**, 87–96. (Interlocking mechanisms and their evolution)

Henry, J.-L. and Nion, J. (1970) Nouvelles observations sur quelques Zeliszkellinae et Phacopidellinae de l'Ordovicien de Bretagne. *Lethaia* **3**, 213–24.

Holm, G. (1898) Über die Organisation des *Eurypterus fischeri* Eichw. *Bulletin of the Academy of Sciences of St Petersbourg* **4**, 369–72.

Horvath, G. and Clarkson, E.N.K. (1993) Computational reconstruction of the probable change of form of the corneal lens and maturation of optics in the post-ecdysial development of the schizochroal eye of the Devonian trilobite *Phacops rana milleri* Stewart 1927. *Journal of Theoretical Biology* **160**, 343–73.

Horvath, G., Clarkson, E.N.K. and Pix, W. (1997) Survey of modern counterparts of schizochroal trilobite eyes: structural and functional similarities and differences. *Historical Biology* **12**, 229–63.

Hughes, C.P., Ingham, J.K. and Addison, R. (1975) The morphology, classification and evolution of the *Trinucleidae* (Trilobita). *Philosophical Transactions of the Royal Society of London B* **272**, 537–607. (Definitive work with descriptions of all trinucleid genera and evolutionary charts)

Jell, P.A. (1978) Trilobite respiration and genal caeca. *Alcheringa* **2**, 251–60.

Jensen, S. (1990) Predation by early Cambrian trilobites on infaunal worms – evidence from the Swedish Mickwitzia Sandstone. *Lethaia* **23**, 29–42.

Jones, T.R. and Woodward, H. (1888–1899) *A Monograph of the British Palaeozoic Phyllopoda (Phyllocarida, Packard)*, Palaeontographical Society, London, pp. 1–211.

Kaufmann, R. (1933) Variations-statistiche Untersuchungen über die Artabwandlung and Artumbildung an der oberkambrischen Trilobitengattung *Olenus* Dalm. *Abhandlung der Geologisches-Paläontologisches Institut der Universität Griefswald* **10**, 1–54. (Classic study of evolution in *Olenus*)

Kjellesvig-Waering, E.N. (1961) The Silurian Eurypterida of the Welsh Borderland. *Journal of Paleontology* **35**, 784–835. (Eurypterid assemblages)

Lindström, G. (1901) Researches on the visual organs of the trilobites. *Kongliga Svensk Vetenskaps Akademie Handlingar* **34**, 1–85. (First work on trilobite eyes, hypostome may have ventral eyes)

Manning, P.L. and Dunlop, J.A. (1995) The respiratory organs of eurypterids. *Palaeontology*. **38**, 287–98.

Manton, S.M. (1973) Arthropod phylogeny – a modern synthesis. *Journal of the Zoological Society of London* **171**, 111–30. (Classification of arthropods)

McKerrow, S. and Scotese, C.R. (eds) (1990) *Palaeozoic Palaeogeography and Biogeography*. Geological Society Memoir No. 12, 240 pp.

McNamara, K.J. (1978) Paedomorphosis in Scottish olenellid trilobites (early Cambrian). *Palaeontology* **21**, 635–56.

McNamara, K.J. (1983) Progenesis in trilobites, in *Trilobites and Other Arthropods* (eds D.E.G. Briggs and P.D. Lane), Special Papers in Palaeontology No. 30, Academic Press, London, pp. 59–68.

McNamara, K.J. and Rudkin, D. (1983) Techniques of trilobite exuviation. *Lethaia* **17**, 153–73.

Miller, J. (1975) Structure and function of trilobite terrace lines. *Fossils and Strata* **4**, 155–78. (Environmental monitoring)

Miller, J. (1976) The sensory fields and life mode of *Phacops rana* (Green 1832) (Trilobita). *Transactions of the Royal Society of Edinburgh* **69**, 337–67. (Functions of cuticular sensory organs)

Miller, J. and Clarkson, E.N.K. (1980) The post-ecdysial development of the cuticle and the eye of the Devonian trilobite *Phacops rana milleri* Stewart 1927. *Philosophical Transactions of the Royal Society of London* **288**, 461–80.

Nathorst, A.G. (1881) Om spar af nagra evertebrerade djur och deras paleontologiska betydelse. *Konglige Svensk Vetenskaps Akademie Handlingar* **18**, 1–104.

Öpik, A. (1960) Alimentary caeca of agnostid and other trilobites. *Palaeontology* **3**, 410–38. (Surface sculpture and its interpretation)

Osgood, R.G. (1970) Trace fossils of the Cincinnati area. *Palaeontographica Americana* **41**, 281–444. (Trilobite trails, burrows and methods of locomotion)

Orlowski, S., Radwanski, A. and Roniewicz, P. (1970) The trilobite ichnocoenoses in the Cambrian sequence of the Holy Cross Mountains Poland, in *Trace Fossils* (eds T.P. Crimes and J.C. Harper) Seel House Press, Liverpool, pp. 345–60. (Description of *Rusophycus* and *Cruziana*)

Palmer, A.R. (1957) Ontogenetic development of two olenellid trilobites. *Journal of Paleontology* **31**, 105–28. (Changes in proportion during ontogeny)

Plotnick, R. (1985) Lift based mechanisms for swimming

in eurypterids and portunid crabs. *Transactions of the Royal Society of Edinburgh: Earth Sciences* **76**, 325–38.

Raasch, G.O. (1939) *Cambrian Merostomata*. Geological Society of America, Specialist Papers No. 19. Geological Society of America, Lawrence, Kan. (Aglaspids)

Raymond, P.E. (1920) The appendages, anatomy and relationships of trilobites. *Connecticut Academy of Arts and Science Memoirs* **7**, 1–169. (Descriptions of trilobite appendages with fine photographs)

Robison, R.A. (1967) Ontogeny of *Bathyuriscus fimbriatus* and its bearing on affinities of corynexochid trilobites. *Journal of Paleontology* **41**, 213–21. (Use of ontogeny in phylogenetic affinities)

Robison, R.A. (1972) Mode of life of agnostid trilobites. *International Geolological Congress 24,* Session 7, 33–40.

Rolfe, W.D.I. (1962) Grosser morphology of the Scottish Silurian phyllocarid crustacean *Ceratiocaris papilio* Salter in Murchison. *Journal of Paleontology* **36**, 912–32. (Description)

Schmalfuss, H. (1981) Structure, patterns and function of cuticular terrces in trilobites. *Lethaia* **14**, 331–41.

Seilacher, A. (1955) Spuren und Lebensweise der Trilobiten; Spuren und Fazies im Unterkambrium, in *Beitrage zur Kenntnis des Kambriums in der Salt Range (Pakistan)* (eds O.H. Schindewolf and A. Seilacher), *Akademie für Wissenschaft und Literatur Mainz Mathematik-Naturwissenschaften Klasse Abhandlung*, 86–143. (First report of sideways-crawling trilobites)

Seilacher, A. (1962) Form and Funktion des Trilobiten – Daktylus. *Paläontologisches Zeitschrifte* (Herta Schmidt Festband), 218–27. (Matching of trilobite trails with appendage structure)

Seilacher, A. (1964) Biogenic sedimentary structures, in *Approaches to Palaeoecology* (eds J. Imbrie and N. Newell), John Wiley, New York, pp. 296–316. (Includes discussion of trilobite tracks)

Selden, P. (1981) Functional morphology of the prosoma of *Baltoeurypterus tetragonophthalmus* (Fischer) (Chelicerata: Eurypterida). *Transactions of the Royal Society of Edinburgh Earth Sciences* **72**, 9–48.

Selden, P.A. (1984) Autecology of Silurian eurypterids, in *Autecology of Silurian Organisms* (eds M.J. Basset and J.D. Lawson), Special Papers in Palaeontology No. 32, Academic Press, London, pp. 39–54.

Sheldon, P.R. (1987) Parallel gradualistic evolution of Ordovician trilobites. *Nature* **330**, 561–3.

Sheldon, P.R. (1988) Trilobite size–frequency distributions, recognition of instars, and phyletic size changes. *Lethaia* **21**, 293–306.

Shergold, J.H. (1971) Late Upper Cambrian trilobites from the Gola Beds, Western Queensland. *Bureau of Mineral Resources Geology and Geophysics Bulletin* **112**, 1–126.

Siveter, D.J. and Selden, P. (1987) A new giant xiphosurid from the Lower Namurian of Weardale, County Durham. *Proceedings of the Yorkshire Geological Society* **46**, 153–68.

Speyer, S. and Brett, C.G. (1987) Clusterid trilobite assemblages in the Middle Devonian Hamilton Group. *Lethaia* **18**, 85–103.

Speyer, S., and Chatterton, B.D.E. (1989) Trilobite larvae and larval ecology. *Historical Biology* **3**, 27–60.

Stitt, J. (1976) Functional morphology and life habits of the Late Cambrian trilobite *Stenopilus pronus* Raymond. *Journal of Paleontology* **50**, 561–76.

Stitt, J. (1983) Enrolled Late Cambrian trilobites from the Davis Formation, southeast Missouri. *Journal of Paleontology* **57**, 93–105.

Stockton, W.L. and Cowen, R. (1976) Stereoscopic vision in one eye – palaeophysiology of the schizochioal eye of trilobites. *Palaeobiology 2,* 304–15. (New interpretation of eye function)

Størmer, L. (1930) Scandinavian Trinucleidae. *Norske Videnskaps Akademie Mathematik-Naturvidenskaps-Klasse Skriften* **4**, 1–111. (Detailed morphology of fringe and sensory organs)

Størmer, L. (1939) Studies on trilobite morphology. Part 1. The thoracic appendages and their phylogenetic significance. *Norsk Geologisk Tidsskrift* **19**, 143–273. (*Ceraurus* appendages; a very detailed study)

Størmer, L. (1970–1974) Arthropods from the Lower Devonian (Lower Emsrain) of Alken an der Mosel Germany. 1970: Part 1, Arachnida; 1972: Part 2, Xiphosura; 1973: Part 3, Eurypterida Hughmilleridae; 1974: Part 4, Eurypterida Drepanopteridae. *Senckenbergiana Lethaea* **51**, 335–69; **53**, 1–29; **54**, 119–295, 359–451. (Classification of eurypterids)

Størmer, L. (1980) Sculpture and microstructure of the exoskeleton in chasmopinid and phacopid trilobites. *Palaeontology* **23**, 237–72.

Stürmer, W. and Bergström (1973) New discoveries on trilobites by X–rays. *Palaontologie* **47**, 104–41. (Appendage morphology)

Stubblefield, J. (1936) Cephalic sutures and their bearing on current classification of trilobites. *Biological Reviews* **11**, 407–40. (Classic paper pointing out limitations of surtures as taxonomic criteria)

Teigler, D.J. and Towe, K.M. (1975) Microstructure and composition of the trilobite exoskeleton. *Fossils and Strata* **4**, 137–49. (Cuticle of several species described)

Thomas, A.T. (1980) Trilobite associations in the British Wenlock, in *The Caledonides of the British Isles – Reviewed* (eds A.L. Harris, C.H. Holland and B.E. Leake), Special Publications of the Geological Society of London, Vol. 8, pp. 447–51.

Thomas, A.T. and Lane, P.D. (1984) Autecology of Silurian trilobites, in *Autecology of Silurian Organisms* (eds M.J. Basset and J.D. Lawson), Special Papers in Palaeontology No. 32, Academic Press, London, pp. 55–69.

Thomas, A.T., Owens, R.M. and Rushton, A.W.A. (1984) Trilobites in British stratigraphy. Geological Society of London Special Report No. 16, pp. 1–78.

Waterston, C.D. (1957) The Scottish Carboniferous Eurypterida. *Transactions of the Royal Society of Edinburgh* **63**, 265–88.

Waterston, C.D. (1960) The median abdominal appendage of the Silurian eurypterid *Slimonia acuminata*. *Palaeontology* **3**, 245–59.

Waterston, C.D. (1975) Gill structures in the Lower Devonian eurypterid *Tarsopterella scotica*. *Fossils and Strata* **4**, 241–54. (Detailed morphology of gills)

Waterston, C.D. (1979) Problems of functional morphology and classification in stylonuroid eurypterids (Chelicerata Merostomata), with observations on the Scottish Silurian Stylonuroidea. *Transactions of the Royal Society of Edinburgh* **70**, 251–322.

Westergård, A. (1922) Sveriges Olenidskiffer. *Sveriges Geologiska Untersokning* **18**,1–205. (Classic taxonomic study of olenid-bearing shales of Scandinavia)

Westrop, S.R. (1983) The life habits of the Ordovician illaenine trilobite *Bumastoides*. *Lethaia* **16**, 15–24.

Whittington, H.B. (1956) Silicified Middle Ordovician trilobites: the Odontopleuridae (Trilobita). *Journal of Paleontology* **30**, 304–520. (Detailed morphology and functional interpretation)

Whittington, H.B. (1957) The ontogeny of trilobites. *Biological Reviews* **32**, 421–69. (Standard reference)

Whittington, H.B. (1966) Phylogeny and distribution of Ordovician trilobites. *Journal of Paleontology* **40**, 696–737. (Evolutionary charts and discussion of faunal provinces)

Whittington, H.B. (1975) Trilobites with appendages from the Middle Cambrian, Burgess Shale, British Columbia. *Fossils and Strata* **4**, 97–136. (Description and functional morphology of *Olenoides* appendages)

Whittington, H.B. (1980) Exoskeleton, moult stage, appendage morphology and habits of the Middle Cambrian trilobite *Olenoides serratus*. *Palaeontology* **23**, 171–204. (Sceptical about *Cruziana* being of trilobite origin)

Whittington, H.B. (1981) Paedomorphosis and cryptogenesis in trilobites. *Geological Magazine* **118**, 591–602.

Whittington, H.B. (1988a) Hypostomes of post-Cambrian trilobites. *New Mexico Bureau of Mines Mineral Research Memoirs* **44**, 321–39.

Whittington, H.B. (1988b) Hypostomes and ventral cephalic sutures in Cambrian trilobites. *Palaeontology* **31**, 577–609.

Whittington, H.B. (1989) Olenelloid trilobites: type species, functional morphology and higher classification. *Philosophical Transactions of the Royal Society of London B* **324**, 111–147.

Whittington, H.B. (1990) Articulation and exuviation in Cambrian trilobites. *Philosophical Transactions of the Royal Society of London B* **329**, 27–46.

Whittington, H.B and Almond, J.E. (1987) Appendages and habits of the Upper Ordovician trilobite *Triarthus eatoni*. *Philosophical Transactions of the Royal Society of London B* **317**, 1–46.

Whittington, H.B. and Hughes, C.P. (1972) Ordovician geography and faunal provinces deduced from trilobite distribution. *Philosophical Transactions of the Royal Society of London B* **263**, 235–78. (With detailed maps; updates Whittington, 1966)

Whittington, H.B. and Hughes, C.P. (1974) Geography and faunal provinces, in *The Tremadoc Epoch* (ed. C.A. Ross), Society of Economic Paleontology and Mineralogy Specialist Publication No. 21, Society of Economic Paleontology and Mineralogy, Tulsa, Okla., pp. 203–18.

Wills, L.J. (1965) A supplement to Gerhard Holm's 'Über die Organisation des *Eurypterus fischeri* Eicher', with special reference to the organs of sight, respiration and reproduction. *Arkiv für Zoologie*, **18**, 93–145.

Wilmot, N.V. (1990) Biomechanics of trilobite exoskeletons. *Palaeontology* **33**, 749–68.

Wilmot, N.V. and Fallick, A. (1989) Original mineralogy of trilobite exoskeletons. *Palaeontology* **32**, 297–304.

Zhang, X.-G. and Clarkson, E.N.K. (1990) The eyes of Lower Cambrian eodiscid trilobites. *Palaeontology* **33**, 911–32.

Exceptional faunas; ichnology

In the final chapter we consider two quite different and contrasting dimensions of palaeontology. The first concerns exceptionally well-preserved fossil faunas, how they come to be preserved and what they have to tell us. The second is ichnology, the study of trace fossils: the tracks, trails, mines and galleries made by long-vanished creatures, on or in the sediment. Whereas the trace makers are not normally themselves preserved, perhaps paradoxically, their behaviour patterns are. These are two end points in palaeontology, each of which illuminates uniquely the history of life on this planet.

12.1 Introduction

'Only a fraction of the myriad creatures that have lived on the Earth have left behind traces of their existence, and only specific parts of those organisms have been preserved' (Briggs, 1991). This statement is indeed true, and it emphasizes that the vast majority of fossil assemblages are composed only of shells or skeletons. Normally we expect to see no more than a narrow band of 'preservable' organisms from an originally much broader biotic spectrum. Accordingly our conception of the course of organic evolution through time is based upon 'a small sample consisting almost entirely of animals with preservable hard parts' (Johnson and Richardson, 1969).

There are, however, rare circumstances where fossils occur either in remarkable concentration or in outstanding preservation. The geological horizons preserving such remains were termed by Seilacher (1970) Fossil-Lagerstätten, which may be roughly translated as 'fossil-bonanzas' or 'rock bodies unusually rich in palaeontological information'.

Of these, first there are **concentration deposits** in which, at a particular horizon, exceptionally large numbers of fossils are preserved. These include the following:

- condensation deposits, where the rate of sedimentation has been so slow, that an extended time period is represented by a thin layer only; this may be crowded, for example, with ammonite shells which had been slowly accumulating for many thousand years;
- placer deposits, such as bonebeds, where fossil material has been concentrated, for example, by tidal lag effects;
- concentration traps such as fissure fillings into which small vertebrates lived or fell and were buried.

These concentration deposits have a greater number, or a more diverse assemblage of fossils than usual, but normally preserve no more than shells or skeletons.

Second, there are **conservation deposits** and it is these that may, and often do, provide information of much higher quality, and can include soft-part preservation. Amongst these are the following:

- stagnation deposits, where anoxic bottom conditions have precluded the activities of scavengers; in black shales thus deposited may be found undisturbed and exquisitely preserved remains of graptolites or fish;
- obrution deposits, where a rich biota may be preserved by rapid burial; in such instances starfish or crinoids, which usually break up very soon after death, remain intact after being smothered by the inrushing sediment and are fossilized;
- conservation traps, a classic example being the trapping of insects in the sticky resin of pine trees, which subsequently hardens as amber;

concretions, where the fossils are preserved within a diagenetic nodule.

If, in these cases, anything other than shells or skeletons is to be preserved, one condition must be fulfilled: that diagenetic changes within the enclosing sediment set in rapidly, within hours of death. Now a freshly deposited sediment may consist of very heterogeneous components, thrown together, and not necessarily in a state of chemical equilibrium. This is particularly true if it contains the remains of living material. The breakdown of this organic material mediated by bacteria may set up reactions which restore chemical balance but may alter the character of the sediment profoundly – this is diagenesis, in which soft-part fossilization involves the growth of new, i.e. authigenic minerals. In certain unusual combinations of circumstance the soft or non-mineralized parts may be preserved as films of phosphate, silicate or kaolinite; alternatively certain organs may be wholly or partially pyritized. Rarely, even whole organs such as muscles may be preserved in phosphate.

In general terms, the rapid burial of fossils, and anoxic conditions (which inhibit scavenging) favour soft-part preservation, but in the vast majority of cases soft parts are preserved only by the growth of authigenetic minerals very soon after the death of the organisms. Very often sediments are oxygenated at the surface and here decay is rapid, but anaerobic conditions are present below. In anoxic layers decay is inhibited but not halted. There may be chemically stratified zones within the sediment where different kinds of bacterial indicators operate (Berner, 1981; Allison, 1988a,b, 1990). Three main parameters: rate of burial, salinity and organic content, control the geochemistry of the sediments, and in different combinations lead to different authigenetic minerals (Fig. 12.11). These minerals are pyrite, carbonates (calcite or siderite), phosphates and silica.

Pyrite preservation of soft parts is favoured by rapid burial, a low organic content and normal or near-normal salinity, and sulphate ions must be available. Carbonate precipitation is likely to take place when burial is rapid and organic content is high. In low-salinity conditions, siderite tends to be deposited whereas normal salinity leads to calcite preservation. The conditions required for phosphatization of soft parts are a low rate of burial and a high organic content. Vivianite is deposited in low salinity, apatite in normal marine conditions.

Calcium phosphate (apatite or fluorapatite) preserves details of structure better than any other authigenetic mineral. However, it will not form where the concentration of HCO_3^- ions is high; such alkaline conditions will activate instead the precipitation of calcium carbonate, and this 'default mechanism' has to be switched off if calcium phosphate is to form (Briggs and Wilby, 1996). Experiments on decaying shrimps in normal 'open' conditions show that $CaCO_3$ crystals begin to precipitate within 3 days of death. Phosphate, however, only precipitates in low-pH conditions, as favoured by 'closed' anoxic conditions, and influenced by microbial activity (Briggs and Kear, 1994). The balance, however, is delicate, and initially phosphatized tissue may be overgrown by, or is found in close association with bundles of $CaCO_3$ crystals – the result of a localized rise of pH.

One remarkable example of sequential authigenic mineralization has been described from the Jurassic of La Voulte sur Rhone, France (Wilby, 1996). Here there is a diverse fauna of marine invertebrates preserved in three dimensions, in a shaly deposit with stratabound minerals, laid down in a restricted basin. In concretions from within the shales, uncrushed animals are uniquely preserved. Apatite formed early as a template upon which other minerals formed in succession: calcite, then ± gypsum, pyrite ± chalcopyrite and finally galena, and different kinds of tissue were preserved by different minerals.

There are three common modes of preservation by authigenic minerals: permineralization of soft tissues, mineral coats and tissue moulds. Permineralization, the replacement of such structure as muscle fibres, is very rare and where found (as in the muscles of Cretaceous fish from Brazil; Martill, 1988) must have taken place within hours of death. Only phosphates, the earliest authigenetic minerals to form, can permineralize in this way, but they may also affect more durable material such as cellulose or chitin.

Mineral coats, the commonest mode of occurrence, may be of phosphate, carbonate or pyrite. The surface of the organism acts as a template for the formation of minerals, whose precipitation is often initiated or mediated by bacteria. These minerals form coats on the surface, and though the soft parts decay, their external coating does not, remaining as a pseudomorph of the soft tissues. Well-known examples such as the Burgess Shale, the

Swedish anthraconites, the Hunsrückschiefer and the Messel Oil Shale are so preserved. At certain periods in the Earth's history, particular minerals seem to have been particularly abundant. There was, for instance, a remarkable amount of phosphate in the early and Mid-Cambrian (Cook and Shergold, 1984), hence phosphate coatings were particularly common at this time. Silica may also form mineral coats, encrusting small trilobites and brachiopods, for example, and fossil plants are sometimes found in three dimensions, with cellular detail precisely replicated by silica coating.

Tissue casts are formed by the stabilization of sediments through diagenesis but before the rock finally lithifies. Faunas found in siliceous or calcareous nodules (e.g. the Mazon Creek and similar biotas) are formed through such processes. Finally the Ediacaran faunas of South Australia (Chapter 3) were preserved by lithification of the containing sand, which underwent very little compaction so that the form of the organisms was retained.

Fossil-Lagerstätten can tell us first much about the anatomy and relationships of animals otherwise known only from hard parts. The discovery of phosphatized 'conodont animals' in the Scottish Carboniferous illustrates this nicely (Briggs *et al.*, 1983); the tooth-like conodonts, the only preservable parts of an otherwise soft-bodied creature, had been known for 125 years previously but not the animal that bore them. Second, they tell us about the nature of entirely soft-bodied or otherwise normally non-preservable animals, and in some cases show the complete spectrum of animals in ancient communities. In this context, the study of Cambrian Lagerstätten has totally altered our perspective on early Phanerozoic life. What has emerged so clearly is that the 'explosion' of life in Cambrian times gave rise to a diversity of forms surpassing that of living creatures today. This concept, nowadays almost taken for granted, would have been inconceivable without the Lagerstätten. The third potential of Fossil-Lagerstätten lies in molecular palaeontology, which will not be pursued further here.

There has been much interest in 'extraordinary' fossil biotas of late (e.g. Whittington and Conway Morris, 1985; Gould, 1990; Simonetta and Conway Morris, 1991; Briggs, 1991; Allison and Briggs, 1991), both in the processes of fossilization and the nature of the biotas themselves. It is surprising how many occurrences there are of fossil assemblages with soft-part preservation; over 60 sites have been documented, and such Lagerstätten occur in virtually all the geological systems. The following list (modified from Briggs, 1991) gives an indication of where some of the more important ones lie. Some sites of greater interest from the palaeobotanical/early terrestrial faunal and vertebrate point of view are included here also.

EOCENE. Grube Messel, Frankfurt, Germany
Lake deposit, plants, vertebrates, insects
(Franzen, 1985)

EOCENE. Green River, Wyoming
Large lake deposit. Fish and other vertebrates
(Grande, 1984)

CRETACEOUS. Sierrra de Montsech, Spain
Web-weaving spiders, insects, crustaceans, vertebrates
(Selden, 1990)

CRETACEOUS. Santana, Brazil
Fish with preserved muscle fibres, pterosaurs with wings
(Martill, 1988, 1993)

U. JURASSIC. Solnhofen, Germany
(see text)

M. JURASSIC. Christian Malford, England
Soft-bodied squids
(Allison, 1988a)

L. JURASSIC. Holzmaden, Germany
Reptiles, crustaceans, cephalopods
(Hauff and Hauff, 1981)

TRIASSIC. Gres à Voltzia, France
Deltaic deposits, plants, insects, and terrestrial organisms are also aquatic fish and crustaceans
(Briggs and Gall, 1990)

U. CARBONIFEROUS. Mazon Creek, Illinois
Deltaic and nearshore marine biotas, and equivalents in England and France (see text)

L. CARBONIFEROUS. Scottish Carboniferous 'shrimp beds'.
Crustaceans, conodont animals, tomopterid worms, fish, chordates
(Briggs and Clarkson, 1983, 1985; Briggs *et al.*, 1991b)

L. CARBONIFEROUS. East Kirkton, Scotland
Hot spring deposits, plants, amphibians, reptiles, scorpions
(Rolfe, 1988)

M. DEVONIAN. Gilboa, New York State
Early terrestrial biota, spiders (with silk spinnerets), pseudoscorpions
(Shear *et al.*, 1984; Selden *et al.*, 1991)

L. DEVONIAN. Hunsrückschiefer, Germany
(see text)

L. DEVONIAN. Rhynie, Scotland
Hot spring deposit, plants, early terrestrial ecosystem
(e.g. Selden and Edwards, 1989)

M. SILURIAN. Herefordshire, England
Arthropods and worms from $CaCO_3$ concretions in a volcanic ash
(Briggs *et al.*, 1996)

M. SILURIAN. Lesmahagow, Scotland
Fish, arthropods, chordates
(Ritchie, 1985)

L. SILURIAN. Waukesha, Wisconsin
Arthropods, conodont animal
(Mikulic *et al.*, 1985)

U. ORDOVICIAN. Soom Shale, South Africa
Conodont animals, other invertebrates
(Gabbott *et al.*, 1995)

L. ORDOVICIAN. 'Beecher's Bed', Utica, New York
Pyritized trilobites with appendages
(Cisné, 1972; Briggs *et al.*, 1991a)

U. CAMBRIAN. Swedish 'Orsten' fauna
(see text)

M. CAMBRIAN. Burgess Shale Fauna of British Columbia and equivalents in Utah (see text)

L. CAMBRIAN. Chengjiang fauna, China
(see text)

PRECAMBRIAN. Ediacara and Avalon faunas
(see Chapter 3)

A number of these are selected for further discussion, beginning with the exhaustively researched Burgess Shale. (N.B. Since the fauna has become well known through the syntheses of Conway Morris (1982), Whittington (1985), Gould (1990) and Briggs *et al.* (1994), references to individual descriptive papers are only given where directly relevant.)

12.2 Burgess Shale fauna (Figs 12.1–12.4)

In 1909 the American geologist Charles D. Walcott was engaged in a reconnaissance survey of the Cambrian geology of the Mount Field area in

(a)

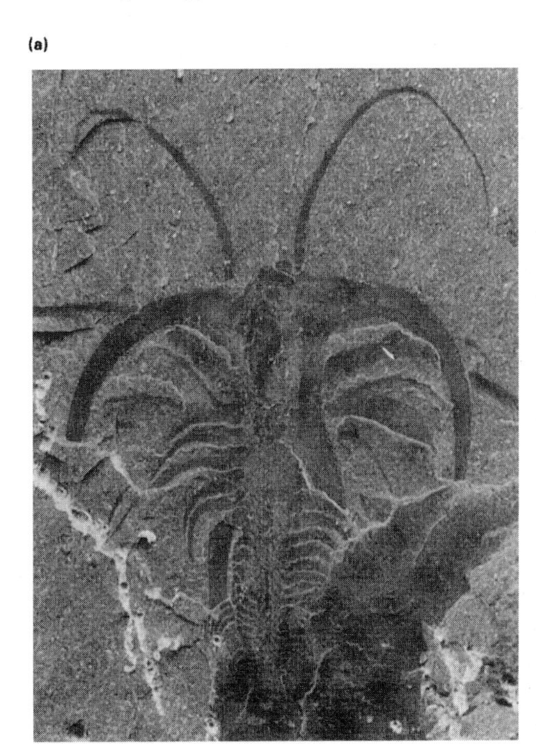

(b)

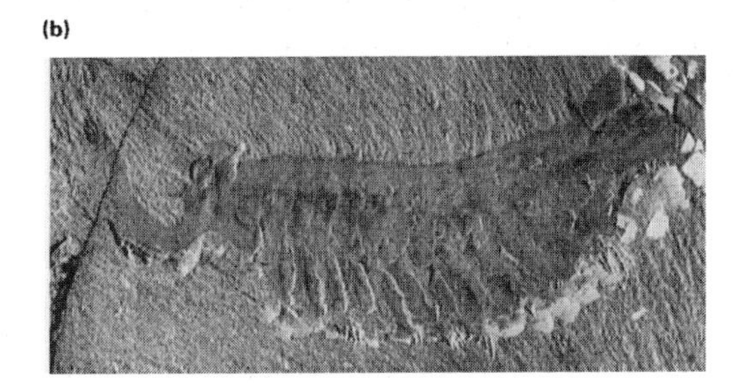

Figure 12.1 (a) *Marrella splendens* (M. Cam.), Burgess Shale, British Columbia, specimen from the Geological Survey of Canada with most of the body preserved though posterior cephalic horns are damaged (×4); (b) *Opabinia regalis* (M. Cam.), Burgess Shale complete specimen in lateral view (×1 approx.). (US National Museum no. 57683.)

British Columbia, Canada. In the process he found a single dislodged slab of rock containing a remarkable assemblage of previously unknown fossils, and he went back the next year to find the horizon from which it came. When the productive sequence in the Middle Cambrian Burgess Shale had been successfully located, he excavated and intensively quarried it, discovering not only trilobites with their appendages preserved but also a very diverse suite of other arthropods, numerous worms, a curious creature like the modern onychophoran *Peripatus*, echinoderms, brachiopods and early molluscs, as well as sponges and algae. There were also many other animals which are not assignable to any living phylum. Walcott's quarrying activities between 1910 and 1917 produced several tens of thousands of specimens from what he called the 'Phyllopod bed' and resulted in diverse publications by himself and other authors. In 1967–1968, under the direction of H.B. Whittington, many thousands of new specimens, including parts and counterparts, were collected from Walcott's phyllopod bed and from a level (the Raymond Quarry) 23 m higher in the section, and these have been intensively studied since.

All the soft-bodied and thin-shelled fossils are flattened, preserved as a dark film which wholly or partly reflects light; pyrite may be associated with the film. This film consists of calcium aluminosilicates with additional magnesium in the more reflective areas. The majority of specimens lie flat upon the bedding planes, but many others lie at an oblique angle to the bedding with their spines and appendages at different levels (Fig. 12.1). In either orientation of preservation, different surfaces of the body separate on part and counterpart, so both are needed for a full interpretation of structure. Since Walcott did not make use of both faces for study, some of his interpretations have been found to be in error.

The Burgess Shale was apparently deposited in relatively deep water off a submarine limestone escarpment on or near a deep-sea fan. The fauna lived on the mud surface below the bank or swam just above it, and it was overwhelmed by a turbidity current of fine suspended sediment which flowed down the slope and transported the animals to an anaerobic environment, not far away, where they were preserved. It is thus, with reference to Seilacher's scheme, an obrution deposit. This explains the attitudes of the animals in the sediment and the completeness of most of the specimens.

Arthropods (Fig. 12.2)

Arthropods account for the largest fraction of the biota (37%). The trilobites, of which *Olenoides* has preserved appendages, are mainly benthic, but there are also some pelagic agnostids and eodiscids. *Naraoia*, which has an unmineralized exoskeleton, is a unique kind of trilobite with a large cephalon and a posterior shield (thoracopygidium). However, it has uniramous antennae and typically trilobitan appendages. *Tegopelte* is a second soft-bodied trilobite and it is very large. It has a cephalon and pygidium, but the three divisions of the thorax embrace three or four body somites.

The non-trilobite arthropods of the Burgess Shale were until recently classified together as Class Trilobitoidea, since their appendages seemed to be generally biramous with leg and gill branches like those of trilobites. However, recent work has shown that this supposed resemblance is evident only in a few genera and even in these is not especially close. The possession of biramous limbs is in any case no more than a sympleisomorphy; a heritage from an original arthropod ancestor with serially uniform limbs, and therefore no guide to affinity.

Marrella splendens, the commonest and perhaps the most elegant of the arthropod fauna, has a wedge-shaped cephalic shield with four long, backwardly directed spines, the posterior pair of which have crenulated margins. On the ventral surface is a two-spined labrum near which are attached the two pairs of antennae: the first pair are long, flexible, multi-jointed rods; the second pair are six-segmented with setose distal joints. Behind the head is the cylindrical segmented body which is devoid of pleurae. Each of the 25 somites has a pair of biramous appendages which possess a jointed walking leg and a feathery gill branch above it. The latter was probably capable of rotation backwards and forwards about a horizontal axis, and the combined rotary effects of all gill branches could have enabled *Marrella* to swim. Rare specimens are preserved with the intestine isolated from the body. The affinity of *Marrella* with other arthropods is still uncertain. The resemblance between the Cambrian *Marrella* and the Devonian

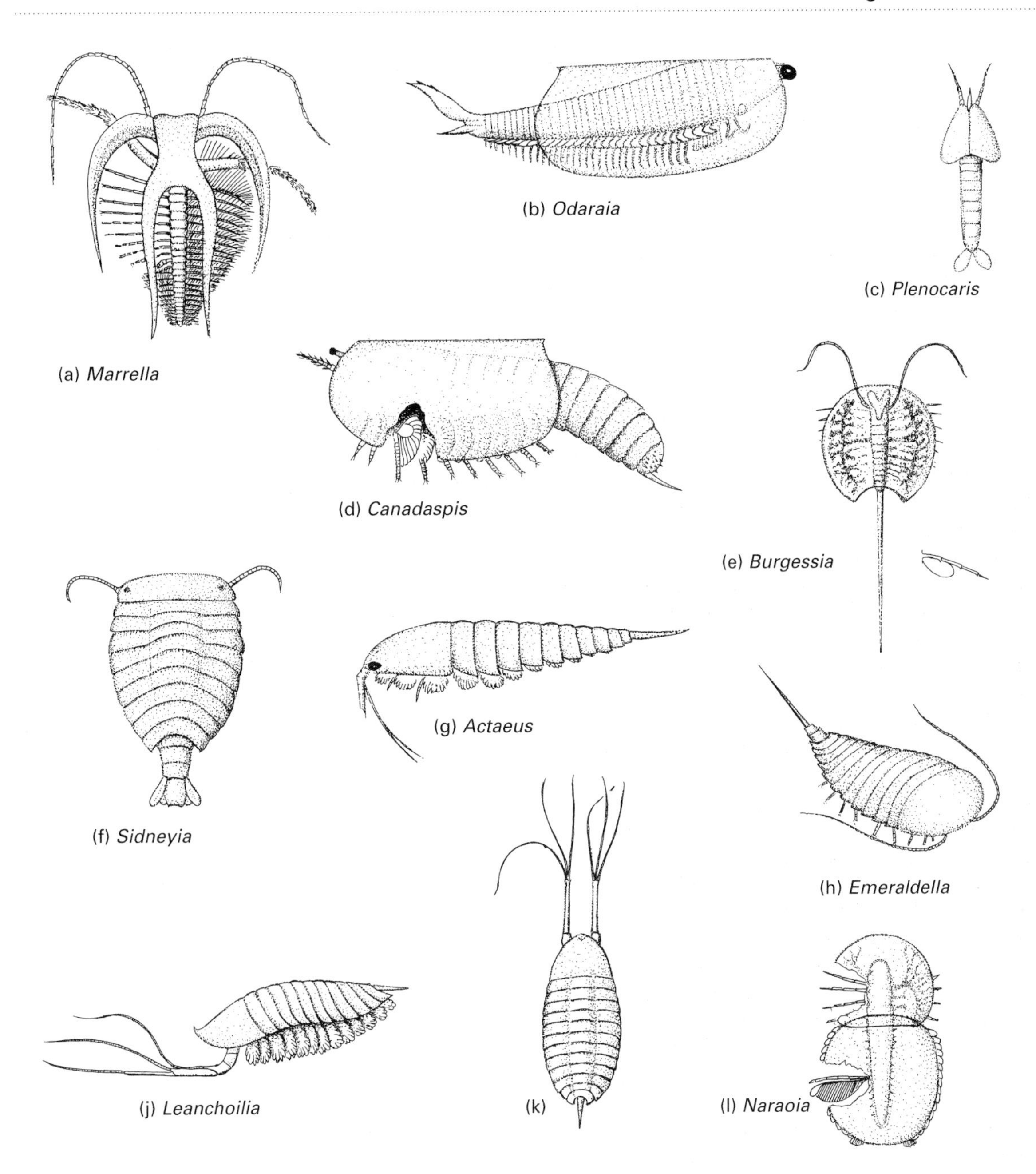

Figure 12.2 Arthropods from the Burgess Shale: (a) *Marrella*, dorsal view (× 2 approx.); (b) *Odaraia*, lateral view (× 0.75); (c) *Plenocaris*, dorsal view (× 1); (d) *Canadaspis*, lateral view, showing structure of the appendages (×1); (e) *Burgessia*, dorsal view and appendage structure (× 1.5); (f) *Sidneyia*, dorsal view (× 0.5); (g) *Actaeus*, lateral view (× 2.5); (h) *Emeraldella*, antero-lateral view (× 1); (j), (k) *Leanchoilia*, dorsal and lateral views (× 1); (l) *Naraoia*, dorsal view, showing appendages (× 0.6). (Redrawn from Briggs, Bruton, Conway Morris, Hughes and Whittington, see text.)

Mimetaster (q.v.) may point to a real affinity and indicate the survival of this stock until much later, but on the other hand the similarities may be symplesiomorphic.

The second commonest species, *Canadaspis perfecta*, is at present regarded as the earliest positively identified crustacean; it is a phyllocarid with a carapace covering most of the body. There are two pairs of antennae, a mandible pair and ten pairs of biramous appendages. It was probably a benthic feeder and is often found in clusters.

Burgessia bella has a large, convex carapace covering the whole body except for the terminal tail spine, which emerges under a posterior indentation and is nearly twice as long as the body. The body is rod-like, and the cephalic region is followed by eight segments and a telson in front of the spine. A pair of very crenulated kidney-shaped organs occupies the lateral parts of the carapace. These have been interpreted as digestive diverticulae. There is a pair of long, multijointed, uniramous antennae projecting in front of the carapace. The cephalic region bears three pairs of biramous, jointed legs. The coxa of each bears an inner branch of six podomeres, identical with the seven pairs of legs on the trunk. The outer branch of these appendages has the form of whip-like flagellae. The walking legs of the trunk bear small, lateral, leaf-like plates which were probably gills, seemingly attached to the coxa.

Plenocaris may be a primitive crustacean. Its bivalved carapace covers the anterior part of the body, which bears one pair of antennae and probably three pairs of indeterminate appendages. Behind this the 12-segmented body terminates in a caudal furca; little is known of the limb morphology. The affinities of *Plenocaris* are unclear. It is not a phyllocarid, even though it superficially appears to be so.

Waptia is a shrimp-like form with a carapace and long antennae. *Leanchoilia* (Fig. 12.2j), which was probably a detritus feeder, has a long trilobed body with a well-defined cephalic part and trunk segments with pleura. The great appendage of the head has a basal joint with four podomeres of which the distal one is clawed and with a long extension. The biramous trunk appendages have a leg and gill branch.

The rare *Odaraia* possesses a long body with up to 45 trunk somites, with biramous limbs and a remarkable carapace so structured that it resembles a cylinder with the appendage series enclosed within it. There is a large anterior pair of eyes and a massive tail with three large flukes which extends posteriorly from the carapace.

Sidneyia is the most abundant large arthropod from the Burgess Shale. Its body is superficially merostome-like and terminates in a fan-like tail, but the presence of antennae and the absence of chelicerae suggest that it is not a merostome. The legs, however, which are provided with gills posteriorly, are in some respects not dissimilar to those of *Limulus*. Gut contents, which include small trilobites and hyolithids, show that this large animal was a predator. The smaller *Emeraldella is* again rather merostome-like, with a long terminal tail spine, but has antennae and biramous limbs. *Actaeus* is of somewhat similar morphology.

There are several other, rare, small arthropod genera, such as *Habelia* and *Molaria*, which extend the range of diversity, and the list given here is by no means exhaustive.

Lobopods

One of the invertebrates originally described by Walcott is *Aysheia* (Fig. 12.3h). It is an annulated, caterpillar-shaped creature, with an anterior pair of branched appendages and ten following pairs of short conical, and likewise annulated 'legs'. It resembles the modern *Peripatus* (Class Onychophora): a terrestrial uniramian which lives in damp soil in the jungles of Minas Gerais and elsewhere in Brazil. The characters of *Peripatus* are in many ways intermediate between those of annelids and arthropods, preserving, for example, the paired segmental excretory organs of the segmented worms, together with an annelid-like eye and a muscular body wall covered by a thin cuticle. Yet the coelom is arthropodan, and the presence of tracheae – ramifying capillaries connecting to the outside and bringing in air to the tissue like those of insects – links *Peripatus* with arthropods. The legs of *Peripatus* move in metachronal rhythm but are unjointed. Their rigidity depends upon the antagonistic operation of muscles against coelomic fluid, so that they can lengthen and be rigid while pressing backwards, thereafter shortening before the next forward stroke. *Peripatus* is highly adapted for terrestrial life and is well defended, being capable of entangling predators with a sticky secretion from glands below the eye.

Aysheaia differs from *Peripatus* in some rather fundamental anatomical respects, and not only in being marine. It is probably not a true onychophoran, but is best considered as a 'lobopod', a term which embraces all such uniramians, including *Peripatus*. *Aysheia* is often found associated with sponges and may have fed upon them.

Opabinia regalis has an elongated, segmented body, in which the head possesses five mushroom-like eyes. From the front of the head extends a long, flexible process terminating in two groups of spines facing each other in a pincer-like fashion. Behind the head is an elongated cylindrical body of 15 segments with a tailpiece having upwardly turned lobes. All the segments bear a pair of appendages, each being a lanceolate gill-blade overlying a flat paddle-like lobe. These were fixed and quite rigid, though possibly capable of movement in an up-and-down plane. The flexible frontal process could reach round to the mouth, which was located ventrally and in the posterior part of the head, and it was probably used to explore for and convey food to the mouth. Budd (1996), when comparing *Opabinia* with new and similar material from Greenland, found limbs below the body. These are ringed, inflatable limbs of lobopod type, and this bizarre animal is reinterpreted as a lobopod, and presumably of benthic habit.

Another Burgess Shale curiosity, *Hallucigenia*, may belong to this same stock. Known from only a very few specimens, it was originally reconstructed as an animal that walked on seven pairs of stiff movable spines, supporting a cylindrical trunk with a globular head, from which seven vertical tentacles arose. The discovery, however, of spiny caterpillar-like animals in the Lower Cambrian Chengjiang Lagerstätte in China, which have many features in common with *Hallucigenia*, suggests an alternative interpretation (Ramsköld and Hou, 1991; Bengtson, 1991). For *Hallucigenia*, turned upside-down is very similar to the Chinese animal. The latter has paired appendages each with a terminal claw; and it now seems most likely that the unpaired appearance of 'tentacles' (in walking legs) of *Hallucigenia* results from imperfect preservation of one set due to the position in which it was buried. The bizarre *Hallucigenia* is therefore probably an 'armoured lobopod', an onychophoran-like animal of the same group as the Chinese fossil. This latter, incidentally, is known as *Microdictyon*. It bears a chain of oval plates along the body, each with a net-like ornament and usually a short spine. Such plates, which often appear in residues of dissolved Lower Cambrian limestone, were described long before the soft-bodied lobopods were known, but since the plates are now understood to be part of the 'lobopod', the whole animal now bears the name. In addition, the 'spines' of *Hallucigenia* (originally thought to be the 'legs') are homologous with those same net-like plates, but with a much more elongated central 'thorn'.

Lobopods were a significant component of the Cambrian radiation. Some of the other peculiar Burgess Shale animals may belong here, including (Budd, 1996) the largest of all, *Anomalocaris*. This has a segmented, flattened, diamond-shaped body with 11 pairs of closely spaced, overlapping lateral fins. These probably undulated in a series of waves, thus propelling the animal forward. Near the mouth are a pair of giant, spiny segmented appendages, the catching apparatus of a predatorial hunter. These structures are sometimes preserved in isolation, and were once believed to represent the abdomen of a crustacean. A circlet of plates surrounding the mouth, and forming a diaphragm with serrated cutting teeth, was previously likewise known only in isolation. For some 70 years this organ, named *Peytoia*, had been believed to be a jellyfish!

Other invertebrates (Figs 12.3, 12.4)

The earliest of all crinoids, *Echmatocrinus* (Fig. 9.51), is found in the Burgess Shale. It has a large conical calyx of irregular plates and plated uniserial arms. There are also eocrinoids and a possible holothurian and edrioasteroid. Molluscs are rare other than hyolithids (if these are indeed molluscs).

The curious plated *Wiwaxia*, with its vertical defensive spines, may well belong to a distinct phylum. Lophophorates indude standard Cambrian inarticulate brachiopods and also the peculiar *Odontogriphus*. This organism is about 6 cm long, and its body is flat and annulated with a poorly defined head of semicircular form. On the head are a pair of lateral palps (sensory organs) of rather indistinct morphology and also a bilaterally symmetrical median structure forming a pair of loops. This apparatus lies at the front end of tubular gut and bears some 25 thorn-like 'teeth'. The teeth were origi-

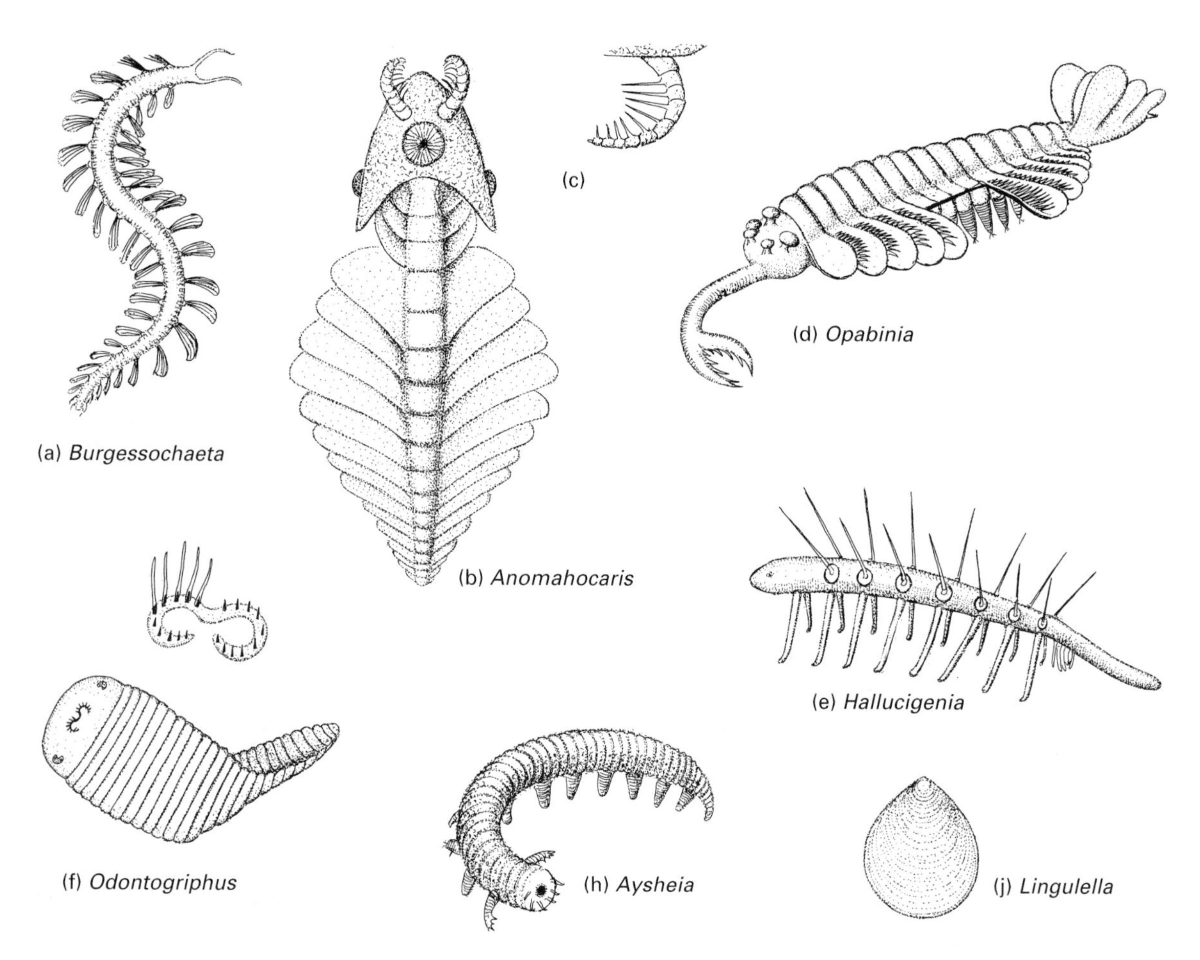

Figure 12.3 Worms, lobopods and other invertebrates from the Burgess Shale: (a) *Burgessochaeta* (× 3); (b), (c) *Anomalocaris*: (b) ventral view and (c) great appendage of same in lateral view (× 0.25); (d) *Opabina*, interpreted as a lobopod, antero-lateral view (× 1); (e) *Hallucigenia*, interpreted as a lobopod (× 1.5); (f) *Odontogriphus* (× 0.5), and (g) reconstruction of its lophophore and tentacles: (h)*Aysheia*, a lobopod (× 1.5); (j) *Lingulella*, a linguliform brachiopod (× 3). [Based on various sources, chiefly Briggs *et al.*, 1994; (d) based on Budd,1996; (e) based on Ramsköld and Hou, 1991.]

nally thought to be possible conodonts, but since the discovery of a true 'conodont animal' (Briggs *et al.*, 1983) of quite different morphology, this now seems unlikely. The 'teeth' of *Odontogriphus* were probably not biting or rasping teeth; they have been interpreted as the supports for a food-gathering apparatus having the form of a 'tentacular lophophore' (Conway Morris, 1979). Such tentacular lophophores are found in modern brachiopods, tube-dwelling phoronid 'worms' and bryozoans, which are all commonly linked together in Superphylum Lophophorata. In these the lophophore, at least in its initial stages of development, is of remarkably constant form and is, incidentally, bilaterally looped like that of *Odontogriphus*. There seems to be a good case for aligning *Odontogriphus* with the Lophophorata as an early derivative of the same superphylum.

The only annelids represented are polychaete worms (e.g. *Burgessochaeta*) and these played a relatively minor role in the fauna. None of these have jaws, for these were not acquired until the Ordovician. *Canadia*, like modern polychaetes, has two paired sets of parapodia (bundles of stiff bristles) for each segment. This is the only genus that might be related to any modern family; none of the others are.

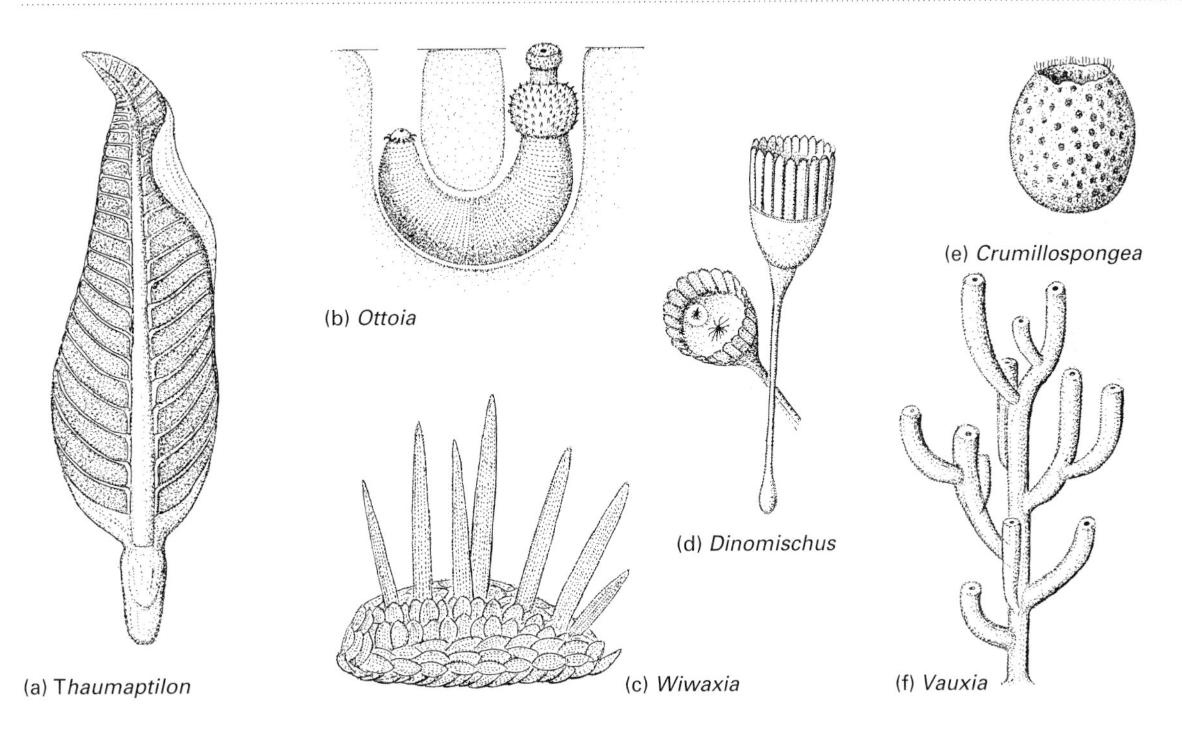

Figure 12.4 Various other Burgess Shale invertebrates: (a) *Thaumaptilon*, a possible Ediacaran survivor (× 3); (b) *Ottoia*, a priapulid worm in its burrow (× 3); (c) *Wiwaxia*, in lateral view (× 1); (d) *Dinomischus* (× 3); (e) *Crumillospongea* (× 1): *Vauxia*, a sponge (× 2). (Redrawn from Briggs *et al.*, 1994; also Conway Morris, Rigby and Whittington.)

There are five species of priapulids whose present-day representatives have a similar retractable spiny proboscis and annulated body to Cambrian priapulids (e.g. *Ottoia*). Modern priapulids are very unimportant, living mainly as cold-water benthos, though one group is endoparasitic. Cambrian forms are more diverse but, as infaunal carnivores, seem to have been displaced from their original environment by predatorial polychaetes later in geological time.

There are a few cnidarians and a very abundant and rich fauna of sponges, (e.g. *Crumillospongia*, *Vauxia*) most of which can be referred to the demosponges and hexactinellids; there are also some lingulate brachiopods. At the other end of the biological spectrum is the primitive and rather fish-like *Pikaia*, which seems to have a notochord and chevron-shaped blocks of muscle (myotomes) typical of chordates.

The position of *Dinomischus* is unclear, but the recently described *Thaumaptilon*, so similar in many ways to the Ediacaran *Charniodiscus*, may prove to be a survivor from the later Proterozoic.

Significance of the Burgess shale faunas

Ecology

The community structure of the Burgess Shale fauna at its type locality has been analysed by Conway Morris (1986). Shelly faunas comprise no more than 20% of the genera (and perhaps no more than 2% of individuals). The soft-bodied or lightly skeletonized faunas are far more abundant and these can generally be analysed in terms of feeding habits, and a food web can be established. In the benthic assemblage there were vagrant benthic deposit feeders (mainly arthropods), infaunal carnivores and scavengers (chiefly priapulids), epifaunal suspension feeders (sponges and brachiopods) and infaunal sessile suspension feeders (hemichordates). There are also large obvious predators such as *Anomalocaris*, and *Ottoia* seems to have been cannibalistic. The diversity of feeding types and their independence shows quite clearly that the fundamental trophic structure of marine metazoans was already established by the Middle Cambrian and possibly earlier.

Geographical distribution

In Middle Cambrian times the Laurentian continent, in which the Burgess Shale occurs, lay isolated and in an equatorial position. It might be expected that other equivalent and contemporaneous faunas would be found in Laurentia, and this has indeed proved to be so. In British Columbia, new discoveries from at least a dozen sites (Collins *et al.*, 1983) show that locally, at least, the Burgess Shale fauna was widespread. The faunas of these localities are broadly similar to those of Walcott's quarry, though some quite new animals, e.g. the early chelicerate *Sanctacaris* ('Santa Claws'; Briggs and Collins, 1988) have been reported.

Further afield there are other Laurentian Middle Cambrian Lagerstätten, which are less well known but of exceptional interest. In Utah (Robison, 1991) there are no less than four, in the Spence, Wheeler and Marjum Formations, where soft-bodied animals also occur with echinoderms, trilobites, brachiopods, hyoliths and sponges. Of the non-trilobite arthropods, worms and sponges, many also occur in the Burgess Shale. The Kinzers Shale in Pennsylvania, poorly exposed and tectonized though it is has likewise clear similarities. The available evidence therefore suggests that, at least in Laurentia, Burgess Shale biotas are only unusual in the way they are preserved and not in their taxonomic content and distribution. Such faunas may indeed have been global in their distribution. The new discoveries of Lower Cambrian but similar faunas in China and elsewhere (q.v.) conduce to the view (Conway Morris, 1989) that Burgess Shale type faunas represent a widely distributed offshore benthic fauna, quite long-ranged (L.–M. Cam.) and with an evolutionarily conservative aspect.

Diversity

The Burgess Shale fauna at its type locality in British Columbia is extraordinarily diverse, with over 140 species in 119 separate genera. It forms a major part of the 'Cambrian evolutionary fauna' (Chapter 3), which otherwise is known only from the hard-shelled trilobites, inarticulate brachiopods, etc. Whereas some genera, such as the crinoid *Echmatocrinus*, the crustacean *Canadaspis* and the chelicerate *Sanctacaris*, are the earliest known representatives of later, important groups, very many others cannot be classified. In terms of the number of different kinds of body plans they show a bewildering diversity. Even though *Hallucigenia* (formerly seen as the most bizarre and peculiar of all the Burgess animals) is now best interpreted as a much more 'respectable' onychophoran-like 'lobopod', the spectrum of diversity is still extraordinary. There thus arises a taxonomic conundrum which led Whittington (1985) to say 'If we knew more of the soft-bodied faunas of the Phanerozoic, our major subdivisions of invertebrates might be viewed differently'. More recently Gould (1990) has eloquently explored this whole concept. In tracing the history of research he shows how Walcott, in his otherwise admirable work, had tried to 'shoehorn' the animals into established systematic categories. He did not realize that the vast diversity which the Burgess Shale animals represented was the key to a new dimension of evolutionary thinking. He 'interpreted them along the path of least resistance' as 'primitive versions of later improvements'. Only when they had been restudied by Whittington and his team and new concepts had taken hold, was the full significance of these Cambrian organisms realized. It is their diversity in terms of body plans that is important; the genera seldom have more than one species and they are separated one from another by wide morphological gaps. Thus early Phanerozoic life shows an explosion of diversity, far greater in variety than today's whole fauna. We do not find a gradual increase of diversity but quite the opposite. The 'tree of life' arising from a root and later flourishing into many branches is unrealistic – a better garden analogy would be a 'grass lawn'. The Early to Middle Cambrian record reveals a great burst of radiation in which marine organisms diversified along many parallel though distinct lines. The elimination of many of these types through increasing competition left the survivors to become the progenitors of today's phyla. There is no indication of a single common ancestor, and many authorities now suggest a polyphyletic organ for the metazoan phyla (i.e. broadly similar kinds of animals arising independently in different parts of the world) rather than a single ancestral form for each phylum.

Persistence

The Burgess Shale fauna was originally known only from the Middle Cambrian of British Columbia. Some remarkable recent discoveries, however, show that many of its elements can be traced right down to the basal Lower Cambrian. Of these, the richest faunas are at Chengjiang in southern China and in Peary Land, Greenland; there are also occurrences in a borehole in northeastern Poland, and in the Sierra Morena of southern Spain.

At Chengjiang a mixed fauna of Lower Cambrian shelly and soft-bodied animals was discovered in 1984 in a mudstone sequence deposited in shallow water. It is the oldest of all Cambrian Lagerstätten, and probably of late Adtabanian age. Stratigraphically below it is a rich assemblage of small shelly fossils. The Chengjiang fauna is dominated by bradoriids, small bivalved arthropods which are superficially similar to, but not closely related to ostracodes. Over 80% of the individuals in this fauna are bradoriids. There are only three genera of trilobites (cf. 40% of the fauna in the Burgess Shale) but many algae, sponges, medusoids, chondrophorans and hyolithids; there are also elements typical of the Burgess Shale. These include priapulids, the scale-bearing onychophoran-like *Microdictyon*, *Dinomischus* and *Anomalocaris*, *Hallucigenia*, *Leanchoilia* and *Naraoia*, but there are also several other non-trilobite arthropods. One of these *Fuxianhuia* looks somewhat like a trilobite, but has a long, parallel-sided, segmented tail. These faunas are still under description: recent references include Conway Morris (1989), Chen and Erdtmann (1991), Hou and Bergström (1990), Ramskold and Hou (1991) and Hou and Bergström (1997). Despite similarities with the Burgess Shale fauna, there are marked differences, probably reflecting an original ecological difference. While it is interesting to compare Cambrian assemblages in sequence, i.e. small shelly fossils, Chengjiang fauna and Burgess Shale faunas, and finally the 'orsten' faunas, it would be misleading to interpret these too directly as a global ecological succession. Each is simply a 'snapshot' of an exceptional fauna in time, specialized for a particular niche, and little is known of contemporaneous, equivalent faunas elsewhere in the world.

A more recently discovered (1989) Lower Cambrian Lagerstätte comes from north Greenland. This fauna (Conway Morris and Peel, 1990; Bengtson, 1991; Budd, 1996) includes many arthropods, polychactes and pirapulids like those of the Burgess Shale and sponges. An unexpected bonus, in addition, solves the problem of the origin of the halkieriids. These are microscopic sclerites of varying form first described in 1967 and previously known only as isolated elements in many Lower Cambrian 'small shelly faunas'. In the Greenland material are complete halkieriid animals (Fig. 12.5).

They are slug-like forms some 3 cm long with a dorsal armour consisting of imbricating rows of the sclerites. Such an organization had actually been predicted, but what was quite unexpected was the presence of a large round shell, one at each end. The affinities of the halkieriids are still very obscure. Regarding these and other such organisms, Bengtson (1990) stated that 'Nature has a way of outshining our most speculative hypotheses'. Who, in considering the Burgess Shale faunas, could possibly disagree?

In broad perspective the Cambrian evolutionary fauna, which included the Burgess Shale forms with large predators such as *Anomalocaris*, spelled the end for the quilted organisms of the 'Garden of Ediacara'. Once it had originated, the Burgess Shale fauna remained very conservative. It may have arisen in shallow waters, and then with the rise of new faunas migrated to deep water, until it too was driven to extinction by later Cambrian and early

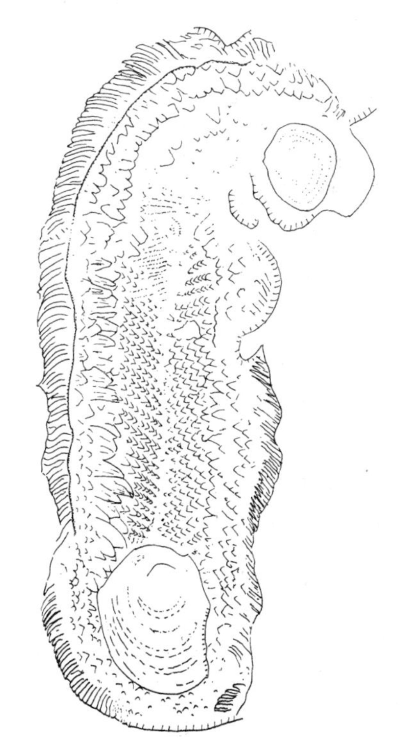

Figure 12.5 A fully preserved halkeriid from Greenland with anterior and posterior shells, and a body covered by dermal sclerites (×1.5). (Redrawn from Conway Morris and Peel, 1990.)

Ordovician competitors. Possibly the peculiar *Mimetaster* from the Devonian Hunsrückschiefer might represent the last survivor of this most extraordinary biota of all time.

12.3 Upper Cambrian of southern Sweden (Figs 12.6, 12.7)

The Upper Cambrian rocks of Sweden are highly fossiliferous and contain abundant olenid trilobites (Chapter 11), usually found in calcareous concretions or lenses known as 'orsten' or 'stinkstones'. These occur over quite a wide area (Skåne, Västergötland, Öland). Some years ago, while searching for conodonts in etched residues of these rocks, Professor K.J. Müller of Bonn made a chance discovery of phosphatized ostracodes and other crustaceans, as well as agnostid trilobites. Many thousands of specimens have been isolated, frequently with the limbs intact. The phosphatic fossils are of two kinds, firstly those with primarily phosphatic hard parts (phosphatocopine ostracodes, inarticulate brachiopods and conodonts) and secondly the arthropods in which the bodies and appendages were preserved by a secondary phosphatic coating which must have formed at the time of burial. In some instances the shell is replaced by phosphate. Preservation is normally exquisite, though in some

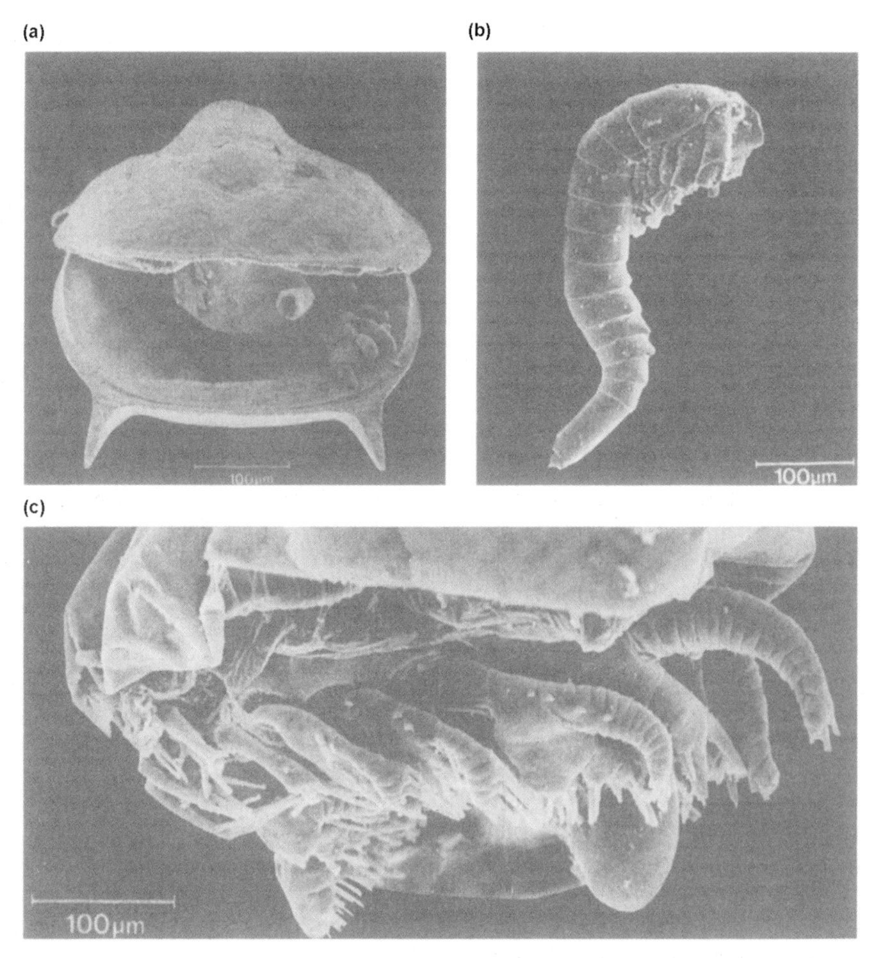

Figure 12.6 Elements of the small phosphatized 'orsten' fauna of the Swedish Upper Cambrian: (a) *Agnostus*, a partially enrolled trilobite showing a phosphatized appendage; (b) *Skara*, a crustacean; (c) phosphatized appendages of a phosphatocopine ostracod. (SEM photographs reproduced by courtesy of Professor Klaus Müller.)

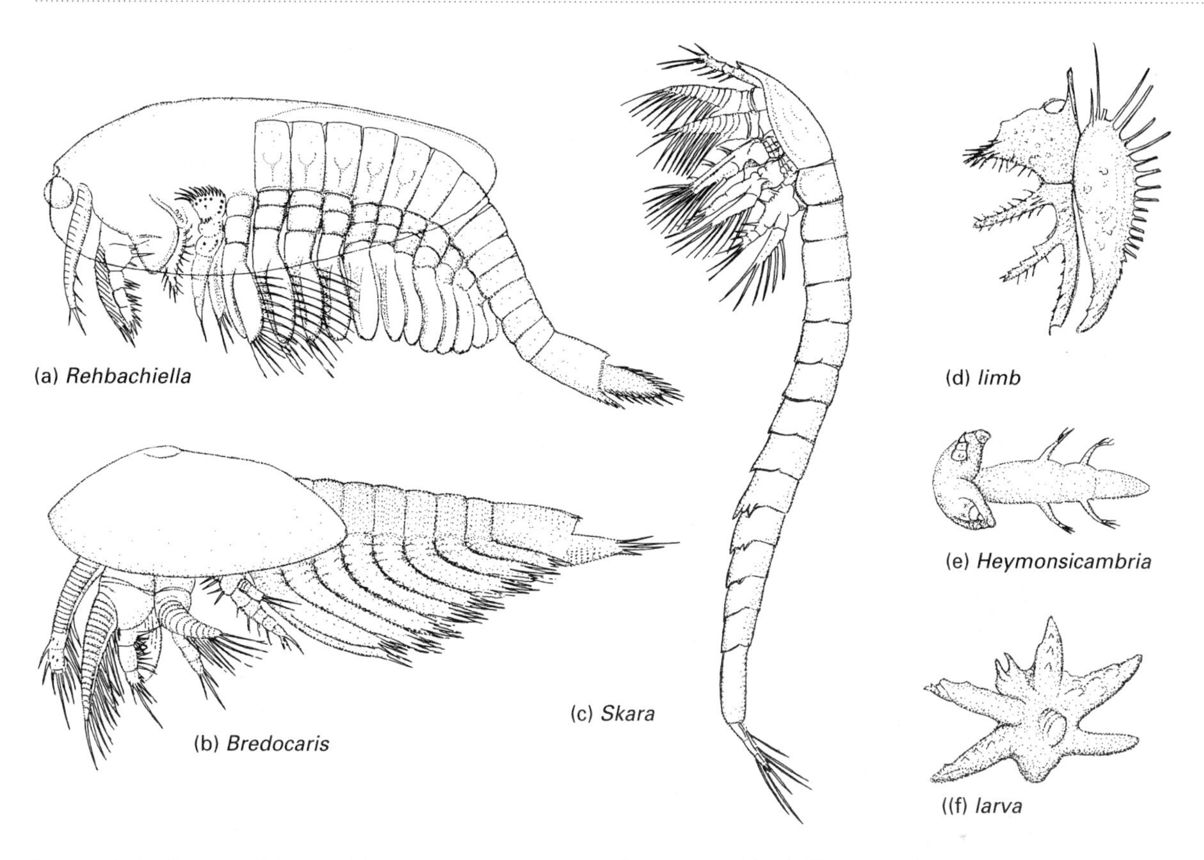

Figure 12.7 Elements of the Swedish Upper Cambrian 'orsten' fauna: (a) *Rehbachiella*, a branchiopod crustacean (× 65); (b) *Bredocaris*, a maxillopod crustacean (× 90); (c) *Skara*, a maxillopod crustacean (× 70); (d) isolated limb of a hesslandonid ostracod (× 70); (e) *Heymonsicambria*, a pentastomid parasite (× 45); (f) a crustacean nauplius larva (× 130). (Redrawn from Müller and Walossek, 1985a,b, 1988; Walossek, 1993; Walossek and Müller, 1994.)

cases, especially at the anterior end of the body, which was in life the most highly phosphatic, the coating is relatively thick and obscures the joints. Where this occurs the spines and small hairs may be diagenetically thickened up to three times their original diameter.

Only small growth stages of the arthropods (<2 mm long) seem to be fossilized in this way. It is quite probable that these inhabited the nutrient-rich flocculent layer on the sea floor, swimming, clambering about and feeding within it. When they died they slid down to the anoxic sea floor just below their habitat and were covered very soon afterwards with a coating of phosphatic bacteria. It is this thin 'shell' of dead, contiguous bacteria that survives when the rest of the animal has decayed. Where the shell is replaced by phosphate it was invaded by the bacteria.

The arthropods were all swimming forms. By far the commonest are the adont primitive ostracodes *Vestrogothia*, *Falites* and *Hesslandona*. These belong to the Order Phosphatocopina, in which the original hard parts were probably phosphate. The abdomen, which may not have been primarily phosphatized, is not usually preserved. In spite of the secondary phosphatic thickening, the information given about the nature of the limbs of these primitive ostracodes is of the highest quality; it is particularly interesting that these phosphatocopine limbs are closely similar to one another within the same animal, and are comparatively unspecialized.

There are now 25 known arthropod genera including 'a large variety of primordial crustaceans whose systematic position is as yet uncertain' (Müller, 1983). Numerous naupliar larval stages of various unidentified crustaceans are likewise present.

Skara has a cephalon with five pairs of appendages and a long flexible trunk and tail. It was a filter feeder and is known only from adults. *Martinssonia* is superficially similar, though it has no specialized cephalic filtering apparatus. It was probably a bottom dweller feeding on detrital particles that it stirred up from the sea floor. The juvenile stages are also known. *Dala* is represented only by thorax and abdomen; it may have been a free swimmer. *Bredocaris* has a large univalved headshield, paired eyes and comparatively unspecialized cephalic limbs. There are seven pairs of thoracic limbs. All larval stages are known. It is unlikely to have been a filter feeder and may have swum just above the flocculent layer. *Walossekia* is not dissimilar, while *Rehbachiella* has a large headshield enclosing much of the body and many limbs. All the early growth stages are known, but the largest specimens are still probably larval. This crustacean is an ancestral branchiopod, and its complex filtering apparatus has some resemblance to that of modern forms. Recently described (Walossek and Müller, 1994) are small pentastomids, and are remarkably similar to their modern counterparts which parasitize the lungs of crocodiles today. It is to be wondered what they fed on in Cambrian times.

The 'orsten' crustaceans are thus very diverse, some being progenitors of modern taxa, others belonging to extinct groups. They were adapted to different microhabitats within or above the flocculent layer.

The trilobite *Agnostus* reveals surprising details of appendage construction. There is one pair of antennae and eight pairs of biramous appendages, but unlike those of 'standard' trilobites these are much diversified. New details have also emerged about the nature of the ventral integument and the stages of early ontogeny. *Agnostus* could not have stretched out fully and it is highly likely that these trilobites lived for much of the time in a state of partial enrollment with a slightly gaping cephalon and pygidium.

The descriptive work on these 'orsten' faunas (Müller, 1979, 1982, 1983, 1985; Müller and Walossek, 1985, 1986, 1987, 1988; Müller and Hinz, 1991; Walossek, 1993) includes perhaps the most precisely documented of all palaeontological studies. It gives a clear insight into the nature and ecology of a Cambrian flocculent-layer community. It provides new information about the limbs of the ancient crustaceans and *Agnostus*. While this gives support to the idea of a monophyletic origin of the crustaceans, there are also similarities between their limbs and those of *Agnostus*, which may (Walossek and Müller, 1990) shed new light on the systematic position of the latter.

Finally, the Swedish faunas, like those of the Burgess Shale, show that the non-trilobite arthropod faunas played a much more important role than has generally been appreciated, in Cambrian times, and the apparent predominance of trilobites must have been accentuated by the higher preservability of their exoskeletons. It is clear also that these Upper Cambrian crustaceans bear little resemblance to those of the Burgess Shale. Although this may, again, be purely due to preservation, the 'morphological and temporal disconnection' of the Burgess Shale fauna referred to by Bergström (1980), in other words its dissimilarity from later faunas may in fact be a real evolutionary phenomenon.

12.4 Hunsrückschiefer fauna (Fig. 12.8)

The Lower Devonian Hunsrückschiefer of the Rhineland has long yielded exquisitely preserved starfish, crinoids, trilobites, crustaceans and cephalopods, mainly preserved in pyrite. In these stagnation deposits the pyrite was formed where organic compounds rich in sulphur were present in decaying material; these combined with iron ions in the water to precipitate FeS and FeS_2. Many such fossils have been known for a long time, but surprisingly complete details of structure have been proved through the application of 'soft' X-rays, often with photographic exposures taking several hours. This technique was begun in the 1930s by W.M. Lehmann, and was resumed in the early 1960s to great effect by Professor W. Stürmer of Erlangen, yielding spectacular results: a fund of information has been obtained, not only on the hard parts but also, in many cases, on the lightly pyritized soft tissues (Stürmer, 1970; Stürmer and Bergström, 1973, 1976, 1978, 1981; Stürmer *et al.*, 1980; Bergström and Brassel, 1984).

Details of limb morphology has been revealed in *Phacops*, *Asteropyge* and other trilobite genera (q.v.) and the detailed structure of the phyllocarid crustacean *Nahecaris* has been made clear (Fig. 12.8a). A

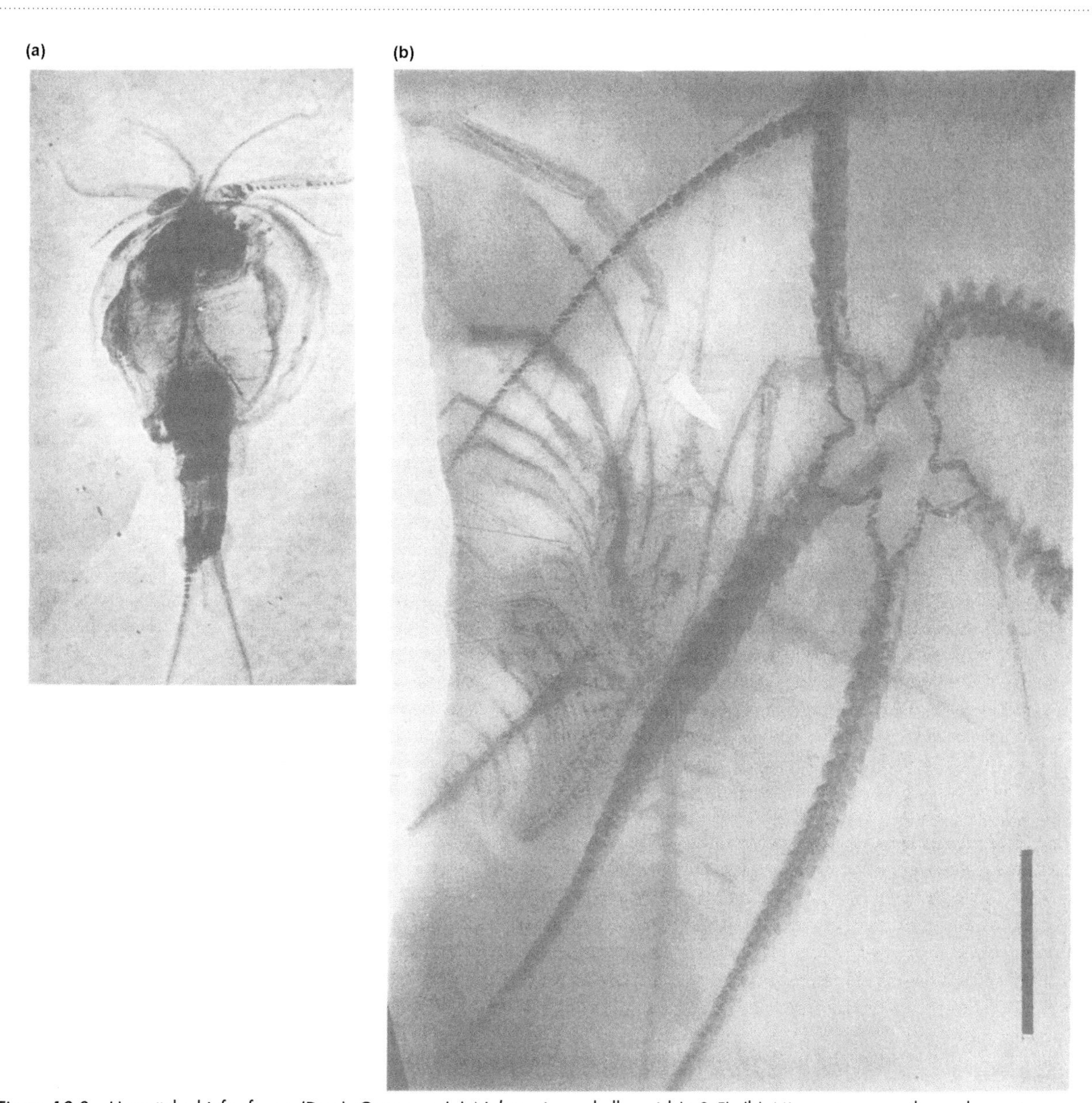

Figure 12.8 Hunsrückschiefer fauna (Dev.), Germany: (a) *Nahecaris*, a phyllocarid (× 0.5); (b) *Mimetaster*, an arthropod associated with an ophiuroid (length of bar is 10 mm). (X-radiographs reproduced by courtesy of Professor Wilhelm Stürmer.)

variety of other arthropods include early representatives of extant groups or members of taxa which are now extinct. *Weinbergina*, a synziphosurid, is unique in having six, rather than five, prosomal limbs behind the chelicerae. It was evidently adapted for life on a soft substrate. Large primitive pycnogonids (sea spiders) belonging to extinct orders also occur, as well as the tiny *Palaeothea*, the most ancient representative of the extant pycnogonid Order Pantopoda.

Some of the arthropods cannot be related to any present-day taxon. The large ovoid *Cheloniellon* has a head with a pair of antennae projecting forwards, nine trunk segments and a curious conical telson. There is one pair of antennae, a second pre-oral pair of appendages, and four pairs of uniramous appendages with gnathobases in the head. Eight pairs of biramous appendages are present in the body. The *Marrella*-like arthropod *Mimetaster*, about half the known specimens of which are found associated with an ophiuroid (Fig 12.8b), was known in gross morphology prior to Stürmer and Bergstrom's work, as was *Vachonisia*, a large arthropod with a bivalved carapace. *Mimetaster*, through X-ray photographs, is now known to have a pair of stalked eyes, stout jointed appendages on the head with pyritized strands of muscle, and about 30 pairs of biramous appendages down the body. *Vachonisia* has turned out to have a surprisingly similar body under the large carapace and is probably related. These two genera could be survivors of the *Marrella* stock, persisting to the Devonian and known only from these exceptional faunas, though some authorities regard their characters as symplesiomorphic.

Other elements in the fauna include early representatives of several groups, e.g. the floating *Velella*-like hydrozoan *Plectodiscus*. Among the molluscs there are thin-shelled ctenodont bivalves, gastropods with soft parts present, ammonoids and bactritids (Chapter 8) which give much information about the early evolution of this group, and straight-shelled cephalopods. Some of these were ectocochlear, with the shell outside the body, whereas others were endocochlear with the outside of the shell invested with soft tissue. Tentacles and other external organs appear in some radiographs, both of the orthocone *Lobobactrites* and in *Goniatites*, and embryonic cephalopods have also appeared unexpectedly on the photographic plates. Coleoids have also been found in the Hunsrückschiefer. *Protoaulacoceras* has some features reminiscent of orthocone cephalopods, but possesses a rostrum, whereas two other genera show marked similarities to Recent teuthids and would seem to be their earliest known representatives. In some beds specimens of the enigmatic conical-shelled tentaculitids are abundant. On X-ray plates their tentacles show up, as does the gut, which tends to support the view of some workers that these septate shells with their long body chambers were an extinct group of cephalopods.

12.5 Mazon Creek fauna (Fig. 12.9)

The Pennsylvanian beds of Illinois have long been strip-mined for coal. Over wide areas the nodular shale overlying the coal has been removed to spoil tips, from which the weathered ironstone nodules or concretions have proved the source of an outstandingly diverse flora and fauna (Nitecki, 1979; Schram, 1974, 1979b; Thompson, 1977, 1979). These nodules provide a valuable record of the soft-part anatomy of many kinds of organisms.

The sediments belong to the Francis Creek Shale, deposited in a generally deltaic environment between a northerly coal swamp and the sea which lay to the south west. Two main floral and faunal groupings can be defined, the non-marine Braidwood and the marine Essex assemblages. All the fossils from these are preserved in ironstone concretions. The non-marine (brackish to freshwater) Braidwood assemblage, which is dominated by exquisitely preserved land plants (350 species) and insects (140 species), was evidently deposited in a swampy lowland region close to the shore. There are also scorpions, spiders, millipedes and centipedes, derived from the shoreward reaches of the coal forest. Small ponds and channels were the home of ostracodes, and of small shrimp-like crustaceans, fish and amphibians. The xiphosuran *Euproops*, often found with the more terrestrial elements, may have spent some of its life subaerially.

The contrasting Essex fauna inhabited the marine facies of the delta front. It is astonishingly diverse, and in many cases the soft parts of the animals are preserved as carbon films. The vertebrate record reveals many fish species as well as hatchling coelacanths with the yolk-sac still preserved, one of the earliest known lampreys, and other jawless fishes. The invertebrate fauna is dominated by large crustaceans (e.g. *Belotelson*), other arthropods (e.g. *Euproops*, *Kottixerxes*) and by epifaunal polychaete worms (e.g. *Levisettius*, *Fossundecima*) of carnivorous habit and complete with bristles and jaws. There are also jellyfish, hydroids (e.g. *Mazohydra*), chaetognaths (e.g. *Paucijaculum*), nemertine worms (e.g. *Archisymplectes*), holothurians, barnacles with chitinous plates and cephalopods with arms and hooks attached. Among other fossils there are specimens of the enigmatic *Tullimonstrum* (Tully monsters), named after their discoverer, and unique to the

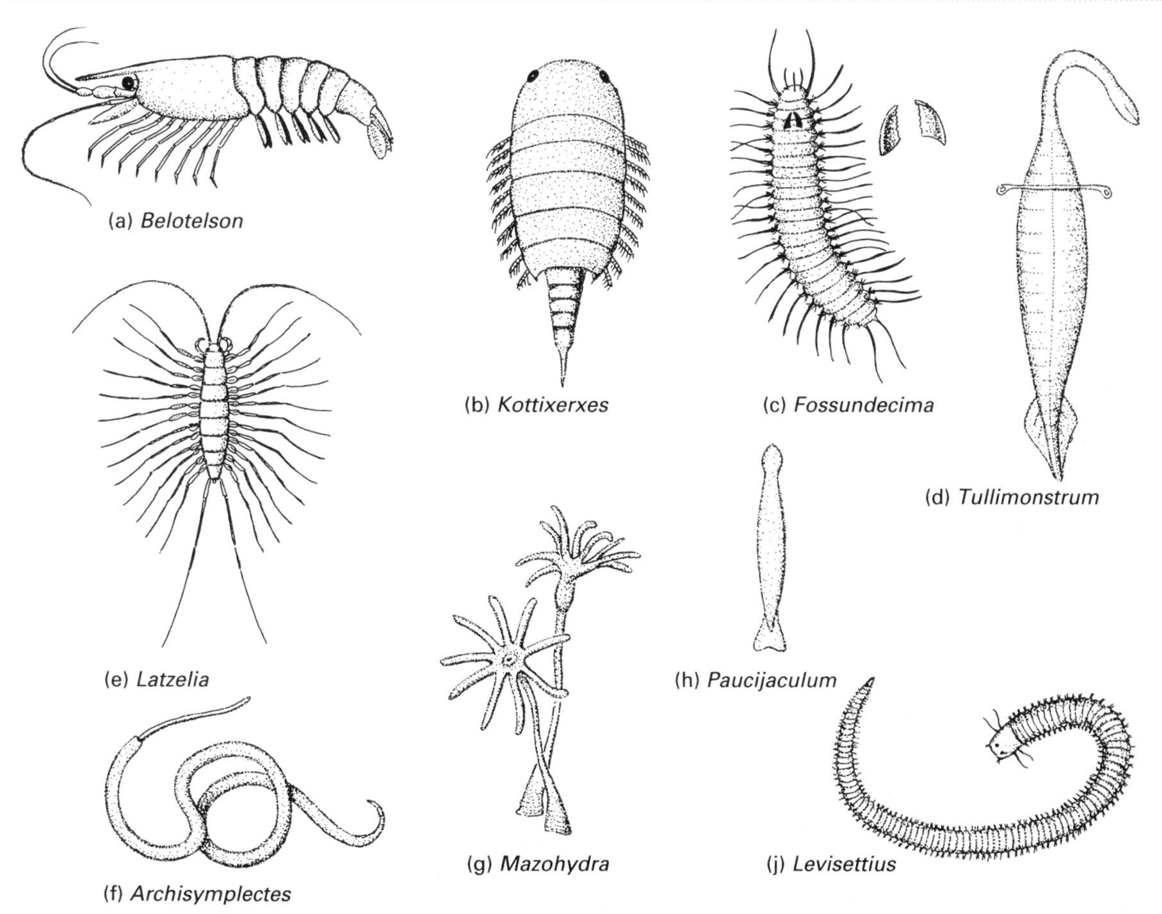

Figure 12.9 Some soft-bodied elements of the Mazon Creek fauna (Carb.), Illinois: (a) *Belotelson*, a malacostracan crustacean (× 2); (b) *Kottixerxes*, a euthycarcinoid arthropod (× 3); (c) *Fossundecima*, a polychaete worm, with jaws (× 1); (d) *Tullimonstrum* (× 1.25); (e) *Latzelia*, a centipede (× 1), (f) *Archisymplectes*, a nemertine worm; (g) *Mazohydra*, a hydrozoan (× 1); (h) chaetognath *Paucijaculum* (× 3); (j) *Levisettius*, a polychaete worm (× 1). [(a), (b), (f), (h) Redrawn from Schram, 1971, 1973; (c), (j), redrawn from Thompson, 1977; others redrawn from Foster, Magel and other authors in Nitecki, 1979.)

Essex fauna. *Tullimonstrum* (Fig 12.9d) was a soft-bodied, bilaterally symmetrical animal, some 8 cm long. Its head tapers to an elongated proboscis with a pair of pincer-like jaws at the end, their inner face armed with minute stylets. Where the head grades into the trunk there is a transverse bar projecting laterally from the body, each end terminating in a hollow, globular bar-organ; these have been variously interpreted as eyes or stabilizers. The trunk is segmented, often showing a median impression, probably the gut. At the rear a spatulate tail is marked with paired fins.

Tullimonstrum was probably a pelagic carnivore which caught its prey with the proboscis. Since the probosics and the transverse bar appear to be unique, most authorities have considered *Tullimonstrum* to be a member of an entirely extinct phylum. Foster (1979), however, points out that *Tullimonstrum* shares many features in common with living heteropods, a group of actively swimming gastropods which have largely lost the shell. These include overall shape, proboscis form, radular teeth and possibly also the eyes. These correspondences are not, however, exact, and most palaeontologists still hold the view that the bizarre *Tullimonstrum* is a member of a long-extinct phylum.

The degree of mixing of the Essex and Braidwood elements varies: sometimes only a few of the more typical Braidwood elements are found with members of the Essex fauna; elsewhere, more truly mixed faunas may have resulted from periodic storm surges which moved the marine animals over the delta, stranding them on its surface. Increased rain during such a storm surge is thought to have swept down masses of sediment to bury the animals almost immediately. The concretions must have formed round the specimens very rapidly after burial, and the preservation itself is rather unusual, for within the siderite concretions calcium carbonate is normally absent while chitin is unaltered, and the soft parts are preserved as carbon films. The nodules might have formed after a sudden influx of silt and fresh water into the area after heavy rains or due to channel diversion. The abundance of Fe^{2+} derived from the coal swamps together with the rapid burial may have proved an ideal environment for the siderite concretion formation.

The Mazon Creek faunas are of unusual composition, for the common Carboniferous marine fossils such as brachiopods are absent, even in the marine Essex facies, and crinoids, gastropods and bivalves are rare. The great bulk of the fossils are, indeed, of uncommon type.

In the faunal assemblage sense, the Essex fauna is unusual. It is not, however, unique, except in the abundance and diversity of fossils found therein. Similar Carboniferous nearshore faunas exist elsewhere, in America, Europe and other places, inhabiting shallow-marine environments and nearshore lagoons between swampy lowlands and the sea. Similar biotic associations are known elsewhere in Illinois (Zangerl and Richardson, 1963), from the Bear Gulch sequence of Montana (Williams, 1983), from Montceau-les-Mines, France (Heyler and Poplin, 1988) and from the Carboniferous of Scotland (Schram, 1979a; Briggs and Clarkson, 1983, 1985) which testify to an ecological continuity over a wide area during the Carboniferous. Indeed a similar biotic association can be recognized as far back as the Silurian in an equivalent nearshore habitat.

Such faunas as these are now beginning to be understood, not just descriptively and taxonomically, but ecologically, as biotic and trophic associations extending through space and time. In the words of Schram (1979b), 'We can no longer be content with treating the Mazon Creek biotas as baroque bric-a-brac; they are in fact part of a combination to unlock the secrets of Carboniferous history'. The understanding of such faunas, which give such a detailed perspective on the whole biota of the time, is a primary task for palaeontology in the years that lie ahead.

12.6 Solnhofen lithographic limestone, Bavaria (Fig. 12.10)

In Bavaria there is a remarkable and widespread limestone of Upper Jurassic (Tithonian) age, formerly quarried as lithographic stone. It is very fine textured, and light grey to buff coloured, and although fossils are rather uncommon therein, they are exceptionally well preserved and very diverse. Some 400 species of animals and plants have been recorded from laminar bedding planes, and nearly all the animal fossils are nektic marine invertebrates and vertebrates. The vertebrate fauna includes many fish species, many aquatic (and a few terrestrial) reptiles, pterodactyls and the early bird *Archaeopteryx* (Barthel, 1978, 1979; Barthel *et al.*, 1990).

Figure 12.10 *Cycleryon propinquus*, a Jurassic crab from the shallow-water Solnhofen Limestone, Germany (× 0.5). The only living eryonids are deep-water species.

The invertebrate fauna is particularly rich in arthropods, some 70 species belonging to 25 genera. They were mainly swimmers, such as the shrimp-like *Aeger*, but there is also a larger macruran, *Cycleryon*, and some small limulines also occur. There are also many species of flying insects, especially dragonflies, brought in during seasonal rain-water floods. Other invertebrates include jellyfish (*Rhizostomites*), cephalopods and the abundant crinoid *Saccocoma*.

There has been much discussion of the environment of deposition of the Solnhofen Limestone. It is clearly a marine sediment, as shown by the presence of coccoliths, cephalopods and crinoids. It is possible that the lime-mud forming the limestone was washed into the lagoon by storms, but it may have been at least partially an algal or chemical precipitate. Some authors have invoked periodic stranding of the fossils on mud flats to explain their preservation, but a more recent interpretation advanced by Barthel (1979) argues for a continuous sea cover throughout deposition. According to Barthel the Solnhofen Limestone was apparently deposited on a sea floor of irregular relief originally consisting of sponge–algal reefs and mounds. Maximum water depth may have been from 30 to 60 m. To the north there were barrier islands, and to the south a coral–hydrozoan barrier reef, which cut off the back-reef lagoonal area in which the limestone was deposited as a carbonate mud in a system of inter-connected basins fringing the deep sponge-reefs. Since the climate was very hot the lagoonal water sometimes became hypersaline through evaporation. The sea floor may have often been stagnant, and both these factors seem to have been responsible for the frequent absence of epi- and infauna [though *Saccocoma* (q.v.) may have been a benthic opportunist], and also for the remarkable paucity of tracks and trails. There are indeed a few trails, short ones made by the decapod *Mecochirus* and longer trails resulting from the movements of limulines. Many specimens of *Mesolimulus walchi* are preserved at the end of their trails. Modern *Limulus* is highly tolerant of change in salinity, temperature and oxygen tension, and the same was probably also true of Jurassic limulines, so that at least some individuals could tolerate to some extent the locally fluctuating and often inimical conditions of the Solnhofen sea floor. And although these same conditions proved too much for some individuals, the circumstances of their demise 140 Ma ago, recorded in detail in the rocks, provide a clearer than normal perspective of the life of that time.

All these exceptional faunas are atypical only in their preservation. It is this alone which distinguishes them from others. What is so important about them is that they provide near-complete pictures of the diversity of animals living at particular times, though they were formed under different conditions. There seems, however, to be no general pattern for the preservation of soft parts, and there is no clear understanding of how decay is inhibited.

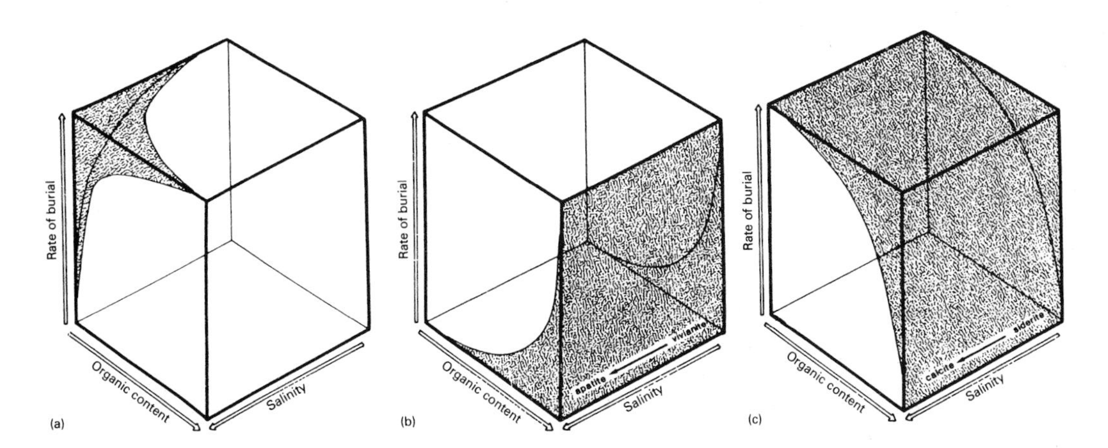

Figure 12.11 Depositional parameters for early diagenesis of soft parts of fossils: (a) pyritization, requiring a very high burial rate and low organic content; (b) phosphatization, requiring a low burial rate and a high organic content; (c) within carbonates, requiring a very high rate of burial and high organic content. (Redrawn from Allison, 1988.)

Much more biogeochemical research (e.g. Fig. 12.11) needs to be undertaken to resolve many current questions.

The further study of Fossil-Lagerstätten would seem to be one of the important tasks for palaeontology in the future.

12.7 Ichnology

Ichnology is the study of the behaviour of once-living animals by examination of the tracks, trails, borings and markings that they made when alive. These are called **trace fossils** or **ichnofossils**. They are usually preserved only at the interface of two types of sediment, for instance where the markings made on a mud surface by a crawling trilobite were filled in by fine sand which later hardened. Ichnology is by no means a recent development in geological science – indeed it can be traced back to the 1820s – but it is only within the last two decades that it has come to be a really powerful tool for understanding certain past sedimentary environments and one of our major clues to the behaviour of the animals that lived within these environments. The different kinds of traces made in shallow-water sediments have been fully documented in the vast mudflats of the North Sea by German workers as summarized by Schäfer (1972), and this knowledge has been applied to the geological record. Since very good summaries of trace fossil work exist elsewhere (Crimes and Harper, 1970; Häntzschel, 1975; Frey, 1976; Donovan, 1994; Bromley, 1996), only a few points will be made here to show something of the scope and methods of ichnology.

In particular I would refer students to Bromley's elegant and biologically based treatment. In this are to be found many illustrated examples of the life habits and traces of modern burrowing animals, and an application of biological principles to trace fossil studies.

Classification of trace fossils

Trace fossils are all sedimentary structures made by the activities of once living animals, mainly invertebrates (Figs 12.12, 12.13).

However, similar traces can be made by quite different kinds of organisms. There are various ways of studying trace fossils and hence different systems of classifying them (Simpson, 1976). Three of these in particular are important: morphological and preservational, behavioural, and phylogenetic.

Morphological and preservational classification

Trace fossils are all given 'form-generic names'; that is, the names are used only for the distinction of various types and do not attempt to identify or suggest their maker. A simplified system of grouping them with interpretation is given here.

tracks or trails on a bedding plane originally made upon the sediment–water interface, e.g. *Cruziana* (trilobite trails), *Nereites*, *Phycosiphon*, *Cosmoraphe* (worm trails), *Gyrochorte* (possible gastropod trail); radially symmetrical horizontal markings, e.g. *Asteriacites* (resting marks of starfish);

tunnels and shafts within the sediment, e.g. *Skolithos* (vertical worm tubes), *Chondrites* (branching galleries probably made by a probing worm);

traces with a **spreite** (a web-like structure, usually with a series of concentric markings) as often found joining two branches of a U-tube and representing former positions of the tube within the sediment, e.g. *Rhizocorallium* (horizontal U-tubes), *Diplocraterion* (vertical U-tubes), *Zoophycos* (inclined spirals) – all of which could have been made by different kinds of animals;

pouch-shaped markings, e.g. *Pelecypodichnus* (bivalve burrows), *Rusophycus* (trilobite resting traces);

others, e.g. *Palaeodictyon* (net-like structure of uncertain origin).

Although this descriptive classification has its usefulness, the characters selected for description must be arbitrary and therefore far from objective. A classification based on preservational features alone is equally hard to apply.

Behavioural classification

Seilacher (1964) designed a very useful scheme based upon the behaviour of the organism that made the traces. In modified form this classification now has at least six categories, but since they overlap to some extent this scheme is also not a perfect system, though it is useful. The categories are as follows.

Crawling traces (Repichnia): these are tracks of moving surface dwellers, e.g. *Cruziana*.

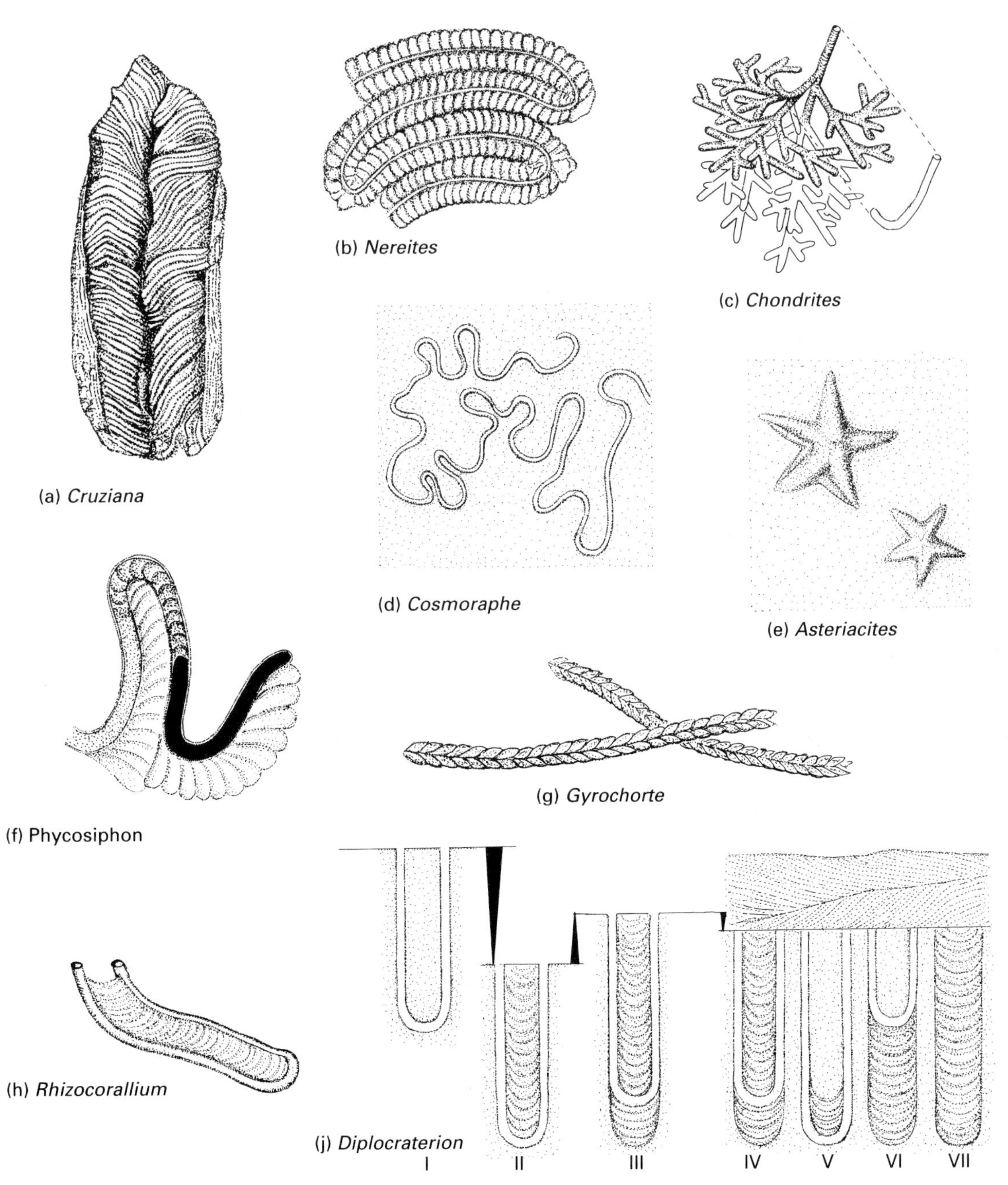

Figure 12.12 Trace fossils: (a) *Cruziana*, made by ploughing trilobites (× 1); (b) *Nereites*, a surface trail (× 0.3); (c) *Chondrites*, mines within sediment, made by a probing worm (× 0.5); (d) *Cosmoraphe*, surface trail (× 0.5); (e) *Asteriacites*, resting trace of starfish (× 1); (f) *Phycosiphon*, surface marking, the originating worm marked in black (× 5); (g) *Gyrochorte*, surface trail (× 1); (h) *Rhizocorallium*, oblique lateral view, as if through transparent sediment (× 0.1); (j) *Diplocraterion*, in vertical view. From left to right: I, normal U-tube; II, U-tube descending as sediment is removed; III, U-tube ascending as sediment is rapidly deposited; IV–VII, U-tubes initiated at various times and showing movement in relation to erosion or deposition of sediments on top; arrows show directions of movement before final plane erosion and deposition of overlying sediment. [Redrawn chiefly from *Treatise of Invertebrate Paleontology*; (a) based on Fillion and Pickerill, *Palaeontographica Canadiana*, 1990; (f) from Bromley, 1996; (j) from Goldring, 1962.]

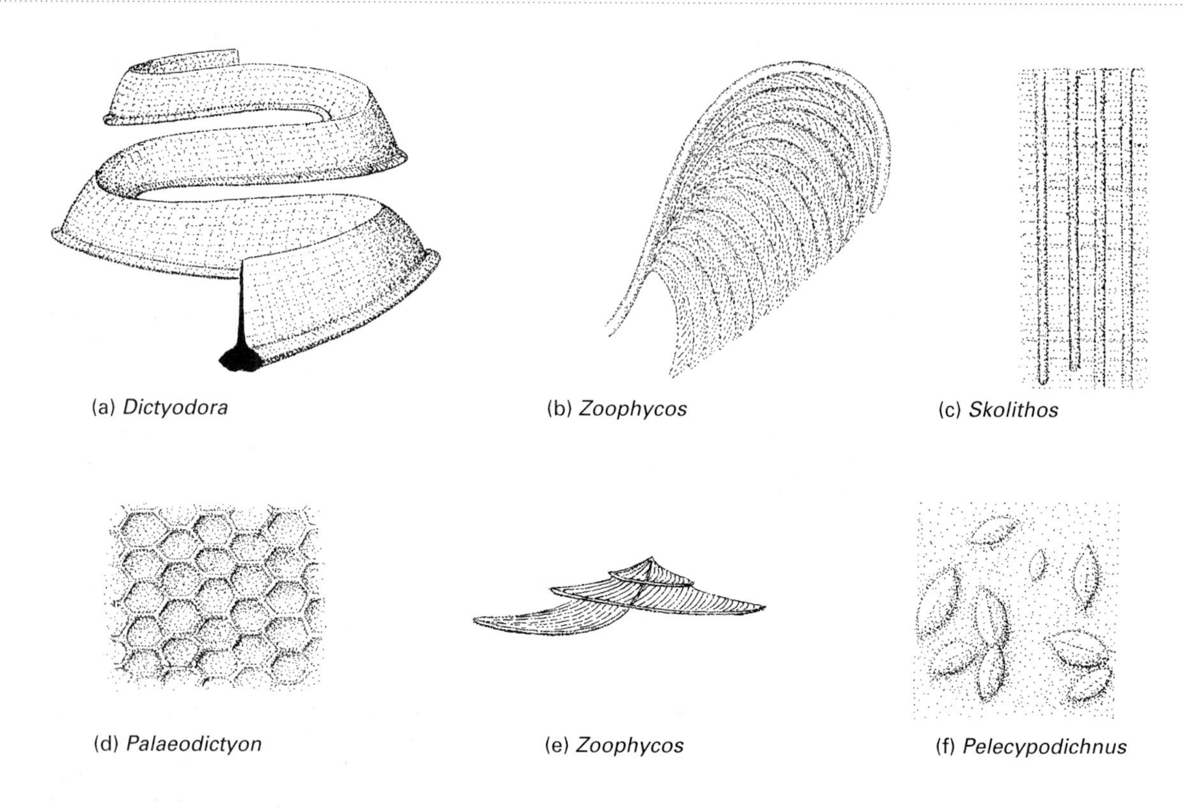

Figure 12.13 (a) *Dictyodora*, reconstructed (× 1); (b), (e) *Zoophycos*, in surface view and in section, respectively (× 0.1); (c) *Skolithos*, in vertical section (× 0.2); (d) *Palaeodictyon* in surface view (× 0.5); (f) *Pelecypodichnus*, surface view (× 0.5).

Resting traces (Cubichnia): these are exemplified by *Rusophycus*, *Asteriacites*.

Grazing or surface feeder traces (Pascichnia): surface feeders make characteristic patterned markings on the sediment; the same sort of patterns may be seen made by Recent snails grazing on algae on stones in rock pools. *Nereites*, for instance, is a trace appearing as a series of parallel lines with lateral lobes, joined by tight loops, each line being equidistant from its neighbours. This is made by a surface feeder economically exploiting a food supply, such as bacteria, upon the sediment surface, covering as much ground as possible but with the minimum of effort and never recrossing its own trail or feeding along previously grazed surfaces. *Phycosiphon* performs the same kind of function in a different way.

Feeding traces (Fodinichnia): these are excavations made by deposit feeders, i.e. animals living on the surface but actively mining within it for their food. They include some U-shaped tubes with internal spreites, which show that the U-burrow has been progressively deepened. There are also various kinds of radial structures, including the many-branched *Chondrites*, which was probably made by a worm exploiting a rich deposit successively, again avoiding any area of previously ingested sediment. An interesting Silurian trace fossil, *Dictyodora*, forms a sinuous ribbon with a flattened T-shaped lower edge, meandering through the rock normal to the bedding (Fig. 12.13). It must have been constructed by an animal moving below the surface as it fed, yet maintaining contact with the water by means of a vertical tube or pipe, and it is the passage of the latter through the sediment which created what is now preserved as the vertical ribbon; the 'worm' made the T-shaped base (Benton and Trewin, 1980).

Dwelling structures (Domichnia): these are permanent burrows and borings of suspension feeders, i.e. animals living within the sediment but straining off particles from the water above. They may be subcylindrical tubes (e.g. *Skolithos*), but there are also some U-shaped burrows, though normally without spreites.

Escape structures (Fugichnia): various kinds of vertical tubes suggest that animals moved upwards to escape from being buried by an influx of sediment, or downwards when the sediment above was eroded. The original tubes are usually therefore modified Domichnia. *Diplocraterion*, for example, is a vertical U-tube found in high-energy environments, with spreite structures either above or below it or both. It is interpreted as the permanent dwelling burrow of an animal that lived below the sediment. When sediment was removed by rapid erosion the animal burrowed down more deeply in order to remain concealed at its habitual depth, leaving a spreite above. Rapid sedimentation caused the organism to move upwards, producing a spreite below of somewhat different morphology since it was no longer linking parallel arms. Such a structure gives information about the environment, suggesting very rapid erosion and deposition, and the related up-and-down movement of the organism that made it led Goldring (1962) to give the organism the appropriately descriptive specific name *D. yoyo*.

Phylogenetic classification

Generally there are few indications as to the identity of a particular trace maker, and often even its phylum can hardly be established since animals of many phyla can make the same kind of traces. In the case of some arthropod ichnofossils, however, the trace maker is known with certainty; trilobites, for instance, have actually been found within *Rusophycus* burrows. However, since this is rare indeed there is no point in trying to erect a classification based upon the identity of the trace maker, and form-generic names have to be given.

No means of classification is therefore entirely satisfactory or complete, and the one that is followed in any particular case depends rather upon the purpose of the study.

Uses of ichnology

Sedimentary environment

By contrast with body fossils, trace fossils are always found in place. If the enclosing sediment had moved they would have been destroyed. They often occur in sedimentary suites where there are no body fossils present and especially in clastic sequences. Rather than being destroyed by diagenesis they are commonly improved by it, since lithological contrast between the fossil and the enclosing sediment may thereby be enhanced. Furthermore they tend to be restricted to a narrow facies range. As guides to the nature of the sedimentary environment in certain types of sequence they have proved invaluable.

If trace fossils are present in a clastic sequence then the sediment must have been well aerated. Conversely, a sequence lacking in traces may well have been anoxic. Thus black shales with abundant traces, even if deep-water in origin, cannot possibly have been anoxic. Graptolitic shales and their Mesozoic equivalents, which contain only thin-shelled planktonic bivalves, are generally lacking in trace fossils, which suggests a euxinic origin.

Rates of sedimentation can often be inferred from the relative abundance of trace fossils, as with *Diplocraterion*, and the presence of definite borings in 'hardgrounds' shows clearly that the sediment must have been indurated when the borings were formed, as may sometimes be confirmed by encrusters upon its surface. Inevitably a temporary break in deposition must be recognized to have allowed such induration. Minor variations in facies can often be inferred from the presence of particular trace fossil assemblages, even if the nature of their makers is not known.

The analysis of trace fossil assemblages is less simple than it appears, since the potential for preservation of 'fossil behaviour' depends upon the depth at which the trace-maker lived within the sediment. Most living infaunal animals are vertically zoned and inhabit different levels or tiers within the sediment (Fig. 12.14).

Within each of these tiers there may be defined ichnoguilds (Bromley, 1996), i.e. groups of ichnospecies which behave in a similar manner, and occupy a similar location within the substrate. In the upper layers there may be highly mobile animals whose life activities totally obliterate other primary structures. They may also sometimes make it difficult for the inhabitants of lower tiers to maintain contact with the surface. While the upper layers of sediment are usually very stirred up (bioturbated), the traces in the lowermost layers (e.g.

Thalassinoides) alone are the ones most likely to be preserved. Bromley's term 'elite trace fossils' applies to these; they may also be enhanced by diagenesis. The trace fossil assemblage that is fossilized is thus not a true reflection of the original biological community. What we see is thus taphonomically 'filtered' and all that remains of the rich and diverse assemblage of the upper layers is simply mixed and bioturbated sediment.

There are also rather generalized trace fossil assemblages which give information of a grosser quality, especially pertaining to the original depth of deposition (Seilacher, 1967). Since physical protection is very important to shallow-water dwellers, partly due to turbulence and partly because the light allows other animals to see them, a high proportion of invertebrates are burrowing suspension feeders (filterers) and the traces are mainly Cubichnia and Domichnia. Farther away in quieter waters, deposit feeders (swallowers) are abundant, and hence Fodinichnia become increasingly common, likewise Repichnia. In the lightless environment of the deep sea there is less need for protection, and because of the rich supply of surface-dwelling bacteria there are very many Fodinichnia and especially Pascichnia. The latter make complex spiral or meandering patterns for systematically grazing from food-rich surface layers. On this basis there were erected four main (Seilacherian) ichnofacies based on bathymetry, the *Skolithos*, *Cruziana*, *Zoophycos* and *Nereites* facies. Other ichnofacies have been added by different authors such as the *Scoyenia* redbed facies. The *Skolithos* facies consists largely of vertical tubes in clean, shallow-water sands; the shallow-water *Cruziana* ichnofacies (deposited above storm wave-base) is of burrows, arthropod tracks and resting traces; the intermediate *Zoophycos* facies is highly bioturbated and was made principally by sediment miners; and the deep-water *Nereites* facies associated with turbidites has grazing trails of complex form.

These ichnofacies are of use in a broad and general sense, but individual traces may occur out of context, and the assemblages themselves are subject to taphonomic filtering.

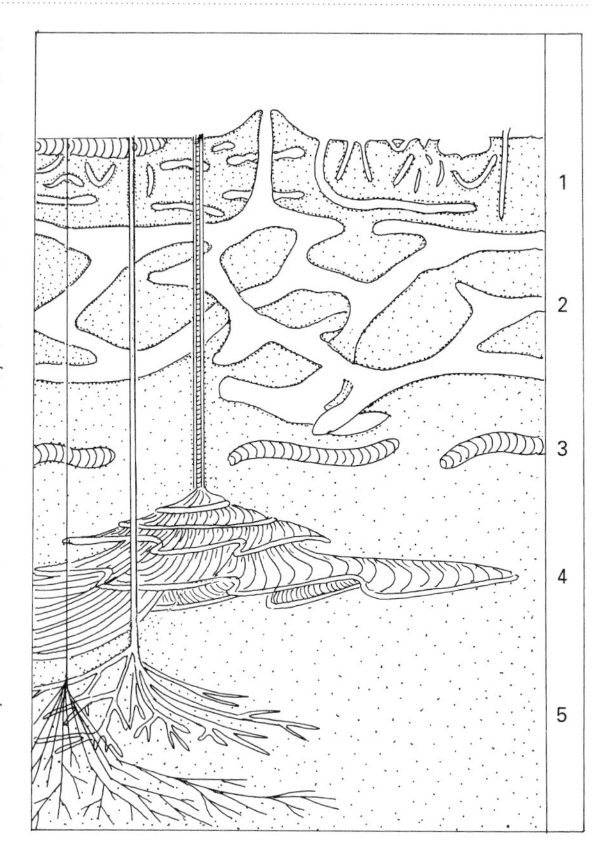

Figure 12.14 Tiering of trace fossils, Cretaceous example: (1) bioturbation from echinoids, gastropods, bivalves, etc.; (2) *Thalassinoides* (crustacean burrow system); (3) *Taenidium* (burrowing worm); (4) *Zoophycos*; (5) *Chondrites*, large-bore and small-bore forms, respectively. (Redrawn from Bromley, 1996.)

Stratigraphy

Generally trace fossils are of very little stratigraphic use. The very characters that make them so valuable environmentally, i.e. long range, strict facies control, etc., are just the opposite of what is needed for good zone fossils. Yet some short-ranged types are useful in local stratigraphy, and most particularly trace fossils have been of immense value in understanding the explosion of life at the Precambrian–Cambrian boundary (Chapter 3), and particularly of metazoan life, for many trace fossils can only be made by an animal with a gut. Other examples have been given by Crimes and Harper (1970).

Fossil behaviour

In a sense, all ichnology relates to the study of fossil behaviour. The complex grazing trails that have already been mentioned testify to a degree of behavioural complexity in their ancient makers like that of many present-day organisms. Trace fossils have been used directly in analysing the life habits and locomotion of trilobites and other arthropods (Chapter 11). Their use in interpreting life habits and past sedimentary environments can but increase in the future.

Bibliography

[N.B. Since references to original descriptions of the Burgess Shale fauna are found in Conway Morris (1982), Whittington (1985) and Gould (1990), they are not included here.]

Books, treatises and symposia (exceptional faunas)

Allison, P.A. and Briggs, D.E.G. (eds) (1991) Taphonomy: releasing the data locked in the fossil record. Plenum Press, New York and London. (Several papers)

Barthel, K.W. (1978) *Solnhofen – ein Blick in die Erdgeschichtce*. Ott Verlag, Thun (Excellent colour plates)

Barthel, K.W., Swinburne, N.H.M. and Conway Morris, S. (1990) *Solnhofen. A study in Mesozoic Palaeontology*. Cambridge. 245 pp.

Briggs, D.E.G., Erwin, D.H. and Collier, F.J. (1994) *The fossils of the Burgess Shale*. Smithsonian Institution Press, Washington.

Conway Morris, S.C. (ed.) (1982) *Atlas of the Burgess Shale*. Palaeontological Association, London. (Folio of photographs and explanation)

Donovan, S.K. (ed.) (1991) *The Processes of Fossilisation*. Belhaven, Boston, Mass. (12 taphomonic papers)

Einsele, G. and Seilacher, A. (eds) (1982) *Cyclic and Event Stratification*. Springer, Berlin.

Gould, S.J. (1990) *Wonderful Life: the Burgess Shale and the Nature of History*. Hutchinson Radius, London, 347 pp. (A mine of data and philosophy)

Hauff, B. and Hauff, R.B. (1981) *Das Holzmadenbuch*. Fellback (Repro-Druck). 136 pp.

Martill, D.A. (1993). *Fossils of the Santana and Crato Formations, Brazil*. Palaeontological Association: Field Guide to Fossils No. 6.

Moore, R.C. (ed.) (1959) *Treatise on Invertebrate Paleontology*, Part O, *Trilobitomorpha*. Geological Society of America and University of Kansas Press, Lawrence, Kan.

Nitecki, M.H. (ed.) (1979) *Mazon Creek Fossils*. Academic Press, New York. (Excellent overview – 16 papers)

Simonetta, A.M. and Conway Morris, S. (eds) (1991) *The Early Evolution of Metazoa and the Significance of Problematic Taxa*. (Proceedings of an International Symposium held at the University of Camerino, 27–31 March 1989). Cambridge University Press, Cambridge, 296 pp.

Whittington, H.B. (1985) *The Burgess Shale*. Yale University Press, 167 pp.

Whittington, H.B. and Conway Morris, S. (1985) *Extraordinary Fossil Biotas: Their Ecological and Evolutionary Significance*. Royal Society, London, 1–92. (19 papers)

Individual papers and other references (exceptional faunas)

Allison, P.A. (1988a) Phosphatised soft-bodied squids from the Jurassic Oxford Clay. *Lethaia* **21**, 403–10.

Allison, P. (1988b) Konservat-Lagerstätten: cause and classification. *Palaeobiology* **14**, 331–44.

Allison, P. (1990) Decay processes, in *Palaeobiology: a Synthesis* (eds D.E.G. Briggs and P.A. Crowther), Blackwell, Oxford, pp. 213–6.

Barthel, K.W. (1979) Solnhofen Formation, in *The Encyclopedia of Paleontology* (eds R.W. Fairbridge and D. Jablonski), Hutchinson, Dowden and Ross, Stroudsburg, Penn., pp. 757–64.

Berner, R.A. (1981) Authigenic mineral formation resulting from organic matter decomposition in modern sediments. *Forstschrifte für Mineralogie* **59**, 117–35.

Bengtson, S. (1991) Oddballs from the Cambrian start to get even. *Nature* **351**, 184–5.

Bergström, J. (1972) Morphology of fossil arthropods as a guide to phylogenetic relationships, in *Arthropod Phylogeny* (ed. A.P. Gupta), Van Nostrand Reinhold, New York, pp. 1–56.

Bergström, J. (1980) Morphology and systematics of early arthropods. *Abhandlung du Naturwissenschafliche Vereinigung Hamburg* **23**, 7–42.

Bergström, J. (1990) Hunsrück Slate, in *Palaeobiology: a Synthesis* (eds D.E.G. Briggs and P.A. Crowther), Blackwell, Oxford, pp. 277–9.

Bergström, J. and Brassel, H. (1984) Legs in the trilobite *Rhenops* from the Lower Devonian Hunsruck Slate. *Lethaia* **17**, 67–72.

Briggs, D.E.G. (1983) Affinities and early evolution of the Crustacea: the evidence of the Cambrian fossils, in *Crustacean Issues 1: Crustacean Phylogeny*, (ed. F.R. Schram), Balkema, Rotterdam, pp. 1–22.

Briggs, D.E.G. (1991) Extraordinary fossils. *American Scientist* **79**, 130–41.

Briggs, D.E.G. and Clarkson, E.N.K. (1983) The Carboniferous Granton shrimp-bed, Edinburgh, in

Trilobites and Other Arthropods (eds D.E.G. Briggs and P.D. Lane), Special Papers in Palaeontology No. 30, Academic Press, London, 161–77. (Crustacean Lagerstätte)

Briggs, D.E.G. and Clarkson, E.N.K. (1985) The Lower Carboniferous shrimp *Tealliocaris* from Gullane, East Lothian, Scotland. *Transactions of the Royal Society of Edinburgh: Earth Sciences* **76**, 173–201.

Briggs, D.E.G. and Collins, D. (1988) A Middle Cambrian chaliccate from Mount Stephen, British Columbia. *Palaeontology* **31**, 779–98.

Briggs, D.E.G. and Gall, J.-C. (1990) The continuum in soft-bodied biotas from transitional environments: a quantitative comparison of Triassic and Carboniferous Konservat-Lagerstätten. *Palaeobiology* **16**, 204–18.

Briggs, D.E.G. and Kear, A. (1994) Decay of *Branchiostoma*: implications for soft-tissue preservation in conodonts and other primitive chordates. *Lethaia* **26**, 275–87.

Briggs, D.E.G. and Wilby P. (1996) The role of the calcium carbonate–calcium phosphate switch in the mineralisation of soft-bodied fossils. *Journal of the Geological Society of London* **153**, 665–8.

Briggs, D.E.G., Clarkson, E.N.K. and Aldridge, R.J. (1983) The conodont animal. *Lethaia* **16**, 1–14. (First known specimen)

Briggs, D.E.G., Bottrell, S.H., and Raiswell, R. (1991a) Pyritisation of soft-bodied fossils: Beecher's Trilobite Bed, Upper Ordovician, New York State. *Geology* **19**, 1221–4.

Briggs, D.E.G., Clark, N.D.L. and Clarkson, E.N.K. (1991b) The Granton 'shrimp-bed', Edinburgh – a Lower Carboniferous Konservat–Lagerstätte. *Transactions of the Royal Society of Edinburgh: Earth Sciences* **82**, 65–85.

Briggs, D.E.G., Siveter, David J., and Siveter, Derek J. (1996) Soft bodied fossils from a Silurian volcanic ash. *Nature* **382**, 248–50.

Budd, G.E. (1996) The morphology of *Opabinia regalis* and the reconstruction of the arthropod stem-group. *Lethaia* **29**, 1–14.

Chen, J.-Y. and Erdtmann, D.-D. (1991) Lower Cambrian Lagerstätte from Chengjiang, China: insights for reconstructing early metazoan life, in *The Early Evolution of Metazoa and the Significance of Problematic Taxa* (eds A.M. Simonetta and S. Conway Morris), Cambridge University Press, Cambridge, pp. 57–76.

Cisné, J.L. (1972). Beecher's Trilobite Bed revisited: ecology of an Ordovician deepwater fauna. *Postilla* **160**, 1–25.

Collins, D., Briggs, D.E.G. and Conway Morris, S. (1983) New Burgess Shale fossil sites reveal Middle Cambrian faunal complex. *Science* **222**, 163–7.

Conway Morris, S.C. (1979) The Burgess Shale (Middle Cambrian) Fauna. *Annual Review of Ecology and Systematics* **10**, 327–49. (Very good overview)

Conway Morris, S. (1986) The community structure of the Middle Cambrian Phyllopod Bed (Burgess Shale). *Palaeontology* **29**, 423–67.

Conway Morris, S. (1989) The persistence of Burgess shale-type faunas: implications for the evolution of deeper water faunas. *Transactions of the Royal Society of Edinburgh: Earth Sciences* **80**, 271–83.

Conway Morris, S. (1990) Late Precambrian and Cambrian soft-bodied faunas. *Annual Review of Earth and Planetary Sciences* **18**, 101–22.

Conway Morris, S. and Peel, J.S. (1990) Articulated halkieriids from the Lower Cambrian of north Greenland. *Nature* **345**, 802–5.

Conway–Morris, S.C. and Whittington, H.B. (1979) The animals of the Burgess Shale. *Scientific American* **241**, 122–33. (Colour plates and reconstructions)

Cook, P. and Shergold, J. (1984) Phosphorus, phosphorites and skeletal evolution at the Cambrian–Precambrian boundary. *Nature* **308**, 231–6.

Foster, M. (1979) A reappraisal of *Tullimonstrum gregarium*, in *Mazon Creek Fossils* (ed. M.H. Nitecki), Academic Press, New York, pp. 269–302. (Was this a gastropod?)

Franzen, J.L. (1985) Exceptional preservation of Eocene vertebrates in the lake deposit of Grube Messel (West Germany), in *Extraordinary Fossil Biotas: their Ecological and Evolutionary Significance* (eds H.B. Whittington and S. Conway Morris), *Philosophical Transactions of the Royal Society of London B* **311**, 181–6.

Gabbott, S.E., Aldridge, R.J. and Theron, J.N. (1995) A giant conodont with preserved muscle tissue from the Upper Ordovician of South Africa. *Nature* **374**, 800–3.

Grande, L. (1984) Palaeontology of the Green River Formation with a review of the fish fauna. *Bulletin of the Geological Survey of Wyoming* **63**, 1–333.

Heyler, D. and Poplin, C. (1988) The fossils of Montaeau-les-Mines. *Scientific American*, Sept., 70–6.

Hou, X.-G. and Bergström, J. (1991) The arthropods of the Lower Cambrian Chengjiang fauna, with relationships and evolutionary significance, in *The Early Evolution of Metazoa and the Significance of Problematic Taxa* (eds A.M. Simonetta and S. Conway Morris), Cambridge University Press, Cambridge, pp. 179–88.

Hou, X.-G. and Bergström, J. (1997) Lower Cambrian arthropods of south-west China. *Fossils and Strata*, in press.

Johnson, R.G. and Richardson, E.S. (1969) Pennsylvanian invertebrates of the Mazon Creek area. Illinois: the morphology and affinities of *Tullimonstrum*. *Fieldiana Geology* **12**, 119–49. (Mazon Creek fauna)

Martill, D.M. (1988) The preservation of fishes in concretions from the Santana Formation (Cretaceous) of Brazil. *Palaeontology* **31**, 1–18. (Phosphatization of muscle)

Martill, D.M. (1993) Fossils of the Santana and Crato Formations, Brazil. *Field Guide to Fossils* **5**, 1–159. Palaeontological Association.

Mikulic, D.G., Briggs, D.E.G. and Kluessendorf, J. (1985) A Silurian soft-bodied biota. *Science* **228**, 715–17.

Müller, K.J. (1979) Phosphatocopine ostracodes with preserved appendages from the Upper Cambrian of Sweden. *Lethaia* **12**, 1–27.

Müller, K.J. (1982) Weichteile von Fossilien aus dem Erdaltertum. *Naturwissenschaften* **69**, 249–54. (Phosphatized faunas)

Müller, K.J. (1983) Crustacea with preserved soft parts from the Upper Cambrian of Sweden. *Lethaia* **16**, 93–109.

Müller, K.J. (1985) Exceptional preservation in calcareous nodules, in *Extraordinary Fossil Biotas: their Ecological and Evolutionary Significance* (eds H.B. Whittington and S. Conway Morris), *Philosophical Transactions of the Royal Society of London B* **311**, 67–74.

Müller, K.J. and I. Hinz (1991) Upper Cambrian conodonts from Sweden. *Fossils and Strata* **28**, 1–152.

Müller, K.J. and Walossek, D. (1985) Skaracarida, a new order of Crustacea from the Upper Cambrian of Västergötland, Sweden. *Fossils and Strata* **17**, 1–65.

Müller, K.J. and Walossek, D. (1986) *Martinssonia elongata* gen. et. sp. nov. a crustacean-like euanthropod from the Upper Cambrian 'Orsten' of Sweden. *Zoologica Scripta* **15**, 73–92.

Müller, K.J. and Walossek, D. (1987) Morphology, ontogeny and life habit of *Agnostus pisiformis* from the Upper Cambrian of Sweden. *Fossils and Strata* **19**, 1–124.

Müller, K.J. and Walossek, D. (1988) External morphology and larval development of the Upper Cambrian maxillopod *Bredocaris admirabilis*. *Fossils and Strata* **23**, 1–70.

Ramsköld, L. and Hou, X. (1991) New early Cambrian animal and onyctophoran affinities of enigmatic metazoans. *Nature* **351**, 225–7.

Ritchie, A. (1985) *Ainiktozoon loganense* Scourfield, a protochordate? from the Silurian of Scotland. *Alcheringa* **9**, 117–42.

Robison, R.A. (1991) Middle Cambrian biotic diversity: examples from four Utah Lagerstätten, in *The Early Evolution of Metazoa and the Significance of Problematic Taxa* (eds A.M. Simonetta and S. Conway Morris), Cambridge University Press, Cambridge, pp. 77–98.

Rolfe, W.D.I. (1988) Early life on land – the East Kirkton discoveries. *Earth Sciences Conservation* **25**, 22–8.

Rolfe, W.D.I., Durant, G.P., Fallick, A.E. *et al.* (1990) An early terrestrial biota preserved by Visean vulcanicity in Scotland, in *Volcanism and Fossil Biotas* (eds M.G. Lockley and A. Rice), Geological Society of America Special Paper No. 244, pp. 13–24.

Schram, F.R. (1971) A strange arthropod from the Mazon Creek of Illinois and the trans-Permo-Triassic Merostomoidea (Trilobitoidea). *Fieldiana Geology* **20**, 85–102.

Schram, F.R. (1973) Pseudocoelomates and a nemertine from the Illinois Pennsylvanian. *Journal of Paleontology*, **47**, 985–9.

Schram, F. (1974) The Mazon Creek Caridoid Crustacea. *Fieldiana Geology* **30**, 9–65.

Schram, F. (1979a) British Carboniferous Malacostraca. *Fieldiana* **40**, 1–129.

Schram, F. (1979b) The Mazon Creek biota in the context of a Carboniferous faunal continuum, in *Mazon Creek Fossils* (ed. M.H. Nitecki), Academic Press, New York, pp. 159–90.

Seilacher, A. (1970) Begriff und Bedeutung der Fossil Lagerstätten. *Neues Jahrbuch für Geologie und Paläontologie Monatshefte* **1**, 34–9. (What are Lagerstatten?)

Seilacher, A., Reif, W.E. and Westphal, F. (1985) Sedimentological, ecological and temporal patterns of fossil-Lagerstätten, in *Extraordinary Fossil Biotas: their Ecological and Evolutionary Significance* (eds H.B. Whittington and S. Conway Morris), *Philosophical Transactions of the Royal Society of London B* **311**, 5–23

Selden, P.A. (1990) Lower Cretaceous spiders from the Sierra de Montsech, North-East Spain. *Palaeontology* **33**, 257–84.

Selden, P. and Edwards, D. (1989) Colonisation of the land, in *Evolution and the Fossil Record* (eds K.C. Allen and D.E.G. Briggs), Belhaven, Boston, Mass., pp. 122–52.

Selden, P., Shear, W.A. and Bonamo, P.M. (1991) A spider and other arachnids from the Devonian of New York, and reinterpretations of Devonian Araneae. *Palaeontology* **34**, 241–81.

Shear, W.A., Bonamo, P.M., Grierson, J.D. *et al.* (1984) Early land animals in North America: evidence from Devonian age arthropods from Gilboa, New York. *Science* **224**, 492–4.

Shear, W.A., Palmer, J.M., Coddington, J.A. and Bonamo, P.M. (1989) A Devonian spinneret: early evidence of spiders and silk use. *Science* **246**, 479–81.

Stürmer, W. (1970) Soft parts of cephalopods and trilobites: some surprising results of X–ray examination of Devonian slates. *Science* **1170**, 1300–2. (First work by Stürmer on X–ray applications; illustrated)

Stürmer, W. (1983) Interdisciplinary palaeontology. *Interdisciplinary Sciences Review* **9**, 1–14. (Use of X–rays and other techniques)

Stürmer, W. and Bergström J. (1973) New discoveries on trilobites by X–rays. *Paläontologisches Zeitschrift* **47**, 104–41.

Stürmer, W. and Bergström, J. (1976) The arthropods *Mimetaster* and *Vachonisia* from the Devonian Hünsruck Shale. *Paläontologische Zeitschrifte* **50**, 78–111.

Stürmer, W. and Bergström, J. (1978) The arthropod *Cheloniellon* from the Devonian Hunsrück Shale. *Paläontologische Zeitung* **52**, 57–81.

Stürmer, W. and Bergström, J. (1981) *Weinbergina*, a xiphosuran arthropod from the Devonian Hunsrück Slate. *Paläontologisches Zeitung* **55**, 237–55.

Stürmer, W., Schaarschmidt, F. and Mittmeyer, H.G. (1980) Versteinertes Leben im Röntgenlicht. *Kleine Senckenberg Reihe* **11**, 1–80. (Excellent illustrations of Hunsrückschiefer fauna)

Thompson, I. (1977) Errant polychaetes (Annelida) from the Pennsylvanian Essex Fauna of Northern Illinois. *Palaeontographica A* **163**, 169–99.

Thompson, I. (1979) Mazon Creek, in *The Encyclopedia of Paleontology* (eds R.W. Fairbridge and D. Jablonski), Dowden, Hutchinson and Ross, Stroudsburg, Penn., pp. 463–9. (Useful summary)

Walossek, D. (1993) The Upper Cambrian *Rehbachiella* and the phylogeny of Branchiopoda and Crustacea. *Fossils and Strata* **32**, 1–202

Walossek, D. and Müller, K.J. (1990) Upper Cambrian stem-lineage crustaceans and their bearing upon the monophyletic origin of Crustacea and the position of *Agnostus*. *Lethaia* **23**, 409–27.

Walossek, D. and Müller, K.J. (1994). Pentastomid parasites from the Lower Palaeozoic of Sweden. *Transactions of the Royal Society of Edinburgh: Earth Sciences* **85**, 1–37.

Wilby, P.R. (1996) Mineralisation of soft-bodied invertebrates in a Jurassic metalliferous deposit. *Geology* **24**, 847–50.

Williams, L.A. (1983) Deposition of the Bear Gulch Limestone: a Carboniferous *Plattenkalk* from central Montana. *Sedimentology* **30**, 843–60.

Zangerl, R. and Richardson, E.S. (1963) Paleoecology of two Pennsylvanian Black Shales. *Fieldiana, Geology Memoirs* **4**, 1–352. (Classic paper)

Books, treatises and symposia (ichnology)

Bromley, R.G. (1996) *Trace Fossils: Biology, Taphonomy, and Applications*, 2nd edn. Chapman & Hall, London (Essential reading, biological slant)

Crimes, T.P. and Harper, J.C. (1970) *Trace Fossils*. Seel House Press, Liverpool. (Proceedings of an international conference; contains 35 papers)

Donovan, S.K. (ed.) (1994) *The Palaeobiology of Trace Fossils*. Wiley, Chichester. (11 papers)

Frey, R.W. (ed.) (1976) *The Study of Trace Fossils*. Springer-Verlag, Berlin. (Invaluable compilation of original papers and introductory chapters)

Häntzschel, W. (1975) Trace fossils and Problematica, in *Treatise on Invertebrate Paleontology*, Part W, *Miscellanea, Supplement* (ed. C. Teichert). Geological Society of America and University of Kansas Press, Lawrence, Kan. (Descriptions and figures of all known genera)

Schäfer, W. (1972) *Ecology and Palaeoecology of Marine Environments*. Oliver & Boyd, Edinburgh.

Individual papers (ichnology)

Benton, M.J. and Trewin, N.H. (1980) *Dictyodora* from the Silurian of Peebleshire, Scotland. *Palaeontology* **23**, 501–13.

Goldring, R. (1962) Trace fossils and the sedimentary surface in shallow water and marine sediments, in *Deltaic and Shallow Marine Deposits* (ed. L.J.M.U. Van Straaten), Elsevier, Amsterdam, pp. 136–43. (Description of *Diplocraterion* and other trace fossils)

Pickerill, R. (1994) Nomenclature and taxonomy of invertebrate trace fossils, in *The Palaeobiology of Trace Fossils* (ed. S.K. Donovan), Wiley, Chichester, pp. 1–42.

Seilacher, A. (1964) Biogenic sedimentary structures, in *Approaches to Palaeoecology* (eds J. Imbrie and N.D. Newell), Wiley, New York, pp. 296–316. (Classification of trace fossils)

Seilacher, A. (1967) Bathymetry of trace fossils. *Marine Geology* **5**, 413–28.

Simpson, S. (1976) Classification of trace fossils, in *The Study of Trace Fossils* (ed. R.W. Frey), Springer-Verlag, Berlin, pp. 39–54.

Systematic index

Numbers in italics refer to figures.

General index

Numbers in italics refer to figures. (Syst.) is a cross-reference to the systematic index.